Silicate Science

VOLUME I

SILICATE SCIENCE

VOLUME I. SILICATE STRUCTURES

Section A. Silicate Crystal Structures
Section B. Clay Mineral Structures
Section C. Silicate Dispersoids

VOLUME II. GLASSES, ENAMELS, SLAGS

Section A. Properties and Constitution of Silicate Glasses
Section B. Industrial Glass and Enamels
Section C. Industrial Slags

VOLUME III. DRY SILICATE SYSTEMS

Section A. Dry Silicate Melt Equilibria, Crystallization, and Polymorphism
Section B. Special Silicate Systems

VOLUME IV. HYDROTHERMAL SILICATE SYSTEMS

Section A. Silicate Systems with Volatiles
Section B. Silicate Hydrates and Zeolites

VOLUME V. CERAMICS AND HYDRAULIC BINDERS

Section A. Solid State Reactions
Section B. Ceramic Silicate Systems
Section C. Portland Cements and Related Hydraulic Binders

SILICATE SCIENCE

BY

WILHELM EITEL

INSTITUTE FOR SILICATE RESEARCH
UNIVERSITY OF TOLEDO
TOLEDO, OHIO

VOLUME I

SILICATE STRUCTURES

1964

ACADEMIC PRESS New York and London

ACADEMIC PRESS INC.
111 Fifth Avenue, New York, New York 10003

United Kingdom Edition published by
ACADEMIC PRESS INC. (LONDON) LTD.
Berkley Square House, London W. 1

LIBRARY OF CONGRESS CATALOG CARD NUMBER: 63-16981

PRINTED IN THE UNITED STATES OF AMERICA.

General Preface

The evolution of Silicate Science has tremendously accelerated in the decade since the publication of "The Physical Chemistry of the Silicates" by the University of Chicago Press in 1954. One is tempted to speak, in the language of our newspapers, of an "explosive" advance in this field of scientific and technological investigation.

The author has decided, in complete accord with the publishers of the 1954 book, and with the new publishers, Academic Press, New York, to restrict this revision exclusively to the progress made in the period between 1952 and 1962.

The new title "Silicate Science" expresses the more extensive fields of investigation covered by the present treatise. The author is very grateful to the publishers for having agreed to issue this new version, which required the more comprehensive form of a series of five encyclopedic volumes.

In this treatise, references cited from "Physical Chemistry of the Silicates" are indicated only by naming the authors and the year of publication. The paragraph system of subdivision, which was welcomed by many readers of the 1954 book, is also used for the new treatise and facilitates cross-references and index quotations. However, the former reference system using numerical keys for periodical, is abandoned. In the footnotes, periodicals are cited by use of the abbreviations now universally known from the "List of Periodicals, Key to Library Files and Other Information," published by *Chemical Abstracts* under the auspices of the American Chemical Society, Washington, D. C., 1961.

The present treatise would not have been possible without the kind help extended by many friends and colleagues all over the world who sent the author their reprints and other publications. This wide correspondence gave the author hope that this treatise will be well received.

With the greatest appreciation the author wishes to express his obligation to Dr. W. S. Carlson, President of the University of Toledo, and the Board of Directors, who granted him multiple facilities for independent work. The National Science Foundation, Washington, D. C., supported his work over the years 1960 to 1962 by important subsidies.

Particular gratitude is owed to Mrs. M. M. Gillham, M. A., and her staff, who helped him master the abundant material through search of the literature and acquisition of rare publications, not to speak of the hospitality the author enjoyed in the office and reading rooms of this University Library.

It would require too much space to give a complete list of those Publishers who granted permission for reproduction of figures. Their names will be cited in special acknowledgments in the individual volumes. The author, however, wishes to express his most sincere appreciation for the spirit of cooperation he met everywhere.

The author would like to include in this general acknowledgment his cordial gratitude to his indefatigable secretary, Miss Arleen Altaffer, who worked with him assiduously and with the greatest accuracy for more than three years, in preparation of the voluminous manuscripts, indexes, and correspondence.

The author also wishes to express his thanks and appreciation to Col. G. Wilfrid Hibbert, B. A., for editorial assistance, comments, and advice. Furthermore, he offers words of personal gratitude to Professor G. E. Pankratz, M. S., M. E., University of Toledo, and Mr. Marvin Johns, B. E., who prepared most of the figures.

Toledo, Ohio W. E.
May 1964

Preface to Volume I

This first volume corresponds to Chapters I and III of the author's previous work, "The Physical Chemistry of the Silicates" (University of Chicago Press, 1954). It comprises the Cystalline State of Silicates in Sections A and B, and the Silicate Dispersoids in Section C. The rapid evolution of clay mineral investigations over two decades has justified the separation of this field from the general section on Silicate Structures.

The reader will find that the Introduction to Crystallochemistry of Silicates is much more detailed than in the previous work. A special chapter on Infrared Investigations describes their importance in structure studies. A more logical quantitative classification of silicate structures than the classical Bragg system of silicates seemed to be necessary; for this reason, the author has adopted in the present text the system proposed by T. Zoltai, based on a classification in "Sharing Coefficients." It is fortunate that Zoltai's system does not entail a basic substitution for the older Bragg system, but only its logical evolution; it also includes the results and proposals made by the school of N. V. Belov. The interlinked groups of Belov's "Second Class" are consistent with the classical principles of chair structures and their derivatives.

The important role of polytypism in silicate structures, particularly for layer structures of the mica and chlorite group, is emphasized much more extensively than in the 1954 book. The author followed here the principles displayed in a pamphlet entitled "Structural Conversions in Crystalline Systems" (The Geological Society of America, Special Paper 66, 1958), in connection with the problems involved with order-disorder conversions as demonstrated in the classical studies of F. Laves and J. R. Goldsmith on the feldspars.

Section B, dealing with the advances in clay mineral structure investigations, discusses many new aspects of mixed layer combinations of different basic structure types. The progress made in this field, particularly by T. Sudo and his school in Japan and the group of investigators in Pennsylvania State University (F. W. Brindley and R. Roy), is highly impressive.

The transition from macroscopic silicates to dispersoid silicate systems is treated in Section C, corresponding to Chapter A. III in the 1954 book. Typical colloid-physical and colloid-chemical reactions are now treated from a viewpoint based on the results of modern X-ray and electron diffraction methods. Entirely new is a subchapter on the nature of alumina-silica gels and the theory of their acidity and catalytic activity, in which the reader will find information on the problematics of reactions which are fundamental in modern cracking processes of hydrocarbons. Also the fascinating field of organic derivatives from silicate layer structures may be mentioned here as an application of modern mineralogical methods combined with classical organic chemical methods.

Toledo, Ohio W. E.
May 1964

Acknowledgments

For this first volume of the series "Silicate Science," the author gratefully acknowledges permission to use text figures from periodicals and books by the following publishers.

Academic Press, Inc., New York and London
Académie royale de Belgique, Brussels
Acta Agriculturae Scandinavica, Stockholm C, Sweden
Akademische Verlagsgesellschaft mbH, Frankfurt a.M., Germany
American Ceramic Society, Columbus, Ohio
American Chemical Society, Washington, D.C.
American Institute of Physics, New York, New York
American Society for Testing Materials, Philadelphia, Pennsylvania
Johann Ambrosius Barth, Leipzig, Germany
Bauverlag GmbH, Wiesbaden, Germany
The British Ceramic Society, Stoke-on-Trent, England
Chemical Society, London, England
Clarendon Press, Oxford, England
Denver Research Institute, Denver, Colorado
Deutsche Keramische Gesellschaft E.V., Bad Honnef/Rh., Germany
Les études des composés siliceux S.A., Brussels, Belgium
The Faraday Society, London, England
Fifth World Petroleum Congress Inc., New York, New York
Gauthier-Villars Imprimeur-Libraire, Paris, France
Geological Society of America, New York, New York
Geologiska Föreningen i Stockholm, Stockholm, Sweden
Geophysical Laboratory, Carnegie Institute of Washington, Washington, D.C.
Hochschule für Schwermaschinenbau, Magdeburg, Germany
Instituto de edafologia y fisiologia vegetal, Madrid, Spain
Kungl. Svenska Vetenskapsakademien, Stockholm, Sweden
Macmillan & Co., Ltd., London, England
Mineralogical Society of America, Ann Arbor, Michigan
Mineralogical Society, London, England
Einar Munksgaard, Copenhagen, Denmark
National Academy of Sciences, National Research Council, Washington, D.C.
National Bureau of Standards, Washington, D.C.
Österreichisch-Amerikanische Magnesit A.G., Radenthein/Carinthia, Austria
Ogden Publishing Group, New York, New York

Pergamon Press, Inc., New York, New York
Dr. Riederer-Verlag, Stuttgart, Germany
E. Schweizerbart'sche Verlagsbuchhandlung, Stuttgart, Germany
Schweizerische Mineralogische und Petrographische Gesellschaft, Basel, Switzerland
Society of Glass Technology, Sheffield, England
Springer-Verlag, Berlin, Göttingen, Heidelberg, Germany
Svenska Forskningsinstitutet för Cement och Betong vid Kungl. Tekniska Högskolan, Stockholm, Sweden
University of Chicago Press, Chicago, Illinois
VEB Verlag für Bauwesen, Berlin, Germany
Verlag von Dr. Dietrich Steinkopff, Darmstadt, Germany
Verlag der Zeitschrift für Naturforschung, Tübingen, Germany
Vulkan-Verlag, Essen, Germany
Wallerstein Laboratories, New York, New York
Westdeutscher Verlag, Köln and Opladen, Germany

Contents

Section A

SILICATE CRYSTAL STRUCTURES

Section B

CLAY MINERALS; STRUCTURES

Section C

SILICATE DISPERSOIDS

Section A

Silicate Crystal Structures

INTRODUCTION TO THE CRYSTALLOCHEMISTRY OF SILICATES

1. The binding mechanism in the molecular unit SiO_2, the tetrahedral unit coordination group $[SiO_4]$, and its tendency toward complex polymerization into chains, networks, and frameworks are fundamental facts on which the structural study of the silicates is based.[1] The ionic and covalent character of the bonding in SiO_2 is in contrast with that of the purely covalent molecule CO_2, as viewed in the light of L. Pauling's resonance theory. SiO_2 and silicates are in an intermediary position between ionic-inorganic and covalent-organic compounds in their physicochemical characteristics, especially in their fusion behavior, hardness, chemical inertia, formation of glass, and polymorphism. The structural principles in quartz, tridymite, and cristobalite are based on variations of frameworks with the same tetrahedral pattern. The same basic principles of coordination and structural interlinkage are assumed to be valid for the structure of the silica and silicate melts, and of glasses. Prompt enantiotropic inversions of the $\alpha \rightleftharpoons \beta$ type in the different mineral groups of SiO_2 may he distinguished from the sluggish, "reconstructive" mutual inversions from one mineral group to the other. These are accompanied by conspicuous changes in the angles of the Si—O—Si bonds (cf. Vol. III, Section A).

2. A *sixfold* coordination of the Si atom to oxygen anions with the formation of $[SiO_6]$ groups was postulated for the structure of SiP_2O_7. From the viewpoints of the *screening theory*, W. A. Weyl and E. A. Hauser[2] discussed the reasons why the

[1] Cf., e.g., A. Badalucco, *Ceramica (Milan)*, **11**, 1956 (3) 41-45; (6) 46-51; (11) 47-49, who gave an excellent introductory discussion on the nature of the Si—O—Si bonds.

[2] *Kolloid-Z.*, **124**, 1951, 72-76. On the relation of the screening theory of the silicic acid hydrates and their polymerization, cf. Section C. Sixfold coordination in a silica modification isostructural with *rutile* is verified in the structure of the recently discovered highest-pressure crystal phase of *stishovite*. Cf. A. Preisinger, *Naturwissenschaften*, **49**, 1962, 345.

complex anion $(SiF_6)^{2-}$ is stable in silicofluorates, and also stable as a structural configuration $[SiF_6]$: four less polarizable F^- anions are unable to saturate the screening tendency of Si^{4+} in a tetrahedral surrounding, and the octahedral surrounding must in this case be more stable. For the more polarizable oxygen anion, corresponding conditions may occur if strongly contrapolarizing cations in the neighborhood of Si^{4+} centers reduce the polarizability. Then, the same conditions as in the formation of $[SiF_6]$ groups may favor the formation of $[SiO_6]$ coordination groups. P^{5+} in SiP_2O_7, or C^{4+} in silicoorganic compounds like silicon acetyl acetonate, act as such strongly contrapolarizing cations. Even in ethyl orthosilicic ester, $Si(OC_2H_5)_4$, the central silicon atom can expand its coordination beyond the number 4, and in silicic acid, $Si(OH)_4$, the presence of $[Si(OH)_6]$ groups is most probable.

3. Great progress has been made in the application of the X-ray Fourier synthesis to the problems of *electron density distribution* in crystalline media,[3] in the determination of the characteristic maxima and minima. The particularly important but difficult location of H^+ ions in *hydrogen bonds* of silicates and related structures was challenged by electron diffraction methods in the investigations of B. K. Vaĭnshteĭn.[4] The superiority of these methods over X-ray diffraction alone is obvious, e.g., for the investigation of heavy metal salt hydrates and of organic compounds. The position of the hydrogen atoms is concluded from zonal projections in a direction perpendicular to the bonds, as B. K. Vaĭnshteĭn and A. N. Lobachev[5] demonstrated for the crystal structure of urotropine (hexamethylene tetramine). The bastion-like prominences in projections on (100) show very clearly the position of the hydrogen atoms. More immediately interesting for silicates is the investigation of the structure of the cubic mineral cryptohalite (ammonium silicofluoride), $(NH_4)_2SiF_6$, by B. K. Vaĭnshteĭn and M. M. Stasova[6] which showed a statistical distribution of the ammonium groups over six possible positions in the elementary cell, with the apexes of the tetrahedra (H atoms) directed towards 4 of 12 F^- anions surrounding them. The Fourier synthesis shows slight prominences of the nitrogen atoms indicating the position of the hydrogen atoms as they are described for urotropine. The bonding N—H is definitely covalent, that of Si^{4+} and F^- ionic.

4. The equal character of the bonding mechanism of "homodesmic" crystal phases

[3] See, e.g., for metals (Cu, Ag, Al): H. Kühne, *Fortschr. Mineral.*, **31**, 1953, 11-13, with critical remarks of H. Jagodzinski and E. Wölfel, on the method of the introduction of an artificial temperature factor in the calculations.

[4] *Trudy Inst. Krist. Akad. Nauk S.S.S.R.*, **10**, 1955, 49-61; 62-70, for paraffin (with Z. G. Pinsker).

[5] *Trudy Inst. Krist. Akad. Nauk S.S.S.R.*, **10**, 1955, 71-75; *Kristallografiya*, **3**, 1958, 452-460; *Doklady Akad. Nauk S.S.S.R.*, **120**, 1958, 523-525, on the distance C—H in paraffins, hexamethylene tetramine, as determined by X-ray, electron and neutron diffraction methods.

[6] *Kristallografiya*, **1**, 1956, 311-320.

($\delta\varepsilon\sigma\mu o\varsigma$ = bond) in contrast with the different bonding mechanism in "heterodesmic" crystals as a basic principle for the classification of minerals, especially for silicates, was discussed by W. Kleber.[7] If p is the quotient of the electrostatic charges, z, of a central cation with its coordination number, n (where $p = z/n$), and y is the charge of the anion, then the relation $p < y/2$ characterizes, e.g., the alkali halides, $p = y/2$ those of SiO_2, and $p > y/2$ those of calcite, perowskite, etc. The alkali halides are called "isodesmic" (simple ionic structures); silicates, borates, metaphosphates, etc. are called "mesodesmic" (ionic structures with complexes interlinking to chains, ribbons, networks, etc.); carbonate, sulfate, and other oxysalt minerals are called "anisodesmic" (ionic structures with noninterlinking complexes). The term "heterodesmic" includes the metalloids, semiconductor compounds, molecule compounds, and van der Waals-bonded structures.

5. The role which is played in silica and silicates by transitions between ionic (heteropolar) and covalent (homopolar) bonding mechanisms is one of the most important problems discussed in recent investigations; e.g., H. Pick[8] emphasized that the combinations and transitions between ionic and valence bonding offer especially the most multiple and attractive aims in modern Cyrstal Physics, as from the viewpoints of structural defects. W. Noll[9] described the Si—O—Si (siloxan) bonds, in polymeric silicate anions, on the one hand, and in silicone molecules, on the other hand, to illustrate this tremendous variability, although in general very simple geometric rules are evident. The tendency of the Si—O—Si grouping *not* to form stretched arrangements with an angle of 180° between the branches but on the average with an angle of 130° is noted. This rule is a powerful principle of selection of a few interlinkages, and the elimination of many others, in tetrahedral $[SiO_4]$ arrangements. This is also the reason why the common structure models for silicates are feasible based on ionic radii, whereas for silicones (e.g., methyl polysiloxene) and other silicoorganic compounds the models of spherical calotte units (after Stuart and Briegleb) are better used, based on action radii and atomic distaces. In the latter case the measured distances are generally smaller than the sum of the bonding radii and of the ionic radii of Si^{4+} and O^{2-}. The mesomeric character of bondings in silicates makes it possible for structural interpretations to be made, often from the ionic as well as the homopolar principle (cf. A. ¶ 225).

[7] *Fortschr. Mineral.*, **26**, 1950, 70-72; the definition of homodesmic and heterodesmic compounds was first given by R. C. Evans, in his book "An Introduction to Crystal Chemistry." Cambridge Univ. Press, London and New York, 1939. See also the recent article of W. Noll, *Naturwissenschaften*, **49**, 1962, 505-512.

[8] *Naturwissenschaften*, **41**, 1954, 346-354, on solid state physics.

[9] *Fortschr. Mineral.*, **34**, 1956, 52. Interesting models of the calotte type are discussed for the urotropine-like structure of (*tert*-$C_4H_9)_4(SiO_{1.5})_4$, by E. Wiberg and W. Simmler, *Z. anorg. u. allgem. Chem.*, **282**, 1956, pp. 334 ff., and for (*iso*-$C_3H_7)_4(SiO_{1.5})_4$. See also for dimethyl polysiloxan chains: W. Noll, *Fortschr. Mineral.*, **34**, 1956, pp. 81 ff.

6. As a new tool for locating hydrogen atoms in solid structures, the *nuclear magnetic resonance* method has great importance; it is characterized by resonance to radio-frequency absorption of the solid structure embedded in a strong magnetic field. The shape and width of resonance lines is determined by the magnetic interactions of the nuclei, based on the theory of Van Vleck by which the second moment of an absorption line for the solid is determined by the internuclear distance and the magnetic properties of the nuclei. H. S. Gutowsky and D. M. Henderson[10] applied this method in the determination of the position of hydrogen in portlandite, $Ca(OH)_2$, a typical layer structure, at the temperature of liquid nitrogen. The distance of separation in the "puckers" of the hexagonal networks of hydrogen nuclei is 0.618 ± 0.04 Å., corresponding to a distance of —O—H of 0.99 ± 0.02 Å., close to the distance in the structures of ice and gypsum. It also agrees very well with the O—H nuclear distance of 0.98 Å. calculated by neutron diffraction of $Ca(OH)_2$, whereas the X-ray determined distance is only 0.87 Å.

7. A basic discussion of the state of oxygen chemically combined in silicates, on the assumption of a *mono*valent O anion action in their structures, was presented by V. I. Lebedev[11] from the ionization potentials and electron affinities of the elements occurring in common rock-forming minerals. Only on the assumption of a singly charged anion, could Lebedev interpret the role of the Si—O bonds when silicon acts as a *di*valent, doubly positive-charged cation to explain the partially ionic, partially covalent character in silicate bonding and the electron density distribution in quartz (cf. R. Brill, K. Hermann, and Cl. Peters, 1942). The ratio of Si : O [in aluminosilicates (Si + Al) : O] varies in igneous rocks from 1 : 4 to 1 : 2. The most striking deviations in the bonding character which are determined by the correlation of ionization potentials of valence electrons in the constituting elements are observed in metamorphic and hydrothermal minerals in the presence of water. Oxygen atoms in special positions in which they cannot exert their tendency to bind silicon in the covalent type (e.g., in amphibole, talc, or the humite minerals) easily bind hydrogen to form OH^- groups. Lebedev's theory claims to have overcome many difficulties of the classical theory of ionic radii, which cannot convincingly explain why aluminum is either structurally equivalent to silicon or to magnesium, but shows different coordinations with oxygen: [AlO_4] and [AlO_6].

8. The *polarization* (deformation) effects of ions by electron configuration and

[10] *Bull. Geol. Soc. Am.*, **67**, 1956, 1703; *Discussions Faraday Soc.*, **19**, 1955, 187-192, and recently *Am. Mineralogist*, **47**, 1962, 1231-1251.

[11] *Vestnik Leningrad. Univ.*, **3**, 1948, 17-30.

[12] For more recent publications of K. Fajans in this field, see the articles "Quanticule Theory" and "Polarization" in the "Encyclopedia of Chemistry" (G. L. Clark and G. G. Hawley, eds.), Reinhold, New York, 1957, pp. 763-767, 803; cf. further reviews in *Vortex*, **15**, February, 1954, 4 pp.; *Physik. Verhandl.*, **5**, 1954, 95; *Angew. Chem.*, **66**, 1954, 608-609.

electrostatic forces in crystalline compounds is the center of the theories of K. Fajans[12] and of W. A. Weyl,[13] in the so-called *screening* effects of electrons and anions surrounding a central cation. These theories are of great importance for the understanding of the constitution of glass and are therefore chiefly discussed in Vol. II, Section A. The screening is a consequence of the tendency of the cations to neutralize their external fields in the smallest possible volume element, and, therefore, to direct the chemical interactions of solid surfaces with their surroundings. The importance of the ion polarizability becomes particularly evident in the discussion of the optical properties of crystalline silicates, their refractive indices, and — as a highly sensitive parameter — the molecular refractivity.

9. J. Verhoogen[14] discussed a very instructive example of these effects for the magnesium-aluminum oxides and silicates, in view of the determination of the degree of ionic and covalent shares in their bonding mechanisms. He continued the previous discussion of W. S. Fyfe,[15] who especially considered the atomic orbital overlap in the linkages with the "overlap integral" $S_{(\psi_A, \psi_B, R)} = \int \psi_A \psi_B \, dv$, in which ψ_A and ψ_B are normalized atomic orbitals of the atoms A and B, and R is the internuclear distance. For purely ionic bonding $S = 0$; for purely covalent bonding $S = 1$. Fyfe gave an extensive tabulation of the overlap integrals in connection with electronegativity for the alkali halides and metal—O bonds, with the important results that for Si—O and Al—O bonds strong overlap occurs, but also in Mg—O and Na—O the covalent contributions to bond energy are rather considerable. Also the double-bonding effects such as in Si—O are indicated by a distinct shortening. In this direction is also explained why there is no $(SiO_3)^{2-}$ group analogous to $(CO_3)^{2-}$, and only the four-coordinated $[SiO_4]$ is stable. Also the fact that coordinations $[BO_3]$ and $[BO_4]$ exist, in lower degree even $[CO_4]$ besides $[CO_3]$ (the first unit with longer bonds) is readily understood in Fyfe's calculations.

10. Such discussions are of considerable interest for the comparison of crystalline silicates, borosilicates, aluminosilicates, and carbonate silicates. Covalently, the four-coordination groups are more stable, and conversely, if a large amount of polarity is introduced, it exerts a greater effect on the C—O groups than on Si—O groups. Verhoogen states in supplementing Fyfe's results that it is not possible by quantum mechanical methods to calculate the bond energies, lattice spacings, or

[13] E. Plumat, *Silicates inds.*, **19**, 1954, 97-107, gave an excellent review of the screening hypothesis and its consequences.

[14] *Am Mineralogist*, **43**, 1959, 552-579.

[15] *Am. Mineralogist*, **39**, 1954, 991-1003. On π-bonding, cf. L. Pauling, *J. Phys. Chem.*, **56**, 1952, 361-365. In the opinion of W. H. Baur, *Naturwissenschaften*, **48**, 1961, 549-550, it is not advisable to conclude from anomalous atomic distances Me—O in silicate structures (see, e.g., cummingtonite, cf. A. ¶ 106) that there is covalent bonding, when Pauling's electrostatic valence rule is apparently not satisfied.

the elasticity coefficients of any structure more complicated than the H_2 molecule. The purely ionic structure models are no doubt much simpler than those with covalent bonding alone. By the important principle of the additivity of entropies,[16] however, it is certain that the bonding type in oxides and especially in magnesium-aluminum silicates must be the same as in the oxide components.

11. We cannot give here in detail the calculations of Verhoogen on the lattice energies, the Cauchy relations for elastic constants, atomic X-ray scattering factors, *f*, the diamagnetic susceptibility, or ionic and electronic (optic) polarizability. But it is important that the electronic polarizability alone does *not* provide clear evidence of covalent bonding in any crystal phases of the system $MgO–Al_2O_3–SiO_2$. Also from the ionic polarizability nothing compels us to abandon the simple ionic representation of these compounds. The degree of covalency calculated from the ionic polarizability does *not* agree with that derived from electronegativity values. Particularly poor indicators of covalent or ionic bonding characteristics are the coordination numbers. Even for SiO_2, which is generally regarded as at least half-covalent, half-ionic, the physical parameters are much more easily explained on the *ionic* than on the covalent basis. Ions, nevertheless, do not behave as rigid spheres; their deformability is clearly indicated in the optical-refractive properties, but this also can be calculated on a purely electrostatic theory, and not easily interpreted from covalency aspects. For practical purposes, Verhoogen believes it fully legitimate to consider compounds like those of the system $MgO–Al_2O_3–SiO_2$ as purely ionic.

12. H. Ramberg[17] examined the polarization of the oxygen anions in silicates as an important principle for calculations of the *stability* of simple silicates. There is a general rule that the larger the size of a noble-gas type cation and the lower its electrostatic charge, the more stable is the silicate relatively to the free oxides. This is due to the asymmetric polarization effects exerted on the O^{2-} anion in the resulting field surrounding the Si^{4+} cations and those of additional metals. The Si^{4+}—O^{2-} bonds are strengthened, the cation —O^{2-} bond is weakened when the oxides react to form the silicate. The thermodynamic calculation shows that the energy release combined with the tightening of the Si—O links is usually larger than the energy absorption connected with the weakening of the cation —O^{2-} bonds. The net exothermic energy release in the formation of the stable silicate from the oxides is closely related to the oxygen polarization; it is increased with increasing polarization and with the stability of the silicate. Ramberg's calculations may also be extended to *polymerized* silicates. For a [SiO-] group in which the central Si^{4+} ion is surrounded by *shared* oxygen ions, the electrons of the shared O^{2-} ions in Si—O—Si bridges are not pulled very close to the Si^{4+} because of the contrapolarization by the

[16] Cf. W. S. Fyfe, F. J. Turner, and J. Verhoogen, *Geol. Soc. Am. Mem.*, **73**, 1958, 259 pp.
[17] *Am Mineralogist*, **39**, 1954, 256-271.

opposite Si^{4+} ions. The effective polarizing power of Si^{4+} increases with an increasing number of shared oxygen anions so that the polarization of each *un*shared O^{2-} by Si^{4+} is the stronger the more shared O^{2-} remain around the central Si^{4+}. These effects in progressive polymerization are well indicated by the heats of formation, ΔH_{298}, for the potassium silicates from the oxides, according to O. Kubaschewski and E. L. Evans[18] for a constant equivalent of the metal oxide:

	K_4SiO_4	K_2SiO_3	$K_2Si_2O_5$	$K_2Si_4O_9$
ΔH_{298} in cal./mole	43,800 ± 300	64,500 ± 700	74,500 ± 700	81,000 ± 700

13. Many attempts have been made to determine the character of the chemical bond in minerals, especially in silicates, by considering the *electronegativity* of the cationic elements in these. A. S. Povarennykh[19] plotted these parameters as a function of the ratio of ionic to covalent bonding on a continuous curve starting from the typically covalent AsH_3, HI, HBr, and HCl; to H_2O and SiO_2 as intermediate bond types; to the ionic alkali halides. The curve is somewhat different from that L. Pauling had given before. How these relations affect the optical properties of minerals of different crystallochemical groups was further demonstrated by Povarennykh.[20] He observed also the screening degree influences on the optical properties; the higher the screening, the higher also the refractive index in analogous crystal groups. Using the differences of electronegativity, $X_O - X_{Si}$, of Si (1.8) and O (3.5), W. Noll[21] made an interesting comparison of the bonding characteristics in silicates and silicones, with their distances Si—O and Si—C. From L. Pauling's approximation formula for the percentage share of heteropolar bonding $= 100 \cdot [1 - e^{-c}(X_O - X_{Si})^2]$ $(c = 0.25)$, one may calculate the ratio 1 : 1 of heteropolar and homopolar bonding in the siloxane group Si—O—Si with the distance Si—O in silicates = 1.60 Å., in silicones = 1.64 Å., whereas for the bond Si—C the covalent distance is = 1.88 Å., and the share of ionic bonding only 12 per cent.

14. H. Krebs[22] discussed the interesting problem of the influence of homopolar bonding on the properties of inorganic salt-like compounds, especially those with a pronounced layer structure like CdI_2, when unoccupied orbitals of the cations overlap occupied orbitals of the anion. Tetrahedral or octahedral coordination is favored if the anion has the electron configurations $s^2 + p^6$, or s^2p^6-hybrid. Krebs

[18] "Metallurgical Thermochemistry," 3rd ed., Pergamon Press, New York, 1958, 426 pp.
[19] *Doklady Akad. Nauk S.S.S.R.*, **109**, 1956, 993-995.
[20] *Zapiski Vsesoyuz. Mineral. Obshchestva*, **87**, 1958, 348-359.
[21] *Fortschr. Mineral.*, **34**, 1956, 63-84; *Naturwissenschaften*, **49**, 1962, 505-512.
[22] *Z. anorg. u. allgem. Chem.*, **278**, 1955, 82-92.

could thus explain the heterotypic crystalline solutions in the fluorite structures, the transitions between CsCl, NaCl, and ZnS structures, and the direction of hydrogen-bonding in hydrargillite, boehmite, and two-layer lattices of the type $(H_2O)_3$ $Co(OH)_3$, as described by W. Feitknecht (1935).

15. The importance of the electronegativity values of the metal and oxygen atoms in the chemical bonds and the distribution of cations in silicate structures was emphasized by H. Ramberg.[23] Differences in the bond types among lithophilic elements cause the formation of different types of silicates in their geochemical differentiation behavior. For simple iron-magnesium silicates, e.g., the distribution is distinctly related to the type of the Si—O skeleton of the minerals concerned. There is an increase in Mg^{2+} ions from the orthosilicates (isle structures) to the layer silicates, and in parallel with a systematic increase in the electronegativity of the oxygen atoms surrounding the metal ions. From the orthosilicates to the framework silicates there is also an increasing capacity to accept Al^{3+} as a substitute for Si^{4+} in the tetrahedral units; the more strongly electropositive Al^{3+} prefers to enter Si^{4+} positions with a maximum number of electronegative neighbors of bridging oxygen atoms. The electronegativity of the oxygen is parallel with the increasing ratio Fe/Mg, as is observed in the crystallization of biotite, anthophyllite, and amphiboles. Alkalies and calcium favor the entrance in layer and framework silicates.

16. Since A. E. Fersman's geochemical ideas (1937) on the paragenesis and succession of silicate minerals, a rule has been generally accepted that the gradually *de*creasing *lattice energies* of the crystal phases during differentiation would control the sequences. L. V. Tauson (1943) critically opposed this assumption and defended the opinion that *in*creasing lattice energies would rule the differentiation courses. An attempt to reconcile these contrasting aspects was made by V. I. Lebedev,[24] going back to the definition of chemical compounds, with ionic, metallic, and molecular binding forces in Haber-Born cycle processes. Evidently, Tauson's conclusions are in general oversimplified since he only considered purely ionic bonding mechanisms, and therefore the calculation of lattice energies, applying A. A. Kapustinskiĭ's formulae must become uncertain. Drastic contradictions concerning the succession mineral phenomena with such a theory are the consequence of these oversimplifications, e.g., in the application of N. L. Bowen's "reaction principle."

[23] *J. Geol.*, **60**, 1952, 331-355. The distribution of major and minor elements in coexisting silicate minerals is in good agreement with the postulates of Ramberg; cf. E. H. Nickel, *Am. Mineralogist*, **39**, 1954, 486-493, as explained by the consideration of the electronegativities. We also emphasize the extensive calculations and tabulations made by M. Vendel, *Acta Geol. Acad. Sci. Hung.*, III, 1955 (1/3), 245-300, on a wide geochemical basis for the substitution of ions and atoms in single minerals and entire geological units.

[24] *Vestnik Leningrad. Univ.*, **7**, 1952 (10) 107-119. Cf. L. V. Tauson, *Doklady Akad. Nauk S.S.S.R.*, **67**, 1949, 687-690.

Thermochemical calculations of Lebedev could demonstrate, not that only an agreement between the theory of Fersman with the natural occurrences is observed if the real bond types are considered, and that the *minimum* lattice energies govern the differentiation reactions, but also that more special problems, like the stability of muscovite and leucite or the incongruent fusion behavior of grossular garnets, can then be solved.

17. The bonding energies of rock-forming silicate minerals were calculated and tabulated by W. D. Keller[25] by adding the bond energies between their constituent cations and oxygen, starting from the elements in the gaseous-ionic state. The standard adjusted to 24 oxygen atoms per unit cell gives a good basis for a comparison of the results of the calculation with the Bowen reaction series in silicate rocks and mineral differentiation, especially for the mica group and metamorphic minerals. A rather good agreement is found between the order of bonding energies and the sequence in the observed mineral occurrences. The energy sequence of the orthosilicates with isle structures (adjusted bonding energy in the order of magnitude of 18,850 kcal. + that of 12 M^{2+} bridges), to the group silicate structures with $[Si_2O_7]$ units (21,500 kcal. + that of $10^2/_7$ M^{2+} bridges), to chain structures (25,000 to 27,000 kcal. + that of 8, or 7, M^{2+} bridges, in the latter case + 2 H^+ bridges), to the layer silicates (30,000 kcal. + that of 6 M^{2+} bridges + 4 H^+ bridges), and finally to framework structures (with 37,300 kcal.) is most instructive and in good analogy with the differentiation principles.

18. More recently, G. B. Bokiĭ and S. S. Batsanov[26] used the electronegativity concepts for a discussion of the nature of the "single" oxygen atoms in *oxysilicates* like disthene, $Al_2(SiO_4).O$, or ramsayite, $Na_2Ti_2(SiO_3)_2.O_3$. The neutral group SiO_2 has an electronegativity of 4.55; that of $Mg_2SiO_4 = 3.32$, with the electronegativity of the $Mg^{2+} = 6.00$, for the atomic Mg = 1.56, and a 40 percent ionization calculated for the structure of forsterite. A totally ionized $[SiO_4]^{4-}$ group would have the electronegativity = 1.45; $[Si_2O_7]^{6-} = 2.02$; $[SiO_3]^{2-} = 2.58$; $[Si_6O_{11}]^{6-} = 2.90$; $[Si_2O_5]^{2-} = 3.25$; $[SiO_2]^0 = 4.55$. Hydroxyl groups reduce the electronegativity values. Simple oxygen atoms also reduce these if the cations of the compound in question are basic, but they increase them if the cations are acidic. Typical examples are $Al_2(SiO_4)(OH_2)$, with an electronegativity of 3.72; $Al_2(SiO_4).O$ with 3.74; fluor-topaz $Al_2(SiO_4)F_2$ with 3.97. The complex fluorosilicate is therefore more acidic than the corresponding hydroxyl-containing compound. Mixed basic and acidic cations appear in ramsayite, with an electronegativity of 3.26; this mineral and disthene are basic, and this fact is important for the genetic conditions of their crystallization and paragenesis.

[25] *Am. Mineralogist*, **39,** 1954, 783-793. For the basic thermochemical data see the previous studies of M. L. Huggins and K. H. Sun (1954).

[26] *Zapiski Vsesoyuz. Mineral. Obshchestva*, **85,** 1956, 137-146.

19. Coming back to the determination of R. Brill *et al.* of the electron density distribution and the bond types in quartz and silicates,[27] we emphasize as a comparison the excellent studies[28] on the structure of MgO by a detailed three-dimensional Fourier analysis which gives evidence that in all parts of the structure, even in positions far away from the atom centers, higher electron densities are found than are characteristic of NaCl and the other alkali halides. Brill *et al.* conclude that this electron distribution indicates a transition state between ionic and homopolar bonding mechanisms. This transition phenomenon is also observed in distinct deviations of the atomar scattering factors for X-rays from the theoretical F-values as a function of (sin ϑ/λ), for completely ionic bonding types. The characteristic temperature of MgO is 660°K.

20. Differences in the degree of covalency in the bonding mechanisms of their crystal structures are also the reason for the striking distinctions observed by A. Dietzel and J. H. Poegel[29] between SiO_2 and $AlPO_4$ which should be expected to express a nearly ideal structural analogy. The same may be expected in their physicochemical properties. They show, however, large differences in their fusion behavior. Whereas silica is a typical glass-forming oxide, $AlPO_4$ behaves like a true salt of prevailingly ionic type, with a strong tendency to crystallization from a fluid melt. Evidently, the very high electrostatic field of the P^{5+} cations contrapolarizes the $[AlO_4]$ groups, widening them to $[AlO_6]$ configurations and making the compound behave as a true aluminum salt. The high fusion point of $AlPO_4$ also lends further support in favor of strongly increased ionic bonding characteristics in $AlPO_4$.

21. That the assumption of a strongly ionic constitution cannot be correct for silica and its glass, especially in view of its anomalies in specific volume as a function of temperature (cf. Vol. II, Section A) was discussed by H. Salmang.[30] This finding was sustained in a general interpretation of the glassy state of silica and of silicates as ionic and/or molecular (covalent) structures, likewise in connection with their high degree of polymerization to chains and complex frameworks, and further in the tendency to form silica molecules with a permanent dipole moment and solvation complexes. Expecially W. Weyl (1933) had emphasized very early the latter facts as a principal basis for interpreting the constitution of colored glasses, and their changes by modifying the electrostatic field conditions by definite additions.

[27] *Ceram. Age*, **61**, 1953 (3) 31, 33 f., 36.

[28] *Z. anorg. u. allgem. Chem.*, **257**, 1948, 151-165 (with C. Hermann and Cl. Peters).

[29] *Naturwissenschaften*, **40**, 1953, 604; *Veröffentl. Max-Planck.Inst. Silikatforsch.*, **13**, 1953, 131; see also H. J. Poegel, Inaug.-Dissertation, Techn. Univ. Berlin, 1955; abstract in *Glastech. Ber.*, **29**, 1956, 266.

[30] *Glastech. Ber.*, **32**, 1959, 137-142; *Chem. Weekblad*, **57**, 1961, 469-474.

A far-reaching reconciliation of the ionic and of the covalent interpretations is therefore possible, opening many ways towards an explanation of glass constitution from both viewpoints, giving a satisfactory approximation of reality.

THE $[SiO_4]$ GROUP AND ITS AGGREGATION INTO COMPLEX TETRAHEDRAL ARRANGEMENTS

22. The character of the Si—O bonds has been extensively discussed in its fundamental importance for the silicate and aluminosilicate structures, therefore also in analogy to Al—O bonds. The interatomic distances of both groupments were especially studied by J. V. Smith,[31] in their variations under the influence of the neighboring cations. A statistical evaluation of the numerous precision determinations in silicate and aluminosilicate structures leads to the average values of Al—O = 1.78 ± 0.02 Å. and of Si—O = 1.60 ± 0.01 Å. used in the location of the Si^{4+} and Al^{3+} ions in feldspars and related structures by accurate measurements of the dimensions of the $[(Si, Al)O_4]$ coordination tetrahedra (cf. A. ¶ 200). For the correct use of the average distances it is also important to know that for high- and low-temperature modifications of silicates and aluminosilicates the tetrahedral bonds are almost independent of temperature. It is therefore permissible to use an elementary linear relation for intermediate ratios of Si/Al in the limits given by the average values indicated above as optimum data.

23. Or further fundamental importance for silicate and aluminosilicate structures is the valence *angle* in the Si—O—Si bondings, and the problem of the existence of stretched configurations, as with an angle of 180°. In this respect some confusion has arisen by an evidently incorrect determination of the angle Si—O—Si for α-cristobalite by R. W. G. Wyckoff (1925). J. D. Mackenzie and J. L. White[32] discussed the angles observed in the structures of the silica modifications and in silica glass, especially emphasizing the fact that J. Zarzycki[33] found for silica and GeO_2 glass (cf. Vol. II, Section A) not only the accurate interatomic distances but also reliable data for the bond angle. For silica glass, the bond angle is 143° ± 17°; for GeO_2 glass at 20°C., it is variable between 125° and 152°, but at 1200°C. the angle is between 137° and possibly 180°. For an entirely disordered structure of the silica glass the angle should be about 150°. In contrast with GeO_2, the angle Si—O—Si remains rather unchanged up to 1600°C., the same as in the α-quartz structure, and it is even preserved in the liquid state, whereas in GeO_2 it is then increased to

[31] *Acta Cryst.*, **7**, 1954, 479-481; especially for the Si—O and Al—O distances in paracelsian and α-carnegieite see also *Carnegie Inst. Wash. Year Books*, **52**, 1952/53, 72-75.

[32] *J. Am. Ceram. Soc.*, **43**, 1960, 170-171.

[33] *Verres et réfractaires*, **11**, 1957, 3-8.

about 180°. Mackenzie and White mention for cristobalite the angle of 142° to 147°, and in Na_2SiO_3 it is 137°. For vitreous silica they give the internuclear distances as shown in Fig. A. 1, based on the neutron diffraction data of W. O. Milligan, H. A. Levy, and S. W. Peterson.[34] Recently also F. Liebau[35] refuted the assumption

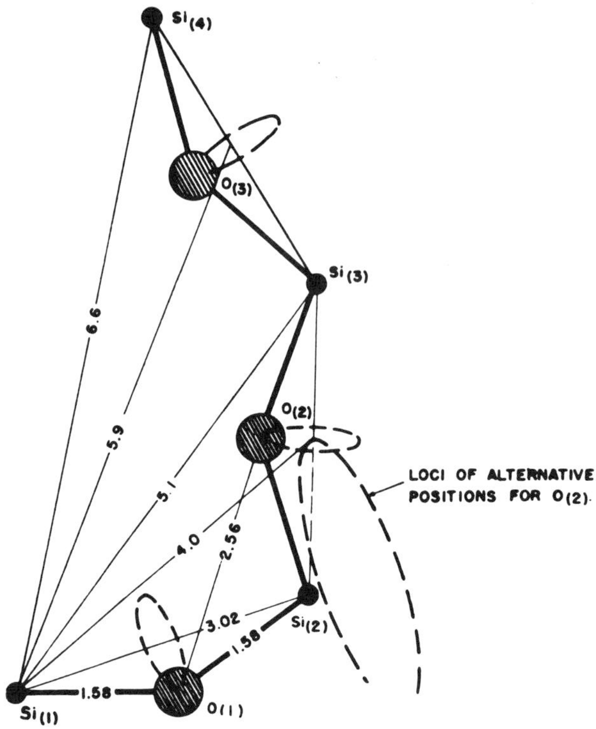

FIG. A. 1. Internuclear distance (in Å.) of the silica glass structure (Milligan, Levy, and Peterson).

of Si—O—Si angles of 180° which in no case have been verified. The largest angle observed was 156° for $Li_2Si_2O_5$, and 163° to 166° for petalite, for the bridging oxygen atoms; on the other hand, it was 125° for the normal oxygen anions in $Li_2Si_2O_5$ and 153° to 150° in petalite. Stretched Si—O—Si bonds are for normal conditions of energy equilibrium in silicate structures evidently highly improbable. Liebau sees in the steady deviation of the bond angles from the stretched groupment a consequence of the strong shares in covalent bonding conditions. Only at high temperatures the increasing ionization may favor an approximation toward stretching in the configuration.

[34] *Phys. Rev.*, **83**, 1951, 226-227.

[35] *Silikat Tech.*, **11**, 1960, 397. See also *Acta Cryst.*, **14**, 1961, 399-406.

24. The problem of the structural premises of the thermal expansion coefficients of monocrystalline silicates and aluminosilicates was again discussed by S. Kumar[36] in a general theoretical examination of oxide structures, based on the previous calculations of H. T. Smyth[37] for silica glass. The agreement between the experimental measurements and the theoretical values of the thermal expansion coefficients over the range from 70° to 700°K. is very satisfactory if the specific heat due to longitudinal vibrations along the Si—O—Si bonds is introduced, whereas large deviations occur if the total specific heat is used to account for the thermal vibrations. The residual discrepancies above 400°K are partly due to decreases in the Si—O distances with increasing amplitude of transverse vibrations, and partly to the (evidently incorrect) assumption of a stretched Si—O—Si alignment.

25. The tendency of $[SiO_4]$ and $[(Si, Al)O_4]$ groups to polymerize to complex arrangements is in no way restricted only to tetrahedral coordination types. If one generalizes the same principles of polymeric interlinkage which are developed in the classical systematic investigations of W. L. Bragg (1930) for silicates and aluminosilicates, for octahedral coordination groups $[RO_6]$ one may also find interesting inorganic "sister structures" which are particularly impressive, such as in the complex fluoroaluminates studied by C. Brosset (cf. Vol. IV, Section A), and also more recently by E. G. Stewart and H. P. Rooksby[38] in the cryolite group, further in phosphates, borates, etc.

26. From the purely geometrical viewpoints, F. Laves[39] gave in 1932 a full derivation of tetrahedral combinations which can be applied to $[SiO_4]$ polymerizations; some of these constructions, however, are not verified in the silicate anions observed in Bragg's systematics. The modern silicoorganic (silicone) chemistry, on the other hand, gives new aspects for a possibility of so-called *cis*- and *trans*-configurations, with equally directed, or oppositely oriented tetrahedral positions, respectively. It is highly significant that also for silicones linear tetrahedral interlinkages of chain or network type are *not* observed. The Si—O—Si angles of 141° are on the other hand, equally observed in the pyroxene group of the silicates, as in the tetrahedral $[(CH_3)_2SiO_{3/2}]$ chains of methyl polysiloxanes of cautchouc-like behavior, studied at low temperatures in the stretched state by P. J. Flory, L. Mandelkern, J. B. Kinsinger, and W. B. Shultz.[40] Only one characteristic apparent exception is to

[36] *Central Glass & Ceram. Research Inst. Bull. (India)*, **7**, 1960, 58-69.

[37] *J. Am. Ceram. Soc.*, **38**, 1955, 140-141.

[38] *Acta Cryst.*, **6**, 1953, 49-52.

[39] *Z. Krist.*, **82**, 1932, 1-14. Cf. on the comparison of silicates and silicone structures under these geometrical viewpoints: W. Noll, *Fortschr. Mineral.*, **34**, 1956, 63-84, especially pp. 72 ff. with many data on valence angles in organic siloxanes, etc.

[40] *J. Am. Chem. Soc.*, **74**, 1952, 3364-3367. On the trigonal, tetragonal, and hexagonal structure of trimeric, tetrameric, and hexameric cyclic ethyl esters of orthosilicic acid, cf. I. Fanku-

be made in the sequence of *alternating* $[SiO_4]$ and $[AlO_4]$ tetrahedra in the polymeric anion of sillimanite (see Fig. A. 2), with the ribbon-shaped arrangement $[SiAlO_5]_n^{3n-}$. It must be emphasized again, that here Si—O—Si bonds are never interlinked, only Al—O—Si bonds. The valence angle 180° is then perpendicular to the ribbon axis,

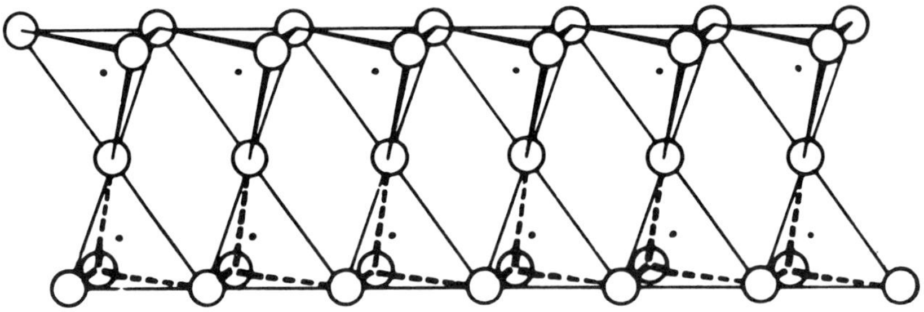

FIG. A. 2. Complex anion $[SiAlO_5]_n^{3n-}$ in the structure of sillimanite. Only $[SiO_4]$ and $[AlO_4]$ tetrahedra are interlinked.

the tetrahedron angle in the direction of this axis. The replacement of one Si^{4+} by Al^{3+} in the siloxane bonding increases the heteropolar (ionic) character of the bonding. Laves also discussed three-dimensional tetrahedra arrangements some of which are not known in silicates (or are very rare), but occur in organic polysiloxanes. The milarite group $[Si_2O_5]_6^{12-} = [RSiO_{3/2}]_{12}$ is highly probable, as in methyl polysiloxane. Urotropine-like steric groups $[RSiO_{3/2}]_4$ are well known in some organic-silicic compounds (cf. Vol. II, Section A).[41]

27. In general, the same principles for a classification of silicate structures as seen by Bragg and Laves were also used by F. Machatschki[42] who chiefly discussed "sister structures," and emphasized the parallel with the complex fluoroaluminates mentioned above, forming chain, layer, and framework arrangements. Especially for the formation of *rings* in the modifications of SiO_2, with four and six tetrahedral units, T. Zoltai and M. J. Buerger[43] made a remarkable attempt to calculate the rel-

chen, H. Steinfink, and B. Post, *Congr. Intern. Cryst.*, 1954; *Acta Cryst.*, **8**, 1955, 420-424 on octamethylcyclotetrasiloxane; G. Peyronel, *Atti accad. nazl. Lincei, Rend. Classe sci. fis. mat. e nat.*, [8] **15**, 1953, 402-408; **16**, 1954, 78-83, 231-236, on hexamethylcyclotrisiloxane.

[41] Cf. E. Wiberg and W. Simmler, *Z. anorg. u. allgem. Chem.*, **282**, 1955, 330-344.

[42] *Periodico Mineral.* (*Rome*), **22**, 1953, 23-34, with interesting proposals for structural nomenclatures, e.g., in the family of the feldspars written in the form ${}^{3}_{\infty}[A^{(10)}B_4^{(4)}C_8]_m$ to indicate the three-dimensional framework with the cations A and B of tenfold and fourfold coordination with oxygen anions, C.

[43] *Z. Krist.*, **114**, 1960, 1-8. In all of the tetrahedral structures (with the exception of isolated groups and single chains) three or more tetrahedra are linked through corners forming a closed ring when separated from other tetrahedra, and loops when linked to neighboring tetrahedra, forming part of a more extensive tetrahedral structure (personal information of T. Zoltai).

ative energies to explain the apparently high stability of such smaller complex units in comparison to larger rings with 8 or 12 tetrahedral units. The calculations were extended up to the number of 100 tetrahedral units. If in the ring plane complete rotation and horizontal mirror plane, symmetry (group $C_{nh} = n/m$) rules, the pentagonal ring is the most stable, and the hexagonal ring has nearly the same stability. Without here discussing the geometrical details, it is highly interesting to see that the sum of the interionic potential energies calculated from the elementary equation $U_{12} = e_1e_2/r_{12}$ (r_{12} is the distance of the ions 1 and 2, and e_1 and e_2 are their electrostatic charges) for the different tetrahedral ring arrangements plotted as a function of the number, n, of the ring members, indicates for the loop (A) with the symmetry n/m (full symmetry) the minimum energy for $n = 5$. For other arrangements (symmetry groups n, $\bar{n}$, $\bar{1}$), shown as (B), (C), and (D) in Fig. A. 3, the corresponding minima are seen from the curves in Fig. A. 4.

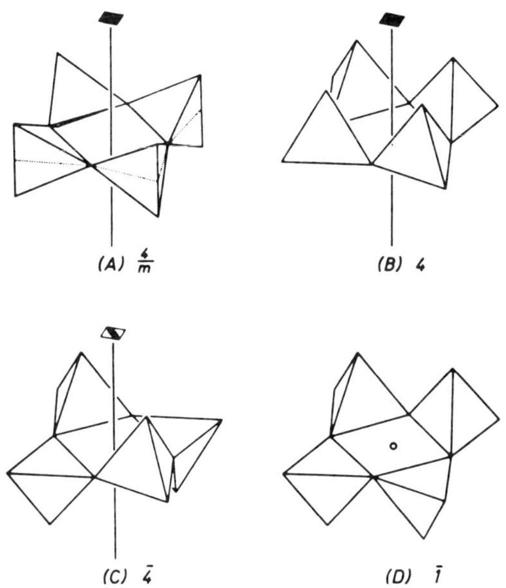

FIG. A. 3. Different tetrahedral ring arrangements (loops) for $n = 4$, in $[SiO_4]$ structures (Zoltai and Buerger). The symmetry symbols are given for the groups $4/m$, 4, $\bar{4}$, $\bar{1}$, C_{4h}, C_4, S_4, and S_2, respectively.

28. It was early recognized that in all the common $[SiO_4]$ tetrahedra complexes the interlinkage is only possible over corners from one tetrahedron to the next, not from edge to edge. That in a high degree of instability, however, this latter arrangement *can* also be verified was shown in the structure of the fluffy fine-fibrous products of silica produced by heterogeneous high-temperature gas reactions (cf. Vol.

IV, Section A), as described by Al. and Arm. Weiss,[44] with the characteristic chains

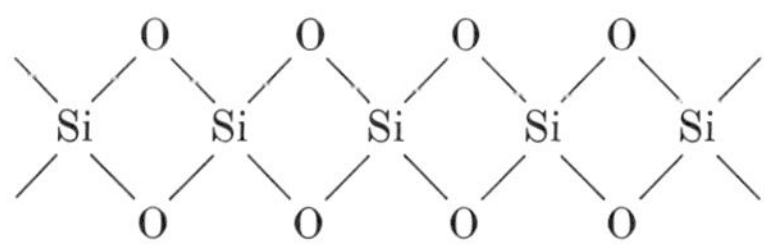

and isostructural with SiS_2 and $SiSe_2$ for which the structural type was previously determined by E. Zintl and K. Loosen (1935), an dby W. Büssem, H. Fischer, and E. Gruner (1935).[45] The extremely fine-fibrous texture of these dichalcogenides and their

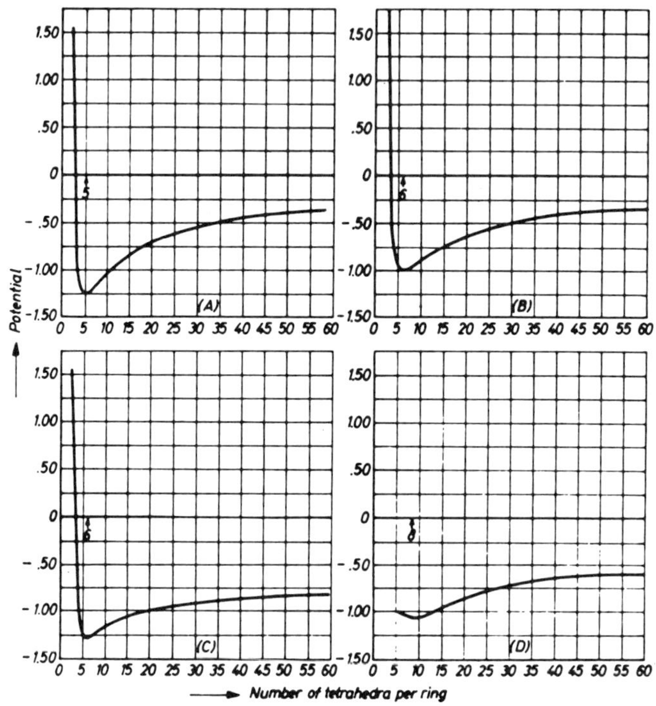

FIG. A. 4. Computed energy potentials for the ring structures shown in Fig. A. 3 (Zoltai and Buerger).

very high reactivity even with traces of water moisture are particularly striking. For $SiSe_2$ which forms a glass, it is characteristic that this vitreous phase is somewhat higher in chemical durability than the crystalline phase.

[44] *Naturwissenschaften,* **41**, 1954, 12; *Z. anorg. u. allgem. Chem.,* **276**, 1954, 95-112.

[45] See further on $SiSe_2$: Al. and. Arm. Weiss, *Z. Naturforsch.,* **7** b, 1952, 483-484, **8** b, 1953, 104-105; *Intern. Union Pure and Appl. Chem.,* **6**, 1954, 41-50. On $SiTe_2$ see *Z. anorg. u. allgem. Chem.,* **273**, 1953, 124-128.

29. *Articulated* connections in layer structures which have a particular importance for the refined interpretation of clay minerals were discussed by W. F. Bradley[46] (cf. B. ¶ 14) for dickite. They may be only shortly mentioned here as important indications for the necessity of abandoning the common assumptions on silicate structures as rigid frameworks. They are able to undergo in reality considerable elastic structural deformation and rearrangements. In another direction the investigation of structures in the disilicate group of lithium, sodium, and silver, further for petalite, by F. Liebau[47] has demonstrated the existence of orthorhombic (or pseudo-orthorhombic) symmetry with corrugated $[Si_2O_5]_\infty$ networks parallel (010).

30. Beyond the Bragg classification of the silicate structures, N. V. Belov[48] introduced a series of additional geometrical principles defining a "second 'Class' of silicate lattice types. Such are the *isomeric* chain groups of the $[SiO_3]$ type, different from that in pyroxenes, namely for wollastonite and $CuGeO_3$, or *condensed ring-chain complexes* of the type $[Si_2O_5]_\infty$ in sillimanite, $[Si_4O_{11}]_\infty$ in amphibole, and $[Si_6O_{17}]_\infty$ in xonotlite. We will discuss these structures more in detail in A. ¶ 109, and therefore give here only some more general viewpoints emphasized by Belov, namely the fact that magnesium, iron, and aluminum silicates prefer the simple (Bragg) chains in pyroxenes and amphiboles, and the sillimanite structure type, whereas calcium silicates appear preferably in the wollastonite or xonotlite chains of Class II. The next step of condensation is that to *networks* with hexagonal $[Si_2O_5]_\infty$ in the magnesium-, iron-, or aluminum-containing layer structures of talc, micas, and pyrophyllite. In calcium silicates of this group appear the Class II polymeric groupings of okenite and apophyllite (with complex octogonal and pentagonal, or tetragonal and octogonal ring combinations, respectively). The distinction between the Bragg Class I and the Belov Class II structures is very important also for the differentiation cycles in nature in which the melanocratic magnesium-iron-aluminum minerals silicates belong to Class I, and the leucocratic calcium and alkali silicates and aluminosilicates belong to group II. The end state of condensation in Class II are the SiO_2-type minerals. It was recently made evident by Belov[49] that even quartz belongs to the Class II type because of the presence of typical xonotlite-type arrangements in the rhombohedron face. The distinction between the two classes of

[46] *Proc. Natl. Conf. on Clays and Clay Minerals*, **6**, 1959, 18-25.

[47] *Acta Cryst.*, **14**, 1961, 389-395 ($Li_2Si_2O_5$); 395-398 (α-$Na_2Si_2O_5$); 398-406 (petalite); *Silikat Tech.*, **11**, 1960, 396.

[48] *Zapiski Vsesoyuz. Mineral. Obshchestva*, **86**, 1957, 239-244; see also *Kristallografiya*, **2**, 1957, 366-370; *Khim. Nauka i Prom.*, **3**, 1958, 46-56; *Silikat Tech*, **11**, 1960, 389-390; **12**, 1961, 329-330.

[49] *Zapiski Vsesoyuz. Mineral. Obschchestva*, **90**, 1961, 168-171. Belov speaks of "okenite" networks in quartz. This is not entirely correct. One may with better justification compare the arrangement of the octogonal nets with those in xonotlite (see below), on a higher degree of condensation.

structures is particularly evident in the comparison[50] shown in the accompanying tabulation.

In Class I (Bragg Structures)			
Pyroxene chains:	Amphibole ribbons:	Mica networks:	
$[Si_{1+1}O_6]_\infty$	$[Si_{2+2}O_{11}]_\infty$	$[Si_2O_5]_\infty^\infty$ with hexagonal rings	
In Class II (Belov Structures)			
Wollastonite chains:	Xonotlite ribbons:	Okenite networks:	Apophyllite networks:
$[Si_{2+1}O_9]_\infty$	$[Si_{4+2}O_{17}]_\infty$	alternating 5- and 8-membered rings	alternating 4- and 8-membered rings

See also Figs. A. 5 and 6.

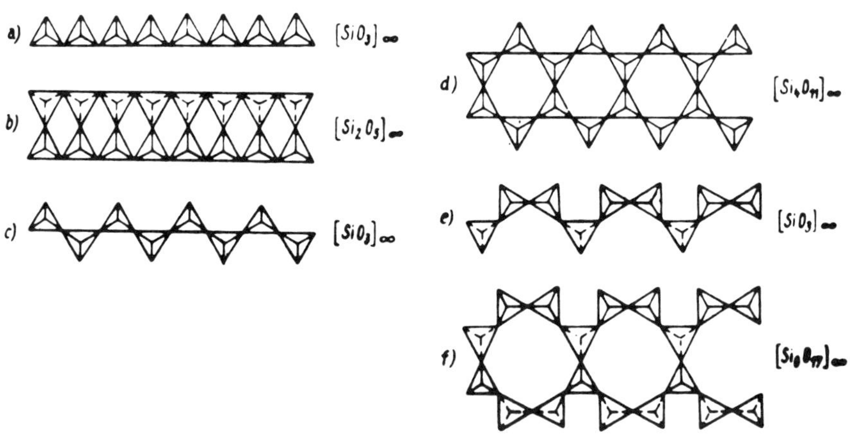

FIG. A. 5. Infinite chains and bands in silicate structures as condensation products of simple chains: (*a*) in $CuGeO_3$; (*b*) of ribbons in sillimanite; (*c*) a pyroxene chain; (*d*) an amphibole ribbon; (*e*) a wollastonite (pyroxenoid chain); and (*f*) a xonotlite ribbon (Belov).

31. Additional more complex condensation products of Class II will be discussed later for the structures of epididymite and narsarsukite (cf. A. ¶ 117). For complex three-dimensional *framework* structures the geometrical principles of Class II are of particularly high importance in the correct interpretation of the zeolites, including

[50] On the xonotlite structure see Kh. S. Mamedov and N. V. Belov, *Zapiski Vsesoyuz. Mineral. Obshchestva*, **85**, 1956, 13-38. A new type silicate radical was found by Yu. A. Pyatenko and A. A. Voronkov, *Doklady Akad. Nauk S.S.S.R.*, **141**, 1961, 958-961, in the zirconosilicate mineral *vlasovite*, $Na_2ZrSi_4O_{11}$, with double chains, $[Si_4O_{11}]_\infty$, which are different from those occurring in amphiboles, cf. similar configurations in the structure of narsarsukite (cf. A. ¶ 117).

the industrially highly useful group of the synthetic "molecular sieves" (see A. ¶¶ 207, 210, 215 ff., and Vol. IV. B. ¶¶ 57 ff., 63 ff.).

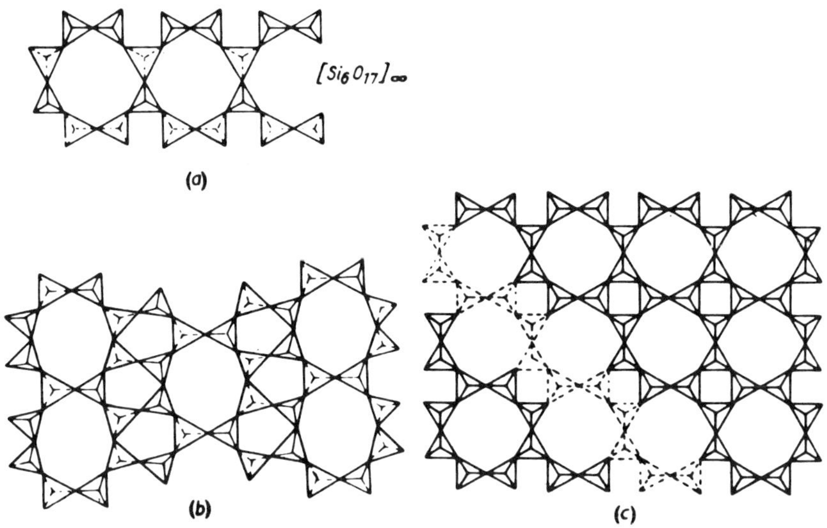

FIG. A. 6. (*a*) Condensation of pyroxenoid chains (of wollastonite-type) in $[Si_6O_{17}]_\infty$ units of xonotlite; (*b*) development of apophyllite network (with additional bridging groups $[Si_2O_7]$; (*c*) development of the okenite networks by condensation of xonotlite ribbons over gliding planes (Belov).

32. Concerning the formation of *aluminosilicates*, which are commonly interpreted by a partial replacement of Si^{4+} cations in a silicate framework by Al^{3+} ions, quite distinct rules are valid which were outlined by M. Balconi,[51] and especially by W. Loewenstein.[52] Starting from the "adjusted standard" of 24 O^{2-} anions per structural unit, Balconi derived empirically some limit ratios for this substitution, namely $Al^{3+}/Si^{4+} = 1/4$ in chain structures; $= 1/2$ in layer structures. In pyroxenes of *magmatic* origin this ratio may even be only 1 : 8.3 (with exceptions in Ti-augites); in amphiboles it may be above 1 : 3, as much as 2.5 : 5.5, and in biotite micas 2.7 : 8. For Ti^{4+} the analogous replacement is much restricted, mostly below 0.4 : 8. In the brittle micas of the *metamorphic* rocks the Al^{3+}/Si^{4+} ratio sometimes reaches 1 : 1, and even 2 : 1, making evident the particular crystallochemical nature of metamorphic mineralization. For the important minerals of the feldspar and feldspathoid groups the ratios Al^{3+}/Si^{4+} are distinctly connected with their chemical stability, as usually characterized in the acid solubility of such structures, including also the zeolites like analcime, pollucite and related minerals.

[51] *Atti accad. nazl. Lincei, Rend. Classe sci. fis. mat. e nat.*, [8] **15**, 1953 (6), 427-432; **16**, 1954 (1), 101-107.

[52] *Am. Mineralogist*, **39**, 1954, 92-96.

33. Of particular importance are the more general structural rules established by W. Loewenstein of which we emphasize the two following:

(1) If two tetrahedra are linked together by one bridging oxygen anion, the center of only one of them can be occupied by Al^{3+} (see above in sillimanite chains); the other center must be occupied by Si^{4+}, or by another related cation, e.g., by P^{5+}.

(2) When two aluminum ions are neighboring one and the same oxygen anion, at least one of them must have a coordination number with oxygen anions higher than four (e.g., 5 or 6, as seen in the structures of the Al_2SiO_5 minerals).

These basic rules illustrate the maximum substitution of 50 per cent of Si^{4+} by Al^{3+} in three-dimensional frameworks and in plane networks of tetrahedra. The 50 per cent maximum of replacement also rigorously postulates the alternation of $[SiO_4]$ and $[AlO_4]$ tetrahedra. The importance of those rules in the structures of the plagioclase feldspars will be discussed in more detail in A. ¶ 195 ff., in connection with the problems of order-disorder phenomena.

STRUCTURE INVESTIGATIONS BY INFRARED SPECTROMETRY METHODS*

34. As one of the most promising tools in structural investigations of silicates and aluminosilicates besides the X-ray diffraction methods, the *infrared absorption* (and reflection) spectrometry was developed rapidly. Fr. Dachille and R. Roy[53] recently demonstrated that by infrared absorption studies one is able to distinguish major changes in coordination, e.g., in the series of the SiO_2 modifications and of related structures, as of GeO_2, of the quartz-type compounds $AlPO_4$, BPO_4, $GaPO_4$, $FePO_4$, $BAsO_4$, $AlAsO_4$, $GaAsO_4$, and the compound $GaSbO_4$ which is isostructural with rutile. In the range from 2 to 25 μ (Fig. A. 7) all of the important cation-anion

[53] *Z. Krist.*, **111**, 1959, 462-470. An analogous examination of the infrared absorption spectra of common B_2O_3, and of the dense modification of B_2O_3 that is stable under very high pressures, however, showed a systematic increase in the average effective coordination from 3.2 to 3.4, and a high variability of the distances B—O, equivalent to the number of different types of distortion of tetrahedra; cf. Fr. Dachille and R. Roy, *J. Am. Ceram. Soc.*, **42**, 1959, 78-80.

* *Note added in proof:* A recent addition to the literature is the valuable collection of 355 charts with reproductions of infrared absorption spectra: H. Moenke, "Mineralspektren, aufgenommen mit dem Jenaer Spektralphotometer UR 10," (published under the auspices of the German Academy of Science, Akademie-Verlag, Berlin, 1962). This thesaurus of spectrographs is very useful in connection with the simultaneously published "Minerals in the Infrared, a Critical Bibliography," by R. J. P. Lyon (76 pp; Stanford Research Institute, Menlo Park, California, 1962); we refer to these works as essential aids in infrared absorption studies of silicate minerals.

stretching, cation-cation stretching, and various cation-anion bending and distorting vibrations are recognized. Highest in frequencies with the strongest absorptions is the cation-anion stretching. When there is a change of primary coordinations of the

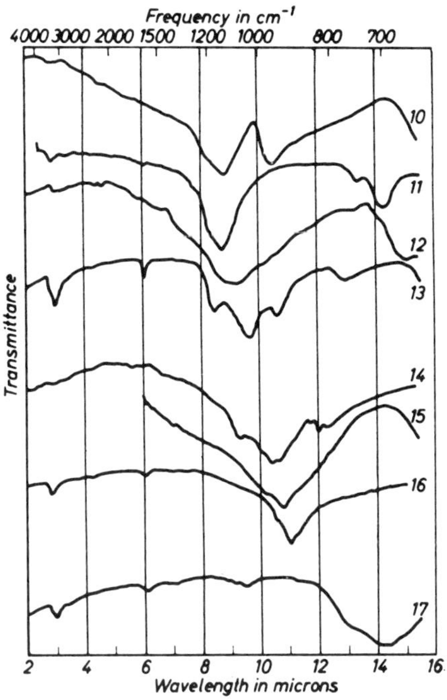

FIG. A. 7. Infrared absorption spectra of the quartz-type forms of BPO_4 (curve 10); $AlPO_4$ (curve 11); $GaPO_4$ (curve 12); $FePO_4$ (curve 13); $BAsO_4$ (curve 14); $AlAsO_4$ (curve 15); $GaAsO_4$ (curve 16); and of the rutile-type of $GaSbO_4$ (curve 17) (Dachille and Roy).

central cation, this process is clearly indicated in the main stretching frequencies. The ratio of the squares of the wave-lengths is, e.g., in the change from $[GeO_4]$ to $[GeO_6]$ (cf. Fig. A. 8) = 1.48 (near to 6/4). The secondary coordinations do not considerably influence on the *main* cation-anion stretching absorption band, even if major changes in physical parameters occur. In Fig. A. 9 the considerable change in density for the transition of quartz into coesite appears only of little influence on the infrared absorption. There is a stray packing effect but the coordination type remains the same (shown also in the BeF_2 "model" structure). The interrelation of coordination (CN), the valence $Z_A . Z_X$ of the cation A and the anion X, a reduced mass term, μ, the cubic root of the sum of the atomic numbers A_A and A_X, and the wavelength of the infrared absorption band, λ, gives in the expression

$$K = \frac{(CN)\mu\,(A_A + A_X)^{1/3}}{Z_A . Z_X . \lambda^2}$$

a valuable characteristic for the special structure type. So $K = 0.168 \pm 002$ for quartz type structures is *in sensu stricto.* These relations are less accurately fulfilled for the derived quartz structures of the compounds ABX_4, as a consequence of higher complication effects, and of superimposed order-disorder relations. It is also in-

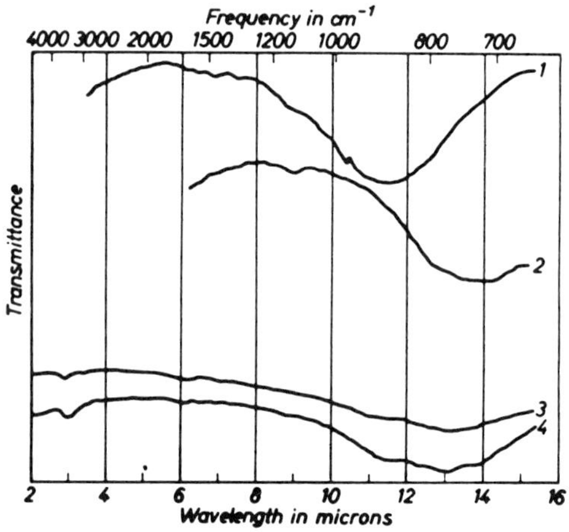

FIG. A. 8. Infrared absorption spectra of GeO_2 (quartz-type) (curve 1); of GeO_2 (rutile-type) (curve 2); BeF_2 (quartz-type) (curve 3); BeF_2 (coesite-type) (curve 4) (Dachille and Roy).

teresting that the refractive index, and the molecular refractivity, R_m, of quartz and rutile type structures follow for the amount of the quotient $R_m/(A.B)X_2$ (for compounds AX_2 and ABX_4 including) a similar approximate constancy with characteristic changes (above a tolerance of 12 per cent) when the coordination changes from 4 to 6 for oxygen anions.

35. Rich experimental data material is available on the determination of infrared absorption and reflection spectra, not only for the silica modifications but also for many aluminosilicate structures. We emphasize the reflection measurements of I. Šimon and H. O. McMahon,[54] using X, Y, and Z cuts (parallel to the c-axis and perpendicular to a diagonal symmetry axis for X; parallel to the c-axis and parallel to a digonal axis for Y; and perpendicular to the x-axis in Z cuts) of standard piezoelectric rock crystal, further at elevated temperatures for α-quartz in Z-cut, for β- and α-cristobalite, and for silica glass, with ordinary and polarized light, in the range from 700 to 1400 cm.$^{-1}$ (14.2 to 7.15 μ), under different angles of incidence. The most

[54] *J. Chem. Phys.*, **21**, 1953, 23-20. On the infrared absorption spectra of tektosilicates (framework structures) see also the investigations and spectral curves presented by R. G. Milkey, *Am. Mineralogist*, **45**, 1960, 990-1007. These data are specifically valuable supplementing the charts of mineral spectra as offered in the compilation by H. Moenke, mentioned in the note added in proof of the present chapter.

important result is the establishment of valence stretching, $\overleftarrow{\mathrm{Si}}$—$\overrightarrow{\mathrm{O}}$—$\overrightarrow{\mathrm{Si}}$ vibration at 1095 cm.$^{-1}$ (9.15 μ), and the functional relation of temperature (between 4° and 880°K.) with intensity and width of the bands. Particularly careful reflection measurements were made by N. A. Sevchenko and V. A. Florinskaya[55] for quartz (par-

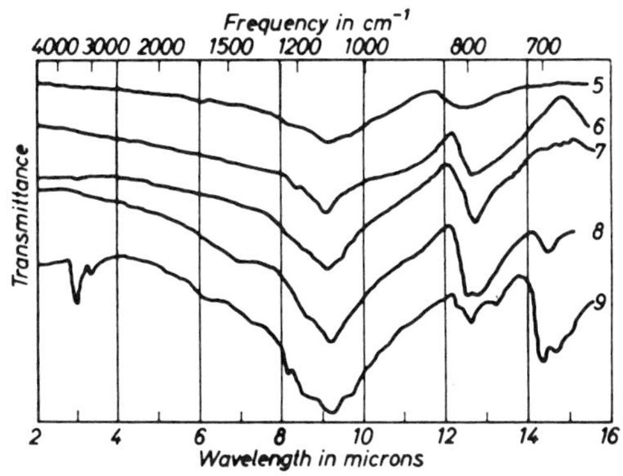

FIG. A. 9. Infrared absorption spectra of silica glass (curve 5); of cristobalite (curve 6); of tridymite (curve 7); of quartz (curve 8); of coesite (curve 9) (Dachille and Roy).

allel and perpendicular to the *c*-axis), cristobalite, and silica glass (Heraeus, Hanau), and also with transmittance curves of the powders of quartz, synthetic tridymite and cristobalite, in a 1 μ thick layer, applied on one face of a KCl prism, and of silica glass (Fig. A. 10).

36. Very precise measurements of transmission spectrograms for β-quartz, β-tridymite, β-cristobalite, coesite, and silica glass, for quartz- and rutile type GeO_2, and GeO_2 glass were made recently in the range from 4000 to 300 cm.$^{-1}$ (25 to 3.35 μ) by E. R. Lippincott, A. Van Valkenburg, Ch. S. Weir, and E. N. Bunting,[56] and further by B. D. Saksena,[57] with the aim of giving a precise assignment on the group frequencies at 1100, 800, and 480 cm.$^{-1}$ (9.10, 12.5, and 21.5 μ), corresponding to a stretching mode involving displacements primarily associated with the oxygen atoms, a stretching mode of similar type for the Si atoms, and a Si—O bending mode. Since these same frequencies are characteristic of coesite equally, the [SiO_4] tetrahedral coordination is certain for this mineral (see above), combined with a high packing density and an angle Si—O—Si of only 120°. The vibrations assigned to

[55] *Doklady Akad. Nauk S.S.S.R.*, **109**, 1956, 1115-1118; see also V. A. Florinskaya and R. S. Pechenkina, *Trudy Soveshchaniya Stroeniya Steklo*, 1953/55, 70-95, especially pp. 70-79.

[56] *J. Research Natl. Bur. Standards*, **61**, 1958, 61-70.

[57] *Proc. Phys. Soc. London*, **72**, 1958, 9-16. Saksena examined besides the X, Y, Z cuts of rock crystal, the V cut of crystal resonators corresponding to the U.S. Patent of Radio Corporation of America, and also quartz powder in pressed pellets with KBr.

GeO_2 (quartz type) are analogous to those in SiO_2 quartz, considering the higher mass factor for the germanium atom. Saksena discussed with great care the range between 12 and 13 μ (835 and 770 cm.$^{-1}$), with finer details of differences between Z and X quartz cuts. It is striking that the 12.32 μ (812 cm.$^{-1}$) band is absent

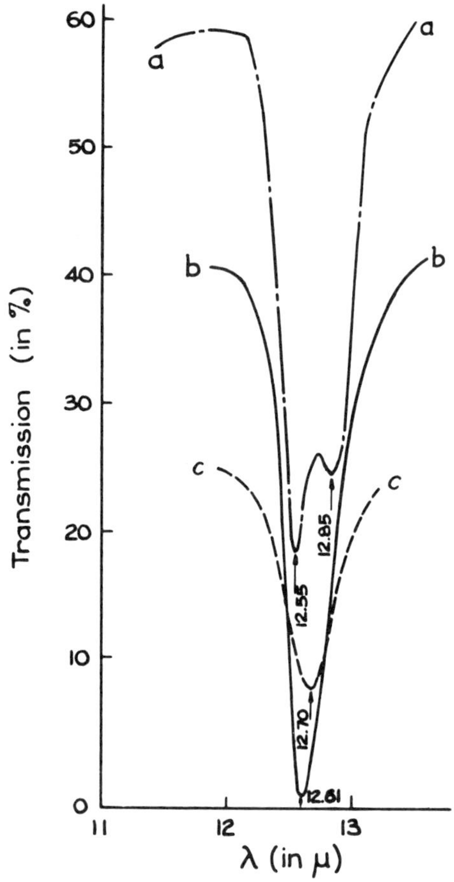

FIG. A. 10. Infrared absorption spectra of quartz (*a*); of cristobalite (*b*) and of tridymite (*c*), in the range from 12 to 13 μ (Florinskaya and Pechenkina).

in the spectra of powdered quartz although the bands at 12.5 μ (800 cm.$^{-1}$) and 12.9 μ (775 cm.$^{-1}$) are present. Interesting is also the disappearance of the band at 14.38 μ (695 cm.$^{-1}$) with the X-cut crystal, independent of the orientation of the sample in relation to the electric vector of the polarized beam. This fact cannot be explained by the previous assignments.

37. Among the framework aluminosilicate structures the *feldspars* were intensely investigated in their infrared characteristics, especially the plagioclases by O. Sh.

Thompson and M. E. Wadsworth[58] to find a connection between the absorption peaks with the chemical composition. Such a functional relation is evident in the albite-anorthite series; even a characteristic shift is observed in the structural type indicated in the critical ranges between 28 and 33, and 67 and 85 per cent anorthite (cf. Vol. III, Section B) of which, however, only the first shift is indicated in infrared characteristics. For diagnostic purposes the range of 15.6 to 16.2 μ may be the optimum for an approximate determination of the plagioclase composition (cf. Fig. A. 11), and of

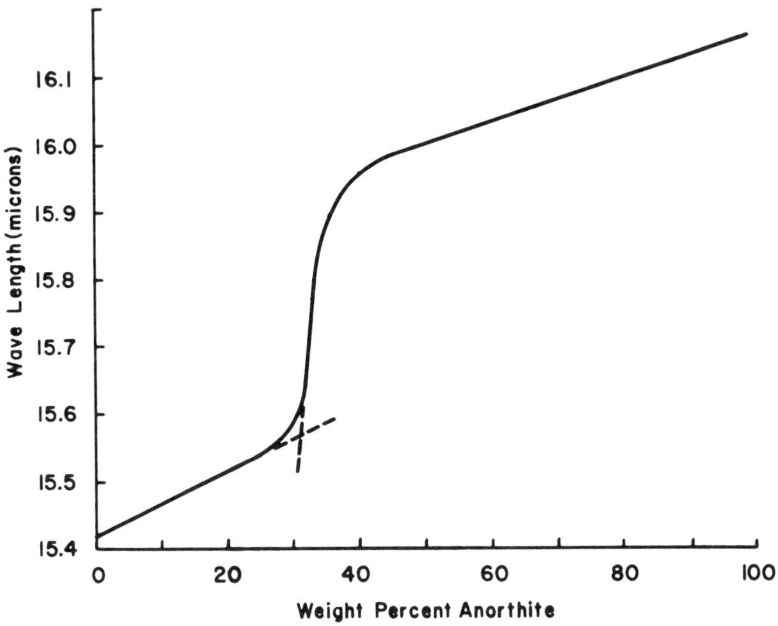

FIG. A. 11. Correlation of the wavelength and composition of plagioclase feldspars for the infrared absorption band in the range from 15.4 to 16.2 μ (Thompson and Wadsworth).

the ratio Si/Al. Thompson and Wadsworth confirmed the interrelations found by Saksena (1940) between the vibrational frequencies of plagioclase and quartz in the range from 8.5 to 11 μ (1180 and 910 cm.$^{-1}$), but not for the 15.4 to 16.2 μ (640 to 618 cm.$^{-1}$) range.

38. Whereas these authors have not been able to find indications of order and disorder effects in the infrared absorption spectra, F. Laves and St. Hafner[59] gave a full explanation for the differences in the low- and high-temperature modifications of feldspars and their Al/Si distribution in the region of 7 to 16 μ of infrared spectro-

[58] *Am. Mineralogist*, **42**, 1957, 334-341.

[59] *Z. Krist.*, **108**, 1956, 52-63; **109**, 1957, 204-225. On the triclinicity of potassium feldspar see A. L. Kosoĭ, *Zapiski Vsesoyuz. Mineral. Obshchestva*, **91**, 1962, 618-620.

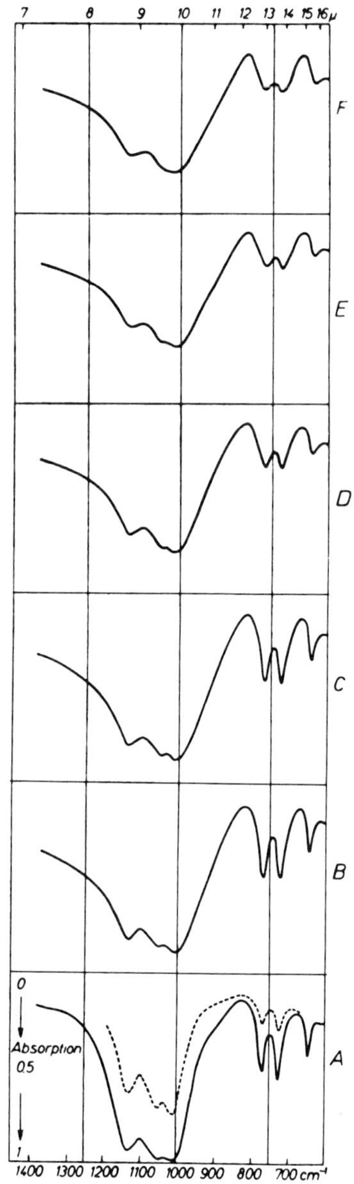

FIG. A. 12. Infrared absorption curves of natural microcline (A), and the same after thermal exposure at 1050°C. for 48 hours (B); 144 hours (C); 240 hours (D); 336 hours (E); 720 hours (F) (Laves and Hafner).

metry (cf. Vol. III, Section B). The studies with alkali feldspars concerned the classical case of the thermal response of microcline to form sanidine by heating, with a characteristic change in the absorption curves (Fig. A. 12) in which especially the disappearance of the 15.4 μ (650 cm.$^{-1}$) band and a shifting towards 15.8 μ (630 cm.$^{-1}$) indicates the increasing degree of structural disorder. Even in sanidine heated for 4 days the peak at 15.4 μ is still perceptible, whereas it definitely disappears in the melt (Fig. A. 13). Adularia is a typical intermediate feldspar and may

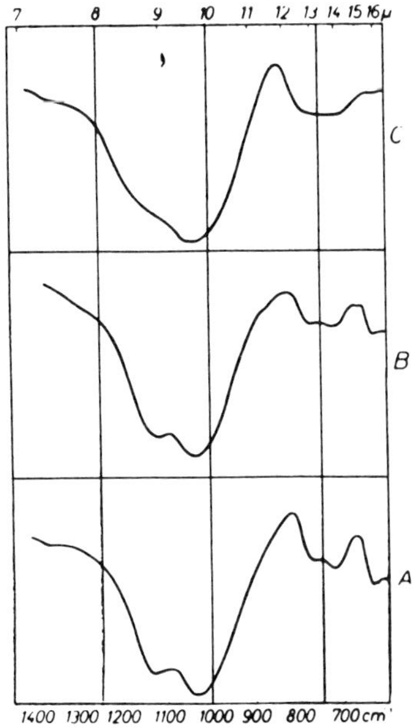

FIG. A. 13. Infrared absorption curves of sanidine from the Kaiserstuhl (*A*); and the same heated at 1050°C. for four days (*B*), in comparison with the curve for the potassium feldspar melt (*C*) (Laves and Hafner).

in natural occurence in the Swiss Alps be closer to microcline, and so fairly well ordered. The low-temperature microcline and albite diagrams are very different, whereas those of sanidine and analbite are very similar. Anorthite is a highly ordered crystal phase, with ordered Al/Si distribution up to high temperatures. Nevertheless, in the plagioclases there are characteristic differences in the low- and high-temperature series, which are well recognized in the infrared absorption curves (Fig. A. 14). They indicate disorder in the high-temperature phases approximating the random distribution in the melts. The effects perceptible in the infrared behavior are in every case larger than those derived from X-ray diffraction methods.

39. More refined investigations of Laves and Hafner deal with variations of the position and intensity of the infrared absorption bands, especially of orthoclase and adularia, as a function of the structural state of order, and of the chemical composition. These are to be distinguished by different degrees of approximation to the thermodynamic equilibrium (cf. Vol. III, Section B), such as by stability problems which are particularly critical in the intermediate phases between microcline and sanidine (orthoclase and adularia as submicroscopically twinned pseudomonoclinic

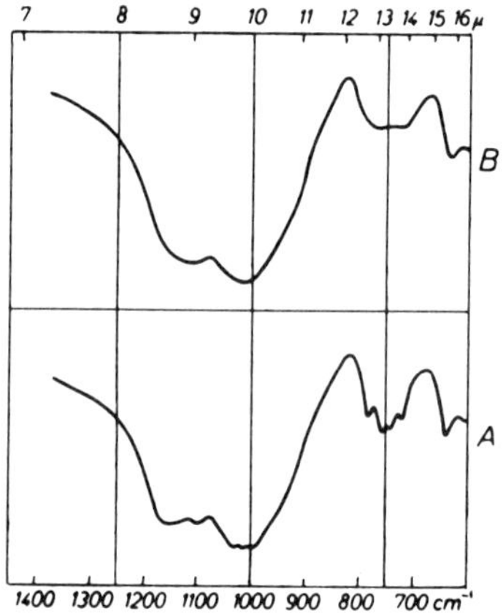

FIG. A. 14. Infrared absorption curve of oligoclase feldspar (with 15 per cent anorthite), in the natural state (A), and after thermal exposure at 1000°C. for six weeks (B) (Laves and Hafner).

crystal phases). For the plagioclases the distinctions between low- and high-temperature modifications are well recognized in the variations of the infrared absorption bands between 15 and 16, and 18 and 19 μ, as a function of the composition (Fig. A. 15), from albite to anorthite from which the relative absorption diagram (Fig. A. 16) is derived. The corresponding diagram for the low- and high-temperature alkali

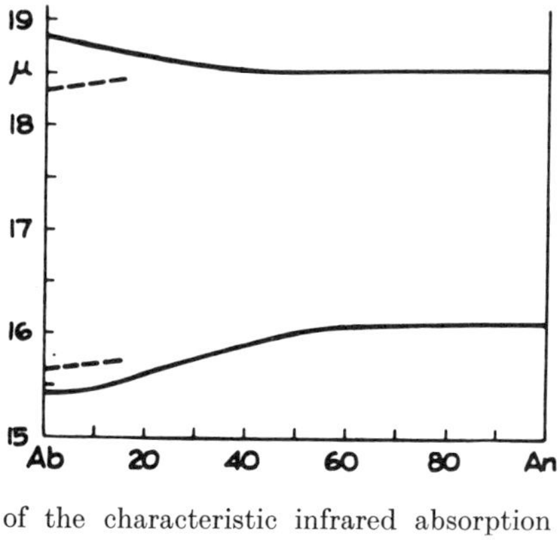

FIG. A. 15. Variation of the characteristic infrared absorption bands of the plagioclases with chemical composition (Hafner and Laves). The curves of the low-temperature modifications are solid lines, those for the high-temperature forms dotted.

feldspars is seen in Fig. A. 17. Fig. A. 18 shows the absorption bands of microcline and albite crystalline solutions as a function of the composition for the ranges between 15.4 and 15.8 μ, 18.2 and 18.9 μ, and 12.7 and 13.9 μ; the splitting of the curves in the latter diagram indicates the transition from microcline toward albite at

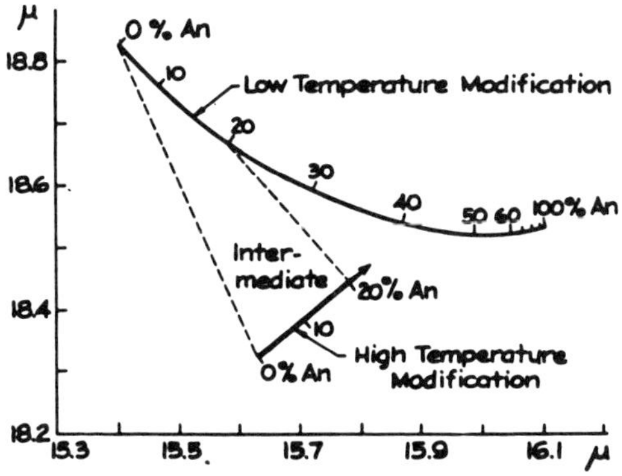

FIG. A. 16. Mutual relation of the infrared absorption bands shown in Fig. A. 15, for low-and high-temperature modifications of the plagioclases (Hafner and Laves).

30 : 70 per cent, in agreement with the analogous splitting in the curves for the angles α^* and γ^* of the reciprocal lattice (cf. J. R. Goldsmith and F. Laves[60]). Among the far-reaching consequences of the results of the investigations by Laves and Hafner perhaps the most important one is the experimentally positive evidence of the pseudomonoclinic nature of intermediate orthoclase. This is in agreement

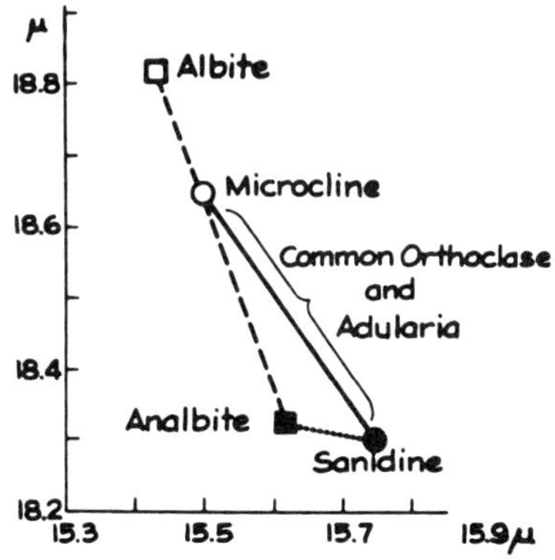

FIG. A. 17. The same as Fig. A. 16 for alkali feldspars in their high- and low-temperature modifications (Hafner and Laves).

[60] *Fortschr. Mineral.*, **35**, 1957, 18.

with the early hypotheses of E. Mallard (1876), and confirms Laves' theories of the degree of triclinicity of the feldspars in its significance for mineral genesis.

40. The particular value of infrared spectrometry for the rapid and reliable identification of silicate minerals in rocks, especially of the sedimentary group, was emphasized in many publications, with a wealth of special numerical data and adsorption (or reflection) diagrams. We mention here especially those of J. M. Hunt and D. S. Turner,[61] and of W. D. Keller and E. E. Pickett[62] in which the clay minerals

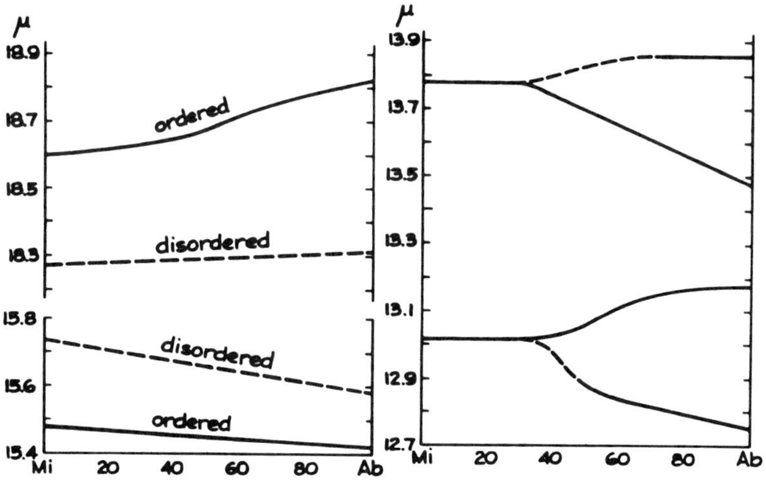

FIG. A. 18. Infrared absorption of microcline-albite crystalline solutions in different wavelength ranges, as a function of the order-disorder state, and showing the splitting by the transition from microcline to albite at 30 : 70 per cent (Hafner and Laves).

[61] *Anal. Chem.*, **25**, 1953, 1169-1174. We also mention in this connection the important identification and accurate quantitative determination of quartz in mine dusts; cf. M. Gade and K. F. Luft, *Naturwissenschaften*, **46**, 1959, 315-316.

[62] *Am. J. Sci.*, **248**, 1950, 264-273; *Ibid.* **250**, 1952, 453-471 (with J. S. Spotts and D. L. Biggs); also E. S. Rudnitskaya, *Zapiski Vsesoyuz. Mineral. Obshchestva*, **85**, 1956, 407-412, with practical details, O. N. Setkina, *Ibid.*, **88**, 1959, 39-47, with extensive tabular data and different applications in ceramics, etc. For the special requirements in ceramics and hydraulic binder investigations see H. Dutz, *Tonind.-Ztg. u. Keram. Rundschau*, **82**, 1959, 219-238, with excellent tabulated data and bibliography up to 1959. For ore and salt mineral deposits H. Moenke gave a series of instructive details of infrared spectrometric practice in *Jenaer Jahrb.*, 1959, I, 154-180 (with P. Görlich and L. Moenke-Blankenburg); *Ibid.*, 1959, II, 361-395; *Ibid.*, 1960, II, 402-466; recently in *Silikat Tech.*, **12**, 1961, 323-327, and *Carl Zeiss Nachr. Jena*, N.F. **9**, 1961 49-81; 82-90. On clay minerals in their infrared characteristics see finally B. M. Bishui and J. Prasad, *Central Glass & Ceram. Research Inst. Bull.* (*India*), **7**, 1960, 97-109. On infrared absorption spectra of typical nesosilicates see H. Moenke, *Silikattechnik*, **13**, 1962, 246-248. The same author (*ibid.* 287-288) makes evident distinguishing infrared peak displacements in minerals containing $[BO_3]$ and $[BO_4]$ coordination groups (cf. A. ¶ 56). This possibility is specifically important for the correct interpretation of the structure of *axinite* (cf. A. ¶ 101).

are particularly considered and which will be discussed more particularly in B. ¶ 29. The powder immersion in Nujol is one of the standard methods for such investigations, although embedding in pressed pellets of KBr has become increasingly popular because there are no disturbing absorption bands of this alkali halide to be considered in the common range of infrared studies. Keller and Hunt's results confirm the previous experience that in infrared absorption the anions are the dominant factors; the heavier the cations present, the more the bands are shifted to longer wavelengths. The replacement of Si^{4+} by Al^{3+} is generally recognized by shifting in the same direction. The classification of the silicates as used by Keller *et al.* strictly follows the Bragg-Strunz (1930, 1937) system progressively discussing the isle, group, chain, layer, and framework structures (neso-, soro-, cyclo-, ino-, phyllo-, and tektosilicates). Regularities in the infrared absorption spectra as an expression of the structural types were also discussed by Ph. J. Launer[63] confirming the general experience that the region of strongest absorption shifts to shorter wavelengths as the ratio of Si/O increases. This is a rule of considerable significance in the differentiation sequence of magmatic minerogenesis (Bowen's principle). With the shifting to higher frequencies is connected a tendency to narrowing wavelength ranges characteristic of a given silicate, as it was observed previously by Cl. Schaefer, F. Matossi, and K. Wirtz (1934).

41. A comparison of (near) infrared absorption with that in the visible range, and in the long-wave ultraviolet for olivine, diopside, and different garnets, was made by S. P. Clark.[64] The high "transparency" of olivine and diopside for heat radiation is in his opinion, a factor of great importance to temperature distribution in the inner layers of the Earth's crust. The role of the *iron* contents in rock-forming minerals must not be defined as "impurities" of the mineral composition, but as "excitons" with specific approximations of semiconductor properties in crystalline solutions. There is an essential widening of the energy levels, either in the valence or in the conduction band at compositions near the iron-rich silicates like fayalite and hedenbergite-grünerite. In iron-rich garnets of almandine and andradite type, such intrinsic semiconduction effects are particularly fascinating. The characteristic peak for the presence of Fe^{II} is that of 8000 cm.$^{-1}$ (1.25 μ) which plays an equally important role in glass colors (cf. Vol. II, Section B). Another peak at 20,000 cm.$^{-1}$ (500 mμ) is assigned to the Fe^{3+} ion which behaves similarly as the Ni^{2+} does in glass (cf. W. A. Weyl, 1950). Also P. Tarte[65] studied the infrared absorption in garnets, especially those containing *titanium* (melanite, schorlomite, jivaarite), with one or two characteristic bands in the range of 350 to 280 cm.$^{-1}$ (28.5 to 36 μ) in the pyral-

[63] *Am. Mineralogist*, **37**, 1952, 764-784.

[64] *Am. Mineralogist*, **42**, 1957, 732-742.

[65] *Nature*, **186**, 1960, 234; extensively discussed in *Silicates inds.*, **25**, 1960, 171-175.

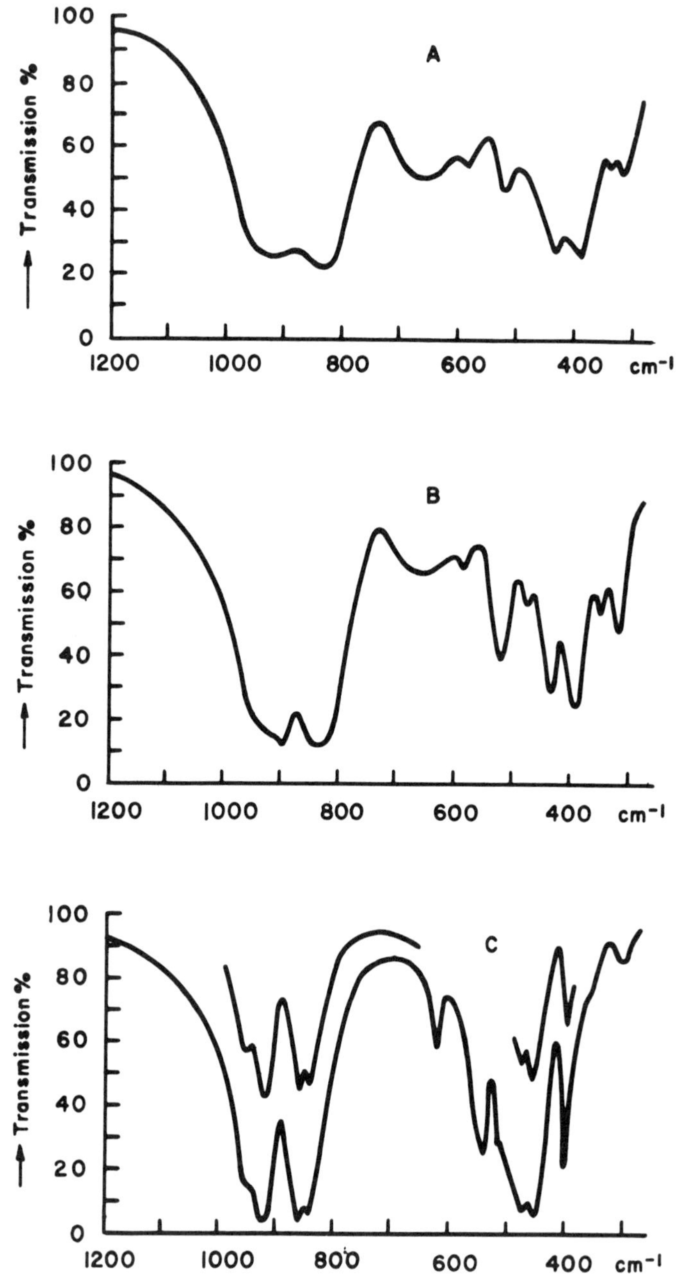

FIG. A. 19. Infrared transmission curves of titanium garnets: jivaarite (*A*), schorlomite (*B*), and grossularite (*C*) (Tarte).

site garnets. The ugrandites do not show this group. They are assigned to divalent cations Fe^{2+}, Mn^{2+}, Mg^{2+}, in $[RO_8]$ coordination whereas Ti^{4+} occurs in $[TiO_4]$.

42. The titanium-free garnets (grossular, spessartine, almandine, and uvarovite) have nearly all the same spectra profiles in spite of considerable variations in the isomorphous substitution of Mg^{2+}, Fe^{2+}, Mn^{2+}, or in Al^{3+}, Fe^{3+}, Cr^{3+} (Fig. A. 19). With increasing titanium contents (in schorlomite up to 5.8 per cent, in jivaarite up to 10.2 per cent) the spectra become remarkably simpler, with a characteristic band at 650 cm.$^{-1}$ (15.4 μ), asisgned to Ti—O vibrations analogous to those in aliphatic orthotitanate esters $Ti(OR)_4$.[66] For *meta*-titanates, $RTiO_3$, the infinite $[TiO_6]_\infty$ chains cause the band at 500 to 600 cm.$^{-1}$ (20 to 16 μ). It is characteristic that the difference of frequencies are the same in *ortho*- and *meta*-titanate, and *ortho*- and *meta*-silicate spectra. In the same way, as was observed by Laves and Hafner for the aluminosilicates of the feldspar group for the effect of a substitution of $[SiO_4]$ by $[AlO_4]$, the titanium garnets show a substitution of $[SiO_4]$ by $[TiO_4]$, with a typical widening of the spectra profiles as seen in the range from 800 to 1000 cm.$^{-1}$ (12.5 to 10 μ). $[TiO_6]$ groups, which may also occur in titanium garnets, do not show independent infrared bands, or these are masked by others. In titaniferous micas with up to 7.7 per cent Ti the substitution is probably also TiO_2 for SiO_2, but in micas with 10 to 12 per cent Ti, it is Ti_2O_3 in $(Al,Fe)_2O_3$. It is highly improbable that Ti_2O_3, however, coexists with Fe_2O_3 in such crystalline solutions since the formation of ilmenite, $FeO.TiO_2$, would prevail as a segregated phase.

43. In the crystalline solutions of the orthosilicates with *ortho*-germanates of magnesium, baryum, and zinc, P. Tarte[67] studied the antisymmetrical valence vibrations in the range from 800 to 1100 cm.$^{-1}$ (12.5 to 9.1 μ), involving four absorption bands corresponding to the ν_3 fundamental oscillations. As soon as the $[SiO_4]$ groups are replaced by $[GeO_4]$ the bands appear widened and their intensities are reduced. In a crystalline solution with 99 Ge^{4+} : 1 Si^{4+} in the tetrahedral groups, the three characteristic bands of a sufficiently deformed $[SiO_4]$ appear for the phases of Mg, Ba, and Zn germanate type, but only one band for the Ni-germanate phase (of spinel structure type) (Figs. A. 20 and 21). The structural disorder of such crystalline solutions causes the typical widening of the absorption bands.

44. In continuation of the previous investigations of Cl. Schaefer, F. Matossi, and K. Wirtz (1934) on the behavior of the different structural groupings of the silicates in the Bragg-Strunz systematics, B. D. Saksena[68] recently developed a

[66] Cf. H. Kriegsmann and K. Light, *Z. Elektrochem.*, **62**, 1958, 1163-1174.

[67] *Bull. classe sci. Acad. roy. Belg.*, [5] **46**, 1960, 169-179. See further P. Tarte and A. E. Ringwood, *Nature*, **193**, 1962, 971-972, and P. Tarte, *Spectrochim. Acta*, **18**, 1962, 467-489, and **19**, 1963, 25-17, 49-71.

[68] *Trans. Faraday Soc.*, **57**, 1961, 242-258; *Proc. Symposium on Raman and Infrared Spectroscopy, Nainital, India*, 1959, pp. 93-102.

much refined discussion of these fundamentally important principles, together with extensive tabulations of the assignments of the absorption bands based on the theoretical deductions of vibration mechanisms by G. Herzberg.[69] Many more details of theoretical-physical interest must here be omitted. We present only a few general hints on the possibilities of finding additional refined distinctions in the infrared characteristics of Class I (Bragg), and Class II (Belov) structure types, such

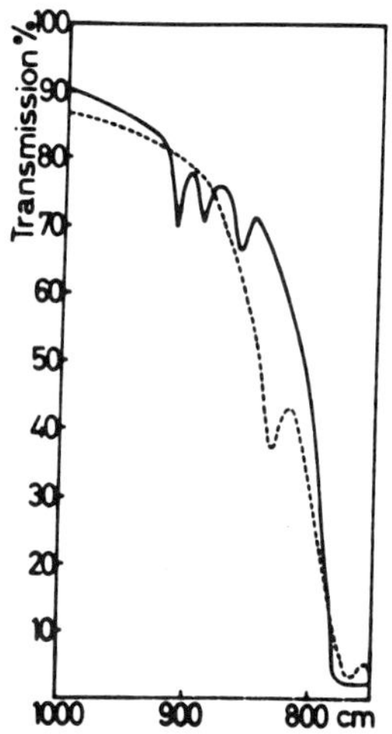

FIG. A. 20. Spectrum of the $[SiO_4]$ ion in dilute crystalline solutions for $Ba_2(Si_{0.01}Ge_{0.99})O_4$ (solid curve), and $Ni_2(Si_{0.01}Ge_{0.99})O$ (dotted curve) (Tarte).

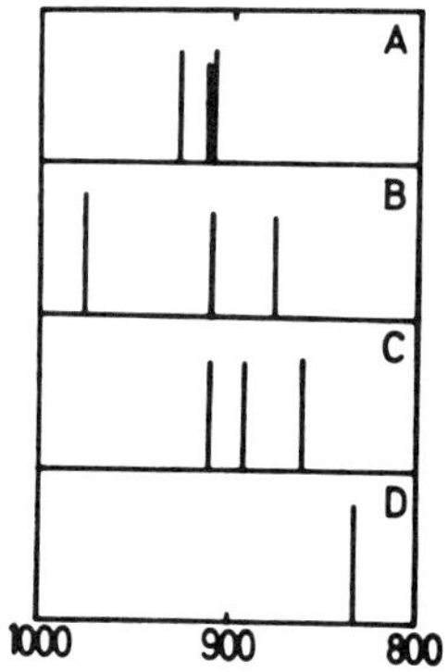

FIG. A. 21. Multiplicity of the fundamental v_3 absorption band for $[SiO_4]$ in silicate-germanate systems of the type $X_2(Si_{0.01}Ge_{0.99})O_4$ (Tarte). (*A*) X = Zn; (*B*) X = Mg; (*C*) X = Ba; (*D*) X = Ni.

[69] "Molecular Spectra and Molecular Structure. II. Infrared and Raman Spectra of Polyatomic Molecules," Van Nostrand, Princeton, New Jersey, 1945, p. 100.

as may have been derived from the investigations of K. A. Wickersheim and R. A. Buchanan[70] on the near-infrared spectra of different beryls, with polarized radiation (cf. A. ¶ 174). Of general interest, however, is the strong dichroism of the absorbing entities relative to the *c*-axis of the beryl crystal, and the strong effect of the water contents, variable from occurrence to occurrence, but clearly made evident by the fundamentals at 3663 and 3700 cm.$^{-1}$ (2.73 and 2.70 μ) (perpendicular, and parallel to the *c*-axis). The vibration at 3598 cm.$^{-1}$ (2.78 μ, polarized parallel *c*-axis) is ascribed to an alkali or alkaline earth hydroxide molecule lying lengthwise in the structural channels of the beryl.

45. This special case leads our attention to the numerous investigations in which the presence of water and generally of *hydroxyl groups* in silicate minerals was intensely discussed (on the analogous problems in the constitution of silicate glasses, see Vol. II, Section A). In this connection we mention first the studies of G. Brunner, H. Wondratschek, and F. Laves[71] on the infrared absorption characteristics of certain quartz and feldspar crystals in the 3 μ-range. A typical spectrum of this kind is shown in Fig. A. 22. An oriented exchange of H atoms by D (deuterium) parallel

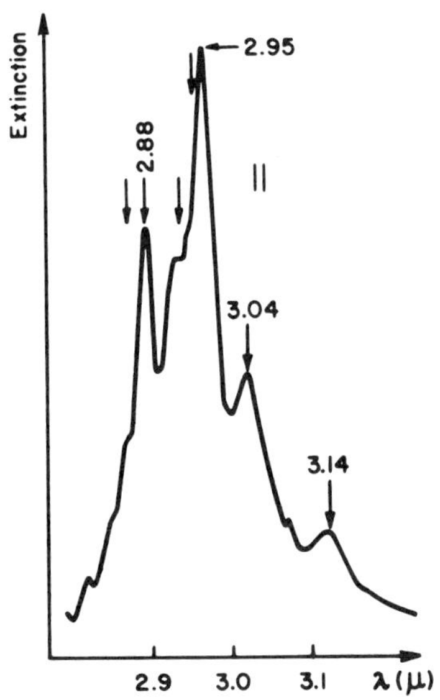

FIG. A. 22. Infrared absorption spectrum in the 3 μ range of a quartz crystal rich in water (Brunner, Wondratschek, and Laves).

[70] *Am. Mineralogist*, **44**, 1959, 440-445.

[71] *Naturwissenschaften*, **46**, 1959, 664; *Fortschr. Mineral.*, **38**, 1960, 125-126. Hafner and Laves used the infrared spectroscopy method in the investigation of *unmixing* phenomena of the system $MgAl_2O_4$—Al_2O_3 as a model of order-disorder problems in cation distribution in crystalline solutions; *Ibid.*, p. 124.

to the c-axis could be achieved either by simple thermal exposure in a D_2O-containing atmosphere, or by electrolysis. The bands 2.88, 2.95, and a part of that at 3.04 μ, polarized perpendicular to the trigonal axis in quartz must be ascribed to water (or hydroxyl groups) because they are shifted by the exchange with D towards the 4 μ range. The concentration in hydroxyl is highly variable in zonar fields of a single crystal, as seen in one example in Fig. A. 23. It can be made directly visible by ir-

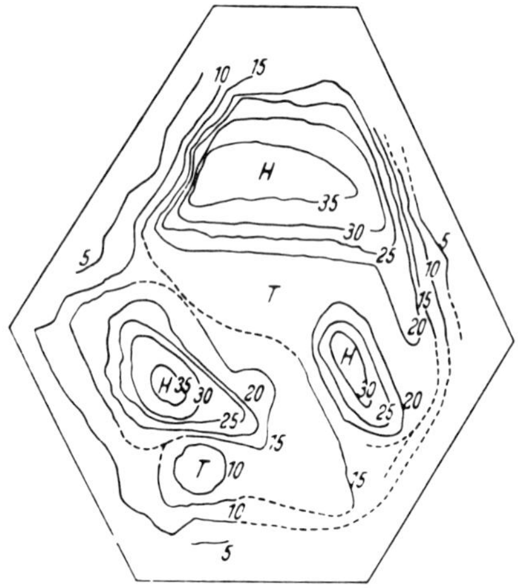

FIG. A. 23. Curves of constant extinction of the 2.95 μ hydroxyl band in a section perpendicular to the c-axis of quartz (Brunner, Wondratschek, and Laves).

radiation with thermal neutrons and γ-rays. Also feldspars show strong differences in infrared absorption, evidently by hydroxyl bands which can be replaced by OD bands. The spectra in the 3 μ range appear the more differentiated, the higher the order of the Al/Si distribution.

46. Very important also is the orientation of the hydroxyl groups with a permanent dipole moment perpendicular to the cleavage of mica crystals in the trioctahedral type, whereas in dioctahedral micas these are oriented under certain oblique angles, as W. A. Bassett[72] observed. The typical infrared absorption band is 3700 cm.$^{-1}$ (2.70 μ) in dioctahedral micas. The orientation effects here described may have an importance in the explanation of the fact that biotite and phlogopite are more susceptible to ion exchange reactions than muscovite, and may also affect the for-

[72] *Bull. Geol. Soc. Amer.*, **71**, 1960, 449-456; *Ibid.*, **69**, 1958, 1532-1533. On the infrared spectra of hydroxyl containing boron minerals and of danburite see H. Moenke, *Jenaer Jahrb.*, 1960, *T*, 191-215. The greatest importance of infrared spectroscopy is in the field of the hydration of basic calcium silicates in hydraulic binders, cf. Vol. IV, Section B, and Vol. V, Section C, cf. recently Ya. U. Ryskin, G. P. Stavitskaya, and N. A. Toropov (see A. ¶ 49).

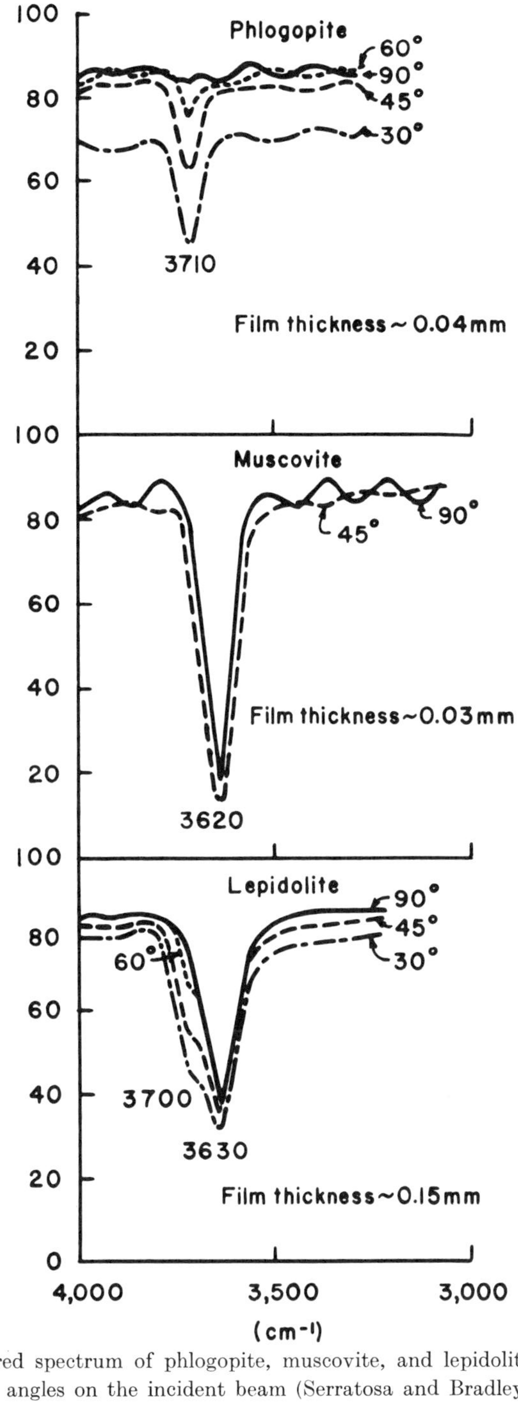

FIG. A. 24. Infrared spectrum of phlogopite, muscovite, and lepidolite for cleavage flakes oriented at different angles on the incident beam (Serratosa and Bradley).

mation of stacking polymorphs (cf. A. ¶ 133 ff.). J. Serratosa and W. F. Bradley[73] refined these observations considerably and extended the infrared examination to lepidolite with the interesting result that at normal incidence, and increased obliquity of it, absorption at the muscovite frequence (3620 cm.$^{-1}$) is observed, but little sensitive to orientation. This is combined with absorption at phlogopite frequency (3700 cm.$^{-1}$), with a sensitivity to orientation as is characteristic of phlogopite (Fig. A. 24). Lepidolite, therefore, exhibits the higher frequency in perpendicular bond axis orientation as for trioctahedral lamellae, the lower frequency for an inclination like the dioctahedral muscovite. Lepidolite is an intimately intergrown mica of muscovite with a lithium-containing biotite. A comparison with clay minerals

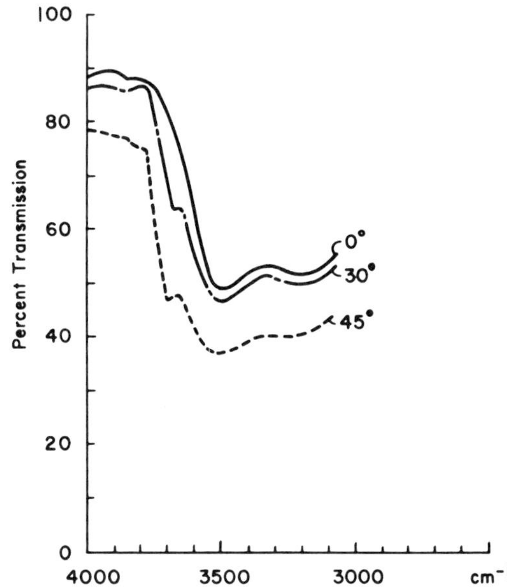

FIG. A. 25. Infrared spectrum of vermiculite (heated at 550°C.), for different angles of incidence (Bradley and Serratosa).

shows Wyoming montmorillonite to be dioctahedral and hectorite to be trioctahedral. The montmorillonite (bentonite paper) has in its aggregates randomly rotated positions within the flakes. Also vermiculite (Fig. A. 25) shows a typical intensity change with varying angles of incidence.

47. As a method for determination of kaolinite in clays, H. W. van der Marel and J. H. L. Zwiers[74] recommend infrared transmission analysis (with KBr pellets)

[73] *Nature*, **181**, 1958, 111; *J. Phys. Chem.* **62**, 1958, 1164-1167.
[74] *Silicates inds.*, **25**, 1960, 76-86, especially pp. 78-80 and 82-83.

for the wavelengths 698, 754, and 795 cm.$^{-1}$ (14.1, 13.2, and 12.6 μ), with characteristic apparently linear functions of the extinction versus the surface area of the kaolinite (in m.^{2}/gm.). This effect is not an elementary Rayleigh scattering but is caused by the decreased order of the stacking in the crystal structure, a decrease in infrared-active material, and a coating of the surface layer with an inactive material in a thickness of 15 to 25 Å. A more detailed study of the same authors on the O—H stretching bonds in kaolinite,[75] identified frequencies at 3698, 3660, 3630, and 3440 cm.$^{-1}$ (2.71, 2.74, 3.03, and 3.21 μ), all of these related to definite positions in the crystal structure. The bands at 1630 cm.$^{-1}$ (6.13 μ) correspond to free water (deformation band of OH) (cf. A. ¶ 45; B. ¶ 29, 98).

48. A very fundamental investigation of the effects of systematic isomorphous substitutions on the infrared absorption characteristics of layer silicates (mostly synthetic) was made over a range from 400 to 5000 cm.$^{-1}$ (25 to 2 μ) by Vl. Stubičan and R. Roy,[76] especially for the substitution of Si^{4+} by Al^{3+}, or of Mg^{2+} by Al^{3+}. In dioctahedral clay minerals of beidellite type the intensity of the bands at 538 and 475 cm.$^{-1}$ (18.5 and 21.8 μ) are correlated with the amount of Al^{3+} in the tetrahedral layers. For the entrance of Mg^{2+} in the octahedral layers of muscovite (in the phengite series) also a correlation could be found by the changes in frequencies in the ranges from 870 to 930, 720 to 830, and 520 to 550 cm.$^{-1}$ (12.75 to 11.8, 13.9 to 12.05, to 19.2 to 18.2 μ). In the trioctahedral clay minerals of the saponite type the substitution of Si^{4+} by Al^{3+} in the tetrahedral layers is quantitatively related to the frequencies of absorption between 650 to 670 cm.$^{-1}$ (15.4 to 14.9 μ) (Fig. A. 26). Also for the chlorite minerals of the 7 Å. (Roy's "septechlorites") (Fig. A. 27) and 14 Å. types characteristic changes of frequency as functions of the aluminum contents are observed.

49. The infrared absorption in *hydrated calcium silicates*, which have a particularly high importance for the investigation of hydration reactions of hydraulic binders, was studied by Ya. I. Ryskin, G. P. Stavitskaya, and N. A. Toropov[77] covering the typical valence oscillation with frequencies of 3408 and 3554 cm.$^{-1}$ (2.92 and 2.80 μ), and the deformation oscillation at 1670 and 1630 cm.$^{-1}$ (6.05 and 6.13 μ) which are usually more or less shifted. These are observed as in tobermorite, $4CaO.5SiO_2.5H_2O$, and also in opal, dioptase, $Cu_6[Si_6O_{18}].6H_2O$, and beryl. For tobermorite the doublet 2922 and 2854 cm.$^{-1}$ (3.40 and 3.50 μ) is particularly characteristic, combined with

[75] *Silicates inds.*, **24**, 1959, 359-369.

[76] *Am. Mineralogist*, **46**, 1961, 32-51; *Z. Krist.*, **115**, 1961, 200-214; *J. Am. Ceram. Soc.* **44**, 1961, 625-627.

[77] *Zhur. Neorg. Khim.*, **5**, 1960, 2727-2734, further on hydrogen bonding in acidic silicates and phosphates: Ya. I. Ryskin and G. P. Stavitskaya, *Optika i Spektroskopiya*, **8**, 1960, 606-613.

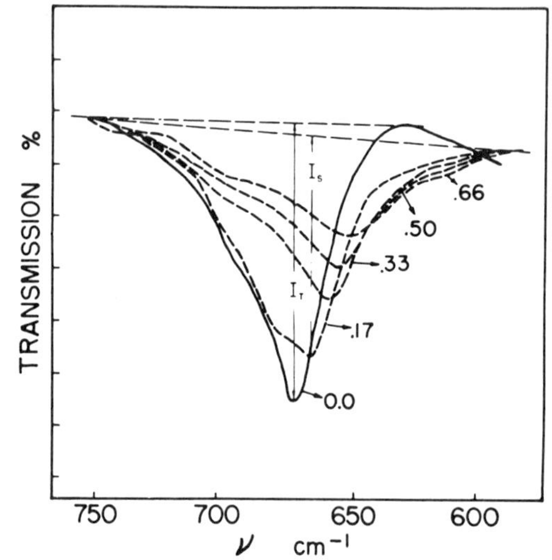

FIG. A. 26. The Si—O band in the infrared spectrum of talc and saponites (the numbers on the curves indicate the amounts of Al^{3+} ions present in tetrahedral coordination) (Stubičan and Roy).

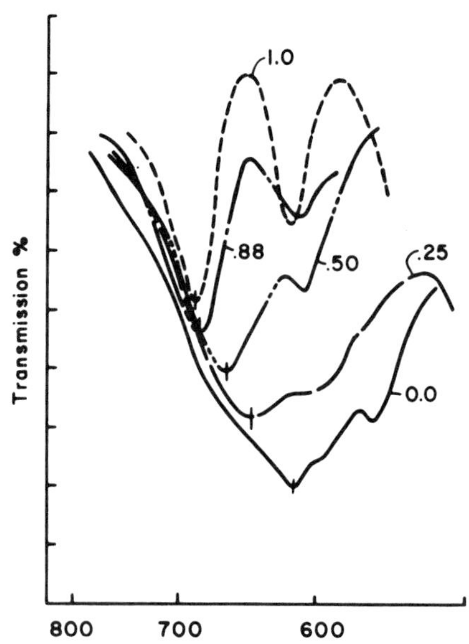

FIG. A. 27. The Si—O band in the infrared spectrum of 7 Å. chlorites (Stubičan and Roy). The numbers on the curves indicate the amounts of Al^{3+} ions in octahedral or tetrahedral coordination. (Stubičan and Roy).

the OH vibrations at 3445 cm.$^{-1}$ (2.90 μ) (Fig. A. 28). In afwillite, $Ca_3[SiO_3(OH)]_2$. $2H_2O$, there is a broad band between 3340 and 3130 cm.$^{-1}$ (3.00 and 3.19 μ), in polarized light at 3270 cm.$^{-1}$ (3.06 μ). The similarity of the infrared absorption in dioptase having sharp bands at 3370 and 3248 cm.$^{-1}$ (2.96 and 3.07 μ) with that of

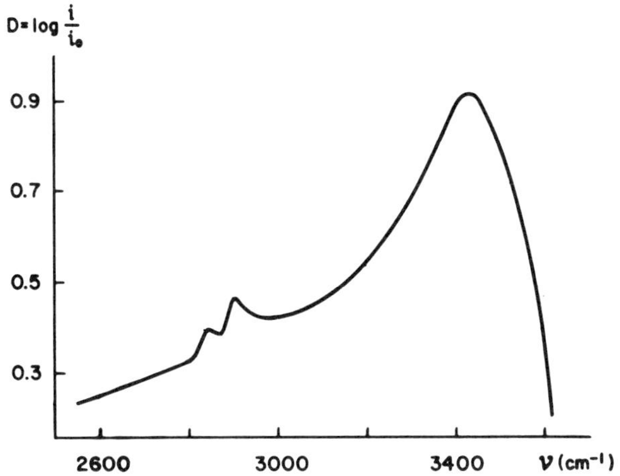

FIG. A. 28. Infrared absorption spectrum of synthetic tobermorite, $4CaO.5SiO_2.5H_2O$ (Ryskin, Stavitskaya, and Toropov).

water dissolved in CCl_4 is striking (Fig. A. 29). The 3370 cm.$^{-1}$ band corresponds to the antisymmetric oscillation of the H_2O molecule; the 3248 cm.$^{-1}$ band, to the symmetric oscillation. The free OH oscillations are those of the band at 3700 cm.$^{-1}$ which is shifted in xonotlite, $Ca_6[Si_6O_{17}](OH)_2$, to 3640 cm.$^{-1}$ (2.74 μ). In hillebrandite, $Ca_{12}[Si_6O_{17}]_2(OH)_4.12Ca(OH)_2$, it appears shifted to 3617 and 3553 cm.$^{-1}$ (2.75 and 2.80 μ), in α-$Ca_2[SiO_3(OH)]OH$ (Fig. A. 30) to 3532 cm.$^{-1}$ (2.83 μ). The frequency of the OH oscillations depends on the more or less covalent character of the bonds Si—O or Al—O. The prevailingly ionic character of the bonds Ca—O and Mg—O in the structures of the free hydroxides reduces the deformation oscillations of OH below 400 cm.$^{-1}$ (25 μ), and also for xonotlite and hillebrandite they are below the limit of observation possible with infrared spectrometers furnished with a NaCl prism. For calamine (hemimorphite, $Zn_4[Si_2O_7](OH)_2.H_2O$), it was observed at 1085 cm.$^{-1}$ (9.2 μ) characteristic of the configuration Zn—OH. If afwillite, which contains $[SiO_3OH]$ groups in its structure, H bonds are active between this configuration, with oscillations at 2400 and 2800 cm.$^{-1}$ (4.17 and 3.57 μ), analogous with that known from infrared characteristics of $KHCO_3$, $NaHCO_3$, and KH_2PO_4. The presence of strong absorption bands in the range from 3000 to 2000 cm.$^{-1}$ (3.33 and 5.00 μ) in silicates indicates the presence of $[SiO_3OH]$ groups. Only in liquid trimethyl silanol the weak proton acceptor nature of the oxygen atom causes a moderate H bonding, and the hydroxyl absorption is observed at 3257 cm.$^{-1}$ (3.05 μ).

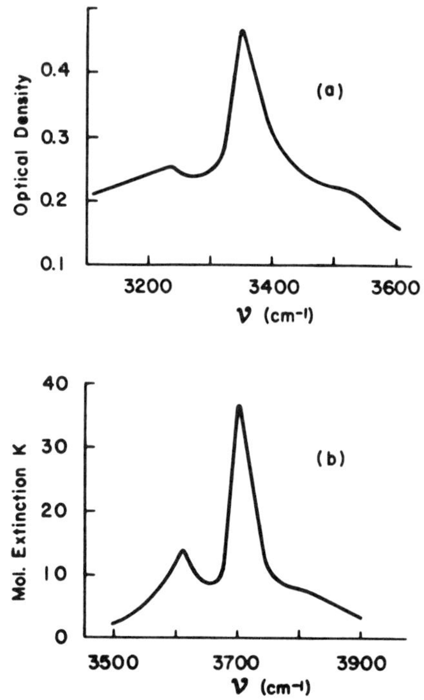

FIG. A. 29. Infrared absorption spectrum of water in dioptase (*a*), and of water dissolved in carbon tetrachloride (*b*), as a comparison (Ryskin, Stavitskaya, and Toropov).

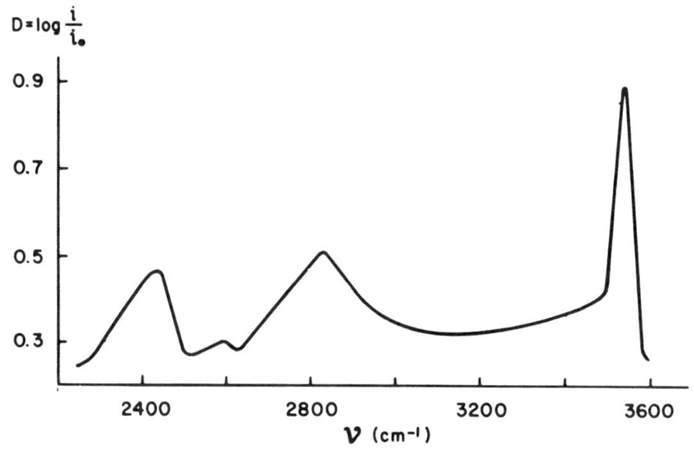

FIG. A. 30. Infrared absorption spectrum of α-Ca_2SiO_4 hydrate (Ryskin, Stavitskaya, and Toropov).

COMPLEX ANIONS IN SILICATES; ALUMINO-, BERYLLO-, AND BOROSILICATES

50. Among the *complex anions* formed with silicates, the *beryllo*silicates (e.g., phenacite) play a quite specific role. An interesting phenomenon is here verified when it is demonstrated how the oxygen anions in coordination not with two but with *three* neighboring cations are able to form groups [BeX_3], as E. Kordes, W. Vogel, and R. Feterowsky[78] emphasized. They could, in an analogous way, explain the structure of beryllium phosphate glasses which contain [BeO_4] and [PO_4] groups at the same time, together with groups [OBe_2P]. By using the apparently anomalous coordination number 3 for O^{2-} in this configuration it becomes possible, however, that in a phenacite-type structure not only the Si^{4+} cations but also the Be^{2+} ions can form a tetrahedral coordination in the trigonal framework.

51. The tremendous importance of the [AlO_4] group in *alumino*silicates is especially evident in the constitution of glasses containing alumina, as A. Silverman[79] postulated. On a much more generalized basis of discussion of the aluminosilicate structures, J. R. Goldsmith[80] developed his efficient principle of "*simplexity*" as a measure of how the complex structures of alumino-silicates respond to the thermodynamic conditions of high entropy in the simplest structure types with a high degree of statistical disorder, and influence the nucleation and crystal growth from melts. This is a petrological principle of great significance in magmatic differentiation. The so-called "ease of crystallization" and prompt synthetic reproduction is characteristic of those aluminosilicates of high simplexity in which [AlO_4] and [SiO_4] are equivalent. Less readily synthesized are those aluminosilicates in which these coordinations are in strict order, or in which [AlO_6] groups appear, both with low degrees of simplexity. The high energy barriers, due to strong bonding in ordered structures of low simplexity at moderate and low temperatures, favor the persistence of first-crystallized or metastable compounds of higher simplexity.

52. Especially in the family of the feldspars and of the feldspathoids, Goldsmith's simplexity principle is of significance as a guiding rule for the crystal phases expected in rock-forming mineral assemblies and the specific role of the framework crystal phases. Many of the [AlO_6] containing aluminum silicates have for a long time defied synthetic attempts, whereas the disordered four-coordinated aluminosilicates generally are promptly accessible to synthesis in their high-temperature modifications (e.g., sanidine, analbite, high-plagioclases, kalsilite, nepheline, carnegieite, leucite, eucryptite), while the ordered forms, because of their high energy barrier (e.g., in microcline and albite), are extremely difficult to produce in synthesis. The rule

[78] *Z. Elektrochem.*, **57**, 1953, 282-288, especially p. 287.
[79] *Am. Ceram. Soc. Bull.*, **17**, 1938, 226-227.
[80] *J. Geol.*, **61**, 1953, 439-451.

especially underlined by F. Laves becomes immediately understood when it is realized that in nature virtually all potassium feldspars, initially crystallized in the disordered monoclinic modification, become ordered (microcline) only by a subsequent conversion in the course of a long period of time. Adularia is a good example of metastable and persistent monoclinic potassium feldspar, crystallized at low temperatures. The simplexity principle is in no way restricted to the typical framework aluminosilicates. It is equally demonstrated by Goldsmith to be valid also for chain, layer, ring, and isolated group silicates.[81]

53. Recently, W. White, H. A. McKinstry, and Th. F. Bates[82] made it evident that the modern method of *X-ray fluorescence* analysis responds to the valence state and to the coordination of a gievn element, as in the case of aluminum with its characteristic Kα radiation. The diagram Fig. A. 31 shows a distinct relationship of the

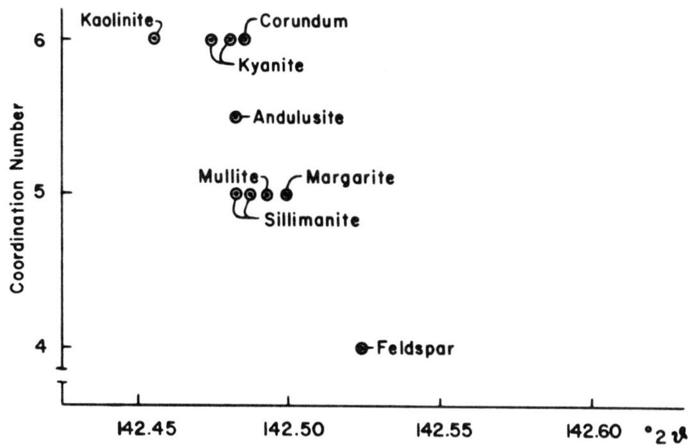

FIG. A. 31. Variation of the X-ray fluorescence peaks of $Al^{3+}K\alpha$, with the average coordination number in various silicates (White, McKinstry, and Bates).

wavelength for Al^{3+} Kα as a function of the angle 2Θ and of the average coordination number, N, in which it occurs; for typical *aluminum* silicates (disthene, andalusite, sillimanite), with $N_{\text{aver.}} = 6$, 5, and 4, whereas among the aluminosilicates feldspar, the only investigated mineral of this class ($N = 4$), bevaves quite specifically. Interesting also is the comparison of kaolinite and corundum in this diagram, both with $N = 6$, but with a characteristic difference in the 2Θ angles, probably due to the fact that the Al—O distance in corundum is smaller than in kaolinite.

[81] On the particular role of the presence of water in the differentiation regularities of silicate minerals with different degrees of simplexity, cf. A. E. Ringwood, *Heidelberger Beitr. Mineral. u. Petrog.*, **6**, 1959, 346-351.

[82] *Proc. Conf. Ind. Appl. X-Ray Anal.*, **7**, 1958, 239-245.

54. Another possibility for study of problems of the aluminosilicates (and of related complex anion structures) is the application of the *magnetic nuclear resonance method.* E. Brun, P. Hartmann, H. H. Staub, St. Hafner, and F. Laves[83] investigated the magnetic resonance behavior of Al^{27} and Na^{23} in feldspars to detect the degree of statistical disorder in the distribution of the Al and Si atoms in the tetrahedral framework of albite, microcline, and sanidine. Very satisfactory, strong resolutions of the resonances were observed for those nuclei in albite and microcline, due to their characteristic quadrupole interactions, whereas for sanidine only a rather unsharp (smeared-out) resonance peak appeared in the spectrum. The effect of a thermal exposure at 1050°C. over a long period of time is distinctly an increasing dis-

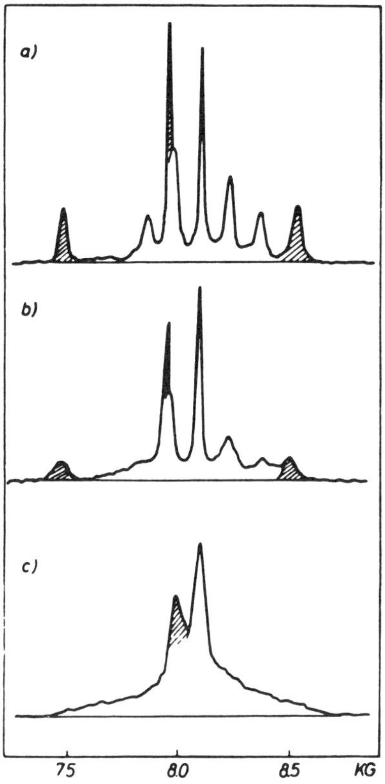

FIG. A. 32. Magnetic nuclear resonance peaks for Al^{27} and Na^{23} (shaded fields) in albite (Brun, Hartmann, Staub, Hafner, and Laves): (*a*) natural mineral; (*b*) heated at 1000°C. for 7 days; (*c*) heated at 1060°C. for 44 days.

[83] *Z. Krist.*, **113**, 1960, 65-76. On the theory of the magnetic nuclear resonance see P. J. Bray and A. H. Silver, *J. Chem. Phys.*, **29**, 1958, 984-990; also "Modern Aspects of the Vitreous State" (J. D. Mackenzie, ed.), Butterworths, London, 1960, pp. 92-119; *Acad. roy. Belg., Classe sci. mém.* **33**, 1961, 289-320, with extensive bibliography.

order of the Al/Si distribution by developing a sanidine-type high-temperature modification from the better ordered low-temperature feldspars (see Figs. A. 32 and 33).

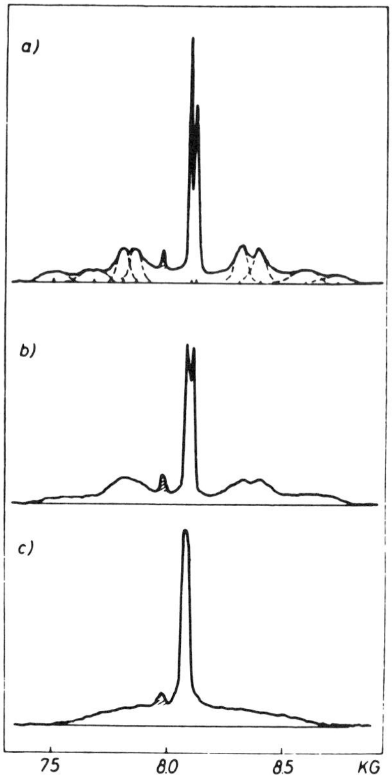

FIG. A. 33. Magnetic nuclear resonance peaks for Al^{27} and Na^{23} (shaded fields) in microcline (Brun, Hartmann, Staub, Hafner, and Laves). (*a*) natural mineral; (*b*) heated at 1050° C. for 1 day; (*c*) at 1050°C, for 18 days.

55. The theoretically interesting question whether there may be $[CrO_4]$ groups replacing $[SiO_4]$ which could form a kind of "*chromino*"-silicates analogous to the aluminosilicates, was answered by M. W. Shafer and R. Roy[84] in the negative. Their conclusions are based on the nature of the compound $CrPO_4$ which is much different from the SiO_2 (quartz) type structures. A. Neuhaus[85] reported the presence of $[CrO_4(OH)_2]$ groups in the green-colored layer structure of the Cr-mica fuchsite (cf. A. ¶ 139), and of $[CrO_2(OH)_4]$ groups in Cr-kaolinite, whereas $[CrO_6]$ groups are characteristic of the green-colored pyroxenes like Cr-diopside or smaragdite.

[84] *J. Am. Chem. Soc.*, **78**, 1956, 1087-1089.

[85] *Z. Krist.*, **113**, 1960, 195-233, especially pp. 224 f.

The red-colored chlorite-type mineral kaemmererite contains [(Mg, Al, Cr)$O_4(OH)_2$] groups, and [(Mg, Al, Cr) $(OH)_6$] groups in the brucite-like octahedral interlayers, with strongly contrapolarizing neighboring cations which are responsible for the red color effect in the absorption spectrum.

56. The incomparably more important complex *borosilicate* structures were studied in a series of investigations of borate minerals in which the boron appears doubtless in fourfold coordination, such as in the complex infinite borate chain structures with the units $[B_3O_4(OH)_3]^{2-}$ in colemanite, consisting of one $[BO_3]$ coupled with two $[BO_4]$ groups as C. L. Christ, J. R. C. Clark, and H. T. Evans[86] found. F. Machatschki[87] developed a systematic classification of complexes of higher boric acids and compounds derived from these. Certain of these natural borate minerals[88] contain both $[BO_3]$ and $[BO_4]$ groups side-by-side. By nuclear quadrupole resonance methods V. Ross and J. O. Edwards[89] confirmed the tetrahedral boron coordination in simple hydroxyl borates (teepleite, $Na_2B(OH)_4Cl$, and bandylite, $CuB(OH)_4Cl$). The structure of B_2O_3 itself was determined by Sv. V. Berger[90] as hexagonal, with 6 units B_2O_3 in the elementary cell, consisting of $[BO_4]$ tetrahedra of two types, linked by the corners to a three-dimensional framework which in a certain degree resembles the quartz structure. W. Loewenstein[91] discussed regularities in the condensation of $[BO_4]$ with $[SiO_4]$ as in the structure of tourmaline and axinite in which the $[BO_3]$ groups are in reality distorted tetrahedra, with the central B^{3+} cation close to the triangular plane of the O^{2-} anions. Sharing of $[BO_4]$ and $[SiO_4]$ tetrahedra over the corners is the rule, but $[AlO_4]$ cannot participate in the same corners because of the strongly weakening effect by the B^{3+} cations in the neighborhood. This is the reason why, in tourmaline, no Al^{3+} can enter the six-membered rings characteristic of this structure (cf. A. ¶ 103).

SPECIAL STRUCTURES OF SILICATES

General Classification Principles

57. Based on his extensive theoretical discussion on the natural state of the oxygen anion in chemical bonding in silicate structures, especially on his aspects of the co-

[86] *Am. Mineralogist*, **40**, 1955, 312.

[87] *Radex Rundschau*, 1952, 77-83, especially pp. 77 f.

[88] Cf. H. Meixner, *Fortschr. Mineral.*, **31**, 1952, 39-42.

[89] *Am. Mineralogist*, **44**, 1959, 875-877.

[90] *Acta Chem. Scand.*, **7**, 1953, 611-622; *Proc. Intern. Symposium on Reactivity of Solids, Gothenburg*, 1952 (pub. 1954), Vol. I, pp. 424-426. See also Bh. K. Banerjee, *J. Am. Ceram. Soc.*, **36**, 1953, 294-298, on complex borates and oriented B_2O_3 crystallites in fibers of B_2O_3 glass.

[91] *Am. Mineralogist*, **41**, 1956, 349-351.

valent character of the Si—O bond, V. I. Lebedev[92] developed a classification with symbols derived from the monovalent scheme used for oxygen, and the divalent scheme for the silicon atom, to account for the partially covalent, and partially ionic character of bonding in silicates (cf. A. ¶7). The classification given by Lebedev is in general analogous to that of W. L. Bragg, and enlarged by H. Strunz. He distinguishes "*A*" groups

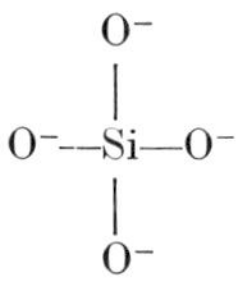

(*A*)

for the orthosilicates, "*B*" groups

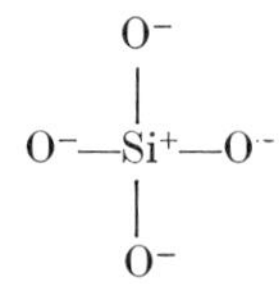

(*B*)

in chain or ring structures, "*C*" groups in the silica modifications

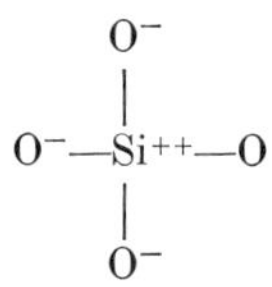

(*C*)

and "*D*" groups

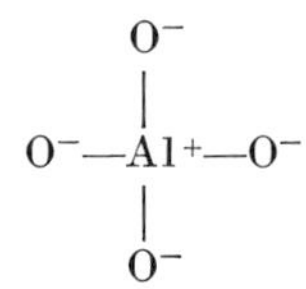

(*D*)

together with "*C*"-groups in aluminosilicates. Combinations of *A* and *B* occur in silicates with diorthosilicate [Si_2O_7] groups, $2B + C$ in ribbons, $B + C$ in layer structures. The increasing ionization of silicon from *A* to *C* corresponds to the systematic decrease in the distances Si—O from 1.70 in garnets and zircon, to 1.61 in diopside, 1.62 in α-quartz, and 1.54Å. in β-quartz. In the feldspars the principle of densest packing is somewhat modified. Lebedev's ideas are chiefly aimed at the mineral- and petrogenetic aspects of magmatic differentiation sequence and postmagmatic hydrothermal and metamorphic conversions.

58. A much more detailed systematic classification also based on the principles given by Bragg was proposed by F. Machatschki[93] improving his previous systematics (1928), which is in its strict consequence and easy readability of the introduced symbols of great didactic value, including the detailed coordination characteristics of the single structural types. A fundamental revision of the classifications of the silicate structures actually used was recently developed by T. Zoltai[94] who strictly postulates that all tetrahedrally coordinated cations should be considered as part of the tetrahedral framework of silicates in question. In this way every ambiguity in doubtful

[92] *Vestnik Leningrad. Univ.*, **3**, 1948 (3) 17-30.

[93] *Fortschr. Mineral.*, **31**, 1953, 47-50.

[94] *Am. Mineralogist*, **45**, 1960, 960-973. See also Fr. Liebau, *Naturwissenschaften*, **49**, 1962, 481-491, on new ideas of silicate classification.

structure classifications is eliminated. There is still a certain objection, especially to the Bragg system, if as in cordierite or gehlenite $[AlO_4]$ groups are disregarded, while $[BO_4]$ groups in danburite, however, are accepted. The justification of the strict postulate of tetrahedral framework structures is evident in the close similarity in cationic radii and in the electronegativity of Si^{4+} with other four-coordinated cations (Al^{3+}, B^{3+}, P^{5+}, Be^{2+}, etc.). The random distribution of tetrahedrally coordinated cations is also accepted in the general use of the crystallochemical formulae given to silicates with framework-type structures. For silicates containing loops of tetrahedra rings, Zoltai suggests that the size of the loops (digon arrangements) may be used as a classification criterion related to the energy of the structures, as was already shown in A. ¶ 27. Further, Zoltai proposes the introduction of the corner sharing of tetrahedra as an additional criterion for systematics; one may introduce a numerical "*sharing coefficient*" (see below) which is related to the cation: anion ratio of the crystallochemical formula for the given silicate.

59. The classification proposed by Zoltai uses the previous geometrical types characteristic of the Bragg-Strunz systematics, namely:

(1) isolated groups of tetrahedra;

(2) one-dimensional infinite tetrahedral arrangements (chains, ribbons);

(3) two-dimensional infinite tetrahedral arrangements (networks, sheets, layers);

(4) three-dimensional infinite tetrahedral arrangements (frameworks).

The possibility of the existence of structure composed of two and more of these four different basic types of tetrahedral arrangements is covered by the addition of group;

(5) mixed tetrahedral structures.

Starting from these basic structural types Zoltai develops a table of subtypes which are characterized by the sharing coefficients and cation: anion ratios as shown in Table I.

60. The consequence in Zoltai's classification principles is worthy of high acknowledgement. It opens a way to a systematics which is practically free from inner contradictions. The only objection may be derived from the (still somewhat enigmatic) introduction of a tetrahedral calcium coordination in the one structure of the orthosilicate Na_2CaSiO_4, which is a derivative of cristobalite according to the very early determination of its structure type (F. W. Barth, 1935). Another somewhat difficult case is axinite in which four-membered rings of $[SiO_4]$ tetrahedra are connected by $[AlO_4]$ tetrahedra, but the Al^{3+} contributes in this structure less than one half of the bond strength to connect the Si tetrahedra, and more than one half to connect Fe octahedra and B triangles. The $[AlO_4]$ groups belong, therefore, not to the framework in axinite but to the rest of the structure. This exceptional behavior of one special structure may not be reason enough to give up the other evident advantages of

TABLE I

SHARING COEFFICIENTS AND CATION-ANION RATIOS OF THE TYPES AND SUBTYPES OF TETRAHEDRAL STRUCTURES

Types	Subtypes	Sharing coefficients[a,b]	Cation-anion ratios
1. Isolated groups of tetrahedra	*a.* Single tetrahedron	1.00	1:4
	b. Pair of tetrahedra	1.25	1:3½
	c. Large groups	1:25–1.50–(1.75)	1:3½–1:3–(1:2½)
	d. Mixed groups	1:00–1.50–(1.75)	1:3½–1:3–(1:2½)
2. One-dimensionally nonterminated structures of tetrahedra	*a.* Single chains	1.50	1:3
	b. Single rings	1.50	1:3
	c. Double chains	1.50–1.75	1:3–1:2½
	d. Double rings	1.50–1.75	1:3–1:2½
	e. Multiple chains	(1.50)–1.75–2.00	(1:3)–1:2½–1:2
	f. Multiple rings	(1.50)–1.75–2.00	(1:3)–1:2½–1:2
	g. Mixed chains and rings	(1.50)–1.75–2.00	(1:3)–1:2½–1:2
3. Two-dimensionally nonterminated structures of tetrahedra	*a.* Single sheets	1.50–1.75	1:3–1:2½
	b. Double sheets	(1.50)–1.75–2.00	(1:3)–1:2½–1:2
	c. Multiple sheets	(1.50)–1.75–2.00	(1:3)–1:2½–1:2
	d. Mixed sheets	(1.50)–1.75–2.00	(1:3)–1:2½–1:2

4. Three-dimensionally nonterminated structures of tetrahedra	*a.* Networks with corners shared	1.75–4.00	1:2$\frac{1}{2}$–1:1
	b. Networks with one or more edges shared	4.00–8.00	1:1–1:$\frac{1}{2}$
	c. Networks with one or more faces shared	8.00–	1:$\frac{1}{2}$–
5. Mixed types			

[a] Sharing coefficients in parentheses indicate theoretically possible but practically improbable ranges.

[b] ANNOTATIONS

The sharing coefficient of a tetrahedral structure is obtained from the subscripts of the tetrahedrally coordinated cation, T, and the anion, A, in the crystallochemical formula of a given silicate. There is a total number of $4T$ tetrahedral corners in a chemical formula unit, x of these joined to anions shared by $n + 1$ tetrahedra, and $4T - x$ to anions shared by n tetrahedra.

The sharing coefficient is therefore $= n + x/4T$.

$$x = 4T(n + 1) - A(n^2 + n);$$

or sharing coefficient $= 2n + 1 - (n^2 + n)A/4T$.

The number n is obtained from the ratio of the number of available tetrahedral corners to the number of shared corners: $n =$ the integer of $4T/A$.

Zoltai's classification, which for so many reasons is adopted in our discussions. This will not bring about too many difficulties in comparing our text with that of our previous one in the book of 1954 in which beryl, cordierite, hemimorphite, phenacite, willemite, bertrandite and melilite were classified in an alternate subclass of Bragg structures. They are now unequivocally three-dimensional framework structures. Zoltai's corner sharing principle is also in agreement with the previous rules of L. Pauling on the correlation of sharing modes and structure stability. Corners should be shared by the least possible number of tetrahedra, thus no corner is shared by three tetrahedra (unless exceptionally there should be no other state of lower energy available). Of considerable importance for the stability of silicate structures is also the occurrence of tetrahedral "loops."[95] One may therefore distinguish between structures containing no loops of tetrahedra, subdivided according to the number of tetrahedra in the periodic unit of the tetrahedral structure, and structures with tetrahedral loops, subclassified according to the number of tetrahedra in the loops. Here also is the possibility of introducing the Class II structures of Belov (cf. A. ¶ 30) which show, in the simplest case of infinite chains, repetitions of only every third tetrahedron as translation equivalents, whereas in Class I (Bragg structures) every second tetrahedron is oriented differently from the first one.

1. ISOLATED GROUPS OF TETRAHEDRA

Subgroup 1a. Tetrahedra (Isle Structures)

61. A most valuable monograph on the minerals of the *humite group* was given by Th. G. Sahama,[96] based on chemical precision analyses and the optical and structural constants of natural occurrences, with definite ranges of variation of the chemical composition including their correlations to powder diffraction data. Particularly important are the correlations to the isomorphous substitution of the anions F^- and OH^-, and also the iron content which was calculated as being present in replacing the Mg^{2+} ions of the olivine core by Fe^{2+}, not the Mg^{2+} in the fluoride-hydroxide slabs of the structure. From this, one would postulate a considerable degree of order in the distribution of iron and magnesium cations. Most interesting is the possibility of a replacement of MgF_2 by TiO_2 in humite minerals which is easily understood from their high degree of isostructural analogy. The prompt synthesis of

[95] This important point was previously emphasized by A. F. Wells, *Acta Cryst.*, **7**, 1954, 545-554, who, however, considered only highly regular tetrahedral frameworks with sharing coefficients of 1.75 and 2.00. Also F. Liebau, *Z. physik. Chem. (Leipzig)*, **206**, 1956, 73-92, considered the subclasses of the simpler tetrahedral structures but not sufficiently those of three-dimensional frameworks (see his most recent proposals in *Fortschr. Mineral.*, **39**, 1961, 38 f.).

[96] *Ann. Acad. Sci. Fennicae, Ser. A*, III, No. 31, 1953, 1-50. On the problems of the iron and titanium content of humite group minerals, see also I. D. Borneman-Starynkevich, *Doklady Akad. Nauk S.S.S.R.*, **71**, 1937, 137-140.

fluorine-humite minerals from melts is discussed in Vol. IV, Section A, whereas the hydroxyl-members of the series could not be prepared in hydrothermal reactions. Concerning the refractive indices and the molecular refraction, the available humite minerals behave as do nearly ideal crystalline solutions.

62. On the other hand, of great interest is the hydrothermal synthesis by E. R. Buckle and H. F. W. Taylor[97] of a *calcio-chondrodite*, $Ca_5(SiO_4)_2.(OH)_2$ (from $3CaO.SiO_2$ at 600° to 700°C.), of monoclinic symmetry. This compound, however, is evidently different from the humite minerals in that no replacement of hydroxyl by fluorine anions is observed in its composition. In contrast with the always fluorine-containing natural and synthetic chondrodite, the calcio-chondrodite is in the same chemical and structural relation to γ-Ca_2SiO_4 as magnesium chondrodite is to forsterite. Unknown are still calcium-norbergite, calcium-humite, and calcium-clinohumite. *Hodgkinsonite*, a manganese-zinc mineral of the norbergite family, occurs of the composition $Zn_2(SiO_4).Mn(OH)_2$, described with its crystallographic constants by A. J. Rentzeperis[98] as being of pneumatolytic origin, and associated with willemite and franklinite.

63. The highly complex structure of *andalusite* found a refinement in investigations of Ch. W. Burnham and M. J. Buerger.[99] In general, the previous interpretation by W. H. Taylor (1929) was confirmed. Chains of octahedral $[AlO_6]_\infty$ chains sharing two edges are arranged parallel to the *c*-axis. These octahedra appreciably deviate from ideality: two of the Al—O bonds have a length of 1.83 Å.; two others, of 1.89 Å.; and two perpendicular to the *c*-axis, of 2.09 Å. The octahedral chains are linked together by alternating single $[SiO_4]$ and $[AlO_5]$ groups (Fig. A. 34). A highly remarkable change in the interpretation of the *sillimanite* structure is its separation from the structures of disthene and andalusite, and its new classification in group 2.*c* of Zoltai's systematics of complex structures, because of the presence of double chains of tetrahedra $[SiO_4]$ and $[AlO_4]$ to form infinite $[AlSiO_5]_\infty$ double chains.

64. In the *garnet* group, the structure of *grossularite* was determined by S. C. Abrahams and S. Geller[100] with a considerable improvement of the previous data going back to G. Menzer (1926) (see Fig. A. 35). This new determination is of particular importance in view of the garnets with Fe^{3+} and Cr^{3+} (andradite and uvarovite), and to the ferrimagnetic silicon-free rare earth ferrite garnets of the type $3Y_2O_3.5Fe_2O_3$ which are of great significance in modern electronics (cf. A. ¶ 244).

[97] *Am. Mineralogist*, **43**, 1958, 818-823. A phase called "*X*" by D. M. Roy, *Ibid.*, **43**, 1958, 1009-1028, is identified with calciochondrodite.

[98] *Acta Cryst.*, **11**, 1958, 448.

[99] *Z. Krist.*, **115**, 1961, 269-290. See also the discussion remarks of W. H. Taylor, *Ibid.*, pp. 314-315, based on studies of P. C. Thornely and W. H. Taylor, *Mem. Proc. Manchester Lit. & Phil. Soc.*, **83**, 1939, 17-30, on the coordination of Al^{3+} in andalusite.

[100] *Acta Cryst.*, **11**, 1958, 437-441.

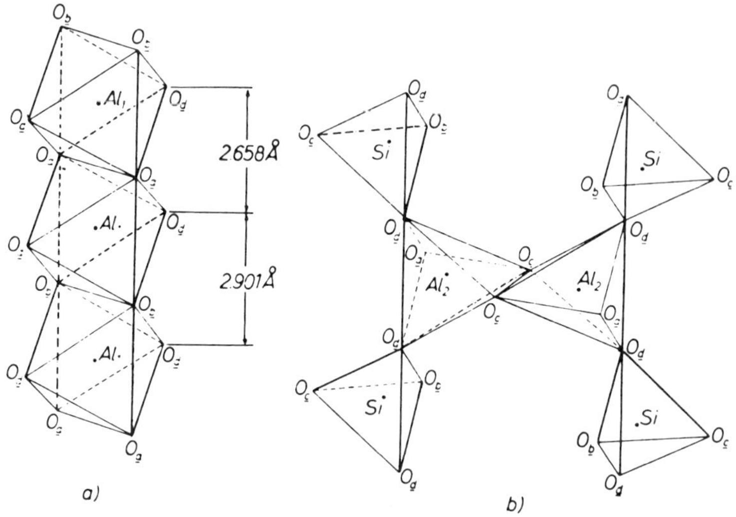

FIG. A. 34. Octahedral chains parallel to *c* in the structure of andalusite (*a*), and geometry of the $[SiO_4]$ and $[AlO_5]$ polyhedra (*b*) (Burnham and Buerger).

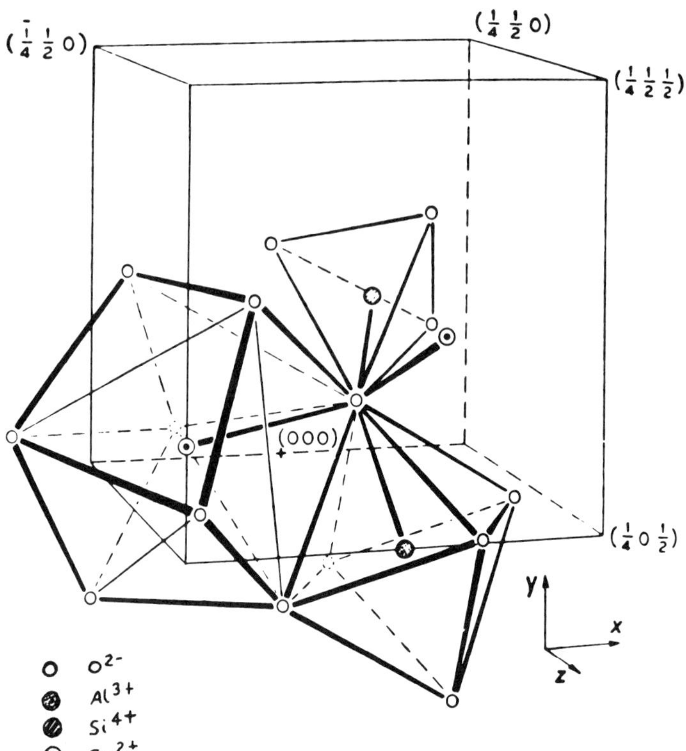

FIG. A. 35. Surrounding of an oxygen anion in the structure of grossularite (Abrahams and Geller).

65. Typical of subgroup 1.*a* with isolated [SiO_4] groups are also *titanite* (sphene), and especially the minerals of the *zircon* family, in which the hydrothermal synthesis of $ZrSiO_4$ (as *thorite*, and *huttonite*) was demonstrated by Cl. Frondel and R. L. Collette.[101] Huttonite is the monoclinic, monazite-like modification of $ThSiO_4$ (cf. Vol. III, Section B). The analogous synthesis of *coffinite*, $USiO_4$, was described by H. R. Hoekstra and L. H. Fuchs[102] and by L. R. Stieff, T. W. Stern, and A. M. Sherwood,[103] with X-ray studies of L. H. Fuchs and E. Gebert.[104] The possibility of a substitution of [SiO_4] by $[OH]_4$ groups in such hydrothermal products analogous to the natural *thorogummite*, $Th(SiO_4)_{1-x}(OH)_{4x}$, is particularly important as it was proposed by Cl. Frondel.[105] The identity of the structure type of coffinite with that of zircon is quite evident as seen from the X-ray data given by Stieff, Stern, and Sherwood, and from infrared absorption spectra.

66. For the highly complex *staurolite* structure there is a monograph of its physical constants as a function of composition, with particular emphasis given to the variation in the FeO and Fe_2O_3, MgO, and H_2O contents, in connection with the structural constants, by A. Juurinen.[106] The structure is described as orthorhombic, as is commonly accepted, with the average formula $H_4Fe_4Al_{18}Si_8O_{48}$. Many irregularities and variabilities which can only be explained by the consideration of the three variable oxide components mentioned above, make an accurate redetermination of the structure highly desirable. In this respect also the crystallographic-geometric discussions of the twin laws for staurolite by V. J. Hurst, J. D. H. Donnay, and G. Donnay[107] are most remarkable. The orthorhombic pseudosymmetry is reduced to the true symmetry without the glide plane *c* of the monoclinic elementary cell, as a twinning-determined "high order merohedry," with penetration twins simulating the orthorhombic symmetry. Precession methods established a series of such actual twin laws.

67. A most remarkable new type of silicate structures with isolated [SiO_4] groups is that of *spurrite*, $Ca_5(SiO_4)_2CO_3$, as was determined by J. V. Smith, I. L. Karle, H. Hauptman, and J. Karle.[108] The [SiO_4] tetrahedra and the equilateral triangles of [CO_3] perpendicular to the *y*-axis are linked together by Ca_2^+ ions in an arrangement of distinctly orthorhombic symmetry which is the reason for the well-known polysynthetic twinning parallel (001) of this interesting contact mineral. The char-

[101] *Am. Mineralogist*, **42**, 1957, 759-765.

[102] *Science*, **123**, 1956, 105.

[103] *Science*, **121**, 1955, 608-609; *Am. Mineralogist*, **41**, 1956, 675-688.

[104] *Am. Mineralogist*, **43**, 1958, 243-248. On a genetically instructive occurrence of coffinite see H. Strunz and E. Seeliger, *Der Aufschluss*, **12**, 1961, 353-359.

[105] *Am. Mineralogist*, **38**, 1953, 1007.

[106] *Ann. Acad. Sci. Fennicae. Ser. A*, III, No. 47, 1956, 1-53.

[107] *Mineral. Mag.*, **31**, 1956, 145-163.

[108] *Acta Cryst.*, **13**, 1960, 454-458.

acteristic linkage of the $[SiO_4]$ groups in different orientations parallel to the y-axis to two sorts of Ca^{2+} ions is seen from Fig. A. 36; three of the Ca^{2+} ions have seven O^{2-} neighbors, and the other Ca^{2+} has eight O^{2-} neighbors.

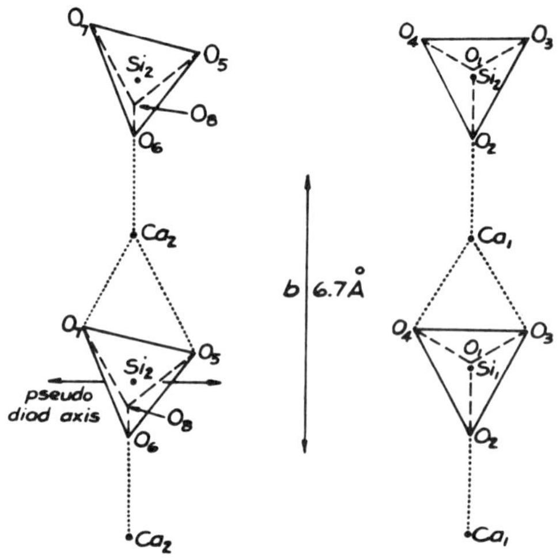

FIG. A. 36. Nature of the linkage between single $[SiO_4]$ tetrahedra and two kinds of Ca^{2+} ions in the structure of spurrite (Smith, Karle, Hauptman, and Karle).

68. *Afwillite*, $Ca_3(SiO_3OH)_2.2H_2O$, which was extensively studied in its structure by H. D. Megaw[109] is particularly important because of its relation to the hydration products of hydraulic binders (cf. Vol. IV, Section A, and Vol. V, Section C), and the presence of hydroxyl groups directly bonded to a central Si^{4+} cation in the tetrahedral groups. The monoclinic crystals are strongly piezoelectric and pyroelectric. The three-dimensional Fourier analysis resulted in the indentification of single tetrahedra $[SiO_3OH]$ interlinked with $[CaO_6]$ and $[CaO_7]$ polyhedra along the y-axis, sharing edges with each other and the tetrahedron (Fig. A. 37), causing the fibrous habitus of afwillite crystals. Characteristic of this highly interesting silicate is also the variability of the Ca—O distances, some much above 2.40 Å., the theoretical distance, such as 2.48 to 2.75 Å. belonging to Ca—H_2O bonds. The distances O—O are distinctly different in ionic bonds with other cations (= 2.80 Å.), but above 3.10 Å. for contacts of van der Waals type. The bond angle of the Si^{4+} cations ranges from 106° to 116° ± 3°. The infrared absorption bands between 3300 and 4000 cm.$^{-1}$ (near 3 μ) is sharp, indi-

[109] *Acta Cryst.*, **5**, 1952, 477-491. On the infrared absorption spectrum of afwillite in unpolarized and polarized radiation, see H. E. Petch, N. Sheppard, and H. D. Megaw, *Ibid.*, **9**, 1956, 29-34. Both short and long hydrogen bonds are confirmed in the structure, and the polarization directions of the absorption bands agree with those derived from the X-ray structure for H_2O and OH groups.

cating the long hydrogen bonds characteristic of such hydrate minerals. There are also short bonds indicated by the frequencies between 2500 and 3000 cm.$^{-1}$ (broad infrared bands of $\lambda = 3.3$ to $4.0\ \mu$). The different types of hydrogen bonds as shown in the afwillite structure given by Megaw could be expected to be of considerable influence on the anisotropic thermal expansion of the crystals. R. Shaw[110] confirmed

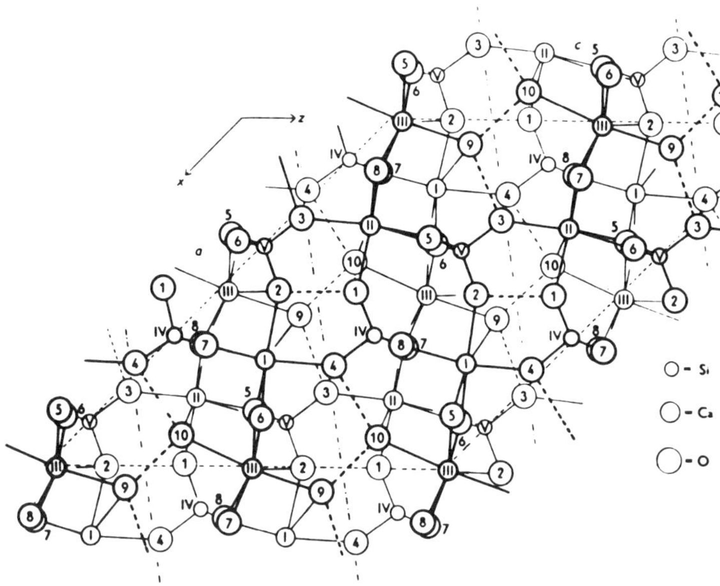

FIG. A. 37. Projection of the afwillite structure on (010) (Megaw). A continuous "Belt" running through the structure in the direction [001] is formed by two [SiO_4] tetrahedra and Ca_I and Ca_{II} polyhedra sharing vertical edges.

this expectation and gave a full explanation for the changes of bond angles with temperature which cause a large expansion in the direction of the cleavage plane (10$\bar{1}$), the trace of which is shown in Fig. A. 37 by dotted-dashed lines, not cutting Si—O bonds in the structure.

69. The afwillite structure further illustrates excellently the fact that by thermal dehydration this phase is directly changed to a *monocrsytalline* pseudomorph of γ-Ca_2SiO_4 (of olivine structure type), as H. F. W. Taylor[111] demonstrated, and is shown in Fig. A. 38, for a half-cell of the hydrated afwillite. Of every four half-cells, three are changed *in situ* to γ-Ca_2SiO_4, and one is destroyed in splitting off the [SiO_4] tetrahedra which are not incorporated in the γ-Ca_2SiO_4 monocrystal but form amorphous silica.

[110] *Acta Cryst.*, **6**, 1953, 428-429.
[111] *Acta Cryst.*, **8**, 1955, 440.

70. Isolated groups $[SiO_3(OH)]$ are also typical for the structure of $2CaO.SiO_2.H_2O(A)$ (cf. Vol. IV, Section A) (*dicalcium silicate α-hydrate*, Taylor), as determined by L. Heller.[112] By heating to 500°C., all water is lost and γ-Ca_2SiO_4 is formed; lacking every evidence to the contrary, Heller assumed that the hydrate must contain discrete tetrahedra with Si^{4+} as the central cation. The projection of the hydrate

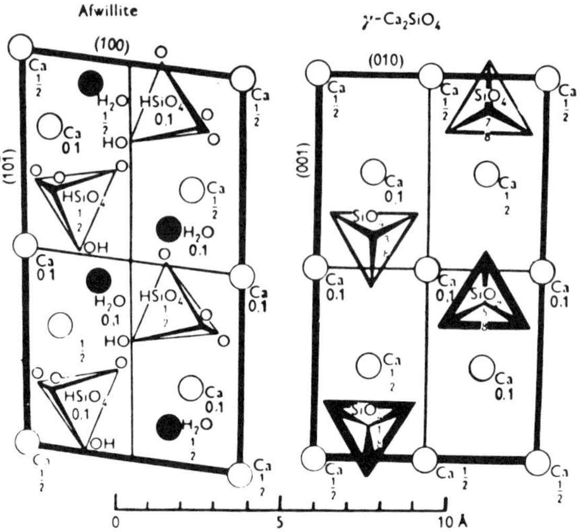

FIG. A. 38. Structure of afwillite, somewhat idealized, in a half-cell, and γ-Ca_2SiO_4 structure projected parallel the *b*-, and the *a*-axis, respectively, to show the oriented conversion by dehydration of the group $[SiO_3OH]$ into $[SiO_4]$ (Taylor).

structure in Fig. A. 39 shows the position of the tetrahedra $[SiO_3(OH)]$ in the pseudocell (space group) $V_h^{15} = P\ bca$, in superposition by projection on the (*ac*) plane.

71. An interesting comparison may be made between afwillite and the mineral *bultfonteinite*, $Ca_4[SiO_2(OH)_2]_2F_2$, the structure of which was studied by H. D. Megaw and C. H. Kelsey.[113] The similarity to afwillite is so great that by simple dehydration at 400°C. a monocrystal of bultfonteinite is changed again to a monocrystal of cuspidine, $Ca_4[Si_2O_7]F_2$ (cf. A. ¶ 78 ff.), with a condensation of two groups $[SiO_2(OH)_2]$ to one $[Si_2O_7]$ group $+ H_2O$. This process is connected with only a slight change in the position of the Ca^{2+} cations. The *c*-axis of bultfonteinite is the *b*-axis of afwillite.

[112] *Acta Cryst.*, **5**, 1952, 724-728. Isolated $[SiO_4]$ tetrahedra are characteristic of the baryum silicate hydrate, $BaO.SiO_2.6_2HO$, as described in its structure by E. Höhne, *Fortschr. Mineral.*, **39**, 1961, 332-333.

[113] *Mineral. Mag.*, **30**, 1955, 569-573. Cf. E. J. McIver and H. D. Megaw, *Silikat Tech.*, **11**, 1960, 395.

72. In this respect the complex silicate *thaumasite*, $Ca_3H_2(CO_3/SO_4)[SiO_4].13H_2O$, which was assigned by H. Strunz (1949) to the olivine-norbergite group is also highly interesting. Besides the occurrence of planar $[CO_3]$ and tetrahedral $[SO_4]$ groups, the silicon appears in two single $[SiO_4]$ groups, but *not* coupled to a $[Si_2O_7]$ group

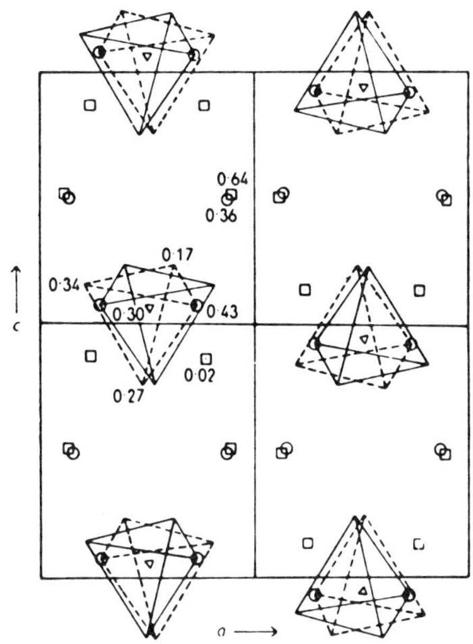

FIG. A. 39. Structure of $Ca_2SiO_4.H_2O$ in projection on plane (*ac*) (Heller). The squares represent Ca^{2+}; the triangles, Si^{4+}; the circles, the probable position of hydroxyl ions.

as E. Welin[114] interpreted from extensive structural studies. The most characteristic feature of the rather complex structure is the interlinkage of chains parallel to the hexagonal *c*-axis, of $[SiO_4]$ tetrahedra with Ca^{2+} coordination polyhedra each containing three molecules of H_2O, as shown in Fig. A. 40 a; this is compared with an analogous interlinkage in γ-Ca_2SiO_4 (Fig. A. 40 b). In a lower degree a similar interlinkage occurs also in monticellite but with the same (slightly distorted) $[CaO_6]$ octahedra. This mutual linkage, however, is not the same in thaumasite and in afwillite, although bonds of the same kind as those found in afwillite also exist in thaumasite. Characteristic are the O—O distances of 2.48 to 2.60 Å. between O^{2-} of the $[SO_4]$ and $[CO_3]$ groups, and oxygen distances in the Ca-polyhedra. Short hydrogen bonds have bond lengths of 2.48 to 2.60 Å., longer ones above 2.60 Å. as mentioned previously also for afwillite. On the analogies between thaumasite and tilleyite see below. Also interesting is a comparison of thaumasite with *gypsum*,

[114] *Arkiv Kemi*, **2**, 1956 (2) 137-147.

$CaSO_4.2H_2O$, with similar distances Ca—H_2O (2.44 Å. in gypsum; 2.50 Å. in thaumasite).

73. An entirely different type of isolated [SiO_4] groups connected with [UO_7] polyhedra, forming uranyl silicate sheets of the unit type $[(UO_2)_2(SiO_4)_2]^{4-}$ occurs in

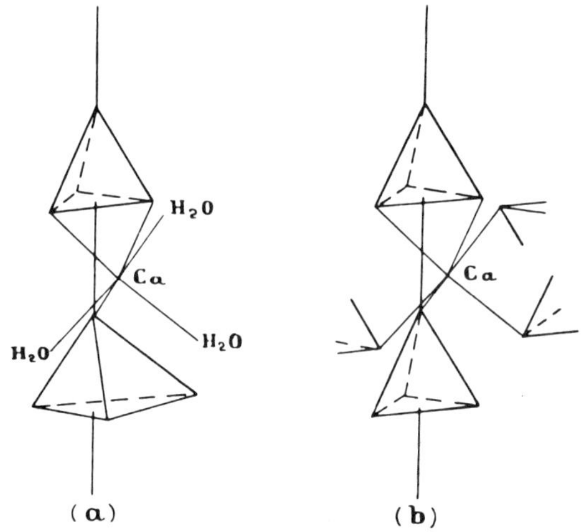

FIG. A. 40. Comparison between the [SiO_4]—Ca^{2+} chains parallel to the *c*-axis in the structure of thaumasite (*a*), and in γ-Ca_2SiO_4 (*b*) (Welin).

the interesting structures of *uranophane* and *sklodovskite*, the mineral first investigated by D. K. Smith, J. W. Gruner, and W. N. Lipscomb,[115] the second by V. I. Mokeeva.[116] The aspect of this unit in uranophane is seen from Fig. A. 41; the [SiO_4] tetrahedra are not distorted, and are the major control on the bond patterns in the unit. This interpretation is the more remarkable since previously uranophane was assumed to be a structure with [Si_2O_7] groups, or even with $[Si_2O_6]_\infty$ chains but there are strong arguments against both interpretations, especially because of intolerable distance conditions for the best possible connection of a double-tetrahedron to an U atom. In sklodovskite, the same unit $[(UO_2)_2(SiO_4)_2]^{4-}$ occurs but with layers shifted parallel to one another by the translation $c/2$. They are interlinked in this mineral by Mg^{2+} cations, in uranophane by Ca^{2+}. The analogy of the sheet unit with that in autunite, $[(UO_2)_2(PO_4)_2]^{2-}$, is most evident. The perfect cleavage of the "uran micas" is caused by these sheet arrangements although uranophane has a more fibrous habitus.

[115] *Am. Mineralogist,* **42,** 1957, 594-618; see also D. H. Gorman and E. W. Nuffield, *Ibid.,* **40,** 1955, 634-645, who distinguish a "β"-crystal phase besides normal uranophane.

[116] *Doklady Akad. Nauk S.S.S.R.,* **124,** 1959, 578-580. See also D. H. Gorman, *Can. Mineralogist,* **6,** 1957, 52-60, and recently E. A. Jumpertz, *Fortschr. Mineral.,* **39,** 1961, 86, who gives evidence for the monoclinic space group of sklodovskite.

74. The rather irregular replacement of Ca^{2+} ions by Mg^{2+} which occurs in orthosilicates of the garnet group is also typically observed in the rare complex fluorocar-

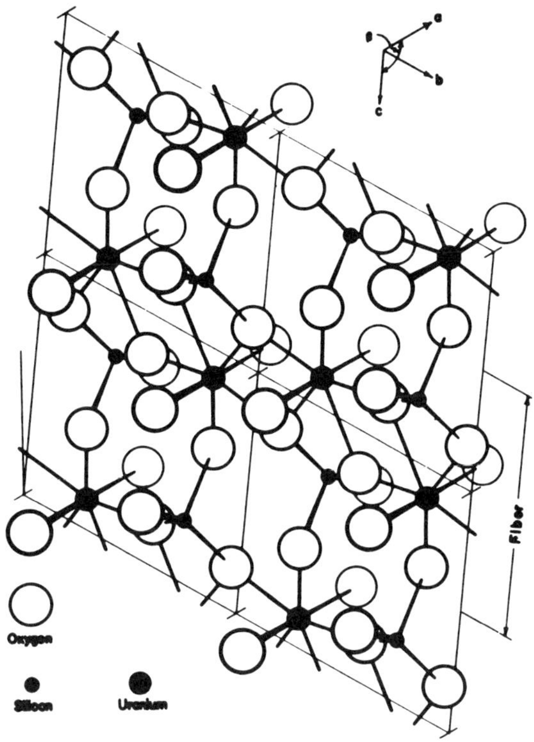

FIG. A. 41. Isometric drawing of the uranyl silicate unit $[(UO_2)_2(SiO_4)_2]^{4-}$ in uranophane (Smith, Gruner, and Lipscomb).

bonate mineral *cerite*, as described by J. J. Glass, H. T. Evans, M. K. Carron, and F. A. Hildebrand.[117] In the proposed crystallochemical formula

$$(Ca,Mg)_2(RE)_8[SiO_4]_{7-x}[FCO_3]_x(OH)_x(H_2O)_{3-x}$$

(observed $x = 0$ to 1.6) the interesting possibility of a complex substitution of the single $[SiO_4]$ tetrahedra by the groups $[FCO_3]$ is obvious; it was also discussed as an explanation of the constitution of *apatite* minerals like *francolite*, if we do not prefer a replacement of $[PO_4]$ by $[CO_3]^{2-}$, or a tetrahedral combination of $F^- + CO_3^{2-}$, or $CO_3^{2-} + OH$. In the apatite minerals *wilkeite* and *ellestadite* the isomorphous substitution of $[PO_4]$ by $[SO_4]$, $[SiO_4]$, and $[CO_3]$ is a rather familiar phenomenon (cf. A. ¶ 234).

[117] *Am. Mineralogist*, **43**, 1958, 460-475.

Subgroup 1b. Pair af Tetrahedra (Sharing Coefficient 1.25)

75. Typical of this Subgroup is the complex $[Si_2O_7]^{6-}$, called by H. Strunz (1949) the "*diorthosilicate*" anion to distinguish it from other isolated complex groups of tetrahedra. We follow Zoltai's classification principles in no longer assigning the name *hemimorphite* to this subgroup but to three-dimensional frameworks (Subgroup 4. a) (cf. A. ¶ 173). According to recent results of G. A. Barclay and E. G. Cox[118] in *thortveitite* ($Sc_2Si_2O_7$) as a particularly characteristic representative of Subgroup 1. b, the symmetry of the monoclinic space group differs from the original determination; especially the angle Si—O—Si is not collinear (180°) as was formerly assumed. A very simple representative of structures with $[Si_2O_7]$ was recently described by I. Náray-Szabó and A. Kálmán[119] in the compound $K_2Pb_2[Si_2O_7]$, as given in Fig. A. 42; it is characteristic of potassium-lead silicate glasses.

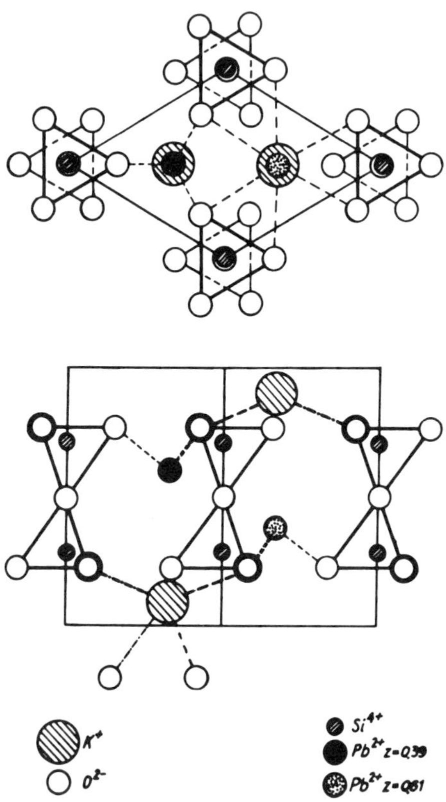

FIG. A. 42. Structure of $K_2Pb_2Si_2O_7$ (Náray-Szabó and Kálmán).

[118] *Z. Krist.*, **113**, 1960, 23-29.

[119] *Silikat Tech.*, **12**, 1961, 316-318. The lead cations are here coordinated to O^{2-} anions in flattened trigonal pyramid configurations.

76. The highly complex structure of *ilvaite* (*lievrite*), $CaFe_2^{2+}Fe^{3+}[Si_2O_7](OH)$, was determined by N. V. Belov and V. I. Mokeeva[120] with the characteristic dior-tho-silicate group, together with $[FeO_6]$ and $[CaO_7]$ coordinations. The $[FeO_6]$ octahedra are arranged in double chains parallel to the *c*-axis (see Fig. A. 43), similar to that of $[AlO_6]_\infty$ in disthene and andalusite, $[CaO_7]$ connecting the double chains $[FeO_6]$. Peculiar is the onesided contact of the $[Si_2O_7]$ groups, ad only one

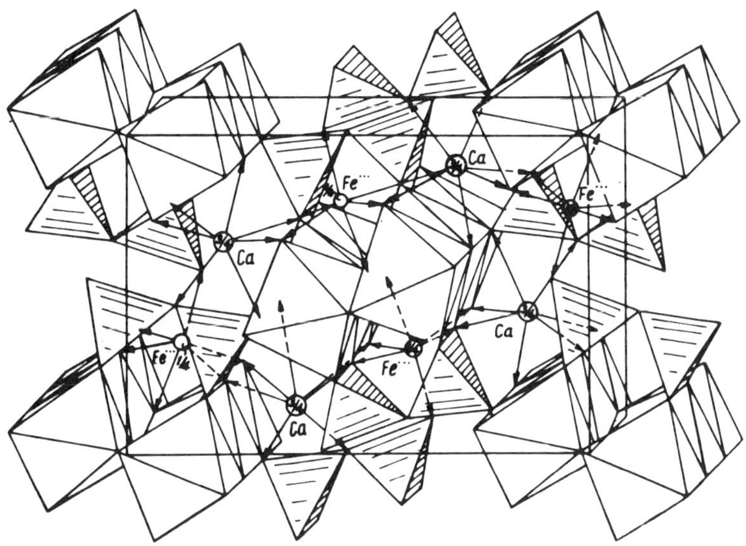

FIG. A. 43. Perspective representation of the unit cell of ilvaite, showing the $[FeO_6]$ double chains and their interlinkage with $[CaO_7]$ groups, and the $[Si_2O_7]$ groups (shaded) (Belov and Mokeeva).

$[FeO_6]$ chain. The $[SiO_4]$ tetrahedra, however, are so situated that they also contact chains of virtual (empty) octahedral groups (Fig. A. 44) as they occur in olivine, disthene, zircon, and apatite. Hydroxyl groups are on the same level as $2[FeO_6]$ and $1[CaO_7]$ group. Belov's and Mokeeva's results differ somewhat from the structure as given by Y. Takeuchi[121] who interpreted ilvaite as an orthosilicate, with single $[SiO_4]$ tetrahedra, but this assumption leads to intolerable distances O—O (1.8 Å. in the place of theoretical 2.5 Å.), and of highly distorted tetrahedra with Si—O distances of 1.93 to 2.8 Å. (usually 1.55-1.79 Å.). Takeuchi's proposal is therefore abandoned. Of high crystallochemical interest, on the other hand, are the close relations of Belov's and Mokeeva's ilvaite structure with that of *lawsonite* (see below), and to *ramsayite*, $Na_2Ti_2O_3[SiO_3]_2$. The angle Si—O—Si is 122°.

[120] *Doklady Akad. Nauk S.S.S.R.*, **81**, 1952, 581-584, 825-827.

[121] *X-Sen*, **5**, 1948, 8-14; abstract *Chem. Abstr.*, **44**, 1950, 5766e.

77. For *lawsonite*, $CaAl_2[Si_2O_7](OH)_2.H_2O$, Fr. E. Wickman (1947) had given an orthorhombic space group with $[Si_2O_7]$ groups and characteristic water "pockets" in the structure (not channels, in distinction from zeolites). A refined re-examination of the structure by I. M. Rumanova and R. K. Skipetrova[122] corrected the proposal of Wickman. The structure is characterized by infinite $[AlO_6]_\infty$ chains or "columns" around screw axes parallel *b*, and every unit cell with four of these columns interlinked with isolated $[Si_2O_7]$ groups. The Si—O and Al—O distances are normal (some-

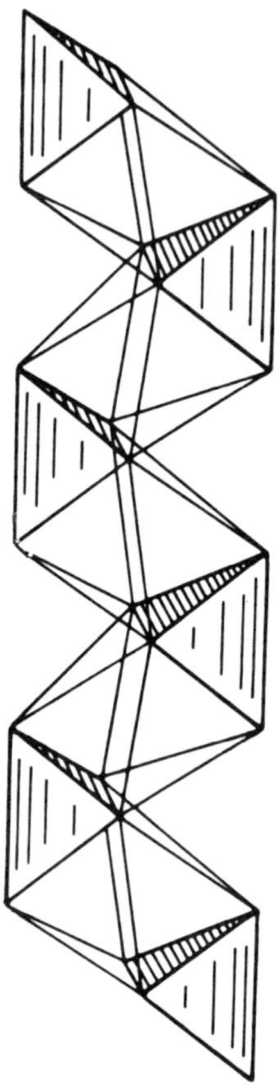

FIG. A. 44. One-sided contacting of $[SiO_4]$ tetrahedra of the $[Si_2O_7]$ groups (shaded), with $[FeO_6]$ columns, not designed, but the empty (virtual) octahedra chain resulting is designed (Belov and Mokeeva).

[122] *Doklady Akad., Nauk S.S.S.R.*, **124**, 1959, 324-327. On the so-called Templeton-effect and the apparent symmetry of lawsonite due to special values of the *y*-parameters of all atoms in general positions of its structure see A. Pabst, *Z. Krist.*, **112**, 1959, 53-59; *Ibid.*, **115**, 1961, 307-309, and Karl Fischer, *Ibid.*, **115**, 1961, 310-313.

what anomalous in Wickman's proposal), the angle Si—O—Si = 135° (Wickman gives 149°). Between the $[Si_2O_7]$ groups and the $[AlO_6]$ columns are arranged coordination groups $[Ca(O, OH)_6]$ (Fig. A. 45).

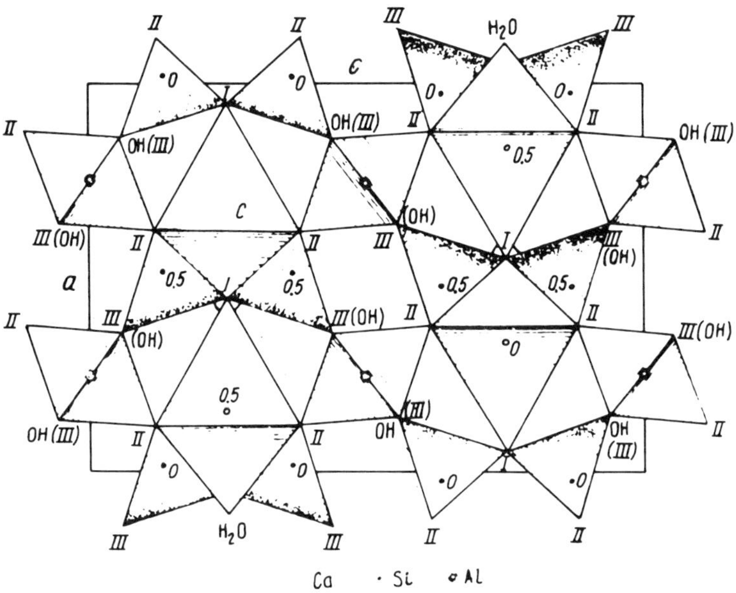

FIG. A. 45. Projection of the structure of lawsonite on the (xy) plane to show the interlinkage of the (shaded) $[Si_2O_7]$ groups with the $[AlO_6]$ columns and the $[Ca(O, OH)_6]$ polyhedra. (Rumanova and Skipetrova).

78. Of great theoretical interest is the structure of *cuspidine*, $Ca_3[Si_2O_7].Ca(OH, F)_2$, which was investigated by R. F. Smirnova, I. M. Rumanova, and N. V. Belov[123] (Fig. A. 46). Similar to the arrangements in ilvaite, cuspidine is built up of $[CaO_6]_\infty$ octahedral columns parallel to the x-axis; bundles of four chains are combined to wider ribbons parallel (012) and $(01\bar{2})$ between which the single $[Si_2O_7]$ groups are inserted, with their elongation parallel x. The angle Si—O—Si for cuspidine = 180° is in contrast with ilvaite (in which it is = 135°), but the interatomic distances are entirely normal, the Ca—O distances variable between 2.22 and 2.59 Å. Characteristic is the pseudoperiod $a/2$ which also occurs in tilleyite. The cuspidine structure explains very well the characteristic polysynthetic twinning in (010).

79. Very closely related to tilleyite and cuspidine is also the *tricalcium silicate hydrate* (TSH) the crystallochemical formula of which was written by N. V. Be-

[123] *Zapiski Vsesoyuz. Mineral. Obshchestva*, **84**, 1955, 159-169.

lov[124] $Ca_4[Si_2O_7](OH)_2.2Ca(OH)_2$. This is evidently the same composition which was already predicted by G. E. Bessey (1936), and discussed in the synthetic studies of E. R. Buckle, J. A. Gard, and H. F. W. Taylor[125] (cf. Vol. IV, Section A).

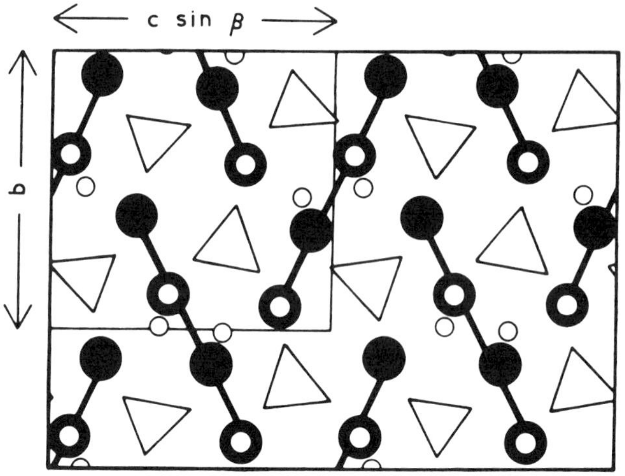

FIG. A. 46. Structure of cuspidine (Smirnova, Rumanova, and Belov). Large full circles represent Ca^{2+} ions at levels 0 and $^1/_2$; large open circles, Ca^{2+} ions at levels $^1/_4$ and $^3/_4$; small circles, F^- anions; triangles, $[Si_2O_7]$ groups, seen from above.

80. The *tilleyite* structure (formula $Ca_5[Si_2O_7](CO_3)_2$), is according to J. V. Smith[126] highly analogous to that of cuspidine, with $[Si_2O_7]$ groups correlated to pairwise $[CO_3]$ triangles as seen in Fig. A. 47 and with a characteristic pseudo-period $a/2$, interlinked by $[CaO_6]$ groups.

81. The determination of the tilleyite structure gave the key also for the structural interpretation of the complex minerals of the *woehlerite-låvenite*, and the *rinkite-mosandrite* families to which generally the orthosilicate formulae $A_3B[SiO_4]_2(F, OH)$ are assigned. However, they should be written in the form of diorthosilicates, because of their close relation to cuspidine and tilleyite, with isolated $[Si_2O_7]$ groups and infinite $[CaO_6]_\infty$ columns. Kh. S. Mamedov, V. I. Simonov, and N. V. Belov[127] speak of a characteristic structural unit in the combination $\{[CaO_6]_\infty.[Si_2O_7]\}$ as the "*tilleyite ribbons*" (Fig. A. 48). These are discrete in tilleyite, connected by $[CaO_6]$ polyhedra and $[CO_3]$ groups as seen above. They are directly interconnected in cuspidine, woehlerite, and låvenite to double $[CaO_6]$ chains. Another principle of

[124] *Fortschr. Mineral.*, **38**, 1960, 5.
[125] *J. Chem. Soc.*, 1958, 1351-1355.
[126] *Acta Cryst.*, **6**, 1953, 9-18.
[127] *Doklady Akad. Nauk S.S.S.R.*, **126**, 1959, 379-381.

interconnection rules in *seĭdozerite.*[128] The somewhat schematic projections in Figs. A. 49-52 show the interlinkage principles in the minerals here discussed, and Table II shows their composition in crystallochemical formulation to illustrate their analogies.

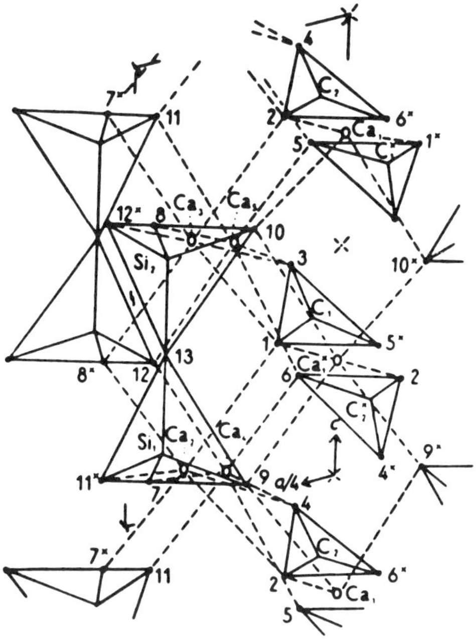

FIG. A. 47. Projection of the tilleyite structure on the (*xz*) plane to show the interlinkage of the $[Si_2O_7]$ groups with the $[CO_3]_2$ pairs, and the $[CaO_6]$ polyhedra (Smith).

82. In the woehlerite-låvenite family zirconium is the characteristic element; in that of rinkite and mosandrite, titanium. Cuspidine is related to the first family, seĭdozerite to the second; both mineral groups being in a relation of structural polytypism, particularly evident in projections on the plane (*ab*). In cuspidine, etc.,

FIG. A. 48. The "tilleyite unit" (Mamedov, Simonov, and Belov). Each octahedron is the basis of an infinite column $[CaO_6]_\infty$ type, perpendicular to the plane of design. The black-filled tetrahedra are the basis of $[SiO_4]$ tetrahedra which pairwise build up the diorthosilicate groups $[Si_2O_7]$.

[128] Cf. V. S. Simonov and N. V. Belov, *Doklady Akad. Nauk S.S.S.R.*, **122**, 1958, 473-476; *Kristallografiya*, **4**, 1959, 163-182. On the isomorphic relations between Zr^{4+} in the Mn containing minerals of the Lovozero Massive see the same authors, *Doklady Akad. Nauk S.S.S.R.*, **125**, 1959, 888-889.

TABLE II

COMPARISON OF THE MINERALS OF THE WOEHLERITE AND RINKITE FAMILIES
(MAMEDOV, SIMONOV, AND BELOV)

A. *Woehlerite–Låvenite Family*							
Woehlerite	$Ca_2NaZrO[Si_2O_7]F$	$a_0 = 10.82$	$b_0 = 10.28$	$c_0 = 7.27$	$\beta = 108°57'$		
Låvenite	$(Na,Ca,Mn)_3(Zr,Ti)O[Si_2O_7]F$	10.95	10.01	7.19	110°18′		
Hellandite	$(Ca,RE,Mn)_3(Al,Fe)O[Si_2O_7].H_2O$				109°45′		
Guarinite	$CaNaZrO[Si_2O_7]F$	10.93	10.31	7.33	$\alpha = 90°29'$	$\beta = 108°50'$	$\gamma = 90°$
Rosenbuschite	$(Na,Ca)_3(Fe,Ti,Zr)O[Si_2O_7]F$	10.14	11.41	7.28	91°21′	99°38′	112°
Hjortdahlite	$NaCa_2ZrO[Si_2O_7]F$	10.91	10.29	7.32	90°29′	108°50′	90°
Niocalite	$Ca_{3.5}Nb_{0.5}O[Si_2O_7](OH,F)$	10.83	10.42	7.38	$\beta = 109°40'$		
Cuspidine	$Ca_4[Si_2O_7](F,OH)_2$	10.35	10.43	7.55	$\beta = 110°04'$		
B. *Rinkite–Mosandrite Family*							
Rinkite	$Na(Ca,Ce)_2(Ti,Ce)O[Si_2O_7]F$	18.51	5.68	7.47	$\beta = 91°13'$		
Johnstrupite	$(Ca,Na)_3(Ti,Ce)O[Si_2O_7](OH,F)$				93°04′		
Mosandrite	$(Ca,Na,RE)_{2-3}(Ti,Zr,Ce)O[Si_2O_7](OH,F)$	18.41	5.64	7.43	93°		
Lamprophyllite	$Na_2Sr_2Ti_3O[Si_2O_7](O,OH,F)_2$	19.09	5.35	7.06			
Seĭdozerite	$Na_2Mn_{0.5}Zr_{0.75}Ti_{0.75}[Si_2O_7]F$	18.30	5.53	7.10	102°43′		

the period b_0 is constant, and a_0 doubled in relation to the tilleyite unit. Inversely, in seĭdozerite, a_0 is constant, and b_0 doubled. Approximately,

$$(a_0 \times b_0)_{\text{woehler.}} = (a_0 \times b_0)_{\text{rink.}}$$

If the $[Si_2O_7]$ groups are differentiated in the positions $z = 0$ and $z = \frac{1}{2}$, irregular distortions occur which make the monoclinic minerals change to triclinic symmetries in the *rosenbuschite* type.

FIG. A. 49. Schematic diagram to show how the tilleyite ribbons develop infinite octahedral columns by extension of the ribbons of one orientation (Mamedov, Simonov, and Belov).

83. A recent special determination of the structure of *låvenite* by V. I. Simonov and N. V. Belov[129] confirmed the existence of octahedral $[Ca, Zr]O_6]_\infty$ chains between which two types of $[Si_2O_7]$ groups are interlinked with the vertical edges of three octahedra surrounding the large cations which occupy positions of the cations

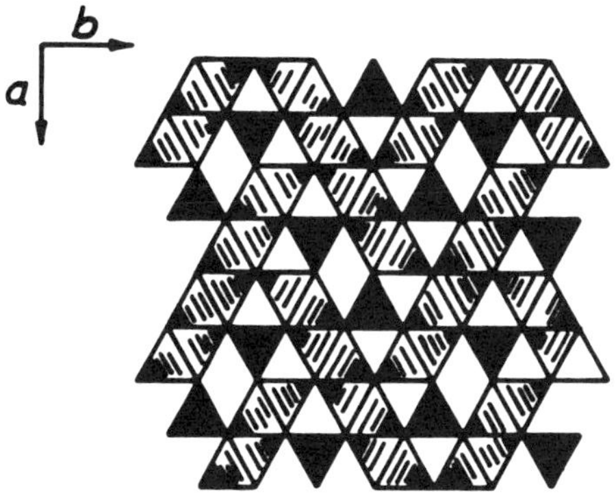

FIG. A. 50. Structure of cuspidine, built up of tilleyite ribbons, to show the fourfold $[CaO_6]_\infty$ chains formed by interlinkage of these units in one direction (Mamedov, Simonov, and Belov).

[129] *Doklady Akad. Nauk S.S.S.R.*, **130**, 1960, 1333-1336; V. I. Simonov, *Silikat Tech.*, **11**, 1960, 392-394.

FIG. A. 51. Structure of seĭdozerite, principle sketch (Mamedov, Simonov, and Belov).

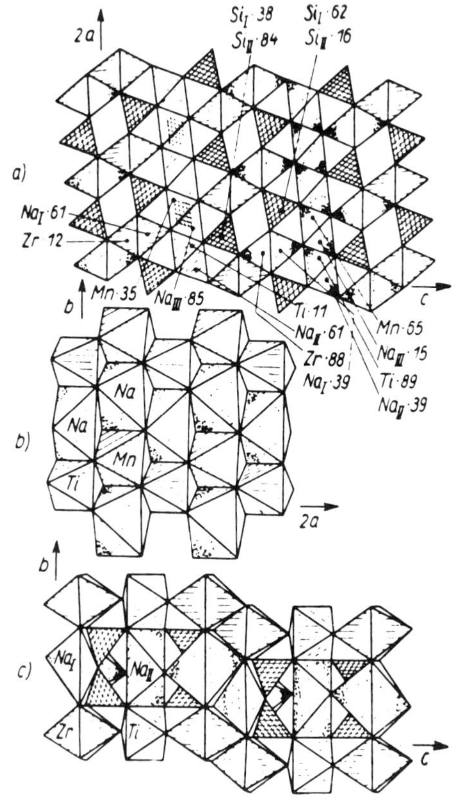

FIG. A. 52. Details of the structure of seĭdozerite (Belov). There are four kinds of columns: two with [ZrO_6] octahedra, one with [TiO_6] octahedra, and one with [MnO_6] octahedra. In each column the octahedra alternate with [NaO_6] octahedra; the [Si_2O_7] groups are inserted between the Na-octahedra. (*a*) Projection on the basis; (*b*) to show the densely packed octahedra "walls"; (*c*) the coherent interlinkage of the [Si_2O_7] groups with the (elongated) Na-octahedra.

in cuspidine by complex ionic exchange reactions such as $Ca^{2+} \rightleftarrows Zr^{4+}$, or $Ti^{4+} \rightleftarrows$ $\rightleftarrows Fe^{2+}$, Mn^{2+}, and also Na^{+} participates in such reactions. Fig. A. 53 shows the structure plan of låvenite in analogy to Figs. 49-52, Fig. A. 54 the mode of the occupation of the octahedral columns with different cations. According to Gl. Gottardi[130] we should also classify *perrierite* together with the structures having $[Si_2O_7]$ groups,

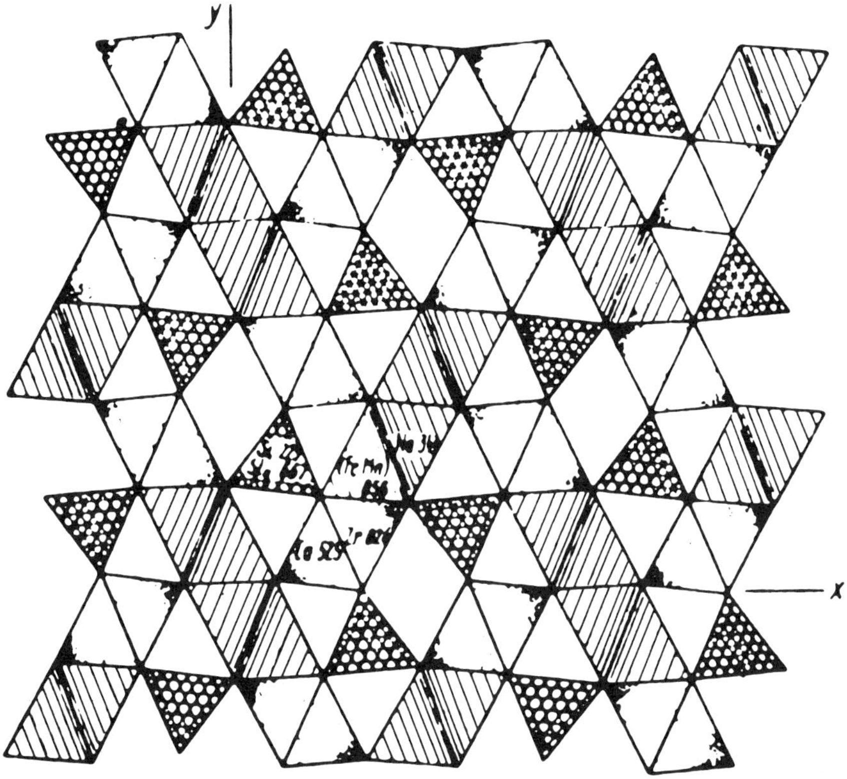

FIG. A. 53. Special structure plan of låvenite to show the tilleyite units in the plane (xy) projection (Simonov and Belov).

interlinked by $[TiO_6]_\infty$ chains parallel to the *b*-axis (Fig. A. 55). Every two octahedra have an edge in common. Actually there are two of these chains, together forming a layer.

Subgroup 1c. Large Groups (Sharing Coefficients = 1.40)

84. No recent structural investigations on the *zunyite* minerals with the complex group $[Si_5O_{16}]^{12-}$ have been known. It is, however, interesting that according to J.

[130] *Am. Mineralogist,* **45**, 1960, 1-14.

Konta and L. Mráz[131] zunyite is the hydroxyl end member and the rare mineral *dillnite* the fluorine end member of the series of crystalline solutions $Al_{15}Si_5O_{20}(OH, F)_{18}Cl$.

Subgroup 1d. Mixed Groups (Sharing Coefficients 1.00 to 1.50)

85. *Vesuvianite* (idrocrase) which was previously described by B. E. Warren and D. J. Modell (1931) as an orthosilicate-isle structure mineral (Subgroup 1a) is classified by Zoltai with mixed groups, i.e., having a sharing coefficient larger than 1.00. A typical representative of the simultaneous presence of single $[SiO_4]$

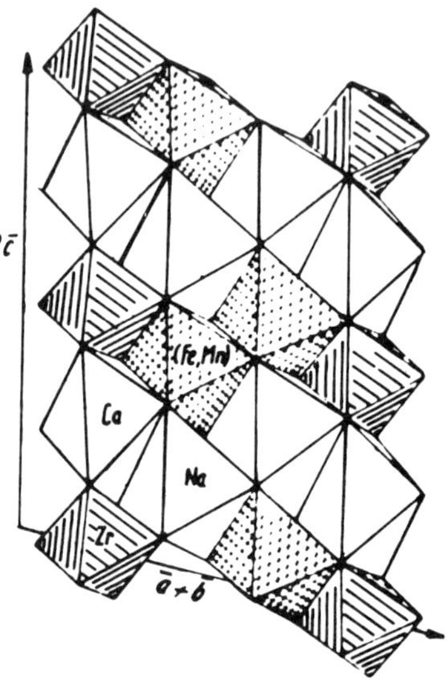

FIG. A. 54. Occupation of the octahedral columns in the låvenite structure by various cations (Simonov and Belov).

and $[Si_2O_7]$ groups in the unit cell is *zoisite*, $Ca_2Al_3[Si_2O_7][SiO_4]O(OH)$, for which E. G. Fesenko, I. M. Rumanova, and N. V. Belov[132] gave a complete structure determination. Interesting is the Si—O—Si angle in the $[Si_2O_7]$ group of 161°. The structure is built up by $[AlO_6]$ columns parallel to the *b*-axis. One group $[Si_2O_7]$ and three groups $[AlO_6]$ form in the plane (xy) a very characteristic pentagonal configuration (Fig. A. 56). The $[CaO_6]$ coordination groups have the shape of trigonal prisms, and the groups $[CaO_6]$ are combined from a trigonal prism and a half-octahedron, similar to those in epidote, titanite, and even in apatite. The previous struc-

[131] *Am. Mineralogist*, **46**, 1961, 629-636.

[132] *Doklady Akad. Nauk S.S.S.R.*, **102**, 1955, 275-278; *Kristallografiya*, **1**, 1956, 171-196. The previous proposals of T. Ito, "X-Ray Studies on Polymorphism," Maruzen, Tokyo, 1950, 50-69, are refuted.

ture proposal of T. Ito (see below) is refuted. The multiple similarity with epidote is particularly evident in the approximate identity of the parameters a_0 (Zoisite) $= 2 \times a_0$ (Epidote) $\times \sin\beta$ (Epid.), and can be interpreted as a structural twinning in zoisite with (100) (Epid.) as the twinning plane. Also the change from the positive optical character of zoisite to the negative character of epidote is understood from the structure.

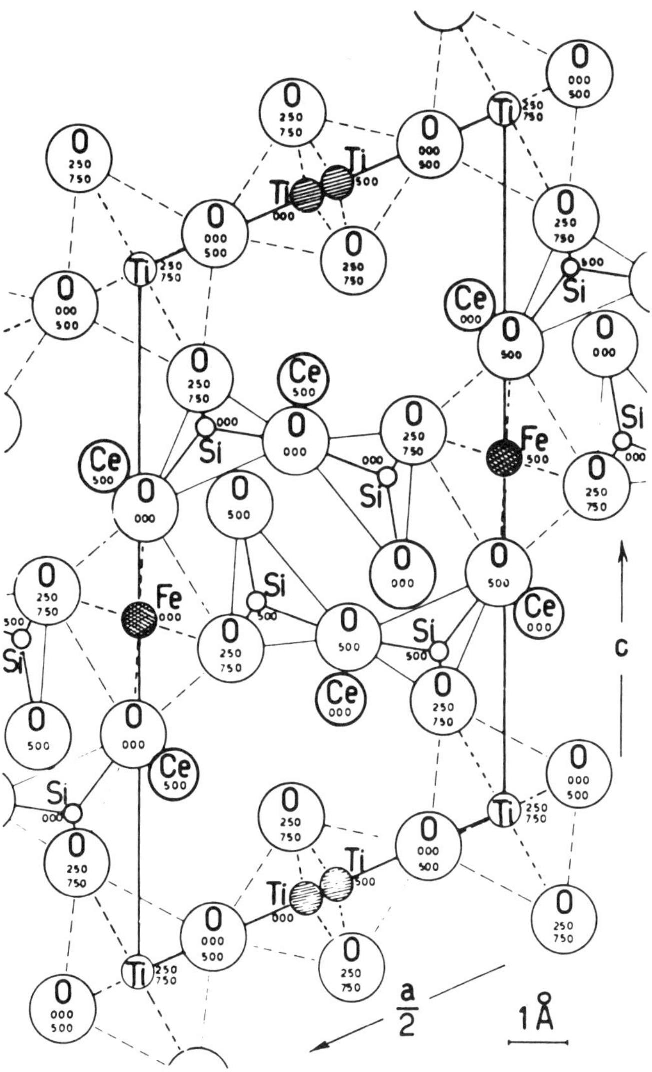

FIG. A. 55. Structure of Perrierite, projected onto (010) (Gottardi).

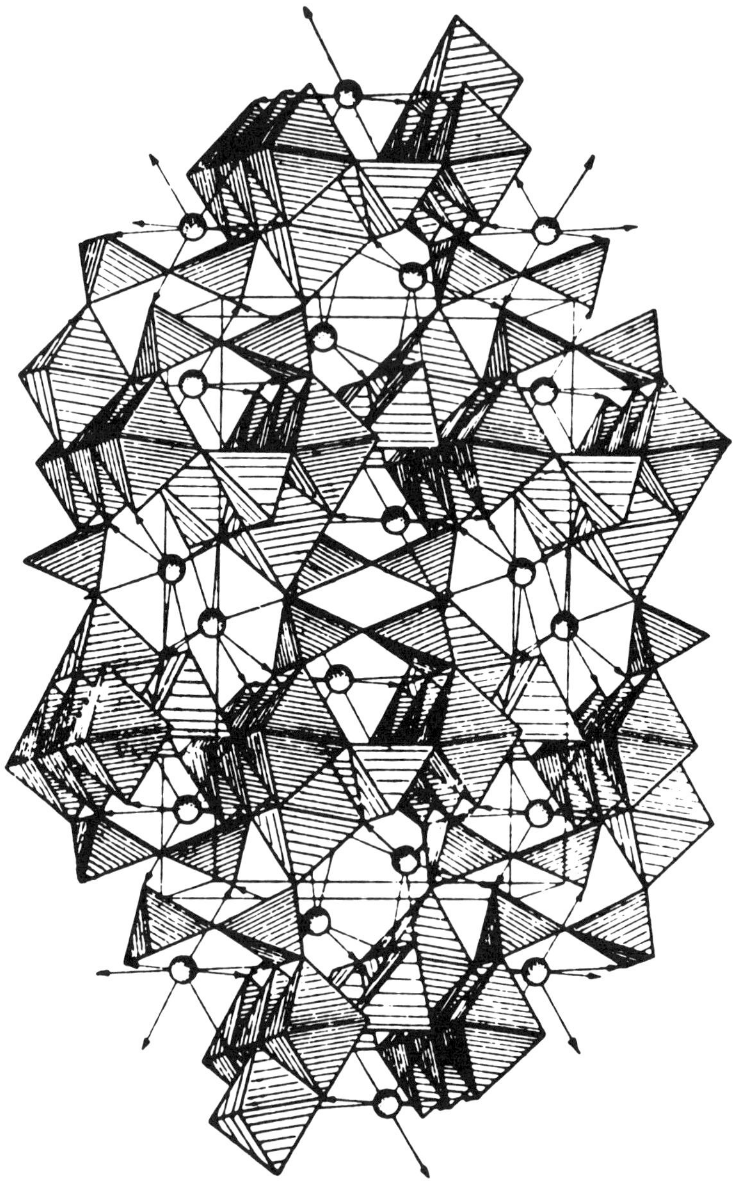

FIG. A. 56. Structural interlinkage of $[Si_2O_6]$ groups with $[AlO_6]$ columns in the structure of zoisite (Fesenko, Rumanova, and Belov).

86. For *epidote* the structure as proposed by T. Ito[133] is refuted; this author had postulated complex fourfold ribbons of the type $[Si_4O_9]_\infty$, but later corrected his first calculations to confirm also for epidote the presence of single $[SiO_4]$ and $[Si_2O_7]$ groups, as seen in Fig. A. 57. Epidote is in a close structural and chemical relationship to the minerals of the rinkite-mosandrite group, but also particularly to *chevkinite* and *perrierite* with the general formula $A_3B_2Si_2O_{12}$. An extensive structural study

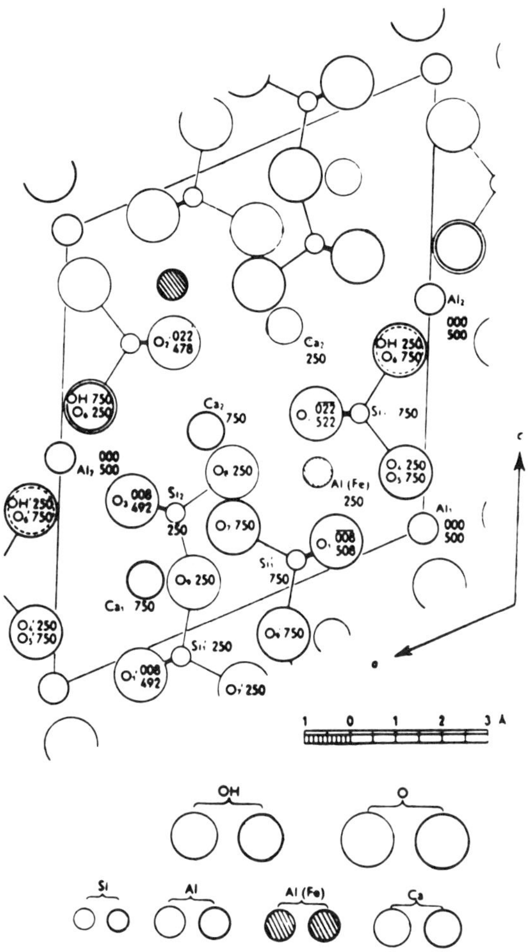

FIG. A. 57. Structure of epidote, projected on (010) (Ito, Morimoto, and Sadanaga). $[SiO_4]$ and $[Si_2O_7]$ groups are indicated.

[133] *Am. Mineralogist*, **32**, 1947, 309-321. For a re-examination, see *Acta Cryst.*, 7, 1954, 53-59 (with N. Morimoto and R. Sadanaga).

by St. Bonatti[134] illustrates these analogies, as in the common characteristics of oxygen anion chains along the monoclinic symmetry axis.

2. One-Dimensionally Infinite Structures of Tetrahedra

Subgroup 2a. Single Chains (Sharing Coefficient 1.50)

87. The classical metasilicates, $RSiO_3$, to which Bragg in his classification ascribed a single chain of $[SiO_3]_\infty$ type, is restricted to the *pyroxenes* and a few related crystal structures. It is of fundamental significance that other unit types than the pyroxene chain have been detected in recent structure investigations with the important result that besides the alternating sequence of chains in the type of Fig. A. 58 b as 2-repeat-units (F. Liebau called them "*Zweierketten*"), written $[Si_{1+1}O_6]_{,\infty}$ other different chains exist in which the *pyroxenoid* 3-repeat-units ("*Dreierketten*") $[Si_{2+1}O_9]_\infty$ shown in Fig. A. 58 c are the ruling coordination units. F. Liebau[135] could even deter-

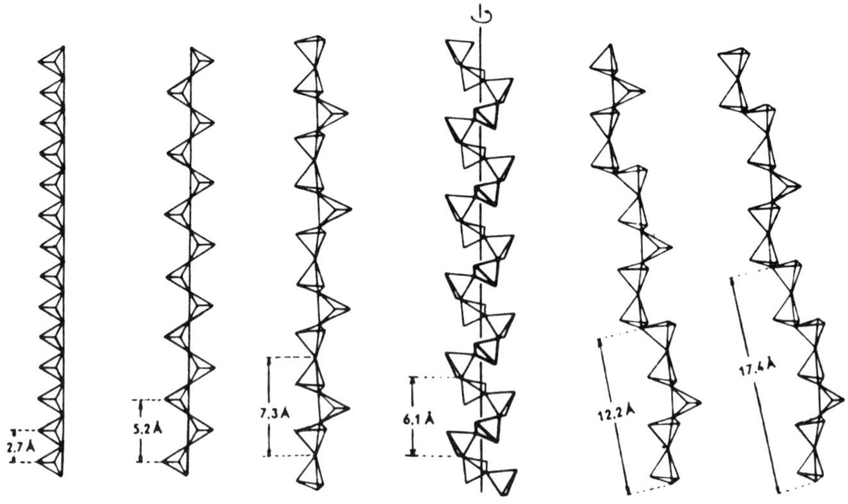

Fig. A. 58. Schematic representation of silicate chains (Liebau): one-repeat chain (*a*); pyroxene (2-repeat-unit) chains (*b*); pyroxenoid (3-repeat-unit) chains (*c*); 4-repeat-unit chains (*d*); rhodonite (5-repeat-unit) chains (*e*); pyroxmangite (7-repeat-unit) chains (*f*).

[134] *Am. Mineralogist*, **44**, 1959, 115-137; see also B. A. Makarochkin, K. A. Gonibesova, and N. S. Makarochkina, *Zapiski Vsesoyuz. Mineral. Obshchestva*, **88**, 1959, 547-553. See further St. Bonatti and G. Gottardi, *Rend. soc. mineral. ital.*, **10**, 1954, 200-225; *Periodico Mineral.* (*Rome*), **23**, 1954, 245-250.

[135] *Naturwissenschaften*, **43**, 1956, 117-178; *Ibid.*, **44**, 1957, 178. In *Acta Cryst.*, **12**, 1959, 177-181. F. Liebau mentions on p. 180 different types of infinite chains, e.g., the 1-repeat units

mine chains in *rhodonite* as 5-repeat-units, $[Si_{2+1+2}O_{12}]_\infty$, and in *pyroxmangite* chains as 7-repeat-units $[Si_{2+1+2+2}O_{22}]_\infty$, as shown in Figs. A. 58e and f. These structural differences explain why previous attempts failed to group in a mineralogical systematics such minerals as wollastonite, rhodonite, and pyroxmangite as members of the pyroxene family, because they have their own independent structural physiognomy, in spite of some similarities with the pyroxenes. It is also particularly interesting that one may interpret a 5-repeat-unit as the sum of a 3- and a 2-repeat-unit which is evident in the additivity of the lattice parameters parallel to the chain elongation, as for rhodonite (5-repeat-unit and parameter 12.20 Å.) proved to be the sum of wollastonite (3-repeat-unit, 7.16 Å.) and the pyroxene type diopside (2-repeat-unit, 5.24 Å.), with a certain tolerance.

88. Single 2-repeat-units occur as in the structure of orthorhombic Na_2SiO_3, which was investigated in its structure by A. Grund and M. Pizy[136]; the $[Si_2O_6]_\infty$ chains are interlinked with sodium cations in the coordination $[NaO_5]$.

89. In the pyroxene family proper, N. Morimoto, D. E. Appleman, and H. T. Evans[137] studied the interrelations of diopside, clinoenstatite, and pigeonite, all characterized by the single chains $[Si_2O_6]_\infty$ (2-repeat-units), in clinoenstatite with all R^{2+} positions occupied by Mg^{2+}, in diopside (cf. B. E. Warren and W. L. Bragg, 1928) one, R_I, by Ca^{2+}, the other, R_{II}, by Mg^{2+} ions, whereas in pigeonite R_I is occupied by $Ca^{2+}_{0.24}Fe^{2+}_{0.76}$ and R_{II} by $Mg^{2+}_{0.52}Fe^{2+}_{0.48}$, in completely ordered substitutions. The chains are in diopside (diallage) crystallographically equivalent, in clinoenstatite and pigeonite, however, in two sets as seen in Fig. A. 59. Changes in shape and relative position are larger in clinoenstatite than in pigeonite, compared with the chains in diopside. It is not yet settled whether all the atoms in R_I in pigeonite are in $[RO_7]$ coordination, or if this coordination may be the result of statistical averaging from $[RO_6]$ and $[RO_8]$ configurations. Diopside-hedenbergite crystalline solutions show displacements of the silicate chains. Distortion of the chains takes place in *spodumene* and in various *augites* (Warren and Biscoe, 1931). Clinopyroxenes of compositions on the join enstatite-ferrosilite ($FeSiO_3$) (cf. Vol. III, Section B) show displacements of the cations and transitions from the monoclinic to the orthopyroxenes, with unmixing of $CaSiO_3$ in the form of diopsidic pyroxenes. Morimoto also investigated the polymorphism of $MgSiO_3$ in the modifications enstatite, protoensta-

in the structure of $CuGeO_3$ and K_2CuCl_3, the "2-repeat-units" in $NaLi[BeF_3]_2$, the "3-repeat-units" in $(NaBeF_3)_x$, the "4-repeat-units" in $(AgPO_3)_4$ (cf. K. H. Jost, *Z. anorg. u. allgem. Chem.*, **296**, 1958, 154-156), and "5-repeat-units" in rhodonite and *babingtonite*, $Ca_4Fe^{II}_2Fe^{III}_2(Si_{10}O_{24})(OH_2)$.

[136] *Acta Cryst.*, **5**, 1952, 837-840.

[137] *Carnegie Inst. Wash. Year Books*, **58**, 1958/59, 193-197; *Z. Krist.*, **114**, 1960, 120-147; 197-198 (on the system $MgSiO_3$ and its polymorphism).

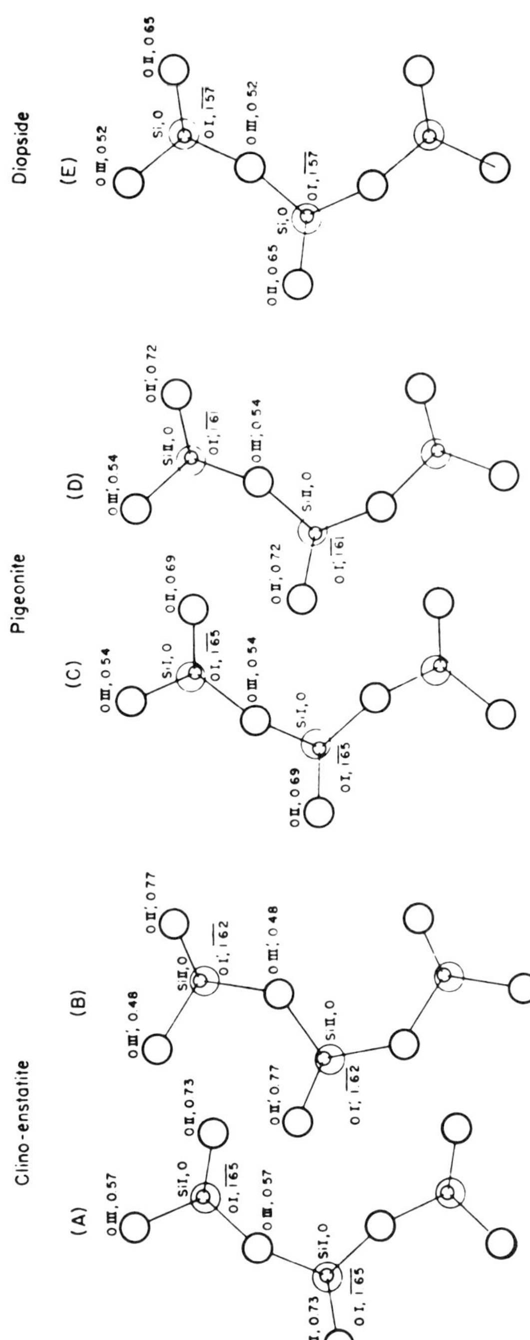

FIG. A. 59. Single $[Si_2O_6]_\infty$ chains in the structures of diopside, clinoenstatite, and pigeonite, projected perpendicular to (100). (Morimoto, Appleman, and Evans).

tite, and clinoenstatite, after the determination of the *protoenstatite* structure by J. V. Smith.[138]

90. All these modifications show in their projections on (001) a very close analogy, with practically the same position and coordination of the cations. Differences are more evident in the projection on (010) (Fig. A. 60), the chains being slightly different from one another. Transitions are connected with displacements of the Mg^{2+} ions in $[MgO_6]$ coordination, without any breaking of the single-chain units, whereas some of the Mg—O bonds are broken in the reaction. Such an interrelation of structurally related modifications makes it highly probable that *intermediate* forms may appear, as they were really observed by J. F. Schairer and N. Morimoto,[139]

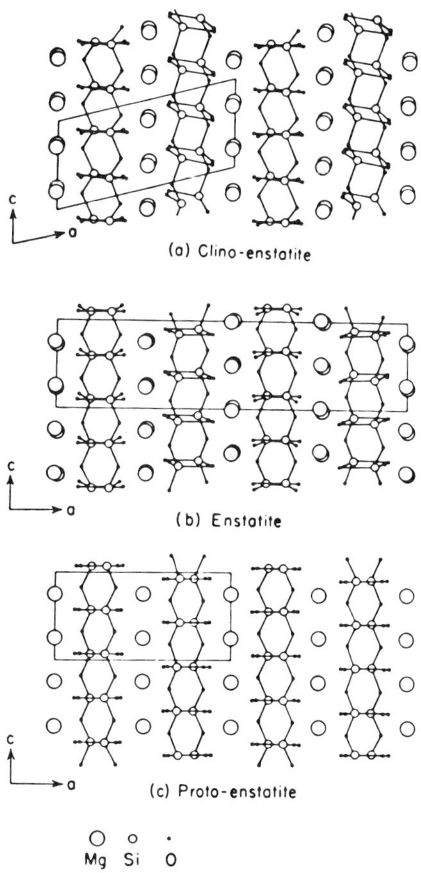

FIG. A. 60. Structure of clinoenstatite (*a*), enstatite (*b*), and protoenstatite (*c*), projected on (010). The unit cells are outlined (Morimoto).

[138] *Acta Cryst.*, **12**, 1959, 515-519.

[139] *Carnegie Inst. Wash. Year Books*, **58**, 1958/59, 113-118.

although these forms are not stable and approach in their powder diagrams that of natural enstatite after thermal exposure during some weeks. A preliminary note of W. Lindemann[140] confirms the presence of pyroxene chains also in the structure of monoclinic γ-$MgSiO_3$, observed in talc after calcination at 1240° for 6 to 24 hours. The microscope shows the lath shape of the single crystals of γ-$MgSiO_3$ which are distinctly different from clinoenstatite. The orthorhombic structure of protoenstatite is seen in a-axis projection in Fig. A. 61.

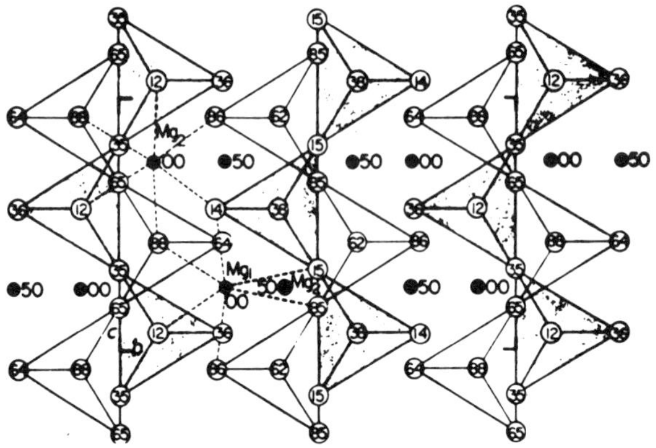

FIG. A. 61. Projection along the c-axis of the structure of protoenstatite (Smith). The two longer Mg—O distances in the Mg_I—O polyhedron are shown in heavy-dashed lines; the others, in light-dashed lines.

91. Structures with 3-repeat-units are typical of wollastonite, pectolite, and related minerals. The previous interpretation given by M. Barnick (1936) is not correct; the ascribed arrangements of $[Si_3O_9]$ annular units, however, are typical of pseudowollastonite (cf. A. ¶ 99) as J. W. Jeffery and L. Heller[141] demonstrated. Natural wollastonite crystals are often composites of monoclinic and triclinic forms as emphasized by M. A. Peacock (1934) to explain the inner twinning effects. The key to the correct interpretation of the chain structure of wollastonite was found by K. Dornberger-Schiff, F. Liebau, and E. Thilo[142] in the determination of the model structure of the compound $(NaAsO_3)_x$ and of the Maddrell salt, the high-temperature-stable salt $(NaPO_3)_x$, with nonbranched chains. Whereas $(NaAsO_3)_x$ does not show structural defects, $(NaPO_3)_x$ and wollastonite have many of these. The skeleton chains in $(NaAsO_3)_x$ and wollastonite are schematically given in Fig. A. 62.

[140] *Naturwissenschaften*, **48**, 1961, 428-429. See also *Sprechsaal*, **94**, 1961, 508-514, on the magnesium pyroxenes as important constituents in electric insulator ceramics (cf. Vol. V, Section B).

[141] *Acta Cryst.*, **6**, 1953, 807-808 (on pseudo-wollastonite), 821-825 (on the inner twinning).

[142] *Naturwissenschaften*, **41**, 1954, 551; *Ibid.*, **43**, 1956, 178; *Acta Cryst.*, **8**, 1955, 752-754.

92. The interpretation of the wollastonite structure with chains of the type $[Si_{2+1}O_9]_\infty$, with interlinking $[Si_2O_7]$ units in the elongation direction, was also given by M. J. Buerger,[143] and extended to the wollastonite family (pectolite, bustamite, see below). The full determination of the wollastonite structure was later done by

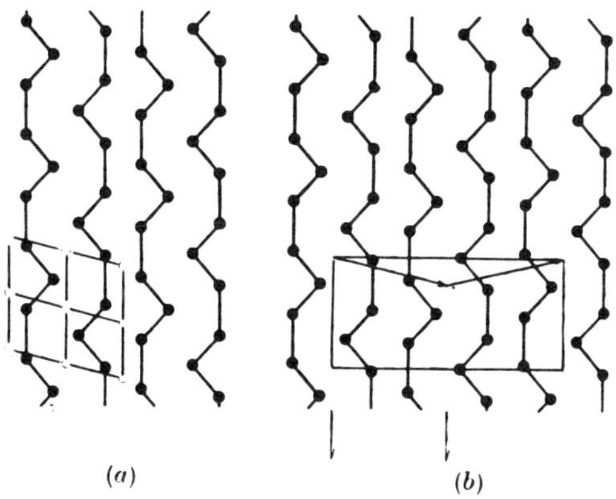

FIG. A. 62. Schematic arrangement of the chains in $(NaAsO_3)_x$ (*a*), in Maddrell's Salt, $(NaPO_3)_x$, and wollastonite (*b*) (Dornberger-Schiff, Liebau, and Thilo).

Kh. S. Mamedov and N. V. Belov[144] following its analogy with xonotlite (see below), and in its ambiguity of monoclinic and/or triclinic symmetry. $[CaO_n]$ polyhedra are interlinked between the chains but in a manner different from that in cuspidine (see above). The better cleavage parallel (100) than that in (010) is also explained by the structure (Fig. A. 63). The distinction of monoclinic para-wollastonite and triclinic wollastonite is further discussed by J. Tolliday[145] who shows their interrelation very clearly in a projection along [001] (Fig. A. 64) in which two wollastonite cells are outlined on one para-wollastonite cell, with equal *c*-axes. The chains I and II are different from the chains III and IV of para-wollastonite, a shift by the period *b*/2 bringing chain IV into a position of wollastonite type. The difference consists only in the packing, explaining the often observed intimate intergrowth of both crystal phases.

93. *Pectolite*, $Ca_2NaH[Si_3O_9]$, is very clearly related to wollastonite, chemically as well as structurally. E. Thilo and H. Funk[146] studied their correlation in connection with the hydrothermal synthesis of this interesting mineral (Vol. IV, Section A), but in the structural formula they adopted for pectolite the previous opinions of M. Barnick (1936) concerning the annular group $[Si_3O_9]$. That this cannot be correct

[143] *Proc. Natl. Acad. Sci. U.S.*, 42, 1956, 113-116.
[144] *Doklady Akad. Nauk S.S.S.R.*, 107, 1956, 463-466.
[145] *Nature*, 182, 1958, 1012-1013.
[146] *Z. anorg. u. allgem. Chem.*, 262, 1950, 185-191.

was demonstrated by M. J. Buerger[147] applying his theory of so-called "sub-structures," and a new method of structure analysis from Patterson projections.[148] The typical 3-repeat-unit of wollastonite type was established for pectolite, and a rather "unorthodox" interlinkage with $[CaO_6]$ groups, namely octahedral configurations sharing three edges to form sheets parallel (101), with two sets of Ca^{2+}, and one set

FIG. A. 63. Pyroxenoid (wollastonite) chain $[Si_3O_9]_\infty$ adjusting itself to the rectilinear chain of $[CaO_6]$ octahedra (Liebau).

of Na^+ ions, in lath-like strips parallel *b* on (101). The Na^+ is close to only three of six O^{2-} anions, and in a distorted octahedral configuration, much farther away from three other O^{2-} ions. This inequality in distance causes rifts in the (101) sheet dissecting it to parallel strips extending infinitely in the *b*-direction with planar surfaces parallel (010). The coordination is more that of a trigonal pyramid form than in octahedra.

94. Although an explicit determination of the structure of *hillebrandite*, $Ca_2SiO_4.H_2O$ is not yet available, an attempt to interpret this important calcium silicate hydrate was made by L. Heller.[149] Kh. S. Mamedow and N. V. Belov[150] classified its structure

[147] A complete determination of the structure in *Z. Krist.*, **108**, 1957, 248-262.

[148] *Acta Cryst.*, **4**, 1951, 531-544; *Anales real soc. españ. fis. y quim.* (*Madrid*), **50**, A, 1954, 221-254, especially pp. 231-237.

[149] *Mineral. Mag.*, **30**, 1953, 150-154.

[150] *Doklady Akad. Nauk S.S.S.R.*, **123**, 1958, 741-743.

as related to that of wollastonite, including the typical 3-repeat-unit, in near relation to other calcium silicate hydrates (like xonotlite, foshagite, tobermorite). Regular intergrowths with $Ca(OH)_2$ parallel (001) are also particularly characteristic of hillebrandite, in spite of its acicular crystal habitus. Whether the interpretation of hillebrandite as a kind of mixed-layer structure of xonotlite and $Ca(OH)_2$ is correct

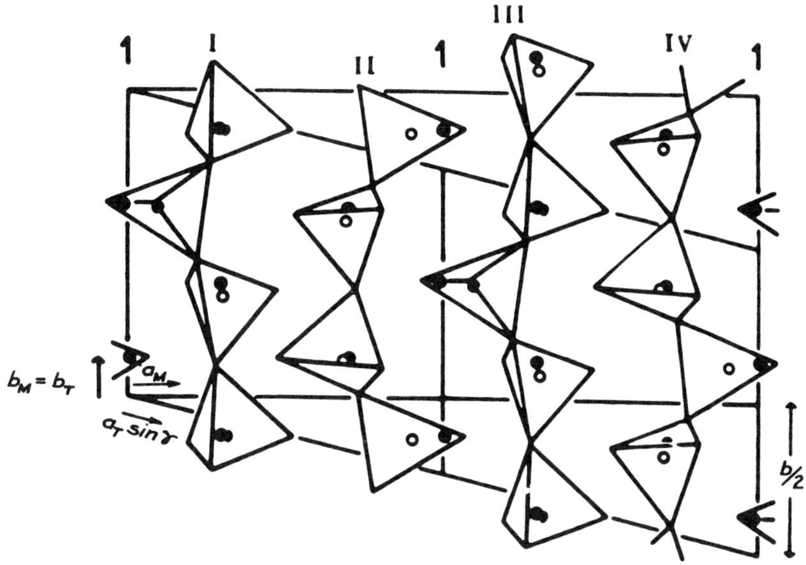

FIG. A. 64. Structural relation of two wollastonite cells (triclinic) to one para-wollastonite (monoclinic) cell to explain their intimate intergrowths (Tolliday).

cannot yet be exactly decided, with Ca^{2+} ions on both sides of $[Si_6O_{17}]_\infty$ chains. In every case, by calcination at 550°C. hillebrandite is destroyed, the alleged $Ca(OH)_2$ layers react with the silicate skeleton, and nonoriented Ca_2SiO_4, not an oriented wollastonite, is formed. The latter is, on the other hand formed, e.g., from okenite and tobermorite by calcination (see below).

95. The 5-repeat-unit chain in *rhodonite* was first described by W. Hilmer, F. Liebau, and G. Lindemann,[151] based on the chemical formula $CaMn_4[Si_5O_{15}]$. The triclinic structure with its characteristic chain anions of a period with five $[SiO_4]$ tetrahedra units shows alternate layers with the cations parallel $(\bar{1}10)$ (Fig. A. 65), and is in many respects similar to, though different from, wollastonite. This similarity is best seen, e.g., in a comparison of rhodonite and $(NaAsO_3)_x$ structures (Fig. A.

[151] *Fortschr. Mineral.*, **36**, 1958, 74; *Naturwissenschaften*, **43**, 1956, 117-118; *Acta Cryst.*, **12**, 1959, 182-187.

66). Also in the structure proposed by Kh. S. Mamedov[152] the difference of the chains from those of the pyroxenes and wollastonite is emphasized, with octahedral coordination of the Mn^{2+} cations arranged in zig-zag chains of the heavy metal-O^{2-} groups.

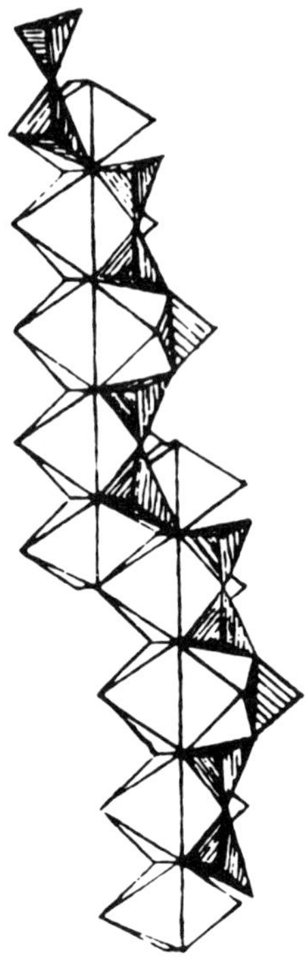

FIG. A. 65. Rhodonite 5-repeat-unit $[Si_5O_{15}]_\infty$ adjusting itself to the agnailed chain of $[MnO_6]$ octahedra (Liebau).

96. Particularly interesting are the trimorphic crystalline solutions of the system $CaSiO_3$-$MnSiO_3$ which still represent the rhodonite structure type up to 20 molecular per cent $CaSiO_3$. Moreover the intermediate crystal phases of *bustamite* type have a structure similar to, but not identical with β-$CaSiO_3$ (wollastonite), and at higher temperatures that of α-$CaSiO_3$ (pseudowollastonite, see below), according to F. Liebau,[153] and more extensively to F. Liebau, M. Sprung, and E. Thilo[154] (see also Vol.

[152] *Doklady Akad. Nauk Azerbaĭdzhan. S.S.R.*, **14**, 1958, 445-450.

[153] *Fortschr. Mineral.*, **37**, 1959, 54.

[154] *Z. anorg. u. allgem. Chem.*, **297**, 1958, 213-225. On p. 215 is an instructive representation of the chains and double chains of Class I (Bragg) and Class II (Belov) structure systematics, with German nomenclature. A complete determination and refinement of the crystal structure of *bustamite* was recently presented by D. R. Peacor and M. J. Buerger, *Z. Krist.*, **117**, 1962, 331-343.

III, Section B). The limited crystalline solubility in the series $CaSiO_3$-$MnSiO_3$ with the break at the ratio 1 : 4, is understood by the incompatibility of 3- and 5-repeat-unit arrangements in bustamite. There is also no completely continuous miscibility of that with β-$CaSiO_3$. At elevated temperatures the pseudowollastonite structure with $[Si_3O_9]$ rings prevails in the crystalline solutions.

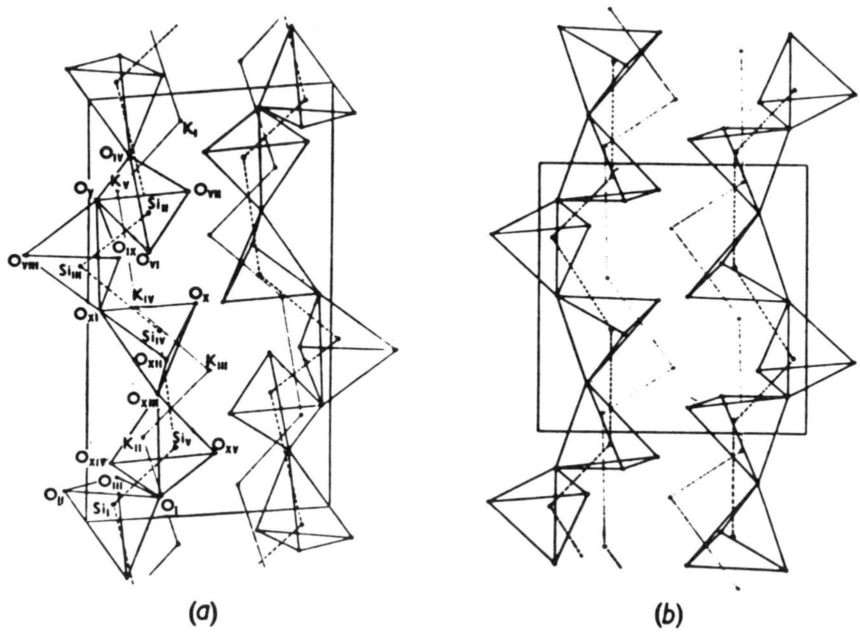

FIG. A. 66. Structures of rhodonite (*a*), projected on the (*xz*) plane, and of $(NaAsO_3)_x$ (*b*), projected on the (*yz*) plane for comparison (Hilmer, Liebau, and Lindemann).

97. Concerning the detection of the existence of 7-repeat-units $[Si_7O_{22}]_\infty$ we once again emphasize the investigation of F. Liebau[155] on *pyroxmangite*, $[(Fe, Mn, Mg, Ca)SiO_3]_n$, which is in many respects similar to rhodonite. Layers of the chains $[Si_7O_{22}]_\infty$ alternate with layers of the cation $[RO_6]$ octahedra parallel (110).

Subgroup 2b. Single Rings (Sharing Coefficient = 1.50)

98. Analogous to *benitoite*, $BaTi[Si_3O_9]$, the three-membered ring structure of the anion $[Si_3O_9]^{6-}$ is typical of *wadeite*, $K_2Zr[Si_3O_9]$, according to D. E. Henshaw.[156] The rings are arranged around hexagonal screw axes, the K^+ and Zr^{4+} cations are in the interstices, K^+ symmetrically above and below the rings, Zr^{4+} in the centers

[155] *Naturwissenschaften*, **44**, 1957, 178; *Acta Cryst.*, **12**, 1959, 177-181.
[156] *Mineral. Mag.*, **30**, 1955, 585-595; *Am. Mineralogist*, **40**, 1955, 775.

of octahedral $[ZrO_6]$ groups. The single $[SiO_4]$ tetrahedra are somewhat distorted causing a position of two thirds of the O^{2-} ions closer to the cations than could be the case for a regular tetrahedron shape of the $[SiO_4]$ units. Rings of $[Ge_3O_9]$ coordination quite analogous to those in benitoite and wadeite have been found by W. Hilmer[157] in the structure of $Sr_3[Ge_3O_9]$. The rings are arranged alternately between layers of the $[SrO_6]$ octahedra, in the sequence *ABCA'B'C'*... There is another modification of the Sr germanate with similar atlernating sequences but the stacking scheme is not yet determined. The structure of *pseudowollastonite* shows the same type of trigonal rings $[Si_3O_9]$ as given in that of $Sr_3[Ge_3O_9]$ (Fig. A. 67).

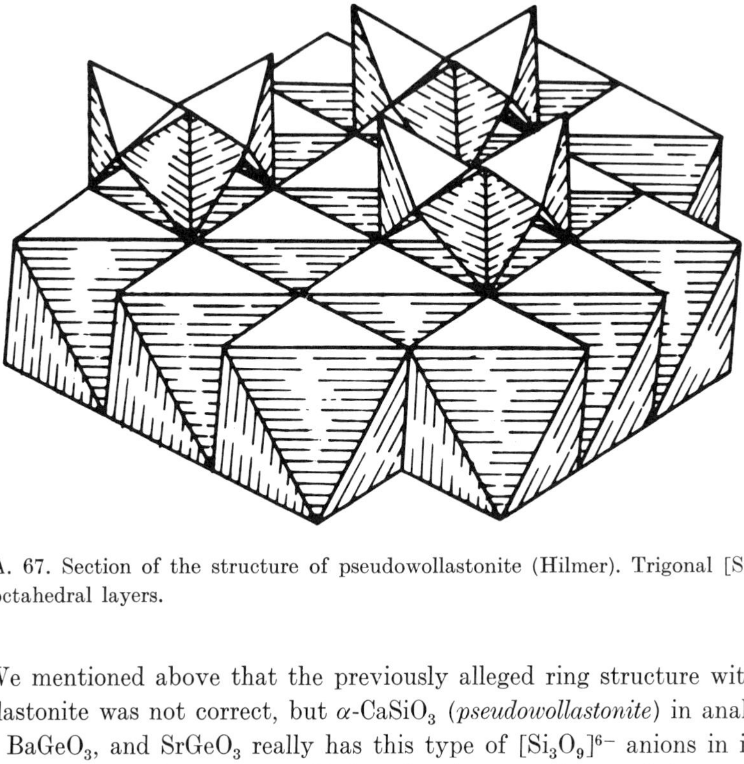

FIG. A. 67. Section of the structure of pseudowollastonite (Hilmer). Trigonal $[Si_3O_9]$ rings on the octahedral layers.

99. We mentioned above that the previously alleged ring structure with $[Si_3O_9]$ for wollastonite was not correct, but α-$CaSiO_3$ (*pseudowollastonite*) in analogy with $SrSiO_3$, $BaGeO_3$, and $SrGeO_3$ really has this type of $[Si_3O_9]^{6-}$ anions in its cell as J. W. Jeffery and L. Heller[158] established for the strongly pseudohexagonal high-temperature compound. The complex polysynthetic twinning indicates its low (triclinic) symmetry with pseudosymmetry planes perpendicular to the crystallographic

[157] *Fortschr. Mineral.*, **37**, 1959, 58; *Naturwissenschaften*, **45**, 1958, 238-239; *Silikat Tech.*, **11**, 1960, 397-398.

[158] *Acta Cryst.*, **6**, 1953, 807-808; see also quite recently K. Boll-Dornberger, *Silikat Tech.*, **12**, 1961, 327-329. $[X_3O_9]$ rings are also characteristic of the $Ca_2BaSi_3O_9$ structure (triclinic), as described by F. P. Glasser and L. S. Dent Glasser, *Z. Krist.*, **116**, 1961, 263-265.

axes. It may be recalled that N. V. Belov[159] ascribed to *catapleiite*, $H_4Na_2Zr[Si_3O_9]O_2$, a structure with the same $[Si_3O_9]$ rings, as determined by B. K. Brunovskiĭ[160] (Fig. A. 68).

100. H. O'Daniel and E. Hellner[161] proposed a structure of $3CaO.SiO_2$ which would contain $[Si_3O_9]$ annular groups. The fundamental importance which this compound has as the most characteristic mineral constituent in Portland cement clinker, in

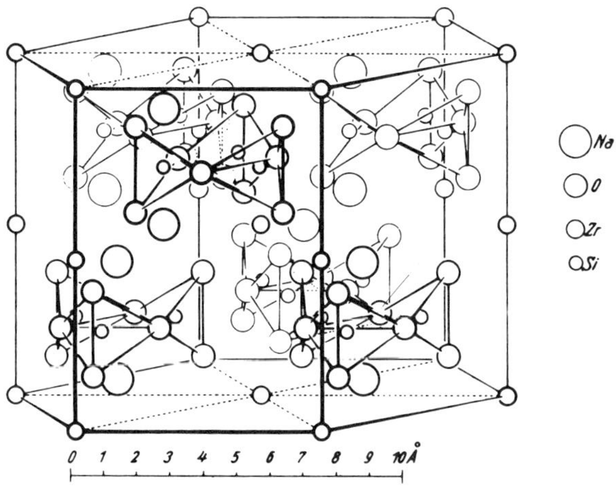

FIG. A. 68. Structure of catapleiite (Brunovskiĭ).

the somewhat modified form of *alite*, will be more extensively discussed in Vol. V, Section C. But it must be emphasized that the assumption of $[Si_3O_9]$ groups in Ca_3SiO_5 must be abandoned after detailed studies of its structure by J. W. Jeffery[162] who assigned to this compound *isolated* $[SiO_4]$ groups. The problem of the interrelation of Ca_3SiO_5 to the industrial alites shall also be discussed in Vol. V, Section C. We may only refer briefly here to recent studies of F. W. Locher[163] on the alite phase and its X-ray characteristics in complex compositions containing MgO and Al_2O_3 in crystalline solution, with a substitution of Ca^{2+} in Ca_3SiO_5 by Mg^{2+} and of Si^{4+} by Al^{3+}; Al_2O_3 (usually assumed to be present in the form of the $3CaO.Al_2O_3$ molecule) has no considerable effect, either on the unit cell dimensions or on the stability

[159] *Compt. rend. acad. sci. U.R.S.S.*, **37**, 1942, 135-138, 139-140.

[160] *Acta Physicochim. U.R.S.S.*, **5**, 1936, 863-892.

[161] *Neues Jahrb. Mineral. Monatsh.*, 1950, 108-111; see also H. O'Daniel, Th. Hahn, and H. Müller, *Ibid.*, 1953, 1-15.

[162] *Acta Cryst.*, **5**, 1952, 26-36; *Proc. Intern. Symposium on Chem. Cements, 3rd, London* 1952, 1954, 30-48. See also the confirmation of Jeffery's structure proposal by Th. Hahn, *Silikat Tech.*, **11**, 1960, 396-397; further *Ibid.*, **12**, 1961, 331-336, especially pp. 334 f.

[163] *Zement-Kalk-Gips*, **13**, 1960, 389-394; *Silikat Tech.*, **11**, 1960, 360-362.

of alite, whereas MgO distinctly decreases the stability and makes the unit cell parameters shrink. The determination of the structure of pure Ca_3SiO_5 as given by G. Yamaguchi and H. Miyabe[164] shows once again the high approximation of this lattice to trigonal symmetry, but it is slightly distorted to triclinic symmetry; above 970°C. a really trigonal modification is observed. Alite is monoclinic and is inverted at 830°C. to an analogous trigonal form.

101. *Four-membered rings* are ascribed to the structure of *axinite*, H(Fe, Mn) $Ca_2Al_2B[Si_4O_{12}]$, by T. Ito and Y. Takeuchi[165] as single structural units $[Si_4O_{12}]$, associated with planar $[BO_3]$ groups as shown convincingly in a projection in the plane perpendicular to the axis b' of the reduced cell unit shown in Fig. A. 69a, b. The groups $[Si_4O_{12}]$ are parallel to one another and lie with their flat side nearly parallel to (010), four of them are interlinked by double octahedra $[Al_2O_{10}]$, and another set of four by double octahedra $[Fe_2O_8(OH)_2]$. Ca^{2+} ions occupy the centers of irregular polyhedra $[CaO_{10}]$. The $[BO_3]$ planar groups are not directly linked to the $[Si_4O_{12}]$, to the double octahedra, or to other$[BO_3]$ groups. Al^{3+} also occurs in single $[AlO_3(OH)]$ tetrahedra. Axinite could therefore also be interpreted as a structure with a three-dimensional network of tetrahedra, in Zoltai's classification. He, however, discarded the Al-tetrahedra since they do not contribute half or more of their bond strength to the framework. Concerning the interlinkage of the $[Si_4O_{12}]$ groups with $[BO_3]$, W. Loewenstein[166] discusses the possibilities of the condensation to one complex group $[BO_3.Si_4O_{12}]$ (Fig. A. 70) in which the B—O coordination is practically tetrahedral, or at least trigonal-pyramidal, with the central B^{3+} cation close to the triangular plane of the basis. We will find similar conditions in tourmaline (see below).

102. *Six-membered rings* $[Si_6O_{18}]$ occur in the structure of *dioptase*, $Cu_6[Si_6O_{18}].6H_2O$, which was determined by N. V. Belov, V. P. Butuzov, and N. I. Golovastikov,[167] with a complete Fourier-Patterson synthesis of the X-ray intesities, using

[164] *J. Am. Ceram. Soc.*, **43**, 1960, 219-224. On the crystalline solutions of $3CaO.SiO_2$ with 0.3 to 0.7 percent Na_2O see the recent studies of G. Yamaguchi and H. Uchikawa, *Zement-Kalk-Gips*, **14**, 1961, 497-504.

[165] *Acta Cryst.*, **5**, 1953, 202-208; see also T. Zoltai, *Am. Mineralogist*, **45**, 1960, 971, especially on the role of the $[AlO_4]$ tetrahedra in the axinite structure. To axinite also belongs *tinzenite*, a manganese axinite, described by C. Milton, F. A. Hildebrand, and A. M. Sherwood, *Ibid.*, **38**, 1953, 1148-1153. Quadruple metasilicate rings of $[Si_4O_{12}]$, and chains of $[(Ti, Nb)O_6]$ octahedra are characteristic of the interesting new mineral *baotite* (*pao-t'ou-k'uang*). $Ba_4Ti_7NbSi_4O_{28}Cl$, the structure of which was determined and discussed by E. I. Semenov, Weng-Hsing Hung, and T. A. Kapitanov, *Doklady Akad. Nauk S.S.S.R.*, **136**, 1961, 915-916; V. I. Simonov, *Kristallografiya*, **5**, 1960, 544-546; E. Wm. Heinrich, Wm. H. Boyer, and F. A. Crowley, *Am. Mineralogist*, **47**, 1962, 987-993.

[166] *Am. Mineralogist*, **41**, 1956, 349-351.

[167] *Doklady Akad. Nauk S.S.S.R.*, **87**, 1953, 953-956.

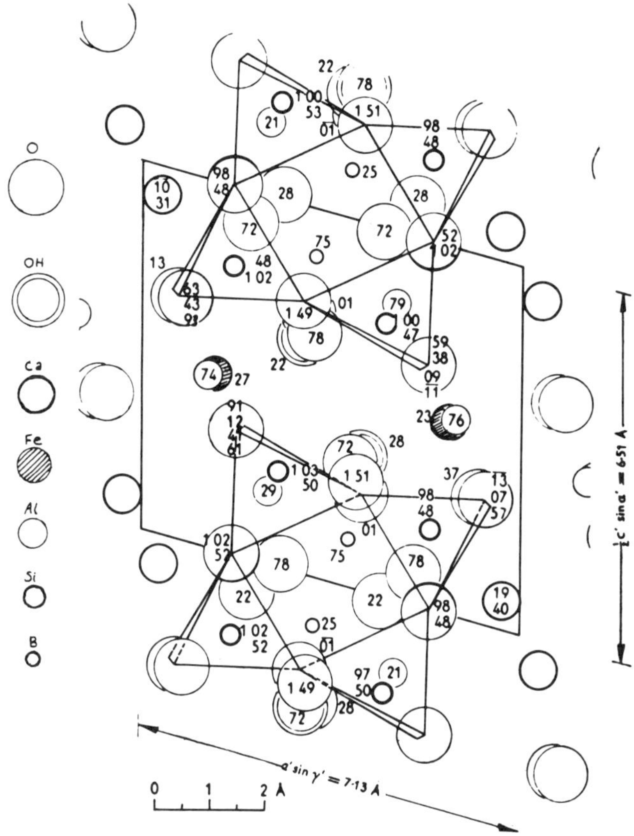

FIG. A. 69. Structure of axinite, projected on plane perpendicular to the b'-axis. Tetrahedra of $[Si_4O_{12}]$ groups are traced by straight lines (Ito and Takeuchi).

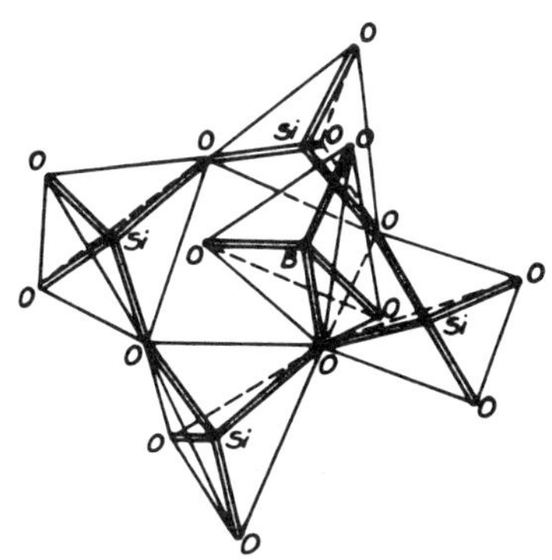

FIG. A. 70. The complex group $[BO_3.Si_4O_{12}]$ in the structure of axinite (Loewenstein).

M. J. Buerger's implication rules for the accurate position of the cations. The $[Si_6O_{18}]$ ring in dioptase has a 6-fold mirror axis, the Cu^{2+} ions being arranged in narrow channels between the rings, in coordination $[CuO_6]$. The water molecules are connected by hydrogen bonds to neighboring O^{2-} anions; to every water molecule belong four neighbors in a somewhat distorted tetrahedral arrangement $[Cu^{2+} + O_{II} + 2(H_2O)]$. The dioptase structure was also determined by H. J. Heide,[168] who emphasized the tube-shaped hollow channels. But whereas the interpretation of Belov *et al.* would make a zeolite-like behavior of water in dioptase expected, Heide emphasizes the relatively firm position of water molecules between the $[Si_6O_{18}]$ rings which impedes the water from free mobility (Fig. A. 71). In this respect there

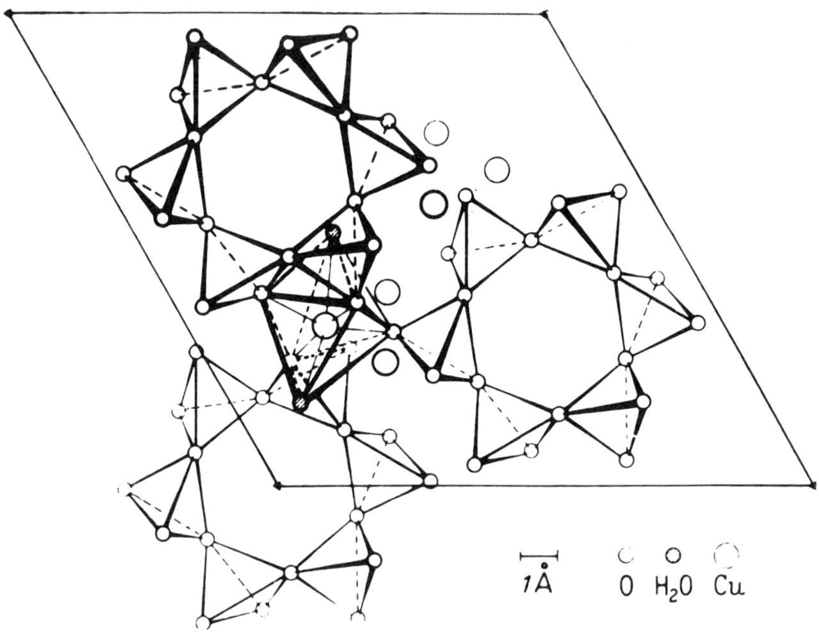

FIG. A. 71. Structure of dioptase, projected on (001) (Heide). Besides the tetrahedral ring $[Si_6O_{18}]$ is indicated one $[CuO_6]$ octahedron.

is no agreement between Belov's and Heide's interpretation. The fact is important that dioptase does not show a zeolite-like reversibility of the dehydration and rehydration cycles, but is decomposed in dehydration to CuO and silica. Heide's structure proposal is therefore more plausible.

103. In the classical structure of *tourmaline*, which contains the $[Si_6O_{18}]$ rings as

[168] *Naturwissenschaften*, **41**, 1954, 402-403; *Acta Cryst.*, **8**, 1955, 425-430, with K. Boll-Dornberger, E. Thilo, and E. M. Thilo.

the principal structural units, the connection of these groups with the planar $[BO_3]$ groups is a problem which was extensively discussed by W. Loewenstein,[169] namely for the possibility of a condensation of B^{3+} and Si^{4+} coordination groups. In the same way mentioned as characteristic of axinite, so too in the case of the groups $[BO_3.Si_4O_{12}]$ these $[BO_3]$ groups are better interpreted as $[BO_3.O]$ pyramids, or as distorted tetrahedra as also seen in the structure of *boracite*.[170] The occurrence of $[AlO_4]$ tetrahedra in the annular complexes sharing common corners with B^{3+} and Si^{4+} tetrahedra, however, is never observed.

104. In the form of single annular groups $[Si_6O_{12}(OH)_6]$ a peculiar modification of a six-membered ring appears in the structure of the rare mineral *lovozerite*, $NaZr[Si_6O_{12}(OH)_6]$ for which V. V. Ilyukhin and N. V. Belov[171] determined the detailed constitution. The cations Na^+ and Zr^{4+} appear in $[NaO_7]$ or $[NaO_8]$, and in $[ZrO_6]$ groups,

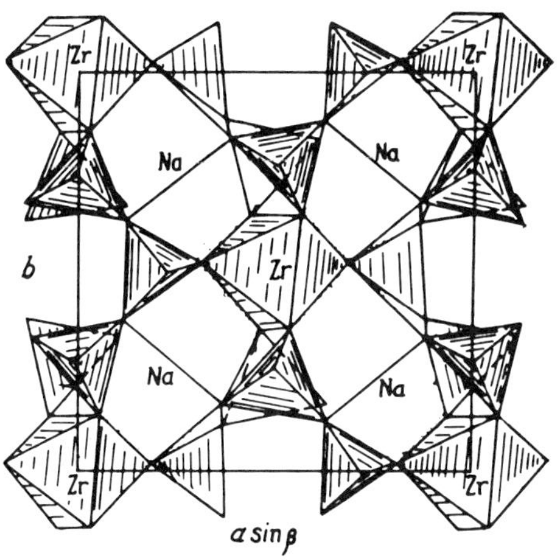

FIG. A. 72. Structure of lovozerite, projected on the pseudotetragonal plane (xy) (Ilyukhin and Belov).

building up with the rings a pseudotetragonal structure, with Zr^{4+} at the corners of the squares, and Na^+ in the centers of the faces. The $[Si_6O_{12}(OH)_6]$ rings are oriented pairwise, but separate, correlated by a twofold symmetry axis. The arrangement of

[169] *Am. Mineralogist*, **41**, 1956, 349-351. A thorough Fourier synthesis of the tourmaline structure was made by T. Ito and R. Sadanaga, *Acta Cryst.*, **4**, 1951, 385-390, as a refinement of the previous structure determinations of M. J. Buerger and G. Hamburger (1948), further M. J. Buerger, *Geol. Soc. Amer., Spec. Paper*, No. 68, 1961, 143, for recent refinements. See also previously N. V. Belov and E. N. Belova, *Doklady Akad. Nauk S.S.S.R.*, **69**, 1949, 185-188; *Ibid.*, **75**, 1950, 807-810.

[170] Cf. T. Ito, N. Morimoto, and R. Sadanaga, *Acta Cryst.*, **4**, 1951, 310-314.

[171] *Doklady Akad. Nauk S.S.S.R.*, **131**, 1960, 176-179.

these rings as seen in Figs. A. 72 and 73 can be understood best as a combination of two $[Si_2O_7]$ groups in which two upper tetrahedra are connected over two single $[SiO_4]$ tetrahedra with two lower tetrahedra of $[Si_2O_7]$ groups. With the rings in dioptase the lovozerite ring has in common the joining of each constituent tetra-

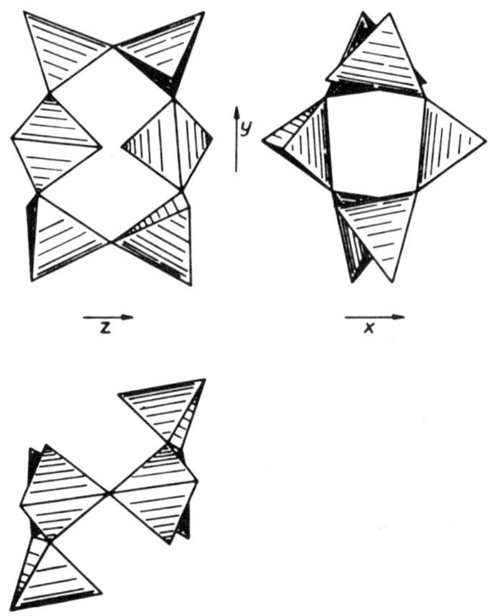

FIG. A. 73. Projection of the single group $[Si_6O_{12}(OH)_6]$ in lovozerite, to show the difference from the normal $[Si_6O_{18}]$ ring (Ilyukhin and Belov).

hedron by two apexes, as in other tetrahedra, but two other edges are free. Each $[SiO_4]$ of the ring has in common one edge with the aluminum polyhedra, one apex with $[ZrO_6]$, one apex being free. Important is the entrance of hydroxyl groups in direct coordination to the Si^{4+} centers of the tetrahedra in the ring, similar to the entrance of hydroxyl in afwillite causing the coordination $[SiO_3OH]$. It is also highly interesting to compare the unit $[Si_6O_{12}(OH)_6]$ with the double ring group $[Si_6O_{15}]$ in *elpidite*, $H_4Na_2Zr[Si_6O_{15}]$, which according to Ilyukhin and Belov originates from two single $[Si_3O_9]$ rings by condensation following the equation

$$2\ [Si_3O_9] - 3\ O^{2-} \rightarrow [Si_6O_{15}]$$

Group 2c. Double Chains (Sharing Coefficient between 1.50 and 1.75)

105. The classical double chains built up from condensed pyroxene 2-repeat-units is the fundamental principle in the structure of the *amphiboles*, with the unit $[Si_4O_{11}]_\infty$ as Bragg defined it (sharing coefficient 1.625). An interesting structural investigation was made by E. Schiebold[172] of natural and synthetic amphibole asbestos, among the synthetic products those of the pneumatolytic process developed by W. Lüdke (1934 to 1944), characterized by highly alkaline amphiboles with high contents in fluorine (cf. Vol. IV, Section A). These amphiboles are based on the structure of *tremolite*, filled with Na^+ ions in the (normally empty) positions 4_g, shown in Fig. A. 74, with coordinations up to $[RO_{10}]$. Beyond these alkali ion positions about two excess Na^+ ions are present per unit cell for which a defined position could not be specified by Schiebold, in spite of an extensive discussion of this fascinating problem. It is not quite excluded that in the reaction products of the Lüdke process under a low pressure and at low temperatures, but grown with a remarkably large rate of crystallization, the crystal phases are not yet completely condensed from the pyroxene chains to the $[Si_4O_{11}]_\infty$ double chains (ribbons) and may have an intermediate position between alkali-saturated tremolite-type amphiboles, and alkaline pyroxenes. If this is conceivable, then it is only a question as to why the pyroxenes are not indicated in the X-ray diffraction diagrams separately for identification as such. In every case, the chemical composition of such extremely alkaline synthetic amphiboles speaks for such a hypothesis.

106. According to precision measurements by S. Ghose[173] the structure of *cummingtonite* is very similar to that of tremolite, except that the curvature of the silicate double chains away from the plane of the metal cations is more pronounced in cummingtonite. Particularly characteristic is the short distance of only 2.04 Å. for the position of M_4 cations and O_4 anions, explained by Ghose as a covalent bonding effect. W. H. Baur[174] is of the opinion that this conclusion is not convincing, and that only the distortion effects in the electrostatic field of these positions are real.

107. *Proto-amphibole*, a new polytype of this group, was described by G. V. Gibbs, F. D. Bloss, and H. R. Shell.[175] It is a synthetic crystal phase of the system Li_2O–MgO–MgF_2–SiO_2 (cf. Vol. IV, Section A). The structural relation of proto-amphibole which is orthorhombic in symmetry, to *anthophyllite*, $Mg_2(Mg, Fe)_5[Si_8O_{22}(OH)_2]$, is given by the approximation a_0 (anthoph.) $= 2 \times a_0$ (proto-amph.), and they correspond to each other in the same way as proto-enstatite does to the orthorhombic pyroxenes. Characteristic is the presence of lithium in the place of calcium in proto-amphibole. This orthorhombic phase is immediately changed to monoclinic amphibole as soon as Ca^{2+} ions are introduced. The postulated structure of proto-amphibole is seen from Fig. A. 75 in projection on (100); it can be derived

[172] *Wiss. Z. Hochsch. Schwermaschinenbau Magdeburg*, **2**, 1958, 51-84.

[173] *Acta Cryst.*, **14**, 1961, 622-627.

[174] *Naturwissenschaften*, **48**, 1961, 549-550, in a general discussion of distortions of heteropolar crystal structures and deviations from Pauling's valence rule.

[175] *Am. Mineralogist*, **45**, 1960, 974-989.

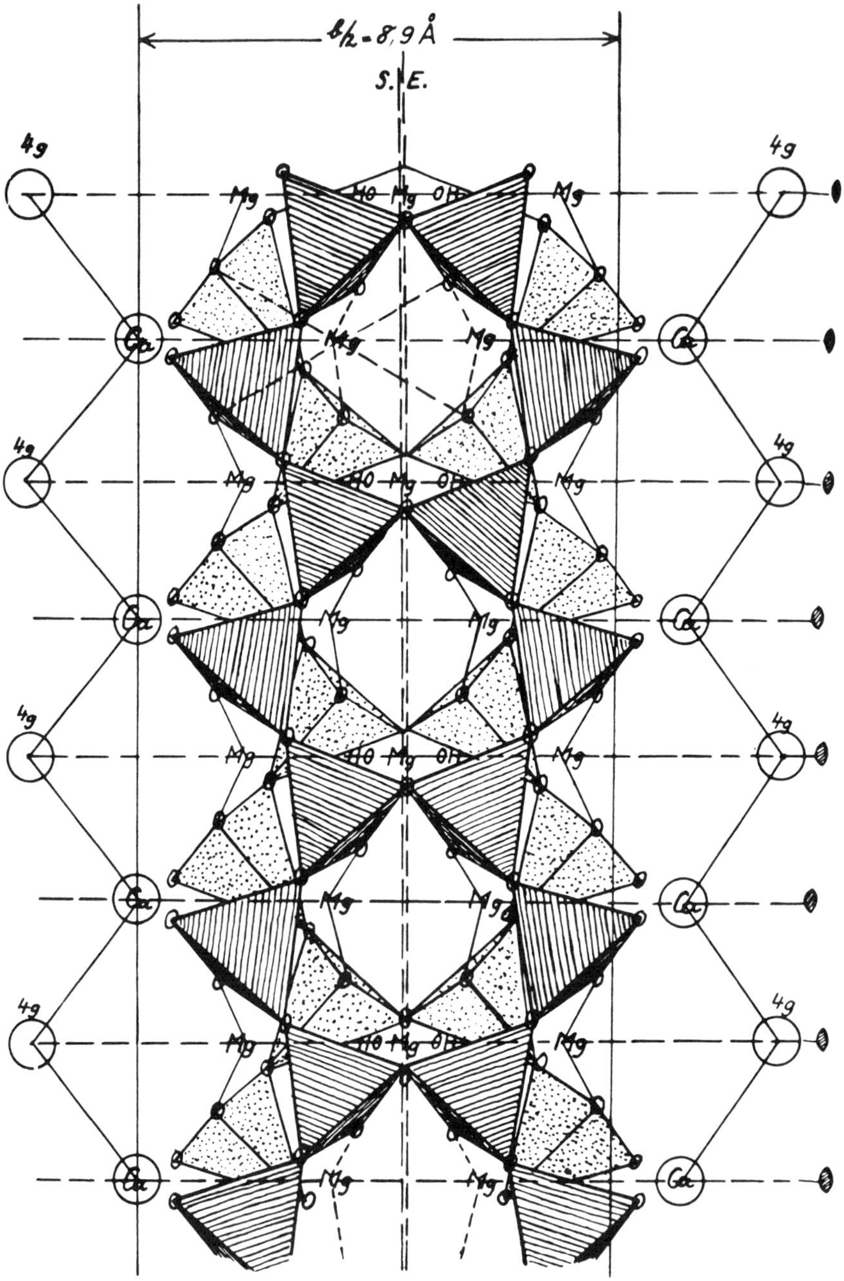

FIG. A. 74. Tremolite-type amphibole structure, projected parallel (100) to show the double chains $[Si_4O_{11}]_\infty$ and the (empty) positions (4g) which may be occupied by Na^+ ions in synthetic sodium-rich amphiboles according to W. Lüdke (Schiebold).

from that of protoenstatite (see above) by placing a mirror plane parallel (010) to double the pyroxene chains $[Si_2O_6]_\infty$ to the $[Si_4O_{11}]_\infty$ ribbons of proto-amphibole, just as B. E. Warren (1929) had done to derive the structure of tremolite from that of diopside. The chains in proto-amphibole are slightly shifted relative to each other along the *c*-axis, in distinction to those in normal amphiboles.

108. The thermal breakdown of tremolite to a pyroxene of diopside-clinoenstatite type by dehydroxylation (cf. Vol. V, Section B) was studied by A. G. Freeman and

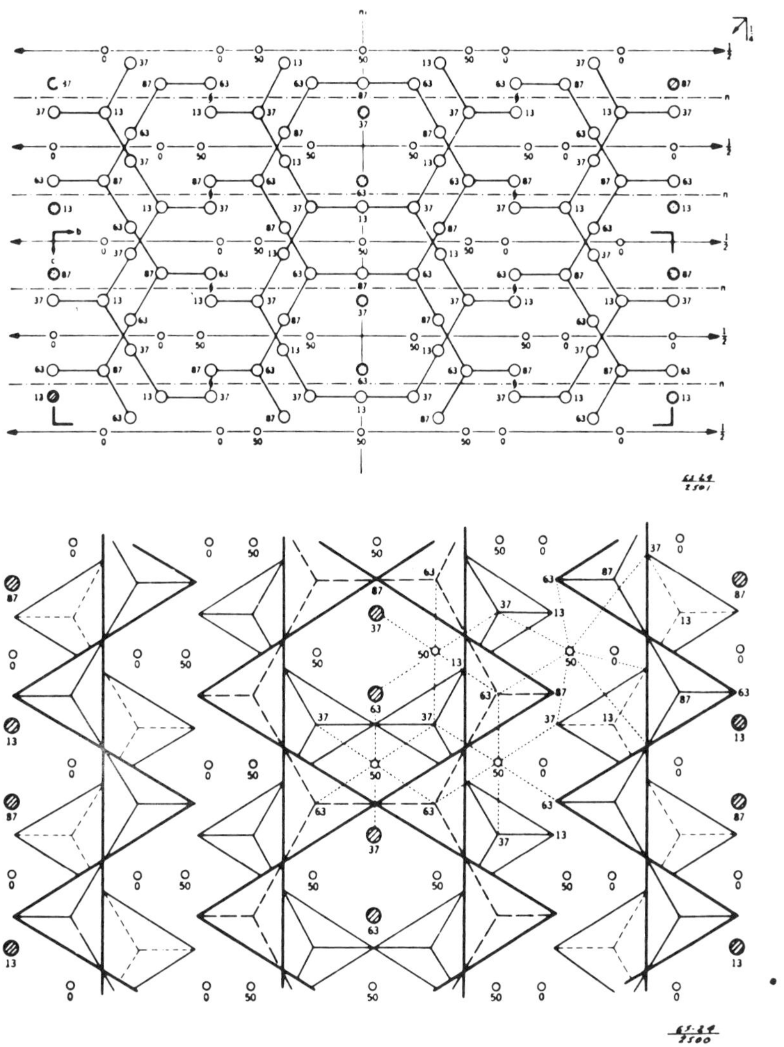

FIG. A. 75. Proto-amphibole-type structure, projected parallel (100), to show the orthorhombic arrangement of the double chains (Gibbs, Bloss, and Shell).

H. F. W. Taylor.[176] The orientation of the crystallographic axes remains completely preserved in this reaction, but the parameters a, b, c of tremolite are changed to a, $2b$, c in the resulting pyroxene; this is, however, not simply diopside, $CaMg[Si_2O_6]$, but more similar to clinoenstatite, of the composition $Ca_2Mg_5[SiO_3]_7$, which is metastable and changed to a mixture of diopside and clinoenstatite. An analogous dehydroxylation mechanism will be discussed for xonotlite in its transition to wollastonite (see below). The structural relations between tremolite and its dehydroxylation products are shown in Fig. A. 76.

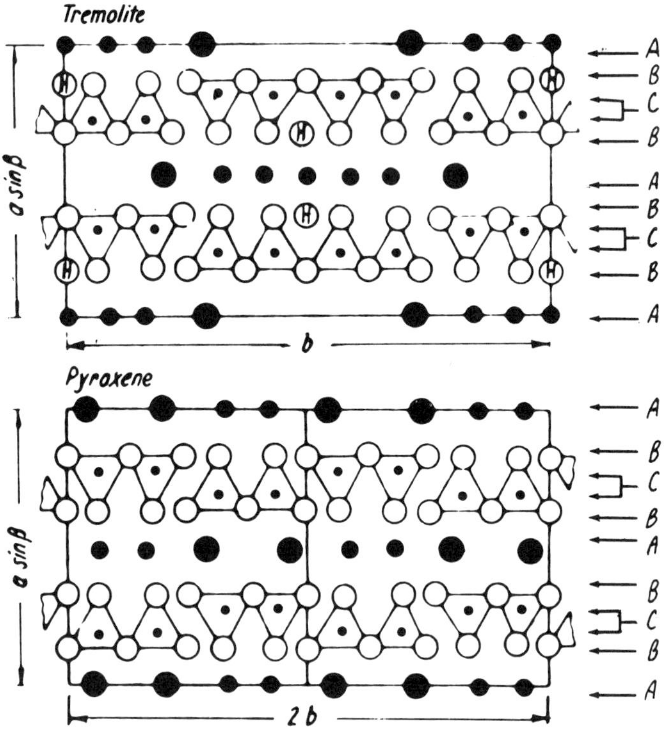

FIG. A. 76. Structures of tremolite and of pyroxene $Ca_2Mg_5(SiO_3)_7$, in its products of dehydroxylation, projected in the direction of the common c-axis (Freeman and Taylor). A are the cation layers; B those of oxygen anions; C those of Si^{4+} parallel (100). H indicates the position of hydroxyl groups in tremolite; the black circles are the cations Ca^{2+}, Mg^{2+}, and Si^{4+}.

109. The same principle which rules the condensation of pyroxene 2-repeat-units to amphibole double 2-repeat-units (Bragg type), applied for the pyroxenoid 3-repeat-units of wollastonite brings about the structure elements of $[Si_6O_{17}]_\infty$ in the

[176] *Silikat Tech.*, 11, 1960, 390-392.

ribbons of *xonotlite* as they were identified by Kh. S. Mamedov and N. V. Belov[177] in close structural relation to cuspidine and tilleyite on the one hand and to wollastonite on the other hand, especially in the occurrence of $[CaO_6]_\infty$ columns. In xonotlite, however, only one-third of these columns are octahedral, two-thirds of them are, in the same coordination $[CaO_6]$, in the shape of trigonal prisms as also observed in apatite. Parallel to the c axis the octahedral columns determine thus the fibrous habitus of xonotlite. The good cleavage parallel (100) is explained by the sheet-wise orientation of the $[Si_6O_{17}]$, groups; a weaker cleavage parallel (001) is caused by the "walls" of the $[CaO_6]_\infty$ columns. A pseudoperiod of $b = 3.66$ Å. is very distinct and typical for the whole xonotlite-derived structures whereas the true period is $b' = 7.32$ Å. with an angle $\beta = 90°$. The characteristic $[Si_6O_{17}]_\infty$ "ribbons"

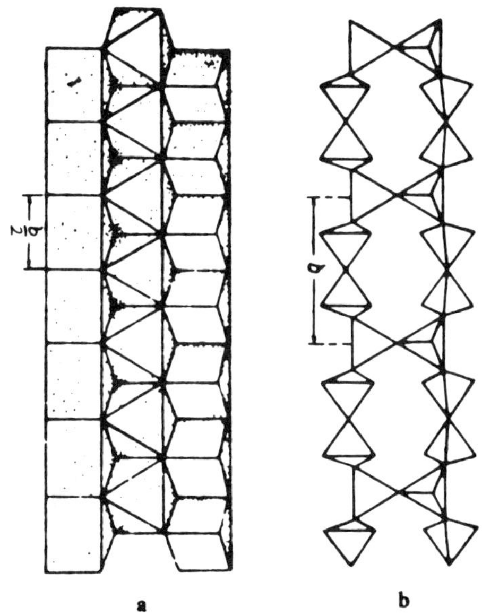

FIG A. 77. Chains in the xonotlite structure: (*a*) $[CaO_6]_\infty$ columns, built up from octahedra and trigonal prisms; (*b*) $[Si_6O_{17}]_\infty$ ribbons (Mamedov and Belov).

are seen in Fig. A. 77, with an angle Si—O—Si of 158°, and a very typical ambiguity of orthorhombic and triclinic arrangements of these chains as discussed by Mamedov and Belov, and shown in Fig. A. 78, with identical cell parameters b_0 and c_0, but different a_0, to explain the polytypism of xonotlite. In this interpretation, xonotlite is, like wollastonite, a typically defective structure.

[177] *Doklady Akad. Nauk S.S.S.R.*, **104**, 1955, 615-618; *Zapiski Vsesoyuz. Mineral. Obshchestva*, **85**, 1956, 13-38.

110. To clearly distinguish the $[Si_6O_{17}]_\infty$ chains (Class II type of Belov) from the $[Si_4O_{11}]_\infty$ ribbons in amphibole (Class I type of Bragg) and the single 2-repeat-units of pyroxenes (Class I), we refer the reader to Fig. A. 79 (in which the ribbons $[SiAlO_5]_\infty$ of sillimanite are added as a comparison). The close relation of xonotlite

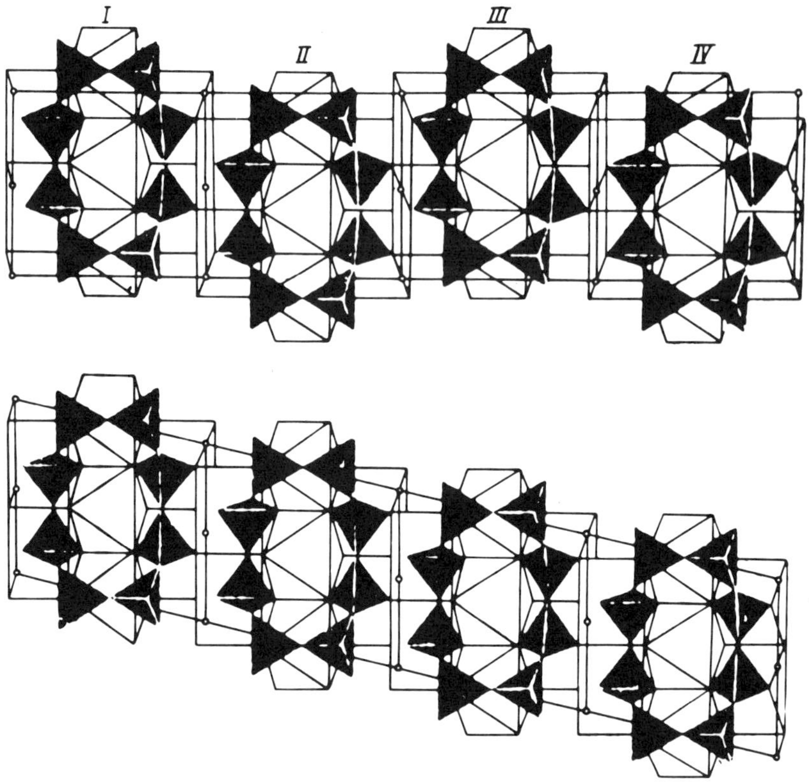

FIG. A. 78. Polytypism of xonotlite (Mamedov and Belov).

to wollastonite becomes evident also in the conversion of single crystals of xonotlite to single crystals of wollastonite by dehydroxylation at 775° to 800°C., structurally studied by L. S. Dent and H. F. W. Taylor.[178] The differences in the orientation are best seen from Fig. A. 80, with preservation of the orientations parallel b and (100) within $\pm 2°$. The double chain is split to two single 3-repeat-units of the wollastonite structure (Fig. A. 81). A slight slip occurs parallel to the good cleavage parallel (100) in xonotlite; it corresponds to the perfect cleavage $(10\bar{1})$ in afwillite during its conversion to γ-Ca_2SiO_4 (see above).

[178] *Acta Cryst.*, **9**, 1956, 1002-1004.

111. For *foshagite* Mamedov and Belov first adopted the dualistic formula $4CaO.3SiO_2.H_2O$ proposed by J. A. Gard and H. F. W. Taylor (see below), and emphasized the conversion into wollastonite by dehydroxylation as the characteristic structural indication, as well as the pseudoperiod $b = 3.68$ Å. which also occurs in

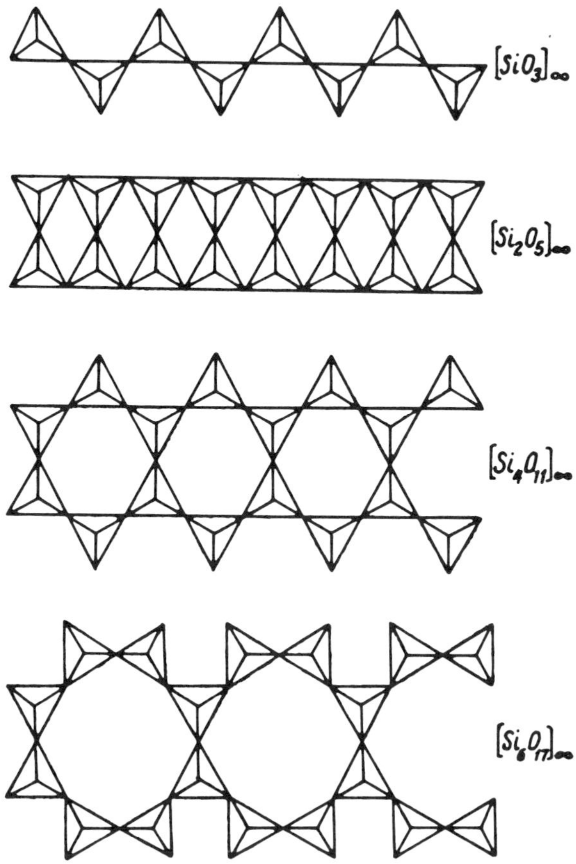

FIG. A. 79. Different interlinkage types of tetrahedral Si—O chains and ribbons (Classes I and II, Bragg and Belov).

xonotlite and hillebrandite (see above). Therefore, Mamedov and Belov[179] wrote the foshagite formula in the form $Ca_8[Si_6O_{18}](OH)_4$, whereas Gard and Taylor wrote it $Ca_8[Si_3O_9]_2(OH)_4$ because at 750°C not only β-$CaSiO_3$ but also β-Ca_2SiO_4 was concluded to be formed, although the latter silicate could not be identified in the X-ray diagrams. The Russian authors assumed that the water content of foshagite would be 1.5 molecules per formula unit, not 1 molecule as Gard and Taylor

[179] *Doklady Akad. Nauk S.S.S.R.*, **121**, 1958, 901-903.

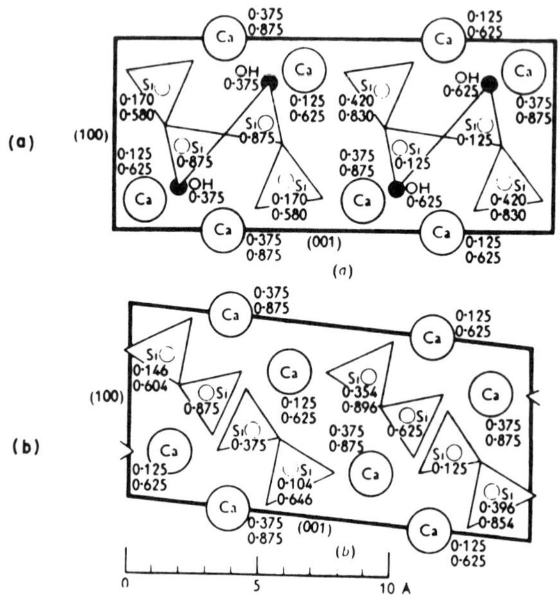

FIG. A. 80. Structure of xonotlite (a), and of monoclinic para-wollastonite (b), in projections on (010) (Dent and Taylor).

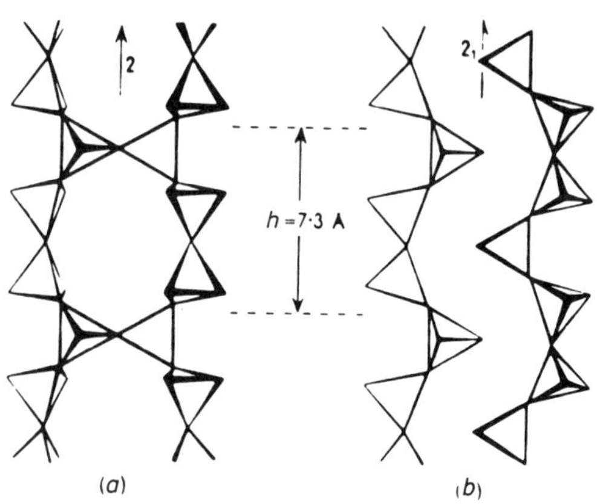

FIG. A. 81. Splitting of the $[Si_6O_{17}]_\infty$ ribbon of xonotlite into two $[Si_3O_9]_\infty$ chains in wollastonite after dehydroxylation (Dent and Taylor).

proposed. The crystallochemical formula of foshagite would then be $Ca_8[Si_6O_{17}](OH)_6$, with the xonotlite ribbon $[Si_6O_{17}]_\infty$ (cf. Fig. A. 82). In this interpretation foshagite would be a condensation product of xonotlite and $Ca(OH)_2$, and the xonotlite unit would be coupled with octahedral columns of $Ca(OH)_2$ composition. The only structural difference from xonotlite would then be that in foshagite *all* of the $[CaO_6]_\infty$ colums are octahedral, while in xonotlite (see above) only one-third of them are octahedral and the others are trigonal-prismatic.

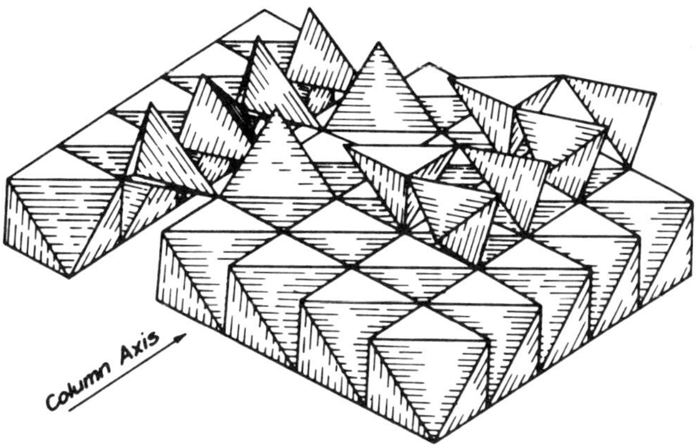

FIG. A. 82. Section of the structure of foshagite to show the four octahedral columns, with Sections of two 3-repeat units of $[SiO_4]$ tetrahedra (Boll-Dornberger, after Mamedov and Belov).

112. J. A. Gard and H. F. W. Taylor[180] are of the opinion that the structure as given by Mamedov and Belov does not seem to be correct. They emphasize the existence exclusively of $[Si_3O_9]_\infty$ (wollastonite) chains in this calcium silicate hydrate. Their measurements, based on electron diffraction data, confirm in their interpretation the monoclinic symmetry of foshagite. The central point of the discussion between both groups of authors may now be seen in the correct formulation of the dehydroxylation products, either as single crystals of β-$CaSiO_3$, or in a mixture of β-$CaSiO_3$ and β-Ca_2SiO_4. Free CaO was previously mentioned by L. Heller, but it is not confirmed. Gard and Taylor believe that random orientation and extremely small crystal size might explain the striking failure of the electron diffraction method to reveal β-Ca_2SiO_4. It would be plausible that the (001) plane of foshagite would have merged into (001) of the pseudocell of β-$CaSiO_3$. The relation of β-$CaSiO_3$ to the initial foshagite is demonstrated by Gard and Taylor in the schematic projection shown in Fig. A. 83.

113. The ribbons of the type $[SiAlO_5]_\infty$ (cf. Fig. A. 2) which were mentioned sev-

[180] *Am. Mineralogist*, **43**, 1958, 1-15; *Nature*, **183**, 1959, 171-173; *Silikat Tech.* **11**, 1960, 396.

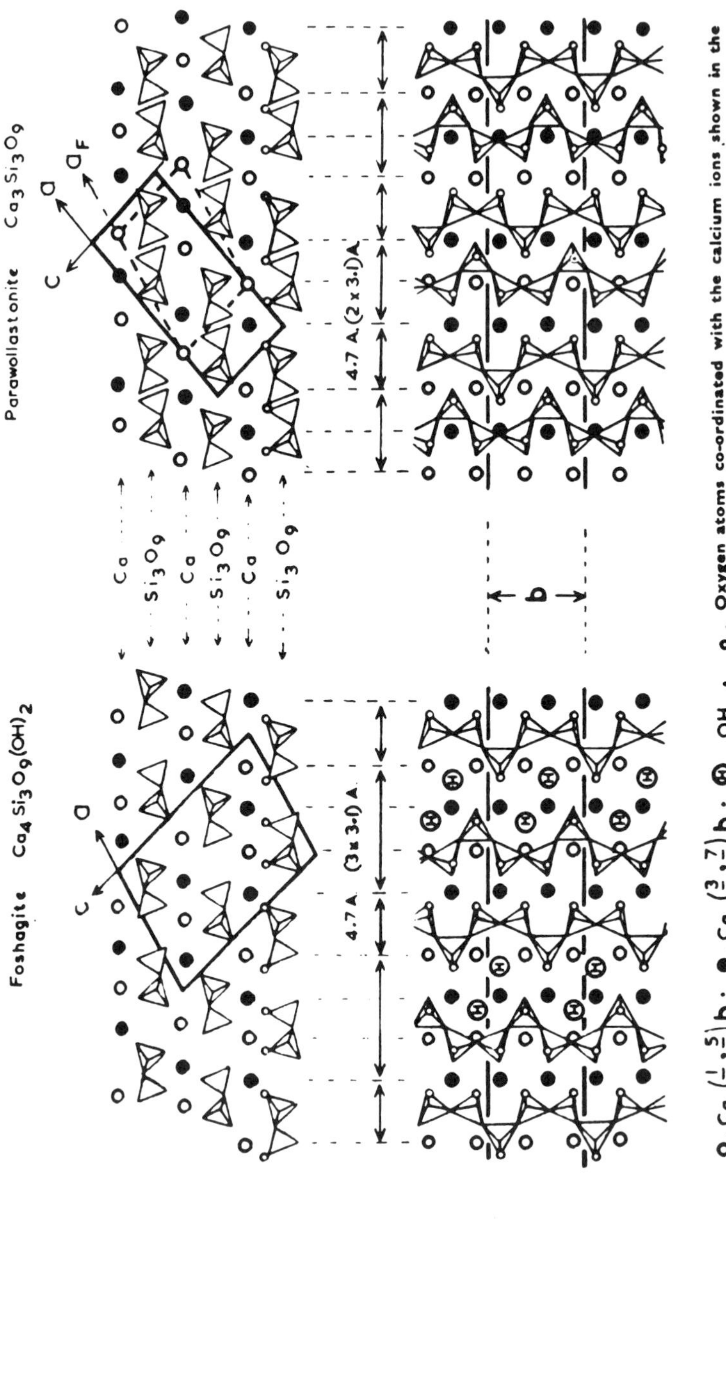

FIG. A. 83. Projections of the structures of foshagite and para-wollastonite (Gard and Taylor). Full lines define unit cells, broken lines the unit cells of foshagite, showing how the Ca^{2+} ions are related in both structures.

eral times above and by Belov in his discussions on structural classifications of silicates, are typical of *sillimanite* (sharing coefficient 1.75) (cf. ¶¶ 26 and 63, and W. Noll in A. ¶ 5, footnote 7). $[SiO_4]$ and $[AlO_4]$ tetrahedra alternate with the characteristic bonding angle Si—O—Al of 180° perpendicular to the length axis and the tetrahedron angle in this direction. In sillimanite two silicon and two aluminum tetrahedra can never be immediate neighbors to each other. The fibrous character of the crystals is equally evident as it is for the amphiboles.

114. The fundamental problem of whether or not *mullite* is in its structure a strict analogue to sillimanite is not yet clearly answered, in spite of many analogies between their physical properties. Furthermore, the principles upon which W. H. Taylor (1932) developed a proposal of the mullite structure still make a very plausible hypothesis, based on the assumption of the extensive replacement of aluminum tetrahedra for silicon tetrahedra, up to the limit concentration $3Al_2O_3.2SiO_2$, and even $2Al_2O_3.SiO_2$. The problem of the mullite crystallization, which is so fundamental for the ceramic reaction from kaolinite to mullite, was recently studied by G. W. Brindley and M. Nakahira[181] from structural viewpoints in connection with the so-called meta-kaolin phases. Single crystal techniques in X-ray diffraction brought about important information on the relative orientation of the typical meta-kaolin in the original kaolinite (cf. Vol. V, Section B), and the new aspects of the formation of a so-called silicon-spinel of highly defective structure indicated by the unit cell composition $Si_8Al_{10^2/_3}\square_{5^1/_3}O_{32}$, as compared with the so-called γ-Al_2O_3 spinel or $Al_{21^1/_3}\square_{2^2/_3}O_{32}$. The transition into mullite is connected with a considerable loss in silica, from the composition $3SiO_2.2Al_2O_3 \rightarrow 3Al_2O_3.2SiO_2$. It is extremely difficult to have some better information of this process by the actual X-ray diffraction methods because of the diffuseness of the lines. The fact, however, is established that mullite also must have the same $[AlO_6]_\infty$ chains, as they are typical of sillimanite, parallel to the *c*-axis, and parallel to the [110] direction, which are already present in the silicon-spinel phase, with right- and left-handed rotations to adjust the cations bonding the chain systems. On the role of the $[SiAlO_5]_\infty$ ribbons Brindley's and Nakahira's discussions do not give exhaustive information.

115. Nevertheless, S. O. Agrell and J. V. Smith[182] made another challenge to the interrelations of sillimanite-mullite, in which a new type ("S"-mullite) is characterized by sharp additional X-ray diffraction lines, in contrast with corresponding highly diffuse lines of common "D"-mullite. If in the latter alumina is enriched in crystalline solution, the parameters a_0 and c_0 are increased, but not b_0. In natural mullites containing some iron and titanium, on the other hand, a slight decrease in a_0 and b_0 is observed, whereas higher contents of Fe_2O_3 and TiO_2 bring about an

[181] *J. Am. Ceram. Soc.*, **42**, 1959, 311-314; 314-318; 319-324.

[182] *J. Am. Ceram. Soc.*, **43**, 1960, 69-76.

increase. The parameter c_0 is uniformly increased for all concentrations in those oxides. The parameters of the unit cells are relatively sensitive indicators of the composition of the crystalline solutions in the series between theoretical mullite, $3Al_2O_3.2SiO_2$, the alumina-enriched mullite, $2Al_2O_3.SiO_2$, and $3(Al, Fe)_2O_3.2SiO_2$, with the extreme ratio of $9\ Al^{3+} : 1\ Fe^{3+}$ cations. For sillimanite the corresponding changes in parameters are relatively subordinate. The single crystal diffraction measurement made by Agrell and Smith, however, do not lead far beyond the theories of W. H. Taylor, with a doubling of the c_0 parameter in mullite. It is still difficult to understand the adjustment of the number of oxygen anions in mullite when Si^{4+} is replaced by Al^{3+} in the tetrahedra. Possibly some distortion of the chains takes place in the alumina-enriched mullites. The accompanying scheme may help to clarify the distinctions between D and S mullites, their composition, and volume effects (densities).

Composition	Structure type	Cell content	Calculated density
$3Al_2O_3.2SiO_2$	Constant (Al + Si)	$Si_6Al_{18}O_{39}$	3.166 D-Mullite
$3Al_2O_3.2SiO_2$	Constant oxygen	$40/39(Si_6Al_{18}O_{39})$	3.247 D-Mullite
$2Al_2O_3.SiO_2$	Constant (Al + Si)	$24/25(Si_5Al_{20}O_{40})$	3.125 S-Mullite; 3.132 D-Mullite
$2Al_2O_3.SiO_2$	Constant oxygen	$Si_5Al_{20}O_{40}$	3.255 S-Mullite; 3.262 D-Mullite

Subgroup 2d. Double Rings

116. We mentioned above that N. V. Belov ascribed to *elpidite* a *trigonal* double-ring unit of the coordination type $[Si_6O_{15}]$, explained as a condensation product of two benitoite rings $[Si_3O_9]$ with the elimination of three O^{2-} anions in the sharing tetrahedron apexes.

The *hexagonal* double rings in the structures of *milarite* and *osumilite* will be discussed among the double sheet structures of these minerals in A. ¶ 171, to follow Zoltai's classification principles.

Subgroup 2e. Multiple Chains

117. In *epididymite* and *eudidymite*, $HNaBeSi_3O_8$, infinite triple chains of the type $[Si_3O_8]_\infty$ were identified by T. Ito (1937), but evidently epididymite should be classified among the network structures because of the highly complex sheets $[NaSi_3O_7]$ and the participation of $[BeO_4]$ tetrahedra in these sheets. Eudidymite was reexamined in its structure by E. A. Pobedinskaya and N. V. Belov.[183] Ito's interpreta-

[183] *Doklady Akad. Nauk S.S.S.R.*, **131**, 1961, 1448-1450. Peculiar double chains are characteristic of the structure of *narsarsukite*, $Na_2(Ti, Fe)[Si_4O_{10}](O, OH)$, according to Yu. A. Pya-

tion of this structure was refuted, and the analogy with the framework structure of the feldspars emphasized to establish the structure according to their framework principles (cf. A. ¶ 195 ff.). In epididymite the analogy to feldspars is indicated only in the crystallochemical formula.

* * *

Representatives of the *Subgroups 2f. Multiple Rings* and *2g. Mixed Chains and Rings* cannot be given.

3. Two-Dimensionally Infinite Networks of Tetrahedra

Subgroup 3a. Single Sheets (Sharing Coefficients 1.75, and higher)

118. *Apophyllite*, $KCa_4[Si_4O_{10}]_2 \cdot 8H_2O$, is the impressive example of the application of the network condensation principle developed by N. V. Belov (see A. ¶ 30) for silicates of the Class II type. The simultaneous presence of tetragonal rings formed of single $[SiO_4]$ tetrahedra, with the $[Si_2O_7]$ groups as the building units is most evident in Fig. A. 84. The analogy of the apophyllite "skeleton" with that of xonotlite and wollastonite chains is striking. There is also a distinct similarity of the apophyllite structure with that of the borosilicate *datolite*, $HCaBSiO_5$. P. V. Pavlov and N. V. Belov[184] determined the structure of datolite in analogy with *herderite*, $CaBePO_4F$, and *gadolinite*, $FeY_2[Be_2Si_2O_{10}]$. The monoclinic cell with $\beta \cong 90^\circ$ improved the previous data given by T. Ito and H. Mori.[185] Most characteristic is the interlinkage in herderite of $[PO_4]$ groups with tetrahedral groups $[BeO_3F]$, and distorted $[CaO_8]$ polyhedra. Analogous sheets of $[SiO_4]$ tetrahedra in datolite are interlinked with $[BO_3H]$ groups. There are no tetrahedral groups between the sheets and datolite, therefore, cannot be a framework structure, as other borosilicates are (cf. A. ¶ 190).

119. A somewhat different condensation principle rules in the structure of *gillespite*, $BaFe[Si_4O_{10}]$, according to A. Pabst (1943), also with a tetragonal network of interlinking $[Si_2O_7]$ groups. The skeleton of the tetragonal ring system is so rigid that it withstands to a high degree acid leaching, as Pabst demonstrated.[186] The tetragonality of the glistering slabs and flakes of leached gillespite, with not consid-

tenko and Z. V. Pudovkina, *Kristallografiya*, 4, 1959, 929. It is tetragonal, with $a_0 = 10.72$ Å., $c_0 = 7.99$ Å., and contains infinite columns of $[TiO_6]$ and $[FeO_6]$ octahedra along the *c*-axis, and infinite $[Si_4O_{10}]_\infty$ chains-columns.

[184] *Doklady Akad. Nauk S.S.S.R.*, 114, 1957, 884-887.

[185] *Acta Cryst.*, 6, 1953, 24-32.

[186] *Am. Mineralogist*, 43, 1958, 970-980. On the compounds of type $RCu[Si_4O_{10}]$ see A. Pabst, *Acta Cryst.*, 12, 1959, 733-739.

erably changed parameters in the (xy) plane, is preserved, whereas the basal spacing is reduced. An impression of this paramorph may be seen in Fig. A. 85, showing the practically persistent tetragonal sheets, whereas all of the Ba^{2+} and Fe^{2+} ions are removed. The hydrated sheets are in part uncoupled (cf. Vol. IV, Section B). Analogous to gillespite are the structures of $CaCu[Si_4O_{10}]$ ("Egyptian Blue"), $SrCu[Si_4O_{10}]$ and $BaCu[Si_4O_{10}]$ (cf. Vol. III, Section B), but these compounds are much more re-

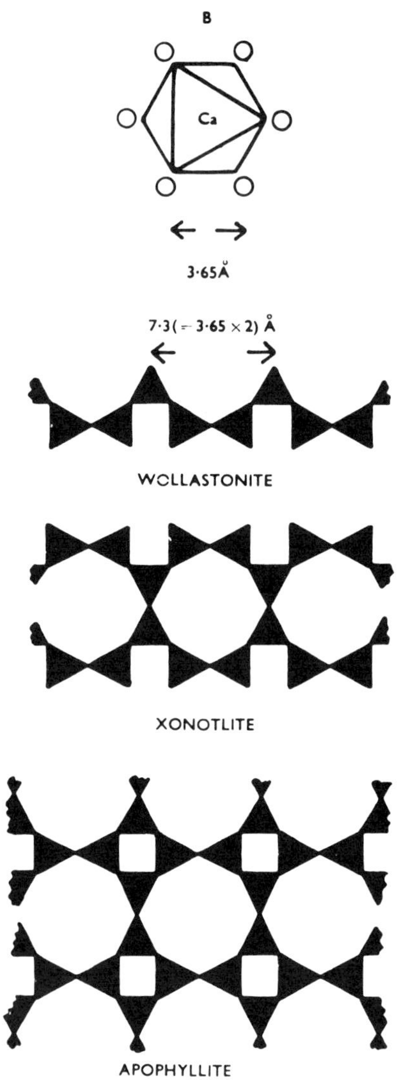

FIG. A. 84. Groupings of $[SiO_4]$ tetrahedra showing the relation between wollastonite, xonotlite, and apophyllite (Taylor)

sistant to acid leaching than gillespite proper. The reason for the high sensitivity of the Fe^{2+} cations in gillespite to the acid reaction is presumably to be seen, according to Pabst, in the incomplete or degenerate coordination of the octahedral group $[FeO_6]$.

120. In spite of a certain analogy with gillespite, the orthorhombic *sanbornite*, $Ba_2[Si_4O_{10}]$, contains other types of $[Si_4O_{10}]_\infty$ sheets, pairwise arranged parallel (001), of six-membered rings, with simple $[SiO_4]$ units, as seen in Fig. A. 86. The arrangements are "undulating" of "corrugated," as determined by R. H. Douglass.[187]

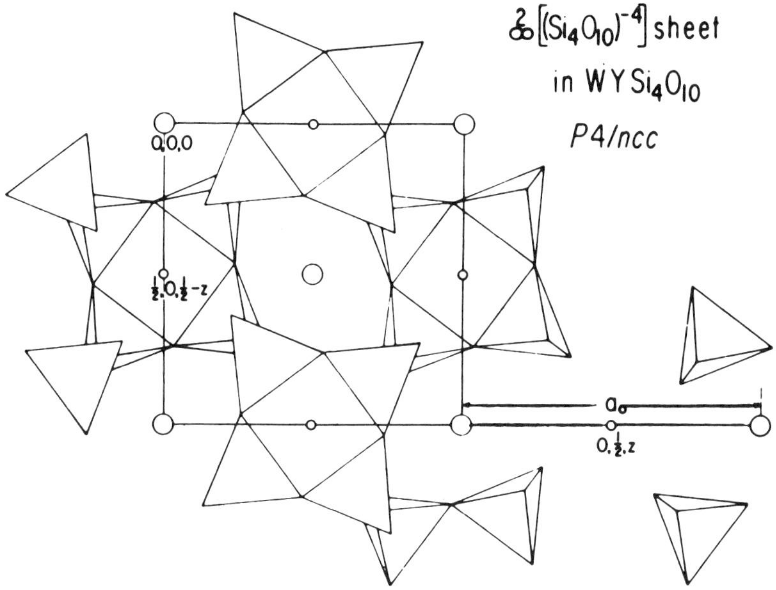

FIG. A. 85. Structure of a tetragonal $[Si_4O_{10}]$ sheet in gillespite, after acid-leaching and elimination of barium and iron cations (Pabst).

[187] *Am. Mineralogist*, **43**, 1958, 517-536. This investigation is particularly interesting by the application of an X-ray-photographic method of M. L. Huggins, *J. Am. Chem. Soc.*, **63**, 1941, 66-69; *J. Chem. Phys.*, **12**, 1944, 520; *Science*, **100**, 1944, 570, for the projection of Patterson patterns. — In relation to sanbornite, we may also mention the structure of *prehnite* recently studied by Peng Sze-Tzung, Chou Kung-Du, and Tang You-Chi, *Acta Chim. Sinica*, **25**, 1959, 56-64, with peculiar layer structural units $[Si_3AlO_{10}]_n^{5n-}$ parallel to (001). These layers are actually frameworks of finite thickness, as mentioned by W. H. Taylor (1934) in his discussion of the heulandite structure. We further note the triple ribbons similar to those in sanbornite which were found by A. N. Lazarev, T. F. Tenisheva, and R. G. Grebenshchikov, *Doklady Akad. Nauk S.S.S.R.*, **140**, 1961, 811-814, for α-$BaSiO_3$ and α-$BaGeO_3$. The β-modifications have a pseudowollastonite-type structure. Infrared absorption spectra confirm these structures. Also $2BaO.3SiO_2$, $5BaO.8SiO_2$, and $3BaO.5SiO_2$ are included in these studies.

The sheets in sanbornite are much less stable than those in gillespite and swell up to X-ray-amorphous shreds by acid treatment. Undulating $[Si_4O_{10}]_\infty$ also occur in the (monoclinic) structure of β-$Na_2Si_2O_5$, according to A. Grund,[188] parallel (001). For such *disilicates* it is characteristic that they also do not contain any hydroxyl groups (in contrast to micas, chlorites, clay minerals, etc.), the successive sheets

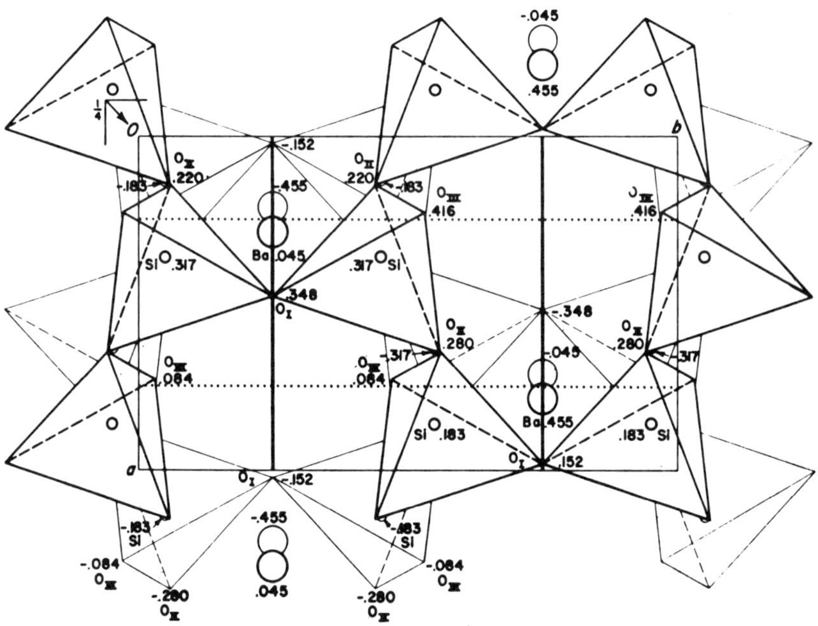

FIG. A. 86. Projection of the sanbornite structure on (001), to show the six-membered $[Si_4O_{10}]$ sheets (Douglass).

being held together by larger cations (as in the micas and apophyllite), the tetrahedral vertices pointing out from both sides of the corrugated sheets (unlike apophyllite which shows the vertices only pointing to one side). F. Liebau[189] confirmed these conclusions also for α-$Na_2Si_2O_5$ (with strongly undulating $[Si_4O_{10}]^{4-}$ sheets), and very similar results for the pseudo-orthorhombic $Li_2Si_2O_5$, with somewhat different mutual arrangements, and angles of the Si—O—Si bonds smaller than 180°.

[188] *Bull. soc. franç. minéral. et crist.*, **77**, 1954, 775-785.

[189] *Fortschr. Mineral.*, **37**, 1959, 99; *Ibid.*, **38**, 1960, 45; extensively in *Acta Cryst.*, **14**, 1961, 389-395, for the room-temperature-stable modification of $Li_2Si_2O_5$, pp. 395-398 for α-$Na_2Si_2O_5$, pp. 399-406 for petalite. On this mineral see also A. Zemann-Hedlick and J. Zemann, *Naturwissenschaften*, **41**, 1954, 476-477. Corrugated disilicate sheets are also characteristic of the phyllosilicic acids (I and II), $H_2Si_2O_5$, as described by F. Liebau and F. Wodtcke, *Fortschr. Mineral.*, **39**, 1961, 333.

Even the lithium aluminosilicate *petalite*, $LiAlSi_4O_{10}$, shows the analogous sheet structure type (Figs. A. 87 and A. 88), in spite of its excellent basal cleavage because of the interlinkage with $[AlO_4]$ tetrahedra, as a framework structure; therefore, it will be discussed in A. ¶ 202.

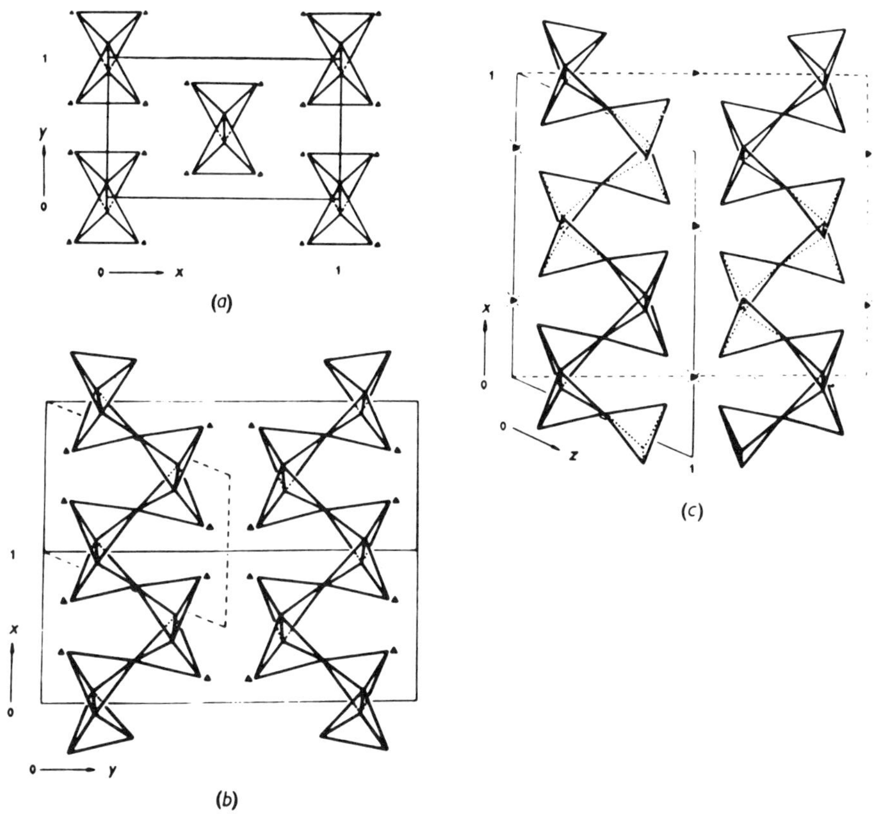

FIG. A. 87. Comparison of the structures of (*a*) Li_2SiO_3, (*b*) $Li_2Si_2O_5$, and (*c*) petalite, (with two elementary cells), in corresponding projections (Liebau).

121. The classical hexagonal networks of $[Si_4O_{10}]_\infty$ type in the talc-prophyllite-mica gruop can only be discussed very briefly here in the multiple cross-relationships existing between these important minerals. We begin with a succinct hint to more recent investigations on *talc*, which is studied extensively as a raw material for modern high-quality electrotechnical ceramic products. J. A. Pask and M. F. Warner[190] discussed the considerable variations in texture and chemical compo-

[190] *J. Am. Ceram. Soc.*, **37**, 1954, 118-128. General classification principles, and schemes for identification of layer silicates were recently elaborated and discussed by Ch. M. Warshaw and R. Roy, *Bull. Geol. Soc. Amer.*, **72**, 1961, 1455-1492 (cf. B. ¶ 1ff.).

sition of different natural occurrences, especially in connection with the formation of crystalline solutions, taking up Al_2O_3, CaO, Fe_2O_3, and even alkalies, influencing the crystal habitus. Relationships to anthophyllite, tremolite, and actinolite are evident, indicating transitions of different degrees of condensation of the $[Si_2O_6]_\infty$

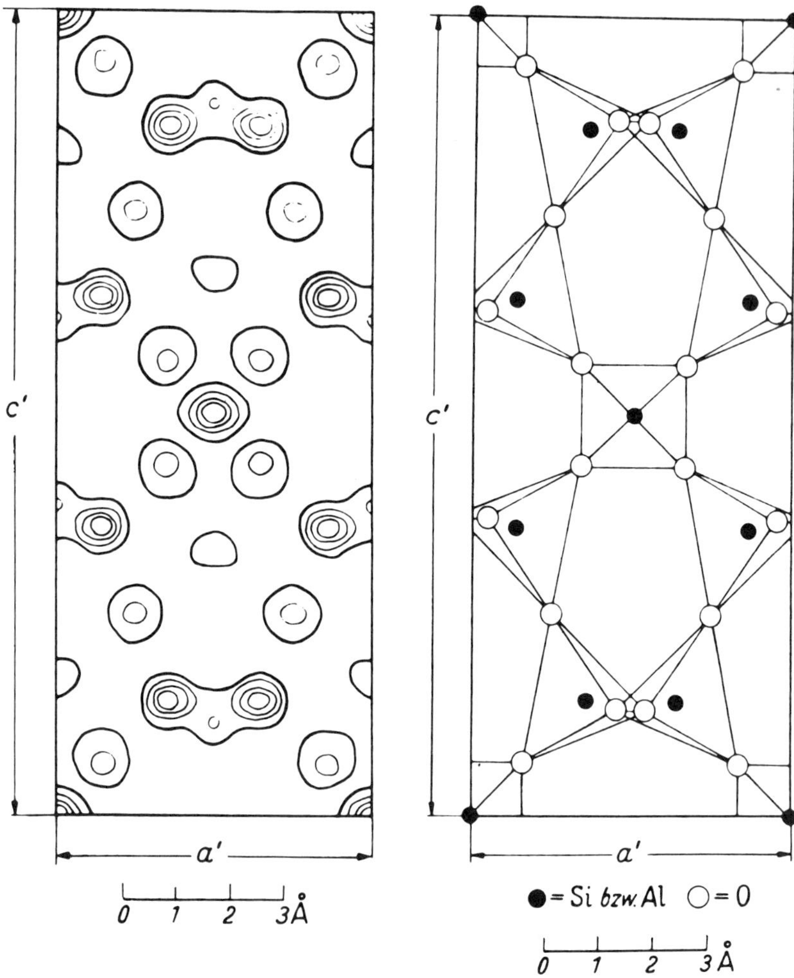

FIG. A. 88. Projection onto (010) of the structure of petalite (Zemann-Hedlick and Zemann).

chains to ribbons, $[Si_4O_{11}]_\infty$, and networks, $[Si_4O_{10}]_\infty$, but also to the chlorites in which additional units like brucite, $Mg(OH)_2$, are taken up into the complex crystal structure, or interlayer mixed types, as concluded by Pask and Warner from powder diffraction studies.

122. The structural relations existing between talc and tremolite were investigated by I. S. Stemple and G. W. Brindley[191] under similar viewpoints, but with not very conclusive results concerning the formation of mixed-layer sequences in talcs. Electron microscopy shows varieties of platy and fibrous forms. The entrance of some alumina into the talc structure is confirmed by synthesis, indicating slight substitutions of Mg^{2+} by Al^{3+}. Fibrous or lath-shaped talc crystals are composites of talc and tremolite; the electron diffraction patterns show excellent regular intergrowths indicating structural superpositions, plausible from the data of Fig. A. 89,

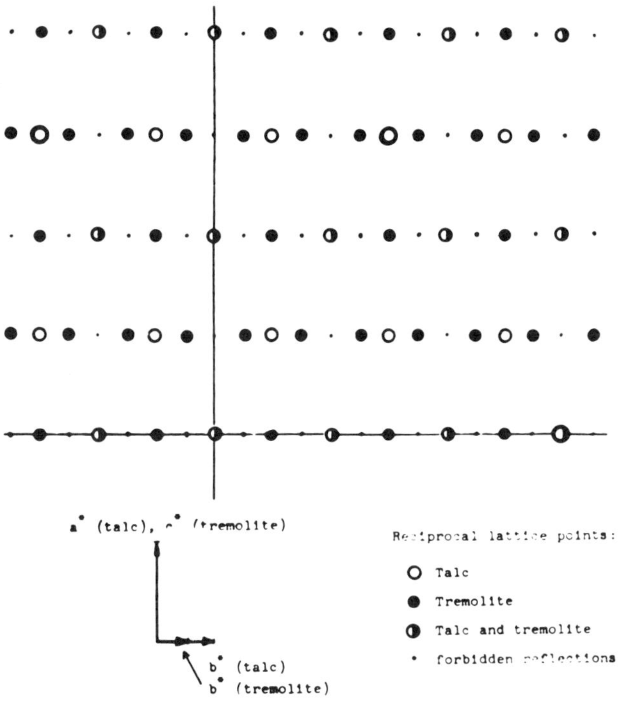

FIG. A. 89. Superposition of a^*b^* from the talc network and c^*b^* of tremolite (Stemple and Brindley).

for the interpretation of the $(hk0)$ reciprocal net of talc, and the $(0kl)$ reciprocal net of tremolite. This intimate relation makes a mutual partial conversion thoroughly plausible, with rearrangements of the Si—O groupings, and a replacement of Ca^{2+} for Mg^{2+} cations, although in talc normally no positions for Ca^{2+} exist. But also Al^{3+} ions may enter the talc structure, with a maximum shrinkage of the basal spac-

[191] *J. Am. Ceram. Soc.*, **43**, 1960, 34-42. The intergrowth observed by Stemple and Brindley is confirmed by optical measurements of H. D. Wright, *Ibid.*, **43**, 1960, 42-43, although on contacts always sharp boundaries between talc and tremolite are observed with no transition zones.

ing up to 0.6 per cent, the b_0 parameter being slightly increased by a substitution of $[AlO_4]$ for $[SiO_4]$ tetrahedra.

123. Fibrous pseudomorphs of a particular talc after amphibole were described by D. P. Serdyuchenko[192] as "*calcio-talc*," containing rather much (up to 0.8 atoms) Ca^{2+} replacing Mg^{2+} (2.0 atoms remaining) in the octahedral layers of normal talc. The calcio-talc is distinctly harder than normal talc, the unit cell parameters are somewhat widened, and the nature of this phase as a transition from tremolite to talc is analogous with that observed by Stemple and Brindley, whereas there is no analogy to the calcium brittle-mica structures, as Serdyuchenko assumes. N. V. Belov,[193] however, is of the opinion that Serdyuchenko's calcio-talc is better interpreted as a margarite with inserted octahedral layers of $[Mg(O, OH)_6]$, or as a derivative of muscovite with a replacement of octahedral groups $[Al(O, OH_6]$ by the same $[Mg(O, OH)_6]$ octahedra, simultaneously of K^+ by Ca^{2+} in the interlayers, to explain the greater hardness. Furthermore, the structural relations to primary amphibole are better understood in this way than by the unusual replacement of Mg^{2+} by Ca^{2+} in a talc structure. The condensation of the $[Si_4O_{11}]_\infty$ ribbons of an amphibole with eight cations in nontetrahedral coordination would explain the transition structure of what Belov called a *magnesio-margarite.*

124. Related to the talcs but characterized by a typical *superstructure* is the mineral *ekmanite*, first discussed by J. Jakob[194] in its chemical composition, with the simplified formula (Fe, Mg, Mn)O.SiO_2.H_2O. Weissenberg and rotation diagrams of single crystals studied by B. Nagy[195] established an anomalously large ortho-hexagonal unit cell, with $a_0 = 66.52$ (12×5.54) Å., $b_0 = 38.40$ (4×9.60) Å., $c_0 = = 96.62$ (8×12.08) Å., or in hexagonal interpretation, $a_0 = b_0 = 38.40$ Å., $c_0 = 96.62$ Å. The interrelation of both cells with the hexagonal network of common layer structures of the talc type (shaded area corresponding to the (001) section of talc) is shown in Fig. A. 90. The ekmanite structure shows signs of disorder related to random-irregular displacement of the layers along the a and b axes. A typical polymorphism comparable to that of the micas along the c-axis is indicated by variable stacking in this direction (Fig. A. 91).

125. Among the hydration products of hydraulic binders, the layer structures of calcium aluminate and silicate hydrates were estensively discussed by K. Spangenberg[196] (cf. Vol. V, Section C), as in the similarities of calcium aluminate hydrates

[192] *Zapiski Vsesoyuz. Mineral. Obshchestva*, **88**, 1959, 298-304.

[193] *Zapiski Vsesoyuz. Mineral. Obshchestva*, **88**, 1959, 305-308.

[194] *Schweiz. mineral. petrog. Mitt.*, **3**, 1923, 237-239.

[195] *Am. Mineralogist*, **39**, 1954, 946-956.

[196] *Tonind.-Ztg. u. Keram. Rundschau*, **75**, 1951, 378-384; *Tagungsber. Zementind.*, No. 4, 1951, 102-156, especially pp. 125, 128, 147.

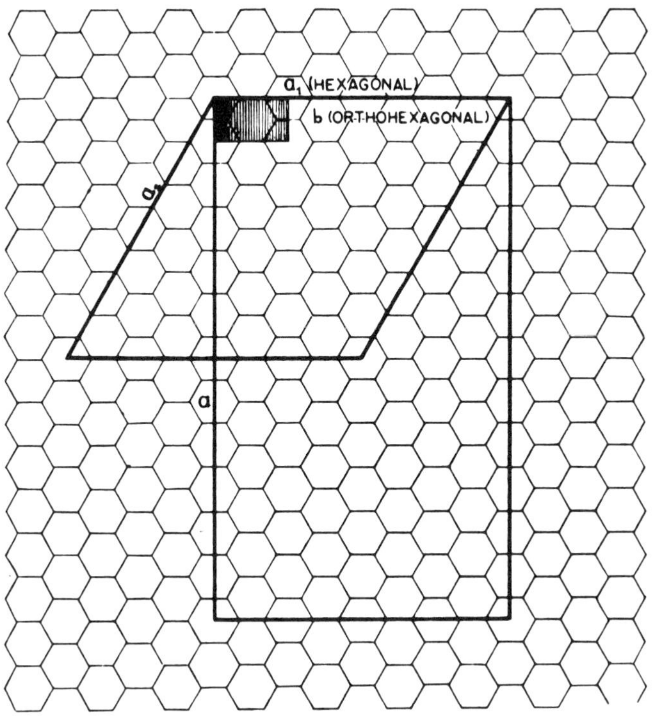

FIG. A. 90. Relation between orthohexagonal and hexagonal unit cell of ekmanite, on the hexagonal network of layer structure silicates. The talc unit cell in the (001) section is shaded (Nagy).

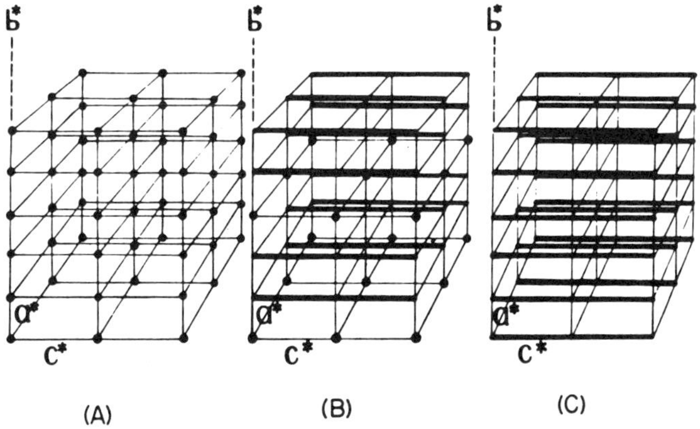

FIG. A. 91. Reciprocal lattice (*a*) of the perfect-order ekmanite structure; (*b*) the same affected by random displacements by $n \cdot b/3$ increments along the *b*-axis; (*c*) affected by random-irregular displacements along the *a* and *b* axis, as occurring in natural ekmanite (Nagy).

with the structure of *hydrargillite*, and in many optical analogies. The important minerals of the *ettringite* family, $3CaO.Al_2O_3.3CaSO_4.32H_2O$, belong to typical layer structures, and several calcium aluminate hydrates show complex intergrowths of $[Ca(OH)_6]_\infty$ layers with those of $[Al(O, OH)_6]_\infty$ behaving as an "anomalous crystalline solution." The phase $3CaO.Al_2O_3.12H_2O$ is thus interpreted as an ordered mixed structure between $2CaO.Al_2O_3.8H_2O$, and $4CaO.Al_2O_3.8H_2O$, and $4CaO.Al_2O_3.13H_2O$.

126. Such discussions also proved to be helpful in the correct interpretation of calcium silicate hydrates of the *tobermorite* family (with the empirical formula $5CaO.6SiO_2.5H_2O$), for which H. D. Megaw and C. H. Kelsey[197] postulated a very particular layer structure. The striking similarity of this silicate hydrate with clay minerals of the montmorillonite group is also evident in the electron-microscopic aspect of typical flakes of extreme thinness and flexibility, and by their swelling properties. Megaw and Kelsey ascribed to tobermorite a pseudo-orthorhombic unit cell containing two identical layers parallel (001), displaced relative to each other, the $[SiO_4]$-tetrahedral portions in "puckered chains" with hydrogen atoms attached to O^{2-} anions of the tetrahedra parallel b. The Ca—O octahedra show four O^{2-} neighbors in complex sheets $[4Ca_2Si_3O_9]$, and two O^{2-} in the sheet of the next level, as schematically shown in Fig. A. 92. The sheets are like double-sided corrugated paper

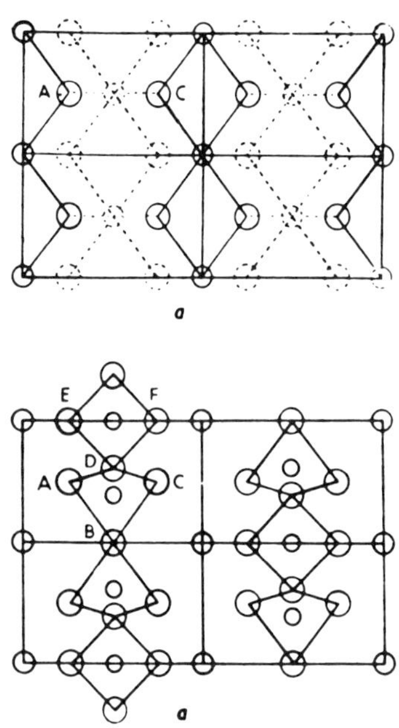

FIG. A. 92. Structural principle of tobermorite (Megaw and Kelsey). The small circles are Si^{4+} ions; the large circles, O^{2-} anions; the intermediate circles, Ca^{2+} cations. (*a*) Two central sheets of layers projected on (001). (*b*) The upper half of layer projected on (001).

[197] *Nature*, 177, 1956, 390-391.

having a strong central sheet reinforced on each surface by longitudinal toothed ridges and grooves between these ridges.

127. The proposed structure accounts well for both the platy and the fibrous habitus of tobermorites observed in the hydration products of basic calcium silicates (cf. Vol. IV, Section A), as described by G. L. Kalousek.[198] The relationships between tobermorite and the clay minerals were especially studied by H. F. W. Taylor and J. W. Howison,[199] emphasizing that of tobermorite with vermiculite, as a 2 : 1 layer type first discussed by J. D. C. McConnell.[200] The variability of the basal spacings with the water content is another analogy with montmorillonite and the wide variability of the degree of crystallinity and composition in the $CaO : SiO_2$ ratio. Previous opinions of an adsorption of $Ca(OH)_2$ on a compound of lower $CaO : SiO_2$ ratio are unsatisfactory. Taylor and Howison even consider the possibilities of a replacement of Si^{4+} in (SiO_2H_2) in the sheets by Ca^{2+}, with a relatively low effect on the weight of the pseudocell, or the unit cell dimensions, and the ratio $H_2O : SiO_2$, as seen from the data presented in the tabulation.

Composition	Ratio $CaO : SiO_2 : H_2O$	Weight of the pseudocell
$Ca_4Si_6O_{18}.CaO_4H_{10}$	0.83 : 1 : 0.83	730
$Ca_4Si_{4^1/_2}O_{15}.Ca_{2^1/_2}O_4H_7$	1.44 : 1 : 0.78	697
$Ca_4Si_4O_{14}.Ca_3O_4H_6$	1.75 : 1 : 0.75	686

128. Characteristic of tobermorite is the distinct formation of three most frequently observed hydration states, with corresponding basal spacings (d_{002}), of 14.0, 11.3, and 9.33 Å., respectively. The latter tobermorite phase has a pseudocell which is face-centered on A, persisting up to 700°C. Complete dehydration at about 800°C.[201] converts the phase into β-$CaSiO_3$, with twinning in two orientations, and the b-axes in common with both silicates. The original faces (101) and $(10\bar{1})$ are changed to (201) in wollastonite. The structural changes are seen from Fig. A. 93, with an approximative preservation of the Ca^{2+} positions, as a relatively stable skeleton. The conversion, however, brings about considerable re-arrangements in the Si—O network, with a migration of the Si^{4+} cations, a partial elimination of silica, and the formation of defective metasilicate chains of wollastonite. This crystal phase has then anomalously low refractive indices and is positive in its elongation zone.

198 *J. Am. Concrete Inst.*, **26**, 1955, 989-1011.
199 *Clay Minerals Bull.*, **2**, 1956, 98-111.
200 *Mineral. Mag.*, **30**, 1954, 293-305.
201 Cf. H. F. W. Taylor, *Proc. Natl. Conf. on Clays and Clay Minerals*, **6**, 1959, 101-109.

129. Kh. S. Mamedov and N. V. Belov,[202] in a re-examination of the tobermorite structure (cf. Fig. A. 94), emphasized the analogies to the micas, vermiculite, and montmorillonite, but with the important difference in the strictly step-wise increasing water contents. They distinguish *plombierite* as the 14 Å. tobermorite (with

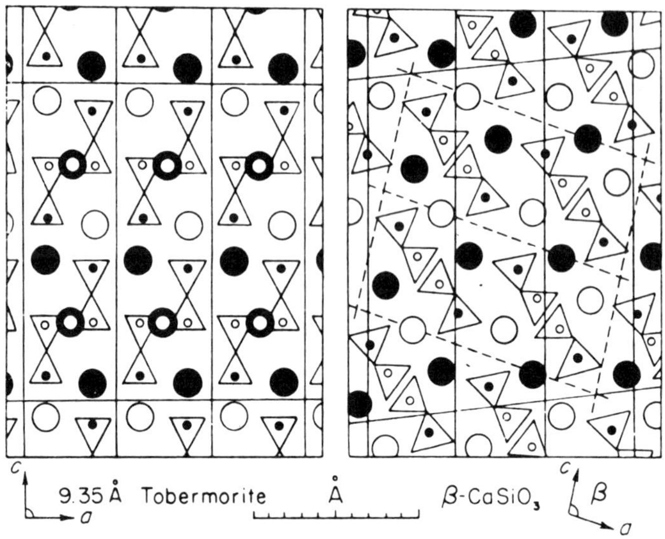

FIG. A. 93. Comparison of the structure of the 9.35 Å. tobermorite with that of β-$CaSiO_3$, as formed after calcination at 800°C. (Taylor).

doubled c_0 parameter = 28 Å.), tobermorite proper (with $c_0 = 2 \times 11.3$ Å.), and *riversideite* (with $c_0 = 2 \times 9.0$ Å.). The unit cell contains four molecules $5CaO.6SiO_2.xH_2O$, the ratio $SiO_2 : H_2O$ being = 0.5 for the 9 Å. tobermorite, $\cong$ 1.0 for the 11 Å. tobermorite, and = 2.0 to 2.5 for the 14 Å. tobermorite. Megaw's and Kelsey's structure proposal concerned only the 11 Å. tobermorite. The condensation from chains of wollastonite type to networks is analogous to that from xonotlite to okenite (see below), to corrugated networks, best seen in the (*ac*) plane, and perpendicular to the nets, showing the xonotlite rings. The period $a_0 = 11.22$ Å. is equal to two-thirds of the corresponding period in xonotlite, but whereas xonotlite does not show the infrared absorption peak at 6.2 μ, this is very clear in tobermorite. The hydroxyl bonding is different in xonotlite and tobermorite, namely with the peak 2.75 μ for xonotlite, and 2.9 μ for tobermorite. The condensation to the network is described by the reaction

$$2\,[Si_6O_{17}]^{10-} - 3\,O^{2-} \rightarrow [Si_{12}O_{31}]^{14-}$$

[202] *Doklady Akad. Nauk S.S.S.R.*, **123**, 1958, 163-165.

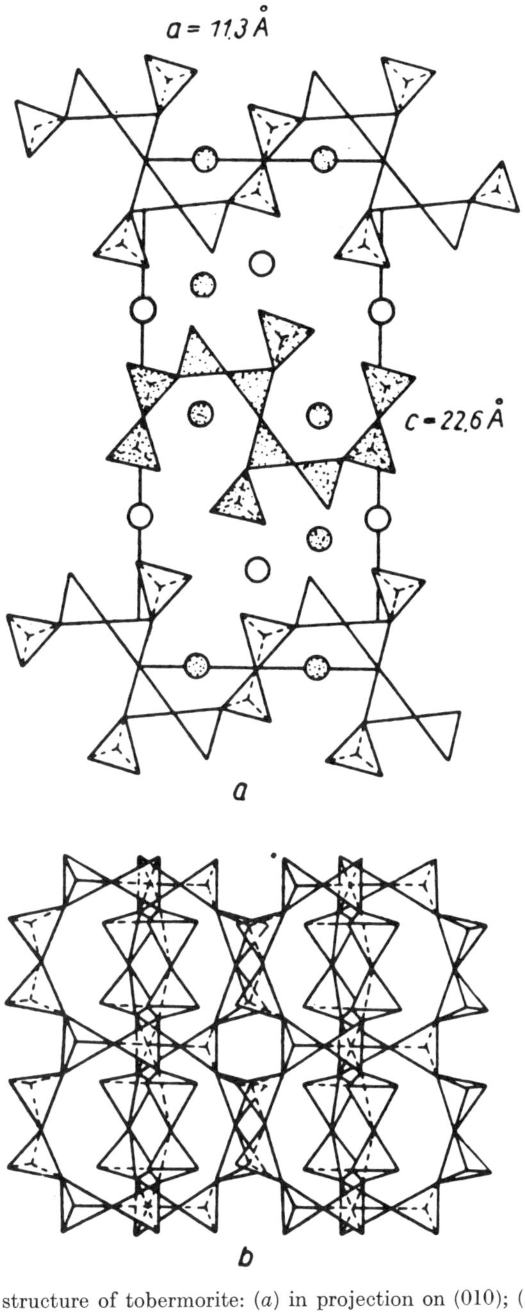

Fig. A. 94. Crystal structure of tobermorite: (*a*) in projection on (010); (*b*) in schematic representation of the network (Mamedov and Belov).

The crystallochemical formula of tobermorite is then 2 $(Ca_{10}[Si_{12}O_{31}](OH)_6.3H_2O)$ for the 9 Å. tobermorite; the corresponding formula for 11 Å., tobermorite is 2 $(Ca_{10}[Si_{12}O_{31}](OH)_6.8H_2O)$, and for the 14 Å. tobermorite the same with 18 molecules of water. Filling the gaps in the networks are 20 Ca^{2+} cations, but the position of the water molecules has not yet been defined.

130. A quite paradoxical role is played by one natural tobermorite from Loch Eynort (Scotland). In the electron diffraction diagrams it shows the a_0 period $3 \times 5.5 = 16.5$ Å., and it is more similar to xonotlite in details of the X-ray diagram, with one-dimensional defects and asterism strokes oriented perpendicular to the reciprocal lattice axis b^*. This tobermorite corresponds to the formula $Ca_4Ca_6[Si_6O_{17}]_2.10H_2O$, as a condensation product from three xonotlite layers from which one molecule $Ca(OH)_2$ was removed and five molecules water added. Mamedov and Belov, therefore, are of the opinion that the Loch Eynort tobermorite is an intermediate crystal phase between common tobermorite and xonotlite. The cleavage which is normally the basis plane is in the Loch Eynort mineral (100). According to J. A. Gard and H. F. W. Taylor[203] the dehydration behavior is also anomalous because the usually observed shrinkage of the basal spacing to 9.35 Å. on heating to 300°C. is absent, the 11.3 Å. spacing persisting up to 800°C. until β-$CaSiO_3$ is formed.

131. The coordination principles assumed from xonotlite units by Mamedov and Belov were extended by the same authors to the structures of the so-called "mica-like" calcium silicate hydrates,[204] with hexagonal $[Si_2O_5]_\infty$ networks, differing from those of the common "phyllosilicates." The presence of ideally eight-membered rings in such networks is contradictory to the crystallographic symmetry laws, but in apophyllite (see above) the combination of eight- and four-membered tetragonal networks is possible. In *okenite*, $3(Ca_3[Si_6O_{15}].2H_2O.4H_2O)$ appears to be a combination of eight- and five-membered rings (cf. Fig. A. 95), similar to that also mentioned for epidote and zoisite (cf. A. ¶ 85 ff.), and gehlenite and melilite (cf. A. ¶ 170). The triclinic mineral *nekoite* is the simplest representative of such a condensation structure, with only one layer $[Si_6O_{15}]$, corresponding to the crystallochemical formula $Ca_3[Si_6O_{15}].2H_2O.4H_2O$.[205] It is triple in okenite as seen above. Another condensation type in the repeat units is observed in *truscottite* (pseudohexagonal), having the formula 4 $(Ca_3[Si_6O_{15}].2H_2O)$, and even the 18-repeat-unit structure of *gyrolite* (pseudohexagonal), with the formula 24 $(Ca_4[Si_6O_{15}(OH)_2].4H_2O)$. All these structures have about the same a_0 parameter (9.86 to 9.72 Å.), and b_0 (7.3 to 7.2 Å.) in the triclinic sheets. The double formulation of the water contents is to distinguish zeolitic and constitution water, as in okenite and nekoite. The

[203] *Mineral. Mag.*, **31**, 1957, 361-370.
[204] *Doklady Akad. Nauk S.S.S.R.*, **121**, 1958, 720-723.
[205] Cf. J. A. Gard and H. F. W. Taylor, *Mineral. Mag.*, **31**, 1956, 5-20.

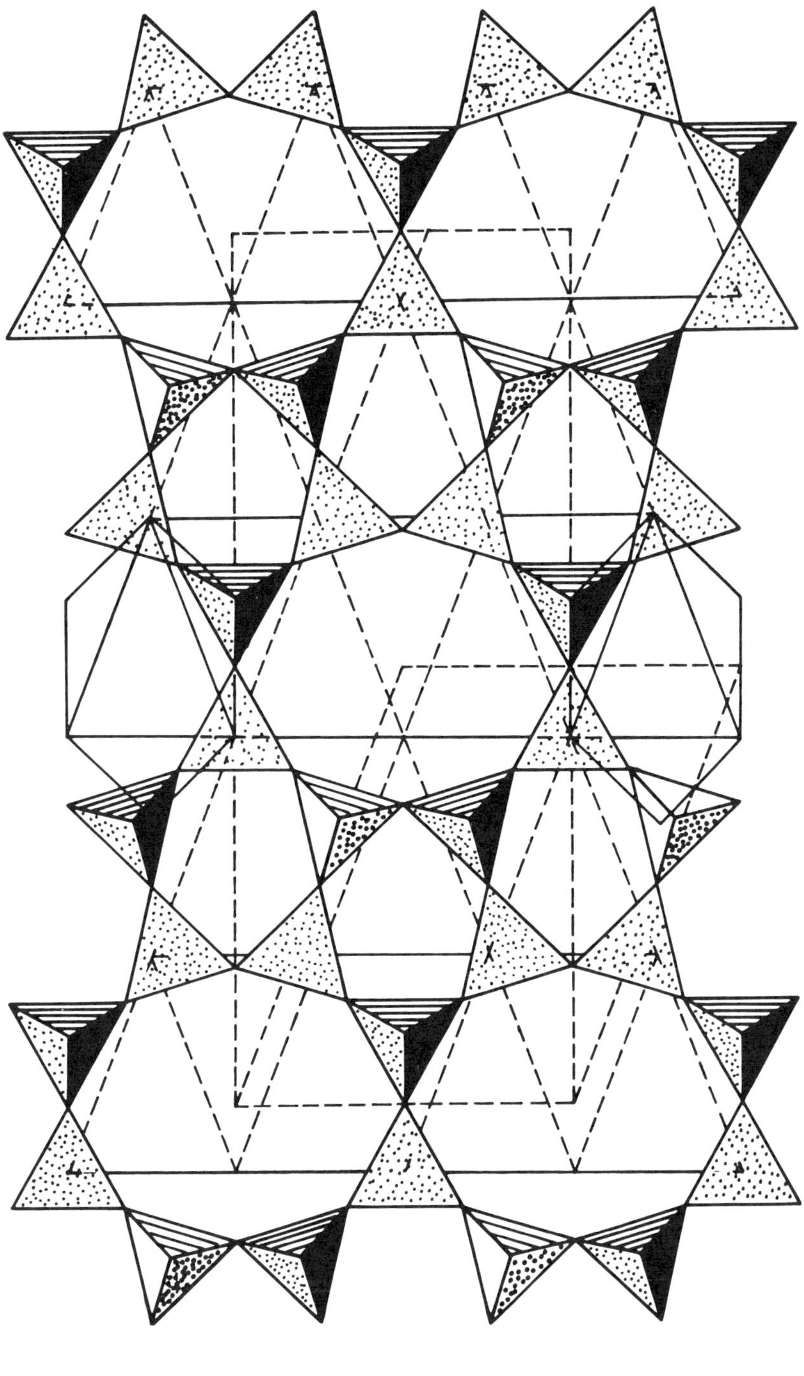

FIG. A. 95. Structure of okenite, showing the network $[Si_6O_{15}]_\infty$ with insertion of $[CaO_6]$ octahedra, two of these designed in position to the network (Mamedov and Belov).

dehydroxylation at 800°C. delivers pseudomorphs of β-$CaSiO_3$ (wollastonite), with the fiber axis b the same as the initial b axis. Truscottite does not contain any zeolitic, water only constitution water, and is also converted to wollastonite. Gyrolite contains a double set of constitution water: one bound to excess $Ca(OH)_2$ columns in the structure, and the other being the normal constitution water.[206].

132. A most particular sheet structure was proposed by R. A. Chalmers, L. S. Dent, and H. F. W. Taylor[207] for the triclinic *zeophyllite*, $Ca_{12}[Si_9O_{24}](OH)_6F_6.6H_2O$, which has a pronouncedly pseudohexagonal sub-lattice with the period $a = 3.53$ Å., corresponding to the a_0 parameter of $Ca(OH)_2(= 3.59$ Å.). Zeophyllite is not only interesting because of its high water content, but also by its conversion into cuspidine at 710°C.; the ratio Ca : F is the same in both minerals. The structure of cuspidine is

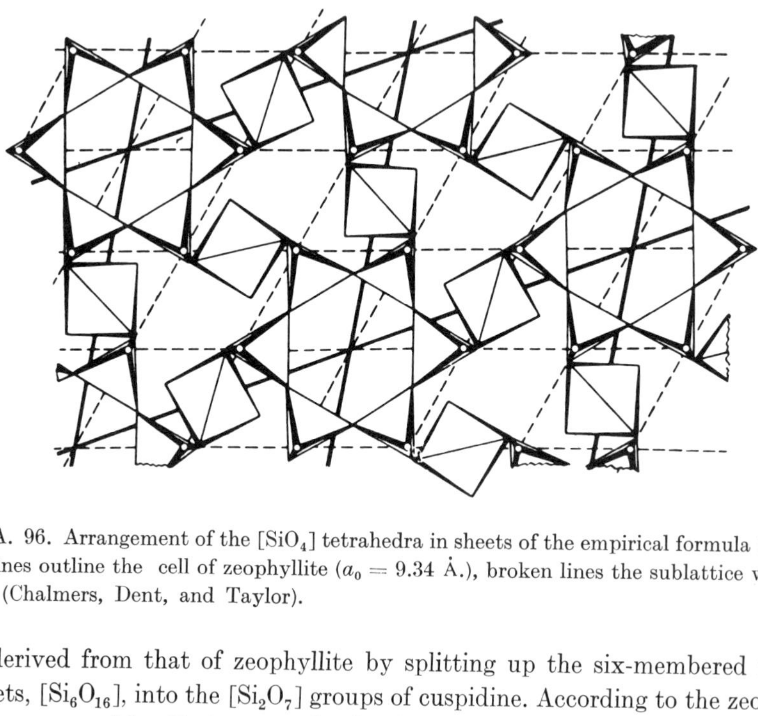

FIG. A. 96. Arrangement of the $[SiO_4]$ tetrahedra in sheets of the empirical formula $[Si_3O_8]^{4-}$. Heavy lines outline the cell of zeophyllite ($a_0 = 9.34$ Å.), broken lines the sublattice with $a_0 = 3.53$ Å. (Chalmers, Dent, and Taylor).

easily derived from that of zeophyllite by splitting up the six-membered rings of the sheets, $[Si_6O_{16}]$, into the $[Si_2O_7]$ groups of cuspidine. According to the zeophyllite structure proposed by Chalmers *et al.*, the rings are in one plane level, joined to one another by bridge tetrahedra at a different level, as is seen in the arrangement in Fig. A. 96. The calcium ions occur in this proposal in positions very similar to those in $Ca(OH)_2$. Characteristic is the pronounced stacking disorder of the basal layers in zeophyllite.

[206] Cf. A. L. Mackay and H. F. W. Taylor, *Mineral, Mag.*, 30, 1953, 80-91.
[207] *Mineral. Mag.*, 31, 1958, 726-735.

133. In the *mica* minerals the investigations on the *polytypism* of this group are of particular importance and genetic significance. Going back to the studies of St. B. Hendricks and M. E. Jefferson (1939) a highly refined reexamination of the previous measurements by W. W. Jackson and J. West (1931, 1933) made polytypism phenomena most evident (commonly called "polymorphism," although this term is not correct), concerning multiple possibilities of adjusting the hexagonal networks $[Si_2O_5]_\infty$ to the different symmetries concluded from morphological and X-ray diffraction determinations. Hendricks and Jefferson[208] solved this problem, showing up seven different ways of shifting the cation layers relative to the $[Si_2O_5]_\infty$ networks, and for alternating the anion layers by rational fractions of the elementary cell periods a_0 and b_0. The different kinds of polytypism are demonstrated by different repetitions of identical positions after every second, third, or sixth layer (the "stacking" sequence). The corresponding symmetries of the whole crystal edifice may then be changed; they may be monoclinic, triclinic, but also trigonal. The most common are the monoclinic stackings (abbreviated symbol *M*) with one (1 *M*), or two (2 *M*)

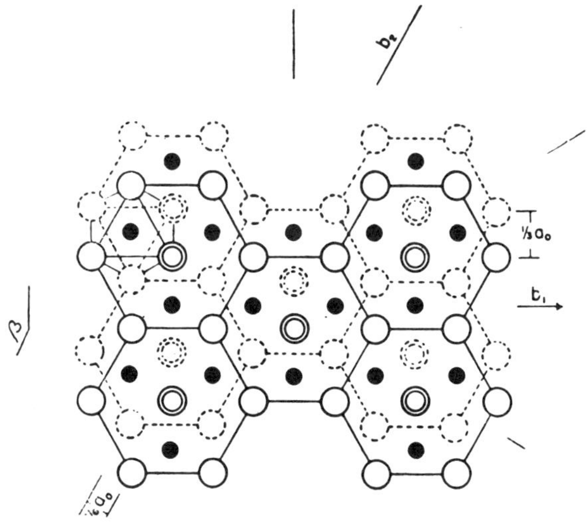

FIG. A. 97. Projection of a phlogopite layer in the 1 *M* polytype structure, on (001) (Hendricks and Jefferson).

[208] *Am. Mineralogist*, **24**, 1939, 729-771. Considerable refinement of the structure as described by Jackson and West was made by W. Radoslovich, *Acta Cryst.*, **13**, 1960, 919-932, indicating essential distortions from the ideal structure. An analogous refinement for the structure of an iron-rich *phlogopite*, by H. Steinfink, *Am. Mineralogist*, **47**, 1962, 886-896, concerns the separate distribution of iron (Fe^{III}) in the tetrahedral sites, of magnesium exclusively in octahedral sites. Complete adjustment of both sheets occurs by a 12° rotation about the normal to the base of the tetrahedra. The highly complex polysynthetic twinning types in micas were extensively studied in their structural basis by R. Sadanaga and Y. Takeuchi, *Z. Krist.*, **116**, 1961, 406-429.

alternating periods. The 1 *M* structure is in nature most common for biotite, phlogopite, and the lithium micas; 2 *M* structures occur also in phlogopite, manganophyllite, and lepidomelane as typical *trioctahedral micas*, i.e., those micas with completely filled [Me(O, $OH_6)_\infty$ layers. If there are only two-thirds of the octahedral positions in those layers filled, the resulting *dioctahedral* micas of the muscovite group prevail in the 2 *M* type (Jackson and West had designated this type generally as "the" mica structure). A three-layer mica with a trigonal symmetry (symbol 3 *T*) is less common for biotite, phlogopite, and lepidomelane. A pseudorhombohedral type 6 *M* was also postulated by Hendricks and Jefferson. For several lepidolites a triclinic six-layer structure (6 *Tr*) was indicated in a few biotites; also a 24-layer triclinic (24 *Tr*) structure was cited as a rarity in one biotite. The projections (Figs. A. 97 and 98) on (001) show a monoclinic phlogopite layer 1 *M*, and a trigonal-enantiomorphic (3 *T*) structure type.

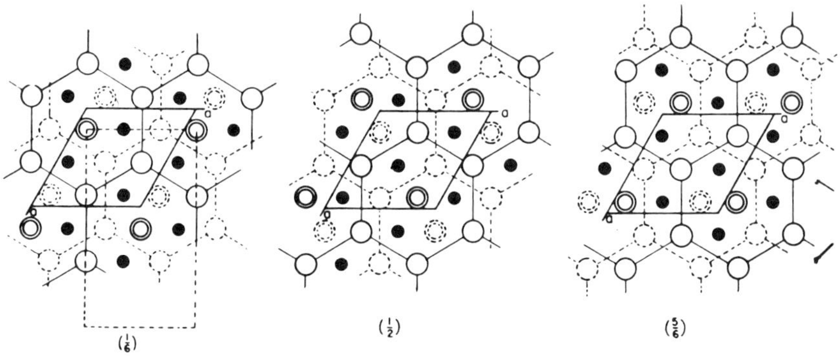

FIG. A. 98. Sequence of the layers in the trigonal-enantiomorphic 3 *T* polytype structure of mica (Hendricks and Jefferson).

134. More recently E. Wm. Heinrich, A. A. Levinson, D. V. Lewandowski, and C. F. Hewitt[209] made extensive studies on the occurrence of the different polytypes of natural micas in connection with their genetic characterization. In pegmatites and effusive rocks, the biotites and related micas show a lower degree of order than in

[209] "Natural Mica Studies," Quarterly Reports of the Engineering Research Institute, University of Michigan, Project No. 978, Signal Corps U.S. Army Contract; Final Report "Studies in Natural History of Micas," 1954, 241 pp., especially pp. 7-11 on the methods, further A. A. Levinson, *Am. Mineralogist*, **38**, 1953, 88-107, on the relations between polymorphism and composition in the series muscovite-lepidolite; Levinson and Heinrich, *Ibid.*, **39**, 1954, 937-945, on single crystals of phlogopite, biotite, and manganophyllite, the polytypes of which are determined by the geological conditions of environment. *Anomite* is identical with 2 *M* biotite, *meroxene* with 1 *M* biotite. The well-known percussion method for a distinction of micas of the first and second class is often unreliable; only the X-ray determination of the (010) plane is unequivocal. Barium-muscovite (*oellacherite*) and *roscoelite* are 1 *M* micas which are not mere varieties of 2 *M* muscovite; cf. *Am. J. Sci.*, **253**, 1955, 39-43.

the low-temperature muscovites, as in vein mineralizations. Here is a wide field of practical application of structural details in minerogenesis and petrogenesis, comparable to the conversions in the feldspar group minerals. We mention only as an interesting example the discovery of a 3 *T* muscovite of nearly uniaxial optical characteristics, described by J. M. Axelrod and F. S. Grimaldi,[210] although according to J. V. Smith and H. S. Yoder[211] the optical symmetry must not in every case be congruent with the X-ray structure symmetry. The results of Heinrich's *et al.* are fundamentally confirmed by H. S. Yoder and H. P. Eugster,[212] although some corrections must be noted, concerning the 6 *Tr* polytype which is now described as a differently interpreted 2 *M* layer type. No further details on the exact atomic positions are available. It is only concluded that the hexagonal layers in muscovite must be buckled, whereas in phlogopite they are planar. Particularly important, however, are the detailed conclusions of Yoder and Eugster concerning the state of order, and the relations of this property with the genesis of the micas in nature. From hydrothermal experiments only low-ordered states of 1 *M* with diffuse interferences are the rule. They are, therefore, called 1 *Md* structures (to indicate their diffuse reflections), but they can be gradually changed to 1 *M* crystals, and in the case of muscovite even reach the fully ordered 2 *M* type. The sequence 1 *Md* → 1 *M* → 2 *M* is characteristic of this progressive change, as from *illite* into muscovite micas in metamorphic sediments. For phlogopite the 1 *M* type is dominant from lava rocks, but they can also be observed in metamorphic sediments.

135. Of the greatest importance for the elucidation of the polytypism in the mica group are the investigations of H. S. Yoder and H. P. Eugster[213] on the hydrothermal synthesis of muscovite, phlogopite, and paragonite, and a rediscussion of the various polytype modifications. The X-ray powder diagrams of natural 1 *M* and 3 *T* structures are identical for synthetic phlogopite. However, a partial conversion of natural 2 *M* phlogopite into 1 *M* and 3 *T* structures by a hydrothermal treatment was possible at 900°C (and 1000 kg./cm.^{2} pressure). In the same way, as phlogopite in effusive rocks, synthetic fluorphlogopite is of the 1 *M* type, as R. A. Hatch and J. A. Kohn[214] established. Hydrothermally prepared hydroxyl phlogopite is highly disordered (1 *Md*) (analogous to "pholidolite" in soils), and gradually may be ordered

[210] *Am. Mineralogist*, **34**, 1949, 559-572.

[211] *Mineral. Mag.*, **31**, 1956, 209-235.

[212] *Carnegie Inst. Wash. Year Books*, **52**, 1952/53, 64-69; *Am. Mineralogist*, **39**, 1954, 343-344. On the composition of the trioctahedral micas see the extensive investigations by M. D. Foster, *Am. Mineralogist*, **45**, 1960, 383-398; and *U.S. Geol. Survey. Profess. Papers*, **354**, *B*, 1960, 11-48.

[213] *Geochim. et Cosmochim. Acta*, **6**, 1954, 157-185; *Ibid.*, **8**, 1956, 225-280; *Carnegie Inst. Wash. Year Books*, **54**, 1954/55, 127-129. A summary in *Proc. Natl. Conf. on Clays and Clay Minerals*, **6**, 1958, 42-60.

[214] *Am. Mineralogist*, **40**, 1955, 10-21.

to the 1 *M* type (see the X-ray diagrams in Fig. A. 99). Completely disordered muscovite, 1 *Md*, does not show any distinct (*hkl*) interferences; in gradual ordering appear (112) and (11$\bar{2}$) reflections of 1 *M*, which may be superseded to 2 *M* by hydrothermal exposure at 250° to 350°C. (Fig. A. 100). *Sericite* is usually 1 *M* or 2 *M*. *Paragonite* behaves very similar in all these conversions. The type 1 *M* is characteristic of low-temperature, poorly developed sericites, *phengites* (high in silica contents), and *mariposite* as Heinrich and Levinson[215] observed, whereas *alurgite* is normally of 2 *M* or even 3 *T* type.

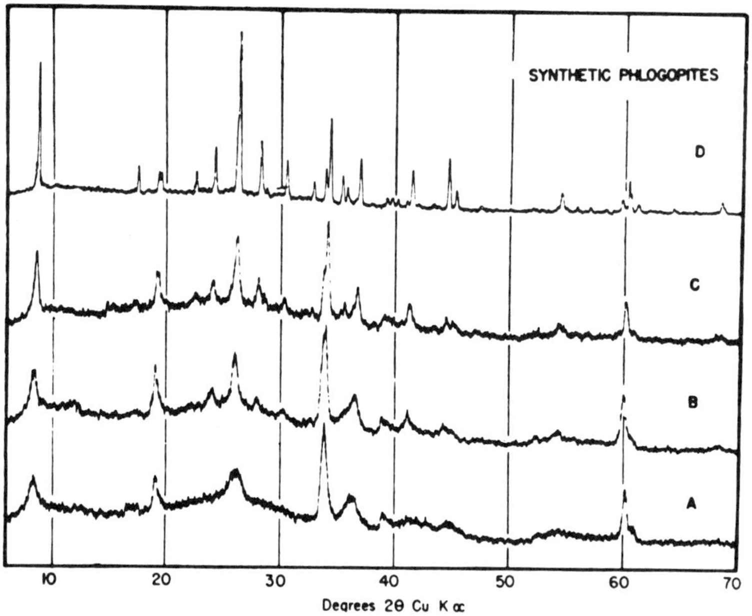

FIG. A. 99. Ordering process 1 *M d* → 1 *M* of synthetic phlogopite by hydrothermal methods (Yoder and Eugster). The lowest temperature used in the synthesis of 1 *M d* mica *A* was below 400°C., the best-ordered phlogopite of 1 *M* type is *D*.

136. Smith and Yoder also revised the geometric premises for the different stackings in the micas since there are in the hexagonal networks angles of 0°, 60°, 120°, 180°, 240°, and 300° between the directions of the stagger of two adjacent layers. The polytypes then possible include those called by Hendricks and Jefferson 1 *M*, 2 *M* (in two slightly different arrangements now called 2 M_1 and 2 M_2), 3 *T*, and the new types 2 *O* and 6 *H*, both not yet observed in nature. We mentioned above that Hendricks' and Jefferson's type 6 *M* is identical with 2 M_2, or with a smaller unit cell in which only two layers are repeated; 6 *Tr*, however, could not be confirmed,

[215] *Am. Mineralogist*, 40, 1955, 983-995; mariposite is a chromium containing phengite.

whereas an 8 *Tr* and an 18 *M* structure with a highly complex stacking are conceivable. A 12 *M* cell was observed in *ephesite* from South Africa. Disordered polytypes, the stacking angles of which occur in random sequence of whole layers, not of single atoms, are possible. There are also transitions from entirely disordered polytypes to completely ordered ones. Random stacking faults bring about twinning around the [310] and [3$\bar{1}$0] axes, with the composition plane (001) in 1 *M* and 2 *M*.

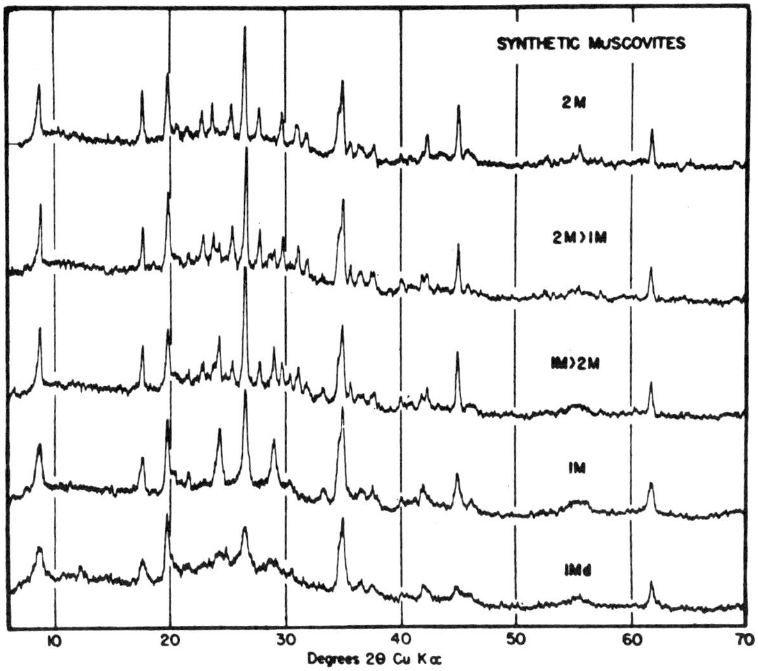

FIG. A. 100. Reordering effects of synthetic 1 *M d* muscovite to a 1 *M* structure, and gradual transition of 1 *M* into 2 *M* muscovite in the temperature range from 350° to 600°C. (Yoder and Eugster).

Another description of the single layer structures in micas was given by A. Pabst[216] for a uniformly holohedral interpretation. For the detection of disorders, spiral growths, etc. the multiple-beam optical interferometric method of S. Tolansky,[217] and more recently of S. Amelinckx and W. DeKeyser,[218] have been of great diagnostic value.

[216] *Am. Mineralogist,* **40,** 1955, 967-974.

[217] *Proc. Roy. Soc.* (*London*), A **184,** 1954, 41-51; *Ibid.,* **186,** 1946, 261-271; *Phil. Mag.,* [7] **37,** 1946, 390-398; 453-462.

[218] *Compt. rend. Congr. Géol. Intern. Alger,* 1952, No. 18, 9-33.

137. An excellent discussion of the properties of micas observed in sandstones and other sediments was given by J. Rimsaite,[219] based on extensive optical and X-ray data on this type of material, especially for the distinction of the muscovite and biotite isotypes. The most frequent one is 2 *M* (less frequently 3 *T*) for muscovite, 1 *Md* for biotite, and 3 *T* very rare for biotite. Sometimes intergrowths of muscovite with paragonite and chlorite also occur. On the weathering products, with their characteristic changes from the initial mica type, cf. Vol. IV, Section A; the octahedral layer in muscovite is much more stable than in biotite in which the iron ions are easily leached away. Potassium is equally leached in amounts up to 20 per cent from muscovite, but to 50 per cent from biotite, with a corresponding widening of the structural spacings.

138. Important are further studies of H. Harder[220] on similar occurrences of muscovite and paragonite, the latter being a distinct and independent mineral, with a $c_0 = 19.28$ Å. (muscovite 20.08 Å.), because of the smaller Na^+ ion. Both minerals form only restricted crystalline solutions; K^+ can be introduced into paragonite only in a narrow range, whereas Na^+ may appear in muscovite in somewhat higher concentrations, and with a characteristic shrinkage of the basal spacing. Paragonite is more restricted in its occurrence to metamorphic rocks (crystalline schists) in paragenesis with pronounced aluminum silicate minerals (andalusite, disthene, chiastolite, staurolite, glaucophane, also corundum). Paragonite was synthesized by Eugster and Yoder[221] in 1 *M* and 2 *M* forms, the natural crystals being 2 *M* (as in ephesite from South Africa), as the stable sodium-mica phase. Characteristic in many occurrences is the formation of muscovite and paragonite, side-by-side, in so-called "white mica."

139. In this connection it may even be surprising to a certain degree that in clays may occur *fuchsite* and other chromium containing micas (and chlorites, see below) as described by P. F. Kerr and P. K. Hamilton,[222] with 1 *M* and 2 *M* polytypes, which differ from the well-crystallized fuchsite and mariposite (cf. B. ¶ 60).

140. For the group of the *brittle micas* (clintonite, margarite, xanthophyllite), it is characteristic that the ratios Al^{3+}/Si^{4+} in the tetrahedral layers are above 1 : 1, which is generally the limit of the aluminum contents in the tetrahedral layers of between 2.20 and 2.33 : 1. The brittle micas do not belong to the magmatic parageneses, but to those in contact-metamorphic assimilation reactions. As M. Balconi[223]

[219] *Heidelberger Beitr. Mineral. u Petrog.*, **6**, 1957, 1-51; see also E. Hellner, *Fortschr. Mineral,.* **34**, 1956, 43-44.

[220] *Heidelberger Beitr. Mineral. u Petrog.*, **5**, 1956, 227-271; *Fortschr. Mineral.*, **34**, 1956, 44.

[221] *Carnegie Inst. Wash. Year Books*, **53**, 1953/54, 111-114.

[222] *Am. Mineralogist*, **43**, 1958, 34-37.

[223] *Atti accad. nazl. Lincei, Rend. classe sci. fis. mat. e nat.* [8], **16**, 1954, 101-107.

demostrated, this ratio varies in xanthophyllite between 2.20 and 2.33 : 1. A complete structural determination was made for *brandisite*, $Ca_{1.0}(Mg_{1.7}Al_{1.1}Ca_{0.2})[(Al_{2.8}Si_{1.2})O_{10}](OH)_2$, by Yu. A. Akhundov, Kh. S. Mamedov, and N. V. Belov.[224] Figure A. 101 shows the structure schematically, projected on the (*yz*) plane, indicating the brucite-like octahedral layers between the tetrahedral sheets. For *xanthophyllite* with the extreme ratio $Al^{3+}/Si^{4+} = 29.5 : 1.05$, Y. Takeuchi and R. Sadanaga[225] gave a similar structure. The trigonal tetrahedral sheets are deformed, with a random distribution of Al^{3+} and Si^{4+} ions, whereas the octahedral sheets have an ordered Mg^{2+}–Al^{3+} distribution. A comparison of the brittle micas is given in Table III.

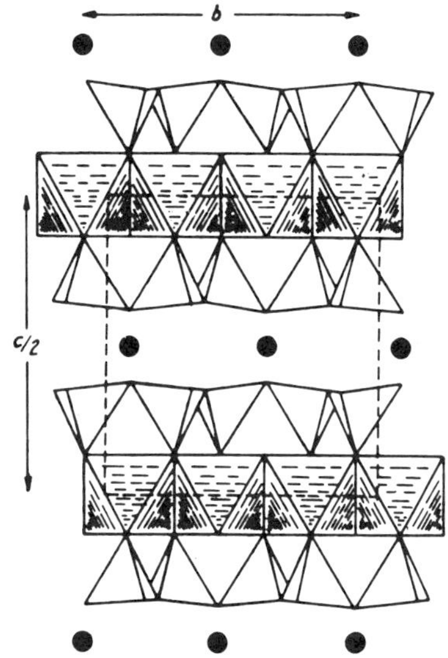

FIG. A. 101. Projection of the structure of brandisite on (*yz*0) (Akhundov, Mamedov, and Belov). Full circles represent Ca^{2+} cations; the buckled tetrahedral sheets and octahedral layers are evident.

141. Among the sheet structures of $[Si_2O_5]_\infty$ network types we may also calculate the minerals of the *pyrosmalite-schallerite-friedelite* group for which Cl. Frondel and L. H. Bauer[226] write crystallochemical formulae containing the group $[Si_6O_{15}]_\infty$ but expressively emphasizing that there is no relation to the gyrolite, truscottite, okenite, and centralassite group discussed above. The habitus of the pyrosmalite minerals is pronouncedly mica-like in the basal cleavage and is also related to the

[224] *Doklady Akad. Nauk S.S.S.R.*, **137**, 1961, 167-170.

[225] *Acta Cryst.*, **12**, 1959, 945-946.

[226] *Am. Mineralogist*, **38**, 1953, 755-760. Critical remarks of D. McConnell, *Ibid.*, **39**, 1954, 929-936, concern difficulties in interpreting schallerite as a layer structure silicate; the stacking as postulated by Frondel and Bauer, however, is confirmed.

micas in the phenomenon of polytypism, with stackings along [0001] with almost exact multiples of the c_0 values as seen in Table IV.

142. Concerning the *chlorite* mineral group, J. A. Pask and M. F. Warner[227] extensively discussed the many relations existing between this group and talc. From the structure determination of chlorite by L. Pauling (1930), the insertion of a brucite-layer of octahedral coordination type, $[Mg(OH)_6]_\infty$ between the "mica parcels" is the characteristic feature of the chlorites, with varying substitutions of Al^{3+} for Fe^{2+}, or Fe^{3+} for Mg^{2+} in the brucite layer, and makes a corresponding number of Si^{4+} ions in the networks of the mica parcel possible. Thus the elementary distinction of

Pennine	$Mg_{11}Al[AlSi_7O_{20}](OH)_{16}$
Clinochlore	$Mg_{10}Al_2[Al_2Si_6O_{20}](OH)_{16}$
Prochlorite	$Mg_9Al_3[Al_3Si_5O_{20}](OH)_{16}$
Amesite	$Mg_8Al_4[Al_4Si_4O_{20}](OH)_{16}$

in a series of crystalline solutions is plausible. Ca^{2+}, however, is not able to enter this structure type. A particular question is the double role we have to consider for the *amesite* structure as an aluminous end member in the chlorite series, and as a magnesium member of the *kaolinite* family, as J. W. Gruner (1944) had demonstrated.[228] We will therefore return to amesite, further on, in the discussion on the structure of clay minerals (Section B).

143. Recent structural aspects on the constitution of the chlorites, especially in their relation to the minerals of the *serpentine* group (see below) were given by R. Roy and B. W. Nelson[229] by an investigation based on hydrothermal syntheses in the system $MgO–Al_2O_3–SiO_2–H_2O$ (cf. Vol. IV, Section A). The most important result of these studies was the discovery of two different types of chlorite structures, one with the normal basal spacing of about 14 Å., and a second one with only 7 Å., but also appearing in hexagonal plates. This second type was called by Roy "septechlorites" and is similar to kaolinite and antigorite (see below). The basal spacing becomes smaller with increasing aluminum contents in both the 7 and 14 Å. phases. The existence of a complete series of crystalline solutions on the

[227] *J. Am. Ceram. Soc.*, **37**, 1954, 118-128.

[228] Cf. *Am. Mineralogist*, **29**, 1944, 422-430; G. W. Brindley, B. M. Oughton, and R. F. Youell, *Acta Cryst.*, **4**, 1951, 552-557. A particularly careful systematic classification of the chlorite minerals in their chemical composition was made by M. H. Hey, *Mineral. Mag.*, **30**, 1954, 277-292.

[229] Roy discussed extensively the postulates of the theory of chlorite constitution of G. Tschermak (1890) and of A. N. Winchell (1928, 1936), in *Proc. Natl. Conf. on Clays and Clay Minerals*, **2**, 1953, 335-348; *Natl. Acad. Sci.–Natl. Research Council, Publ.*, No. 327, 1954, 335-348. With the establishment of 7 Å.-chlorites, Roy and Nelson refute the opinion of G. W. Brindley and K. Robinson (1951) that all chlorite-like minerals giving a 7 Å. series of basal interferences should be divorced from the chlorite group.

TABLE III

UNIT CELL CHARACTERISTICS OF THE BRITTLE MICA GROUP (TAKEUCHI AND SADANAGA)

	a_0	b_0	c_0	β	Space group	$d(001)$	Unit cell contents
Xanthophyllite	5.19 Å.	9.00 Å.	9.74 Å.	100°1′	C 2/m	9.59 Å.	2 CaMg $Al[SiAl_3O_{10}](OH)_2$
Margarite	5.13 Å.	8.89 Å.	19.40 Å.		C 2/c	2×9.56 Å.	4 Ca Al_2 $[Si_2Al_2O_{10}](OH)_2$
Seybertite	5.19 Å.	8.99 Å.	28.76 Å.		$P\,3_1\,12$ or $P\,3_2\,12$	3×9.59 Å.	6 $Ca(Mg, Al, Fe)_3[(Al, Si)_4O_{10}](OH)_2$

TABLE IV

UNIT CELL CHARACTERISTICS OF THE PYROSMALITE MINERAL GROUP (FRONDEL AND BAUER)

	a_0	c_0	Lattice type	Unit cell contents
Pyrosmalite	13.36 Å.	7.15 Å. = 1	Hexagonal	2 $(Fe, Mn)_8[Si_6O_{15}](OH, Cl)_{10}$
Schallerite	13.43 Å.	14.31 Å. = 2	Hexagonal	4 $(Mn, Fe)_8[Si_6O_{15}](OH, Cl)_{10}$
Friedelite	13.40 Å.	21.43 Å. = 3	Rhombohedral	6 $(Mn, Fe)_8[Si_6O_{15}](OH, Cl)_{10}$

join serpentine-amesite for the 7 Å. phases, and a partial crystalline solubility in the 14 Å. chlorites of the same composition type supports G. Tschermak's concepts of isomorphism of the "orthochlorites." The transition of the 7 into the 14 Å. chlorite structures explains the existence of the natural 14 Å. chlorites with intermediate compositions, and the existence of mixed 7 and 14 Å. minerals, especially of the amesite composition. Also the ferrous analogues in the chlorite group are characterized by 7 and 14 Å. polytypes, as in the natural 7 and 14 Å. *chamosite*, and mixed 7 and 14 Å. types in *daphnite*.

144. Summing up, we have to distinguish between the four-layer (mica-parcel + brucite layer) type of the perfectly and more coarsely crystallized chlorites, and the two-layer structures of trioctahedral type which are characterized by fine-granular, less perfectly crystallized minerals. The fundamental substitutions in both 7 Å. and 14 Å. structures are characterized by the simultaneous replacement of tetrahe-

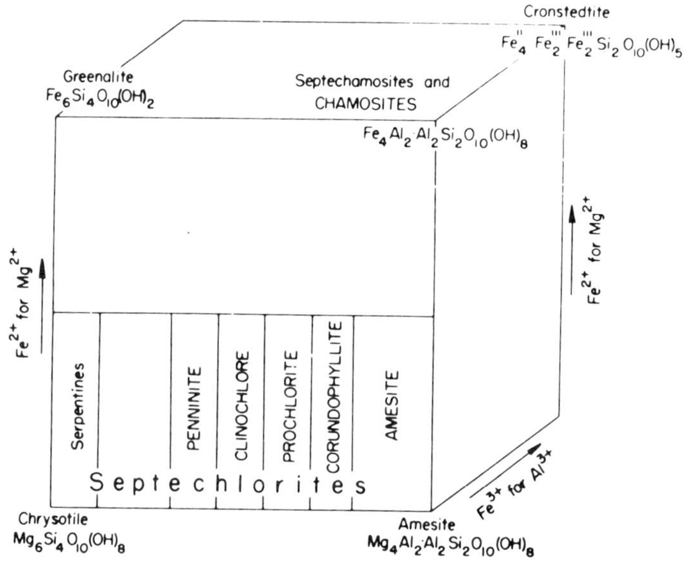

FIG. A. 102. Relationships of the chemical composition of the minerals in the chlorite group. The "septechlorites" are low-temperature polytypes to the more stable 14 Å. chlorites (Nelson and Roy).

dral Si^{4+} and octahedral Mg^{2+} ions by trivalent cations. It is also not necessary to separate the structures of antigorite and chamosite from the chlorite group as was often done in mineralogical systematics. Roy expresses the "consanguinity" of all these minerals in the diagram (Fig. A.1 02) which excludes, however, kaolinite as a 7 Å. layer mineral, which is not trioctahedral. It is recommended by Roy to avoid speaking of the "septechlorites" as being of "kaolin type." The structural "misfit" which impedes a perfect crystallization of 7 Å. chlorites is not valid for a 7 Å. magnesium-germanium serpentine which is synthesized in larger crystals than is usually observed in the magnesium-silicon analogue. Apparently, the 7 Å. chlorites are of lower stability than the 14 Å. types.

145. A thoroughly satisfactory system of the chlorite structure minerals, based on synthetic experience, was more recently given by B. W. Nelson and R. Roy.[230] The polytypes of the 14 and 7 Å. chlorites are able to show an extensive isomorphous coupled replacement of the type $2Al^{3+} \rightleftharpoons Si^{4+} + Mg^{2+}$, with the limit composition in the amesite molecule as mentioned above. The series of the 7 Å. chlorites synthesized extends from the amesite composition, $4MgO.2Al_2O_3.2SiO_2.4H_2O$, to that of serpentine, $6MgO.4SiO_2.2H_2O$, the series of normal 14 Å. chlorites from amesite to pennine, $21MgO.3Al_2O_3.13SiO_2.16H_2O$, with the monotropic transition of 7 → 14 Å.

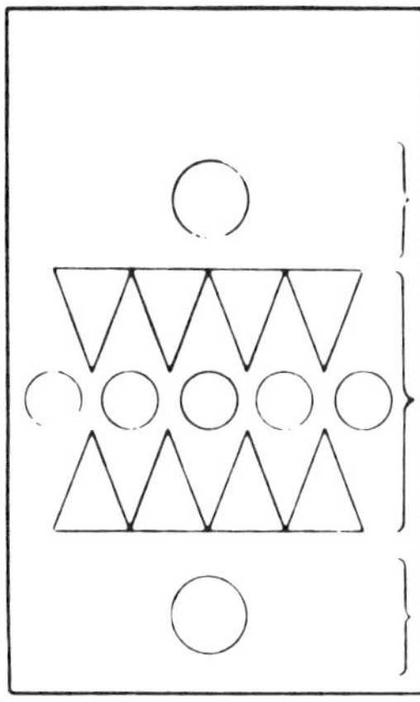

	PHLOGOPITE		HYPOTHETICAL 14 Å CHRYSOTILE		"NEUTRAL" CLINOCHLORE		MAX^M CHARGE PENNINITE		MAX^M CHARGE CLINOCHLORE	
	K	+ 1	3Mg(OH)$_2$	0	3Mg(OH)$_2$	0	2 5 Mg, 0 5 Al	+ 5	2Mg Al	+
	3Si Al		4 Si		3Si. Al		3 5 Si, 0 5 Al		3Si Al	
	6Mg	- 2	6Mg	0	4Mg 2Al	0	6Mg	- .	6Mg	- 2
	3Si Al		4 Si		3Si Al		3 5 Si, 0 5 Al		3Si Al	
	K	+ 1	3Mg(OH)$_2$	0	3Mg(OH)$_2$	0	2 5 Mg, 0 5 Al	+ 5	2Mg Al	+ 1

FIG. A. 103. Various possibilities of filling the different layers in pennine and clinochlore structures. (Nelson and Roy).

types under proper temperature and pressure conditions (cf. Vol. IV, Section A). The structural conditions of the possible substitutions in chlorites are illustrated in Fig. A. 103. This, and Fig. A. 102, also include the *leptochlorites* of the iron chlorite group (cronstedtite, greenalite, etc.). Another most elaborate classification of the chlorites on the basis of their cationic replacement regularities was given by G. W.

[230] *Am. Mineralogist,* **43**, 1958, 707-725, with an etymology of the word "septechlorites." Regular and semi-random one-layer structure polytypism in chlorite minerals is recently discussed by B. E. Brown and S. W. Bailey, *Ibid.*, **47**, 1962, 819-850. A. L. Albee, *Ibid.*, 851-870, demonstrates relations between chemical composition, physical (particularly optical) properties, and mineral association conditions of chlorites.

Brindley and F. W. Gilley,[231] especially for the practical X-ray identification from diffraction data, chiefly the basal spacings, the (060) reflection for the determination of the b_0 parameter, and the structure factors of the lower order interferences for which calculated data are given. The relations of chlorites to the kaolinite structure are demonstrated by Brindley and Gillery[232] in a description of intergrowths of both in mixed-layer types (called *daphnite*) which in their basal spacings appeared to be incompatible with a normal chlorite structure. Such a modified chlorite structure could be explained from a Fourier analysis by a modified distribution of the Si and O atoms. Some Si are displaced in their tetrahedral layers, and interlinked to the adjacent brucite layers, bringing about kaolinite-like portions within the chlorite structure. The required movements of the Si atoms are seen from Fig. A. 104.

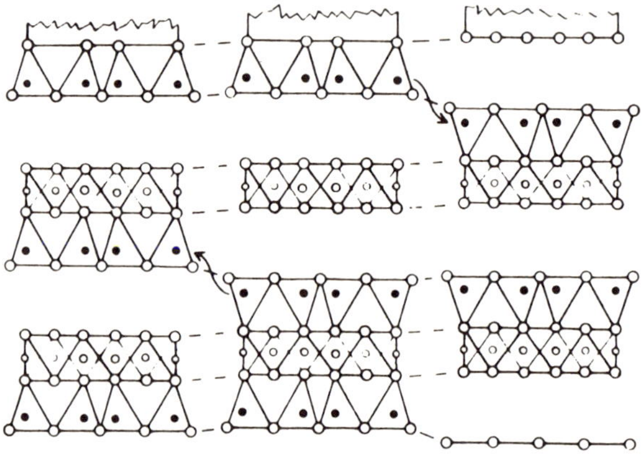

(a) (b) (c)

FIG. A. 104. Relation of a chlorite structure (*b*), to that of kaolinite, (*a*) and (*c*). The arrows indicate the displacements of Si-atoms required to change (*b*) into (*a*) or (*c*) (Brindley and Gillery).

146. A surprising exterior similarity of an *asbestos-like* (cross-fibrous) chlorite with chrysotile was described by V. P. Ivanova and N. A. Kronilov,[233] with a larger than normal basal spacing of $c_0 \sin \beta = 14.35$ Å. (normal 14.20 Å.). Besides the fibrous texture, the chemical composition of the chlorite is perfectly normal (with Si : Al in the tetrahedral layers = 3.22 : 0.78).

[231] *Can. J. Research,* **41**, 1956, 169-186. As leptochlorites the "oxidized chlorites" are defined by M. H. Hey, *Mineral. Mag.,* **30**, 1954, 277-292.

[232] *Proc. Natl. Conf. on Clays and Clay Minerals,* **2**, 1953, 349-353; *Natl. Acad. Sci.-Natl. Research Council, Publ.,* No. 327, 1954, 349-353.

[233] *Doklady Akad. Nauk S.S.S.R.,* **119**, 1958, 154-157. N. A. Kornilov, *Zapiski Vsesoyuz. Mineral. Obshchestva,* **89**, 1960, 160-165, gives valuable X-ray data for the diagnostic distinction of iron-magnesium chlorites from serpentines, to be compared with the data of V. I. Mikheev, *Zapiski Vsesoyuz. Mineral. Obshchestva,* **82**, 1953, 214-217.

147. The many possibilities of the isomorphous introduction of different metal cations into the structure of chlorites, which were demonstrated by the synthetic experiments of Nelson and Roy (see above), are in agreement with the existence of interesting *chromium chlorites* which were described by D. M. Lapham.[234] The Cr^{3+} ion replaces Al^{3+} or Mg^{2+} in the octahedral layers, although it is not excluded that at low temperatures Cr^{3+} may go into tetrahedral positions, too. The extensive X-ray studies of chromium chlorites chiefly consisted in precision determinations of the changes in the basal spacings, to distinguish both possibilities in their effects on the unit cell parameters. In *kotschubeite* with more than 2 per cent Cr_2O_3 content, the tetrahedral substitution prevails; in *kaemmererite*, the octahedral position of chromium is predominant.

148. According to Cl. Frondel,[235] the iron chlorite *cronstedtite*, $Fe_4^{2+}Fe_2^{3+}[Si_2Fe_2^{3+}O_{10}](OH)_5$, is highly remarkable in its polytypism. St. B. Hendricks (1939) described its rhombohedral structure (type 3 R) (with $c_0 = 21.25$ Å.; $a_0 = 5.48$ Å.) with three layers, but Frondel added a one-layer hexagonal polytype (1 H) (with $a_0 = 5.49$ Å.; $c_0 = 7.08$ Å.), a two-layer hexagonal type (2 H) (with $a_0 = 5.49$ Å.; $c_0 = 14.16$ Å.), and a nine-layer rhombohedral polytype (9 R) (with $a_0 = 9.54$ Å.; $c_0 = 63.72$ Å.). This latter crystal phase shows much stacking disorder. All polytypes have in common a distinct pseudocell with $a_0 = 3.18$ Å., and c_0 as given above for the single types, by ordered alternating distribution of Si^{4+} and Fe^{3+} ions in the tetrahedral layers.

149. The interesting lithium chlorite, *cookeite*, $LiAl_4[Si_3AlO_{10}](OH)_8$, was described by A. I. Ginzburg[236] as an analogue to kaolinite, $Al_4[Si_4O_{10}](OH)_8$, but it is characterized as a true chlorite by the coupled substitution of 2 Mg^{2+} for $Li^+ + Al^{3+}$, with smaller spacings, namely $a_0 = 5.13$ Å., $b_0 = 8.93$ Å., $c_0 = 28.3$ Å., and $\beta = 90^o$. Also B. B. Zvyagin and E. I. Nefedov[237] described cookeite, replacing lepidolite in lithium pegmatite, although with $c_0 = 14.3$ KX, and $\beta = 98^o45' \pm 30'$, determined from electron diffraction diagrams, as a typical chlorite with "parcels" $[AlSi_3AlO_{10}(OH)]^{1-}$ and the inserted brucite-hydrargillite layer $[LiAl_2(OH)_6]^{1+}$.

150. That the infrared absorption spectra of layer structure silicates may give important indications in polytypism problems was made evident in a recent in-

[234] *Am. Mineralogist*, **43**, 1958, 921-956. Lapham calls "comorphism" the appearance of one and the same cation in two different structural positions as in tetrahedral and octahedral layers of chlorites. See also A. Neuhaus, *Z. Krist.*, **113**, 1960, 195-233, especially pp. 225-227, who, however, concludes that in kaemmererite the chromium is *only* octahedral, as is also the case in chromium mica (fuchsite).

[235] *Am. Mineralogist*, **47**, 1962; 781-783.

[236] *Doklady Akad. Nauk S.S.S.R.*, **90**, 1953, 871-874.

[237] *Doklady Akad. Nauk S.S.S.R.*, **95**, 1954, 1305-1308.

vestigation of Vl. Stubičan and R. Roy.[238] We may here only mention briefly the discussion in A. ¶ 48, and refer to Figs. A. 105 and 106, showing some spectra of septechlorites and 14 Å. chlorites. The general similarity of these spectra is evident, but in the range from $\nu = 750$ to 850 cm.$^{-1}$ there are characteristic distinctions. With 14 Å. chlorites there is no resolved band at 910 cm.$^{-1}$, regardless of the aluminum content. Such a band, however, occurs in the spectra of the 7 Å. chlorites. The main difference is that the bands between 750 and 850 cm.$^{-1}$ are not so well resolved with normal chlorites as they are with 7 Å. chlorites. The strong band for

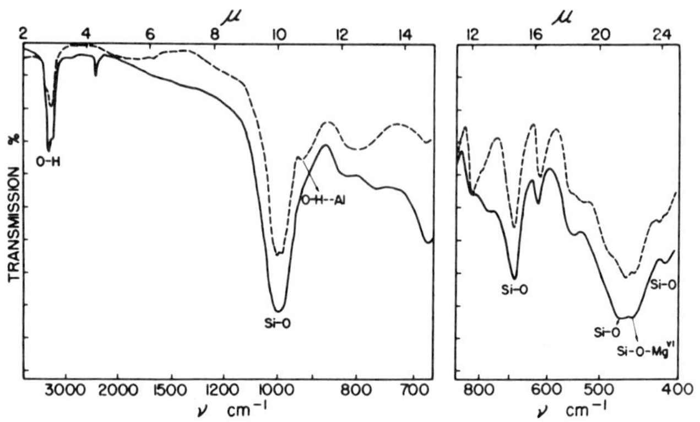

FIG. A. 105. Infrared absorption spectrum of 14 Å. chlorites (full curve) and 7 Å. chlorites (dashed curve) (Stubičan and Roy).

Si—O between 665 and 685 cm.$^{-1}$ can be used for the quantitative determination of the amount of tetrahedral Al^{3+} with 7 Å. chlorites. It is practically in the same position also for 14 Å. chlorites, and the shifts of the bands are analogous. The relatively small shift in this band from clinochlore to amesite composition is perhaps related to a disproportionation of the Al^{3+} ions between the two octahedral layers.

151. An interesting intermediate position between *pyrophyllite* and chlorites is ascribed to *donbassite* (called "pseudo-pyrophyllite" by F. A. Loewinson-Lessing, 1935) by E. K. Lazarenko,[239] projected in the composition triangle kaolinite—talc—amesite. The X-ray diffraction pattern of donbassite is distinctly different from that of pyrophyllite. We further mention in this connection the interesting hydrous iron-manganese silicate *ekmanite* with its layer structure, related to *stilpnomelane*, and the fascinating polytypism studied by B. Nagy[240] in relation to the disorder in the superlattice discussed above (cf. A. ¶ 124).

[238] *Am. Mineralogist,* **46,** 1961, 32-51, especially pp. 42-46.
[239] *Doklady Akad. Nauk S.S.S.R.,* **84,** 1952, 781-784.
[240] *Am. Mineralogist,* **39,** 1954, 338; 946-956.

152. In connection with the very sensitive changes in infrared absorption spectra which indicate changes in layer silicate structures of micas, chlorites, etc., M. S. Crowley and R. Roy[241] discussed the Al^{3+}—Si^{4+} ordering degree in the tetrahedral layers, as indicated by the *b* and *c* parameters of the unit cell, as a function of hydrostatic *pressure* during the crystallization. *Contraction* effects are observed in the range from 0.1 to 0.5 per cent under experimental pressures in the range from 350 to 3500 kg./cm.². The greatest contraction effect was observed in muscovite; smaller effects occur in chlorites of the clinochlore type. At higher pressures there is evi-

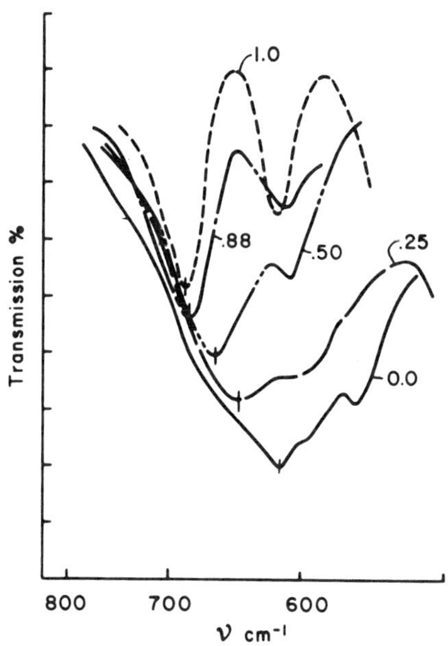

FIG. A. 106. Infrared absorption spectrum, with band in the region 550 to 800 cm.$^{-1}$, for septechlorite (Stubičan and Roy). The numbers on the curves refer to the amount of Al^{3+} ions in octahedral and tetrahedral coordination. (The spectrum illustrated here is identical to the spectrum in Fig. A. 27).

dently a progressive tendency to ordering in the distribution of the Al^{3+} and Si^{4+} ions. This tendency is more pronounced in the dioctahedral than in the trioctahedral layer silicate structures because vacancies and substitutional defects in the octahedral layers in the dioctahedral structures are more efficient for displacements of neighboring atom positions. In muscovite the initial 1 *Md* type is, therefore, relatively promptly changed to 1 *M* and later to 2 *M*; in phengite, even to 3 *T* (see above). Attempts to synthesize under pressure a "condensed" phlogopite have always been

[241] *Geochim. et Cosmochim. Acta,* 18, 1960, 94-100.

unsuccessful. It is also not possible to observe under high pressure a contraction in pyrophyllite, in spite of the vacancies in the octahedral layer, because no mobilization for ordering is possible in this mineral as is the case in muscovite and also in chlorites.

153. We also wish to mention briefly the extensive studies of Ch. Kiefer[242] on the stability of layer silicates in general, presenting data on a great amount of experimental material, mostly for practical application of those silicates in the manufacture of Portland cement, but also confirming the general conclusions made especially by Nelson and Roy.

154. For the very interesting structure of the mica-like minerals of the *chloritoid* group, G. W. Brindley and F. W. Harrison[243] developed a peculiar layer model which is, in spite of the many relations to real phyllosilicates, basically different in the presence of continuously linked octahedral oxide-hydroxide sheets, cross-linked by single, discrete $[SiO_4]$ tetrahedra. The crystallochemical formula is therefore not,

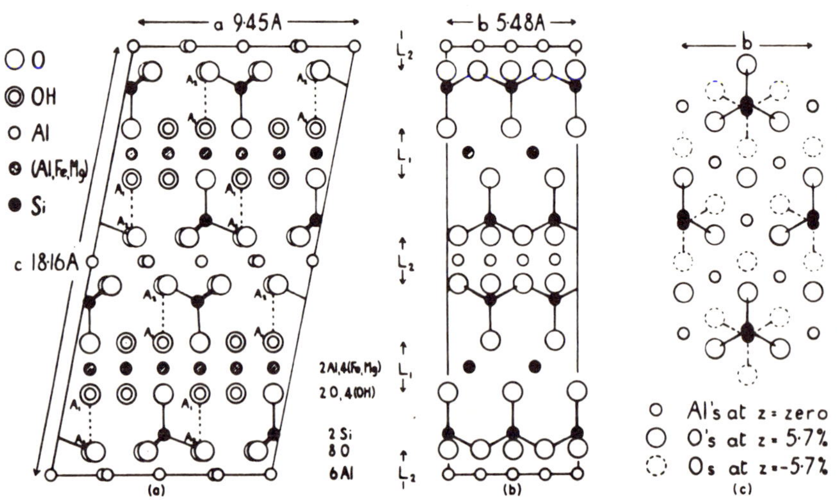

FIG. A. 107. Structure of chloritoid, projected on (010) (*a*); the same, viewed along the *a*-axis (*b*); layer L_2, viewed perpendicular to the (*ab*) plane (*c*) (Brindley and Harrison).

[242] *Rev. Matériaux construct. et trav. publ.*, **424**, 1951, 19-23; *Ibid.* **425**, 1951, 53-58; *Ibid.*, **426**, 1951, 84-86; *Ibid.*, **427**, 1951, 109-114; *Ibid.*, **428**, 141-146; *Ibid.*, **429**, 1951, 191-194; *Ibid.*, **430**, 1951, 225-232; *Ibid.*, **431**, **432**, 1951, 258-263. Thesis, Faculté des Sciences, Université de Paris, No. 288, 1951, 55 pp.

[243] *Natl. Acad. Sci.–Natl. Research Council, Publ.*, **327**, 1954, 360-363; *Acta Cryst.*, **5**, 1952, 698-699; *Ibid.*, **10**, 1957, 77-82. The structural analogy of the chloritoids with chlorites and micas is here chosen as a principle of classification; cf. H. Strunz, "Mineralogische Tabellen," Akademische Verlagsges., Leipzig, 1957, p. 319.

as one would expect, from the mica-analogies, $(Fe,Mg)_2Al_2[Al_2Si_2O_{10}](OH)_4$, but it is written by Brindley and Harrison either $[(Fe,Mg)_2Al](OH)_4Al_3(O_2[SiO_4]_2)$, to show the isolated $[SiO_4]$ groups, or alternately $[Al_3O_8]Si_2[(Fe,Mg)_2AlO_2(OH)_4]$, to demostrate the oxide and hydroxide layers linked by the Si atoms. Figure A. 107 gives a projection on (010) viewed along the *a* axis. Interesting is the difference between two sets of the octahedral layers: that indicated as L_1 is of trioctahedral (brucite) type; the second layer L_2 resembles *corundum* (Fig. A. 108). It is this cor-

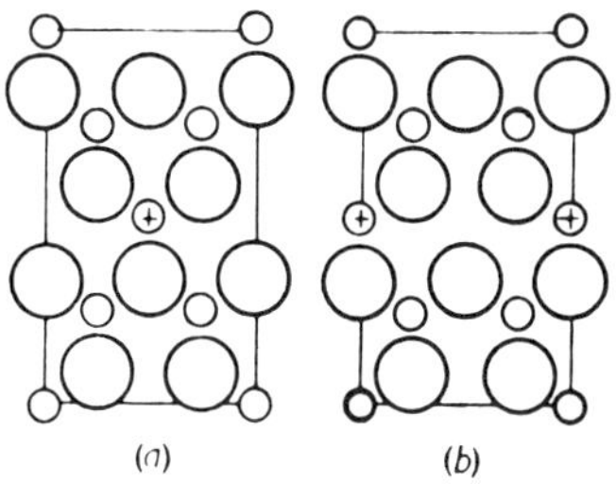

FIG. A. 108. The Al—O layer (*a*) in the structure of chloritoid; (*b*) in corundum (Brindley and Harrison).

undum-type layer which sets the scale of the unit cell dimensions in chloritoid, just as the continuous Si-tetrahedral network does in the micas. Chloritoid contains only half as many silicon atoms as there are in the (Si, Al) positions in micas and brittle micas. Hydrogen bonds occur between the $(OH)^-$ in L_1 and the O^{2-} anions above and below those in L_2, with H more closely attached to the (OH). The infrared absorption spectrum shows the sharp band at 2.86 μ. The octahedral arrangements are far from being regular.

155. L. B. Halferdahl[244] distinguished two polytype modifications of *ottrelite*, written $FeAl_2(SiO_5)(OH)_2$, one monoclinic with $c_0 = 18.19$ Å., the other, triclinic, with $c_0 = 9.16$ Å. In the latter case the unit cell is related to the monoclinic cell given by Brindley and Harrison by a zigzag arrangement with about $2 \times c_0$ (Fig. A. 109). It is interesting that the monoclinic polytype is preferably associated with almandine garnet in metamorphic rocks. By dehydration, chloritoid is changed to an anhydride structure characterized by H. G. Bachmann[245] as closely related to the initial chloritoid (Fig. A. 110), but with a considerably shorter c_0 parameter, although in a simple numerical relation (cf. Vol. IV, Section B). Chloritoids are minerals of

[244] *J. Petrol.*, **2**, 1961, 49-134.

[245] *Z. Krist.*, **108**, 1956, 145-156. L. B. Halferdahl (*Ibid.*, pp. 90 ff.) also discussed the dehydration and decomposition of chloritoid, especially in connection with the studies of J. F. Schairer and K. Yagi (1952) of the system $FeO–Al_2O_3–SiO_2$ and the hydrothermal stability conditions (cf. Vol. IV, Section A).

a high degree of structural disorder, and inclined to undergo intergrowths, as with sericite, and many similarities also exist such as in the dehydration behavior with clay minerals of the montmorillonite type.

156. Since the investigations of E. Aruja (1944) on the structure of *chrysotile*, it has been established that the early interpretation of this mineral as a chain structure, similar to the amphiboles (as given by B. E. Warren and W. L. Bragg, 1930), could not be correct. Aruja based his conclusions on the discussion of the structure type of a magnesium kaolinite, $Mg_6[Si_4O_{10}](OH)_8$,[246] in contrast with the ribbon structure type indicated by the formula $Mg_6[Si_4O_{11}](OH)_6.H_2O$. In spite of the im-

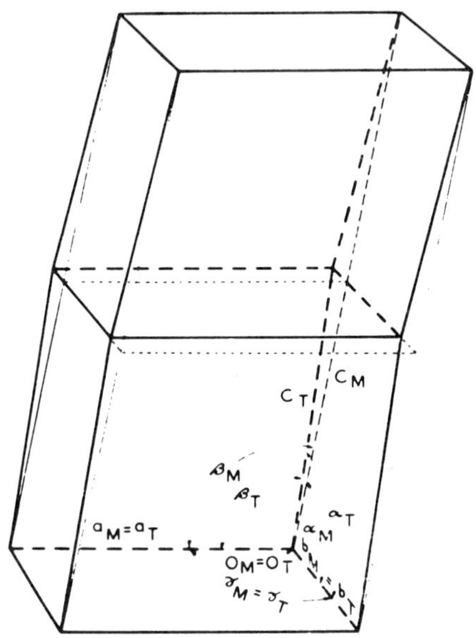

FIG. A. 109. Dimensional relation between the unit cells of triclinic and monoclinic chloritoid (Halferdahl).

pressive plausibility of the earlier proposal, it was difficult to explain on this basis the relations to *antigorite* with its pronounced tabular crystal habitus, and the analogies with amesite. The idea of the structure type of a magnesium kaolinite arose for the first time in the discussion of the structural interrelation between chrysotile (B. E. Warren and K. W. Hering, 1941), amesite (J. W. Gruner, 1944), and antigorite (E. Aruja, 1945). Decisive is the principle of all the *serpentine*-type minerals, that they are two-layer silicates built up by parcels of a tetrahedral $[Si_2O_5]_\infty$ layer and a brucite-like octahedral layer, giving to this complex a polar character.

[246] A short description of the Mg-Kaolin structure see by Å. Grudemo, *Svenska Forskningsinst. Cement Betong vid Kgl. Tek. Högskol. Stockholm, Handl.*, No. 22, 1951, p. 17-18.

157. The relation between antigorite and chrysotile, which Aruja found, is that of a typical *superstructure* in the dimensions of the a-axis of the unit cell, namely with A_0 (antigorite) $= 8 \times a_0$ (chrysotile) $= 8 \times 5.42$ KX $= 43.36$ KX, and 16 molecules $Mg_3Si_2O_9H_4$ in the cell. The superstructure is indicated by clusters of interference spots in the a^* row of the reciprocal lattice reflections where chrysotile shows only single spots. Inner tensions arising from the randomness of stacking cause asterism phenomena in antigorite, and in chrysotile a *curvature* of the magnesium kaolinite layer sheets. In addition, a polytype of antigorite was found in an orthohexagonal antigorite from Unst (Shetland Islands), and G. W. Brindley and O. von Knorring[247] called this type ortho-antigorite to distinguish it from the common monoclinic antigorite (clino-antigorite). Also ortho-antigorite is a superstructure to chrysotile, with the period $C_0 = 3 \times 14.52$ Å. $= 43.59$ Å., and a sixfold layer structure.

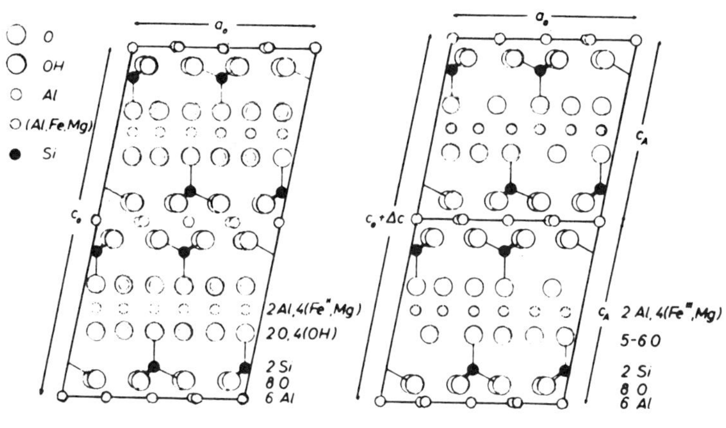

Fig. A. 110. Projection on (010) of the structure of chloritoid (a), and of the dehydration (anhydride) phase (b) (Bachmann).

158. Of tremendous importance in the investigation of the chrysotile and antigorite structure was the advanced knowledge of their morphology for which the electron microscopy was a great help, combined with electron diffraction methods of minute single crystals (cf. B. ¶ 7 ff.). The presumption of the effects of inner tensions influencing the shape of the crystals of the chrysotile-antigorite group was fully confirmed by the analogous observation of curvature and *tube*-formations in halloysite (meta-halloysite) clay minerals by T. F. Bates, F. A. Hillebrand, and A. Swineford,[248]

[247] *Am. Mineralogist*, **39**, 1954, 794-804. On the superlattice problems see an interesting discussion of J. Zussman and G. W. Brindley, *Ibid.*, **41**, 1956, 148-151 and 151-152.

[248] *Am. Mineralogist*, **35**, 1950, 463-484; T. F. Bates, L. B. Sand, and J. F. Mink, *Science*, **111**, 1950, 512-513. The hollow fibers were also observed by W. Noll and H. Kircher, *Neues*

applying L. Pauling's ideas (1930) on structural "misfits" and distortions. Excellent electron microscopic replica pictures of chrysotile were given by T. F. Bates and J. J. Comer[249] more recently, in comparison with those of kaolinite and halloysite. All these observations confirm the correctness of the tube-shaped habitus of natural chrysotile and of many serpentines which according to B. Nagy and G. T. Faust[250] are mixtures of chrysotile fibers and antigorite flakes.

159. W. F. Bradley[251] compared the crystal structure of magnesium silicate hydrates (chrysotile, antigorite, sepiolite) with that of magnesium-containing clay minerals (attapulgite, vermiculite, montmorillonite) (cf. B. ¶ 76 ff.) from the viewpoints of the evidently temperature-sensitive, more or less developed perfection of articulation between tetrahedral- and octahedral-coordinated layers, and the different crystallization habitus and defectiveness which reflect the degree to which difficulty of articulation limits the growth of particular compositions. This is equally valid also for *chlorite* structures in which the tetrahedral networks are at or near the limit of their lateral extensibility, and the octahedral layers at or near their limit of compressiblity. Therefore, the crystals under strain have a low durability to weathering actions, as seen especially for the highly ferriferous chlorites (leptochlorites). Bradley demonstrated, as applied to chrysotile, the high misfit in this mineral structure. This is also emphasized in the very elaborate discussions of H. Jagodzinski *et al.*,[252] who definitely accepted the structure proposal of Warren and Hering as the basis for the accurate determination of the constitution of chrysotile and antigorite, using the low-angle X-ray diffraction method. The tube-shaped growth of the chrysotile fibers is then the consequence of radial and axial displacements, perhaps also of edge-dislocations, as shown in Figs. A. 111 a-b. These dislocations also influence

Jahrb. Mineral. Monatsh., 1951, 219-240; *Fortschr. Mineral.*, **29/30**, 1952, 391-392. For synthetic serpentine see N. L. Bowen and O. F. Tuttle (1949), and D. M. Roy and R. Roy, *Am. Mineralogist*, **39**, 1954, 957-975. See further a collection of instructive electron micrographs by Ch. W. Huggins, *U. S. Bur. Mines, Rept. Invest.*, No. 5551, 1959, 14 pp.; for chrysotile, also *Am. Mineralogist*, **44**, 1959, 78-114. Definite evidence for the tube shape of chrysotile fibers was given by M. Maser, R. V. Rice, and H. P. Klug, *Ibid.*, **45**, 1960, 680-688, using the diamond-knife ultra-microtome method (cf. C. ¶ 112). Very interesting electron-optical fringe interferences were observed by G. W. Brindley, J. J. Comer, R. Uyeda, and J. Zussman, *Acta Cryst.*, **11**, 1958, 99-102, with antigorite (Yu-Yen stone), with a superstructure parameter $A = 100$ Å.

[249] *Proc. Natl. Conf. on Clays and Clay Minerals*, **6**, 1959, 237-248.

[250] *Am. Mineralogist*, **41**, 1956, 817-838. G. L. Kalousek and L. E. Muttart, *Ibid.*, **42**, 1957, 1-22, extensively discuss the serpentine problems, in connection with synthetic observations (cf. Vol. IV, Sectin B). For criticism of Nagy and Faust's opinion, see E. J. W. Whittaker and J. Zussman, *Ibid.*, **43**, 1958, 917-920.

[251] *Natl. Acad. Sci.-Natl. Research Council, Publ.*, No. 395, 1955, 94-102.

[252] *Fortschr. Mineral.*, **32**, 1954, 86-87 (with G. Kunze and S. N. Bagchi). For low-angle diffraction measurements on chrysotile the special camera developed by H. Jagodzinski and K. Wohlleben, *Z. Elektrochem.*, **64**, 1960, 212-216, proved to be very helpful.

the intensity in the calculations of the structure. The mathematical principles for such calculations for cylindrically curved atomic arrangements in networks was advanced by E. J. W. Whittaker[253] who also gave important details on the chrysotile structure.[254] A valuable general idea on the possibilities in special structural arrangements caused by inner tensions was indicated by L. Onsager.[255]

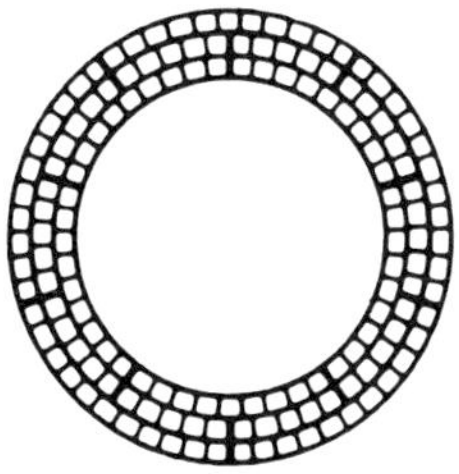

FIG. A. 111. *a*. Edge displacements in chrysotile; the network planes of the inner $[Si_2O_5]_\infty$ layer remain constant, whereas those of the brucite-like layer vary (Jagodzinski and Kunze).

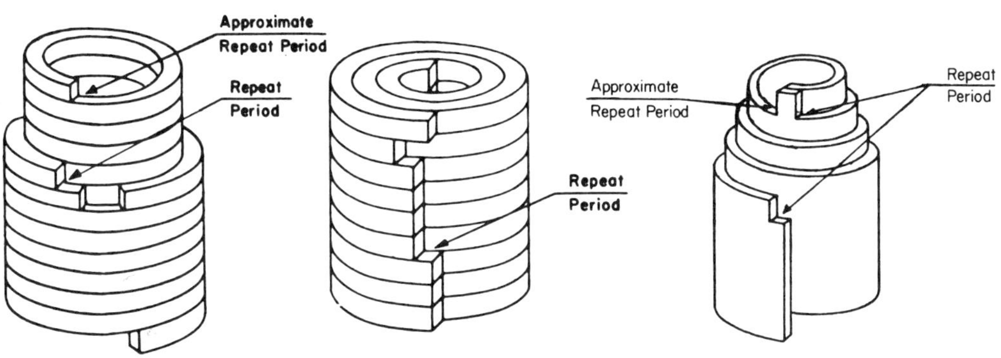

FIG. A. 111. *b*. Axial and radial displacement in chrysotile; *c*. Coupled displacement in chrysotile (Jagodzinski and Kunze).

160. A distortion with a superstructure as was observed by Aruja (see above) for antigorite may be combined with Pauling's curvature principle in structural misfits. An arrangement satisfying the postulates of Onsager's theory would be a *corrugated sheet* structure, with an inversion from one superperiod (e. g., in antigorite that of 43.4 Å.) to the next superperiod. There can be either an alternating wave structure in which convex and concave sides of curved sheets alternate by every

[253] *Acta Cryst.*, **7**, 1954, 827-832; *Ibid.*, **8**, 1955, 261-265; 265-271; 571-574 (a classification of cylindrical lattices); 726-729.

[254] *Acta Cryst.*, **4**, 1951, 187-188; *Ibid.*, **6**, 1953, 747-748; *Ibid.*, **9**, 1956, 855-862, 862-864, 865-867 (see below).

[255] *Brit. J. Appl. Phys.*, **3**, 1952, 281-282, as a discussion remark to a conference on silicate structures by K. Robinson and E. R. S. Shaw.

half-wave, or a rectified-wave structure in which the convex and concave sides have the same orientation. The latter idea, however, implies difficulties in the transition of lattice configurations near the inversion line or "suture" (Fig. A. 112) as demonstrated in the elaborate theoretical and experimental investigations of the chrysotile structure by H. Jagodzinski, S. N. Bagchi, and G. Kunze.[256]

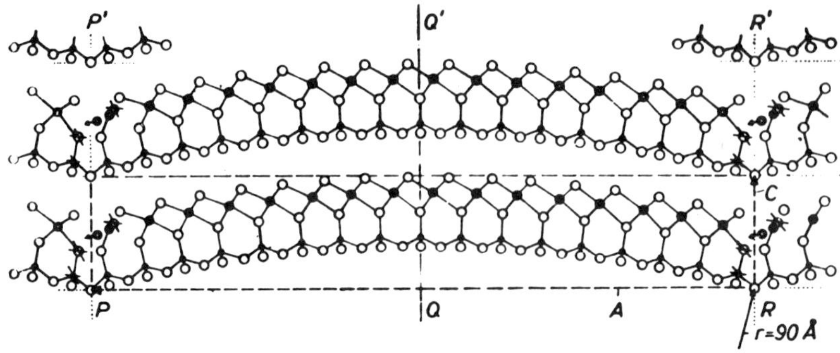

FIG. A. 112. One-sided rectified wave-structure (Kunze). The cylinder axis is the crystallographic *b*-axis; the superperiod is $A = 43.3$ Å.; the radius of curvature in the oxygen anion network is 90 Å. The atomic positions in the suture *P* and *R* imply great difficulties.

161. J. Zussman[257] gave evidence that in the structure of antigorite (from Mikonui) as described by Aruja the curvature about the *b* axis is the reason for the superperiod observed in the multiple of the a_0 parameter. These problems are definitely solved in favor of the *alternate wave* structures with curvature, also for clay minerals, or even for nonsilicates. The coincidence conditions of the atoms in the suture (inversion lines) are decisive, and must not be in contradiction with the continuity and completeness of the $[Si_2O_5]_\infty$ layers, although certain "coincidence losses" by an elimination of Mg^{2+} and OH^- ions in the suture must occur. Starting from the double-half-wave models, the sinusoidal structure type seen in Fig. A. 113 for the most plausible interpretation of the interference phenomena was given by Kunze, in the continuous symmetrical curvature structure, analogous to the model ortho-antigorite (see above) in which the sheets have a sequence of the type

[256] *Neues Jahrb. Mineral. Monatsh.*, 1953, 97-100; *Ibid.*, 1954, 95-108, 113-130, 137-150; *Veröffentl. Max-Planck-Inst. Silikatforsch.*, **14**, 1954, 63-109. For further detailed discussions of the wave structure of antigorite by G. Kunze, see *Fortschr. Mineral.*, **35**, 1957, 46-48; *Z. Krist.*, **108**, 1956, 82-107; *Ibid.*, **110**, 1958, 282-320; and on the defects in antigorite see *Fortschr. Mineral.*, **37**, 1959, 50-54; *Z. Krist.*, **111**, 1959, 190-212. We regret particularly that it is not possible to give in the frame of the present work an exhaustive appreciation of these most fascinating refinements of the suture-problems in the wave structure of antigorite. In general we refer to the detailed monograph by H. Kunze on antigorite problems, in *Fortschr. Mineral.*, **39**, 1961, 206-324.

[257] *Mineral. Mag.*, **30**, 1954, 498-512.

ABABAB..., and a monoclinic model for the sequence AAAAA... (in clino-chrysotile), to follow Whittaker.

162. Kunze's structure determination is in harmony with the results of electron microscopy by J. Zussman, G. W. Brindley, and J. J. Comer[258] explaining the chrysotile-antigorite group by isotypism and superstructure relations as shown in Table V, which also includes the six-layer ortho-serpentine (Brindley's and von Knorring's "ortho-antigorite," a designation which was later dropped by the authors), with the superperiod C'. In *lizardite*, on the other hand, there is a subperiod $C' = c_0/2$ of ortho-chrysotile; it includes the *bastite* pseudomorphs after ortho-pyroxenes. In

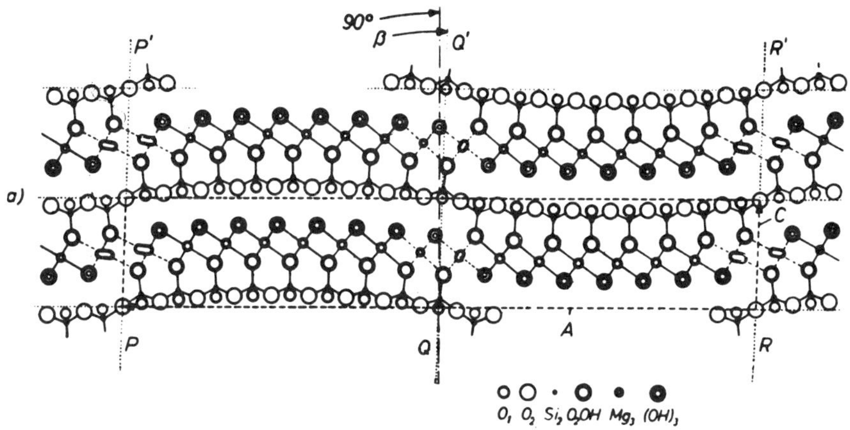

FIG. A. 113. Double-half wave-structure of antigorite (Jagodzinski and Kunze). Shows the symmetrical curvature and the characteristic suture distortion, with coincidence losses along the inversion line Q-Q' (Kunze).

nature ortho- and clino-chrysotiles are mixed in variable ratios, as in silky chrysotile, whereas massive serpentines are lizardites, often mistaken for antigorite because of their scaly habitus, but clearly distinguished from it by the hexagonal arrangement of the electron diffraction spots. Typical antigorites differ by clusters of spots spaced at regular intervals, the superperiods A varying between 34 and 90, and even 110 Å., mostly around 43.5 Å. To the six-layer ortho type also belongs a synthetic magnesium-germanium serpentine described by D. M. Roy and R. Roy,[259] similar to lizardite, with the superperiod $C' = 44.67 = 6 \times 7.44$ Å., and a high order in the structure. The absence of structure misfit in this case is made clear

[258] *Am. Mineralogist,* **42**, 1957, 133-153.

[259] *Am. Mineralogist,* **39**, 1954, 957-975, especially p. 967. On the six-layer serpentines see J. Zussman and G. W. Brindley, *Ibid.*, **42**, 1957, 666-670, and quite recently E. J. Olsen, *Ibid.*, **46**, 1961, 434-438.

TABLE V

UNIT CELL PARAMETERS OF THE SERPENTINE MINERALS

	a_0 (Å.)	b_0 (Å.)	c_0 (Å.)	β	Reference
Ortho-chrysotile	5.32	9.2	14.6	90°	Whittaker (1953)
	14.63	9.2	5.34	90°	Whittaker (1956)
Clino-chrysotile	5.32	9.2	14.6	93°12′	Whittaker (1951)
	14.65	9.25	5.34	93°16′	Whittaker (1956)
Para-chrysotile *	14.7 ± 0.1	9.24 ± 0.2	5.3	90°	Whittaker (1956)
Antigorite, including picrolite	43.5	9.2	7.28	91°24′	Aruja (1945)
Lizardite	5.31	9.2	7.31	90°	Whittaker and Zussman (1956)
Six-layer ortho-serpentine	5.32	9.22	43.6	90°	Brindley and von Knorring (1954)
Magnesium-germanium serpentine	5.44	9.42	44.7	90°	Zussman, Brindley, and Comer (1957)

* Para-chrysotile is a component of silky fiber bundles, and was brifly mentioned by Whittaker and Zussman.

by the schematic sketch Fig. A. 114, in comparison with common serpentine and with halloysite.

163. How the polytypism in the serpentine minerals is connected with their genetic conditions is still a rather difficult problem. A substitution of Mg^{2+} and Al^{3+} cations in various ratios may influence the cylindrical arrangements. Kunze emphasizes that the superperiods of antigorites of variable aluminum contents are highly sensitive to slight changes in the chemical composition. The antigorite from the

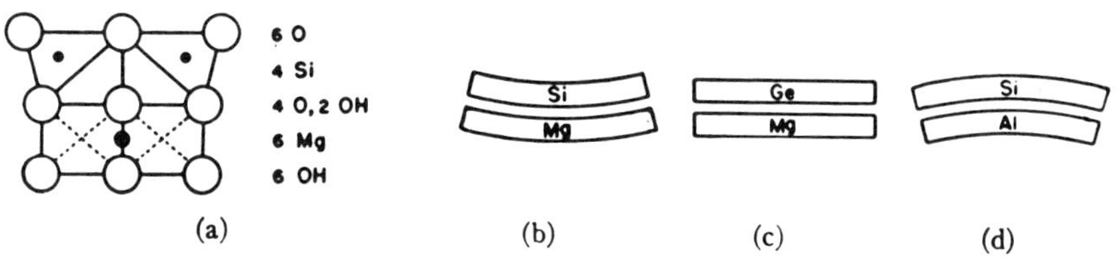

FIG. A. 114. Schematic interpretation of structural misfits in serpentine and halloysite (Roy and Roy).

original occurrence of Antigorio is high in alumina and low in magnesia, whereas that of Mikonui studied by Aruja is just of the opposite type. Highly defective is the structure of a "pure" antigorite from Caracas, Venezuela, studied by Brindley,[260] parlicularly low in hydroxyl (3.73 in the place of 4.00 in the crystallochemical formula) and in R^{2+} cations (2.84 in the place of 3.00). This corresponds nearly ideally to the coincidence losses predicted by Kunze in Fig. A. 113. We note the interesting observation of Kunze (1951) that there are fibrous antigorites on veins which contain chrysotile tube-shaped fibers inserted parallel to the texture axis, and, on the other hand, that there are chrysotiles with antigorite fibers inserted parallel to the texture axis. As a matter of fact, serpentine rocks which underwent strong tectonic displacements show preferably, or exclusively, antigorite; serpentinites of a low degree of dynamo-metamorphism show chrysotile besides antigorite.

164. We repeatedly emphasized how much the knowledge of the structure of the serpentine group minerals owes to synthetic investigations, chiefly in the system $MgO–Al_2O_3–SiO_2–H_2O$, by N. L. Bowen and O. F. Tuttle (1949), and of D. M. Roy and R. Roy[261] on the influence of cation substitutions in the tetrahedral (Si by Ge) as in the octahedral (Mg by Al, Fe, Co, Ni, etc.) portions. The *nickel serpentines*, $Ni_6[Si_4O_{10}](OH)_8$ and $Ni_6[Ge_4H_{10}](OH)_8$, are examples of this kind, the first with a tubular texture, the second in well-developed hexagonal tabular crystals. Whereas

[260] *Am. Mineralogist*, 39, 1954, 391-393.
[261] *Am. Mineralogist*, 39, 1954, 957-975.

these authors could not prepare a cobalt serpentine, recent experiments of W. Noll, H. Kircher, and W. Sybertz[262] describe a *cobalt chrysotile* as a typically tubular phase (called a soleno-silicate, from the Greek word $\sigma\omega\lambda\acute{\epsilon}\nu$ = tube). The calculated diameters of the tubes are in full agreement with the postulates of the structural misfit concept. In comparison with the cations Mg^{2+}, Ni^{2+}, Co^{2+}, and Al^{3+} in the curved layers, a thoroughly regular functional relation is derived between the curvature and the sizes of these cations. In this respect, it becomes understandable why Mn^{2+} or even Fe^{2+} cannot form serpentine-type layer structures. For Z^2n^+, which might fulfill suitable conditions of ionic size, the stability of the orthosilicate, willemite, Zn_2SiO_4, is so much greater that this silicate and magnesium chrysotile are formed in all the compositions from which the Zn-serpentine would have been expected. The particular stability of the $[ZnO_4]$ configuration is of great importance (see below).

165. A particular problem is the introduction of Al^{3+} into the serpentine structure. We may only repeat what was discussed also in the chlorite structures, namely, that a simple replacement $2\ Al^{3+} \rightleftharpoons 3\ Mg^{2+}$ in the octahedral layers is unlikely. It must always be of the coupled type; a part of the Al^{3+} must simultaneously replace Si^{4+} in the tetrahedral layers. This reaction, which is of great importance for the double role the amesite composition plays in chlorites and serpentines, was discussed by F. H. Gillery,[263] who found, by hydrothermal-experimental synthesis, the limit conditions between both mineral families (cf. Vol. IV, Section A). Temperature is the chief controlling variable, whereas the composition of the mixes determines the polytype phase appearing. Gillery finds the limit concentration of Al_2O_3 between fibrous and tabular natural serpentines of the system $MgO–Al_2O_3–SiO_2–H_2O$ at 4.5 per cent (or $x = 0.2$ atoms Al^{3+} in the general formula $[(Si_{4-x}Al_x)_4O_{10}](Mg_{6-x}Al_x)(OH)_8$, in agreement with the limit composition of natural serpentines as given by Nagy and Faust (see above). We note, however, that E. J. Olsen[264] is inclined to accept a much lower limit, of about 0.6 per cent Al_2O_3.

166. That Cr^{3+} may enter the octahedral layer in antigorite was demonstrated by J. J. Glass, A. C. Vlisidis, and N. C. Pearre,[265] combined with a remarkably high alumina content (more than 7 per cent). The electron micrograph shows angular plates and laths of such a chromium antigorite.

167. In view of the rather strong differences in the structure of the serpentine minerals, the *infrared absorption* spectra studied by G. W. Brindley and J. Zussman[266]

[262] *Naturwissenschaften*, **45**, 1958, 489; *Beitr. Mineral. u Petrog.*, **7**, 1960, 232-241; *Silika Tech.*, **11**, 1960, 395-396.

[263] *Am. Mineralogist*, **44**, 1959, 143-152.

[264] *Am. Mineralogist*, **46**, 1961, 434-438.

[265] *Am. Mineralogist*, **44**, 1959, 650-656.

[266] *Am. Mineralogist*, **44**, 1959, 185-188.

did not show any particular distinctions in the curves for chrysotile with a tubular habitus, or for one- and six-layer ortho-serpentine, with the exception of some differentiation of the water content retained in the tubular channels, thus giving rise to some differences in the 2.7 to 3.0 μ range, but without any obvious connection with the structure type. The very accurate recent infrared measurements of Vl. Stubičan and R. Roy[267] in the 12 to 24 μ range show for tabular and fibrous antigorite and chrysotile some slight differences due to the bonding movements affected by differences in stacking, as seen from the strong bands near 700 cm.$^{-1}$, or in the intensity of the bands at 645 and 625 cm.$^{-1}$ (Fig. A. 115).

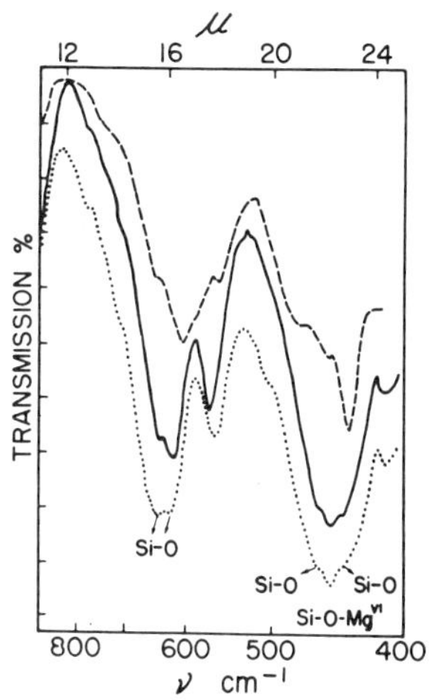

FIG. A. 115. Infrared absorption spectra of fibrous antigorite (full curve), chrysotile (dashed curve), and tabular antigorite (dotted curve) (Stubičan and Roy).

168. Although they belong to the clay minerals and are, therefore, discussed more in detail there, the interesting structures of *sepiolite* and *attapulgite* may be here mentioned in their analogies to those of the minerals of the serpentine group. Sepiolite, $Mg_8[Si_{12}O_{30}]H_6(OH)_{10}$, + aq., was studied by B. Nagy and W. F. Bradley[268] as a composite of three linked chains (cf. Bradley,[269]) (in attapulgite there are very similar triple chains), enclosing a second central array of channels in which the water molecules are disposed. The condensation of the chains is the same as in the tetra-

[267] *Am. Mineralogist*, **46**, 1961, 32-51. The assignment of definite infrared absorption bands for layer-structure silicates is considerably promoted by the replacement of characteristic cations, e.g., of Si^{4+} by Ge^{4+}, Al^{3+} by Ga^{3+}, Mg^{2+} by Fe^{2+} or Ni^{2+} in hydrothermal synthesis, and also by the replacement of OH^- groups by OD^-, as recently demonstrated by Vl. Stubičan and R. Roy, *Z. Krist.*, **115**, 1961, 200-214.

[268] *Am. Mineralogist*, **40**, 1955, 329, 885-892.

[269] Cf. W. F. Bradley, *Am. Mineralogist*, **25**, 1940, 405-410.

hedral layers of mica and chlorites. The fibrous character is also shown in the electron micrographs with distinct laths, well explained by the ribbon structure indicated in the triple chain complexes. The structure is best seen in the projection onto (001) (Fig. A. 116) either of orthorhombic, or of monoclinic symmetry. Very par-

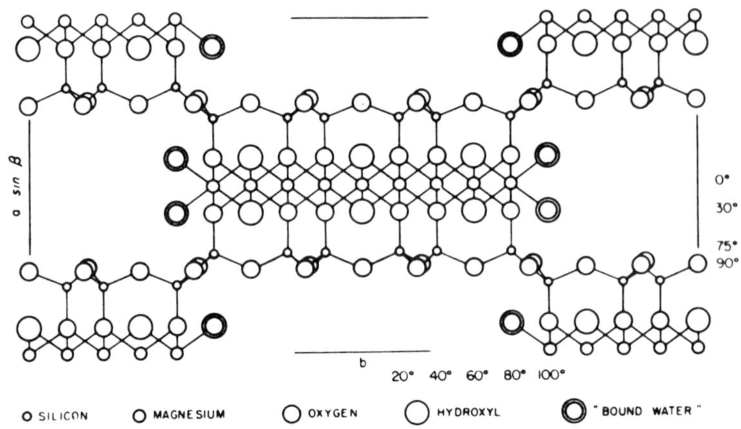

Fig. A. 116. Structure of sepiolite, projected on (001) (Nagy and Bradley).

ticular is the fourfold role of the water content in its bonding energy, namely as molecular water, water bound on brucite ribbon edges, hydroxyl groups analogous to those in micas, chlorites, etc. and balancing "extra" positions analogous with exchange ions, in the ratios 6 : 4 : 3 : 1, respectively. The oxygen anions joining the chains are defined as *double* links, whereas in attapulgite they are assumed to be *single* oxygen positions, given for comparison in the structure proposal of Bradley (Fig. A. 117).

169. Further studies of K. Brauner and A. Preisinger[270] confirmed the structure as given by Nagy and Bradley, emphasizing especially the layers in (100) in which the tetrahedral apexes alternate in the three chains. Like in micas, $[MgO_6]$ interlinks with these layers. Characteristic are the wide cavities with the water molecules next to the exterior $[MgO_6]$ groups. Brauner and Preisinger write the crystallochemical formula $\{Mg_8(H_2O)_4(OH)_4[Si_{12}O_{30}]\}.n(H_2O)$, with $n = 8$. The characteristic difference of this structure proposal is the *single* oxygen linkage with the chains, as in attapulgite (Fig. A. 118). A. G. Kulbicki[271] remarks that the transition of sepiolite into enstatite at 800°C. can be particularly well understood from these structures (Fig. A. 119). The three "basic units" in the sepiolite ribbons are separated in the

[270] *Rend. soc. mineral. ital.*, **12**, 1956, 69-71; *Mineral. u. petrog. Mitt.*, **6**, 1956, 120-140.

[271] *Am. Mineralogist*, **44**, 1959, 752-764; see also A. Preisinger, *Proc. Natl. Conf. on Clays and Clay Minerals*, **6**, 1957, 61-67.

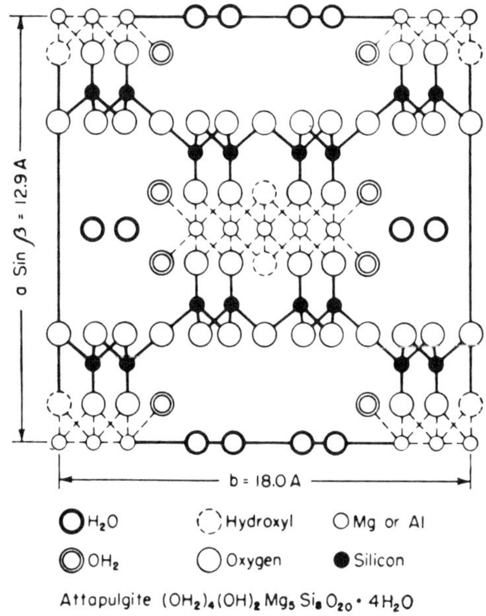

FIG. A. 117. Structure proposal of attapulgite, projected on (001) (Bradley).

thermal dissociation reaction, then regrouped to the enstatite chain structures, but the ease of this reaction is claimed to be greater for the Brauner-Preisinger proposal with a *single* oxygen bond. It may be that both structural variations occur in nature, thus explaining the differences observed in the transition rate from sepiolite into enstatite. Still slower is the transition for the layer structure of *saponite* (cf. B. ¶ 78).

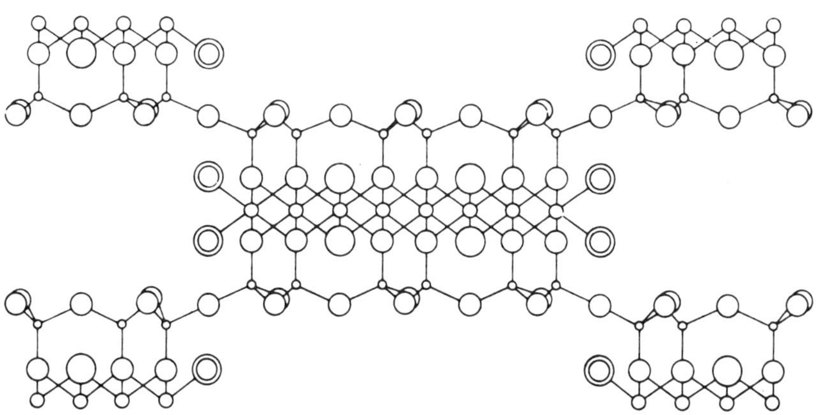

FIG. A. 118. Structure of sepiolite, projected on (001), as given by Brauner and Preisinger.

170. G. W. Brindley[272] studied the electron diffraction of orthorhombic sepiolite in single crystals, confirming the (100) layer position of the laths. The oxygen layers are schematically seen from the sketch in Fig. A. 120, as a projection on (001). J. V. Smith[273] carried out a precise re-examination of the structure of *melilite* and

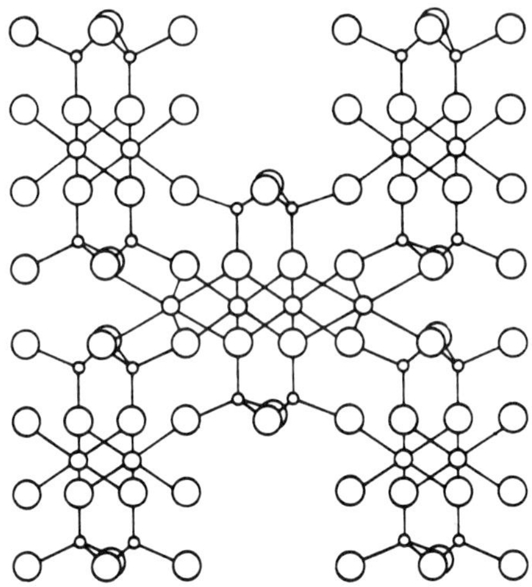

Fig. A. 119. Idealized structure of enstatite, projected onto (001), showing the relationship with sepiolite structure (Kulbicki).

its interesting crystalline solution series gehlenite-akermanite. In comparison with the previous structural interpretations by B. E. Warren (1930) and F. Raaz (1930, for gehlenite), especially a revision of the $[CaO_n]$ polyhedron was required. This configuration is highly distorted and very different from what is normally observed,

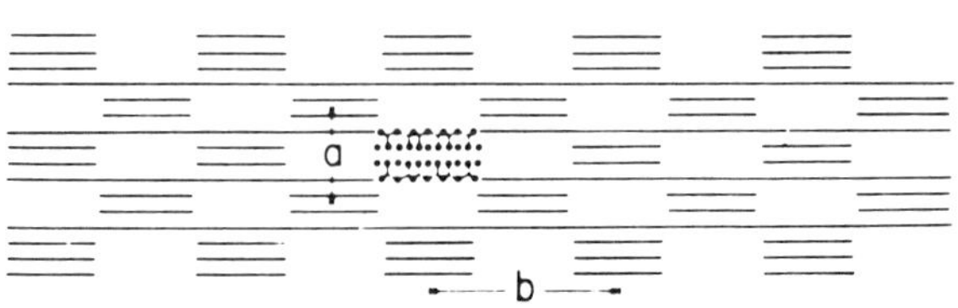

Fig. A. 120. Schematic cross-section of the lath structure of sepiolite showing the more accentuated development in the *b* than in the *a* direction (Brindley).

causing strong deviations from Pauling's rule of electrostatic valence bonds, also made evident in the O—O distances explained by the sharing of polyhedron edges. The $[CaO_n]$ groups are described as square antiprisms, with two groups of short

[272] *Am. Mineralogist,* **44**, 1959, 495-500.
[273] *Am. Mineralogist,* **38**, 1953, 643-661; *Ibid.*, **39**, 1954, 139.

(2.4 Å.) and long (2.7 Å.) distances Ca—O seen in stereographic projection in Fig. A. 121. The *c*-axis projection of the melilite structure is given in Fig. A. 122. The bonds Ca—O are probably not truyly ionic, but to a considerable portion homopolar.

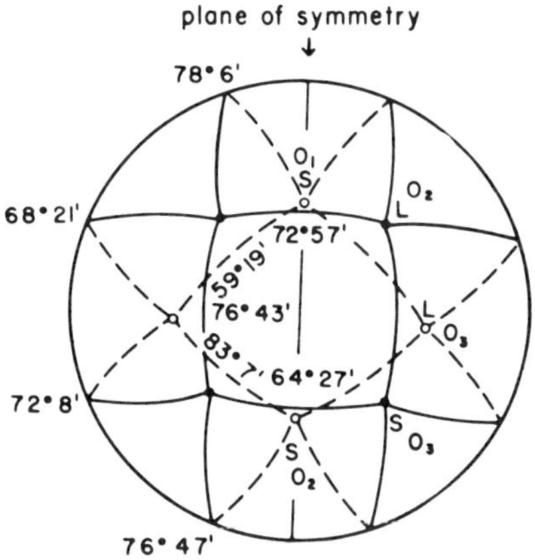

FIG. A. 121. Stereographic projection of the $[CaO_8]$ polyhedron in the structure of melilite (Smith).

Subgroup 3b. Double Sheets (Sharing Coefficient 2.00)

171. Double hexagonal ring structures are known either as single double-ring systems (in milarite, osumilite), or as hexagonal networks of two-dimensionally infinite extension (as in hexagonal $BaAl_2Si_2O_8$ and hexagonal $CaAl_2Si_2O_8$). We first discuss the interesting structure of *osumilite*, $(K,Na,Ca)(Mg,Fe^{2+})_2(Al,Fe^{2+},Fe^{3+})_3[(Si,Al)_{12}O_{30}].H_2O$, according to A. Miyashiro[274] who, in connection with his studies on cordierite crystal phases (see below), discussed this high-temperature mineral occurring in volcanic rocks. The condensed double rings have the type $[(Si, Al)_{12}O_{30}]$, as a somewhat isolated unit in the hexagonal structure (Fig. A. 123), connected to similar double rings by Al—Si tetrahedra, the same as previously described by N. V. Belov and P. N. Tarkhova[275] for *milarite*, $KCa_2(Be,Al,Si)_3[(Si,Be)_{12}O_{30}].H_2O$. In osumilite the constitution water is "stuffed" inside the double rings. The remarkable contents in alkalies, calcium, and divalent and trivalent iron are particular for its magmatic crystallization, somewhat similar to that of tridymite and cristobalite in effusive rocks.

[274] *Proc. Japan Acad.*, **29**, 1953, 321-323; *Am. Mineralogist*, **41**, 1956, 104-116. See on metastability of osumilite: W. Schreyer and J. F. Schairer, *Ibid.*, **47**, 1962, 90-104 (cf. Vol. III, B. ¶¶ 219f.).

[275] *Doklady Akad. Nauk S.S.S.R.*, **69**, 1949, 365-368; *Trudy Inst. Krist. Akad. Nauk S.S.S.R.*, **6**, 1951, 83-140.

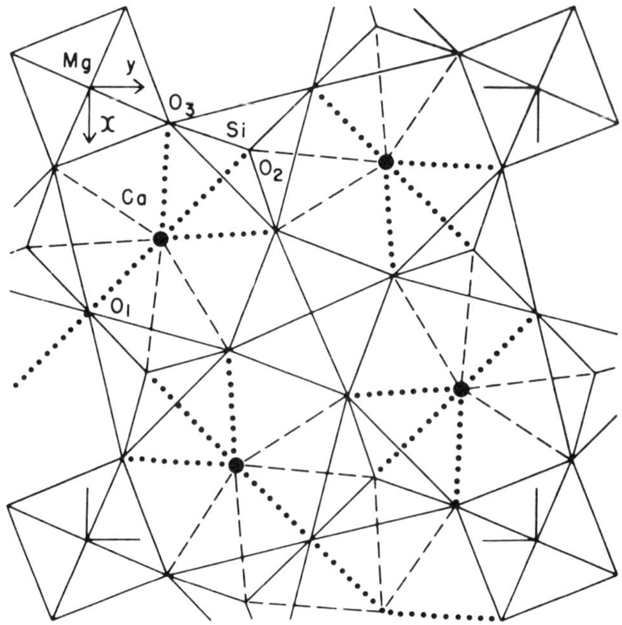

FIG. A. 122. *Z*-axis projection of the structure of melilite (Smith).

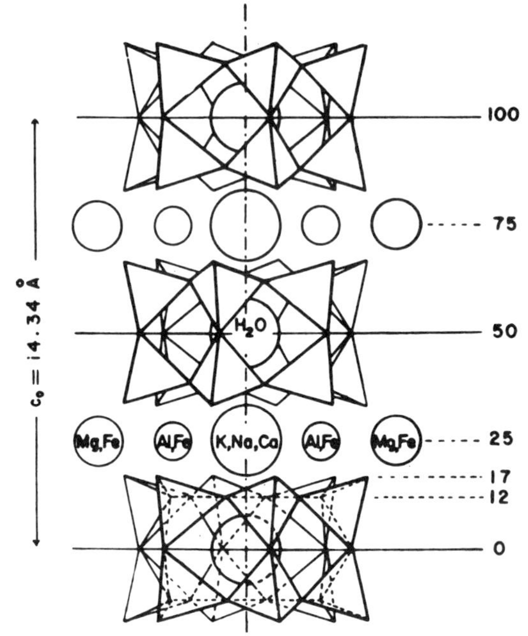

FIG. A. 123. Structure of osumilite, with double rings $[(Si,Al)_{12}O_{30})]$ (Miyashiro).

172. The structure of *hexagonal* $CaAl_2Si_2O_8$ is similar to that previously determined by T. Ito, B. Yoshiki, and K. Matsumoto (1951) for "α-celsian," hexagonal $BaAl_2Si_2O_8$, only with that distinction that the single rings in the basal double-layers structure are not strictly hexagonal, but ditrigonal, as Y. Takeuchi and G. Donnay[276] established. $[AlO_4]$ tetrahedra in the lower half of the leyers point upward, those in the

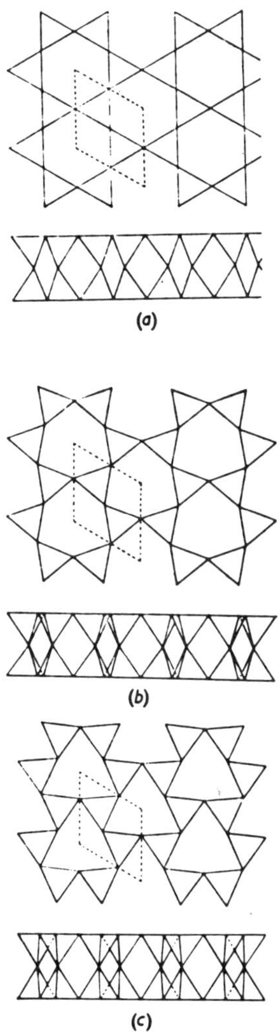

FIG. A. 124. Layers $[(Si_2Al_2O_8)]_\infty$: (*a*) in the hexagonal high-temperature modification of $BaAl_2Si_2O_8$; (*b*) in the hexagonal low-temperature $BaAl_2Si_2O_8$; (*c*) in hexagonal $CaAl_2Si_2O_8$ (Takeuchi and Donnay).

[276] *Carnegie Inst. Wash. Year Books*, **58**, 1958/59, 198-199; *Acta Cryst.*, **12**, 1959, 465-470.

upper half point downward. The apical bridge oxygen shared between two opposite tetrahedra is in a bond angle Si—O—Si = 180°, and all corners of the tetrahedra are shared, as in a framework structure. This fact is important for a 50 and 50 per cent share of Al^{3+} in the centers of the tetrahedra. The $[CaO_6]$ polyhedra are of two kinds, a rather regular octahedron ideally ordered, and a trigonal prism form for faults of the frequent layer stacking. In the corresponding $BaAl_2Si_2O_8$ there is also a characteristic difference between the really hexagonal high-temperature modification with regular hexagonal, and a low-temperature form with pseudohexagonal, ditrigonal layer units. No corresponding really hexagonal modification of $CaAl_2Si_2O_8$ was observed[277] (see Figs. A. 124 and 125).

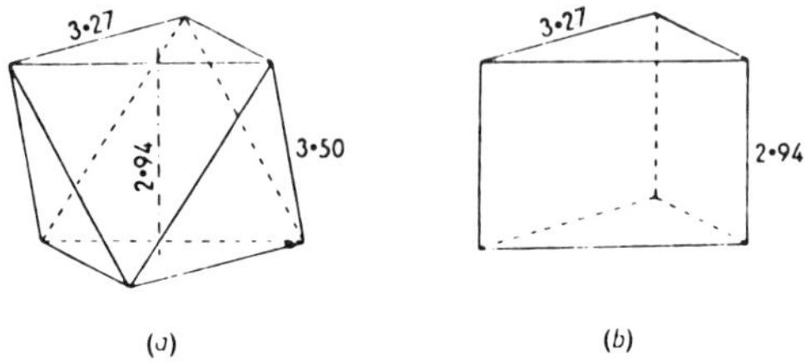

FIG. A. 125. $[CaO_6]$ polyhedra in hexagonal $CaAl_2Si_2O_8$: (*a*) trigonal antiprism (almost regular octahedron); (*b*) trigonal prism (Takeuchi and Donnay).

4. Three-Dimensionally Infinite Tetrahedral Frameworks (Sharing Coefficient 2.00 and Higher)

173. *Hemimorphite* (calamine) which was previously classified as an "di-orthosilicate" structure with $[Si_2O_7]$ groups (cf. T. Ito and J. West, 1932) must be in Zoltai's definition a typical framework silicate. G. A. Barclay and E. G. Cox[278] made a complete redetermination in which the existence of $[Si_2O_7]$ groups with a shared corner is confirmed, the Si—O—Si bond angle being 133°. The distortion of the tetrahedra is quite small. The $[ZnO_4]$ tetrahedra interlink between the $[Si_2O_7]$ groups, the water molecules lying in channels parallel to the *c*-axis, held by two bonds to hydroxyl groups, with an angle of 78°.

174. In the same way, we group *phenacite*, Be_2SiO_4, and *willemite*, Zn_2SiO_4, definitely to the framework structures, because $[BeO_4]$ and $[ZnO_4]$ tetrahedra interlink

[277] We refer in this connection to the fact that Ito and Sadanaga (1951) in lithium-aluminum tourmalines found a six-membered (single) ring; Donnay and Buerger (1950) in magnesium tourmalines found a ditrigonal configuration of this unit.

[278] *Z. Krist.*, **113**, 1960, 23-29. The complex crystallochemical reactions on dehydration of hemimorphite were recently investigated by H. F. W. Taylor, *Am. Mineralogist*, **47**, 1962, 932-944. An intermediate formation of trigonal "β-"Zn_2SiO_4 is observed which probably has a distorted tridymite- or cristobalite-like structure, in oriented conversion intergrowths.

between rings of $[SiO_4]$ groups around the trigonal axes. The principles of this interlinking between two kinds of tetrahedra were correctly formulated by W. L. Bragg (1937). Most convincing is also the analogous structural interpretation of *beryl* as a framework structure, with $[BeO_4]$ tetrahedra interlinking $[Si_6O_{18}]$ rings. To this important group of beryllosilicates we may also add, according to W. Nowacki, G. Bergerhoff, and H. Huttenlocher,[279] the rare mineral *bazzite*, a scandium beryl in which two thirds of the Al^{3+} ions in $[AlO_6]$ coordination are replaced by Sc^{3+} ions, and a widened beryl structure results. Both minerals contain additional atoms (chiefly alkalies), "stuffed" in the channels of the wide hexagonal ring units, and water in a nonstoichiometric, variable proportion. Nowacki *et al.*, emphasize that in the silicate structures of beryl type the high variability of the total composition is a characteristic property. The Al^{3+} ions are partly replaced by Sc^{3+} (in bazzite), and replace in their turn Si^{4+} in the tetrahedral rings. The Fourier projection (Fig. A. 126) parallel to the *c*-axis shows very impressively the manner in which the additional atoms and the water molecules increase the central maxima from beryl to bazzite. Even Be^{2+} in tetrahedral coordination can be partially replaced by Al^{3+}.

175. Alkaliės (especially lithium, sodium, and cesium) containing beryls were studied by T. A. Sosedko[280] by examining the systematic replacement of beryllium by the alkalies, calculated in the ratio $Be^{2+} : R^+$, variable between 2.8 to 0.35 in the extremes: a green beryl (with 1.8 per cent R_2O) and a pinkish beryl (with 7.2 per cent R_2O, not less than 4.1 per cent being Cs_2O), in different zones of the pegmatite in which these varieties occur. Sosedko and V. A. Frank-Kamenetskiĭ[281] followed these variations in changes of the c_0 parameters. The planar arrangement of $[AlO_6]$ and $[BeO_4]$ groups perpendicular to the *c*-axis is changed in the isomorphous replacement of Be^{2+} by alkali ions occupying positions in the centers of the empty channels of the $[Si_6O_{18}]$ rings, with corresponding defects in the $[BeO_4]$ groups.

176. *Bertrandite*, $(Be_4Si_2O_7(OH)_2)$, is in certain respects similar to hemimorphite, both now being grouped as framework structure silicates. The structure previously described by T. Ito and J. West (1932) was recently reexamined by L. L. Solov'eva and N. V. Belov[282]; characteristic of the new interpretation is the interlinkage of

[279] *Schweiz. mineral. petrog. Mitt.*, **34**, 1954, 501-504; *Ibid.*, **35**, 1955, 410-421; *Fortschr. Mineral.*, **35**, 1957, 43-44. For detailed studies of the excess helium and argon in beryl, trapped in the channels of the structure as magmatic gases, and by far in excess over that which would be accounted for by radioactive decay, see P. E. Damon and J. L. Kulp, *Am. Mineralogist*, **43**, 1958, 433-459. The same is valid for the rare gases in cordierite and tourmaline. K. A. Wickersheim and R. A. Buchanan, *Ibid.*, **44**, 1959, 440-445, determined the water content in beryl by infrared absorption measurements, with remarkable dichroism effects resulting from a high degree of orientation of the water molecules relative to the *c*-axis (cf. A. ¶ 44).

[280] *Zapiski Vsesoyuz. Mineral. Obshchestva*, **86**, 1957, 495-499.

[281] *Doklady Akad. Nauk S.S.S.R.*, **118**, 1958, 815-817.

[282] *Doklady Akad. Nauk S.S.S.R.*, **140**, 1961, 685-688.

$[Si_2O_7]$ groups (as in hemimorphite) oriented parallel to the x-axis, with two kinds of $[BeO_4]$ tetrahedra in ditrigonal (pseudo-hexagonal) rings. The bertrandite structure is pronouncedly *polar*, the tetrahedra being oriented with their apexes to one side parallel c.

177. Many interesting new aspects on the structure of *cordierite*, $Al_3Mg_2[Si_5AlO_{18}]$, in analogy to that of beryl, were found by the more recent investigations of A. Miyashiro[283] *et alt.*, especially concerning the discovery of the high-temperature phase

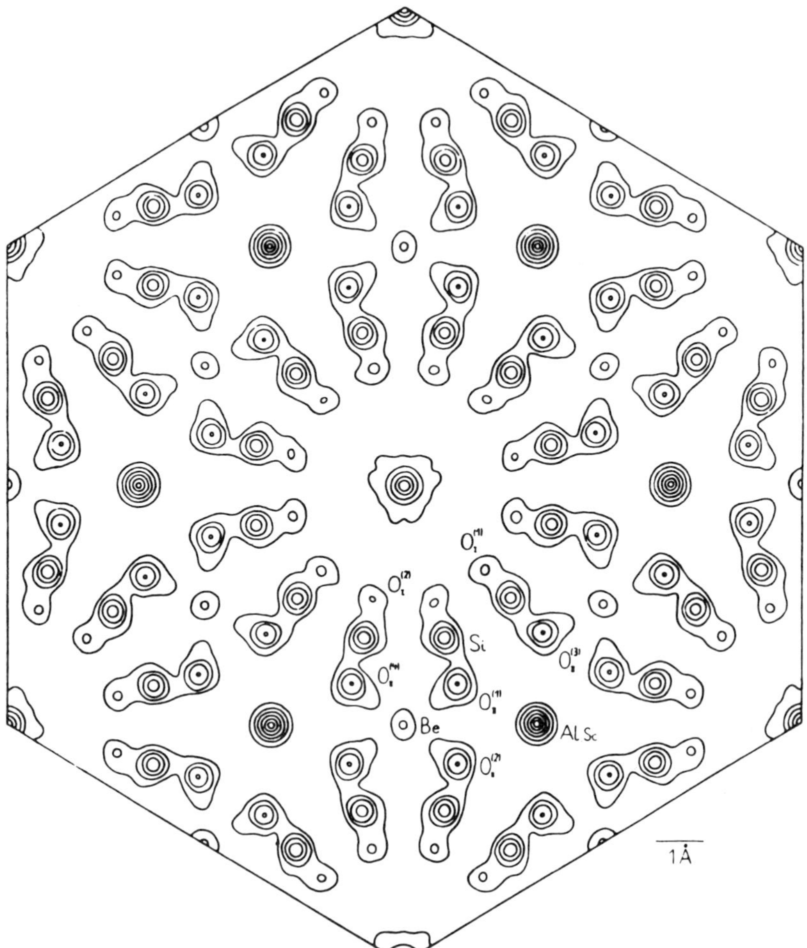

FIG. A. 126. Fourier projection of the bazzite structure on (0001), with the contour lines of the | F | values (Nowacki and Bergerhoff).

[283] *Am. J. Sci.*, **253**, 1955, 185-208 (with T. Iiyama, M. Yamasaki, and T. Miyashiro); *Proc. Japan Acad.*, **30**, 1954, 746-750.

of indialite and its relations to cordierite and beryl. Since these mineralogical-petrological problems are extensively discussed in Vol. III, Section B, we restrict discussion here to a few general remarks on the polytypism and the polymorphic phenomena, as observed in the truly hexagonal space groups of α-(high temperature), and β(low temperature) indialite, pseudohexagonal (orthorhombic) high-temperature cordierite, and pseudohexagonal low-temperature cordierite; furthermore, we deal with the isolated phase called μ-cordierite by G. A. Rankin and H. E. Merwin (1918), and also in its relation to hexagonal *osumilite* (see above), in spite of its different structure type. Indialite is doubtless a highly disordered high-temperature modification, the orthorhombic cordierite a better-ordered low-temperature form. The iron-cordierite synthesized by J. F. Schairer and K. Yagi (1952) is hexagonal, therefore, a high-temperature iron-indialite.

178. The indialites are true analogues of beryl, but the distribution of $[AlO_4]$ and $[SiO_4]$ tetrahedra in the annular rings is at random. All the polytypes and polymorphs of the group are in disorder $\rightleftharpoons$ order reaction relations, and there are crystalline solutions with the coupled exchange mechanism $Mg^{2+} + Si^{4+} \rightleftharpoons 2\ Al^{3+}$, with $Mg_2Al_3[Si_5AlO_{18}]$ as one end member, and the "magnesium beryl" $Mg_3Al_2[Si_6O_{18}]$ as the other, as T. Iiyama[284] defined it. As a measure of the structural disorder Miyashiro defined an index, Δ, for suitably chosen interference lines in the X-ray powder diagrams of the orthorhombic modifications, with $2\,\Theta$ between 29° and 30° for CuKα radiation, for peaks called A, B, D (cf. Fig. A. 127):

$$\Delta = 2\,\Theta_D - \frac{2\,\Theta_A + 2\,\Theta_B}{2}$$

($\Delta = 0$ for indialite), being proportional to the difference $a_0 - b_0\sqrt{3}$, of the orthorhombic cell unit. "Perdistorted" cordierite has $\Delta = 0.31$. Intermediate crystal phases show changes in Δ with a thermal treatment in both directions of the conversions, highly analogous to those in alkali feldspars with a great sensitivity of optical constants (e.g., 2 V) to changes in Δ, but also with metastable anomalies of crystallization, as below 700°C.

179. Among the silica structures, a revision of the unit cell dimensions of *tridymite* and *cristobalite* was made by J. V. Smith[285] with precision data: for α-cristobalite $a_0 = 7.1382 \pm 0.0010$ Å. at 405°C.; for α-tridymite $a_0 = 5.0463 \pm 0.0020$ Å.; $c_0 = 8.2563 \pm 0.0030$ Å. at 405°C. The unit cell volumes of both modifications are 363.72 ± 0.2 Å.^{3} and 363.98 ± 0.4 Å.^{3}, respectively; these are nearly the same, in view of the general similar structural principles in both crystal phazes, differing only in the long-range parameters. This similarity, although with a different stacking in

[284] *Proc. Japan Acad.*, **31**, 1955, 166-168.
[285] *Carnegie Inst. Wash. Year Books*, **52**, 1952/53, 61-62.

two layers for tridymite (with the sequence ABABAB...) and in cristobalite with three layers (sequence ABCABCABC...) is the starting point of the basically important postulates of O. W. Flörke[286] who concluded from widely extended analytical and physicochemical data that tridymite is a "stuffed" phase. It always contains alkalies, calcium and aluminum, even in low concentrations, and is, therefore, not a true polymorph of pure silica. This type of contamination rules the occurrence of one-dimensional dislocation phenomena, the formation of multiple polytypes with

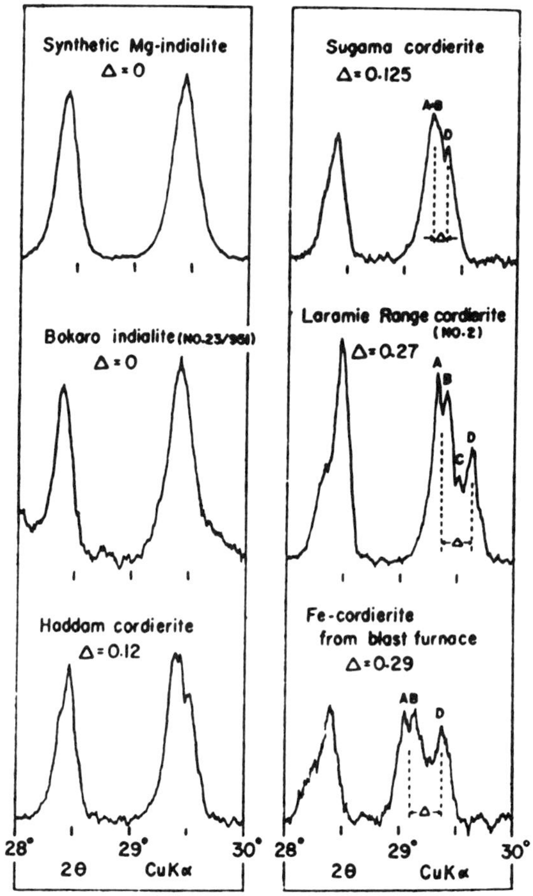

FIG. A. 127. X-ray powder diagrams of indialite and cordierites (Miyashiro). The characteristic peaks are A, B, D, for the determination of the distortion index, Δ.

[286] *Naturwissenschaften*, **41**, 1954, 371-372; *Fortschr. Mineral.*, **32**, 1954, 33-35; *Ibid.*, **33**, 1955; 151-152; and particularly *Ber. deut. keram. Ges.*, **32**, 1955, 369-381; a short summary was given by W. Eitel, *Am. Ceram. Soc. Bull.*, **36**, 1957, 142-148, and recently by O. W. Flörke, *Silikat Tech.*, **12**, 1961, 304-305.

stacking anomalies, and the occurrence of superstructures. Tridymite-type phases with ten and even twenty or more layers stacked may also be formed, although never in the pure state, but always intergrown with others, as sequences of the ABC... type of cristobalite.

180. Flörke's ideas, here indicated only in a few words, have developed to be an important basis for explaining the quite variable thermal expansion behavior of silica as a function of the thermal history, either in the pure system SiO_2 or in complex systems, such as those encountered in refractories used in glass melting or metallurgical furnaces (discussed in Vol. III, Section B and Vol. V, Section B). From the structural viewpoint, the most important result in Flörke's discussions is that the theory of one-dimensional structural defects (H. Jagodzinski) can be applied in its full significance for the X-ray structural determination of silica crystal phases from multicomponent systems, for which the original phase equilibrium diagram of SiO_2 as given by C. N. Fenner (1913) is not now sufficient. Important to an understanding of the specific mineralizing properties of alkali salts is a consideration of possibilities of alkali ions entering the hollow channels parallel c in tridymite. Flörke has calculated the radii of the gaps in α-quartz, α-cristobalite, and α-tridymite. This affords a possibility for discussing the coordination of these alkali ions to O_2 anions (in theory mostly between 6 and 9), and explains, in addition, why in quartz the ions of lithium and sodium can enter the structure, why ions of sodium and potassium can enter in tridymite, and why ions of rubidium and cesium can enter in cristobalite.

181. The structure of *β-tridymite*, in comparison with the incomplete and idealized determination for α-tridymite by R. E. Gibbs (1927), was quite recently calculated by J. E. Fleming and H. Lynton,[287] from a single crystal found in a refractory furnace lining. This monocrystal is a polytype containing not less than 864 units (a_0 = 30.08 Å., c_0 = 49.08 Å.). The strong interference spots indicating the hexagonal subcell with a'_0 = 5.01 Å., c'_0 = 8.18 Å., which is similar to that proposed by Gibbs for α-tridymite. The new determinations also definitely correct the idealized and unsatisfactory conclusions from Gibbs' proposal that the angle Si—O—Si would be 180°; it is much smaller. Nevertheless, the hexagonal network in the α-tridymite structure is also present as an approximation in β-tridymite as seen from the Fourier projection for (hk0) (Fig. A. 128).

182. The fundamental difference of the structure of *coesite*, the high-pressure modification of silica, from the classical modifications (quartz, tridymite, cristobalite) becomes evident in the results of the lattice investigation by L. S. Ramsdell[288] and

[287] *Phys. Chem. Glass*, **1**, 1960, 148-154.
[288] *Am. Mineralogist*, **40**, 1955, 975-982.

of the structure determination by T. Zoltai and M. J. Buerger.[289] Ramsdell described a monoclinic structure for coesite, although it is dimensionally hexagonal with $a_0 = c_0 = 7.23$ Å., $b_0 = 12.52$ Å., and $\beta = 120°$, explaining the characteristic twinning with (100) and (021) as derived from the X-ray data (Weissenberg diagrams). Zoltai and Buerger confirmed in general the structure as proposed by Ramsdell, only exchanging the b and c axes, and derived the structure from three-dimensional Patterson analysis. The coesite structure has the same tetrahedral coordination $[SiO_4]$ as the normal silica modifications in a framework configuration which is

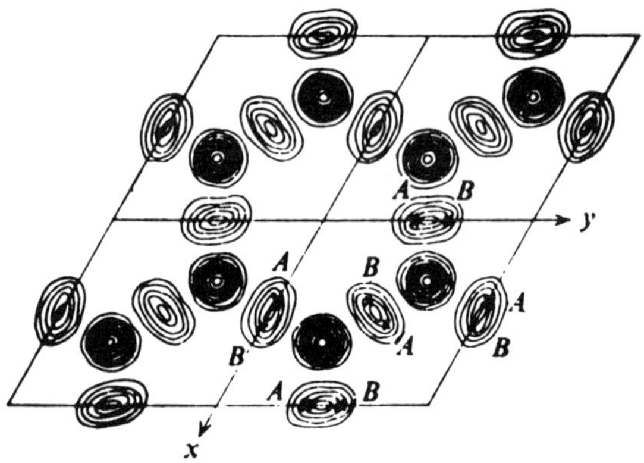

FIG. A. 128. Fourier synthesis for (hk0) projection for low-temperature (β)-tridymite (Fleming and Lynton).

somewhat similar to the skeleton of $[(Si, Al)O_4]$ tetrahedra in the feldspars and in paracelsian (see below), and with characteristic fourfold ring groupings parallel (010) and (001) in a very dense arrangement. Figure A. 129 shows a perspective aspect described as a framework of tetragonal rings, each ring having eight external connections with two sets of criss-crossed chains parallel (101) and (10$\bar{1}$) which are isolated in the one plane (010). Equivalent with this description is a configuration of the rings symmetrical with respect to twofold axes, parallel (001) and forming chains parallel (010) (Fig. A. 130). The chains are joined by the rings of glide-equivalent chains in the levels below and above the plane (ab) (Fig. A. 131).

183. Concerning the tetragonal structure of *keatite*, the other nonclassical silica modification, intermediate between cristobalite and quartz in its density, between

[289] *Z. Krist.*, **111**, 1959, 129-141. The existence of a silica modification which is characterized by the absence of $[SiO_4]$ groups, but consisting, as an isostructure to *rutile*, exclusively of $[SiO_6]$ octahedral coordination units, is made evident in the recently discovered highest-pressure mineral *stishovite* (cf. Vol. III, Sections A and B). On the space group of stishovite ($D_{4h}^{14} = P\,4_2/mnm$), its density (4.28), and the distance Si—O $= 1.77_8$ Å., see A. Preisinger, *Naturwissenschaften*, **49**, 1962, 345.

the low-pressure silica modifications, and coesite, J. Shropshire, P. P. Keat, and P. A. Vaughan[290] determined a system of four [SiO_4] tetrahedral spirals, interlinked by additional [SiO_4] groups, sharing all corners with the four spirals. A projection on (001) makes this interlinking evident (Fig. A. 132), with slight deviations

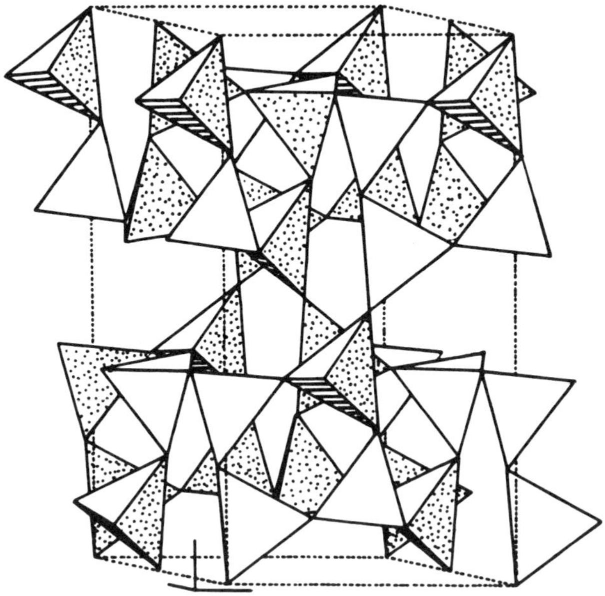

FIG. A. 129. Perspective aspect of the structure of coesite (Zoltai and Buerger).

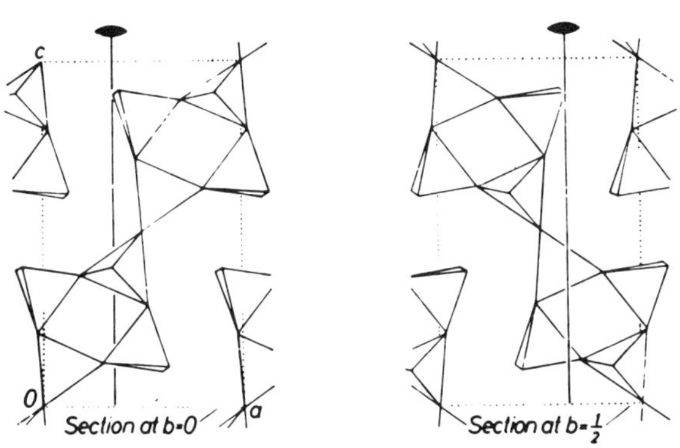

FIG. A. 130. Criss-crossed chains of fourfold rings parallel to the twofold symmetry axis of coesite (Zoltai and Buerger).

[290] *Z. Krist.*, **112**, 1959, 409-413. The previous assumption of P. P. Keat, *Science*, **120**, 1954, 328-330, that keatite would be similar in structure to α'-$LiAlSi_2O_6$ and μ-cordierite is not confirmed.

from regularity in the tetrahedra which are somewhat larger than Zoltai and Buerger found them in coesite. Keatite contains five- and seven-membered rings, in contrast with the six-membered rings in tridymite and cristobalite. In a comparison of all the structures observed for the silica modifications, Zoltai and Buerger[291] found characteristic six-membered "loops" in quartz, tridymite, and cristobalite, five-

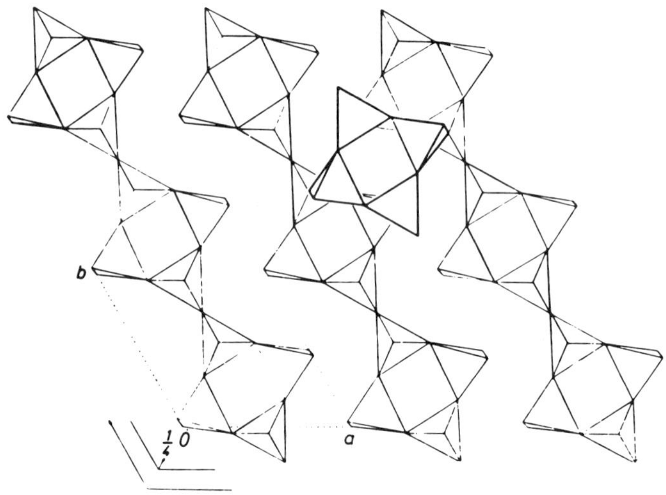

FIG. A. 131. Projection onto the (ab) plane of coesite, to show the chains parallel to the b-axis (Zoltai and Buerger).

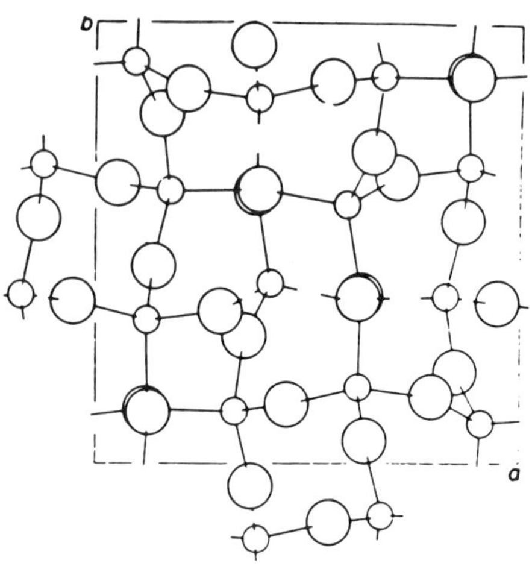

FIG. A. 132. Projection on the (001) plane of keatite (Shropshire, Keat, and Vaughan).

[291] *Z. Krist.*, **114**, 1960, 1-8.

membered loops in keatite, and four-membered loops in coesite. The stability of these loop configurations is best judged from a calculation of the relative energies (cf. A. ¶ 27), for the central ion in the tetrahedral units of $+4$, the vertex ions of -2, if shared, or -1 if not shared. The minimum energy is that for five-membered rings. A higher energy for six and four-membered rings is in agreement with the experimental stability determinations, namely quartz being the stable modification for normal pressure, keatite being less stable, and coesite the most unstable form.

184. Among the *derivatives* of the silica structures we first mention the structure of *chkalovite*, $[Na_2BeSi_2O_6]$, which was described in its pseudotetragonal, orthorhombic type by Yu. A. Pyatenko, G. B. Bokiĭ, and N. V. Belov[292] as a framework of $[SiO_4]$ and $[BeO_4]$ tetrahedra with $(Si^{4+} + Be^{2+}) : O = 1 : 2$, and very similar to aluminosilicates derived from the structure of α-cristobalite: c (chkal.) (6.87 Å.) $= a$ (cristob.); $a = b$ (chkal.) ($=$ 21.1 Å.) $= 3 \times a$ (cristob.). This relation is seen in a replacement of two thirds of the Si^{4+} in cristobalite by Be^{2+}; the Na^+ ions are arranged in polyhedral cavities of the diamond-like cell, in coordinations $[NaO_{10}]$ and $[NaO_{12}]$. The empty interstitial polyhedra parallel (100) cause the cleavage. The $[BeO_4]$ and $[SiO_4]$ tetrahedra have the same infinitely extended bands as in cristobalite.

185. A remarkable attempt to explain the highly complex phenomena of the formation of Na_2O-enriched phases of *α-carnegieite*, $NaAlSiO_4$, from the structural viewpoints was made by W. Borchert and J. Keidel,[293] derived from the type of α-cristobalite, with an extensive statistical distribution of the excess Na^+ ions in a highly defective structure, the occurrence of superstructures and rotational inversions. In connection with the important problem of the influence of statistical distributions of the $[AlO_4]$ and $[SiO_4]$ tetrahedra in aluminosilicate frameworks to which α-carnegieite belongs as well as, e.g., *paracelsian*, $BaAl_2Si_2O_8$, with the average intermediate distances Al—O and Si—O, J. V. Smith[294] gave the valuable diagram shown in Fig. A. 133. Smith also determined the structure of α-carnegieite to complete the preliminary data of T. F. W. Barth (1932). His results confirm the α-cristobalite type structure with a statistically random arrangement of the O^{2-} anions so that they appear to rotate in a circle of 0.5 Å. radius, indicating the very high degree of disorder of this high-temperature phase. The inversion into β-carnegieite at 690°C. is structurally described either as a collapse of the high-temperature arrangement around the holes, or as a cessation of the rotations, or both phenomena combined.

186. As a derivative of the α-tridymite structure, that of *nepheline* is the most re-

[292] *Doklady Akad. Nauk S.S.S.R.*, **108**, 1956, 1077-1080.

[293] *Heidelberger Beitr. Mineral. u. Petrog.*, **1**, 1947, 17-30; *Fortschr. Mineral.* **27**, 1950, 27-30, with particular emphasis given to analogous phenomena in ore minerals.

[294] *Carnegie Inst. Wash. Year Books*, **52**, 1952/53, 72-75; *Acta Cryst.*, **7**, 1954, 479-481.

markable example, as determined by M. J. Buerger, G. E. Klein, and G. Donnay.[295] Like tridymite, nepheline is also a typical stuffed structure. All of the $[SiO_4]$ tetrahedra in the framework pointing toward one end of the *c*-axis are replaced by $[AlO_4]$, giving a polar character to the hexagonal axis. The valence-compensating Na^+ ions fill the structural voids. The most important result in this respect, however,

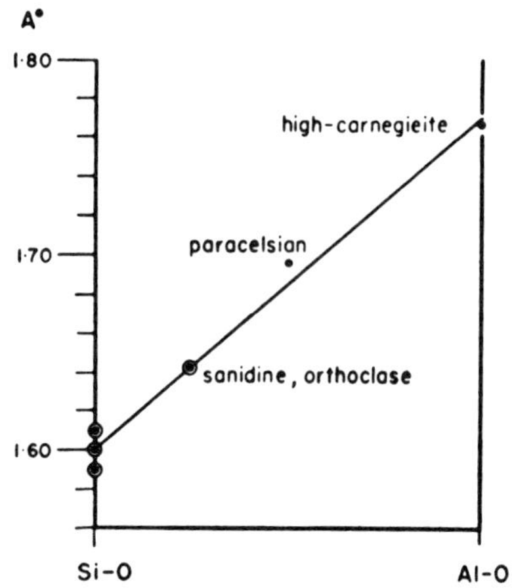

FIG. A. 133. Variation of the distances Si—O and Al—O in tetrahedral coordination in aluminosilicates (Smith). For Si—O the values are for quartz, cristobalite (in α- and β-modifications), diopside, afwillite, tilleyite, and α-carnegieite.

is that the natural nepheline is practically never the pure $NaAlSiO_4$, but $KNa_3Al_4Si_4O_{16}$. One fourth of the ideally present Na^+ ions are replaced by the larger K^+ ions with a coordination $[KO_9]$ which occupy larger cavities than those filled with the smaller Na^+ ions in coordination $[NaO_8]$. These latter cavities are produced by a twisting of the tridymite framework about the sets of the trigonal symmetry axes. On these axes the O^{2+} anions probably rotate at elevated temperatures contacting the surrounding Na^+ ions only part of the time. From the electron density projection onto the plane (*xy*) the interlinkage of the tetrahedra in relative position to Na^+ and K^+ ions is seen in Fig. A. 134.

187. The more detailed structure of $KNa_3Al_4Si_4O_{16}$ given by Th. Hahn and M. J. Buerger[296] indicates in natural nepheline a small deficiency in alkalies, probably due to a ratio Si^{4+}/Al^{3+} above 1.0. The O^{2-} anions are considerably deformed to ellipsoids elongated in directions which do not change the Si—O and Al—O bond distances, and indicating strong thermal motion. Concerning the order principles

[295] *Am. Mineralogist*, **39**, 1954, 805-818.
[296] *Z. Krist.*, **106**, 1955, 308-338.

valid in the nepheline structure only four of the sixteen tetrahedra per unit cell have ordered (Al, Si) positions, the others are disordered. The projection on the (xz) plane (Fig. A. 135) shows the elevation of the structure. We further mention the interpretation of the nepheline structure as given by N. V. Belov[297] who attempted to introduce double rings of the type $[Si_5AlO_{18}]$ as in cordierite (see above), and Na^+ grouped as in ramsayite in $[NaO_6]$ units. We prefer the structure as given by Buerger *et al.* A direct structural analysis of natural nephelines with variable compo-

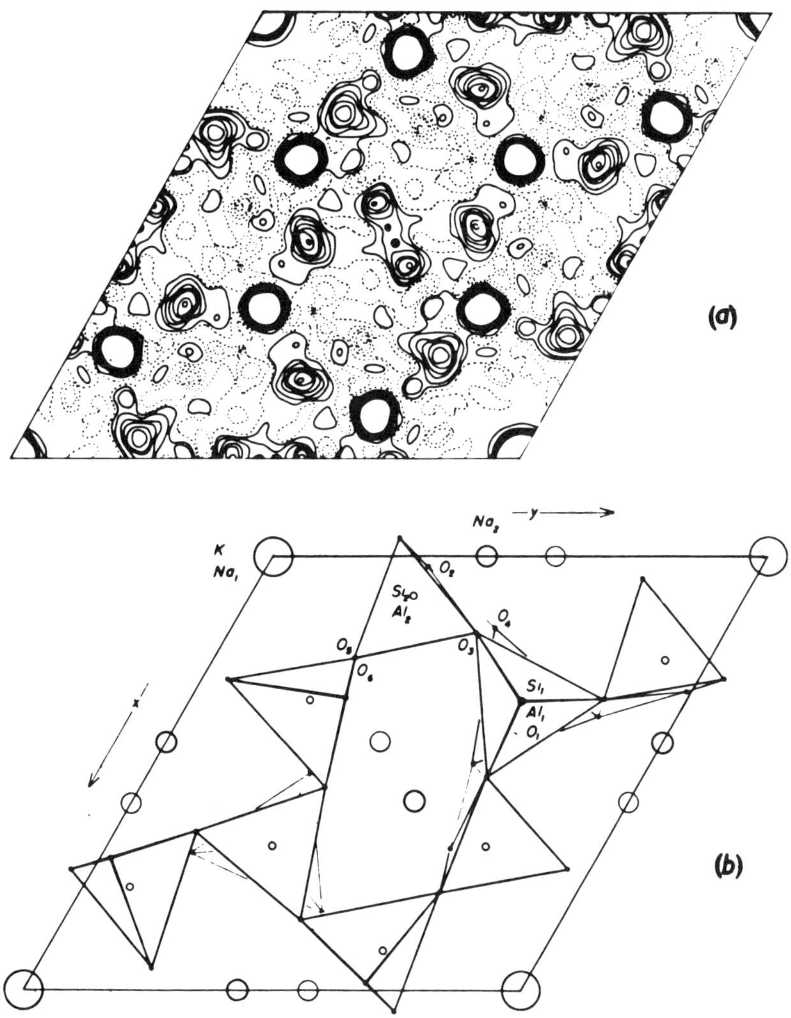

FIG. A. 134. Structure of nepheline (b), derived from electron density projection (a) onto the (xy) plane (Buerger, Klein, and Donnay).

[297] *Trudy Inst. Krist. Akad. Nauk S.S.S.R.*, 10, 1954, 6-9.

sitions (atomic per cent ratios K/(K + Na + Ca)) was made by J. V. Smith and Th. G. Sahama[298] with the remarkable result that there is a sharp discontinuity of 2Θ for (210) as a function of this ratio, at 25 per cent, confirming the singularity of the composition $KNa_3Al_4Si_4O_{16}$, as given by Buerger *et al.* On the highly complex polymorphism and polytypism in the series $NaAlSiO_4$–$KAlSiO_4$, with *kalsilite*, *kaliophilite*, and related phases see Vol. III, Section B.

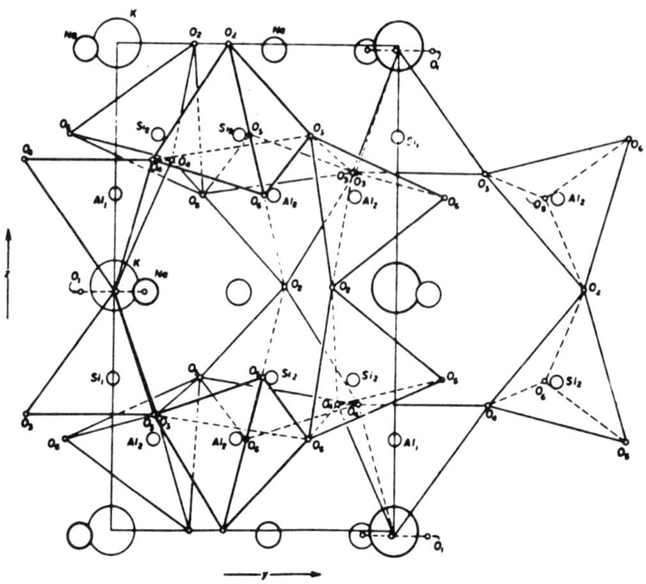

FIG. A. 135. Elevation of the nepheline structure interpreted from electron density ϱ (xz) plan (Hahn and Buerger).

188. In the *sodalite* group of complex sodium aluminosilicates H. Saalfeld[299] writes the composition of *nosean*, $Na_2[SO_4][Al_6Si_6O_{24}]$, and refines the previous structure proposals of L. Pauling (1930), T. F. W. Barth and F. Machatschki (1932), with a statistical distribution of the $[SO_4]$ groups in the cavities of the tetrahedral framework. In connection with his studies of the reactions between kaolinite and sodium salt melts, Saalfeld prepared a compound $Na_2[WO_4]$ $[Si_6Al_6O_{24}]$, which unequivocally confirmed the structural proposal of Barth based on the space group $T_d^1 = P\,\bar{4}\,3\,m$, but Machatschki's definition of a statistical distribution of the $[SO]_4$ groups in nosean is also correct. The $[SO_4]$ groups are larger than both authors had assumed (1.58 and 1.60 Å. for the distance S—O), causing a widening of the nosean

[298] *Mineral. Mag.*, 30, 1954, 439-449.

[299] *Fortschr. Mineral.*, 37, 1959, 48; *Neues Jahrb. Mineral. Monatsh.*, 1959, 38-46; *Veröffentl. Max-Planck-Inst. Silikatforsch.*, 19, 1959, 73-82.

structure relative to that of sodalite, Na[Cl][$Si_6Al_6O_{24}$], although the cavities in sodalite would be large enough to be filled with [SO_4] units. The widening by [WO_4] is so considerable that also the Si^{4+} and Al^{3+} ions are somewhat dislocated from their ideal positions. Typical one-dimensional displacements are indicated by crosswise diffuse diffraction tails in the X-ray monocrystal diagrams.

189. For *hauyinite*, $(Ca,Na)_{4-8}[SO_4]_{1-2}[Si_6Al_6O_{24}]$ Saalfeld,[300] observed besides the normal diffraction spots of the space group $T_d^4 = P\,\bar{4}\,3\,n$, superstructure effects with additional "satellites" arranged along the [110] directions, combined with diffuse scattering indicating one-dimensional disorder. It is remarkable that some natural hauyinite crystals from Mount Vesuvius do *not* show any superstructure of the kind mentioned here. The superstructure is caused by the [SO_4] groups, and also partly by the Ca^{2+} and Na^+ ions, but not by the tetrahedral framework.

190. *Borosilicates* of the danburite group are of great importance because they serve as "models" which provide the key for the deeper interpretation of the structure of the feldspars (see below), especially concerning the order of distribution of the Al^{3+} and B^{3+} cations in tetrahedral coordination, replacing [SiO_4] groups. H. Kramer and R. D. Allen[301] made a comparison of the structure of *bakerite*, $8CaO.6SiO_2.5B_2O_3.6H_2O$, with the borate *priceite*, $4CaO.5B_2O_3.7H_2O$, and *veatchite*, $3(Sr,Ca)O.8B_2O_3.5H_2O$, among which bakerite is the most interesting because of its near similarity in the X-ray diffraction diagrams with datolite, $Ca_8B_8[SiO_4]_2[SiO_4]_6(OH)_8$. For bakerite, therefore, a crystallochemical formula, $Ca_8B_8[BO_4]_2[SiO_4]_6(OH)_6.3H_2O$, may be considered. A determination of the structure, however, on this basis was not yet given.

191. The structure of *danburite*, Ca[$B_2Si_2O_8$], and of *hurlbutite*, Ca[$Be_2P_2O_8$], are correlated by similar analogies and are of considerable importance for the correct interpretation of complex boro-, alumino-, and beryllosilicates. V. V. Bakakin, V. B. Kravchenko, and N. V. Belov[302] found similar tetragonal and octagonal rings of [SiO_4], [BO_4], and [BeO_4] tetrahedra in these minerals as they are observed in anorthite, with [Si_2O_7] and [B_2O_7] groups, caused by symmetry planes parallel to the plane (xy), interlinked by [CaO_7] polyhedra arranged in columns in the midst of the pseudoquadratic (xy) plane. The angles Si—O—Si and B—O—B are 133° and 129°, respectively. Hurlbutite differs only in having no planes of symmetry, every [PO_4] group being connected with four [BeO_4] groups, and vice-versa (Figs. A.

[300] *Z. Krist.*, **115**, 1961, 132-140.

[301] *Am. Mineralogist*, **41**, 1956, 689-700; cf. Cl. Frondel, H. Berman, and C. Palache in Dana's "System of Mineralogy," 7th ed., Vol. II, 1951, pp. 341-343, 348-349, 363. On the crystallographic forms of bakerite see also J. Murdoch, *Am. Mineralogist*, **47**, 1962, 919-923.

[302] *Doklady Akad. Nauk S.S.S.R.*, **129**, 1959, 420-423; refinements of the hurlbutite structure in *Ibid.*, **135**, 1960, 587-590.

136 and 137). The similarity with anorthite is particularly striking in the pseudo-tetragonal projection with the characteristic alternations of four- and eight-membered rings. There is also a great similarity with paracelsian (Fig. A. 138).

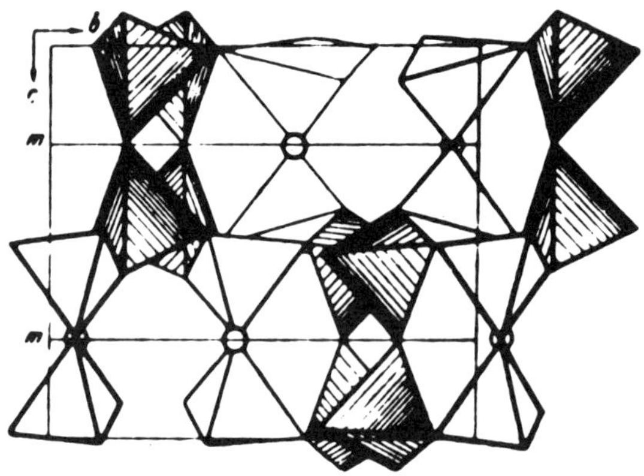

FIG. A. 136. Projection of the structure of danburite on the (xy) plane. (Bakakin, Kravchenko, and Belov).

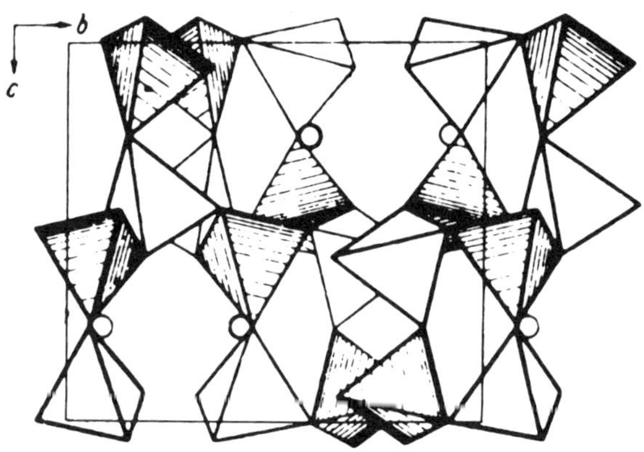

FIG. A. 137. Projection of the structure of hurlbutite onto the (bc) plane (Bakakin, Kravchenko, and Belov).

192. Borosilicate analogues to the alkali feldspars and feldspathoids are the synthetic compounds $KBSiO_4$, $KBSi_3O_8$, and $NaBSi_3O_8$, the latter known as the mineral *reedmergnerite*, which was discussed in this analogy by H. P. Eugster and N. L. McIver.[303] In this group we may also include monoclinic *searlesite*, $NaBSi_2O_6.H_2O$, the analogue to analcime, although it differs much in the crystallographic parame-

[303] *Bull. Geol. Soc. Amer.*, **70**, 1959, 1598-1599.

ters.[304] For reedmergnerite a limited crystalline solubility with albite is observed up to the composition 80 albite : 20 reedmergnerite. The crystallographic similarity between both is very striking, although natural reedmergnerite is never twinned, as described by Ch. Milton, E. C. T. Chao, J. M. Axelrod, and Fr. S. Grimaldi.[305] J. P. Clark and D. E. Appleman[306] gave some refinement of the reedmergnerite structure, especially concerning the introduction of B^{3+} in the tetrahedra, analogous to that of Al^3 in low-temperature albite, as determihed by R. B. Ferguson *et al.* (see below).

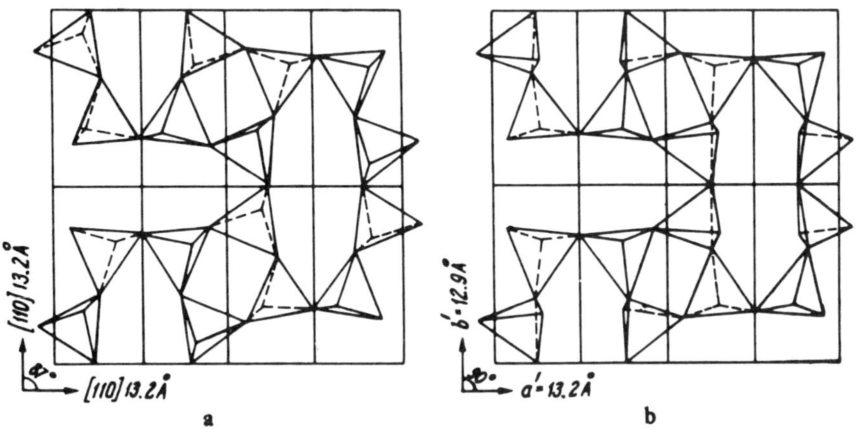

FIG. A. 138. Comparison of the structures of (*a*) hurlbutite and (*b*) anorthite, projected in the direction of the pseudotetragonal axis (Bakakin and Belov).

The framework is somewhat distorted as compared to that of albite in that the smaller $[BO_4]$ group dimensions are considered; the average reduction of the parameters relative to albite is about 4 per cent. The Na^+ ions are either coordinated in $[NaO_5]$ or in $[NaO_7]$ units, while in albite they are $[NaO_6]$ groups.

193. Also similar to feldspar is *eudidymite*, $Na[BeSi_3O_7(OH)]$, for which T. Ito (1943) had described a triple-chain type structure related to epididymite. A redetermination of the structure by E. A. Pobedimskaya and N. V. Belov[307] shows spiral chains of $[NaO_7]$ polyhedra elongated parallel *b*, each coordination unit being a trigonal pyramid with a quadratic half-octahedron. The $[SiO_4]$ groups are arranged in wollastonite-like chains $[Si_{1+2}O_9]_\infty$ similar to those of epididymite, not in pairwise symmetric arrangement but linked by O^{2-} in centers of symmetry positions. Corru-

[304] Cf. J. J. Fahey and J. M. Axelrod, *Am. Mineralogist,* **35**, 1950, 1014-1020, especially pp. 1019 f.

[305] *Am. Mineralogist,* **45**, 1960, 188-199.

[306] *Bull. Geol. Soc. Amer.*, **71**, 1960, 1843-1844, *Science*, **132**, 1960, 1837-1838.

[307] *Doklady Akad. Nauk S.S.S.R.*, **136**, 1961, 1448-1450.

gated layers of $[SiO_4]$ tetrahedra parallel (010) form a threefold framework with $[BeO_4]$ tetrahedra, of the composition $[BeSi_3O_7](OH)$ as seen in Figs. A. 139 and 140.

194. The structure of *hexagonal* $CaAl_2Si_2O_8$ was studied by Y. Takeuchi and G. Donnay[308] as a typical double-sheet, and not as a framework structure, whereas *para-celsian*, $BaAl_2Si_2O_8$, in distinction from α-hexagonal $BaAl_2Si_2O_8$, is a framework structure, similar to that of danburite, as J. V. Smith[309] confirmed, with a monoclinic, although nearly orthorhombic symmetry (β about 90°). The zigzag of the tetrahedral

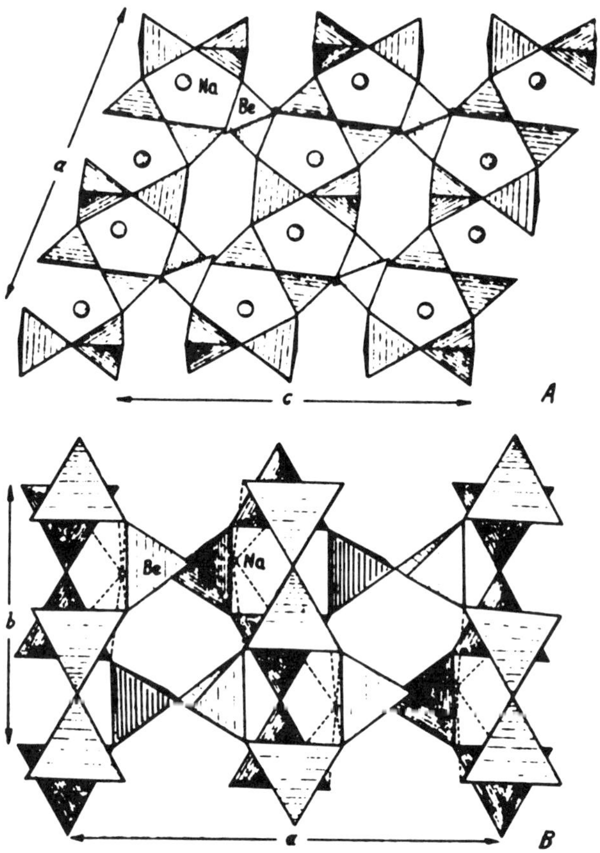

FIG. A. 139. Structure of eudidymite (Pobedimskaya and Belov).

[308] *Carnegie Inst. Wash. Year Books*, **58**, 1958/59, 198-199; *Acta Cryst.*, **12**, 1959, 465-470. It is our opinion that one should avoid calling this compound "*hexagonal anorthite*." Anorthite as a classical mineral is triclinic calcium feldspar, and nothing else, just as α'-$LiAlSi_2O_6$ is never a "β-spodumene," but an entirely different crystal phase than the mineral spodumene. Such arbitrary inconsequences in nomenclature are troublesome.

[309] *Am. J. Sci.*, Bowen-Volume, 1952, 513-515; *Acta Cryst.*, **6**, 1953, 613-620.

arrangement is identical with that in the feldspars (cf. W. H. Taylor, 1933), but must be rotated by 90° to the orientation in the feldspar structure as the only essential difference (Fig. A. 141).

195. In the structure of the *feldspar minerals* the introduction of the principles of conversions as a function of the thermal history and of the degree of order-disorder in the distribution of the Al^{3+} and Si^{4+} cations in the tetrahedral units has become a

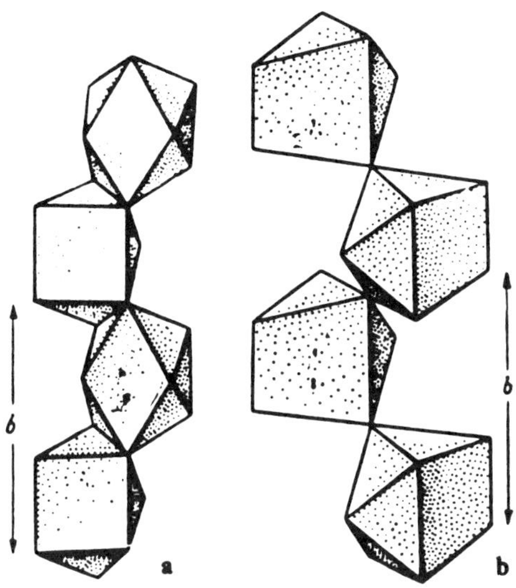

FIG. A. 140. Comparison of the $[NaO_7]$ polyhedra columns (*a*) in orthorhombic epididymite, (*b*) in monoclinic eudidymite (Pobedimskaya and Belov).

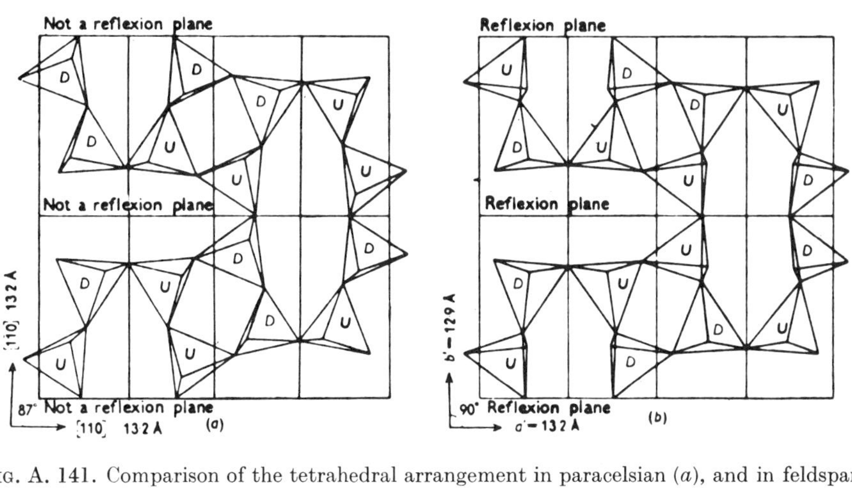

FIG. A. 141. Comparison of the tetrahedral arrangement in paracelsian (*a*), and in feldspar (*b*) (Smith).

fundamentally important factor, especially in view of the genetic conditions of these crystal phases (cf. Vol. III, Section B).[310] The detailed studies of F. Laves and J. R. Goldsmith,[311] partly based on the previous work of E. Spencer (1937) and A. Köhler (1941), definitely established the polymorphism in high- and low-temperature structures of the feldspars, as well in the alkali feldspars as in the plagioclases. Concerning the rather complex nomenclature of the different minerals and crystal phases here observed, the reader is referred to the more detailed classification given in Vol. III, Section B; we will emphasize here only the central problems of the order-disorder relations of the Al—Si distribution in the tetrahedral frameworks, remarking that in general the fundamental structures as given by W. H. Taylor *et al.* (1937, 1940) were not essentially changed.

196. The superstructure and short-range order in *anorthite* is expressed in the characteristic doubling of the *c*-axis relative to the K—Na feldspars; also the base-centering of the (*ab*) plane is lost, and additional reflections (first observed by Taylor) appear as compared with other feldspars. The degree of the diffuseness of such additional interference spots depends on the thermal history, and there are two different kinds of spots, diffuseness of which is affected differently by heat treatment. It is concluded, therefore, by Laves and Goldsmith that the Si^{4+} and Al^{3+} ions control in their order-disorder relations the first, the Ca^{2+} ions the second type of the interference spots. An appropriate analysis of the degree of diffuseness for the latter spots is, therefore, of great importance for the use of calcic plagioclases in geological thermometry[312].

197. For the *alkali feldspars* the study of the interrelations between the observed thermal history effects and the order-disorder distribution in the structures goes back to the observations of W. H. Taylor, S. H. Chao, and D. L. Smare (1939), and of O. F. Tuttle and N. L. Bowen (1949/50) on the structure of *perthites* and granitic-pegmatitic feldspars. Laves *et al.* more extensively observed the transitions between the different type phases (microcline, orthoclase, adularia, sanidine, albite, analbite) in their degrees of disorder, increasing with the temperature, and the delayed reversibility of these conversions, even over geological periods. The modern experience of structural dislocations and one-dimensional imperfections in crystalline structures (as taken from metal alloys) could be applied to these conversions

[310] Cf. the short introduction into this field, W. Eitel, "Structural Conversions," *Geol. Soc. Am. Spec. Pepers*, **66**, 1958, 80-119.

[311] We emphasize here especially *Acta Cryst.*, **7**, 1954, 465-472, on the long-range and short-range order in the structure of calcic plagioclases as a continuous and reversible function of temperature, and on the superstructure in anorthite, *Ibid.*, **7**, 1954, 131-132. Comments on the anorthite structure were contributed by H. D. Megaw, F. Laves, and J. R. Goldsmith, *Cursillos y confers. inst Lucas Mallada*, No. 8, 1961, 81-96; cf. Vol. III, B. ¶ 306.

[312] *J. Geol.*, **62**, 1954, 405-408.

with great success. The refined X-ray diffraction methods, especially the Buerger precession method, became an important help in establishing the phase equilibrium diagrams of the systems of the K-Na feldspars and also of the ternary feldspars, combined with their optical characteristics for diagnostic and genetic investigations. An example may be found in correlating the parameters and unit cell (reciprocal lattice) constants to the composition of the crystalline solutions in natural and synthetic feldspars, their unmixing reactions, and similar basic problems of physicochemical petrology.

198. For a more detailed study of the principles ruling the distribution of the $[SiO_4]$ and $[AlO_4]$ in the feldspars, W. Loewenstein's discussions[313] are essential, and even more especially those of G. W. DeVore,[314] based on terms of relative bond energies. The structural conditions of given feldspars require the distinction of polymerized chains of $[SiO_4]$ and $[AlO_4]$ tetrahedra, abbreviated here by the symbols Si—Si—Si... for pure $[SiO_4]$ unit ("quartz") chains, and mixed ("anorthite") chains Si—Al—Si—Al... In albite both kinds of chains occur parallel to the *b*-axis, bonded to one another, and parallel the *c*-axis an Al—Si-ordered array of the configurations

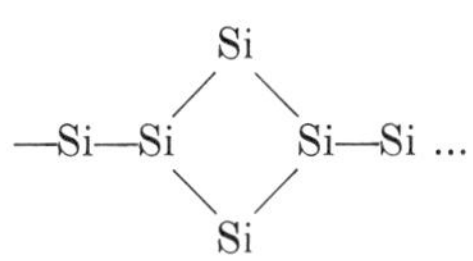

which alternate along the *b*-axis with modified anorthite configurations

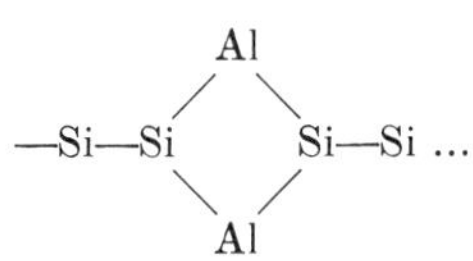

whereas for anorthite the ordering in the *c*-axis the Al—O—Al is characteristic, indicating the lack of disorder in the Al—Si positions. In the *a*-axis such bonds, however, are avoided.

199. A considerable progress in the structure determination of the alkali feldspars was recently possible by high-precision X-ray studies combined with 2- and 3-dimensional Fourier synthesis, first for a triclinic potassium feldspar (an intermediate microcline) by S. W. Bailey and W. H. Taylor,[315] and then for an essential refinement

[313] *Am. Mineralogist*, **39**, 1954, 92-96; this theoretical investigation concerns especially the problems in aluminosilicates like the feldspars and feldspathoids, but also problems of aluminates, the aluminum silicates, etc. and illustrates very well the reasons of the superstructure in anorthite.

[314] *Science*, **121**, 1955, 707-708; *Z. Krist.*, **107**, 1956, 247-264.

[315] *Acta Cryst.*, **8**, 1955, 621-632.

of the earlier structural data for sanidine. The microcline examined showed a partial order in the Al—Si distribution over the tetrahedral units, entirely confirming the previous assumptions and deductions of Laves and Goldsmith on the polymorphism in the system $KAlSi_3O_8$. Also, concerning the *covalent* share in the bonding mechanism in feldspars as aluminosilicates, the results of Bailey and Taylor give an important hint, with about a 50 per cent share of covalency and ionization, for the electron contents of Si, Al, and O atoms.

200. Further investigations of R. B. Ferguson, R. J. Traill, and W. H. Taylor[316] on the low- and high-temperature modifications of *albite* were also based on a refinement of the previous results of W. H. Taylor, J. A. Darbyshire, and H. Strunz (1934), by two- and three-dimensional Fourier synthesis. Concerning the Al—Si distribution over tetrahedral positions, it is important that in low-temperature albite one site $[Si_I(O)]$ contains all the Al^{3+} ions, whereas in high-temperature albite the Al and Si ions are distributed at random over all four different types of tetrahedral sites of the structure. Omitting here the detailed discussions on the temperature factor of the atoms, we emphasize concerning the Na^+ ions their distribution at random through the structure, in large cavities which in high-temperature albite are effectively larger than the sodium atom can fill. In low-temperature albite, on the other hand, the cavities are small and nearly enclose the Na^+. In microcline the large K^+ ions are fixed immovably in cavities of exactly adequate sizes.

201. Taylor's and Ferguson's discussions on the electrostatic charge balance in the albite structures, and in microcline or sanidine, brought about some items which do not appear in parallel with what had been provided in the theories of Laves *et al.*, on the maximum stability of fully ordered microcline. Against these consequences, W. S. Mackenzie and J. V. Smith[317] made considerable objections. They restrict the validity of charge balance factors to decide which tetrahedra are able to contain Al^{3+} in "maximum microcline," but not to determine the thermodynamic stability which is ruled by disorder entropy factors and also by inner strains. The incomplete filling of the cavities in high-temperature albite by Na^+ is the reason for the striking anisotropy by an elongation of the contour curves around the sodium center as seen in Fig. A. 142.

202. The monoclinic structure of *petalite*, $Li[AlSi_4O_{10}]$, was defined by A. Zemann-Hedlick and J. Zemann[318] as a layer type with $[Si_2O_5]_\infty$ tetrahedral sheets, of six-membered rings in double layers. The tetrahedra are distorted only a little. We pre-

[316] *Acta Cryst.*, 11, 1958, 331-348.

[317] *Acta Cryst.*, 12, 1959, 73-74. On the sodium content of microcline and the microcline-albite series see also recently F. Laves and J. R. Goldsmith, *Cursillos y confers. inst. Lucas Mallada*, No. 8, 1961, 81-96; cf. Vol. III. B. ¶ 306.

[318] *Naturwissenschaften*, 41, 1954, 477-478.

fer the classification of petalite as a framework structure in the definition of Zoltai, because of the interlinking by bridging tetrahedra between the sheets and chains. The angle Si—O—Si is variable between 150° and 160°, as a compromise between theoretically fully ionic bonding and covalent bonding requiring 180° and 140°, respectively. A projection on (010) is seen in Fig. A. 143. F. Liebau[319] refined the proposal of Zemann-Hedlick and Zemann, to angles Si—O—Si of 166° for one kind of O^{2-} atoms, and 150° to 153° for the other oxygen atoms.

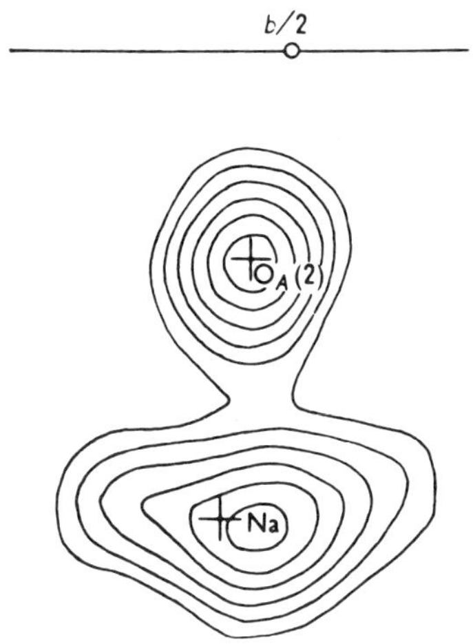

FIG. A. 142. Contour lines indicating the electron density distribution around Na atoms in the structure of high-temperature albite, projected along the [001] direction, with pronounced anisotropy along the y-axis (Ferguson, Traill, and Taylor). In low-temperature albite the contour curves are nearly circular.

203. The complex structure of the *scapolites*, especially of the carbonate containing *mejonite*, $Ca_4(CO_3)[Si_2Al_2O_8]_3$, includes the problem of the presence of the anion $(CO_3)^{2-}$, either in planar $[CO_3]$, or in tetrahedral $[CO_4]$ coordination. The analogous problem in *carbonate apatite*, $Ca_{10}[PO_4]_6(CO_3)$, was discussed by Th. Geiger,[320] by W. Kleber and B. Winkhaus,[321] and recently by D. McDonnel[322] for *dahllite*. Whereas Geiger and some authors with him deny entirely the possibility of the introduction of carbonate groups in phosphate (and equally in silicate) structures, and, therefore, explain the carbonate apatite as an intimate intergrowth of common (e.g., hydroxyl apatite) with calcite, McConnell defends the direct substitution of $[CO_3]$ (not $[CO_3OH]$) for $[PO_4]$, as in francolite and dahllite, further in fossilized dental enamel.

[319] *Acta Cryst.*, **14**, 1961, 399-406.
[320] *Schweiz. mineral. petrog. Mitt.*, **30**, 1950, 161-181.
[321] *Fortschr. Mineral.*, **28**, 1951, 175-202, especially p. 184.
[322] *Am. Mineralogist*, **45**, 1960, 209-216.

Carbonate anions cannot substitute for hydroxyl or fluorine in normal apatite. Applied for scapolites, McConnell's intepretation may speak for a possible introduction of $[CO_3]$ replacing $[SiO_4]$ in mejonite.

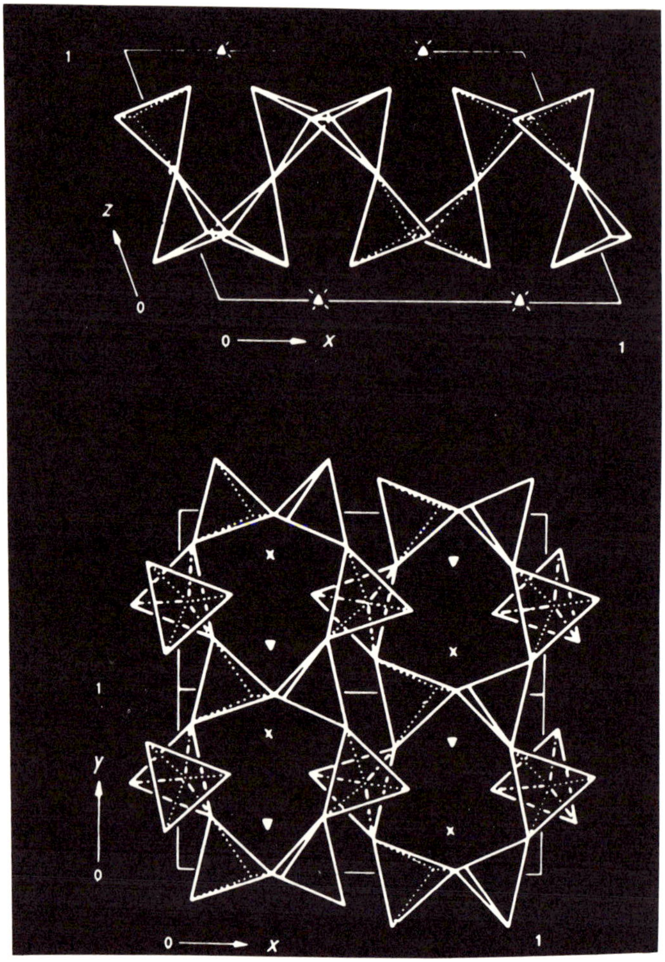

FIG. A. 143. Structure of Petalite, as a Framework Type, schematically projected on Planes (xz) and (xy) (Lieban).

204. Among the feldspathoid structures, the rare mineral *pollucite*, $Cs[AlSi_2O_6]$, is particularly interesting. Although the analyses of the natural occurrences mostly show a certain content in water which makes pollucite comparable to analcime and simple zeolites, the synthesis of pure anhydrous pollucite was confirmed by R. M. Barrer and N. McCallum.[323] They emphasized that a typical "sieve effect" of the

[323] *Nature*, 167, 1951, 1071.

large Cs^+ cations in the cavities of the framework structure is possible and impedes it from "vagabonding"; this phenomenon is observed, however, for the smaller cations Na^+ and K^+, and even for the large Rb^+, Tl^+, and NH_4^+ ions by an ionic mobility and exchangeability. This phenomenon is of the greatest importance in the zeolites which have similar specific cavity size structures (cf. Vol. IV, Section B). The volume conditions in the pollucite cavities make an exchange of the type $Cs^+ \rightleftarrows Na^+ + (H_2O)$ possible. An exchange of high-temperature $Li[AlSi_2O_6]$ with Cs_2SO_4 in aqueous solution, however, is always incomplete as V. E. Plyushchev[324] observed, whereas in the melt of Cs_2SO_4 the exchange at 950° to 1000°C is prompt. The synthetic pollucite has the parameter $a_0 = 13.646$ KX. The analogy of the pollucite structure with that of the zeolite analcime was fully described by St. Náray-Szabó (1938, 1942), as a tetrahedral framework of $[Si_4O_{12}]_\infty$ type, combined with $[AlO_4]$ groups.

205. In the large group of the *zeolites*, which all have typical framework structures, that of *analcime*, $Na[AlSi_2O_6].H_2O$, was early determined by W. H. Taylor (1930, 1938), who assumed a statistical distribution of the 16 Na^+ cations in the unit cell over 24-fold sites of the lattice. Against this assumption, I. R. Beattie[325] objected that other distributions are also justifiable, such as that of only 8 of the Na^+ ions in random distribution, the remaining 8 ions near the 2 Al^{3+} ions. Beattie also underlines the close relation of the analcime structure to that of leucite and pollucite, and the possibility of considerable variation of the crystallographic constants of analcime by ion exchange reactions. More recently, Pr. Saha[326] extended his structural studies to synthetic analcimes of varying chemical compositions and physical properties, as a function of the silica and water contents. The main structural features of all of the modified analcimes are seen in the presence of diagonal channels, surrounded by six-membered rings of $[SiO_4]$ tetrahedra not crossing one another, in a pseudocubic framework. The exchangeability of Na^+ by large cations like K^+, Rb^+, and Tl^+ as observed by R. M. Barrer[327] (cf. Vol. IV, Section B) is not easily compatible with Taylor's structural proposal, and as Beattie observed the exchange-modified analcime crystals are not cubic but more or less strongly anisotropic. Saha gives for the variability of the chemical compositions of analcimes the limits shown in the accompanying tabulation.

[324] *Zapiski Vsesoyuz. Mineral. Obshchestva*, **88**, 1959, 152-156; *Nauch. Doklady Vysshei Shkoly, Khim. i Khim. Tekhnol.*, 1959 (2) 284-288; *Chem. Abstr.*, **53**, 1959, 18708 d. — The analogous synthesis of $Rb[AlSi_2O_6]$ was successful, but the compound is tetragonal, with $a_0 = 13.36$ KX, $c_0 = 13.72$ KX, and a characteristic inversion at about 1060°C. Cf. V. E. Plyushchev, Yu. P. Simanov, and I. V. Shakimo, *Zhur. Neorg. Khim.*, **3**, 1958, 2133-2137.

[325] *Acta Cryst.*, **7**, 1954, 357-359.

[326] *Am. Mineralogist*, **44**, 1959, 300-313; see also D. S. Coombs, *Mineral. Mag.*, **30**, 1955, 699-708.

[327] *J. Chem. Soc.*, 1950, 2342-2350.

	Natrolite-related analcime $Na[AlSi_{1.5}O_5].0.75H_2O$	Ideal analcime $Na[AlSi_2O_6].H_2O$	Albite-related analcime $Na[AlSi_3O_8].1.5H_2O$
Ratio $Na:Al:Si:H_2O$	1 : 1 : 1.5 : 0.75	1 : 1 : 2 : 1	1 : 1 : 3 : 1.5
Distribution over 48 tetrahedral units	19.2 Al^{3+} 26.8 Si^{4+}	16 Al^{3+} 32 Si^{4+}	12 Al^{3+} 36 Si^{4+}
Distribution over 24-fold sites	19.2 Na^+	16 Na^+	12 Na^+
Distribution over 16-fold sites	14.4 (H_2O)	16 (H_2O)	18 (H_2O)

206. It is, therefore, evident that the natrolite-related extreme composition is compatible with a position of the Na^+ ions and water molecules in 24- and 16-fold positions, as Taylor's structure postulates, but it is impossible to locate the 18 water molecules of the albite-related composition in the 16-fold positions alone. This isotropic, fully cubic modification of analcime can only be understood if the two excess water molecules enter vacant positions normally occupied by Na^+ ions. The distortions of the structure by the larger water molecules is probably too small to be detected by X-ray diffraction.

207. The structure of *mordenite*, $Na[Si_5AlO_{12}].3H_2O$, is orthorhombic, with eight molecules in the unit cell ($a_0 = 18.13$ Å., $b_0 = 20.49$ Å., $c_0 = 7.52$ Å.), according to recent determinations by W. M. Meier,[328] using three-dimensional Patterson analysis. The framework of the zeolite is particularly remarkable because it is built up from five-membered tetrahedral rings in infinite columns, as shown in Figs. A. 144 and 145. The numbers of 4-, 5-, 6-, and 8-membered loops of tetrahedra per unit cell are 4, 48, 15, and 16, respectively, corresponding to the minimum for $n = 5$ or 6 for the most stable rings (of lowest relative energy in the isolated ring models of Zoltai and Buerger, cf. A. ¶ 27). Mordenite, therefore, can be expected to have a high stability. The fibrous habitus is well illustrated in this structure principle. Remarkable are the wide structural channels with a diameter of 6.6 Å. which show the suitability of such a structure as a two-dimensional molecular sieve (Vol. IV, Section B). The channels are oriented parallel [001] and are interconnected by smaller openings of 2.8 Å. oriented parallel [010]. Meier demonstrated that the ideal molecular sieve action of the 6.6 Å. channels is considerably reduced by structural imperfections (stacking faults as shown in Fig. A. 146), which interrupt the full aperture of the

[328] *Z. Krist.*, **115**, 1961, 439-450. The fibrous material is often called *ptilolite*, a word which should be eliminated from current use. A confusion of mordenite and *clinoptilolite* or silica-rich heulandite is often found in the literature and should be considered as a possibility in every case.

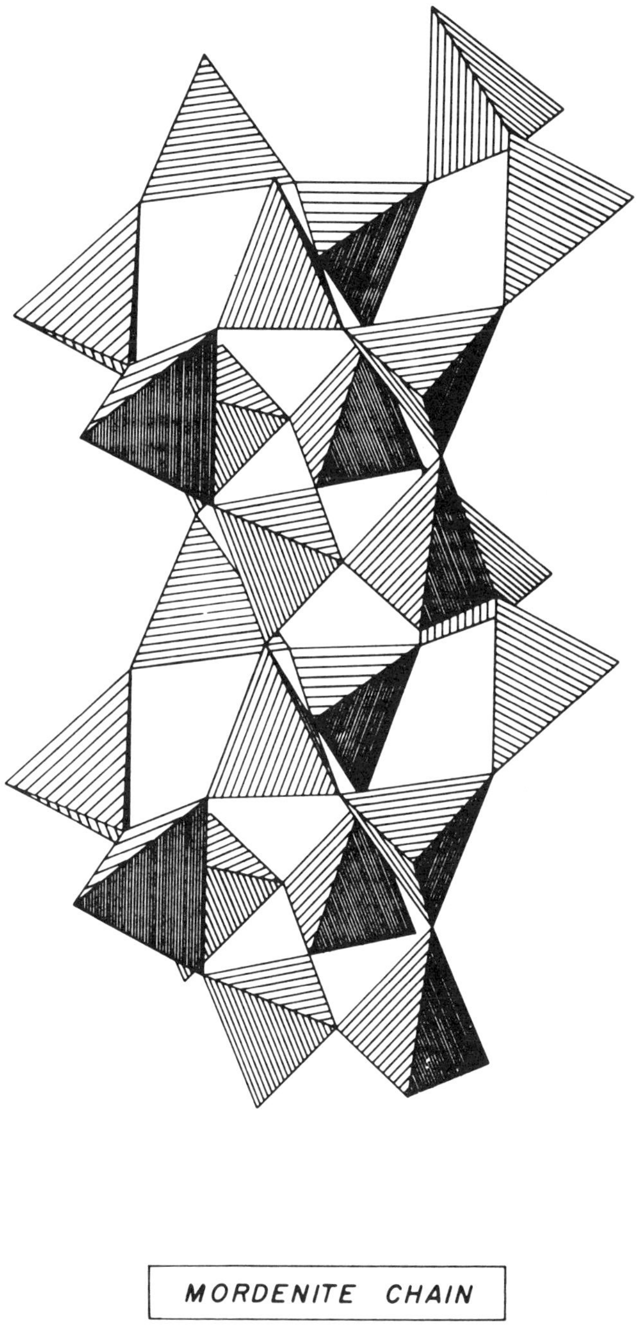

FIG. A. 144. Tetrahedral columns parallel to the c-axis in the structure of mordenite (Meier).

channels, lowering it to an effective pore diameter of only 4 Å. The distribution of the Al^{3+} and Si^{4+} cations in the framework is with a high degree of probability ordered; four of the Na^+-ions per unit cell are located in fixed positions, the others at random. The similarity of the mordenite structure with that of heulandite is striking. Both zeolites are often associated in nature.

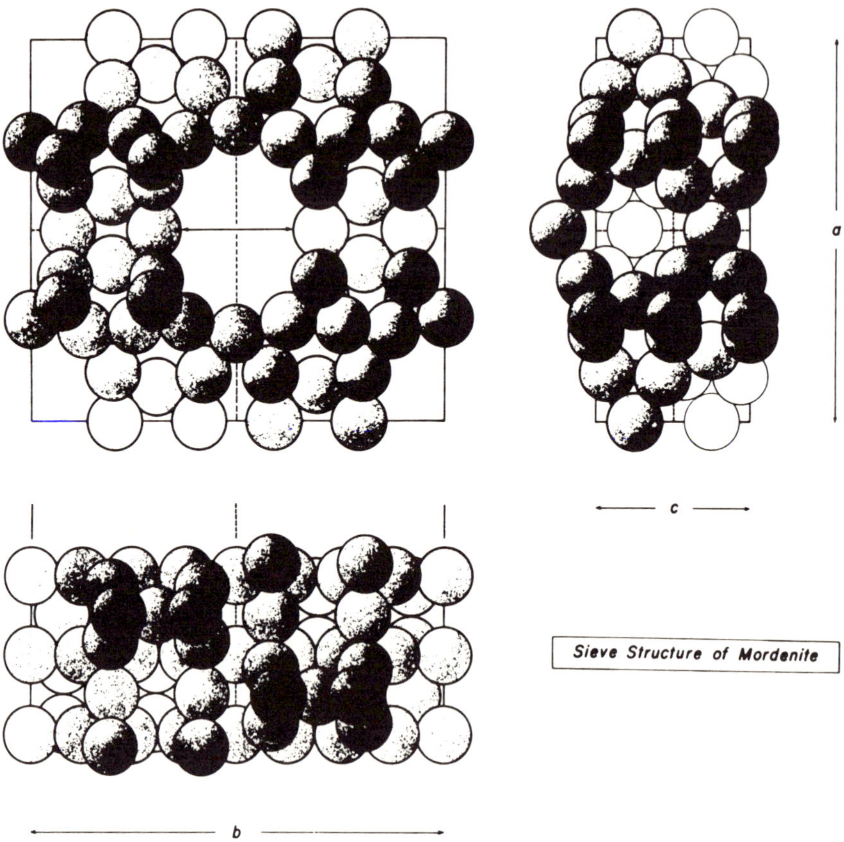

FIG. A. 145. Sieve structure of mordenite (Meier). The small cations Na^+, Al^{3+}, and Si^{4+} are not shown.

208. In close relation to mordenite is the monoclinic zeolite *dachiardite*, $(K_2, Na_2Ca)_{2.5}[Si_{19}Al_5O_{48}].14H_2O$, as described in its structure by Gl. Gottardi.[329] These

[329] *Periodico Mineral.* (*Rome*), **29**, 1960, 183-191; a preliminary note in *Ibid.*, **29**, 1960, 103-108 (S. Bonatti and Gl. Gottardi); determination of the dachiardite structure by G. Gottardi and W. M. Meier, *Proceed. Third. General Meetg., Internation. Mineral. Association, Washington D. C.*, April 1962, 4 pp.

minerals are related as polytypes by the identity of z (morden.) = y (dachiard.). In projection on (010) is a (dachiard.) = 18.75 Å., parallel to a (morden.) = 17.95 Å., and b (morden.) = 20.35 Å. Dachiardite is a structural twin phase with individuals of (001) as twin face. This polytypism is, therefore, in principle the same as that is of zoisite and clinozoisite (T. Ito, 1950).

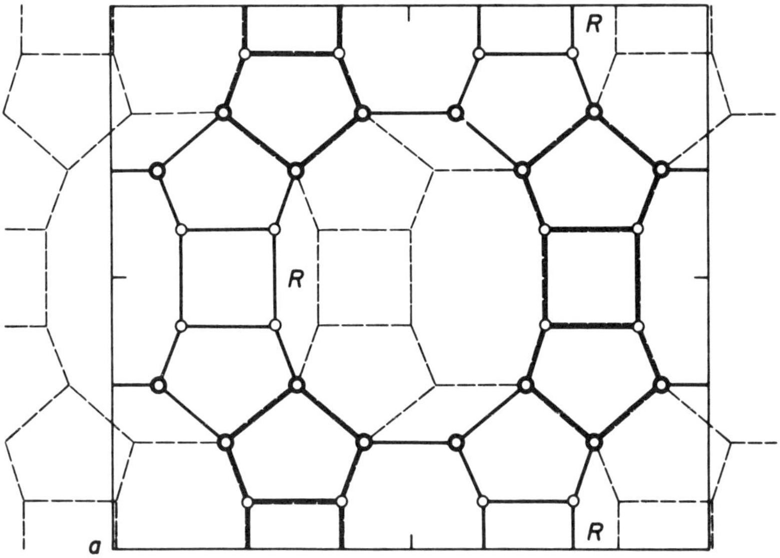

FIG. 146. Probable stacking disorder in the structure of mordenite, projected onto (001), showing exclusively the positions of the Si^{4+} and Al^{3+} ions (Meier). The displaced portion of the structure which reduces the free diameter of the channels parallel c is outlined by broken lines.

209. The apparent monoclinic structure of *heulandite*, $Ca[Al_2Si_7O_{18}].6H_2O$, is only a pseudosymmetry according to U. Ventriglia;[330] the mineral is triclinic-pedial, and without any centers of symmetry. In (001) appear characteristic groupings of rings with five $[SiO_4]$ tetrahedra, elongated parallel [010], in (100) chains of $[SiO_4]$ tetrahedra with alternating apexes in the plane of perfect cleavage (010). In (010) appear rings of six $[SiO_4]$ tetrahedra, elongated parallel [001]. This principle of the framework aspects distinguishes heulandite from chabazite in which quadratic configurations occur which are absent in heulandite. Interlinking tetrahedra form hexagonal rings in (110), and pentagonal and octogonal rings in (001). The framework shows channels of hexagonal cross-section parallel (001). Nothing is known on the degree of order of the Al^{3+}–Si^{4+} distribution, but the Ca^{2+} ions (and substituting Na^{+}) are assumed to be at random in the channels like the water molecules.

[330] *Periodico Mineral.* (*Rome*), **24**, 1955, 49-83.

210. W. M. Meier[331] re-examined the structure of *natrolite*, $Na_2[Si_2Al_3O_{10}].2H_2O$, as first proposed by L. Pauling (1930) and W. H. Taylor (1934). In general the previous interpretation could be confirmed by Fourier synthesis concerning the tetrahedral framework skeleton. Characteristic is the formation of hydrogen bonds of 2.85 and 2.99 Å. in length, bridging the water molecules with O^{2-} in the Al–Si tetrahedral framework. Natrolite has rather wide open channels in a complex intersecting system. The characteristic base exchange and dehydration reactions take place with a high degree of probability, however, *not* in the much narrower (diameter 2.08 Å.) channels parallel to the fiber axis, as Taylor assumed, but in the internal crossing systems of communicating channels with a minimum free diameter of 2.60 Å. These dimensions are too small to develop typical molecular sieve qualities, especially since the tetrahedral chains are able to rotate somewhat around the *c*-axis.

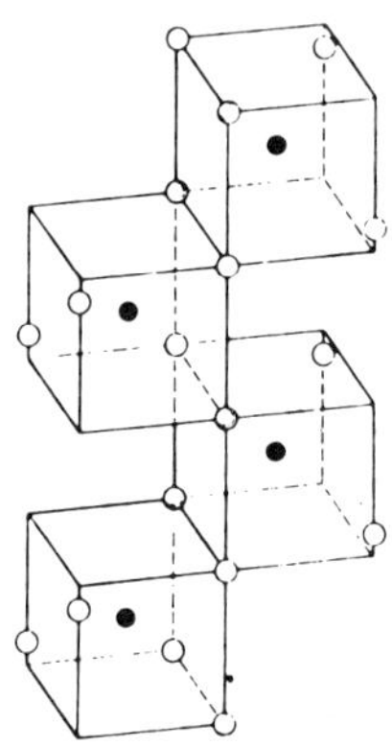

FIG. A. 147. Arrangement of the Na^+ ions relative to O^{2-} anions and water molecules in the structure of natrolite (Meier).

The distribution of the Al^{3+} and Si^{4+} cations in the framework is ordered, as Pauling and Taylor had assumed; the ratio Al/Si = 2 : 3 is fixed; the tetrahedra are slightly distorted. For the arrangement of the Na^+ ions and water molecules Meier gives the Fig. A. 147.

211. For *gismondite*, $Ca[Al_2Si_2O_8].H_2O$, as first described in its unit cell by O. Kraus (1939), K. Fischer and H. Kuzel[332] propose a different orientation of the crystallographic axes in a pseudotetragonal, monoclinic symmetry, but no structural details are given. The synthesis of a Na-free *thomsonite*, $Ca_3[Al_6Si_4O_{20}].6H_2O$, by J. R.

[331] *Z. Krist.*, **113**, 1960, 430-444.

[332] *Naturwissenschaften*, **45**, 1958, 488.

Goldsmith[333] is remarkable because here the unusual ratio $Al^{3+}/Si^{4+} = 1.5:1$ appears. The synthetic material was too fine-grained for a thorough structure determination.

212. Perhaps, the structurally best-investigated group of zeolites is that of the *chabazite* family ("rhombohedral" or less justifiable "cube-shaped" zeolites). K. Fischer and H. O'Daniel[334] gave fundamental data for faujasite, gmelinite, and herschelite. It is particularly important that intimate intergrowths occur between chabazite and gmelinite, because of their great similarity in the Si–Al framework structure, which can be related by gliding in the (0001) plane. Its effect can be observed in the characteristic rotation interference diagrams. More detailed data on chabazite, gmelinite, and levynite were also given by W. Nowacki, M. Aellen, and H. Koyama[335] who confirmed the frequent intergrowths with chabazite. We may also briefly refer to the similarity of the structures of the chabazite-gmelinite family with that of *cancrinite.* This latter mineral has single hexagonal rings in its structure, whereas chabazite and gmelinite have double rings.[336]

213. I. R. Beattie[337] rediscussed the *chabazite* structure as a defective phase because there is no space free to accomodate larger cations to be exchanged for the original smaller cations (Na^+ and Ca^{2+}) exactly in the theoretical positions, and a displacement of the ions must occur. The distortion by larger and smaller monovalent and divalent cations underlies the conditions of minimum potential energy requirements. Such complex structures as ion-exchanged forms of chabazite are under Coulombian cation-cation repulsion actions to reach final equilibria of distribution. A definite proposal for the structure of chabazite, $(Ca,Na_2)[Al_2Si_4O_{12}].6H_2O$, was made by W. Nowacki, H. Koyama, and M. H. Mladeck.[338] The preliminary structure as first indicated by J. Wyart (1933), postulates unlikely Si—O and Al—O distances and channels with molecular sieve qualities of only 3 Å. The structure derived by three-dimensional Fourier analysis is rhombohedral ($a_{rhb.} = 9.35$ Å., and $\alpha = 94°37'$). Two kinds of framework "cages" are described, one in hexagonal prism form, the other as a large polyhedron limited by two hexagons, six squares, and six octagons. Such a polyhedron contacts six equal surrounding polyhedra. The very loose framework model makes fully understood a possible molecular dif-

333 *Mineral. Mag.*, **29**, 1952, 952-954.

334 *Naturwissenschaften*, **43**, 1956, 348. Br. Mason, *Am. Mineralogist*, **47**, 1962, 985-987, emphasizes that *herschelite* is a valid mineral species, namely sodium chabazite, $Na_2[Al_2Si_4O_{12}].6H_2O$.

335 *Schweiz. mineral. petrog. Mitt.*, **38**, 1958, 53-60. On the intracrystalline channels in the structure of *levynite* see R. M. Barrer and I. S. Kerr, *Trans. Faraday Soc.*, **55**, 1959, 1915-1923.

336 Cf. Sh. Kozu and K. Takane, *Proc. Imp. Acad. (Tokyo)*, **9**, 1933, 56-59, 105-108, on the structure of cancrinite.

337 *Nature*, **172**, 1953, 999-1000.

338 *Experientia*, **14**, 1958, 396.

fusion through the structure in all directions, especially through the octagons with an opening of 4 Å. as the "sieve mesh."

214. These results of Nowacki *et al.* are in good agreement with the determination of the chabazite structure by L. S. Dent and J. V. Smith[339] who give in Fig. A. 148 an illustration of the two sets of "cages." The positions of the Ca^{2+} ions and the water molecules are irregular, making large vibrations of both possible. In spite of its surprisingly regular aspect, the chabazite structure is highly disordered, consistent with the ease of ion exchange and molecular sieve (sorption) qualities.

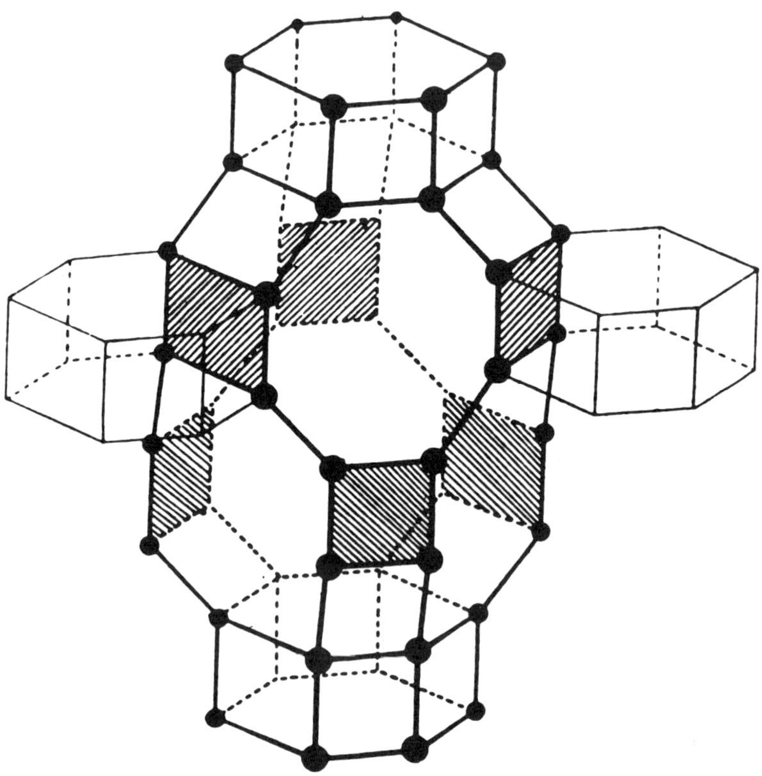

FIG. A. 148. Positions of the Al^{3+} and Si^{4+} ions in the framework skeleton of chabazite (Smith and Dent). The shaded polygons are the points of attachment of the hexagonal-prismatic "cages" of which four are indicated by broken outlines.

215. Particularly fascinating is the structure of the rare zeolite *faujasite*, which in its natural crystals has the composition $NaCa_{0.5}[Al_2Si_5O_{14}].10H_2O$, with not less

[339] *Nature*, **181**, 1958, 1794-1796; *Bull. Geol. Soc. Am.*, **70**, 1959, 1678; *Ibid.*, **71**, 1960, 1980.

than 192/7 formulae units in its cell, as W. Nowacki and G. Bergerhoff[340] established. The Si—Al skeleton is characterized by "cages" of cubooctahedral shape as these are also known from the ultramarines, and similar to those of chabazite. The cages are arranged like the carbon atoms in the diamond lattice, interlinked by bridging O^{2-} anions (192 tetrahedra corner ions). The alkali ions and water molecules are situated in the wide cavities and channels of 6 to 7 Å. width which run through the structure parallel to the directions [110]. This structure fully explains the cation exchange and molecular sieve properties assigned to natural (and to synthetic) faujasite, as it was studied by F. M. Barrer, W. Buser, and W. F. Grütter[341] (cf. Vol. IV, Section B). G. Bergerhoff, W. H. Baur, and W. Nowacki[342] emphasize that silver-faujasite prepared by cation exchange with sodium-faujasite is only isostructural with normal faujasite concerning the Si–Al framework, but that the silver ions are able to occupy other positions than those of Na^+. The characteristic channels parallel [110] are surrounded by rings of twelve Si^{4+} and O^{2-} ions. They have an elliptical cross section, the large axis of which in $[1\bar{1}0]$ is 10.4 Å., the short axis parallel [001] is 8.9 Å. The opening of 10.4 Å. can be followed on a zigzag path alternately in [111] and $[11\bar{1}]$.

216. As a synthetic "Linde A" zeolite, with highly developed molecular sieve properties, faujasite was prepared and determined in its structure by D. W. Breck, W. G. Eversole, R. M. Milton, T. B. Reed, and T. L. Thomas,[343] with the chemical compositions $Na_{12}[(AlO_2)_{12}(SiO_2)_{12}].27H_2O$, and the unit cell dimension $a_0 = 12.32$ Å. We discuss this important industrial product more extensively in Vol. IV, Section B. The diameter of the cavities is given by those authors as 11.4 Å., that of the circular openings as 4.2 Å. The dehydrated crystals have voids of 45 volume per cent. The crystal habit is for individuals of the 1 to 5 μ size ((110)); for larger crystals also ((100)), with penetration twinnings. Characteristic is the ample possibility of "ion exchange tailoring," i.e., a systematic variation of the geometric constants by cation replacement. The apparent pore size of 4.2 to 4.75 Å. and its variation up to 5 to 5.8 Å. makes the synthetic faujasites highly useful in the selectivity of their molecular sieve qualities for practical purposes.

[340] *Verhandl. schweiz. naturforsch. Ges.*, Basel 1956, 115-116. An explicit comparison of the structures of chabazite, faujasite, and gmelinite was given by G. Bergerhoff, H. Koyama, and W. Nowacki, *Experientia*, **12**, 1956, 418; see also *Fortschr. Mineral.*, **35**, 1957, 44. On synthetic faujasite see also R. M. Milton, British Patent No. 777,233 (December 21, 1954); U.S. Patents 2,882,243 and 2,882,244 (April 14, 1959).

[341] *Helv. Chim. Acta*, **39**, 1956, 518-530.

[342] *Neues Jahrb. Mineral. Monatsh.*, 1958, 193-200.

[343] *J. Am. Chem. Soc.*, **78**, 1956, 2338; 5963-5971; for the structure, cf. pp. 5972-5977 (T. B. Reed and D. W. Breck, with complete Fourier analysis). See also D. W. Breck and J. V. Smith, *Sci. American*, **200**, 1959, 85-94.

217. In the same way as the mineral faujasite can vary in rather wide limits in its relative silica contents, as Barrer, Buser, and Grütter (see above) had emphasized, it is also possible to raise the silica content of synthetic faujasites. "Linde X," a synthetic zeolite was prepared by D. W. Breck, E. M. Flanigen, and T. B. Reed[344] with a composition of the type $Me_{56+x}[(AlO_2)_{56+x}(SiO_2)_{136-x}.M\ H_2O$ ($M = 260$-264; $x = 0$ to 38), with $a_0 = 24.65$ to 24.95 Å., cavities of 13 Å. (therefore also called "zeolite 13 A"), and channels with rings of 8 to 9 Å. opening. The dehydrated material has voids of 51 volume per cent, with a corresponding high adsorption and molecular sieve activity. The extreme composition $Na_{94}[(AlO_2)_{94}(SiO_2)_{98}].M\ H_2O$ (for $x = 38$) is nearly a "nepheline hydrate," $NaAlSiO_4.nH_2O$, whereas in $Na_{56}[(AlO_2)_{56}(SiO_2)_{136}].M\ H_2O$ ($x = 0$) the ratio $Si^{4+} : Al^{3+}$ is $= 2.38 : 1$ (the highest for natural faujasite is 2.33 to 2.50 : 1). Linde X zeolite is, therefore, very near to natural faujasite.

218. Also L. Broussard and D. P. Shoemaker[345] determined with a three-dimensional Fourier synthesis molecular sieves of the type "4A" ($NaAlSiO_4.xH_2O$), "5A" ($Na_{1/3}Ca_{2/3}AlSi_4.xH_2O$), and "13X" ($NaAlSi_{2.8}.xH_2O$ in their hydrated forms, with 25 to 33 weight per cent water. All three types have in common the "sodalite" block unit containing 24 (Si, Al) ions, interconnected with 36 O^{2-}. In "4A" and "5A" these are arranged to simple cubic arrays, with each sodalite unit connected to its neighbor by four bridging O^{2-} anions. In "13X" the diamond array of faujasite is typical, each sodalite unit being connected with its neighbors by six bridging O^{2-} atoms. The cations are mostly near the centers of six-membered rings of O^{2-} in the cavity channels (cf. Fig. A. 149, and those in Vol. IV, Section B).

219. In close relation to the chabazite family zeolites is the rare hexagonal mineral *erionite* for which H. Strunz[346] assigned the crystallochemical formula $(Na,K,Ca_{1/2})_3[Al_3Si_9O_{24}].8H_2O$, in comparison with chabazite, $(Ca,Na_2)_3[Al_4Si_4O_{24}].12H_2O$, and levynite, $Ca_6[Al_4Si_8O_{24}].12H_2O$. Peculiar is the occurrence of erionite in wool-like filaments and bundles, with extreme elongation parallel to [0001], $a_0 = 13.26$ Å.,

[344] Am. Chem. Soc., Div. of Phys. Chem., Abstracts of Papers presented, Chicago, Illinois, Sept. 7-12, 1958, 5S-6S.

[345] *J. Am. Chem. Soc.*, **82**, 1960, 1041-1052. Identical results concerning Linde A and Linde X zeolites as described above were also found by M. M. Dubinin, *Doklady Akad. Nauk S.S.S.R.*, **138**, 1961, 866-869, with extensive calculations of the volume conditions, in agreement with the data of R. M. Barrer, F. W. Bultitude, and T. W. Sutherland, *Trans. Faraday Soc.*, **53**, 1957 1111-1123. The interpretation of the structure of *phillipsite*, $(K_xNa_{1-x})_5[Si_{11}Al_5O_{32}].10H_2O$, as given by R. M. Barrer, F. W. Bultitude, and I. S. Kerr, *J. Chem. Soc.* 1959, 1521-1528, is not correct. It was revised by H. Steinfink, *Acta Cryst.*, **15**, 1962, 644-651; cfr. Vol. IV, Section B), with large channels parallel *a* and *b*, the first set octagonally shaped with an open passage of 12 Å². cross section, the second set of rectangular section of about 9 Å². Barrer *et al.* included in their studies also the structures of *harmotome* and *wellsite*.

[346] *Naturwissenschaften*, **47**, 1960, 59.

and $c_0 = 15.12$ Å. Ll. W. Staples and J. A. Gard[347] gave the formula $(Ca,Mg,Na_2,K_2)_{4.5}[Al\text{-}Si_{27}O_{72}].27H_2O$ for erionite and found the same unit cell dimensions from electron diffraction single crystal diagrams. The orthorhombic interpretation of these and of X-ray rotation patterns (with $a_0 = 13.3$ Å., $b_0 = 22.4$ Å., $c_0 = 15.1$ Å.)

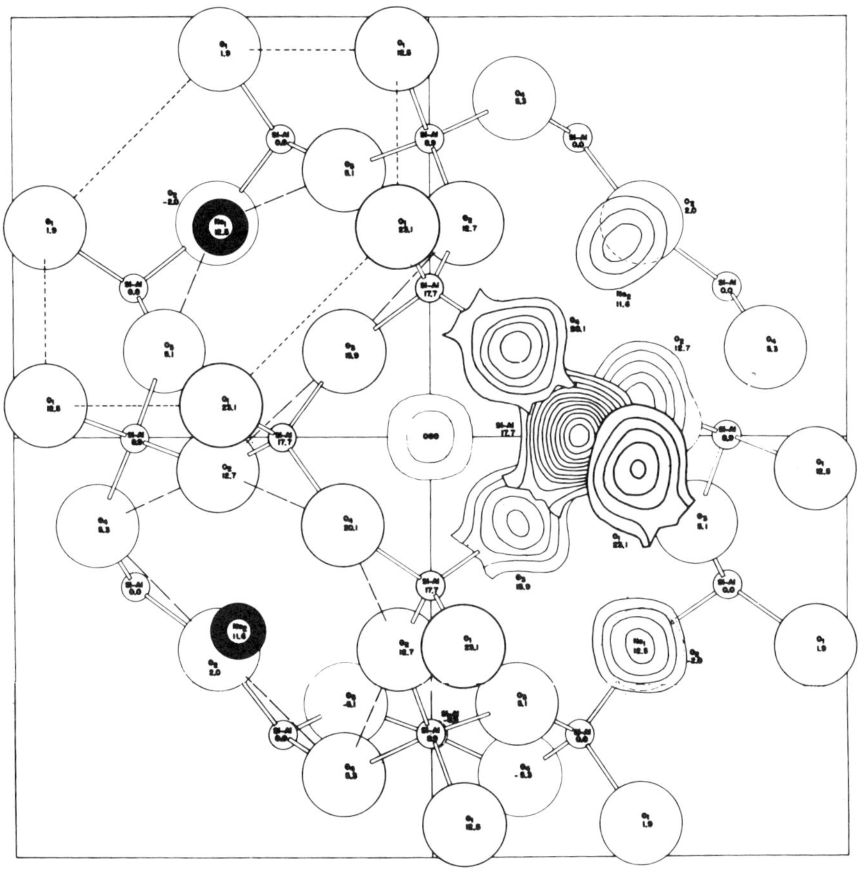

FIG. A. 149. Sodalite unit of the synthetic "13 X" molecular sieve zeolite (faujasite type) showing sections of three-dimensional Fourier synthesis (Broussard and Shoemaker).

is geometrically equivalent to the hexagonal cell units mentioned above, but the single crystal data are unequivocal, speaking for the hexagonal structure with 36 $[(Si, Al)O_4]$ tetrahedra in the cell. Characteristic of the structure is the occurrence of typical 3-repeat-unit columns $[Si_{2+1}O_9]_\infty$ parallel to c, indicating the pseudo-

[347] *Mineral. Mag.*, **32**, 1959, 261-281. The presence of 0.46 per cent NH_4^+ ions in the composition of erionite is very remarkable, characterizing this mineral as a natural ammonium zeolite.

period $c/2 = 7.56$ Å. as observed, further six-membered rings of tetrahedra stacked either in the mode *A* or *B* of Fig. A. 150, with *B* as the most probable arrangement. The (001) projection is given in Fig. A. 151. There is a "window" of 2.5 Å. in diameter as the free space at the center of the cavities, and other elliptical openings of 3.5 and 4.7 Å. free dimensions as seen in the $(10\bar{1}0)$ projection, with continuous channels in the direction $[11\bar{2}0]$, and others in $[11\bar{2}1]$. In levynite similar six-membered ring arrangements may occur. The erionite type structure is definitely more highly complex than that of chabazite, a fact which may explain its relative rarity in comparison with the more simple chabazite.

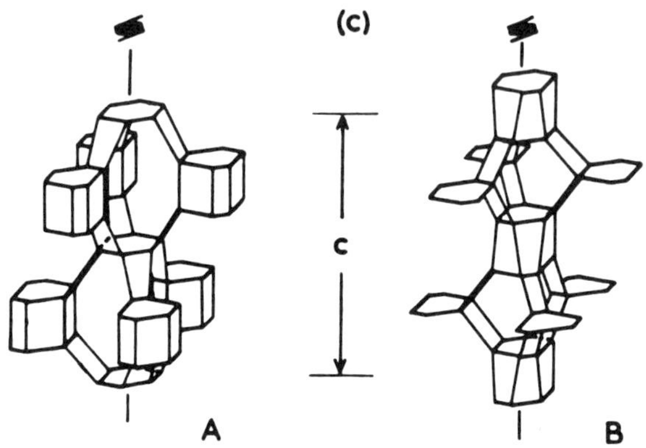

FIG. A. 150. Stacking of six-membered tetrahedral rings in the framework structure of erionite (Staples and Gard).

220. W. B. Kamb and W. C. Oke[348] found a new zeolite called *paulingite*, in rhombic dodecahedron forms, with the unusually large unit cell parameter $a_0 = 35.10$ Å. The chemical composition is that of a calcium-potassium zeolite with subordinate Ba^{2+} and Na^{+} contents, the ratio $Si^{4+} : Al^{3+}$ being about 3 : 1. The space group is probably $O_h^9 = I\,m3m$. The cell is the largest ever observed for an inorganic compound including intermetallic complex phases, sulfides, or sulfosalts, and is surpassed only by unit cells of some proteins. It contains not less than 580 Si and Al atoms, 1160 atoms of oxygen, 690 molecules of water, 120 cations of alkalies and alkaline earths. Paulingite occurs associated with erionite in acicular prismatic crystals.

221. Another very remarkable zeolite mineral, unique in its content in magnesium besides sodium and potassium, is the orthorhombic *ferrierite* with the crystallochemical formula $(Na,K)_4Mg_2[Al_6Si_{30}O_{72}](OH)_2.18H_2O$, which was characterized by L. W. Staples[349] as a tektosilicate with a typical framework structure, in spite of

[348] *Am. Mineralogist*, **45**, 1960, 79-91.

[349] *Am. Mineralogist*, **40**, 1955, 1095-1099.

its tabular habitus and cleavage parallel (100). Particularly interesting is the presence of additional hydroxyl groups, comparable to those of Cl^- and other anions in scapolite and cancrinite.

ERIONITE

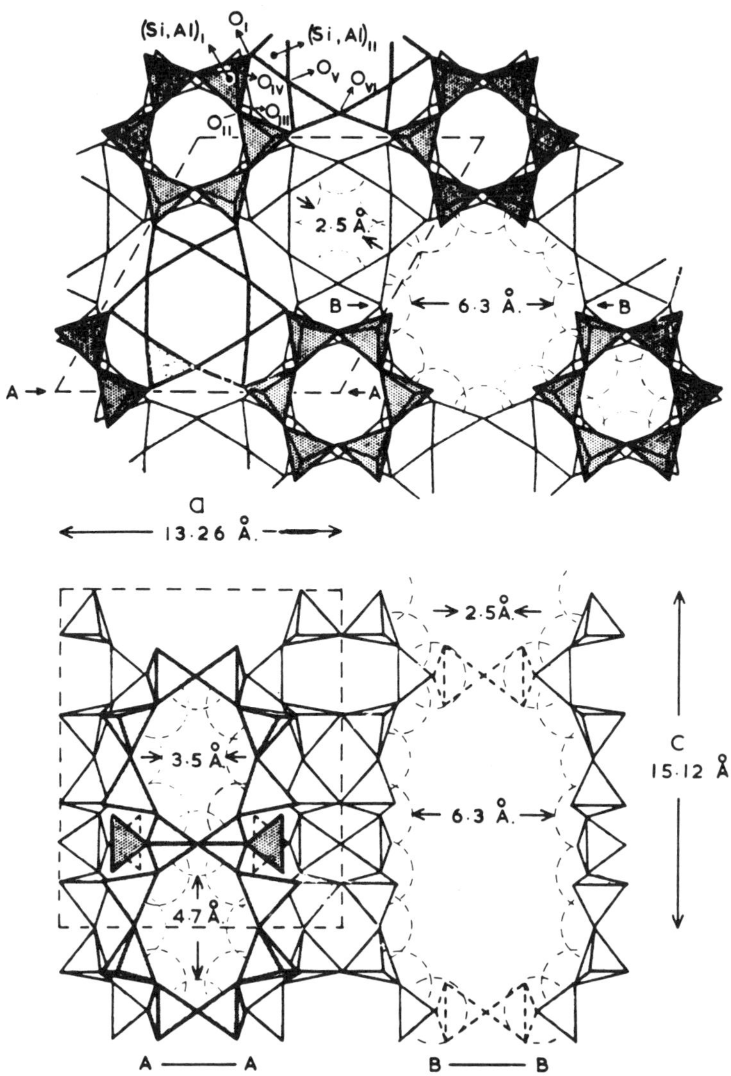

FIG. A. 151. Projections (0001) and (10$\bar{1}$0) of the structure of erionite, to show the characteristic arrangement of the tetrahedral frameworks (Staples and Gard).

CRYSTALLOCHEMICAL ASPECTS OF ISOMORPHISM AND ISOSTRUCTURAL RELATIONS OF SILICATES WITH OTHER COMPOUNDS

222. The fundamental importance of the phenomenon of isomorphism for *geochemical* investigations in the occurrence and distribution of elements in rocks and silicate minerals, which was so impressively discussed in the classical work of V. M. Goldschmidt, was emphasized anew by M. Vendel[350] who studied the atomic and ionic substitution in minerals and especially in silicates from the general viewpoints of the cationic radii relations, the electronegativity, and the ionization potentials. Vendel developed a "substitution index," H, to characterize the ability of a given cation to substitute for other cations, in the expression

$$H = (r_n^3/r_k^3) \times (I_n/I_k) \times (E_n/E_k)$$

(where r is the ionic radii, I the ionization energies, E the electronegativity, of the larger cation, n, and of the smaller cation, k, respectively), in a pair of atoms relating to each other. If the covalent bonding (cf. A. ¶ 9 ff.) prevails, the atomic volume must be referred to the equation for the index H. Extensive tabulations of this index for the most important cation groups illustrate the application of this parameter for the characterization of the substitution rules of rock-forming minerals, sulfide ores, etc., including the Goldschmidt rules of admission and capturing, as for the ions of different valence numbers, and of coupled substitutions.

223. In continuation of his previous theoretical calculations and deductions of the ionic radii, the molecular and ionic refraction, and polarizability, E. Kordes[351] discussed the ions of noble gas configuration in detail, especially for alkali halides and alkaline earth chalcogenides, for the case of reciprocal relations, and under the assumption of rigid radii values, but also for SiO_2 (with purely ionic radii, without any correction for covalent shares in bonding) and grossularite, $Ca_3Al_2Si_3O_{12}$. Kordes' results are in close relation to the discussions of Th. Neugebauer[352] on the mutual polarization effects of ions in coordination structures, not only of simple alkali halides, but also for the ZnS and NiAs types. Problematic factors in the use of ionic radii calculations are critically discussed by Br. Eckstein,[353] concerning the deviations

[350] *Acta Geol. Acad. Sci. Hung.*, III, 1955 (1-3) 245-300; the calculations are based on the lattice energy formulae of A. A. Kapustinskiĭ and the EK values of A. E. Fersman, in the theory of the so-called energy index of the elements (cf. *Compt. rend. acad. sci. U.R.S.S.*, **2**, 1935, 266-269, 564-566).

[351] *Naturwissenschaften*, **39**, 1952, 488-489; *Z. Elektrochem.*, **59**, 1955, 551-560, 927-932, 932-938; *Fortschr. Mineral.*, **34**, 1956, 51.

[352] *Naturwissenschaften*, **40**, 1953, 18.

[353] *Fortschr. Mineral.*, **36**, 1958, 62-63.

from the rule of additivity of the radii which are interpreted not only as deformation phenomena but as limits of the applicability of elementary models for the quasi-geometric arrangements of atomar units in structures and the very questionable concept of the "shape" of such units.

224. New data for the ionic refractions were given by G. B. Bokiĭ and S. S. Batsanov,[354] including ions of non-noble gas configuration, illustrated for carbonates, sulfates, wolframates, molybdates, and the pair Pb_2SiO_4/Ca_2SiO_4, in good agreement with the postulates of K. Fajans and E. Kordes for the ionic refractions and their deformation effects. The strongly polarizing Mg^{2+} cation, however, brings about distinct anomalies from the common additivity rules of the calculation of molecular refractions for salts with complex anions. The application for silicates was demonstrated by the same authors.[355] Deviations, ΔR, are particularly striking if the strongly polarizing B^{3+} ion in $[BO_4]$ groups is introduced in silicates, whereas for aluminosilicates the refractometric measurements are reliable and may even be used as a control of the coordination state of the Al^{3+} ion in such compounds. An exception is observed in celsian, $BaAl_2Si_2O_8$, which shows with $\Delta R = 0.90$, a surprisingly high deviation for both coordination principles, $[AlO_4]$ and $[AlO_6]$, from the theoretical values. For sillimanite ΔR even $= 0.93$ if $[AlO_4]$ alone is calculated; $\Delta R = 1.87$ if $[AlO_6]$ alone is calculated. Other remarkable examples are given for $LiAlSiO_4$ and $LiAlSi_2O_6$; the latter shows prevailing $[AlO_4]$ in the α-(high-) modification and prevailing $[AlO_6]$ in the α' (β)- (low-) modification. The possibility of distinguishing (OH^-) groups and water molecules is also remarkable. Thus, hillebrandite is characterized as an orthosilicate with OH—O hydrogen bonds to $[SiO_4]$ as in gypsum.

225. A. S. Povarennykh[356] gave highly valuable tabulations of the effective ionic radii of the most important elements in rock-forming minerals, in comparison with the classical data of Goldschmidt and L. Pauling for the same, with corrections for covalent bonding shares. W. Noll[357] recommends in such and similar calculations that one not overestimate the influence of covalent bonding. In the calculation of the interatomic distance of siloxane bonds in silicate anions, silicoorganic compounds, or in silicones it is rather irrelevant, whether or not purely ionic, or "corrected" covalent-ionic conditions of bonding are taken as a basis. The siloxanes of both groups are mesomeric in their bonding mechanisms and may therefore be interpreted equally well from both viewpoints (cf. A. ¶ 9 ff.).

226. The fundamental relations of *electronegativity* (in kcal/gram atom) of the cationic elements in inorganic compounds with the ratio of covalent bonds to ionic

[354] *Kristallografiya*, **1**, 1956, 81-89.
[355] *Zapiski Vsesoyuz. Mineral. Obshchestva*, **86**, 1957, 421-431.
[356] *Doklady Akad. Nauk S.S.S.R.*, **109**, 1956, 1167-1170.
[357] *Fortschr. Mineral.*, **34**, 1956, 52 with important discussion remarks of E. Kordes.

bonds were plotted by A. S. Povarennykh[358] in a curve which also well illustrates the intermediate bond type from the potentials of the constituent ions in silica and silicates. We refer in this respect to the discussions of W. S. Fyfe[359] (cf. A. ¶ 9).

227. In numerous publications[360] E. Szádeczky-Kardoss combined theoretical deductions on the *anionic* potentials, the additive potentials of bonding, ψ, of chemical compounds, the ionic characteristics (e.g., their specific gravity, $3A/4\pi r^3$), with the geochemical role and appearance of different elements in minerals, including their migration phenomena.

228. For the geochemistry of silicates the isomorphic relations of Si^{4+} and Be^{2+} in the corresponding groups $[SiO_4]$ and $[BeO_4]$ are of great theoretical and practical importance. A. A. Beus[361] applied Fersman's energy coefficients EK which are for $[SiO_4] = 2.75$ and $[BeO_4] = 6.0$. They play an essential role in the mineral association of alkaline rocks. These concepts are also applicable to the analogy of $[BeO_4]$ and $[ZnO_4]$ in the willemite-phenacite group, occurring together as in the zinc ore paragenesis of Franklin Furnace, New Jersey, whereas between $[BeO_4]$ and $[AlO_4]$ there is practically no such corresponding analogy although beryl and cordierite contain those tetrahedral $[RO_4]$ groups in corresponding positions. In rock-forming silicates of the alkaline pegmatites (nepheline syenite) and their pneumatolytic sequences, Be is always associated with Ti, Zr, and rare earth elements. The Be concentration in alkali rocks is in general about six times larger than in granitoids, most of the Be being "camouflaged" in aluminosilicates, and particular enrichment in characteristic of hybridic granitoid pegmatites crossing ultrabasic rocks (enrichment by 5 to 10 times), or in contact skarns. Characteristic ionic replacements are the coupled association of $(K, Na)^+ + [SiO_4]$ = rare earth elements + $(BeO_4]$, or $2Ca^{2+} + [SiO_4]$ by two rare earth element ions and $[BeO_4]$, in grossularite, spodumene, tourmaline, gilbertite, apatite, muscovite, lepidolite, as well as in pneumatolytic-pegmatitic formations margarite, phlogopite with Ti, Cr, Mg, which are foreign to the granite pegmatites of the "pure line." Also the fluorine contents of

[358] *Doklady Akad. Nauk S.S.S.R.*, **109**, 1956, 993-995.

[359] *Am. Mineralogist*, **39**, 1954, 991-1004.

[360] A complete list of these unfortunately not easily accessible articles is given by P. Eskola, *Compt. rend. soc. géol. Finlande*, **29**, 1957, 85-100; *Bull. comm. géol. Finlande*, No. 172, 1956, in an investigation of the postmagmatic potassium metasomatism of granite, (cf. Vol. IV, Section A). Abstracts see in *Chem. Abstr.*, **47**, 1953, 6823 c, e; 6834 c; 7267 i; *Ibid.*, **48**, 1954, 12633 a, g; *Ibid.*, **49**, 1955, 8747 g, h.

[361] *Doklady Akad. Nauk S.S.S.R.*, **90**, 1953, 425-428; *Ibid.*, **104**, 1955, 108-111 (with S. N. Fedorchuk); *Geokhimiya*, 1956 (1), 67-80, with interesting data of Be in titanium clinohumite, vesuvianite, orthite, etc. In *bavenite* a possibility is seen for the generally uncommon analogy of $[BeO_4]$ and $[AlO_4]$. Bavenite with a characteristic constant content of 6.5 to 8.0 per cent BeO was described by M. Fleischer and G. Switzer, *Am. Mineralogist*, **38**, 1953, 988-993.

such formations are coupled with the appearance of Be by the replacement of Si^{4+} + $2(OH)^-$ by Be^{2+} and 2 F^-.

229. The classical isomorphous consanguinity of zirconium and *hafnium* was discussed by M. Fleischer[362] in a thorough compilation of the analytical data and of the ratios Hf/Zr in zirconium minerals, from the geochemical and economic viewpoints. The average ratio Hf/Zr may be equal to 0.02 for the earth's crust. Minerals from alkali rocks (nepheline syenite) have in general a lower ratio Hf/Zr than those from granitoids. It is highest in granite pegmatites, in alvite, cyrtolite, and naegite (varieties of zircon). In thortveitite Hf may be even higher than Zr (ratio 1.8 to 1.9 of both elements).

230. The very important isomorphism between silicates and *germanates*, but also of the complex salts with central Si and Ge cations, such as fluorosilicates and fluorogermanates of the type $R^I_2(Si,Ge)F_6$, were discussed by C. Cipriani,[363] with a complete crystalline solubility of the sodium salts, whereas the potassium salts have different crystal structures, and only a very limited crystalline solubility. As a matter of fact, all of the pneumatolytic-hydrothermal silicates contain germanium in isomorphous substitution for silicon, including the fluorosilicates. The pure oxides SiO_2 and GeO_2, on the other hand, are merely in isostructural relation in their quartz-type modifications. The isomorphous miscibility in crystalline solutions is much limited and has the character of a typical accessorial (trace) isomorphism[364] in the important effects caused by the formation of color centers after X-ray irradiation (cf. C. ¶ 77 ff.). The pressure-temperature diagrams of both systems, however, have much in common as J. F. Sarver and F. A. Hummel[365] demonstrated, especially in the $\alpha \rightleftharpoons \beta$ inversion type of the quartz-like modifications of GeO_2 with true quartz. J. F. White, E. R. Shaw, and J. F. Corwin[366] described a synthetic fibrous chalcedony-like form of GeO_2, with the same optical anomalies known from common chalcedony.

231. H. Strunz[367] developed a systematic classification of germanate structures based on the isostructural relations with the silicates. These include the typical

[362] *U. S. Geol. Survey. Bull.*, No. 1021-A, 1955, 13 pp.

[363] *Fortschr. Mineral.*, 32, 1954, 87.

[364] V. A. Frank-Kamenetskiĭ, *Silikat Tech.*, **12**, 1961, 306-308, speaks of "micro-isomorphism" of GeO_2 in SiO_2 (quartz); cf. Volume V, Section B..

[365] *J. Am. Ceram. Soc.*, **43**, 1960, 336.

[366] *Am. Mineralogist*, **43**, 1958, 580-584.

[367] *Naturwissenschaften*, **47**, 1960, 1954-155; *Silikat Tech.*, **11**, 1960, 397; *Rend. soc. mineral. ital.*, **17**, 1961, 537-542. On germanates with tektosilicate structures see H. Strunz and E. Ritter, *Neues Jahrb. f. Mineral. Monatsh.*, 1961, 22-24. The complete analogy of calcium germanates and silicates is demonstrated by S. M. Royak and I. A. Prokhvatilova, *Doklady Akad. Nauk S.S.S.R.*, **141**, 1961, 880-883. On synthetic *germanium melilites* see C. Brisi and F. Abbatista,

isle (neso-), group (soro-), ring (cyclo-), chain (ino-), layer (phyllo-), and framework (tekto-) structures, with numerous synthetic crystal phases of both series of compounds. H. Strunz and P. Jacob[368] described in detail germanates with the structure of phenacite, olivine, monticellite, and a Ge eulytine, $Bi_4(GeO_4)_3$, synthesized by A. Durif-Varambon.[369] Germanates with apatite and garnet structures were synthesized by H. Strunz, H. Freigang, and B. Contag,[370] as well as nepheline, celsian, sodalite, and analcime-natrolite-type compounds. Germanate garnets of the type $Mn_3Cr_2Ge_3O_{12}$ have interesting paramagnetic properties in the range from 90° to 500°K. if the magnetic cations occupy fourfold positions in the crystal structure, as A. Tauber, E. Banks, and H. H. Kedesdy[371] observed. In hydrated germanate minerals the isostructural relation to arsenates is much more pronounced than that to silicates, as in the secondary oxidation zone minerals of the Tsumeb Mine, South-West Africa, according to Cl. Frondel and J. Ito.[372] Hemimorphite is the only silicate hydrate mineral which is able to take up $[GeO_4]$ in crystalline solution. It is absent in the Tsumeb mineral association. E. Schroll[373] described such Ge-containing hemimorphites from the lead-zinc ore deposits of Bleiberg-Kreuth, Karinthia.

232. In many cases we have already noted isostructural relations with *phosphates* and *arsenates* in the systematic classification and description of silicate structures, in the present Section. As a matter of fact, the replacement of $[SiO_4]$ by $[PO_4]$ or $[AsO_4]$ in the crystallochemical formulae of minerals illustrates a great number of such structural analogies of which we will here only speak generally, with a selection of a few characteristic examples among typical well-known silicate-phosphate structures.

233. F. Machatschki,[374] in his studies on comparative crystal chemistry of silicates, gave a well-selected tabulation of such analogies, with special reference to the quartz-like structures of (RPO_4) compounds (see below). In the rare mineral

Ann. chim. (Rome), **50**, 1960, 1786-1790. For steric reasons, it is not possible to prepare pure calcium germanate melilites, but those containing Sr^{2+} or Ba^{2+} in the place of Ca^{2+} are stable.

[368] *Neues Jahrb. Mineral. Monatsh.*, 1960, 78-79.

[369] *Bull. soc. franç. minéral. et crist.*, **82**, 1959 285-314, including germanium spinels, scheelites, etc.

[370] *Neues Jahrb. Mineral. Monatsh.*, 1960, 47-48.

[371] *J. Appl. Phys.*, **29**, 1958, 358-387.

[372] *Am. Mineralogist*, **42**, 1957, 743-753.

[373] *Mitt. österr. mineral. Ges.*, Sonderheft 2, 1953, 1-60. We further emphasize the occurrence of $FeGe(OH)_6$ as the new mineral *stottite* which was described by H. Strunz and M. Giglio, *Fortschr. Mineral.*, **38**, 1960, 40-41; *Naturwissenschaften* **46**, 1959, 489-490. The existence of the hexahydroxo-germanate group $[Ge(OH)_6]$ is particularly interesting in this crystal phase as a structural coordination unit.

[374] *Periodico Mineral. (Rome)*, **22**, 1953, 23-24. See also valuable tables and bibliography given by H. Brasseur, *Silicates inds.*, **26**, 1961, 169-177.

dixenite, $(Mn,Fe,Cu,As^{3+})_{20}[(Si,As^{5+})_3O_{12}](O,OH)_{20}$, Fr. E. Wickman[375] sees one of the most striking examples of isostructural and even isomorphic analogies of $[SiO_4]$ and $[AsO_4]$ groups, indicating in a certain range a mutual replacement in crystalline solutions. An extensive discussion of the isotypism and the isostructural relations of phosphate minerals was given by W. Kleber and B. Winkhaus.[376] Phosphates are anisodesmic and, in isostructural relation to silicates, distinctly mesodesmic, as in the tendency to form $[P_2O_7]$ groups in analogy to $[Si_2O_7]$. In the ratio of the central cation radius to that of O^{2-}, r_A/r_0, is $[SiO_4]$ characterized by this quotient = 0.292, $[PO_4]$ by 0.243, whereas $[AsO_4]$ is more remote with 0.336, $[SO_4]$ with 0.207, however, compatible, $[VO_4]$ with 0.422 already beyond a normal tolerance. $[SO_4]$ on the other hand, does not show mesodesmic tendencies as $[PO_4]$ does, but can replace $[PO_4]$ in monazite.[377]

234. Particularly typical are the isostructural relations in the wagnerite-rhenanite family (cf. R. Klement and Fr. Steckenreiter, 1940), further in the rich material of observations in the apatite family, previously studied by D. McConnell (1937/39), with the classical example of *ellestadite*, $Ca_{10}[SiSO_8]_3F_2$, *wilkeite*, $Ca_5[(P,S,Si)O_4]_3(F,OH)$, and *britholite*, $(Na,Ce,Ca)_5[(Si,P)O_4]_3(F,OH)$.[378] The isostructural possibilities of modifying apatites and *pyromorphites*, not only with $[SiO_4]$, $[AsO_4]$, and $[VO_4]$ but also with $[GeO_4]$ replacing $[SiO_4]$ and $[PO_4]$, likewise were considerably enlarged by the excellent studies of L. Merker and H. Wondratschek.[379] Those com-

[375] *Geol. Fören. i Stockholm Förh.*, **73**, 1952, 637-638 (from notes of the late K. Johansson on minerals from the Långban deposits). A quite exceptional structure is that of *mcgovernite*, which is the largest unit cell (rhombohedral) yet discovered for an inorganic compound, and not a stacking polytype, with the parameters $a_0 = 8.22$ Å.; $c_0 = 205.5$ Å., in close relationship to dixenite; cf. B. J. Wuensch, *Am. Mineralogist*, **45**, 1960, 937-945. Space group is $D_{3d}^6 = R\ 32/c$, with six molecules $Mn_9Mg_4Zn_2As_2Si_2O_{17}(OH)_{14}$.

[376] *Fortschr. Mineral.*, **28**, 1951, 175-202.

[377] On an interesting sulfate monazite occurring in the carbonatites of the Kola Peninsula, see A. A. Kukharenko, A. G. Bulakh, and K. A. Baklanova, *Zapiski Vsesoyuz. Mineral. Obshchestva*, **90**, 1961, 373-381.

[378] For the problems of the britholite composition see E. A. Nechaeva and I. D. Borneman-Starinkevich, *Zapiski Vsesoyuz. Mineral. Obshchestva*, **85**, 1956, 509-514, on a skarn occurrence in Transbaĭkaliya. Similar to britholite is also the new mineral *belovite*, described by L. S. Borodin and M. E. Kazakova, *Doklady Akad. Nauk S.S.S.R.*, **96**, 1954, 613-616, with some $[SO_4]$, but only subordinate $[SiO_4]$, occurring in nepheline syenite pegmatite. On *synthetic* britholite cf. G. Trömel and W. Eitel, *Z. Krist.*, **109**, 1957, 231-239, including britholite proper, $Ca_2Ce_3[SiO_4]_3(OH,F)$, and *abukumalite*, $Ca_2Y_3[SiO_4]_3(F,OH)$, or other rare earth cations substituting for Ce^{3+} in britholite. The possibility of introducing $[CO_3]$ groups in the apatite structures (the carbonate apatite problem) is recently also discussed by D. McConnell, *Naturwissenschaften*, **45**, 1958, 83-84.

[379] *Naturwissenschaften*, **43**, 1956, 495-496, and *Fortschr. Mineral.*, **35**, 1957, 45-46. Characteristic is the occurrence of such modified pyromorphite structures in which the (F,OH) positions are empty.

pounds are probably truly isomorphic since nothing indicates their limited crystalline miscibility in synthetic compositions of this class, with strict hexagonality, but with a statistical distribution of the anion tetrahedra. The least stable are the vanadium compounds, in conformity with the criteria given above on the ratio r_A/r_O. Some of the arsenates are not equally as stable as the phosphate-silicate complexes, and only incongruently fusing (cf. Vol. III, Section B).

235. An enumeration of the observed isostructural relations of silicates and phosphates is not necessary since all essentials are excellently given in the afore mentioned report of W. Kleber and B. Winkhaus;[380] this is equally true for the families of the olivine, andalusite, garnet, zircon, titanite, thortveitite, trimerite, datolite, and the silica modifications. We may only add the remarks of R. Blix and Fr. E. Wickman[381] on berzeliite, $Ca_2Na(Mg,Mn)_2[AsO_4]_3$, in isostructural relation to grossularite, a slight substitution of $[SiO_4]$, $[PO_4]$, and $[SO_4]$ for $[AsO_4]$, and some hydroxyl groups like those occurring in calcium garnets. To the zircon type of rare earth phosphates (*xenotime*, YPO_4), M. K. Carron, M. E. Mrose, and K. J. Murata[382] assign observations on synthetic compounds like $SmAsO_4$, $PrVO_4$, $ScVO_4$, $CeVO_4$, $TbPO_4$, etc., whereas $NdAsO_4$ and $LaVO_4$ have a *monazite* structure. The possibility of the presence of $[PO_4]$ replacing $[SiO_4]$ in *thorite* and *thorogummite*, coupled with the introduction of hydroxyl groups in the tetrahedral coordination $[(OH)_4]$ is discussed by Cl. Frondel,[383] as seen from analytical data (cf. Vol. III, Section B). An isostructural relation between *amblygonite*, $LiAlPO_4(F,OH)$, and titanite, $CaTiSiO_5$, is discussed by V. I. Simonov and N. V. Belov,[384] particularly in the analogies of $[AlO_6]$ and $[TiO_6]$ chains, on the one hand, and of $[PO_4]$ and $[SiO_4]$, on the other hand. The analogy to *herderite*, $CaBePO_4F$, previously defended by W. R. Richmond[385] is not confirmed. Therefore, the analogy to datolite (see below) also is unconfirmed.

236. The rare pegmatitic mineral *morinite*, $(Ca,Na)_{4-8}Al_3[PO_4]_3(OH,F)_{5-9}.2H_2O$, shows a distinct isostructural relation to *epidote*, as D. J. Fisher and J. J. Runner[386] observed, especially when one plots the zero level planes of the lattices of both minerals normal to the *b*-axis, the axes *c* (morinite) and *a* (epidote) coinciding. If one,

[380] For an excellent discussion on the isostructural relations of compounds of the general type A_2BX_4 and A_3BX_5, with tabulations of the olivine and phenacite analogues, see Th. Hahn, *Silikat Tech.*, **12**, 1961, 331-336.

[381] *Arkiv Kemi*, **2**, 1959 (33) 417-424.

[382] *Am. Mineralogist*, **43**, 1958, 985-989, with a discussion of the limits between xenotime-zircon, and monazite-huttonite type compounds.

[383] *Am. Mineralogist*, **38**, 1953, 1007-1018.

[384] *Doklady Akad. Nauk S.S.S.R.*, **119**, 1958, 354-356.

[385] *Am. Mineralogist*, **25**, 1940, 441-479.

[386] *Am. Mineralogist*, **43**, 1958, 585-594.

however, considers the presence of four double tetrahedra [Si_2O_7], and two single [SiO_4] tetrahedra in epidote, morinite contains only four single [PO_4] groups. Morinite is, therefore, interpreted by Fisher and Runner as a neso-phosphate with isolated tetrahedral groups, unless the Al^{3+} ions are assumed to be in [AlO_4] coordination, whereas epidote is a mixed neso-soro-silicate (group 1*d* of Zoltai).

237. The well-known analogy of *datolite*, CaB[SiO_4](OH), with *herderite*, CaBe[PO_4](F,OH), and with *gadolinite*, $FeY_3[B_2Si_2O_{10}]$, was discussed by P. V. Pavlov and N. V. Belov.[387] The atomic positions in all three minerals are highly analogous, with distinction in the parameters caused by the ionic replacements of different radii and charges. For herderite the development of layers of [PO_4] groups interlinked with [BeO_3F] tetrahedra and [CaO_8] coordinations in distorted cubes is typical. The evident similarities of the diffraction patterns of datolite with those of *bakerite*, $Ca_4B_4[BO_3(OH)]_2[SiO_4]_3(OH)_4$, and *homilite*, $(Ca,Fe)_4B_4[SiO_4][SiO_4]_3(OH)_4$, further with *garrelsite*, $(Ba,Ca,Mg)_4B_4[SiO_4]_2[BO_3(OH)]_2(OH)$, were emphasized by C. L. Christ,[388] with a substitution of [SiO_4] by [$BO_3(OH)$] rather than by [BO_4]. The infinite sheets of the composition $[BSiO_4(OH)]_n^{2n-}$, contain corner-interlinked [SiO_4] and [$BO_3(OH)$] groups, each [SiO_4] sharing three corners and having one corner unshared, the [$BO_3(OH)$] groups also sharing three corners, and the unshared (OH) at the fourth corner. The relation to herderite is evident in the analogy between the [PO_4] and [SiO_4] groups, and the [$BeO_3(F,OH)$] and [$BO_3(OH)$] groups,

238. A very convincing isostructural relation also exists between *danburite*, $CaB_2[SiO_4]_2$, and *hurlbutite*, $CaBe_2[PO_4]_2$, the structure of which was determined by V. V. Bakakin and N. V. Belov,[389] indicated already by F. Machatschki and E. Stradner.[390] There are remarkable genetic correlations of hurlbutite also with *beryllonite*, $NaBePO_4$, and herderite. Hurlbutite, however, does not contain [R_2O_7] groups; every [PO_4] is connected with four [BeO_4] tetrahedra, and viceversa.

239. Extensive studies of the structural analogy of the *silica modifications* with compounds of the type ABX_4 have been made, going back to the data given by M. Strada[391] on the phosphates and arsenates of aluminum, $AlPO_4$, and $AlAsO_4$, and of V. Caglioti[392] on ferriphosphate, $FePO_4$, which were continued by B. Wink-

[387] *Doklady Akad. Nauk S.S.S.R.*, **114**, 1957, 884-887.

[388] *Am. Mineralogist*, **44**, 1959, 176-177; on garrelsite cf. Ch. Milton, J. M. Axelrod, and F. S. Grimaldi, *Bull. Geol. Soc. Am.*, **66**, 1957, 1597.

[389] *Doklady Akad. Nauk S.S.S.R.*, **125**, 1959, 383-385. More in details see the same authors and V. B. Kravchenko, *Doklady Akad. Nauk S.S.S.R.*, **129**, 1959, 420-423, and *Ibid.*, **135**, 1960, 587-590.

[390] *Osterr. Akad. Wiss. Math. naturw. Kl. Anz.*, **90**, 1953, 26-29.

[391] *Gazz. chim. ital.*, **64**, 1934, 653-662.

[392] *Atti accad. nazl. Lincei. Rend. Classe sci. fis. mat. e nat.*, [6] **22**, 1935, 146-149.

haus.[393] Detailed synthetic-hydrothermal experiments for the systematic isostructural relations of quartz to ABX_4 compounds were enterprised by R. Roy with R. C. DeVries, D. E. Rase, M. W. Shafer, F. W. Harrison, and E. F. Osborn.[394] The detailed discussions of the analogies and distinctions of the many crystal phases with quartz or cristobalite structure, on the one hand, and with a rutile and other dioxide structures, on the other hand, cannot be given here. We mention, however, the tabular collection of characteristic relations given by Shafer and Roy (see Table VI), which shows the important correlation of the structural types with the average cation radii $(A + B)/2$, and the ratio $\frac{1}{2}(A + B)/O^{2-}$. The exceptional position of $CrPO_4$ is immediately evident as was also discussed by M. W. Shafer and R. Roy.[395] The interesting $GaPO_4$ compound was also studied in extenso by A. Perloff.[396] It is isostructural with the *berlinite* (low-temperature quartz) form of $AlPO_4$, and some other modifications. $GaPO_4$ forms modifications of α- and β-cristobalite type, but no analogue to *tridymite*. Higher pressure modifications of BPO_4 and $BAsO_4$ were synthetically prepared by Fr. Dachille and L.S.D. Glasser[397] as quartz analogues, except that the c-axis parameter of the unit cell is doubled (see also Vol. III, Section A, on the quartz → coesite inversions), and particularly in the observations of Fr. Dachille and R. Roy.[398]

240. In connection with derivative structures we also have to consider the crystal phases which M. J. Buerger[399] characterized as "stuffed" structures. Of particular

[393] *Fortschr. Mineral.*, **28**, 1951, 194-197; *Neues Jahrb. Mineral. Abhandl.*, **83**, 1951, 1-22; B. Winkhaus, Inaug.-Dissertation, Univ. of Bonn, 1951. More recently see O. W. Flörke, *Fortschr. Mineral.*, **37**, 1959, 2, concerning a *tridymite*-like $AlPO_4$ form, and the structural disorder in these phases. See further G. Trömel and K.-H. Obst, *Arch. Eisenhüttenw.*, **26**, 1955, 307-318, who recommended the introduction of $AlPO_4$ into silica bricks for an improvement of their refractory properties (cf. Vol. V, Section B).

[394] *Penna State Univ. Mineral Inds. Expt. Sta. Tech. Paper*, "Crystal Chemistry Studies," Quarterly Reports, 1953-1955; *Z. Krist.*, **108**, 1956, 263-275; *J. Am. Ceram. Soc.*, **39**, 1956, 330-336 (with E. Camilli-Shafer, on $GaPO_4$, $GaAsO_4$, $GaSbO_4$). New data on $AlPO_4$ in *Z. physik. Chem. (Leipzig)*, **11**, 1957, 30-40.

[395] *J. Am. Chem. Soc.*, **78**, 1956, 1087-1089.

[396] *J. Am. Ceram. Soc.*, **39**, 1956, 83-88.

[397] *Acta Cryst.*, **12**, 1959, 820-821.

[398] *Z. Krist.*, **111**, 1959, 451-461. See also J. D. Mackenzie, W. L. Roth, and R. H. Wentorf, *Acta. Cryst.*, **12**, 1959, 79.

[399] *Am. Mineralogist*, **39**, 1954, 600-614, with valuable tabulations of comparable stuffed structures, extending even to non-silicates, e.g., aluminates and ferrites. On the fundamental structure determination of *nepheline*, see M. J. Buerger, G. E. Klein, and G. Donnay, *Am. Mineralogist*, **39**, 1954, 805-818, and especially Th. Hahn and M. J. Buerger, *Z. Krist.*, **106**, 1955, 308-338. On crystals with a "*sub-structure*" see M. J. Buerger, *Proc. Natl. Acad. Sci. U.S.*, **40**, 1954, 125-128, concerning sets of atoms arranged in a pattern with *simpler* geometrical characteristics than the entire pattern indicates. Such structures are characterized by close packings of the larger atoms, e.g., of the oxygen anions.

TABLE VI

$A^{III}B^{V}O_4$	Average radius	Radius ratio	Structures formed					
	$\frac{A+B}{2}$	$\frac{1}{2}\left(\frac{A+B}{O^{2-}}\right)$	Quartz		Tridymite	Cristobalite		Other
			Low	High		Low	High	
BPO_4	0.29	0.21	+			+		
$BAsO_4$	0.34	0.24				+		
BVO_4	0.41	0.29						Unknown Structure
$SiSiO_4$	0.42	0.30	+	+	+	+	+	at least 2 other forms
$AlPO_4$	0.43	0.31	+	+	+	+	+	
$AlAsO_4$	0.48	0.34	+					
$GaPO_4$	0.48	0.34	+			+	+	
$FePO_4$	0.49	0.35	+	+				
$CrPO_4$	0.49	0.35						Unknown
$MnPO_4$	0.50	0.36	+			+	+	
$GaAsO_4$	0.54	0.38	+					
$FeAsO_4$	0.55	0.39						Unknown
$CrAsO_4$	0.55	0.39						Same as $CrPO_4$
$AlVO_4$	0.55	0.39						Unknown
$AlSbO_4$	0.56	0.40						Rutile
$MnAsO_4$	0.56	0.40						Unknown
$GaVO_4$	0.60	0.43						Unknown
$GaSbO_4$	0.62	0.44						Rutile

importance among these are the phases derived from the silica modifications as the "basic" structures, either by a distortion of the latter, or by cation replacements by other ions of different chemical type. Compensation of electrostatic charges, in such additions of other atoms "stuffing" (filling up) the structure, is the fundamental postulate to maintain the isostructural lattice also in the derivative product, and the added ions must fit in the available cavities of the basic structure. Buerger discusses especially the derivatives of *α-quartz* which have only restricted possibilities to fill up cavities of suitable sizes, illustrated by the small cation Li^+ in α-$LiAlSiO_4$ and α-$LiAlSi_2O_6$,[400] or by Be^{2+} and even by B^{3+}. *α-Tridymite* and *α-Cristobalite* have larger cavities into which the ions of Na^+, K^+, or Ca^{2+} may be introduced. Tridymite as a typical stuffed phase was extensively discussed especially by O. W. Flörke[401] in connection with the important phenomena of one-dimensional structural anomalies in the stacking of the tridymite- and cristobalite-units (cf. Vol. III, Section B). The structure of *nepheline*, $KNa_3Al_4Si_4O_{16}$, could be correctly interpreted only by an analogous stuffing mechanism, and similar phenomena are valid for $KAlSiO_4$ phases, the magnesium-potassium compounds $K_2MgSi_2O_8$ and $K_2Mg_5Si_{12}O_{30}$, for $K_2FeSi_3O_8$, and perhaps also for Na_2MgSiO_4 and Na_2BeSiO_4. As stuffed structures derived from cristobalite Buerger includes *carnegieite*, $NaAlSiO_4$, and the somewhat problematic Na_2CaSiO_4.

241. A particular question is the isostructural relation of *forsterite* to Li_3PO_4, as natural mineral called *lithiophosphate*,[402] which according to F. Zambonini and F. Laves (1932) is a "filled" olivine structure, or *triphylite*, $FeLiPO_4$, with all of the lithium ions in octahedral cavities. A re-examination of this proposal by J. Zemann[403] confirmed the analogy of the position of the $[SiO_4]$ and $[PO_4]$ groups in both structures. The Li^+ ions, however, are not in octahedral, but in tetrahedral gaps (with the reliability coefficient $R = 0.26$ to 0.30 for the octahedral positions, and $R = 0.18$ for the tetrahedral positions). The correct crystallochemical formula of Li_3PO_4 is then written $Li_2^{(4)}Li^{(6)}[PO_4]$. Also Ag_4SiO_4 is a "filled" olivine type structure, according to E. Thilo and Fr. Wodtcke,[404] with the silver ions assumed to be in the octahedral gaps, but there are indications that this silicate has a lower symmetry than that of the olivine group. The deep-cinnabar-red color which is not changed by quenching in liquid air, indicates a lattice which cannot be bonded prevailingly by ionic forces, but must be considerably covalent.

[400] See also R. Roy, *Z. Krist.*, **111**, 1959, 185 ff.

[401] *Ber. deut. keram. Ges.*, **32**, 1955, 369-381.

[402] V. V. Matias and A. M. Bondareva, *Doklady Akad. Nauk S.S.S.R.*, **112**, 1957, 124-126, describe lithiophosphate from lithium pegmatites; see also D. J. Fisher, *Am. Mineralogist*, **43**, 1958, 761-762.

[403] *Naturwissenschaften*, **45**, 1958, 182.

[404] *Z. anorg. u. allgem. Chem.*, **295**, 1958, 247-262, especially pp. 248-250.

242. The important phenomena of *coupled* isostructural and isomorphic cation replacement, which was often mentioned in our text on special silicate structures, may once again be illustrated in a few chosen examples. A. I. Ginzburg and S. I. Berkhin[405] discussed the role of metasomatic exchange reactions in *micas*, particularly on the formation of *lepidolite* from muscovite and biotite by the exchange of 3 Al^{3+} ions by $Li^{+} + 2\ Si^{4+}$, or of 2 Mg^{2+} ions by $Li^{+} + Al^{3+}$, in connection with a very characteristic zoning in the lithium micas. Often the crystallochemical formulae of phlogopite-biotite micas show deficits in K^{+} (plus Na^{+}, Ca^{2+}, and Ba^{2+}), which were explained by D. P. Serdyuchenko[406] as a possible entrance of Na^{+} into the octahedral layers replacing Mg^{2+} and Li^{+} by a coupled exchange of $2\ Mg^{2+} \rightleftharpoons Al^{3+} + Na^{+}$, especially if the sum of the alkali ions in the formula is higher than 1.00.

243. The geochemical role of *beryllium*, and especially its distribution and specific enrichment, which was extensively studied by A. A. Beus,[407] is particularly characteristic for an ion exchange in rock-forming minerals for which the energy coefficient EK of A. E. Fersman (1935) could be applied. The coupling of Si^{4+} (EK 8.60) with Sc^{3+} (EK 4.65), and also of Be^{2+} (EK 2.65) with 2 Zr^{4+} (EK 7.85) in thortveitite are typical. The coupling of Mg^{2+} with Si^{4+}, on the one hand, of Be^{2+} with Ti^{4+}, on the other hand, in clinohumite, pyroxenes, amphiboles, and Fe–Mg micas are other characteristic groupings. Furthermore, in alkali-rock minerals groupings of K^{+} and Mg^{2+} with Si^{4+} occur, both being replaced by rare earth elements (e.g., La^{3+}). The occurrence of beryllium in feldspars, nepheline, and aegirite, with the grouping $2\ Li^{+} + Si^{4+}$ replaced by $2\ Mg^{2} + Be^{2+}$, is observed as in lepidolite and tourmaline. Beryllium, as a rule, is associated with Ti, Zr, and rare earth elements in alkali rocks, most of the Be^{2+} being "*camouflaged*" (Goldschmidt) in the aluminosilicates. The "capturing" is of the type K^{+} or $Na^{+} + Si^{4+} \rightarrow$ rare earth elements $^{3+} + Be^{2+}$, or $2Ca^{2+} + Si^{4+} \rightarrow$ rare earth elements^{3+}. In seĭdozerite, $Na_4TiMn(Zr_{1.5}Ti_{0.5})O_2[Si_2O_7]_2(F,OH)$, V. I. Simonov and N. V. Belov[408] observed the striking reaction $Mn^{2+} + Ti^{4+} \rightleftharpoons Mn^{3+} + Ti^{3+}$.

244. Recently, the practical (electrotechnical) interest in the isomorphism phenomena of the *garnet* mineral family and their synthetic analogues has been concen-

[405] *Trudy Mineral. Muzeya, Akad. Nauk S.S.S.R.*, No. 5, 1953, 90-131. The simple coupled exchange of 2 Mg^{2+} for $Li^{+} + Al^{3+}$ is., e.g., observed in the orthorhombic anthophyllite-like *holmquistite* which was discussed in its occurrence and chemical composition by Th. Vogt, O. Bastiansen, and P. Skancke, *Am. Mineralogist*, **43**, 1958, 981-982, and by N. Sundius, *Am. Mineralogist*, **44**, 1959, 669-670. The similarity of the relations of holmquistite with anthophyllite in the amphibole family, and of spodumene with clinoestatite in the pyroxene family, is evident.

[406] *Doklady Akad. Nauk S.S.S.R.*, **97**, 1954, 315-318.

[407] *Doklady Akad. Nauk S.S.S.R.*, **90**, 1953, 425-428; *Ibid.*, **104**, 1955, 108-111 (with S. N. Fedorchuk).

[408] *Doklady Akad. Nauk S.S.S.R.*, **122**, 1958, 473-476; see also *Ibid.*, **125**, 1959, 888-889.

trated on the occurrence of *rare earth elements* in garnet structures. S. Geller and C. E Miller[409] studied the series of crystalline solutions between grossularite, $Ca_3Al_2[SiO_4]_3$, and $Y_3Fe_2[FeO_4]_3$, over the range with 10 to 100 per cent of the latter compound. The analogous series of *spessartine*, $Mn_3Al_2[SiO_4]_3$, with the same yttrium-ferrite garnet is not complete, as only homogeneous crystalline solutions are formed on the extremes. The range of crystalline miscibility is even smaller in the system $Ca_3Al_2[SiO_4]_3$–$Gd_3Fe_2[FeO_4]_3$. A. L. Gentile and R. Roy[410] demonstrated in an extensive investigation on the synthesis and crystalline solubility in isomorphous garnet-type compounds how the ionic radii are the dominant factor in the more or less ideal miscibility phenomena, with proper corrections for simultaneous changes in electrostatic charges. The high degree of isomorphism between grossularite and the rare earth ferrite garnets is also paralleled by that of the *cadmium garnet*, $Cd_3Al_2[SiO_4]_3$, with spessartine, $Mn_3Al_2[SiO_4]_3$, and uvarovite, $Ca_3Cr_2[SiO_4]_3$. The analogy is most evident if one considers the crystallochemical analogy between silicate and germanate garnets on one hand, and aluminate-ferrite garnets on the other hand, as of $Ca_3Al_2[SiO_4]_3$ with $Y_3^{(8)}YAl^{(6)}[AlO_4]_3$, or with substitutions in octahedral and tetrahedral sites, and coupled VIII–IV, or VIII–VI–IV substitutions (cf. Vol. III, Section B).

245. We must also consider the suggestive concepts of G. W. De Vore[411] on interrelations between the crystal growth and habitus as ruled by the presence of "*foreign*" elements in a crystal phase. These are often assumed to be mere "contaminations," but doubtless must not be underestimated in their ionic effects, of which De Vore discussed especially the polarizing and contrapolarizing actions. In olivine, garnet, orthopyroxenes, diopside, hornblende, biotite, muscovite, chlorite, pyrophyllite, and talc the fractionation and distribution of different elements between coexisting minerals of a given paragenesis could be tabulated. For coexisting orthopyroxenes and calcium pyroxenes, the distribution of Mg^{2+}/Fe^{2+} is strongly determined by the relative distribution of the Fe^{3+}, Ti^{4+}, and Al^{3+} ions in octahedral coordination. If these elements are present, iron-rich orthopyroxene prevails, and in analogous coexistence of amphibole and pyroxene, the hornblende is the iron-rich phase. Sim-

[409] *Am. Mineralogist*, **44**, 1956, 1115-1120.

[410] *Am. Mineralogist*, **45**, 1960, 701-711. In the rapidly increasing special literature on the synthetic magnetoelectrical properties of the garnet phases here discussed, we shall cite only the results of J. W. Nielsen, *J. Appl. Phys.*, **29**, 1958, 390-391; B. A. Calhoun, W. V. Smith, and J. Overmeyer, *Ibid.*, pp. 427-428; R. Pauthenet, *Ibid.*, pp. 253-255; M. A. Gilleo and S. Geller, *Ibid.*, pp. 380-381; C. S. Porter, E. G. Spencer, and P. C. Le Craw, *Ibid.* pp. 495-496. Especially interesting are the methods of growing large monocrystals.

[411] *Am. Mineralogist*, **38**, 1953, 334-335, for the examples of calcium amphiboles, from tremolite to iron-hastingsite in the ratio of the lengths in the directions of the *b* and *c* axes (varying from 0.05 in tremolite to 0.78 for hastingsite). More extensively in *J. Geol.*, **63**, 1955, 471-494; *Ibid.*, **65**, 1957, 178-196.

ilar conditions are indicated in the coexistence of biotite and hornblende relative to pyroxenes. A high degree of association is observed for the strongly polarizing ions, namely of Fe^{3+}, Ti^{4+}, Al^{3+} in $[AlO_6]$, and the weakly polarizing ions Na^+, K^+, Ca^{2+}. If the presence of true *trace elements* (e.g., Cr, Ni, V, Cu) in silicate crystal phases (e.g., in hornblende, biotite, garnet) is concerned, De Vore[412] introduces, besides the physical factor of temperature, the *dispersoid*-active forces of *adsorption* during the crystallization process. The trace elements do not occupy regular structure sites but they are concentrated on surfaces, imperfections, dislocations, and interfaces in the crystal, as is seen from the results of detailed spectrochemical examinations. A characteristic example is the occurrence of such interlayered foreign trace elements in plagioclase during their crystallization periods. It may even become difficult in some cases to determine a sharp boundary between such adsorption and camouflage contents in foreign elements.

246. Nevertheless, true isomorphic replacements are evident in the regular accessorial contents of the potassium-sodium feldspars in Li, Rb, Cs, Ba, Sr, Mn, Pb, Ga, Ti, as are shown in the precision spectral-analytical determinations by E. D. Belyankina,[413] whereas Be, V, Ni, Co, and Cr are absent in spite of their distinct presence in the surrounding rocks. *Amazonite* has a higher than common content in Rb, Cs, and Ga, but less Sr and Ba. These results are in good agreement with the studies of H. P. Eugster[414] especially on the distribution of Cs, Rb, and Tl in sanidine, and on its equilibrium with water, with definite fractionation regularities.

247. Doubtless isomorphous, and not only "trapped," is also the trace element distribution of the rare earth elements in granite and granite pegmatite minerals, as established by R. H. Jahns.[415] This isomorphism and camouflage of the rare earth elements, either in a selective or a complex mechanism of fractionation, was studied by E. I. Semenov,[416] who used the coordination number as an indicator of these mechanisms. The high coordination of 10 or 12 in cerium minerals is characteristic for selective compounds, whereas Y with a coordination number of 6, or other rare earth elements with 7 or 8, occur in complex minerals. If the logarithm of the concentration is plotted as a function of the atomic number of the elements concerned, the slope of the resulting two lines (for even and odd numbers) increases with increasing cerium contents for the different minerals examined. The selectivity is the more developed, the greater this slope. For pegmatite minerals of the Urals,

[412] *Am. Mineralogist*, **39**, 1954, 323; extensively in *J. Geol.*, **63**, 1955, 159-190; *Ibid.*, **64**, 1956, 31-55.

[413] *Voprosy Petrog. i Mineral., Akad. Nauk S.S.S.R.*, **2**, 1953, 77-87.

[414] *Carnegie Inst. Wash. Year Books*, **53**, 1953/1954, 102-104; *Ibid.*, **54**, 1954/1955, 112-114; *Bull. Geol. Soc. Am.*, **66**, 1955, 1555.

[415] *Am. Mineralogist*, **39**, 1954, 314.

[416] *Geokhimiya*, 1957 (7) 626-637.

D. A. Mineev[417] studied the contents of the epidote crystals in rare earth elements, among which the exceptionally high content in erbium, ytterbium, and dysprosium is most remarkable whereas cerium and related elements are relatively low.

EPITAXIS PHENOMENA

248. Before we enter the fascinating field of regular intergrowths of different crystal phases by epitaxis as a consequence of one- or two-dimensional analogies in their structures, we wish to mention the interesting studies of G. and J. D. H. Donnay[418] on what they called "*polycrystals*," namely syntaxic crystalline edifices simulating a single crystal but composed of two or more chemically distinct species intergrown in parallel orientations. Neither chemical nor spectrographic analyses can distinguish such a polycrystal from crystalline solutions, they are *superimposed* reciprocal lattices. Donnay and Donnay studied such phases in the fluorocarbonate group of *bastnaesite*, $CeFCO_3$, *parisite*, $2CaFCO_3.CaCO_3$, *roentgenite*, $3CeFCO_3.2CaCO_3$, and *synchisite*, $CeFCO_3.CaCO_3$, differing in the sequence of layers and the tilt of the carbonate groups. They have a pseudocell (with a' of about 4.1 Å.; c''', systematically variable between 4.7 and 4.9 Å.) (see Figs. A. 152-154) in common. The optical prop-

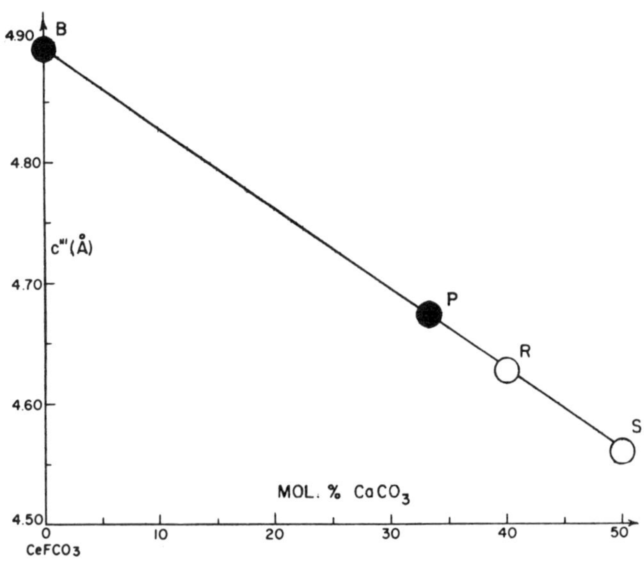

FIG. A. 152. Linear change of the pseudocell parameter c''' for the syntactic intergrowths in the fluorocarbonates of the bastnaesite group (Donnay and Donnay).

[417] *Doklady Akad. Nauk S.S.S.R.*, **127**, 1959, 865-868.

[418] *Am. Mineralogist*, **38**, 1953, 932-963. On the stacking disorders by one-dimensional dislocation cf. H. Jagodzinski, *Acta Cryst.*, **2**, 1949, 201-207.

erties can be calculated after the formulae given by O. Wiener (1921) for "mixed bodies" (plates or rods) with dimensions smaller than the wavelength of visible light (form birefringence), with all the characteristics of disorders by stacking of chemically different slabs (called *B* for $CeFCO_3$, *S* for $CeFCO_3.CaCO_3$, of 4.89 and 9.12 Å., respectively, in Fig. A. 154). The formation of true crystalline solutions in the system $CeFO_3$-$CaCO_3$ is unlikely, and was never observed.

249. In the study of oriented intergrowths (and overgrowths) = *epitaxy*, of chemically different species, A. Neuhaus[419] investigated especially the kinetic condi-

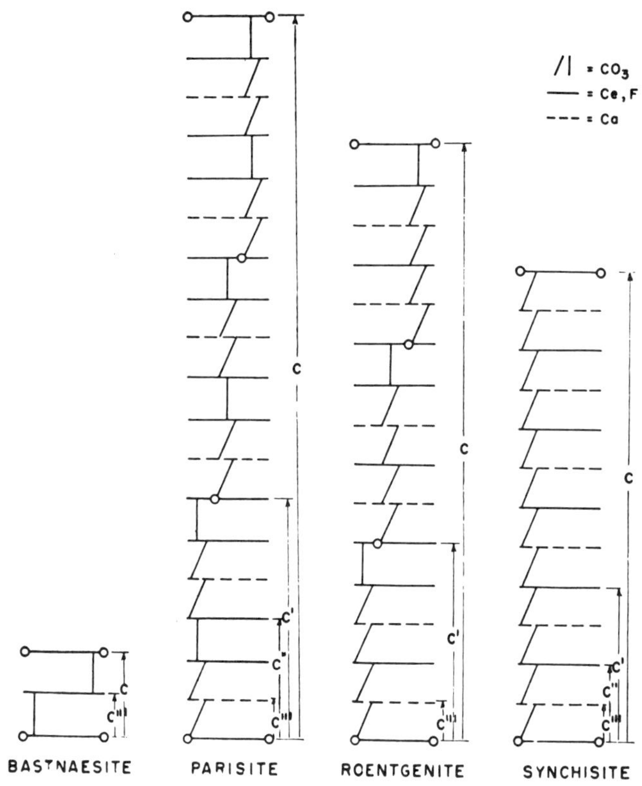

FIG. A. 153. Schematic representation of the stacking in the bastnaesite group of fluorocarbonates (Donnay and Donnay). The rhombohedral lattice points are indicated along the diagonal of the rectangle.

[419] *Naturwissenschaften*, **37**, 1950, 374-375 (with W. Noll); *Fortschr. Mineral.*, **28** 1951, 58-63; *Ibid.*, **29/30**, 1951, 18-23. See also J. Willems, *Naturwissenschaften*, **41**, 1954, 302, who confirmed Neuhaus' conclusions on the influence of raised temperatures on the production of intergrowths, but observed much lower temperatures for those on quartz, feldspar, and disthene, than those expected by Neuhaus; cf. also J. C. Monier, *Compt. rend.*, **237**, 1953, 342-343; Thèse, Faculté

tions of the interconnection between the participating lattices. It is emphasized in a statistical enumeration of the observed intergrowths that most of these take place on a supporting crystal substratum with relatively soft surfaces (Mohs hardness degree commonly 1 to 3), very few on harder substrata (hardness up to 5), as on fluorite, siderite, sphalerite, periclase. On diamond no intergrowths at all could be produced by L. Royer (1928) in his classical experiments. Nevertheless, Neuhaus was able to demonstrate that even orthoclase (hardness 6) orients NaCl (or other halide) crystals on (001) if the feldspar silicate is heated, say to 500°C., and the NaCl condensed from its vapor at 650°C. In this case the "tolerance" (i.e., the permissible difference of the commensurable lattice distances) in only ± 5 per cent, for an in-

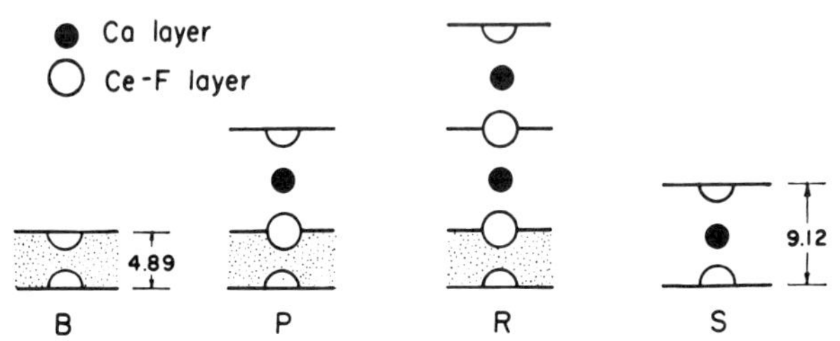

FIG. A. 154. Stacking of $CeFCO_3$ (*B*) slabs, and of $CeFCO_3.CaCO_3$ (*S*) slabs in the crystal structure of the bastnaesite group (Donnay and Donnay).

tergrowth over the K^+ ions by ion-ion interchange. The process is not a mere London-van der Waals adsorption, but a real crystallochemical reaction between the two participating lattices, as though a two-dimensional chemical compound were formed by the action of chemical valences. The hardness of the supporting substratum lattice which is only an empirical measure of the *rigidity* in the structure requires a certain threshold activation energy of the atoms in the surface of the crystal to make it act as an acceptor. Corresponding intergrowths of the vapor-deposit of NaCl were observed on (110) of hornblende, and on (100) of wollastonite, but not on disthene, which does not contain easily activated mono- or divalent cations. Other conditions of intergrowths are valid if the participating lattices are not ionogenic, but homopolar in their bonding characteristics.

250. The threshold temperatures of activation systematically increase with the lattice energies, and with the molecular bonding energies of the halide condensate, as from KBr (threshold temperature 340°C.) to KCl (360°C.), NaBr (420°C.) to NaCl (500°C.), but slightly different for (100), (110), or (111) as the planes of the intergrowth of the deposit.

des Sciences, Université de Paris, 1173-1228. This latter author comes to results in contradiction to Neuhaus' rigidity-temperature rule.

251. Neuhaus[420] also gave an extensive tabulation and discussion of epitaxy in general, with a detailed characterization of the structural premises, and the systematics of the intergrowth phenomena as a function of different valence and bonding types of the participating lattices. Among the intergrowths on silicates are those on *micas*, the classical paradigma. The majority are "partial-isomorphous" intergrowths of organic substances on halides, oxides, sulfides, etc., together with the influences of rigidity as mentioned above in the examples of the system NaCl–orthoclase. Special intergrowths of minerals are tabulated by J. von Vultée.[421] We further emphasize the report given by H. Seifert[422] on epitaxy in general, based on structural considerations, especially in connection with the problems of nucleation and adsorption. The phenomenon of a *selectivity* in epitaxy is of great importance in view of reactions influenced by heterogeneous catalysis, and in biochemical reactions (see below).

252. The influence of the experimental conditions on the orientation of KCl and KBr on (001) of muscovite was studied by N. E. Kamentsev.[423] A complete epitaxy is possible even with relatively high divergences of the structural lattice parameters (by 11.4 per cent for KBr, by 15 per cent for KCl). A decrease in the difference of the parameters of the growing planar lattices influences the probability of *nucleation*, w, and $\ln w$ is proportional to

$$\frac{1}{T^3 \ln (c/c_\infty)^2},$$

in which c is the equilibrium concentration in the halide solution, c_∞ the concentration at the start of the crystal deposition. The epitactic growth of LiF, NaF, CaF_2, BaF_2, and MgF_2 on cleavage faces of mica (in a high vacuum) were examined by E. Bauer[424] using electron microscopic replica methods, and electron diffraction. For the dynamics of epitaxis the nucleation theory gave a full explanation, although quantitative conclusions could not be derived. The theoretical discussion also makes evident that corresponding crystal structures must not necessarily postulate an equal orientation behavior. For the mechanism of the development of combined forms the conditions are in every case much more intricate than was generally assumed (e.g., in the comparison of CaF_2 and BaF_2).

[420] *Fortschr. Mineral.*, **29/30**, 1952, 136-296.

[421] *Fortschr. Mineral.*, **29/30**, 1952, 297-378 (on silicate intergrowths see pp. 338-348).

[422] In "Structure and Properties of Solid Surfaces," by R. Gomer and C. St. Smith, Univ. of Chicago Press, Chicago, 1953, Chapter IX, pp. 318-383. On the "hour-glass" structures observed in intergrowths of organic dyestuffs with inorganic salts, cf. pp. 362 ff.; cf. also the intergrowths of methylene blue on salts, and with clay minerals, C. ¶ 193. Concerning the epitaxy of aminoacids on quartz cf. *Z. Elektrochem.*, **59**, 1955, 409-412.

[423] *Kristallografiya*, **1**, 1956, 240-243. See further N. D. Lisgarten, *Trans. Faraday Soc.*, **50**, 1956, 684-690.

[424] *Z. Krist.*, **107**, 1956, 265-289; 289-317.

253. Highly interesting experiments were made by W. Kleber and D. Weis[425] on the intergrowth and orientation of ice I on fresh cleavage planes (001) of biotite. In spite of the lack of evident morphological and structural interrelations between both crystal phases, there are dendritic epitactic forms. W. Kleber[426] also developed a complete correspondence between the laws of adsorption (Stranski's theory and Langmuir's adsorption isotherm), introducing the "*hypomorphism*" to explain discrepancies between morphological and structural symmetry, and in general a reduc-

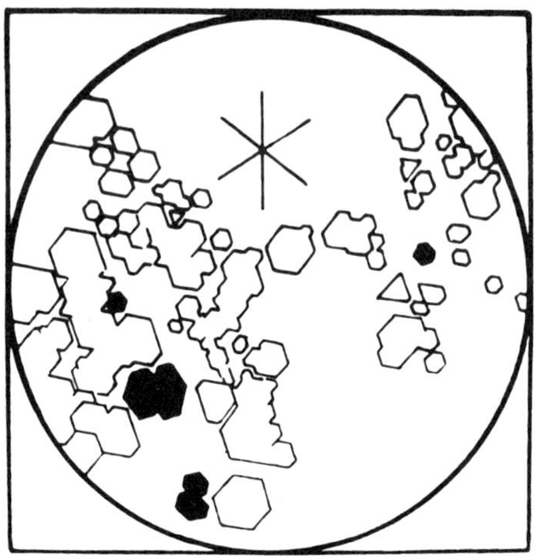

FIG. A. 155. Oriented intergrowth of *d*-camphora on (001) of biotite (Kleber). The two enantiomorphic crystal positions are distinguished in black and white.

tion of the exogenic symmetry degree (symmetry defects, such as the development of enantiomorphic types by the suppression of mirror symmetry). In the case of "enantio-hypomorphism" of $PbCl_2$ by the addition of optically active agents, a characteristic example is given inducing a face-specific adsorption phenomenon and an evident change in the crystal habitus. The epitaxy of *d*-camphora on (001) of biotite in mirror-imaged positions is another important example of enantiomorphic adsorption (Figs. A. 155 and 156).

254. *Overgrowths* of different micas, often with characteristic changes in color, such as rose and green-colored muscovite and lithium micas, were described and

[425] *Fortschr. Mineral.*, **36**, 1958, 66, with important studies on ice I intergrowing with AgI; PbI_2; CdI_2; from the kinetic viewpoint. See further *Naturwissenschaften*, **44**, 1957, 231; extensively in *Z. Krist.*, **110**, 1958, 30-46.

[426] *Fortschr. Mineral.*, **36**, 1958, 38-51. On the theory of oriented crystallization cf. G. Bliznakow, *Izvest. Bulgar. Akad. Nauk. Otdel. Fiz.-Mat. i Tekh. Nauki, Ser. Fiz.*, **3**, 1952, 23-41; *Ibid.*, **4**, 1953, 135-152; *Godishnik Sofiiskiya Univ. Fiz.-Mat. Fak. Kniga* 2. *Khim.*, **49**, 1956, 65-71; *Z. physik. Chem.* (*Leipzig*), **206**, 1957, 271-280 (with E. Kirkowa).

characterized in their differences of chemical composition by E. W. Heinrich and A. A. Levinson[427] by variations of the lithium and manganese contents. The pegmatitic mica crystallization is rich in excellent examples of regular intergrowths with quartz, feldspars (orthoclase and oligoclase), apatite, etc., of which we mention here the descriptions given by V. A. Frank-Kamenetskiĭ,[428] N. Z. Evzikova,[429] and V. F. Barabanov.[430] The planes of intergrowth are for mica (001), for quartz ($10\bar{1}1$), for oligoclase (010). The edges of the growth planes in common are for muscovite [100] and [110]; for quartz [$12\bar{1}3$] and [$\bar{2}110$], for oligoclase [100] and [001]. Analogous parameters are for muscovite $3 \times 5.18 = 15.54$, and $3 \times 5.21 = 15.63$ KX; for

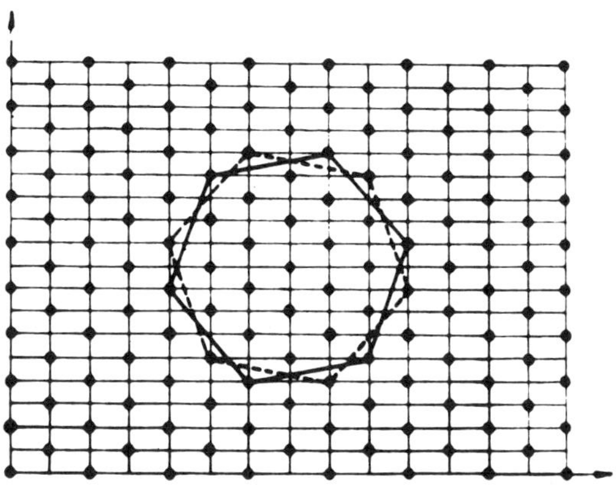

FIG. A. 156. Lattice-geometrical correlation of the (001) biotite, and (0001) camphora meshes of the structure (right- and left-handed positions distinguished in solid and dotted lines) (Kleber).

quartz $2 \times 7.42 = 14.84$, and $3 \times 5.01 = 15.03$ KX, for oligoclase $2 \times 8.16 = 16.32$ and $2 \times 7.26 = 14.52$ KX. There is a complete parallelism of the [SiO_4] chains in all three minerals, especially in the period of the multiple repetition of the edge lengths of the tetrahedra in common (with 2.5 KX as unit). The complex quartz-oligoclase intergrowth is basic also in the phenomenon of graphic granite structures of pegmatites.[431] In the phlogopite–apatite intergrowths there is an analogous orientation

[427] "Natural Mica Studies," Quarterly Reports Univ. of Michigan Engineering Research Inst., Project 978, No. 7, 1953, 12 pp.; *Am. Mineralogist*, **38**, 1953, 25-49.

[428] *Uchenye Zapiski Leningrad. Gosudarst. Univ. im. A. A. Zhdanova, Ser. Geol. Nauk*, No. 178, Geol. Ser. No. 4, 1954, 137-153.

[429] *Zapiski Vsesoyuz. Mineral. Obshchestva*, **83**, 1954, 255-260.

[430] *Zapiski Vsesoyuz. Mineral. Obshchestva*, **83**, 1954, 123-133; concerning the [PO_4] containing phlogopite, and the [SiO_4] containing apatite intergrown in the mica deposits of Sludyanka, see also Vol. IV, Section A.

[431] On the effects of interface energies in mineral contacts of this kind cf. G. W. De Vore, *J. Geol.*, **64**, 1956, 3-55, especially pp. 44-52.

of $[PO_4]$ and $[SiO_4]$ groups in both crystal phases, best observed in projections on $(10\bar{1}0)$ of apatite and on (001) of phlogopite along the edges [110] and $[1\bar{1}0]$, with ionic distances of 9.36 Å. in apatite, of 9.24 in phlogopite, in the percussion figures.

255. E. Ferroni and M. Cocchi[432] describe the interesting case that two polymorphic modifications of one and the same compound (e.g., $CsNO_3$) undergo regular intergrowths on muscovite as the substratum. The epitaxy of the trigonal β- and of the cubic α-modification is different, namely with (100) [010] of α (from the melt), and $(11\bar{2}1)$ $[10\bar{1}0]$ of β (from the saturated solution), in parallelism with (001) [010] of the mica. The very close relation of the two structures of cesium nitrate is illustrated in the identical arrangement of the Cs^+ ions in $(11\bar{2}1)$ of β, and in (100) of α, with a contact of the identical structural planes on (001) of the mica. Of somewhat different type are the *apparent* intergrowths of chrysotile with other minerals of the serpentine family and with calcite, as studied by H. Wondratschek.[433] He especially applied monocrystal rotation diagrams for the elucidation of such highly intricate phenomena of a planar epitaxy but with a considerable share of the fibrous habitus of the chrysotile by a selective nucleation ("Keimauslese"), as for the crystallization of calcite between the fibers, with a minimum of free energy of the surface or boundary faces.

256. For a refined study of regular intergrowths the use of the *field electron emission* may become a remarkable help[434] because the presence of atomar and molecular subjects on a given crystal face can be made visible by this extremely sensitive method. E. Schmid and F. Stangler[435] demonstrated the examination of metal surfaces with field electron microscopy for films of oxides or carbides, illustrating directly oriented crystalline adsorption of materials with cubic space- and face-centered translation lattices, on (001) of a space-centered substratum, as formed by chemical reactions and vapor deposition.

257. By the extensive investigations of J. Willems,[436] A. Neuhaus (see above), and especially of H. Seifert, the knowledge of oriented intergrowths of organic compounds on inorganic substrata has been much advanced and basically interpreted;

[432] *Z. Krist.*, **111**, 1959, 154-158; cf. *Nature*, **181**, 1958, 903.

[433] *Neues Jahrb. Mineral. Monatsh.*, 1957, 135-140; *Veröflentl. Max-Planck-Inst. Silikatforsch.*, **17**, 1957, 131-136.

[434] See the report by E. W. Müller, *Ergeb. exakt. Naturw.*, **27**, 1953, 290-360; on adsorbed layers and single crystalline units in field emission microscopy especially pp. 329 to 346. See further, by the same author, in *Natl. Bur. Standards (U.S.) Circ.*, **527**, 1954, 345-354, on point projection microscopy.

[435] *Radex Rundschau*, 1956, 173-188.

[436] Cf. *Fortschr. Mineral.*, **26**, 1950, 126-130; *Naturwissenschaften*, **42**, 1955, 176-177. M. Brandstätter and A. Martinek, *Ibid.*, **42**, 1955, 369-370, concluded that regular intergrowths of *organic* substances on *organic* substrata are rather non-specific, with wide tolerances in the metric relations.

this work has been continued in recent times by K.-F. Seifert.[437] Starting from systematic observations of epitaxy phenomena of *amino acids* (glycocoll, alanine, glutaminic acid, etc.) on quartz, and later also of oligopeptides, such as L-oxyproline, and other related protein-derivatives, Seifert developed an elaborate theory of the specificity of the interaction of such organic, more or less high-molecular compounds with quartz, and the principle of *matrix*-action underlying the evolution of the silicosis disease phenomena. This matrix theory is not yet universally accepted as the basis of silicosis pathology but it explains thoroughly the fact that true silicosis is necessarily connected to the presence of quartz, and of no other form of silica or of silicates; the reactions of the organism forming the well-known "callosities" in the lungs occur only in the presence of quartz. It is, however, important to know that a regular intergrowth of the same amino acids as used by Seifert on $AlPO_4$, which is isostructural to quartz, is *not* observed, evidently because the electronic configuration of the structure in $AlPO_4$ is different. The isostructural geometry is *not* sufficient for an epitaxy effect in this case. Concerning *enantiomorphy* relations in epitaxis there is a high degree of specificity. L-Oxyproline is equally oriented on L and R quartz, whereas glycocoll is able to intergrow (from the vapor phase) with its enantiomorphic faces (120) and ($1\bar{2}0$) on the same asymmetric quartz face, as well as on L and R quartz, in identical orientation. One-dimensional structural analogy therefore prevails (cf. Fig. A. 157).

258. K.-F. Seifert assumes the presence of surface-bound hydroxyl groups on quartz and of water molecules, irreversibly adsorbed. The dipoles rule the adsorption phenomena (*p–p*, dipole–dipole attraction, according to Neuhaus). On micas the epitaxy is a ion-dipole attraction effect, whereas on the cation-free surfaces of the talc and pyrophyllite layer structure the oxygen anions are the acceptors, with the possibility of hydrogen-bonding actions.

[437] We mention here especially the studies in *Naturwissenschaften*, **42**, 1955, 13; *Ibid.*, **43**, 1956, 156-157; *Trabajos reunión intern. reactividad sólidos*, III, 1956, 93-113; *Z. Krist.*, **114**, 1960, 287-309; *Staublungenerkrankungen*, **3**, 1959, 93-113; *Neues Jahrb. Mineral. Monatsh.*, 1958, 145-151. Further the Dissertation of K.-F. Seifert, Univ. of Münster, Westphalia, No. 13, 1958, 26-27; *Z. Krist.*, **114**, 1960, 361-392; *Naturwissenschaften*, **46**, 1959, 261, and special contributions to the problems of silicosis, e.g. in *Z. Aerosol-Forsch. u. Therap.*, **2**, 1953, (4) 11 pp.; *Staublungenerkrankungen*, **2**, 1954, 199-227; *Beitr. Silikose-Forsch.*, Sonderband "Grundfragen der Silikoseforschung," Bochum 1956, 85-97; 1958, 291-300. Further *Z. Elektrochem.*, **60**, 1956, 848-854; *Fortsch. Mineral.*, **34**, 1956, 24-26; *Staublungenerkrankungen*, **3**, *Wiss. Forschungsber. Naturw. Reihe*, **66**, 1958, 3-26. See also W. Lüdke, *Jenaer Jahrb.*, 1959, II, 496,512, on biological problems under the viewpoints of isomorphism and epitaxy, and H. Seifert, *Naturw. Rundschau*, **14**, 1961, 7-11, 60-65; *Chem. Ingr. Tech.*, **33**, 1961, 210-219, on "biocrystallography". The principle of the matrix theory dates back to H. Friedrich-Freska, *Naturwissenschaften*, **28**, 1940, 376-379. See also on recent experiments of epitactic intergrowths of α-lactose (as an oligosaccharide) and of Vitamins B_1 and B_{12} on quartz see H. Seifert, *Ibid.*, **48**, 1961, 713-714; on the structure of α-lactose monohydrate, cf. H. Seifert and G. Labrot, *Naturwissenschaften*, **48**, 1961, p. 691.

259. Concerning the general importance of Seifert's investigations for the validity of the principles of matrix theories in biochemistry and pathology, the high degree of specificity in the analogies between amino acids and quartz is striking. This is not only observed in the lower-molecular amino acids but also according to the recent results extends to di-, tri- and tetrapeptides. We are not able here to discuss the

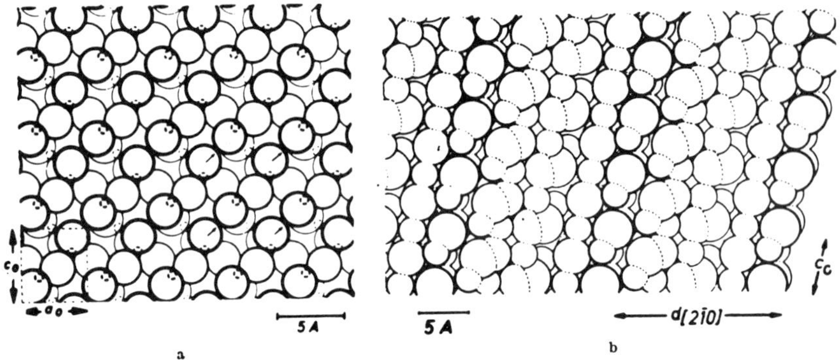

FIG. A. 157. Structure of the substratum [L-quartz (10$\bar{1}$0)] and of the deposit [glycine, (120)], correlated by regular intergrowths (Seifert).

finer details of structural intergrowths as examined by accurate X-ray diffraction goniometry or of the bonding mechanisms, nor the consequences for the pathology of silicosis. But we want to emphasize the high degree of probability that the matrix reactions are of basic significance in this case and for analogous biochemical phenomena (cf. Vol. II, Section A).

EFFECTS OF ISOMORPHISM ON THE PHYSICAL PROPERTIES OF CRYSTALLINE SOLUTIONS

260. Especially for diagnostic purposes by a rapid and sufficiently accurate determination of the composition of crystalline solutions, as in rock-forming silicate minerals, currently investigations are made to examine the additivity of the physical parameters involved. Among these the rapid determination of the density (or of the specific volume) is one of the routine methods in the mineralogical-petrographic laboratory. F. D. Bloss[438] derived a general equation for the specific volume as a

[438] *Am. Mineralogist*, 37, 1952, 966-981. An investigation of the infrared absorption regularities in the series forsterite-fayalite was recently made by Ḥ. Lehmann, H. Dutz and M. Koltermann, *Ber. deut. keram. Ges.*, **38**, 1961, 512-514. A typical case of non-ideal isomorphism is that of the system åkermanite-gehlenite (*melilites*) studied by K. J. Neuvonen, *Bull. comm. géol. Finlande*, **158**, 1952, 5-50, using the thermochemical method of heats of solution. The unit cell parameters a_0 and c_0 nevertheless, follow in this case a linear law of miscibility.

function of the molecular composition, as in the elementary series of the *olivines* (forsterite–fayalite). For such a practically ideal crystalline solubility as a paradigm, the thermochemical measurements of Th. G. Sahama and D. R. Torgeson (1949) had already confirmed the elementary additivity rule to be valid. The ratio of the molecular volume of the end member with the higher molecular weight to the molecular volume of the other component, $K = V_a/V_b$, is a strictly linear function as shown in the density versus composition diagram for the ideal case ($K = 1$), with an upward concave quadratic curve for K below 1, and a convex curve for K above 1. Strictly speaking, the olivines belong to the latter case, with a slightly convex curve represented by a quadratic equation for the molecular percentage in fayalite ($q_{(\varrho)}$), as a function of the density, ϱ, of the crystalline solution. For example $q_{(\varrho)} = -207.754 + 47.68521\,\varrho + 5.25529\,\varrho^2$ (density of forsterite $= 3.217 \pm 0.003$; that of fayalite $= 4.36 \pm 0.01$).

261. The application of the basic Gladstone-Dale rule for a determination of the composition of crystalline minerals from refractive index measurements was demonstrated by H. W. Jaffe,[439] to give a review of the accuracy of such determinations based on the classical data of E. S. Larsen's table work (1921, 1934).[440]

262. The use of X-ray diffraction methods for a rapid determination of the basic unit cell dimensions of silicate crystalline solutions was recommended by H. S. Yoder[441] for the olivines, with an accuracy of the results of ± 2 molar per cent. In the same way, A. Shiramadas[442] gave eight triangular diagrams for the determination of ternary crystalline solutions in the family of the *garnets*, based on the unit cell edges, a_0, and the refractive indices, as a function of the ternary compositions, in analogy with previous correlations given by S. B. Levin (1949). The linear variations in these diagrams are very remarkable and fit well to the measurements. Especially for the garnets, M. A. Gnevushev, A. I. Kalinin, V. I. Mikheev, and G. I. Smirnov[443] gave the results of precision measurements of accurately analyzed garnets from which calculations could be made for the pure end members, and diagrams and equations developed for the determination of crystalline solutions. The series pyrope–almandine–spessartine (pyralspites) follows ideal additivity rules of mixing. Another linear series is that of grossularite–andradite. Polynary garnets can be approximated in their unit cell dimensions as a function of the cationic radii

[439] *Am. Mineralogist*, **41**, 1956, 757-777.

[440] Cf. *U. S. Geol. Survey. Bull.*, **679**, 1921; 294 pp.; *Ibid.*, **848**, 1934, 266 pp.; cf. especially pp. 30-32.

[441] *Carnegie Inst. Wash. Year Books*, **51**, 1951/52, 45; *Ibid.*, **52**, 1952/53, 43, 69.

[442] *Am. Mineralogist*, **42**, 1957, 294-298; cf. S. B. Levin, *Ibid.*, **34**, 1949, 279.

[443] *Zapiski Vsesoyuz. Mineral. Obshchestva*, **85**, 1956, 472-490. See also the excellent data of V. I. Mikheev in *Kristallografiya*, **2**, 1957, 466-469, and *Zapiski Vsesoyuz. Mineral. Obshchestva*, **86**, 1957, 245-253.

of the divalent and trivalent metals, in the form $a_0 = 9.125 + 1.560\, r^{\mathrm{II}} + 2.000\, r^{\mathrm{III}}$, with an accuracy of 3 to 5 per cent. Only the presence of Cr_2O_3 and of MnO brings about slight deviations from strict additivity.

263. The petrographically fundamental determination of the *plagioclase* crystalline solutions from simple optical measurements was recently revised by F. Chayes[444] by the development of new charts for the correlations between the refractive indices and the chemical compositions. This revision is particularly conclusive since it confirmed one or two breaks in the elementary series of the parameters and in the diagrams by curves indicating and emphasizing the role of polytypism in the albite–anorthite feldspars (cf. on high- and low-temperature plagioclases, Vol. III, Section B). The unit cell parameters of natural low- and high-temperature plagioclases, in comparison with those of synthetic feldspars, were measured by J. V. Smith and C. R. Robbins,[445] with a criticism of the higher accuracy of the data for the low-temperature series. It is rather inadequate to determine the composition accurately enough from the high-temperature forms of the plagioclases. Instead of the somewhat poorly resolved reflections $(1\bar{3}1)$ and (131) for albite-rich plagioclases, the difference of the angles $2\,\Theta$ of $(24\bar{1})$ and $(\bar{1}32)$ is more suitable because of the uniformly good resolution over the entire range. The determinative charts were therefore based on the latter difference.

264. The classical universal stage (Fedorow) methods for the determination of the plagioclases from the orientation of the optical indicatrix ellipsoid as a function of the chemical composition are commonly based on the previous stereographic diagrams of V. V. Nikitin and M. A. Ussow (1913). V. Dolivo-Dobrovol'skiĭ[446] gave a very convenient new form of such diagrams with homogeneous triangular coordinates, and the indices α, β, and γ as the apexes at the corners, in which the meridians of the Nikitin projection are straight lines. Also the well-known variations of the rhombus section as a function of the composition, and the most important twin laws are projected in Dolivo-Dovrovol'skiĭ's diagram. A thorough revision of the basic optical data by new, accurate measurements, with a particular consideration of the thermal history (high- and low-temperature) changes in the plagioclases was made by A. S. Marfunin,[447] including the anomalies, as for the Linosa anemousite feldspars, and the order-disorder effects of the Al–Si distribution in the framework. The new Marfunin diagram shows the two-dimensional variation fields of the optical parameters as a function of the chemical composition and of the ther-

[444] *Am. J. Sci.*, Bowen-Volume, 1952, 85-105. On the determination of the composition from plagioclase *glasses* and their refractive indices see W. R. Foster, *Am. Mineralogist*, **39**, 1954, 329.

[445] *Am. Mineralogist*, **39**, 1954, 343.

[446] *Zapiski Vsesoyuz. Minerál. Obshchestva*, **82**, 1953, 122-124.

[447] *Doklady Akad. Nauk S.S.S.R.*, **118**, 1958, 1183-1186.

mal history, at the same time. An excellent example for the application of this diagram and of the thermal history factors in particular was recently given by A. A. Lebedev-Zinov'ev[448] in the oligoclase feldspars of granite-porphyries of the Northern Caucasus. Some of the optical characteristics of such feldspars correspond nearly completely to those of (high-temperature) synthetic plagioclases of corresponding anorthite contents. The presence of potassium in such feldspars does not considerably influence the optical orientation parameters, but markedly decreases the data for the angle 2 *V*.

265. Especially for layer structure silicates, V. Stubičan and R. Roy[449] found in the methods of *infrared absorption* spectrometry a powerful help in the study of the effects of isomorphic replacements, as well as in the investigation of dioctahedral and trioctahedral synthetic clay minerals, micas, chlorites, etc., by quantitative correlations existing between the frequencies and intensities of characteristic absorption bands. Likewise the stacking disorder phenomena are indicated by differences in the intensities (cf. A. ¶ 133 ff.). Infrared absorption methods are also helpful in the distinction of the isomorphous replacement of hydroxyl by fluorine anions, and their quantitative determination, as in the apatite family, demonstrated by R. B. Fischer and Ch. E. Ring.[450] The common chemical-analytical methods are in such cases most difficult, or even not suitable with the desirable degree of accuracy. The hydroxyl and fluorine bands measured are in the range from 16 to 18 μ.

266. Of much practical interest are the studies of the anisotropic vectorial *hardness* parameters of synthetic fluorine micas of the phlogopite group (cf. Vol. IV, Section A) as given by F. D. Bloss, E. Shekarchi, and H. R. Shell,[451] with measurements of the standard Knoop indentation test. It is striking that such pure fluorine micas are harder than the natural prevailingly hydroxyl-containing micas of analogous chemical composition. The difference is evidently a consequence of the higher degree of polarizability of the OH^- ion as compared to F^-. Particularly striking is the influence of *cationic* replacements in synthetic micas. The substitution of Ba^{2+} for K^+ causes a marked increase in hardness, whereas that of Co^{2+} for Mg^{2+} decreases the rigidity. The replacement of Al^{3+} in tetrahedral coordination by B^{3+} produces a softer mica, whereas that of Al^{3+} by Fe^{3+} is only of low effect. Interesting measure-

[448] *Zapiski Vsesoyuz. Mineral. Obshchestva*, **88**, 1959, 172-177.

[449] *Am. Mineralogist*, **46**, 1961, 32-51. For the correct understanding of disorder phenomena and their identification from X-ray interference methods, cf. F. Chayes, *Carnegie Inst. Wash. Year Books*, **56**, 1956/57, 151-156; *Ibid.*, **57**, 1957/58, 240-243.

[450] *Anal. Chem.*, **29**, 1957, 431-434.

[451] *Am. Mineralogist*, **44**, 1959, 33-48. On experiments to exchange in synthetic phlogopite the fluorine by hydroxyl anions by a hydrothermal treatment cf. T. Noda and R. Roy, *Ibid.*, **41**, 1956, 1956, 929-931, further from the practical viewpoint T. Noda, H. Saito, I. Tate, Ts. Fukusawa, M. Fukase, and E. Sekine, *Bull. Chem. Soc. Japan*, **31**, 1958, 508-514.

ments of the magnetic susceptibility and of the Bohr magneton paramagnetism with synthetic fluorine phlogopite, and natural muscovite for a comparison, were made by J. T. Kendall and D. Yeo.[452] The susceptibility varies linearly with the (very low) iron content, the paramagnetism anisotropy being proportional to the content in Fe^{2+}.

MODEL STRUCTURES FOR SILICATES

267. In continuation of the fundamental investigations on *fluoroberyllates* as models of silicate structures in the definition of V. M. Goldschmidt (1926), E. Thilo and H. Schröder[453] studied the system NaF–BeF_2 as the model of the silicate system

[452] *Nature*, **161**, 1948, 476-477; *Proc. Phys. Soc. London*, **64**, B, 1951, 135-142.

[453] *Z. physik. Chem. (Leipzig)*, **197**, 1951, 39-62.

CaO–SiO_2. In Vol. IV, Section A, the melt-equilibria of both (and other) groups of systems will be compared in their striking analogy, with the binary compounds Na_2BeF_4, $NaBeF_3$, and $Na_3Be_2F_7$ and their polymorphic inversions; also in parallelism with those of the calcium silicates, as the consequence of their intimate isostructural correlation. This is also the case for the systems LiF–BeF_2, and MgO–SiO_2, with a complete analogy of the crystal structures of the corresponding models and silicates, as E. Thilo and H.-A. Lehmann[454] observed. The compound $NaLiBe_2F_6$ is the ideal model of diopside, $CaMgSi_2O_6$.

268. That it is advisable to apply Goldschmidt's model hypothesis with some criticism was already recommended by H. O'Daniel (1941) in the discussion of the berylloflurate structures in comparison with more complex silicates, and especially with aluminosilicates. In the same way, D. M. Roy, R. Roy, and E. F. Osborn[455] criticized some of the conclusions made by Thilo, although simple ratios of the fusion points for the beryllofluorates, and the corresponding silicates were alleged in the systems LiF–BeF_2, NaF–BeF_2, and PbF_2–BeF_2. In the latter case there is no analogy to the system TiO_2–SiO_2. The full validity of the model concepts, however, is only evident for the polymorphic relations of the simple system BeF_2 and for the elementary binary beryllofluorates. The quartz form of BeF_2 has an $\alpha \rightleftharpoons \beta$ inversion at 220°C., but the cristobalite form appears only as a metastable phase. A tridymite analogue probably does not exist. $LiBeF_3$ has the structure of β-$CaSiO_3$ (wollastonite) below 343°C., but the α-modification is different from that of α-$CaSiO_3$

[454] *Z. anorg. u. allgem. Chem.*, **258**, 1949, 332-355; see also *Silikat Tech.*, **3**, 1952, 545-548.

[455] *J. Am. Ceram. Soc.*, **36**, 1953, 185-190; *Ibid.*, **37**, 1954, 300-305. Also other details in Thilo's results are not confirmed in those investigations.

(pseudowollastonite). Na_2BeF_4 in its four phases α, β, β', and γ (β is inconsequently often called α') is completely analogous to Ca_2SiO_4. W. Jahn and E. Thilo,[456] however, also observed a monoclinic so-called γ' modification of Na_2BeF_4 which is not known in the system Ca_2SiO_4, whereas the inversion of $\gamma \rightarrow \gamma'$ (called α' by these authors) at 210°C. has the same type of structural change as that of $\gamma \rightarrow \beta'$ in Ca_2SiO_4. The inversion $\gamma \rightarrow \gamma'$ Na_2BeF_4 is irreversible and, followed by that of $\gamma \rightarrow \beta$ Na_2BeF_4. The γ' Na_2BeF_4 appears as a stable phase intermediate between γ and β, between 210° and 270°C. in the schematic $p - T$ diagram given by Jahn and Thilo.

269. The analogy of BeF_2 with quartz was also studied by A. V. Novoselova, O. N. Breusov, D. F. Kirkina, and Yu. P. Simanov.[457] N. A. Toropov and R. G. Grebenenschchikov[458] described the system RbF–BeF_2 as the analogue to the system BaO–SiO_2. Interesting structural relations to *perowskite*, $CaTiO_3$, are evident in the phase $KMgF_3$ of the system KF–MgF_2, as described by H. Remy and Fr. Hansen,[459] with very particular crystalline solutions of KF and $KMgF_3$, which show exsolution phenomena in the composition range of K_2MgF_4, but can also convert to a homogeneous crystal phase of this composition, depending on thermal conditions.

270. Elaborate experiments and discussions of the model relations between fluoroberyllates and silicates were further made by Th. Hahn[460] who found a full analogy between both structural types in the case of compounds with isolated groups $[RX_4]$, or with complex groups such as $[R_2X_7]$, etc., and also with chains $[RX_3]_\infty$, but *not* for layer and framework structures with $[R_2X_5]_\infty$ or $[RX_2]_\infty$. Hahn also emphasizes the very close analogy of the systems Na_2BeF_4 and Ca_2SiO_4, and in addition that of the systems Na_3BeF_5 and Ca_3SiO_5. These structures do not contain annular $[R_3X_9]$ groups, but isolated tetrahedra. Hahn synthesized not only many of the

[456] *Z. anorg. u. allgem. Chem.*, **274**, 1953, 72-80; see also W. Jahn, *Ibid.*, **276**, 1954, 113-127 on $NaLi(BeF_4)$ as the model to monticellite, $CaMg(SiO_4)$, and *Ibid.*, **277**, 1954, 274-286, on $Na_3Li(BeF_4)_2$, as the model of merwinite, $Ca_3Mg(SiO_4)_2$.

[457] *Zhur. Neorg. Khim.*, **1**, 1956, 1619-1628, 2670-2675; *Doklady Akad. Nauk S.S.S.R.*, **107**, 1956, 837-838.

[458] *Zhur. Neorg. Khim.*, **1**, 1956, 2686-2699; *Ibid*, **6**, 1961, 920-927; *Doklady Akad. Nauk S.S.S.R.*, **114**, 1957, 316-319; *Silikat Tech.*, **13**, 1962, 350-353, with bibliography.

[459] *Z. anorg. u. allgem. Chem.*, **283**, 1956, 277-286.

[460] Dissertation University Frankfurt am Main, November 1952; *Neues Jahrb. Mineral. Abhandl.*, **86**, 1953, 1-65; on the Ca_3SiO_5 structure also *Neues Jahrb. Mineral. Monatsh.*, 1953, 1-15. Most recently see *Silikat Tech.*, **12**, 1961, 331-336, chiefly concerning the technological alite problems (cf. Vol. V, Section C). According to recent investigations of V. V. Ilyukhin and N. V. Belov, *Doklady Akad. Nauk S.S.S.R.*, **140**, 1961, 1066-1069, the structure of $RbBe_2F_5$, in spite of some similarities to that of sanbornite, $BaSi_2O_5$, is *not* a true model.

binary beryllofluorates mentioned above in agreement with previous authors, but also $CaBeF_4$ which is strictly analogous to zircon, $ZrSiO_4$, as its model.

271. We do not recommend extending the model concepts to germanates, in spite of their often evident analogies to silicates, since the common aspects of isomorphic replacements of Ge^{4+} for Si^{4+} ions (cf. A. ¶ 230 ff.) in the structures of silicates are fully sufficient. These analogies are, of course, not only of the greatest geochemical significance but may often be used for finer details in the determination of structure investigations. J. R. Goldsmith and F. Laves[461] showed this possibility in the study of the disorder–order relations of the feldspars, by a systematic replacement of Si^{4+} by Ge^{4+}, and/or that of Al^{3+} by Ga^{3+} in tetrahedral coordination in the framework of synthetic feldspar crystals (cf. Vol. III, Section B).

[461] *Acta Cryst.*, **7**, 1954, 131-132; *Z. Krist.*, **106**, 1955, 1955, 213-226. See also on the isostructural relations of $MgGeO_3$ with enstatite, of $SrGeO_3$ and $BaGeO_3$ (both hexagonal) with α-$CaSiO_3$, according to R. S. Roth, *Am. Mineralogist*, **40**, 1955, 332. On the interesting system $Li[AlSi_2O_6]$ - $Li[GaSi_2O_6]$, with *gallium spodumene*, cf. N. A. Toropov and Lin'Tzu-Syan (Tsu-Hsiang Ling), *Zhur. Neorg. Khim.*, **6**, 1961, 928-932.

Section B

Clay Minerals; Structures

INTRODUCTORY REMARKS

1. For the fundamental study of clay minerals in their crystallochemical, mineralogical, and physicochemical behavior a strict *nomenclature* and system is highly desirable. It is, therefore, important that G. Brown in the "Clay Minerals Sub-Committee for the Nomenclature of Clay Minerals"[1] has tabulated seven "family trees" of such mineral types with a strict classification of 2 : 1 structural layer types defined as "*triphormic*," whereas the term "*diphormic*" is suggested for minerals of the 1 : 1 layer types (cf. the Greek word $\varphi o \varrho \mu \acute{o} \varsigma$ = network). Entirely new are the classifications of the kaolinite type minerals as the "*kandite*" family, and in the place of the bulky words "montmorillonitic clay minerals," or "montmorines" and "morillonoids" etc., the very characteristic word "*smectite*" is used, in a broader meaning of the word than formerly. We may further mention that the chlorite-type clay minerals include the chamosites proper and that in the place of the kaolinite-type minerals of the chamosite group the older word "*berthierine*" is proposed again. We will accept in our text, to a certain degree, the designation of the kandites in the meaning as suggested by the Sub-Committee, further the word smectite for the characterization of many montmorillonite-type minerals. The logical principles, after which the proposal of G. Brown has been developed, were also adopted by J. S. Hosking[2] in soil science regulations, strictly applying these principles throughout the whole group of clay minerals.

[1] *Clay Minerals Bull.*, **2**, 1955, 294-302. One may remark that the mineral *berthierite* is a sulfosalt mineral, $FeS.Sb_2S_3$.

[2] Second Australian Conference in Soil Science, Melbourne, Victoria, August, 1957, Vol. I, 1957, 6.1-6.26. Ch. M. Warshaw and R. Roy, *Bull. Geol. Soc. Amer.*, **72**, 1961, 1455-1492, recently developed a structural classification of layer silicates (cf. A. ¶ 121*), including the clay minerals, with highly useful schemes for identification.

2. Another very complete classification of this kind was developed independently by Th. Ernst, W. Forkel, and K. von Gehlen[3] with a widely specialized system of *symbols* to characterize the different clay minerals in natural sedimentary rocks, chiefly the clays. These authors differentiate between a principal group symbol in one Capital letter, a subgroup symbol in two Capital letters, and a species symbol with two letters in lower case. The principal groups are *K* (kaolinite and kandites), *M* (montmorillonites = smectites), *I* (illite-type and mica-like minerals), and a residual group, *X*, in which also mixed-layer minerals (*ML*), the vermiculites (*VM*), chlorites (*CH*), and serpentine minerals (*SP*) are included. For our purposes we suppress in general the nonsilicate and accessorial minerals of the Groups *A* (alumina and aluminum hydroxide), *Q* (quartz), *F* (feldspars), *S* (salts), *C* (carbonates), *Y* (other accessorials), and *O* (organic materials), which, however, may become essential for the characterization of clay rocks in sedimentary petrography, or for the pratical use of clays as raw materials in ceramics. For the latter purpose it is easily possible to develop quantitative formulae for the mineral classification of the composition of clays of great usefulness, especially if, in addition, numerical indications are given in the symbols characterizing the most important particle size distribution parameters from histograms (cf. C. ¶ 7).[4]

3. Quantitative systems of this kind should be based on thoroughly reliable experimental methods for the identification of the single clay minerals. There is a rich literature on the use of *X-ray diffraction* methods for the qualitative and *quantitative* determination of these mineral species. We mention here first the very useful collection of powder diffraction data and charts given by H. Urban,[5] and the report of W. Parrish[6] on advances in the X-ray diffractometry of clay minerals. These especially include the use of modern Geiger, proportional, and scintillation counter tubes, and focusing crystal monochromator arrangements, as supplements to the commercial standard reflecting specimen diffractometers.

4. For the quantitative phase analysis of clay minerals by X-rays, W. von Engelhardt[7] gave basic instructions for the requirements of orientation, and grain size

[3] *Ber. deut. keram. Ges.*, **35**, 1958, 53-56; *Ibid.*, **36**, 1959, 11-18; see also *Sprechsaal*, **91**, 1958, 263-266.

[4] See e.g. R. Mosebach, *Erdöl u. Kohle*, **7**, 1954, 421-425.

[5] "Röntgenkartei zur Bestimmung von Ton- und Sedimentmineralien," Verlag von G. Feller-Nottuln, Kettwig/Ruhr, Vol. 5, of "Opuscula Mineralogica et Geologica," Univ. of Münster, Westphalia, 1954, 37 pp., with 50 diagrams and charts.

[6] *Proc. Natl. Conf. on Clays and Clay Minerals*, **7**, 1959, 230-259. Rapid clay mineral determination methods were recently recommended by T. Sudo, K. Oinuma, and K. Kobayashi, *Acta Univ. Carolinae, Geol. Supplem.* I, 1961, 189-219.

[7] *Z. Krist.*, **106**, 1955, 430-459; *Geol. Fören. i Stockholm Förh.*, **81**, 1959, 374-388. On phase analysis from thin sections, based on X-ray absorption and mass absorption coefficients, see the attractive method developed by H. Haussühl, *Neues Jahrb. Mineral., Monatsh.* 1960, 204-215.

of the samples, their dimensions, and the important effects of crystal defects and of the contents in X-ray-amorphic contaminations. Especially important is the correct application of the *mass absorption coefficients*, μ, as well for the clay mineral to be determined as for the diluent minerals in artificial mixtures for calibration. These coefficients should be determined directly by methods described and recommended by von Engelhardt. We further emphasize the experience given by L. E. Copeland and R. H. Bragg[8] on the use of the Geiger counter diffractometer, with valuable proposals for sampling in order to have reproducible intensity measurements, especially if the powder sample contains constituents of a pronounced tabular texture. An optimum degree of randomness is most essential in such cases which are very frequent in clay mineral analysis. D. H. Porrenga[9] recommended a multiple Guinier camera (modified by de Wolff, see below), with a single focusing monochromator and a single film, applying fine parallel ribs onto the surface of the clay sample to develop favorable orientation to the X-ray beam (cf. Fig. B. 1).

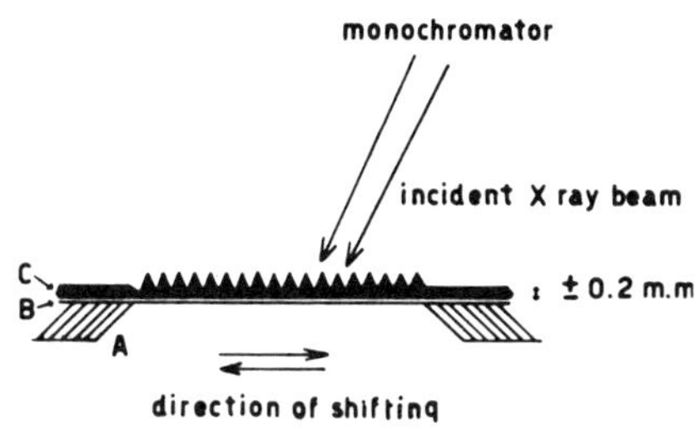

FIG. B. 1. Clay surface (*C*) with ribs, applied for developing optimum orientation conditions in *X*-ray diffractometry with the Guinier — von Wolff monochromator camera (Porrenga).

5. A highly elaborate study, covering a comparison of the different methods for the quantitative mineral analysis of clays and kaolins (for ceramic purposes), was made by U. Hofmann, Th. Ernst, and A. Zwetsch,[10] including a critical discussion of the results from chemical bulk and rational analyses, of dilatometry, differential-thermal analysis, dehydration curves, light- and electron-microscopic examination, and of the qualitative and quantitative X-ray determinations of a carefully chosen selection of well-defined, chiefly Middle-European occurrences. This fundamental

[8] *ASTM Bull.*, **228**, Feb. 1958, 56-59; *Research and Develop. Labs. Portland Cement Assoc. Research Dept. Bull.*, **92**, 1958, 10 pp.

[9] *Am. Mineralogist*, **43**, 1958, 770-774; on the modified Guinier camera mentioned above see P. M. de Wolff, *Acta Cryst.*, **1**, 1948, 207-211; see further Fr. Lippmann, *Heidelberger Beitr. Mineral. u. Petrog.*, **4**, 1954, 251-254.

[10] *Fachausschuss-Berichte der Deutschen Keramischen Gesellschaft*, No. 12, November, 1958, 47 pp., with valuable comparative data tables on pp. 15-47. We further recommend the very useful routine methods as given by K. Oinuma, K. Kobayashi, and T. Sudo, *Clay Science (Japan)*, **1**, 1961, 23-28.

report gives a full account of the advantages and disadvantages of conventional laboratory practices for identification of the clay minerals; it particularly recommends that such methods always be used in suitable combinations. The basic importance of optimum sedimentation and elutriation methods (cf. C. ¶ 16 ff.) for the sampling is most evident.

6. We further emphasize the extensive studies of H. E. Schwiete and R. Vinkeloe[11] on the possibilities of the quantitative X-ray diffractometry in the determination of clay minerals. These authors recommend not only using X-ray methods for this purpose, but in combination with suitable auxiliary determinations, e.g., differential-thermal analysis. Concerning sampling and control of the texture orientation of the particles, the methods of K. Jasmund[12] or those of O. W. Flörke and H. Saalfeld[13] are most adequate. For practical purposes, we also mention the methods derived by I. Pascal,[14] who emphasized the experimental determination of von Engelhardt's mass absorption coefficients. The internal standard method developed by H. P. Klug and L. E. Alexander[15] is often impeded by serious difficulties if the powder diagrams are highly complex and rich in diffraction lines. Especially for kaolinite in soils, H. W. van der Marel[16] gave excellent methods of combined X-ray diffraction, infrared absorption, and differential-thermal analysis. It was observed that an *amorphization* of kaolinite by prolonged grinding is a great hindrance in the quantitative X-ray determinations. Particularly interesting is the recommendation of van der Marel that synthetic boehmite, AlO(OH), be added as a suitable standard for comparison with kaolinite, because the former substance has nearly the same mass absorption coefficient, μ, as kaolinite, and this coefficient is also similar to that of quartz and mica (for Co Kα radiation). Also studied were the effects of structural defects and randomness of the layers, as in the structure of fireclay minerals (see below), which reduce the intensity of the reflections. Mosaic structures in the shape of small crystal blocks inclined at small angles towards one another on the surface of the particles, on the other hand, improve the reflection intensity. It is evident

[11] *Forschungsber. Wirtsch. u. Verkehrsministeriums Nordrhein-Westfalen*, No. 399, 1958, 88 pp.

[12] *Neues Jahrb. Mineral. Monatsh.*, 1950, 63-72; *Fortschr. Mineral.*, **28**, 1950, 34; *Ber. deut. keram. Ges.*, **28**, 1951, 2-13; on the Guinier-von Wolff camera used by K. Jasmund, cf. *Geol. Fören. i Stockholm Förh.*, **78**, 1956, 156-170; *Naturwissenschaften*, **43**, 1956, 275. See also W. D. Keller, *J. Sediment. Petrol.*, **23**, 1953, 93-105, especially p. 94 f.

[13] *Z. Krist.*, **106**, 1955, 460-466.

[14] *Silikat Tech.*, **10**, 1959, 409; *Ibid.*, **11**, 1960, 121-122 (with J. Heinze and H. Ludwig), on the clay minerals in the Mansfeld copper slates.

[15] "X-Ray Diffraction Procedures," Wiley, New York, 1954, 716 pp.; especially pp. 410-439.

[16] *Silicates inds.*, **25**, 1960, 23-31, 76-86. See also numerical data in *Soil Sci.*, **70**, 1950, 109-136, and quite recently J. Wiegmann and G. Kranz, *Ber. deut. keram. Ges.*, **38**, 1961, 294-302; G. W. Brindley and S. S. Kartossy, *Am. Mineralogist*, **46**, 1961, 1205-1215; **47**, 1962, 1213-1215, and H. W. van der Marel, ibid. 1209-1213.

that such factors very much influence the accuracy of the results in quantitative X-ray determinations and must be well considered in their evaluation. In this respect it is remarkable that the infrared O—H stretching bands, as of kaolin minerals, are also much affected by the different degrees of crystallinity, and, in the case of free hydroxyl groups, by the surface of the crystallites, as van der Marel (with J. H. L. Zwiers)[17] observed (cf. A. ¶ 47).

THE KANDITE MINERALS

7. B. B. Zvyagin[18] studied the minerals of the kaolinite group by electron diffraction methods; they included true kaolinite, dickite, and halloysite equally, of which only relatively few occurrences showed constant c periods with a regular sequence of the layers, with interference maxima for the index conditions $k \neq 3k'$. Many of the minerals studied had rather variable c periods, with a limited regularity of the stacking, and diffuse reflections for $k \neq 3k'$, but sharp interferences for this condition on the elliptic curves of the diffraction patterns. Most frequently observed were highly diffuse $k \neq 3k'$ reflections, and a regularity only for the direction a. Triple layers with $c_0 = 21.6$ KX ($\beta = 92^o$) indicate triclinic symmetry.

8. The pseudomonoclinicity of very well developed *kaolinite* (from Turbovsk) was further studied by Zvyagin in electron diffraction methods,[19] with the triclinic unit cell constants $a_0 = 5.13$ KX, $b_0 = 8.89$ KX, $c_0 = 7.25$ KX, $\alpha = 91^o40'$, $\beta = 104^o40'$, and $\gamma = 90^o$, in good agreement with the previous results of G. W. Brindley and K. Robinson (1946). The basic motive of the kaolinite structure is ditrigonal in the arrangement of the octahedral and tetrahedral groups. The layers are oriented with the octahedral edges in the negative, with the tetrahedral edges in the positive direction of the b-axis. The basal network of the tetrahedra is shifted relative to that of the octahedra from the pseudosymmetry plane by $-b/3$, eliminating the pseudomonoclinicity. The octahedra are turned by 3° and 5° from the densest packing positions in the upper and lower layers, respectively; the tetrahedra, by 20° (Fig. B. 2). The structure as given by Zvyagin is about the same as that which G. W. Brindley and M. Nakahira (see below) gave for dickite. Quite recently O. E. Radczewski[20] described some excellently developed kaolinite crystals in Schletta kaolin, with a complete indexing of the reciprocal lattice in electron diffraction

[17] *Silicates inds.*, **24**, 1959, 359-369.

[18] *Doklady Akad. Nauk S.S.S.R.*, **96**, 1954, 809-812. On monoclinic kaolinite see I. Krstanovič and St. Radošević, *Am. Mineralogist*, **46**, 1961, 1198.

[19] *Doklady Akad. Nauk S.S.S.R.*, **130**, 1959, 1023-1026. Cf. G. W. Brindley and M. Nakahira, *Mineral. Mag.*, **31**, 1958, 781-786, and quite recently G. W. Brindley and C. de Kimpe, *Am. Mineralogist*, **46**, 1961, 1005-1016.

patterns, as monocrystal interference diagrams. The indexes include a series up to (190) and (460), and with a clear distinction of ($hk0$) and ($h\bar{k}0$) spots.

9. In the ratio of the intensities for the indexes ($02\bar{1}$) and (060) of the diffraction diagrams of kaolinite, W. D. Johns and H. H. Murray[21] found a measure of the degree of crystallinity, which shows two groups of kaolinite development in different kaolins, bauxitic kaolins and flint clays, in the ranges from 1.00 to 0.80 for the former, and from 0.50 to 0.00 for the latter. The best crystallinity was near 0.50. There is, remarkably, *no* consistent correlation between particle sizes and crystallinity, not even for different grain size fractions for one and the same clay.

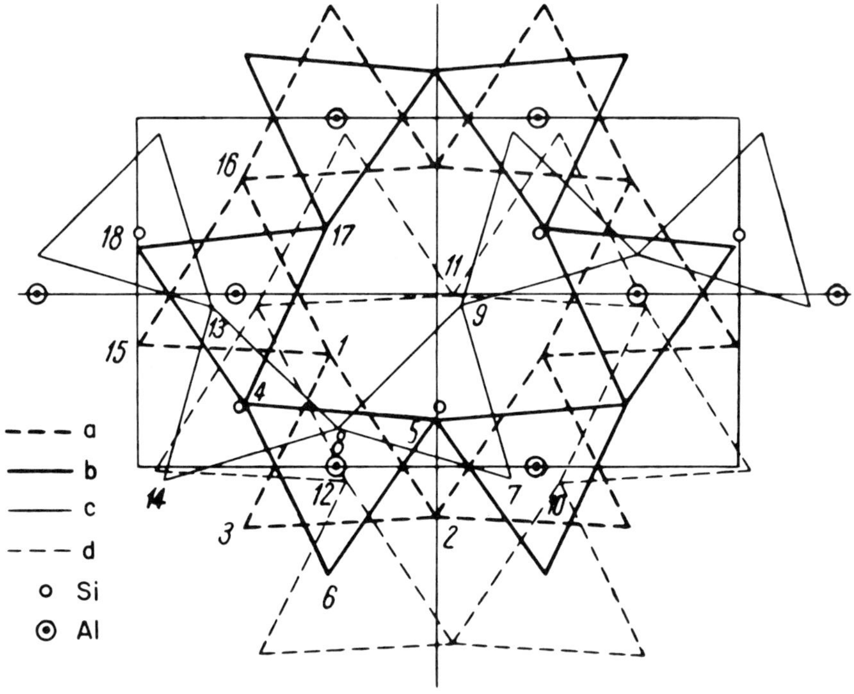

FIG. B. 2. Basis projection of the kaolinite structure to demonstrate the distortion of the octahedral and tetrahedral groups from ideal hexagonal network positions (Zvyagin).

10. The occurrence of *tubular* shapes in kaolinite, differing from those of halloysite, with triclinic layer structures, was described by G. Honjo, N. Kitamura, and K.

[20] *Sprechsaal*, **93**, 1960, 155-159. Excellent electron micrographs of kaolinite crystals and aggregates were described also by H. J. Koenig and S. C. Lyons, *Ceram. Age*, **66**, 1955 (1), 8, 10-14.

[21] *Bull. Geol. Soc. Am.*, **70**, 1959, 1624.

Mihama[22] from different localities in Japan, and from Brazil (Minas Girais, São Paolo, Rio) described by Y. Stourdze Visconti, B. N. F. Nicot, E. G. de Andrade, and H. C. Villanova,[23] showing a great variation of shapes among the particles. There is, however, the typical cracking of the tube walls and a more or less complete unrolling, with elongated hexagonal end forms, similar to those observed in the kaolinite of Les Eyziès, France.

11. *Pholerite* and *Anauxite* as derivatives of kaolinite are discussed by U. Ventriglia,[24] both with the full structural characteristics of kaolinite, but the first mineral showed a partial substitution of silica by alumina, the second a deficit in Al^{3+} from octahedral positions, or a replacement of Al^{3+} by hydroxyl groups. No intercalated silica occurs in anauxite.

12. From the minerogenetic viewpoint we note most remarkable observations made by T. Sudo[25] on the specific occurrence of kaolinite (in association with sericite) in tellurium-gold-silver ores of epithermal type in Japan, whereas in copper ores chlorites prevail (besides sericite). This is in contrast with the abundant occurrence of kaolinite (and halloysite) with nonmetallic deposits. Kaolinite(disordered)–alunite, kaolinite–dickite, and nacrite–alunite associations with gold ores are typical. Clay minerals are in general evidently very sensitively affected by their environmental conditions, i.e., by temperature, pressure, pH, chemical composition of the parent rocks, and by the place in which the mineralization occurred (on corresponding regularities with montmorillonite, cf. B. ¶ 35).

13. The presence of *dickite* (called pholerite) besides diaspore was established in the refractory clay slates of Neurode, Silesia, in well-developed pseudohexagonal tabular crystals, by O. E. Radczewski,[26] with the edges [001][101] and [$\bar{1}$01], as seen in the electron microscope. The electron diffraction patterns normally do not show the (001) interferences because of the strong texture effects found by depositing the scales parallel to the supporting collodion film. R. D. Hill[27] observed a char-

[22] *Clay Minerals Bull.*, **2**, 1954, 133-141.

[23] *Am. Mineralogist*, **41**, 1956, 67-75; *Cerâmica (Brazil)*, **2**, 1956 (6) 59-71; *Ibid.*, **5**, 1959 (20) 2-10. See also H. Urban, *Tonind.-Ztg. u. Keram. Rundschau*, **82**, 1958, 41-43, who emphasized the possibility of transitions of primary kaolinite into halloysite, e.g., in the Brazilian tube kaolins.

[24] *Clay Minerals Bull.*, **2**, 1954, 179-183.

[25] *Sci. Repts. Tokyo Kyoiku Daigaku*, Ser. C, No. 23, July, 1954, 173-197. We very much recommend, for a detailed study of the rich material of observation made by Sudo and his coworkers, the excellent book of T. Sudo, "Mineralogical Study on Clays of Japan," Maruzen, Tokyo, 1959, 328 pp.

[26] *Fortschr. Mineral.*, **28**, 1951, 68-69; extensively in *Ber. deut. keram. Ges.*, **28**, 1951, 119-138. Dickite was also observed (besides ribbon-shaped illite) in Triassic sediments of the northern Eifel, by K. Jasmund and D. Riedel, *Bull. Geol. Inst. Univ. Upsala*, **40**, 1961, 246-257.

[27] *Acta Cryst.*, **8**, 1955, 120. Cf. G. Honjo, N. Kitamura, and K. Mihama, *Ibid.*, **7**, 1954, 684.

acteristic distinction between natural, and calcined, then rehydrated dickite. The original mineral does not show any spacings above 7 Å., whereas the calcined material shows 14 Å. interferences. The structure of dickite, therefore, indicates a peak for 14 Å. spacings owing to the superposition of kaolinite units in the c direction, with alternate shifts of $+ b_0/2$ and $- b_0/2$ in the b direction, whereas in kaolinite such shifts are absent.

14. The accurate structure of dickite was determined by R. E. Newnham and G. W. Brindley[28] by a two-dimensional Fourier analysis, confirming the presence of two polar kaolinite layers in the unit cell. The [SiO_4] and [Al(O, $OH)_6$] units in the layers are strongly distorted (see Fig. B. 3); the "oversized" silica sheets are

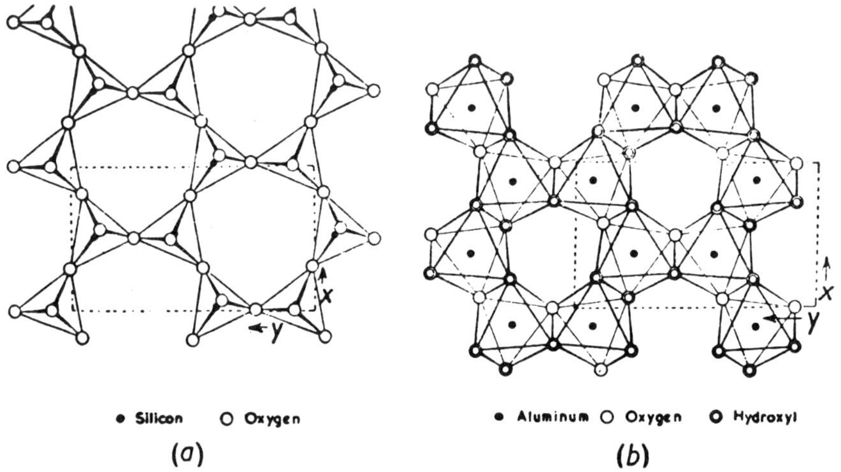

FIG. B. 3. Projection on (001) of the Si—O tetrahedra layer (a), and of the Al—OH octahedra layers (b) in the structure of dickite (Newnham and Brindley).

compressed mainly by articulated rotations of the tetrahedra which tend to a trigonal array which is less open than the hexagonal. On the other hand, the alumina layers are not so widely expanded as the ideal hexagonal arrangement would require. The principal distinction is a shortening of the shared octahedral edges which eliminates the "misfits" and the deviations from Pauling's rules. The O—OH interatomic distances of 2.89 Å. indicate the presence of hydroxyl-bonding between the kaolinite layers. The displacements in the b-direction are cancelled out in the whole crystal and the over-all symmetry is monoclinic, with the unit cell dimensions $a_0 = 5.15$ Å., $b_0 = 8.95$ Å., $c_0 = 14.42$ Å., and $\beta = 96^o48'$.

[28] *Acta Cryst.*, **9**, 1956, 759-764; further G. W. Brindley and M. Nakahira, *Mineral. Mag.*, **31**, 1958, 781-786, and W. F. Bradley, *Proc. Natl. Conf. on Clays and Clay Minerals*, **6**, 1958, 18-25.

15. In the *halloysite*-metahalloysite group, problems of nomenclature were discussed by G. T. Faust.[29] We should state once and for all that names like endellite are not needed if halloysite is clearly the mineral with $Al_2(OH)_4[Si_2O_5].2H_2O$ composition and metahalloysite is $Al_2(OH)_4[Si_2O_5]$ alone. Confusions should be avoided by calling metahalloysite "halloysite"; similarly, it is superfluous to call halloysite "hydrated halloysite." Since the discovery of the tubular shape of natural halloysite which was, of course, changed in the high vacuum of the electron microscope to the metahalloysite form (F. T. Bates, 1949) one was, perhaps, too commonly of the opinion that this morphological property would be the *one* characteristic of the mineral. Halloysite, however, can also occur in totally different forms as was demonstrated by T. Sudo[30] for halloysite in alteration products of volcanic tuffs from Japan, mostly as rounded grains, but also as nearly radial or sugar-plum-like aggregates of fine fibers. These halloysites are always in a very low degree of crystallinity, and usually associated with amorphous coaly, or nearly amorphous allophane, although the typical halloysite X-ray diffraction pattern is distinguished and superimposed on the background.

16. Halloysite is often an incipient stage of crystallization and this conclusion is important for the general problem of the genetic position of halloysite in the kandite group of the clay minerals as it was discussed for clay sediments and the laterite formation by T. F. Bates.[31] That there are transitions is not doubted any more. The clay mineral filling the gap between the structures of optically distinctive dickite and morphologically distinctive halloysite shows a wide range of characteristics, the kaolinite type being most typical in "china clays," the transition mineral type in flint clays, fireclays, and in soils, as G. W. Brindley and K. Robinson recognized before (1946). Randomness increases from kaolinite to halloysite and metahalloysite, with random shifts in both the a and b direction, towards allophane with complete randomness. If, on the other hand, kaolinite is formed from halloysite, it must be by a process more involving solution and recrystallization than simply an increase in order of the same fragments.

17. The purely structural problems of the kaolinite minerals in general were

[29] *Am. Mineralogist*, **40**, 1955, 1110-1118. New data on tubular metahalloysite see by Th. Nemetschek and U. Hofmann, *Z. Naturforsch.*, **16**, b, 620-621. The cross sections of the tubes are often not circular but polygonal.

[30] *Mineral. J. (Sapporo)*, **1**, 1953, 66-68; *Clay Minerals Bull.*, **2**, 1954, 96-106; *Proc. Natl. Conf. on Clays and Clay Minerals*, **4**, 1955, 67-69, *Natl. Acad. Sci.-Natl. Research Council, Publ.*, **456**, 1955, 67-69 (with H. Takahashi). A corresponding occurrence is that described by A. Alietti, *Atti e mem. accad. sci. lettere ed arti Modena*, [VI], **1**, 1959, 3-11, from ignimbrites of Monte Amiata, Italy, originating from allophane. See also K. S. Birrell, M. Fieldes, and K. I. Williamson, *Am. Mineralogist*, **40**, 1955, 122-124 for halloysite derived from volcanic ashes of New Zealand.

[31] Symposium on the Problems of Clays and Laterite Genesis, Amer. Inst. Mining and Metallurg. Eng., 1952, 144-153.

studied by G. A. Kovalev and Yu. S. Dyakonov[32] in respect to the disorder phenomena, the transitions as mentioned above in the discussions of Bates, and with distinct phases showing variable intensities of X-ray diffraction lines, with broadened peaks in the diagrams. It is, however, criticized by Kovalev and Dyakonov that Brindley's and Robinson's definition of "fireclay minerals" also covers several clay type minerals occurring in sediments which have nothing to do with ceramic fireclays. H. H. Murray[33] discussed the same problem of structural variations between kaolinite and halloysite from the viewpoint of step-wise transitions with variable degrees of crystallinity. The X-ray studies are here best combined with differential-thermal examinations which indicate more the inner variations, while the X-ray interferences indicate more the stacking anomalies, superimposed on these. Also the base-exchange methods may become helpful in this respect. Metahalloysite shows a lower, rather than a relatively high base exchange capacity which should be expected from the X-ray behavior. On the other hand the "fireclay minerals" have considerably raised base exchange data (e.g., 18 meq./100 gm., in comparison with 5 for halloysite, 6 for flint clay, and well-crystallized Georgia kaolin only 3 meq./100 gm.).

18. K. Wada[34] investigated the possibility of an *oriented penetration* of ionic compounds (e.g., alkali chlorides, nitrates, sulfates, metaphosphates, and acetates) into the structure of halloysite, with reversible expansions of the lattice by 5 to 45 per cent (from 10.1 to 13.3 and even 14.4 Å.) of the basal spacing. The large ions of potassium, ammonium, rubidium, and cesium are typical of this effect whereas lithium and sodium ions gave no specific expansion reaction of this kind, nor did magnesium and calcium or barium. The *anions* determine the measure of the structure-widening effects and the strength of the bonds between the halloysite and the oriented salt with a weak polarizability, in ammonium salts with NH—O hydrogen bonds. About 200 to 300 moles of the salts can be retained per 100 gm. of the air-dried halloysite in a monomolecular interlayering process, replacing the interlayer water. It is highly characteristic that by heating at 100°C. no transition from halloysite to the metahalloysite type, with a reduction of the spacing from 10 to 7 Å., is observed. If, on the other hand, the two-dimensional "interlayer compounds', are rinsed in excess water, the normal halloysite with the 10.1 Å. spacing is restored. The widening effect is observed exclusively for the (00*l*) reflections, not for the (*hk*) bands.

19. Particularly interesting are the changes in the relative intensities of the in-

[32] *Zapiski Vsesoyuz. Mineral. Obshchestva*, **88**, 1959, 467-473. On the fireclay type minerals, see also H. Saalfeld, *Ber. deut. keram. Ges.*, **32**, 1955, 150-152.

[33] *Am. Mineralogist*, **39**, 1954, 97-108.

[34] *Am. Mineralogist*, **44**, 1959, 153-165; 1237-1242. On the corresponding complexes with phosphates, see *Soil Sci.*, **87**, 1956, 125-150.

terference peaks of the (*hk*) bands shown by the NH_4Cl complex on halloysite (basal spacing 10.5 Å.). In distinction from the simple salt, the diffraction diagram is specific, no polymorphic inversion occurs at 185°C., and a marked shifting of the thermal dissociation of NH_4Cl by about 30° to 50°C. to higher temperatures takes place. The maximum orientation effect is observed for 325 to 330 mmoles/100 gm. of the air-dried halloysite, corresponding to two molecules NH_4Cl per unit cell of the halloysite. The replacement of the two molecules of interlayer water by one molecule NH_4Cl occurs in the first stage of the complex formation. Very characteristic also is the observation that even in the dry state, by grinding halloysite with crystals of NH_4Cl, the complex with 10.5 Å. is formed. NH_4Cl is therefore able to get dispersed from the crystals directly into the interlayers, and a stickiness develops on grinding. The bonding between halloysite and NH_4Cl must be stronger than that of halloysite to water. The geometrical fitting is understood as that of NH_4^+ ions in the cavity of the O^{2-} hexagon in the tetrahedral structural sheets.

20. If, in the place of halloysite, a well-crystallized Georgia kaolinite is ground with *potassium* acetate a widening to 14 Å. basal spacing is observed.[35] From aqueous solution, however, no potassium acetate is interlayered in the structure. At 100°C. the widened structure is collapsed to 11.4 Å. These effects are surprisingly specific since talc, pyrophyllite, antigorite, and chlorites behave totally negative to potassium or ammonium acetate. Rinsing with water reduces the 14 Å. complex to common kaolinite with 7 Å., but the halloysite complex to a basal spacing of 10 Å. The dehydration of halloysite by complex interlayering of salts is therefore completely reversible. A structure cross section of the 14 Å. complex is given in Fig. B. 4. It is not thinkable that the lattice expansion in the complex may be a mere mechanism

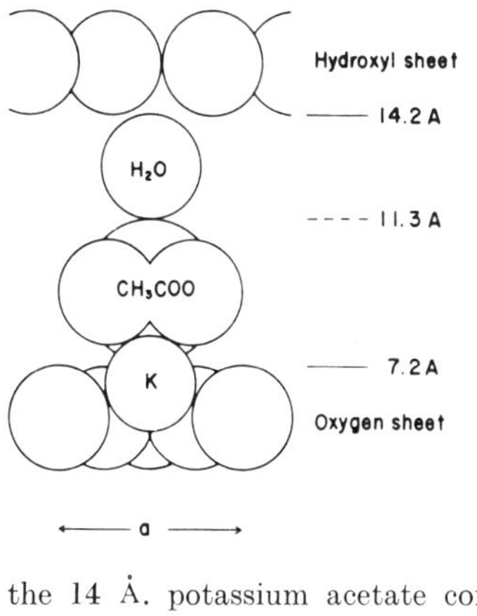

FIG. B. 4. Cross section of the 14 Å. potassium acetate complex compound of halloysite and kaolinite (Wada).

[35] G. K. Wada, *Am. Mineralogist*, **46**, 1961, 78-91.

of disruption of the kaolinite structure to form halloysite, as such a disruption distinctly occurs in dry-grinding of kaolinite for 30 minutes. Figure 4 also shows the reduction of the 14 Å. basal spacing to 11.3 Å. When the monomolecular water layer of the full complex is removed it becomes evident that the clearing space is about equal to the sum of the van der Waals radii for the methyl group of the acetate anion, and of the H_2O molecule.

21. The formation of interlayer complexes is a potential characteristic common to the minerals of the kandite series, but not to the serpentines, either with or without interlayer charges. The reason for the preference of potassium acetate on the other hand is not yet clear. Whereas the water interlayer complexes of halloysite observed by rinsing are relatively stable those with kaolinite are unstable. The break-down of the interlayers depends to a great part on the degree of randomness of the stacking in the structural layers of the kandite minerals. In this respect, a "kaolin" from Spruce Pine, North Carolina, shows a characteristic intermediate position between kaolinite and halloysite, and is higher in crystallinity than common halloysite.

22. We emphasize in this connection also the previous studies of A. Weiss, A. Mehler, G. Koch, and U. Hofmann[36] who described the interlayering of NH_4F in the place of water in halloysite (from Djebel Debar, Algeria) (cf. C. ¶ 183), and those more recently of W. G. Garrett and G. F. Walker.[37] Although no exchangeable interlayer cations could be found, the formation of loose "intercalation compounds" with different inorganic salts and with potassium trimethylacetate was concluded in this case, with basal spacings from 12 to 19.8 Å. in single and double sheets over different portions of the lattice. The potassium ion is characterized as a typical cation of poor hydration.

23. Coming back to the *morphological* factors in the kandite mineral group, we mention here the studies of W. L. De Keyser and L. Degueldre[38] who emphasize that the tubular development of halloysite or metahalloysite is *not* a necessary characteristic, and that kaolinitic clays also may show tubular shapes in the electron microscope (see above), although the X-ray powder diagram is entirely normal for this mineral. The well-known clays of Provins (France), on the other hand, are according to W. von Engelhardt and H. Goldschmidt[39] intermediate phases between the "fireclay" type and halloysite.

[36] *Z. anorg. u. allgem. Chem.*, **284**, 1956, 247-271, especially pp. 268 ff.

[37] *Clay Minerals Bull.*, 4, 1959, 75-80.

[38] *Bull. soc. belge géol. paléontol. et hydrol.*, **63**, 1954, 100-110. We mention in this connection recent electron-microscopic studies on tubular halloysite by W. Behne and W. Müller, *Naturwissenschaften*, **41**, 1954, 138.

[39] *Heidelberger Beitr. Mineral. u. Petrog.*, **4**, 1954, 319-324.

24. In continuation of his first basic studies on the tubular halloysites, T. F. Bates, in cooperation with J. J. Comer[40] refined the electron microscopic methods considerably by a special replica technique, combined with preshadowing with platinum and backing this with a layer of carbon (cf. C. ¶ 119 ff.). For kaolinite, it is often characteristic that the pseudohexagonal plates are cleavage fragments, not primary forms. The tubes of halloysite are beautifully evident in surface replicas and exist as such in the bulk sample, with circular or elliptical cross-section where the tubes intersect the surface at angles approaching 90°. Dickite appears mostly in massive, "wall-like" aggregations.

25. Electron diffraction combined with electron microscopy was applied by J. J. Comer and L. B. Sand[41] for tubular halloysite-glycerine complexes, with the characteristic basal spacing of 10.7 Å. It may, therefore, be concluded that unstrained halloysite with a full water content is also *primarily* tubular, and not just curled up in the high vacuum of the electron microscope after removal of the interstitial water. Brindley and Comer[42] also studied the interesting kaolin clay of Eyziès (France) by such refined methods, with about equal amounts of platy kaolinite and rolled or tubular halloysite. Oriented tubes or rolled sheets may here have been formed from well-aligned platy material after fracturing, and subsequently could have curled. Nevertheless, on the rolled forms hexagonal features appear. These arguments point in the same direction, namely a transition of kaolinite into halloysite, geologically best understood by an acid attack on kaolinite leading to gel formation and a subsequent recrystallization to halloysite.

26. The role of halloysite in red soils from Portugal, Africa, and other countries as a phase of typical crystal imperfection was discussed by L. Bramao, J. G. Cady, and S. B. Hendricks,[43] as an important constituent of terra rossa. Special methods of X-ray diffraction combined with differential-thermal analysis are in such complex systems indispensable for an accurate identification of the kandites, and the thermal methods have been refined to a high degree of accuracy for this purpose. Only a few hints in this respect may be sufficient.[44] Quantitative determinations of the kao-

[40] *Natl. Acad. Sci.-Natl. Research Council, Publ.*, **395**, 1955, 1-9; *Proc. Natl. Conf. on Clays and Clay Minerals*, **3**, 1955, 1-9.

[41] *Natl. Acad. Sci.-Natl. Research Council, Publ.*, **395**, 1955, 26-30; *Proc. Natl. Conf. on Clays and Clay Minerals*, **3**, 1955, 26-30. See also J. J. Comer in *Am. Soc. Testing Materials, Spec. Tech. Pub.*, **257**, 1959, 94-120, with many experimental details and micrographs. See further G. S. Gritsaenko, A. I. Goroshkov and K. E. Frolova, *Zapiski Vsesoyuz. Mineral. Obshchestva*, **87**, 1958, 269-276, and replica micrographs of single crystals, *Ibid.*, **90**, 1961, 266-269.

[42] *Natl. Acad. Sci.-Natl. Research Council, Publ.*, **456**, 1956, 61-66; *Proc. Natl. Conf. on Clays and Clay Minerals*, **4**, 1956, 61-66.

[43] *Trans. Intern. Congr. Soil Sci.*, **4**, Amsterdam 1950, Vol. I, 313; *Soil Sci.*, **73**, 1952, 273-287 (with M. Swerdlow).

[44] Cf. the investigations of H. Lehmann, *Tonind.-Ztg. u. Keram. Rundschau*, Beiheft, No. 1,

linite and halloysite contents of clays on the basis of differential-thermal analysis were further given by L. B. Sand and T. F. Bates.[45] The halloysite share can be distinguished by an endothermic reaction at 500°C. in the ethylene-glycol-saturated form, whereas kaolinite shows its characteristic peak at 575°C. The doublet peaks are the basis for the quantitative evaluation in mixes. Another instructive example is the kaolin from Pagu, Tanganyika, studied by R. H. S. Robertson, G. W. Brindley, and R. C. Mackenzie,[46] with well crystallized kaolinite, and with another *b*-axis-disordered kandite mineral. This latter material shows in the electron microscope well-formed hexagonal plates, with cation exchange properties both in tetrahedral and octahedral positions. It is also interesting because of a certain anatase content which is characteristic of fireclays.

27. Halloysite deposits in the Southern Appalachian region are mostly derived from weathered feldspars, as typically observed in Spruce Pine, North Carolina, under specific environmental conditions, described by L. B. Sand.[47] Kaolinite is developed independently, without any evidence of transition forms. The vermicular kaolinite is generally derived from muscovite-like micas as an intermediate formation between the feldspar and the kaolinite in this complex weathering process from pegmatites. In no sample, even though mica appeared embedded in a matrix of halloysite, was this latter mineral intergrown with mica, whereas kaolinite pseudomorphs after mica orient with the basal plane parallel to that of the mica. On interesting Indian clay deposits with kaolinite and halloysite, we refer to recent investigations of B. M. Bishui and J. Prasad[48] based on electron microscopy and differential-thermal analysis.

28. Particularly remarkable are the observations of R. Pellizzer[49] on halloysite clays (of Capalbio, Orbetello, Italy) with kandites intermediate between kaolinite and halloysite (besides montmorillonite in this case). The genetic conditions are very similar to those in the Utah deposits, the constitution and ceramic characteristics of which were examined by Br. F. Hampel and I. B. Cutler.[50] The genesis of such kandites as described by Pellizzer is particularly fascinating by the occurrence of the clay minerals in calcareous surroundings. In such cases L. L. Ames and L. B. Sand[51] assumed the formation of the halloysite structure from the closely related

1954, 55 pp., with extensive bibliography; *Ibid.* **78**, 1954, 309-310 (with P. Fischer); H. M. Köster and H. Nordmeyer, *Ber. deut. keram. Ges.*, **38**, 1961, 98-106, with examples of "fireclay minerals" in clays of Saxony and the Upper Palatinate.

[45] *Am. Mineralogist*, **38**, 1953, 271-278.

[46] *Am. Mineralogist*, **39**, 1954, 118-138.

[47] *Am. Mineralogist*, **41**, 1956, 28-40.

[48] *Central Glass & Ceram. Research Inst. Bull. (India)*, **6**, 1959, 76-85; *Ibid.*, **7**, 1960, 11-28.

[49] *Atti accad. sci. ist. Bologna Classe sci. fis. Rend.*, [11] **6**, 1959, 12 pp.

[50] *J. Am. Ceram. Soc.*, **36**, 1953, 30-34.

[51] *Proc. Natl. Conf. on Clays and Clay Minerals*, **6**, 1959, 378-385.

structure of a tetracalcium aluminate hydrate, such as $4CaO.Al_2O_3.nH_2O$, which is a synthetic well-known compound with layer structure (cf. Vol. V, Section C). It may also have been an intermediate phase before the calcium ions were removed by low-temperature-hydrothermal waters, with the introduction of silica to crystallize halloysite, under quite specific pH conditions (pH = 8 to 10), in the system $CaO–Al_2O_3–SiO_2–H_2O$ (cf. Vol. IV, Section A), as in the alteration of siliceous volcanic tuffs by calcareous carbonates in hot spring waters.

29. *Infrared absorption* spectrometry must become a powerful new tool in the investigation of the water content of halloysite and of other clay minerals, as is evident from the earlier studies of W. D. Keller and E. E. Pickett,[52] and of J. M. Hunt with D. S. Turner.[53] The free hydroxyl groups in halloysite are indicated by the band at 2.8 μ, the hydrogen-bonded hydroxyl by those at 3.0 and 3.2 μ, in agreement with the bonding mechanism of molecular interlayer water in halloysite sheets as postulated by S. B. Hendricks (1938). B. Nagy[54] applied a polynominal method for the theoretical derivation of the vibrations and frequencies of water adsorbed on kaolinite, with an excellent agreement of his results to experimental determinations of the absorption lines, at $\nu = 3690$ and 3619.3 cm.$^{-1}$ ($\lambda = 2.72$ and 2.77 μ), indicating the O—H—O bonding type on one end of the molecule, whereas the other end is in a state of torsional rotation about a line at a certain angle to the crystal surface.

30. We further mention P. Görlich, H. Moenke, and L. Moenke-Blankenburg[55] on practical uses of infrared spectrometry in mineral diagnostics, and those of E. S. Rudnitskaya[56] and O. N. Setkina[57] on the bands of the clay minerals for testing the purity of clays. A handy introduction on the usefulness of infrared spectromety for investigation of clays as raw materials in silicate technology was further given by H. Dutz.[58] B. M. Bishui and J. Prasad[59] described the practical characterization of

[52] *Am. J. Sci.*, **248**, 1950, 264-273.

[53] *Anal. Chem.*, **25**, 1953, 1169-1174.

[54] *Bull. Geol. Soc. Am.*, **67**, 1956, 1722-1723. On the wave-mechanical theory of the water molecule see the important studies of E. Forslind, *Svenska Forskningsinst. Cement Betong vid Kgl. Tek. Högskol. Stockholm, Handl.*, No. 16, 1952, 43 pp.

[55] *Jenaer Jahrb.*, 1951, I, 154-180. See also very recently H. Moenke, *Carl Zeiss Nachr. Jena*, N.F., **9**, 1961, 49-81. Successful identification of kaolin minerals by infrared absorption spectrophotometry is also described by H. Kodama and K. Oinuma, *Clay Science* (*Jap.*), **1**, 1962, 113-119.

[56] *Zapiski Vsesoyuz. Mineral. Obshchestva*, **85**, 1956, 407-412.

[57] *Zapiski Vsesoyuz. Mineral. Obshchestva*, **88**, 1959, 39-47; cf. H. Scholze and A. Dietzel, *Naturwissenschaften*, **42**, 1955, 575, for kaolinite and montmorillonite.

[58] H. Dutz, Dissertation Bergakademie Clausthal, Germany, 1959, 23 pp.; *Tonind.-Ztg. u. Keram. Rundschau*, **83**, 1959, 219-238 (with H. Lehmann).

[59] *Central Glass & Ceram. Research Inst. Bull.* (*India*), **7**, 1960, 97-109.

clay minerals by infrared absorption measurements, especially based on American and Indian ceramic clays.

THE SMECTITE CLAY MINERALS (MONTMORILLONITE AND ITS DERIVATIVES)

31. For the structural characterization of the montmorillonites as typical 2 : 1 clay minerals, of dioctahedral type with pyrophyllite parcels, or of trioctahedral type with talc parcels, T. F. Bates (1961) emphasized that the differentiation of expandable (swelling) montmorilloids and non-swellable illites or other mica-like clay minerals with fixed alkali, preferably potassium ions in the layers between the parcels, should not be overestimated. According to Ch. E. Weaver,[60] however, real transitions are possible by base exchange reactions. In spite of that, we may discuss in general the typical montmorillonites or smectites as swellable clay minerals in a particular group of system; we come back to these relations in the example of the illites.

32. The basic and simple relations in the series montmorillonite–beidellite were demonstrated by E. Gutierrez Rios and P. Gonzales Garcia[61] in the example of bentonites from Spanish Morocco, with the maximum replacement degree of Si^{4+} in tetrahedral coordination by Al^{3+} in the ratio 3 : 1, as the beidellite extreme. Similar beidellite clays with differential-thermal curves were described by M. A. Rateev[62] from the Maĭkop Region and the Chernaya Balka. Methylene blue staining differentiates such beidellites from montmorillonites proper by a change of blue-violet tints into green or even yellowish-green (cf. C. ¶ 194), whereas benzidine does not stain at all before a 4 to 6 hours treatment (montmorillonite is promptly stained blue).

33. D. P. Serdyuchenko[63] classified the montmorillonite type minerals according to their Si^{4+} positions in the tetrahedral layers. In montmorillonite proper 3.8 to 4.0 Si^{4+} are contained per unit cell, in beidellite 3.8 to 3.6, in saponites 3.6 to 3.4, and in "parahalloysites" 3.4 to 3.1 Si^{4+}, such as those from the Jurassic sediments of the Northern Caucasus. Typical *nontronites* and magnesium-parahalloysites or even iron-magnesium parahalloysites are related to *saponites*. To this parahalloysite group also belong occasional iron-chromium parahalloysites (from the Gedmyshkh Valley) which previously were called "chromium hisingerites," with a very low degree

[60] *Am. Mineralogist*, **38**, 1953, 698-706.

[61] *Anales real soc. espan. fis. y quim.* (*Madrid*), B, **45**, 1949, 1047-1066.

[62] *Doklady Akad. Nauk S.S.S.R.*, **96**, 1954, 813-816.

[63] *Voprosy Petrog. i Mineral., Akad. Nauk S.S.S.R.*, **2**, 1953, 100-122. On the correct calculation of the crystallochemical formula of smectites, see R. C. Mackenzie, *Silicates inds.*, **25**, 1960, 12-18; 71-75. A. H. Weir and R. Greene-Kelly, *Am. Mineralogist*, **47**, 1962, 137-146, re-examined the present experience on beidellite as an independent smectite mineral, based on the previous definition proposed by Cl. S. Ross and S. B. Hendricks (1945).

of crystallinity. Extensive chemical studies especially on the changes in composition of smectites by base exchange and the swelling effects in their geochemical significance, were made by M. D. Foster.[64] Swelling decreases with increasing octahedral replacement. In this respect the introduction of Fe^{2+} into the octahedral layers has a greater depressing effect than that of Fe^{3+} which in turn appeared to have the same depressing effect as Mg^{2+}. The swelling volume may vary between 66 and 21, a more than threefold difference. In a study of the numerical relations of the lattice parameters of smectites to their composition, G. T. Faust[65] demonstrated the characteristic differences between dioctahedral and trioctahedral clay minerals of this type, in the first subtype starting from pyrophyllite to *galapektite*(with $Si^{4+} : Al^{3+} = 3.88 : 0.12$), in the second subtype from talc to *stevensite* (Si : Al = 4 : 0), *saponite* (with $Si : Al : Fe^{2+} = 3.19 : 0.75 : 0.06$), and *griffithite* (with Si : Al = 3.19 : 0.81).

34. Characteristic of the smectites is their very high degree of variability in the chemical composition, which makes a strict nomenclature extremely difficult and often even contradictory in the application of one species name. As an example we mention the interesting montmorillonite-type cementing clay minerals of andesitic detritus of the Northern Sierra Nevada, Central California, described by J. F. Lerbekmo.[66] The rock has a striking blue color caused by a translucent coating on the dominant dark-colored detritus particles, which are extremely fine laths and fibers oriented perpendicular to the surface of the clastic grains. In its mineralogical characteristics this clay mineral is intermediate between beidellite and nontronite, as a relatively rare combination.

35. Japanese clay mineralogists have contributed much to the characterization of complex smectites, either swelling or non-swelling. Among those studies we shall mention Ch. Kato[67] on "acid clays" swelling in boiling basic salt solutions (sulfates, oxalates, citrates), either completely reversible or irreversible, and stable in the dispersed state or unstable, respectively. Also T. Sudo[68] described smectites rich in Fe^{3+}, especially from the Hanaoka ore deposits (Akita Prefecture), in very fine

[64] *Am. Mineralogist*, **38**, 1953, 994-1006.

[65] *J. Wash. Acad. Sci.*, **47**, 1957, 146-147. A redifinition of *stevensite* by G. T. Faust and K. J. Murata, *Am. Mineralogist*, **38**, 1953, 973-987, characterizes this mineral as different from saponite in lacking essential aluminum, from hectorite in lacking lithium and fluorine. It has an anomalously low base exchange capacity. See also G. T. Faust, J. C. Hathaway, and G. Millot, *Am. Mineralogist*, **44**, 1959, 342-370, with many other related species. See also B. ¶ 120.

[66] *J. Sediment. Petrol.*, **27**, 1957, 298-305.

[67] Resumés, Department Applied Chemistry, Waseda Univ., Japan, 1957, 5 pp.

[68] *Proc. Japan Acad.*, **26**, 1950, 91-95. *J. Geol. Soc. Japan*, **58**, 1952, 487-491 (with S. Ota); *Clay Minerals Bull.*, **2**, 1954, 96-106. See also rich details given by T. Sudo in his book "Mineralogical Studies on Clays of Japan," Maruzen, Tokyo, 1959, 328 pp., especially pp. 227-268. See also T. Sudo and H. Hayashi, *Congr. Geol. Intern.*, *XXa Sesión*, Mexico City 1956, Comité Internacional para el Estudio de las Arcillas, 1959, 43-52.

fibers, and a similar iron smectite from "oyaishi," a vitreous tuffaceous rock, as a mineral intermediate between montmorillonite and nontronite (with Si : Al = 3.69 to 0.31). The number of atoms in the octahedral layers is 2.18; the Fe^{3+} content is lower than in nontronite (0.86), and the Fe^{2+} content is very low (only 0.09). The divalent iron in oyaishi is very unstable and easily oxidized to the trivalent ion, even if merely exposed to daylight (changing the color from green to black within about one hour, and into brown in a few weeks).

36. The variability of the tetrahedral Si^{4+} and Al^{3+} ratio in smectites from Texas was studied by H. E. Roberson,[69] using K^+ or Li^+ ion saturation, heating in glycerine, solvation, and using also the temperatures of dehydroxylation in the differential-thermal curves as indications of the different degree of this substitution. The *griffithite* material mentioned above is according to G. T. Faust[70] an iron-rich saponite, and not a chlorite as which it was first described by E. S. Larsen and G. Steiger (1917). X-ray powder diagrams of griffithite and of saponite are in good agreement.

37. Iron-rich saponites were further identified from Tertiary iron sand beds in Japan by T. Sudo.[71] The mineral is very similar to that described by H. Lemberg already in 1877 from Predazzo (simply called "lembergite"), as an alteration product of augite. It is enriched in dark-green aggregates of a tuffaceous sandstone, with the atomic ratio $Fe^{3+} : Fe^{2+} = 7.95 : 1$, and $Si^{3+} : Al^{3+}$ (tetrahedral) $= 3.62 : 0.38$. The basal reflection (001) is shifted after treatment with ethylene glycol from 15.7 to 16.9 ± 0.3 KX. Characteristic is the high degree of chemical instability of such iron-rich saponites which are easily decomposed by dilute HCl even without heating, and the divalent iron is easily oxidized in open air. By simple X-ray diffraction diagrams it is in most cases impossible to distinguish such an iron saponite, which is trioctahedral, from nontronite which is dioctahedral.

38. Saponites very *low* in iron, practically only magnesian in composition (with Si : Al = 3.63 : 0.37), were found in South Africa (Krugersdorp, Transvaal) by E. R. Schmidt and H. Heystek,[72] of high purity, and in association with chrysotile

[69] *Bull. Geol. Soc. Am.*, **70**, 1959, 1661-1662.

[70] *J. Wash. Acad. Sci.*, **45**, 1955, 66-70.

[71] *J. Geol. Soc. Japan*, **59**, 1954, 18-27; see also an iron saponite discussed by N. Miyamoto, *Mineral. J. (Sapporo)*, **2**, 1957, 193-195, occurring especially in altered basaltic tuff brecciae on amygdules. On the electron-microscope habitus of nontronite see O. E. Radczewski, *Fortschr. Mineral.*, **31**, 1952, 59-60; most of the samples examined showed lath-like forms with fringed ends, e.g., in those from St. Andreasberg, Harz Mountains.

[72] *Mineral. Mag.*, **30**, 1953, 201-210. Another saponite derived from serpentines is very similar to that of Krugersdorp, as described by A. Alietti, *Rend. soc. mineral. ital.*, **12**, 1956, 31-41, and *Periodico Mineral. (Rome)*, **28**, 1959, 65-110, especially pp. 95 ff. This occurrence is interesting because here is an example given for the transition of a typical two-layer silicate (serpentine) into a three-layer (and trioctahedral) smectite mineral.

asbestos in serpentinites. It is very similar to another very pure saponite described by H. P. Cahoon[73] from Milford, Utah, as a product of hydrothermal replacement of a dolomitic limestone, with excellent book-like sheets (sometimes folded, crinkled, or bent) as observed in the electron microscope. In a high degree of dispersion, however, this material shows the "wavy" appearance which is so common in montmorillonites.

39. Another highly pure saponite was discussed by A. Weiss, G. Koch, and U. Hofmann[74] from Groschlattengrün, Fichtelgebirge. This saponite is interesting because of the occurrence of lithium-containing samples which show a transition to *hectorite*, the typical lithium smectite. The X-ray diffraction behavior of this saponite is particularly remarkable because, in spite of the extreme thinness of the book-like sheets visible in the electron micrographs, (*hkl*) interferences are distinctly observed, and the maxima, as for (131) and (20*l*), etc., on the slope of the photometer curve beyond (130) to higher 2Θ values are well developed, whereas common montmorillonites show on the same slope only a uniform continuous course. In this respect the saponite from the Fichtelgebirge is also similar to *batavite* in the vermiculite group (cf. B. ¶ 94). It is further rather striking from the data given by Weiss, Koch, and Hofmann that the appearance of (*hkl*) interferences coincides with the presence of more than 0.55 eq./$(Si,Al)_4O_{10}$ in the chemical composition as the number of exchangeable cations. In a corresponding correlation the innercrystalline swelling is, for Na^+ as the exchangeable cation, the maximum basal spacing of the water-cured saponites = 15.8 Å., whereas, for K^+ as exchangeable ion the swelling is reduced and may even be suppressed. The 0.55 electrons/$(Si,Al)_4O_{10}$ clays represent a characteristic threshold value above which the corresponding smectites show a transition to the properties of vermiculites. In the differential-thermal characteristics the saponite from the Fichtelgebirge is similar to Sudo's iron saponites.

40. A *nickel* containing saponite (previously described as an "ideal beidellite") and related to *garnierite*, occurs in laterite-like crusts on serpentine of Kremnitz, Bohemia. E. Slanský[75] called this mineral "*β-kerolite*" and classes it among the smectites between saponite and hectorite, analogous to the Transvaal saponite (see above).

[73] *Am. Mineralogist,* **39**, 1954, 222-230.

[74] *Ber. deut. keram. Ges.,* **32**, 1955, 12-17. See also very recently a saponite from the Keewanawan basalt flows on the North shore of Lake Superior as described by J. A. Whelan and H. Lepp, *Am. Mineralogist,* **47**, 1961, 430-433. The fibrous habitus in the electron micrographs is interpreted in this case as an alteration phenomenon after fibrous zeolites.

[75] *Zapiski Vsesoyuz. Mineral. Obshchestva,* **84**, 1955, 43-46. On the nature of α- and β-"*kerolite,*" cf. I. I. Ginzburg and A. Rukavishnikova, *Ibid.,* **79**, 1950, 33-44. Because of the analogies between the tubular forms of garnierite and chrysotile which were confirmed by W. Noll and H. Kircher, *Naturwissenschaften,* **39**, 1952, 233-234, we prefer to classify this mineral with chrysotile (cf. Vol. IV, Section A). Also the kerolites are transitions to the chrysotile group.

In its differential-thermal curve this material is similar to garnierite and the saponite mentioned, with endothermic effects at 130° and 835°C., and an exothermic reaction at about 600°C. (cf. Vol. V, Section B). On the other hand, *sauconite* is a real zinc smectite. J. A. Kornfeld[76] described the effects of a replacement of magnesium by zinc in sauconite, esclusivelý in the octahedral layers ranging in the atomic numbers 1.23 to 2.89, or from 44 to 94 per cent of the octahedral positions, and with a linear relation between the amounts in Zn^{2+} and (Al^{3+} + Fe^{3+}) also in the octahedral layers, with very low contents in interlayer cations (chiefly Na^{+}, Ca^{2+}). Copper can enter in the place of Zn^{2+} only in much restricted amounts.

41. *Chrysocolla* is the typical copper smectite mineral, as described from the oxidation zone of copper-zinc-lead ore deposits by L. K. Yakhontova[77] from different Russian occurrences, and in variable degrees of crystallinity, isotropic or birefringent, with different degrees of aging and diagenesis. Although the X-ray diffraction diagrams and the differential-thermal analysis show in general the smectite type of chrysocolla, there are some specific distinctions such as the leachability of copper from the octahedral layers, and a somewhat anomalous ratio $Si^{4+} : O^{2-} = 4 : 12$ (not the normal ratio 4 : 10), caused by the presence of brucite-like $Cu(OH)_2$ interlayers. Si^{4+} is replaceable by Al^{3+} as in normal montmorillonites, combined with the presence of alkalies and calcium in exterior layers, and with distinct water interlayers.

42. *Volkonskoite* as the characteristic chromium smectite is described by D. McConnell[78] in Upper-Jurassic sediments of the Morrison Formation in Utah, together with the well-known particular vanadium and uranium minerals. It is enriched on fractures which cut the strata at high angles, thus indicating a deposition from circulating ground waters. The electron micrograph shows typical fibers and laths; the X-ray diagram shows the basic montmorillonite-type structure. A. Weiss, G. Koch, and U. Hofmann[79] identified this interesting chromium mineral also in the basalt of Groschlattengrün, Fichtelgebirge (on saponite occurring in the same locality, see above) with X-ray and differential-thermal analysis, and in the electron microscope with lath-shaped crystallites. The striking similarity with nontronite was early recognized by Cl. S. Ross and St. B. Hendricks (1945); a distinctive mark, however, is found by observing the rapid oxidation of nontronite in the moist atmosphere, whereas volkonskoite remains unchanged. The reaction is much accelerated, to a few minutes, if a 1 N ammonia solution is applied.

[76] *Bull. Geol. Soc. Am.*, **69**, 1958, 1598-1599.

[77] *Vestnik Moskov. Univ.*, 7, 1952, No. 6, *Ser. Fiz. Mat. i Estestven. Nauk*, No. 4, 123-130.

[78] *Natl. Acad. Sci.-Natl. Research Council, Publ.*, **327**, 1954, 152-157; *Proc. Natl. Conf. on Clays and Clay Minerals*, **2**, 1954, 152-157.

[79] *Ber. deut. keram. Ges.*, **31**, 1954, 301-305; cf. D. P. Serdyuchenko, 1933.

43. One cannot say that the knowledge of the crystal structure of the smectites in general, and of montmorillonite in particular, was much advanced in the last decade. The proposal made by U. Hofmann *et al.* (1933) still lacks its definite confirmation by structural details, but also the very fascinating ideas of H. Edelman and J. Ch. L. Favejee (1940) still have only the rank of more or less plausible hypothesis. It is extremely difficult to have sufficiently developed interference diagrams to which the structure calculation methods based on monocrystals may be applied, if one depends entirely on powder interference diagrams with their many possibilities of error. Most authors, therefore, preferred indirect ways of making conclusions by trial and error from the proposals presented by different authors for the best possible approximations of the real physicochemical and structural behavior.

44. G. Brown[80] tried to apply a Fourier investigation on montmorillonite under the assumption that monovalent cations in the montmorillonite-glycerine complex may occupy positions midway between the inorganic layer sheets. Brown prefers in this interpretation the Hofmann proposal for the montmorillonite structure. From the experimental study of electron diffraction by pronounced layer structures in the crystallization of BiI_3, and of the double salt $CdBr_2.CdI_2$, Z. G. Pinsker[81] comes to the conclusion that these are not two-dimensional defect structures but show the "forbidden" interferences of (*hkl*) sharp, as they also occur in montmorillonite. Hofmann's ideas on two-dimensional cross-lattice diffraction patterns are, therefore, criticized. The actual X-ray methods, however, are not sufficiently effective in such cases.

45. A new way of finding out a solution of the structure problems of montmorillonite was developed by H. E. Schwiete[82] starting from the idea that the basic "pyrophyllite parcel" with the octahedral layers "sandwiched" between two tetrahedral layers does not correspond to the optimum arrangement but to another arrangement in which the parcel should be built up from two tetrahedral layers combined, with one subsequent octahedral layer. This idea is derived from the observation of intermediary spinel-like phases in the calcination products of montmorillonite (cf. Vol. V, Section B), with a more or less regular oxygen-packing structure in which the cations are distributed over octahedral and tetrahedral cavities. This very typical transition phase is better understood by two subsequent tetrahedral layers with an α-quartz-like interlinkage, above which the octahedral layer is super-imposed (cf. Fig. B. 5). The water interlayer of montmorillonite appears in this proposal arranged between one tetrahedral and one octahedral boundary layer of subsequent parcels.

[80] *Clay Minerals Bull.*, **1** 1950, 109-110.

[81] *Trudy Inst. Krist. Akad. Nauk S.S.S.R.*, **7**, 1952, 35-48.

[82] Cf. *Zement-Kalk-Gips*, **9**, 1956, 351-357, especially p. 352; *Naturwissenschaften* **42**, 1955, 176 (with G. Franzen and H. Müller-Hesse).

The extensive discussion of this structural possibility by H. E. Schwiete and G. Franzen[83] is based chiefly on elaborate experiments on the calcination behavior of montmorillonite and its interpretation from X-ray interference phenomena before the details of the consequences of the new structure proposal were given. The appearance of (*hkl*) interferences from montmorillonite is again emphasized (see above).

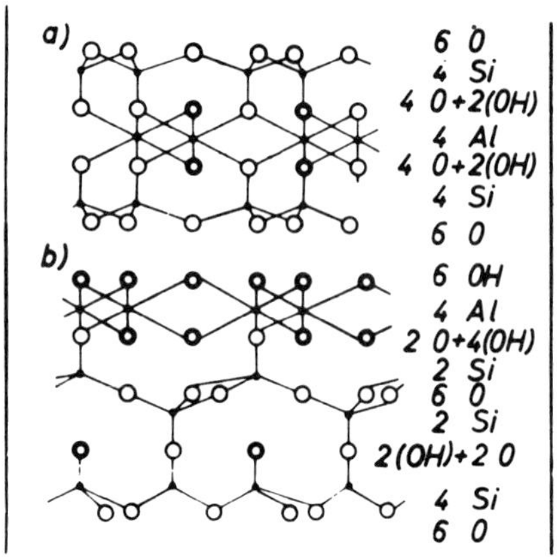

FIG. B. 5. Structure of montmorillonite according to the proposal of Schwiete, Franzen, and Müller-Hesse (*b*), in comparison with that of Hofmann, Endell, and Wilm (*a*).

46. For all the reasons considered, Schwiete and Franzen abandoned the assumption of simple pyrophyllite parcels of the symmetric type, and adopted a *polar* structure. An "ideal" montmorillonite (without any cation substitution) should behave in certain respects as does halloysite. Replacements of Si^{4+} by Al^{3+} should not be restricted to the one exterior tetrahedral layer but might also occur in the central tetrahedral sheet. The resulting negative electrostatic charges can be *irreversibly* compensated by the entrance of *non*-exchangeable cations. If the neutralization is caused by protons, these may react with neighboring hydroxyl groups forming water which is split off. The resulting bridging oxygen anion belongs only by coordination to the $[AlO_4]$ group (cf. Fig. B. 6). It is interlinked to $[SiO_4]$ groups at the same time.

47. Schwiete *et al.* claim to give by this mechanism an interpretation of the base exchange capacity of real montmorillonites which is always considerably lower than that theoretically calculated, because of the reduction of the number of exchangeable

[83] *Forschungsber. Wirtsch. u. Verkehrsministeriums Nordrhein-Westfalen*, **484**, 1958, 70 pp., with extensive critical remarks on the previous proposals; on the dehydration behavior of montmorillonite in its correlation with the new structure model cf. H. E. Schwiete, G. Ziegler, and Chr. Kliesch, *Ibid.*, **545**, 1958, 48 pp.

ions when a part of these enters the central tetrahedral layers. Montmorillonite and beidellite are chiefly different in this respect by the fact that in the first mineral the tetrahedral substitution of Si^{4+} by Al^{3+} occurs in the exterior tetrahedral layer but in beidellite in the central layers. The cations are sorbed on montmorillonite chiefly on octahedral layers, and a strong hydration of the cations is made possible, whereas

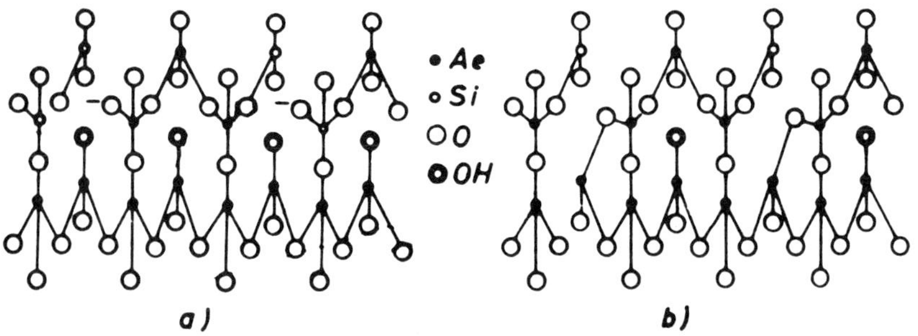

FIG. B. 6. Substitution of the central tetrahedra layer of montmorillonite, and neutralization of the electrostatic charges caused by this process (Schwiete and Franzen).

in beidellite the bridging of the cations impedes the swelling capacity. In trioctahedral montmorillonite-type minerals, the substitution in the octahedral layers brings about positive electrostatic charges; in the tetrahedral layers, negative electrostatic charges. The resulting coulombian attraction between the parcels suppresses the swelling considerably, or even completely.

48. Besides the consequences of Schwiete's structure model for the water content of the montmorillonite "molecule," which are not discussed here, this proposal gives a fascinating illustration of the formation of α-quartz in a distinct calcination phase at 1000°C., previously observed by W. F. Bradley and R. E. Grim.[84] The tetrahedral layers contain the unit cell of α-quartz preformed, as in the tetrahedra I, II, III, and IV, shown in Fig. B. 7 (the oxygen atoms in each tetrahedron are numbered by 1, 2, 3, 4). In every case, the tetrahedra of the quartz unit cell remain coherent by "structural inheritance," the disruption of the structure being restricted to the tetrahedral faces only in those positions where the change into α-quartz requires an intense torsion.

49. It may speak in favor of the Edelman-Favejee structure proposal that E. Forslind,[85] in his studies on the water adsorption and base exchange in montmorilli-

[84] *Am. Mineralogist*, **36**, 1951, 182-201, especially pp. 191 ff.

[85] *Svenska Forskningsinst. Cement Betong vid Kgl. Tek. Högskol. Stockholm, Handl.*, No. 17, 1952, 1-12; No. 18, 1953, 14 pp. Proc. Second Intern. Congr. on Rheology, Oxford, 1953, 50-63,

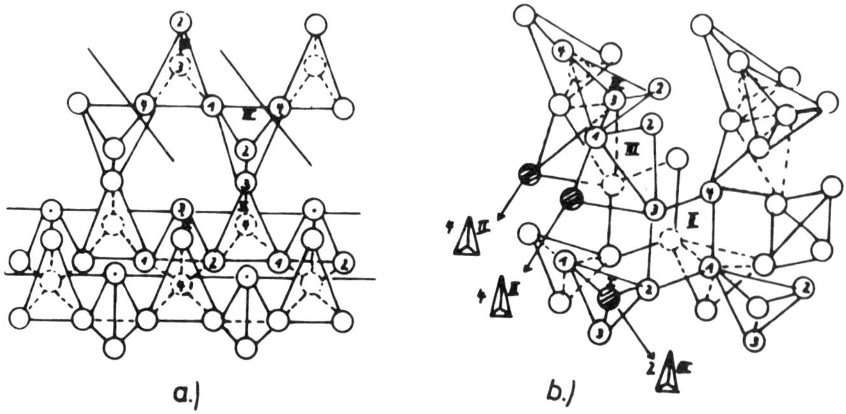

FIG. B. 7. Formation of α-quartz by the thermal decomposition of the montmorillonite structure at 1000°C. (Schwiete and Franzen).

nite, started from an ice-like adsorption layer of water on the clay mineral surface over bridging hydrogen bonds (mechanism of "charge transfer"), and superimposed on the hydroxyl groups of the montmorillonite (as shown in Fig. B. 8) with an excellent fit. The typical "*trans*" configuration in the silica tetrahedra, as given in the Edelman-Favejee model, implies the existence of hydroxyl groups protruding on the surface of the clay mineral sheet. Although this is not yet firmly proved, the interpretation

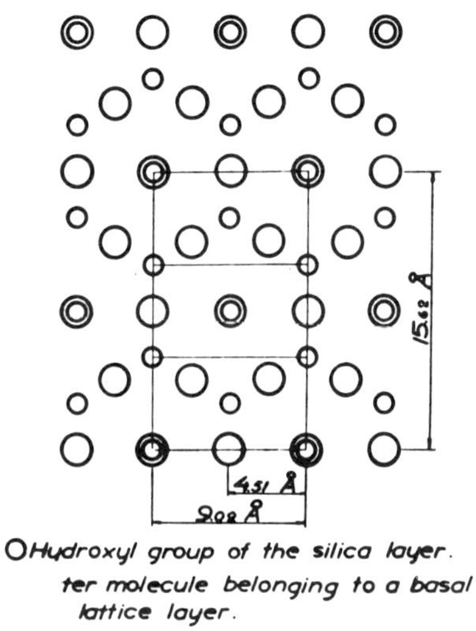

FIG. B. 8. Basis position of ice lattice molecules superimposed onto the hydroxyl groups on (001) of montmorillonite (Forslind).

especially pp. 58 ff. Cf. also Å. Grudemo, *Svenska Forskningsinst. Cement Betong vid Kgl. Tek. Högskol. Stockholm, Handl.*, No. 22, 1954, 56 pp., especially pp. 4 f. and 16.

by the Edelman-Favejee model is in better agreement than is the alternative "*cis*" configuration as given in Hofmann's model. The "*trans*" configuration of the definitely acidic hydroxyl groups is confirmed by the diazomethan reaction as studied by G. Berger (cf. C. ¶ 161), and Forslind is of the opinion that the Edelman-Favejee structure model is the most reasonable working hypothesis available, with $[SiO_3(OH)]$ groups as the basis of the adsorption. The presence of highly reactive hydroxyl groups on the surface of different clay minerals (kaolinite, halloysite, montmorillonite), and on chabazite, on the other hand, is made probable by J. A. Faucher and H. C. Thomas[86] in studies on the exchange of H by D from heavy water. However, these experiments do not completely rule out a modified Edelman-Favejee structure for montmorillonite in which a part of the hydroxyl groups *is* exchangeable with OD.

50. U. Hofmann[87] objected to the Edelman-Favejee hypothesis especially because it postulates a much higher content in thermally easily removed OH groups than ever would be calculated from carefully conducted analytical determinations. Also the presence of *infrared absorption* bands at 2.75 μ must not necessarily give evidence of surface hydroxyl groups directly bonded to Si^{4+} ions, but may be caused by hydroxyl groups of the deeper lying ("sandwiched") octahedral layers as they are typical of the pyrophyllite parcels and in this mineral proper.[88] Concerning the proposal of Schwiete and Franzen, Hofmann makes some preliminary remarks stating why he prefers the symmetrical arrangement of the sequence tetrahedral-octahedral-tetrahedral layers, and not a more irregular one. From a study of *organic* clay mineral derivatives, D. J. Greenland and E. W. Russell[89] state the objection to the Edelman-Favejee proposal that one ought not to speak of covalent bonding of organic groups to the surface of silicate minerals if adsorption phenomena have not been definitely excluded. This criticism chiefly concerns the discussions on clay mineral chlorides and acetylates as claimed in the experiments of E. Deuel (C. ¶ 304 ff.). In every case, Greenland and Russell refute such assumptions as valid arguments and main evidence for the Edelman-Favejee hypothesis.

[86] *J. Phys. Chem.*, **59**, 1955, 189-191; cf. previously C. D. McAuliffe, N. S. Hall, L.A. Dean, and St. B. Hendricks, *Soil Sci. Soc. Am. Proc.*, **12**, 1947, 119-123.

[87] *Angew. Chem.*, **68**, 1956, 53-61.

[88] See, regarding this important argument, P. F. Kerr, *Am. Petrol. Inst. Project* 49, *Clay Mineral Standards*, Report No. 8, 1951, pp. 104, 121, 139. A revision of Edelman's and Favejee's proposal given by Edelman himself in *Verre et silicates inds.*, **12**, 1947 (6), 3-6, opens a possibility that the number of hydroxyl groups in tetrahedra turned outside with their apexes may be considerably reduced, differing from the original proposal. M. Oberlin, R. Glaeser, and J. Méring, *Compt. rend. Congr. Géol. Intern., Algeria*, **19**, 1952, No. 18, 1953, 105-115, even decide against these reduced hydroxyl concentrations, after careful measurements of the water contents in montmorillonite using the K. Fischer reagent method.

[89] *Trans. Faraday Soc.*, **51**, 1955, 1300-1307. Also E. Gruner and R. E. Vogel, *Kolloid-Z.*, **118**, 1950, 100-110, come to the result that the Edelman-Favejee proposal is not able to explain aging processes in calcium bentonites by a chemosorption as this can be done in the Hofmann model.

51. The *swelling* phenomena of the smectites constitute perhaps that physico-chemical property of these fascinating minerals which is the most striking, in correlation with structure. They were early investigated, especially by U. Hofmann and his school, first in analogy with the reactions in the formation of so-called graphite oxides and corresponding "addition compounds" with anhydrous hydrofluoric acid, and even with alkali metals. We mention in this respect the excellent structural studies of W. Rüdorff,[90] and the more recent results of A. Clauss, R. Plass, H.-P. Boehm, and U. Hofmann.[91] The leading principle of these "models" of the smectic swelling reactions is the formation of intercalated interlayers of the structure, with regular stepwise widening in the direction perpendicular to (001) as observed in the X-ray powder diagrams. After the classical investigations of U. Hofmann and his school on the structural effects of the intercrystalline swelling by the introduction of water and other liquid agents, we may emphasize some of the industrial applications seen in this field, as demonstrated hy J. W. Earley, B. B. Osthaus, and I. H. Milne,[92] on the one hand, and U. Hofmann, A. Weiss, G. Koch, and A. Mehler,[93] on the other hand. These studies chiefly concern the *bentonitic rocks*, commonly containing besides montmorillonite more or less cristobalite, hydrous micas, and others as accessories. Considerable amounts of cristobalite were found by G. W. Brindley[94] in a fuller's earth of Georgia. The apparent α-cristobalite in this occurrence is in reality a highly disordered β-cristobalite (cf. C. ¶ 149). The electron microscope shows in this material the fine-flaky or lath-shaped montmorillonite together with globular aggregates of cristobalite.

52. We further emphasize the report of U. Hofmann *et al.*[95] on the interrelations existing between the intracrystalline swelling of montmorillonite and its cation-exchange mechanism, also with its anion exchange, namely of hydroxyl ions as by F^- (cf. C. ¶ 183). The essential functional relation of the swelling phenomenon is that to the surface density of the exchangeable cations between the silicate sheets. It is characteristic of the anion exchange that the exchange capacity is localized within hydroxyl positions on the boundaries of octahedral layers. In nontronite the anion exchange OH^-–F^- is increased if the divalent iron is oxidized to trivalent ions.

[90] *Z. anorg. u. allgem. Chem.*, **254**, 1947, 319-328; *Ibid.*, **277**, 1954, 156-171 (with E. Schulze).

[91] *Z. anorg. u. allgem. Chem.*, **291**, 1957, 205-220.

[92] *Am. Mineralogist*, **38**, 1953, 707-724, with differential-thermal curves, X-ray diagrams, etc.

[93] *Silikat Tech.*, **6**, 1955, 524.

[94] *Clay Minerals Bull.*, **3**, 1957, 167-169. An interesting study of E. J. Weiss and R. L. Stone, *Am. Ceram. Soc. Bull.*, **33**, 1954, 51-54, concerns the presence of "hidden" smectites as typical three-layer clay minerals in Texas and Florida kaolins, and Tennessee ballclays as seen from detailed thermal-differential analysis and X-ray diffraction diagrams.

[95] *Natl. Acad. Sci.-Natl. Research Council, Publ.*, **456**, 1956, 273-287; *Proc. Natl. Conf. on Clays and Clay Minerals*, **4**, 1956, 273-287 (with A. Weiss, G. Koch, A. Mehler, and A. Scholz).

53. The formation of quite definite montmorillonite *hydrates*, complexes of stoichiometrically constant multiples of the ratios between the silicate network and the water interlayers in their structures (cf. W. F. Bradley, R. E. Grim, and G. L. Clark, 1937), was rediscussed by R. S. Roth[96] for bentonite montmorillonites from a more practical viewpoint. One kind of these complex minerals has two water molecules, the second only one water molecule layer, and the third no water at all adsorbed on the inner surface. There is a characteristic correlation to the contents in exchange cations and, therefore, to the practically important properties of bentonites. The double-water-layer particles contain Ca^{2+} as exchangeable cations, the single-water-layer particles contain Na^+, and the zero-water-layer minerals are probably electrostatically neutral. For an estimate of the drilling mud yield values in water suspension (cf. C. ¶ 332), one may use the behavior of the (001) interference in the powder diagram. Best yield values are observed with bentonites having basal spacings between 12.85 and 12.5 Å. A certain exchangeable Ca^{2+} content is needed for a mud with an optimum drilling behavior in order to create sufficient structural disorder in the clay-water system to allow an easy dispersion.

54. On the basis of well-characterized *synthetic* smectite (montmorillonite, beidellite, saponite types) with controlled contents in exchangeable cations, F. H. Gillery[97] investigated systematically the adsorption and desorption characteristics in humid atmosphere. The saturating cations have the predominant influence on the dehydration behavior. Starting from the general formula for beidellites,

$$\underset{\substack{|\\ M_y}}{Al_2[(Si_{4-x}Al_x)O_{10}](OH)_2}$$

and for saponites

$$\underset{\substack{|\\ M_y}}{Mg_3[(Si_{4-x}Al_x)O_{10}](OH)_2}$$

the cation M^+ is monovalent ($x = y$) or divalent ($y = \frac{1}{2}x$). For most natural smectites and the synthetic samples of Koizumi and Roy $x = 0.33$ (normal type with symbol N), and for $x = 0.17$ and $x = 0.67$ the corresponding symbols of montmorillonite are $N/2$ and $2N$, respectively. The $2N$ saponite approaches the composition of vermiculite (cf. G. F. Walker,[98] and B. ¶ 83 ff.), and the accompanying tabulation shows this close relation, with a constant difference of spacings:

[96] *Am. Mineralogist*, **39**, 1954, 340.

[97] *Am. Mineralogist*, **44**, 1959, 806-818. On the synthesis of the samples, cf. M. Koizumi and R. Roy, *Ibid.*, **44**, 1959, 788-805.

[98] *Natl. Acad. Sci.-Natl. Research Council, Publ.*, **327**, 1955, 101-115.

Synthetic Ca–*N* Saponite (Koizumi and Roy) (Å.)	Vermiculite (Walker) (Å.)	Difference (Å.)
15.5 (two layers of water)	14.81	+ 0.7
15.0	14.56	+ 0.6
13.1	13.82	— 0.7
12.4 (one layer of water)	11.59	+ 0.8
11.7	20.6 (irregular mixed layers of 11.59 and 9.02)	
9.8	9.02	+ 5.8
	Average	+ 0.7

55. It is further evident that the smectites of the 1 *N* substitution type are more stable than those of the *N*/2 and the 2 *N* composition, equally thermally stable, and in the development of the maximum spacing 15.5 Å. for the double-water-layer type, if one considers the number of interlayer cations increasing from the implicit pyrophyllite or talc structures. The first few cations expand the structure very little, with distorted layers surrounding the cation positions. If more cations are introduced, progressively more distortion will take place and the crystallinity becomes poor as it is observed for the *N*/2 specimens. With cation concentrations approaching the *N* type (corresponding to 1/3 of the cation number in micas) the associated water molecules extend the layers completely without considerable distortion, in a phase with a minimum of free energy. Further added cations again cause an increase in bonding between the layers which are drawn closer together, decreasing the spacing. Water will then find difficulty in entering between the silicate layers, and the work of hydration will increase with increasing free energy to form the 2 *N* type, and so on.

56. R. C. Mackenzie[99] considered the interactions of the cations in a single-water interlayer of montmorillonite, forming flattened spheres of vibration limited by the interlayer distances. There are evidently differences between the loosely retained cations which are characteristic for the substitution of Mg^{2+} for Al^{3+} in the octahedral structure layer, and the more firmly retained cations if Al^{3+} substitutes Si^{4+} in the tetrahedral layers. In the first case the possible hydration state of the clay mineral will be higher. These conclusions are entirely in agreement with Gillery's observations that the exchangeable cations predominantly determine the water sorption in smectites. The hydration energies of different hydration shells can be estimated from the endothermal peak temperatures in the differential-thermal analysis curves.

[99] *Clay Minerals Bull.*, **1**, 1950, 115-119.

The energy of hydration of the sheet surface can be calculated to be about 10,000 cal./gm. ion. The hydration conditions for the Li^+ ion in smectites, however, show anomalies characteristic of this ion of highest hydration degree.

57. The interlayer water is very distinctly indicated in the broad 2.95 μ infrared absorption band described by H. Scholze and A. Dietzel,[100] the same band which is also characteristic of adsorbed water on silicate surfaces in general. The sharp 2.75 μ band, however, is according to Hofmann's proposal of the montmorillonite structure characteristic of the group Al—O··H····O in the inner of the lattice, and may be interpreted by the $[SiO_3(OH)]$ groups of the tetrahedral sheets on the surface of the montmorillonite structure as proposed by Edelman and Favejee. In the latter case a coincidence of both kinds of hydroxyl groups in their infrared absorption properties would be possible.

MICA-LIKE CLAY MINERALS

58. Typical three-layer minerals of non-swelling type, both of dioctahedral structure (*illites*) or trioctahedral structure (*ledikites*, according to G. Brown's nomenclature),[101] are connected with the smectite family by distinct transitions, as in the so-called K-(or meta-)bentonites described by Ch. E. Weaver and T. F. Bates[102] from Ordovician limestones of Pennsylvania. Such minerals can even be prepared experimentally by a *fixation* of potassium ions on smectites offered from solutions. These classification groups are not so sharply separated as it may appear. In such cases as the "meta-bentonite" complex, mixed-layer stacking may also be observed, namely of a mineral species which is expandable in ethylene glycol, with another non-swellable material. We characterize these minerals in B. ¶ 121. Their non-swellable portion, however, would be a typical dioctahedral illite. Trioctahedral, on the other hand, is the biotite-derived mineral which G. F. Walker[103] described

[100] *Naturwissenschaften*, **42**, 1955, 575.

[101] *Clay Minerals Bull.*, **2**, 1955, 294-302. See also the classification principles discussed by Ch. E. Weaver, *Am. Mineralogist*, **38**, 1953, 698-706.

[102] *Clay Minerals Bull.*, **1**, 1952, 258-261; *Bull. Geol. Soc. Am.*, **64**, 1953, 921-943. The natural K-bentonite was formed according to Weaver from a 2 : 1 montmorillonite (derived from volcanic glass material) which was changed into illite by the adsorption of potassium ions from circulating solutions (with up to 80 per cent of not-expanding shares in the end product).

[103] *Mineral. Mag.*, **29**, 1950, 72-84. The name *pholidolite* proposed by S. B. Hendricks, *J. Geol.*, **50**, 1942, 276-290, is not to be recommended. On the use of the spacing of (060) for the distinction of dioctahedral and trioctahedral mica-like clay minerals cf. also C. I. Rich, *Am. Mineralogist*, **42**, 1957, 569-570. For a distinction of illite from kaolinite it is important to note both minerals are dioctahedral and, therefore, have the spacing for (060) = 1.49 to 1.50 Å. in common; but after calcination at 550°C. the interference for kaolinite disappears, that of illite remains in its full intensity.

from soils of northeastern Scotland. This author also assigned in the powder diagrams of mica-like minerals the interference line (060) to be indicative either for dioctahedral or for trioctahedral types, namely by a spacing of 1.49 to 1.5 Å. in the first case and 1.53 to 1.55 Å. in the second case. This is what Walker called a "montmorillonite product (*b*)" of weathering if clay biotite in soils is a material which takes up glycerine in a double layer, with a spacing of 17.7 Å. The more frequent weathering product, nevertheless, in the Scotland soils, is a vermiculite (cf. B. ¶ 86 ff.).

59. Besides the trioctahedral type, dioctahedral illite may also occur in soil clays of this type, together with kaolinite, and may make up (e.g., in Ordley, Aberdeenshire) 70 to 80 per cent. From a soil-genetic viewpoint, it is therefore often very difficult strictly to use the structural distinction marks in discussing these minerals if both types of illites and ledikites (Brown) are intimately mixed, in addition to vermiculite. Overemphasizing of a single property, such as the expandibility alone, may also lead to much confusion. We agree with Weaver in his classification proposal that it is often preferable to state that all 2 : 1 mica-type clay minerals may be treated as one group, only with a subdivision into varieties of illites or ledikites, if these can be distinguished with certainty.

60. For the illites derived from muscovite, H. S. Yoder and H. P. Eugster[104] also discussed, in their fundamental synthetic studies of the micas, the hydrothermal-genetic conditions (cf. Vol. IV, Section A) in connection with the problems of the genesis of muscovite in different polytypes of various degrees of crystallinity and ordering (cf. A. ¶ 134 ff.). Illite, together with varieties like pinite, gieseckite, groppite, hydromica, mariposite, Maegdefrau's and Hofmann's "sarospatakite," potassium-bentonite (see above), and Hofmann's "glimmerartiger Ton" are here concerned. They are also viewed in relation to magnesium and iron containing phyllosilicates like *celadonite*, nontronite, and the montmorillonites in general, as seen in Fig. B. 9.[105]

61. What Yoder and Eugster called metamorphic 1 M d muscovites may be for a great part *sericite*, which we do not include as such with the clay minerals proper, in spite of its rich abundance in low-metamorphic sedimentary rocks. In their lists, however, are also included the true illites of the occurrence in Vermilion Co., Illinois, potash-bentonites from Tennessee and Georgia, sarospatakite from Hungary, all of the 1 M d type, but not higher in structural order (e.g., 2 M). Yoder and Eugster even go so far as to characterize "most illites in fact as mixed-layer structures." In every case, the illites are on the representative join-line from mont-

[104] *Carnegie Inst. Wash. Year Books,* **52**, 1952/53, 64-69; *Ibid.*, **54**, 1954/55, 127-129; *Geochim. et Cosmochim. Acta,* **8**, 1955, 225-280, especially, pp. 249 ff.

[105] It may be remarked regarding this diagram that not all of the large fields of the indicated minerals correspond to isomorphic-crystalline solutions. They also include merely physical mixtures and mixed-layer assemblages with which we are here not especially concerned.

morillonite to muscovite in the diagram referred to above. "Hydromica" from Montrose, Colorado, is characterized as a mica in which the K^+ ions are replaced by the oxonium ion, H_3O^+. Some of the synthetic muscovites are near the ideal muscovite composition on the join oxonium muscovite to potassium muscovite rather than on the join muscovite-montmorillonite. More recently Yoder[106] included in his discussions of clay minerals also the *di*octahedral *glauconite* with a 1 M d structure, whereas Walker's clay mica from Scotland is a *trioctahedral* 1 M d structure (mostly in mixed layer relation with montmorillonite) and related to synthetic phlogopite (in nature with biotite).

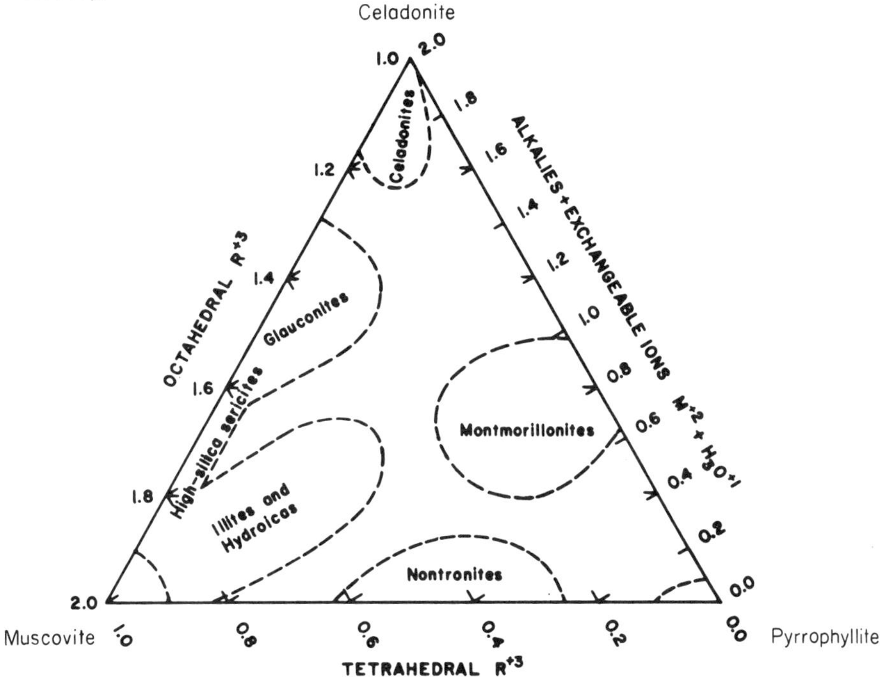

FIG. B. 9. Tetrahedral and octahedral cations R^{3+}, plotted in atomic proportions for dioctahedral micas and derived minerals (Yoder and Eugster).

62. That the slow alteration of micas by hydrothermal metasomatosis not only brings about kaolinite and halloysite clay deposits but sometimes also those of the so-called "*hydrous mica*" (later called "illidromica" in the Italian literature) was demonstrated by C. Andreatta[107] in elaborate studies of clays of Capalbio (Orbetello).

[106] *Proc. Natl. Conf. on Clays and Clay Minerals*, **6**, 1958, 42-60. On an interesting stacked-layer structure of Japanese magnesium containing sericite of great randomness, see M. Nakahira, *Clay Minerals Bull.*, **2**, 1954, 141-146.

[107] *Rend. soc. mineral. ital.*, **6**, 1949, 39-47; *Periodico Mineral.* (*Rome*), **18**, 1949, 11-31; *Clay Minerals Bull.*, **1**, 1949, 96-99.

All of these contain much reduced alkali contents, in comparison with the original mica, until by a complete removal of the alkalies a beidellite of dioctahedral type may have developed. If the original mica was trioctahedral, the resulting illidromica is also usually of the trioctahedral (ledikitic) type. At the same time when K^+ is reduced, water is introduced, probably in interstitial structural layers, and all transitions to real montmorillonite are observed, the intermediate products showing wide systematic changes in optical properties and composition. The chemical relations between the original dioctahedral micas and the hydrous micas were carefully studied by M. D. Forster.[108] In the series from muscovite to dioctahedral phengites to glauconite and celadonite, another series analogous to the hydromuscovites (illites) is noted, in which almost all of the changes in the metasomatic reactions occur not in the tetrahedral but in the octahedral layers. Thus, in the second series the introduction of unusual constituents like chromium, vanadium, or manganese, etc. is even possible.[109] Following the formulation of G. Brown and K. Norrish,[110] one could describe the general crystallochemical composition of hydrous micas and illites as $(Ca,Na,K,(H_3O))_2(Al,Mg,Fe,Ti)_4[(Si,Al)_8O_{20}](OH)_4$.

63. In Japan, Y. Kubo[111] identified mica-like alteration products of muscovite in the so-called "kira" deposits of Gairome which are used for ceramic purposes. T. Sudo[112] described from ore deposits of epithermal character in Japan very common "sericites" of low crystallinity, intermediate between muscovite and illites. They showed reduced contents in K^+, but with increased water contents, and an endothermal peak at 600°C. on the differential-thermal curve. On the other hand, Ch. E. Weaver[113] characterized from grauwackes a lath-shaped, not expanding dioctahedral clay mineral with less Fe^{2+} and Mg^{2+} than illites usually contain. In the X-ray diagram several interference lines appear, which do not occur in common illite but do so in muscovite. No expandibility is observed on treatment with $CaCl_2$ solution, or with ethylene glycol. The endothermic peak characteristic of dehydroxylation for this material is also higher than in common illites, near 600°C., with a material of relatively high crystallinity.

[108] *Am. Mineralogist*, **39**, 1954, 328-329.

[109] An interesting material of similar kind is the "chrome mica clay" described by P. F. Kerr and P. K. Hamilton, *Am. Mineralogist*, **43**, 1958, 34-47, which demonstrates the appearance of a transition between muscovite and illite in differential-thermal behavior, with the typical dehydroxylation at about 600°C., although this species is better crystallized than 1 M d (1 M or even 2 M).

[110] *Mineral. Mag.*, **29**, 1952, 929-932. On the alkali contents of hydromuscovites see recently H. Kodama, *Clay Science (Jap.)*. **1**, 1962, 89-99.

[111] *Tojiki Shikenjo Kenkyu Jiho*, **6**, 1952, 7-12.

[112] *Sci. Repts. Tokyo Kyoiku Daigaku*, Section C, No. 23, 1954, 173-197. See also T. Sudo's book "Mineralogical Studies on Clay of Japan," Maruzen, Tokyo, 1959, 328 pp., especially pp. 73-87.

[113] *Am. Mineralogist*, **38**, 1953, 279-289.

64. In the great confusion arising especially in the classification of ceramic raw materials, it occurs that important illite clay formations from the Vicentinian Pre-Alps in northern Italy are called "kaolins." They are of siallic composition clearly derived from effusive and vein rocks of porphyry or porphyrite type, changed to sericite, montmorillonite, and illite. The deposits of Laghi and Tretto (Arsiero) were accurately analyzed by G. Peco[114] with illite (and sericite) as the principal clay minerals. On the other hand, the Mediterranean "terra rossa" (red earth) from Istria, Friaul, Umbria, Apulia, and Cyrenaica contains abundant illite, as S. Cecconi[115] established. This result is very important in comparison with the clay minerals in lateritic soils (e.g., of Angola and Paraguay) which contain kaolinite as the predominant constituent. I. D. Sedletskiĭ[116] examined sediments of the carboniferous formations of the Donets Basin, with much illite, little kaolinite, and perhaps some montmorillonite. The illite shows a *higher* water content than the usual minerals of this family, but also a higher birefringence. Even the particularly practically important clays of Chassov-Yar show in the electron microscope a complex mixture of a mica-like mineral and kaolinite. The "*monothermite*" (cf. Belyankin, 1932; Sedletskiĭ, 1949) is distinguished from normal illite by its lower contents in alkalies and alkaline earths, but higher in water, and the (001) interference line has a spacing of 10.34 Å. (illite normally 9.98 Å.).

65. An extensive investigation of the weathering of micas (and of chlorites) including the hydromicas, hydromuscovites, and illites, with their transitional phenomena was made by I. I. Ginzburg.[117] On jefferisite of the vermiculite group see B. ¶ 85. Hydromuscovites are defined as the first stage of the changes from muscovite in which essential optical, thermal, and structural changes are not yet observed. When later the alkalies are progressively leached away, lower birefringence is measured, and typical endothermic effects appear which approach the products more and more to kaolinite as the end product. In *bravaisite* just half of the potassium content of the normal muscovite is removed. The mineral is interpreted as a mixed-layer structure of montmorillonite and illite, just as "monothermite" is called a mixed-layer product of hydromica and kaolinite, nearer to the latter clay mineral. The important Chassov-Yar mineral is a hydromica with 70 per cent of the potassium replaced by the hydroxonium ion, and in the monothermite from the Ural as much as 97 per cent of the potassium is replaced. The sum ($H_3O^+ + OH^-$) increases in such transitions from 2.5 to 4.0, as in the monothermites. The heterogeneity of such products is visible in the electron microscope.

[114] *Ceramica* (Milan), **8**, (4) 53-57; (5) 37-42.

[115] *Ricerca sci.*, **28**, 1953, 1645.

[116] *Voprosy Petrog. i Mineral., Akad. Nauk S.S.S.R.*, **2**, 1953, 123-134.

[117] *Voprosy Petrog. i Mineral., Akad. Nauk S.S.S.R.*, **2**, 1953, 139-160.

66. A. A. Levinson[118] called "polymorphism" of the illites and of hydrous micas the various 1 M d disordered structures derived from 1 M, 2 M, and 3 T micas (cf. A. ¶ 133 ff.). It is in his opinion not justifiable to speak of illites only as disordered 2 M-derived muscovites, and great care must be recommended in the correct assignment of the 10.0 Å. interference line to different polytypes of the micas, if these are in transition to illite-type clay minerals. It is also very difficult to correlate the different polytypes and their products to specific conditions of the chemical and geological environment of the micas in their changes to illite and hydrous micas. The well-known illite from Fithian, Illinois, is derived from 2 M and 1 M mica in sedimentary environments. In a non-sedimentary facies they can originate from 1 M and 3 T, as in granites. Levinson could demonstrate how some illites preserve the marks of the mica type from which they are derived, as from the 3 T polytype in illite of Balleter, Aberdeenshire, from 2 M in hydromica of Yorkshire fireclay, from 1 M in illite from Aberdeenshire, and so on.

67. A review of the elementary cell constants of hydromicas, hydromuscovites, and "illidromica," calculated from electron diffraction data, was given by B. B. Zvyagin,[119] with variations for a_0 between 5.16 to 5.29 KX, for b_0 between 8.90 and 9.20 KX, and for c_0 between 10.1 to 10.3 KX, in monolayer structures, and between 20.00 and 20.6 KX, in double-layer minerals, the angle β varying between 99.5° to 101.5°, and 93.5° and 96°, respectively. Classifications can also be made according to the intensity of the interferences (021), ($11l$) and ($\bar{1}1l$). Of course, the structures are here given after the collapse of water interlayers under the high vacuum of the electron diffraction camera.

68. A "*trioctahedral illite*" (ledikite, Brown) is described by A. Weiss, A. Scholz, and U. Hofmann,[120] with a high magnesium content (2.3 atoms per octahedral for-

[118] *Am. Mineralogist,* **40**, 1955, 41-49. Cf. especially R. E. Grim, W. F. Bradley, and G. Brown, in Chapter V on the "Mica Minerals," in the monograph "X-Ray Identification and Crystal Structure of Clay Minerals," Mineralogical Society of Great Britain, 1951, pp. 138-172. A typical 2 M_2 hydromuscovite is described by I. M. Threadgold, *Am. Mineralogist,* **44**, 1959, 488-494, probably of metasomatic origin, from Tasmania. 2 M and 1 M derivatives are also described as "white mica" in the flysch of the Appenine Mts. of Modena, Italy, by P. Gallitelli and R. Olivieri, *Atti e mem. accad. sci. lettere ed arti Modena,* [6] **2**, 1960, 172-178, besides dioctahedral montmorillonite and mixed-layer minerals.

[119] *Kristallografiya,* **1**, 1956, 214-217. On similar problems in the Palaeozoic shales of Illinois, cf. R. E. Grim, W. F. Bradley, and W. A. White, *Illinois State Geol. Survey, Rept. Invest.,* **203**, 1957, 35 pp. Such shales contain besides illite (30 to 80 per cent) also chlorite (0 to 40 per cent), kaolinite (0 to 60 per cent), and mixed-layer minerals. In general a high content in K_2O corresponds to a high content in illite. A much recommended review on the clay minerals of the clays used in ceramic industry for making roof tiles is that of C. W. Correns, *Trans. Intern. Ceram. Congr.,* VI, 1958, 183-194.

[120] *Z. Naturforsch.,* **11**, b, 1956, 429-430.

mula unit), with a pronounced superstructure in the direction of the *a* axis (from 5.32 to 15.0 Å.), and a *regular* distribution of the potassium ions (0.66 atoms per layer unit). A corresponding trioctahedral illite results if swellable vermiculites and batavites (cf. B. ¶ 94 ff.) are treated to exhaustive base exchange in potassium salt solutions. An illite results which swells no more, but has in addition a superstructure with $C_0 = 3 \times 10.44$ Å. If alkyl ammonium ions are introduced, this illite becomes organophilic (C. ¶ 298). Another unusually well-developed trioctahedral illite (with crystals of 1 to 4 mm. in size) in pseudomorphs after biotite, was described by J. Green and P. F. Kerr[121] from Utah (associated with intermixed chlorite). From Bashkyrian horizons E. A. Shabaeva and L. I. Gorbunova[122] described a very peculiar replacement of montmorillonite (with positive methylene blue staining reaction) by illite or hydromicas, characterized by their X-ray and differential-thermal analysis diagrams, with contents in K_2O up to 5.7 per cent. The original montmorillonite is still recognized by the not negligible content of magnesia, and the characteristic ratio Si : Al. The replacement reaction is a common diagenetic phenomenon, widespread in the marine clays of the Saratov District.

69. Another hydromica (illite), very similar in its striking Fe^{2+} and Fe^{3+} content which makes it comparable to nontronite, occurs in the ore deposits of Madzharovo, Bulgaria, as described by T. G. Radonova.[123] In its exterior aspect this hydromica gives the impression of a chlorite-type (chamosite-like) mineral although in the X-ray diagram the typical chlorite interferences are absent. In the electron microscope this very interesting material shows the scaly habitus of hydromicas, though similar to excellently developed illite of Sárospatak.[124] Under favorable conditions one may even observe monocrystalline diffraction diagrams of such illites in the electron microscope examination; this was demonstrated by O. E. Radczewski[125] from the ceramic Schletta clay. The diagrams were evaluated on the basis of the 2 M mica structure, although it is not possible to give a detailed structure analysis. Nevertheless, with a high degree of approximation, the dimensions of the unit cell are calculated to be $a_0 = 5.18$ Å., $b_0 = a_0 \cdot \sqrt{3}$, and $c_0 = 19.84$ Å.

70. An isolated occurrence of halloysite in phosphorite-bearing clayish sediments of the Dniestr Basin was discussed by V. P. Kurochka;[126] here a subsequent change of illite into halloysite was assumed, caused by a diagenetic reaction of specific pH

[121] *Am. Mineralogist*, **39**, 1954, 331.

[122] *Doklady Akad. Nauk S.S.S.R.*, **116**, 1957, 484-487.

[123] *Zapiski Vsesoyuz. Mineral. Obshchestva*, **87**, 1958, 224-230.

[124] See, e.g., A. Weiss, *Z. anorg. u. allgem. Chem.*, **297**, 1958, 257-286, especially p. 275, in which the base exchange behavior is discussed in relation to that of real micas, montmorillonites, and vermiculites.

[125] *Sprechsaal*, **93**, 1960, 155-159.

[126] *Doklady Akad. Nauk S.S.S.R.*, **119**, 1958, 351-353.

conditions during the sedimentation, especially in the Min'kovtsy-Ushitsa series. Such particular conditions correspond to those in the formation of C. Andreatta's "illidromica" (see above) in the Italian Middle Mountains, which were sometimes interpreted as mixed-layer types of clay minerals. R. Pellizzer and G. Guidetti[127] give evidence that "andreattite" is the same as "illidromica," with the characteristic endothermic effect of dehydroxylation at 550° to 600°C. shown on the differential-thermal curves.

71. Although *leverrierite* belongs to the mixed-layer type, we may here insert a few remarks on this problematic mineral which according to H. Saalfeld[128] is an oriented intergrowth of dioctahedral (2 M) illite with a 14 Å. mineral (cf. B. ¶ 86, 108*), with a pronounced one-dimensional dislocation, and the elementary cell dimensions $a_0 = 5.13$ Å., $b_0 = 8.90$ Å., $c_0 = 20.33$ Å., and $\beta = 95°30'$, non-swellable. The (060) spacing of 1.49 Å. makes most probable the identity of the 14 Å. component with a kaolinite-type structure. Leverrierite belongs to the illites, and it is recommended that the species name leverrierite be dropped once and for all for an intergrowth mineral of this sort. The illite share in it is very low in alkalies, and the ratio Si : Al in the tetrahedral layer is high.

72. The morphological characteristics of *glauconite* were studied by W. Müller and W. Behne[129] in the electron microscope. In addition to the typical small-scaly material, more lath-like shapes also occur. J. F. Burst[130] gave a detailed classification of glauconites. We will only discuss here the strictly mineralogical aspects of glauconite, with its characteristic potassium and trivalent iron contents in mica-like positions. Burst finds in the majority of cases more than two thirds of the K^+ positions occupied, although varieties with deficits in potassium occur, with a low degree of cristallinity (1 M d in Yoder and Eugster's definition, or in maximum 1 M), and many even contain interlayered material, such as illite and montmorillonite. All of the glauconites described by Burst are *di*octahedral, in complete analogy with the degrees of disorder

[127] *Rend. soc. mineral. ital.*, **15**, 1959, 215-232. On the name andreattite and its definition, cf. R. C. Mackenzie, G. F. Walker, and R. Hart, *Mineral. Mag.*, **28**, 1949, 704-713, and Mackenzie, *Anales edafol. y fisiol. vegetal (Madrid)*, **13**, 1954, 111-139. The interpretation of andreattite as a mixed layer mineral is given by H. S. Yoder and H. P. Eugster, *Geochim. et Cosmochim. Acta*, **8**, 1955, especially on p. 273.

[128] *Fortschr. Mineral.*, **38**, 1960, 161; *Beitr. Mineral. u. Petrog.*, 1960, 63-70; *Veröffentl. Max-Planck-Inst. Silikatforsch.*, **20**, 1960, 141-148. The material described was from the hard coal seams of Dobrilugk, Germany and from a deep-sea sediment of the Tyrrhenian Sea. See also A. Schüller, *Heidelberger Beitr. Mineral. u. Petrog.*, **2**, 1951, 269-278; 413-427; *Neues Jahrb. Mineral. Monatsh.*, 1951, 156-168; *Geologie (Berlin)*, **2**, 1953, 150-152.

[129] *Naturwissenschaften*, **41**, 1954, 575.

[130] *Am. Mineralogist*, **43**, 1958, 481-497; *Bull. Am. Assoc. Petrol. Geologists*, **42**, 1958, 310-327. We here expressively refute the use of the word glauconite as a sedimentary rock formation in which "green earths" of chlorite, montmorillonite, etc. are sometimes mixed.

which are known among the illites, intermediate between true micas and swelling clay minerals. Biotite, which is high in iron and trioctahedral, can be changed to dioctahedral glauconite, as was early observed by E. W. Galliher.[131] The accompanying scheme, which may illustrate these correlations between illites and glauconites, with mica and montmorillonite as the extremes, is demonstrated by Burst as a function of the potassium content (in atomic-equivalent numbers of the unit cell), varying during the diagenesis, as follows (the arrows indicating the potassium interchanges possible).

Muscovite ($K^+ = 2$) ⇄ Illite ($K^+ = 1.0$ to 1.4)
Muscovite ($K^+ = 2$) ⇄ Glauconite ($K^+ = 1.4$ to 1.8)
Biotite ($K^+ = 2.0$) ⇄ Glauconite ($K^+ = 1.4$ to 1.8)

Illite ($K^+ = 1.0$ to 1.4) ⇅ Degraded illite (K^+ below 1.0) ⇅ Mixed-layer illite ($K^+ = 0.2$ to 1.0) ⇄ Montmorillonite (K^+ below 0.2)

Glauconite ($K^+ = 1.4$ to 1.8) ⇅ Disordered glauconite ($K^+ = 0.8$ to 1.4) ⇅ Mixed-layer glauconite ($K^+ = 0.2$ to 0.8) ⇄ Montmorillonite (K^+ below 0.2)

73. The mica-like minerals of green tuffs derived from volcanic glass in Japan were studied by T. Sudo[132] in which both K^+ and Fe^{3+} are lower than in normal glauconite or celadonite (see below), but silica is higher than in chlorophaeite. The X-ray diffraction diagrams, however, are those of disordered glauconite in the definition of Burst, although Sudo prefers to characterize the material of the green tuffs as a distinct (often fibrous) variety, related to palagonite and chlorophaeite, called "hydrous celadonites."

74. The ideas of Burst were supplemented from the genetic viewpoints by J. Hower[133] and confirmed for the degraded glauconite with a mica-like structure and a low sedimentation process. In this, a gradual substitution of Al^{3+} in octahedral

[131] *Bull. Am. Assoc. Petrol. Geologists,* **23**, 1939, 513-515.

[132] *Proceedings of the Seventh Pacific Science Congress, Pacific Science Association, New Zealand,* 1949, Vol. 2, 1953, 249-255. See also T. Sudo in his book "Mineralogical Study on Clays in Japan," Maruzen, Tokyo, 1959, 83-84, 269-280. We imagine that the simple word "green earth" may bring about much uncertainty, if, e.g., green earths derived from the Deccan traps are defined by N. Sen, *Quart. J. Geol. Mining Met. Soc. India,* **26**, 1954, 1-13, as built up from nontronite and montmorillonite although they were previously called "celadonite" and "delessite." On the relations of chlorophaeite to the saponites, see also the previous work of D. P. Serdyuchenko, *Zapiski Vsesoyuz. Mineral. Obshchestva,* **76**, 1947, 177-189.

[133] *Am. Mineralogist,* **46**, 1961, 313-334. L. G. Toler and J. Hower, *Ibid.,* **44**, 1959, 1314-1318, used a linear relation between the refractive index and the percentage of randomly interstratified expandable layers (of glycol-solvated and K-salt treated glauconite) for diagnostic purposes.

positions by Fe^{3+} takes place, and a consequent increase in interlayer K^+ to form the non-expandable lattice of higher order, independent of the tetrahedral charges. Hower especially emphasizes the distinction of glauconites, according to their geological age in shales: young glauconites are low in K^+, contain more expandable layers, and are also often heterogeneous; on the other hand, aged glauconites with higher potassium contents show the higher order type and a loss in expandability, as mentioned before. These environmental diagenesis factors influence essentially the multiple forms of appearance of glauconite in more or less crystalline pellets and aggregates. H. Urban[134] made a study of the glauconite occurrence in the Lower Cretaceous sediments of Westphalia. He distinguished celadonite not as an independent mineral, but as a magnesium variety of glauconite, a proposal which may simplify the complex chemical and geological discussion of the glauconite problems here involved.

75. For *celadonite* B. B. Zvyagin[135] used electron diffraction methods to determine the unit cell dimensions: $a_0 = 5.20$ KX, $b_0 = 9.00$ KX, $c_0 = 10.25$ KX, and $\beta = 100.1°$. The structure is distorted as well in the octahedral as in the tetrahedral coordination polyhedra. A perspective impression of the structure is given in Fig. B.10 in which characteristic six-membered rings are observed. If a powder of celadonite is deposited from its suspension on a highly polished surface of stainless steel, the electron reflection diagrams show preferably the interferences from $(0kl)$ and $(hk0)$ This means that the plane (a^*b^*) of the reciprocal lattice is projected in the diagrams, pairwise as $(0h0)$ and $(0h1)$, or $(hk0)$ and $(hk1)$, as are reflections belonging to textures with the c^* axis, $(00l)$, and on lines parallel to c^* $(02l)$ $(\pm 11l)$, $(\pm 20l)$, $(\pm 13l)$. The reason for this complex diffraction is the ribbon-shaped habitus of the celadonite.

76. Of particular interest are the magnesium silicate hydrate minerals which either occur in clay deposits or are closely related to them, as independent formations. Many of those magnesium minerals are striking by their *fibrous* habitus, and were discussed by S. Caillère, A. Oberlin Mathieu-Sicaud, and St. Hénin[136] as "fibrous montmorillonites" from Madagascar. Important members of the *sepiolite* family, with the ratios MgO : SiO_2 varying between 1.0 and 0.60, that is from $MgO.SiO_2.2H_2O$ to $6MgO.10SiO_2.10H_2O$, are *palygorskite* and *pilolite*, discussed by N. Efremov.[137] They are not pure magnesium silicate hydrates but variable in their isomorphous replacements of Mg^{2+} and Al^{3+} ions.

77. An excellent new occurrence of palygorskite amid sandstones and marls with rounded inclusions of the fibrous magnesium silicate hydrate mineral, surrounded

[134] *Tonind.-Ztg. u. Keram. Rundschau*, **81**, 1957, 363-371.
[135] *Kristallografiya*, 2, 1957, 181-183 (with R. A. Shakhova); 393-399.
[136] *Compt. rend.*, **236**, 1953, 1581-1583.
[137] *Am. Mineralogist*, **39**, 1954, 324-325.

by montmorillonite, from the Aruk-Tau Ranges (Tadzhik S.S.R.), was described by S. Yusupova and V. Bosov.[138] The formula $2.5MgO.Al_2O_3.8.6SiO_2.12.6H_2O$ is given for this interesting palygorskite of epigenetic origin, probably from colloidal suspensions deposited in cavities, and as a silification product of wood, associated with dolomite, thus indicating the high magnesia content of the solutions. On the other hand, J. B. Droste and G. R. Gates[139] found in lacustrine sediments of southern Cal-

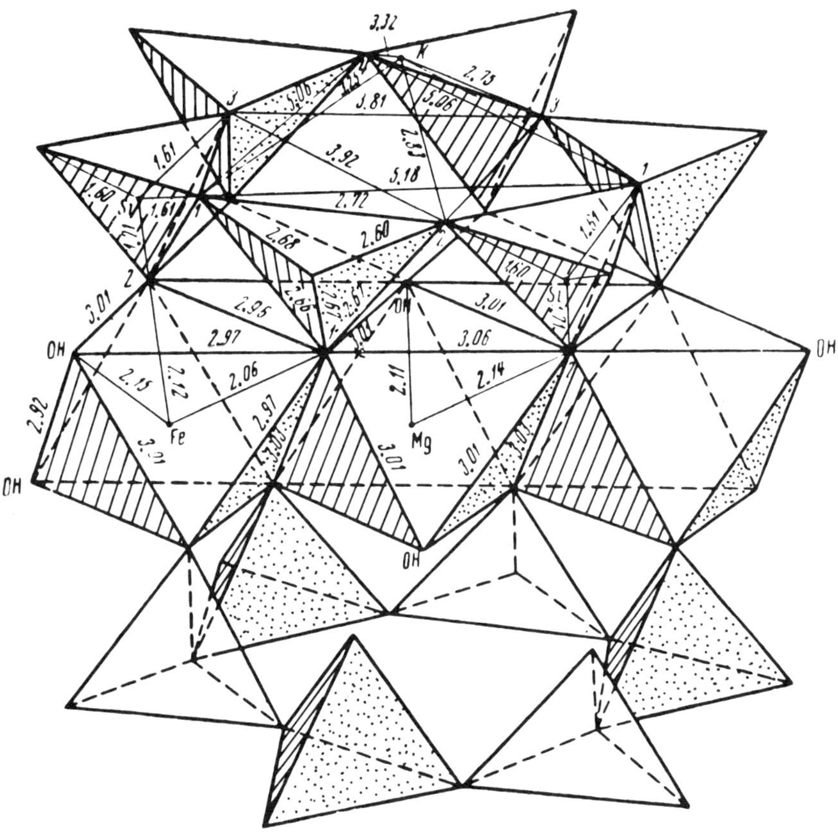

FIG. B. 10. Perspective representation of the structure of celadonite (magnesium glauconite) (Zvyagin and Shakhova).

[138] *Voprosy Petrog. i Mineral., Akad. Nauk S.S.S.R.*, **2**, 1953, 135-138. Palygorskite, not in the usual form of "mineral leather," but in normal clays, marls, etc. was described from the Moscow-Kaluga-Rzhev Basin by I. D. Zkhus, *Doklady Akad. Nauk S.S.S.R.*, **107**, 1956, 734-736. See also from Cherkasskoe (Ukraine) large attapulgite deposits recently described by E. G. Kukovskiĭ and A. B. Ostrovskaya, *Zapiski Vsesoyuz. Mineral. Obshchestva*, **90**, 1961, 598-601. Highly interesting occurrences of palygorskite and sepiolite in young sediments of Hadramaut, southern Arabia, are described by G. Müller, *Neues Jahrb. Mineral., Abhandl.* **97**, 1961, 275-288.

[139] *Bull. Geol. Soc. Am.*, **70**, 1959, 1593.

ifornia, veins filled with *attapulgite*, as a dense-fibrous material, with subordinate montmorillonite and dolomite, suggesting a hydrothermal origin of the attapulgite. This formation together with that of true sepiolite is made highly probable by the studies of Fr. A. Mumpton and R. Roy[140] in the system $MgO–Al_2O_3–SiO_2–H_2O$, although it was not yet possible to synthesize both minerals at low temperatures (below 200°C.). They are decomposed forming smectites (saponite). It is, therefore, very probable that under the experimental conditions attapulgite would be metastable (cf. Vol. IV, Section A). Also interesting is the fact that natural attapulgites contain some lime. The excellent fiber or lath shape of single crystallites of attapulgite, as well as that of sepiolite, is explained by the structural scheme which was proposed by B. Nagy and W. F. Bradley[141] (cf. Fig. A. 116), and confirmed by electron micrographs.

78. For all the structural details on palygorskite and attapulgite we refer to A. ¶ 169, and the studies of K. Brauner and A. Preisinger.[142] We also refer to the investigations of G. Kulbicki[143] on the structural changes by a high-temperature treatment of attapulgite and sepiolite and the orientation of enstatite formed, combined with characteristic effects on the differential-thermal curves. The multiple technological uses of sepiolite (Meerschaum), especially that of Vallecas in Spain, was discussed by R. H. S. Robertson[144] and by G. P. C. Chambers;[145] it is employed in drilling muds, as filter aids, and for other uses because of the extremely fine-fibrous dispersion. We further mention the electron and X-ray diffraction studies of G. W. Brindley[146] for sepiolite. Interesting metasomatic occurrences of sepiolite in silicified serpentinites of the northern Caucasus, showing the primary structure of the original basic rocks in metacolloidal forms, were described by D.P. Serdyuchenko.[147] The fibrous material found in dolomites of many carboniferous sediments of the Russian platform and in the Odessa and Stauropol Regions was called "α-" sepiolite by M. A. Rateev and D. D. Kotel'nikov.[148] It has high magnesia contents (21 to 26 per cent), distinct from palygorskites, which are more frequently found in marine marls.

79. From the important sediments of the Green River formation in Wyoming,

[140] *Natl. Acad. Sci.-Natl. Research Council, Publ.*, **566**, 1958, 136-143; *Proc. Natl. Conf. on Clays and Clay Minerals*, **5**, 1958, 136-143.

[141] *Am. Mineralogist*, **40**, 1955, 329; 885-892.

[142] *Rend. soc. mineral. ital.*, **12**, 1956, 69-71; *Mineral. u. petrog. Mitt.*, N.F., **6**, 1956, 120-140; *Proc. Natl. Conf. on Clays and Clay Minerals*, **6**, 1957, 61-67.

[143] *Am. Mineralogist*, **44**, 1959, 752-764.

[144] *Chem. & Ind. (London)*, 1953, 1492-1495.

[145] *Silicates inds.*, **24**, 1959, 181-189.

[146] *Am. Mineralogist*, **44**, 1959, 495-500.

[147] *Doklady Akad. Nauk S.S.S.R.*, **69**, 1949, 577-580.

[148] *Doklady Akad. Nauk S.S.S.R.*, **109**, 1956, 191-194.

J. J. Fahey, M. Ross, and J. M. Axelrod[149] described "*loughlinite*" as a new sepiolite-like mineral, a hydrous sodium-magnesium silicate of the composition $Na_2O.3MgO.6SiO_2.8H_2O$, occurring in dolomites and oil shales as a fibrous mineral, with about 8 per cent Na_2O exchangeable in the structure with magnesia. An immersion in a 1 per cent $MgCl_2$ solution for 7 days is sufficient to convert loughlinite into normal sepiolite. This change is not reversible. The electron micrograph shows loughlinite as extremely fine fibers, very similar to those of sepiolite, such as that from Madagascar.

80. The relatively scarce occurrence of minerals of the *serpentine* group as clay minerals has been discussed prevailingly from the morphological viewpoint, stressing the characteristic tube-shaped development which this group has in common with metahalloysite. The morphology of the chrysotile fibers was extensively discussed in A. ¶ 156 ff. We, therefore, shall here only mention briefly the investigations with the electron microscope, such as those of W. Noll and H. Kircher[150] which emphasized the great similarity of the chrysotile and halloysite tubes, in the light of L. Pauling's theory of structural misfits, as the cause of curvature in layer-type silicate structures. Concerning the same problems, we further mention recent studies of T. F. Bates and J. J. Comer,[151] especially the excellent replica electron micrographs given by these authors. W. F. Bradley[152] discussed in detail the structural irregularities in magnesium silicate hydrates, such as in chrysotile, antigorite, sepiolite, montmorillonite, and vermiculite. To explain the high dispersity of corresponding "clay minerals," he proposes that perfection of articulation between the tetrahedral and octahedral layers is temperature-sensitive, and that the different development of crystallization reflects the degree to which difficulty of articulation limits the growth of specific compositions. We also recall the peculiar occurrence of tubular kaolinite (cf. B. ¶ 10) under corresponding aspects. G. Honjo, N. Kitamura, and K. Mihama[153] described such tubular kaolins from North Carolina and Indiana; Y. Stourdze Visconti, B. N. F. Nicot, and E. Goulart De Andrade[154] described those from Brazil. We further refer to the rich collection of crystallochemical data given on 1 : 1 layer silicates by T. F. Bates,[155] in connection with D. McConnell's theory of the structural equiv-

[149] *Am. Mineralogist*, **45**, 1960, 270-281.

[150] *Naturwissenschaften*, **37**, 1950, 540-541; *Fortschr. Mineral.*, **29/30**, 1952, 391-392, more extensively in *Neues Jahrb. Mineral. Monatsh.*, 1951, 219-240.

[151] *Proc. Natl. Conf. on Clays and Clay Minerals*, **6**, 1958, 237-248; see the particularly beautiful electron micrographs in *Mineral Inds. Exp. Sta. Circ.*, No. 51, 1958, 58 pp.

[152] *Natl. Acad. Sci.-Natl. Research Council, Publ.*, **395**, 1955, 94-102; *Proc. Natl. Conf. on Clays and Clay Minerals*, **3**, 1955, 94-102.

[153] *Clay Minerals Bull.*, **2**, 1954, 133-141.

[154] *Am. Mineralogist*, **41**, 1956; 67-75; *Cerâmica (Brazil)*, **2**, 1956 (6) 59-71; *Ibid.*, **5**, 1959 (20) 2-10; see also H. Urban, *Tonind.-Ztg. u. Keram. Rundschau*, **82**, 1958, 41-43, who explains these tubes as halloysite formed from primary kaolinite.

[155] *Am. Mineralogist*, **44**, 1959, 78-114.

alency of [SiO_4] and [$(OH)_4$] groups,[156] as in antigorite, and also the studies of G. L. Kalousek and L. E. Muttart[157] on the serpentine family. The theory of McConnell (1950) proved to be as helpful in studying the minerals here discussed as it was in studies of the high-dispersed smectites for which it was originally developed.

81. The close relations existing between the serpentine and *chlorite* mineral families were especially elucidated by the extensive synthetic investigations of R. Roy *et al.* in the system $MgO–Al_2O_3–SiO_2–H_2O$ (cf. Vol. IV, Section A). For clay minerals, the importance of chlorites is especially evident in the field of mixed-layer structures. We therefore shall discuss the chlorites more in detail. B. W. Nelson and R. Roy[158] discovered the low-ordered "*septechlorites*" series with the characteristic (001) spacing interference of 7 Å. in comparison with that of 14 Å. in normal chlorites (cf. ¶ A. 143). These 7 Å. chlorites are the typical high-dispersed phases occurring as clay minerals. The correlations of montmorillonites, serpentines, and chlorites are particularly discussed by D. M. Roy and R. Roy[159] on the basis of the hydrothermal-synthetic data, from which it is evident that the montmorillonites cover a large field of composition in the siliceous ranges whereas aluminum containing serpentines are in the more magnesian ranges. They are changed to clinochlore at higher temperatures.

82. One would expect from equilibria in the system $MgO–Al_2O_3–SiO_2–H_2O$ a rather common occurrence of montmorillonite together with 7 Å. chlorites, and even with 14 Å. chlorites (clinochlore) covering a large field of compositions. The mixed-layer structures to be discussed later, in which the interlayering of montmorillonite and chlorite increments in combination is frequent, evidently do not correspond to equilibria and are typical metastable formations. W. F. Bradley,[160] who emphasizes that the chlorites more frequently occur in argillaceous sediments than was previously assumed, explains why the direct X-ray identification of single chlorite compositions is rather difficult. The natural chlorites are usually complex in their chemical constitution, but in general the observation of samples with preferred orientation (on settled slides) is sufficient for a distinction of chlorite from kaolinite, montmorillonite, mica, and vermiculite in powder diagrams. An important help in such cases is a systematic study of the powder diagrams, after heating the samples to defined temperatures, by the characteristic modifications in intensity and position of definite interference peaks. To the distinction from vermiculite, we will refer later.

[156] *Am. Mineralogist*, **39**, 1954, 830-831.

[157] *Am. Mineralogist*, **42**, 1957, 1-22.

[158] *Natl. Acad. Sci.-Natl. Research Council, Publ.*, **327**, 1953/54, 335-348; *Proc. Natl. Conf. on Clays and Clay Minerals*, **2**, 1953/54, 335-348.

[159] *Am. Mineralogist*, **40**, 1955, 147-178.

[160] *Natl. Acad. Sci.-Natl. Research Council, Publ.*, **327**, 1953/54, 324-334; *Proc. Natl. Conf. on Clays and Clay Minerals*, **2**, 1953/54, 324-334.

VERMICULITES

83. Among the clay minerals occurring in common weathering soils, the vermiculite types were increasingly recognized in their fundamental importance. They are so intimately connected with the formation of mixed-layer minerals (see the following chapter), especially in the combination with montmorillonite, that it is difficult to find a reasonable line of demarcation between both groups. Nevertheless, it is advisable to discuss first the vermiculites proper, and to characterize their most typical physicochemical and structural behavior. F. Ficai[161] gave a very worthwhile review of these factors, especially in view of the previous investigations of J. W. Gruner (1934, 1939), G. F. Walker (1949), and I. Barshad (1950), concerning the structure, the base exchange reactivity, and the exfoliation phenomena. The high water imbibition capacity of the exfoliated vermiculite is practically important for soil melioration. In this respect, U. Ventriglia[162] also emphasized the structural relations of the vermiculites to the micas, with an ample possibility of compensating for deficiencies or excess in electrostatic charges, on the tetrahedral sheets of the pyrophyllite or talc-like core parcels, by cation exchange and water interlayers. Most of the vermiculites are *trioctahedral* and easily derived from biotite by weathering. They may, however, also react as imperfectly crystallized chlorites. The multiple intermediate position of the vermiculites between micas, chlorites, smectites, talc, and pyrophyllite is particularly instructive. That there are also *dioctahedral* analogues to the common trioctahedral vermiculites was demonstrated by G. Brown[163] from soils of Lancashire, England, as is concluded from the low spacing of the (060) interference (below 1.50 Å.) characteristic of dioctahedral layer structures. Such vermiculites are probably derived from dioctahedral micas.

84. A very remarkable attempt to establish the structure of normal (trioctahedral) vermiculite by X-ray diffraction analysis was made by Å. Grudemo,[164] with a combined Patterson-Fourier synthesis. The structure proposed by S. B. Hendricks and M. E. Jefferson (1938) is in general confirmed. Characteristic of the new trials of Grudemo is the introduction of $[SiO_3(OH)]$ groups analogous to those in the Edelman-Favejee model of montmorillonite (cf. B. ¶ 49), with "*trans*-" oriented Si^{4+} ions, to explain the particularly high cation exchange capacity and water adsorption qualities of vermiculites. A high degree of randomness in the pseudo-hexagonal

[161] *Ceramica (Milan)*, **8**, 1953 (6), 46-54; from the viewpoints of economic geology see O. S. North and H. P. Chandler, *U. S. Bur. Mines Inform. Circ.*, **7668**, October, 1953, 27 pp.

[162] *Clay Minerals Bull.*, **2**, 1954, 179-183.

[163] *Clay Minerals Bull.*, **2**, 1953, 64-69.

[164] *Svenska Forskningsinst. Cement Betong vid Kgl. Tek. Högskol. Stockholm, Handl.*, No. 22, 1954, 56 pp. The similarity of the vermiculite structures with those of some calcium silicate hydrates is particularly interesting.

cross-gratings is strongly characteristic, although with a preservation of the hexagonal orientation within the stackings. Figure B. 11 shows a projection of the atoms on (001) of a single talc-type layer, with adjacent brucite positions. The Edelman-Favejee model differs from this arrangement only because every second Si^{4+} ion is placed in an "outside" position with respect to the adjacent oxygen layer. The structure so developed is ortho-hexagonal, with rhombohedral symmetry. The mono-

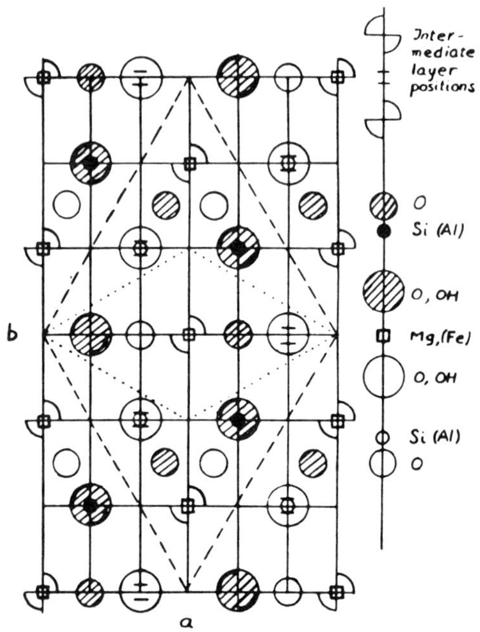

FIG. B. 11. Projection on (001) of a single talc layer and an adjacent brucite layer in the pseudohexagonal structure of vermiculite (Grudemo). The outlines of the larger cell are dashed; those of the smaller cell are dotted.

clinic character of projections of the $(h0l)$ zone only is probably apparent. The unit cell dimensions are $a_0 = 5.338$ KX, $c_0 = 14.37$ KX. In another sample, c_0 is only equal to 7.30 KX.

85. The symmetry discussion gives the remarkable result that this smaller unit is the *magnesium kaolinite* which was previously considered by L. Pauling (1930) and G. W. Brindley (1951), and shown in the ideal (001) projection (Fig. B. 12). The extensive calculations based on intensity measurements of the interferences speak in favor of this kaolinite-type interpretation; also the *trans*-configuration in a chlorite-like structure with brucite layer intermediate between the parcels, and with reversed mutual positions of silica and hydroxide layers, is considered to be favorable evidence. The fact that of the three samples of vermiculite examined by

Grudemo (also among these is jefferisite) only *one* may be representative in its structure of true vermiculite shows most impressively the protean nature of this most remarkable clay mineral. The water interlayers postulated by Hendricks and Jefferson cause in great portion the degree of structural disorder and the characteristic reaction of easy removal of water from the structure by elevated temperatures. More than half of the total water is lost by heating to only 110°C.

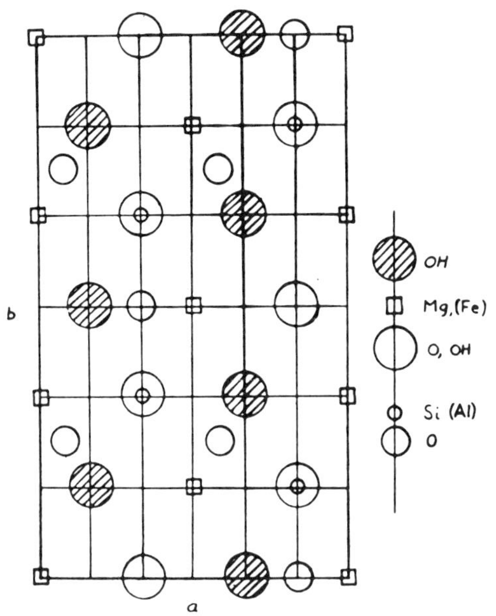

FIG. B. 12. Basis projection of the ideal magnesium kaolinite structure (Grudemo).

86. In a most fascinating series of investigations, G. F. Walker[165] studied the physicochemical characteristics of vermiculites as surface-active structures, such as in the development of water layers and the formation of organic complexes with glycerin, codeine, brucine sulfate, *p*-phenylene diamine, benzidine hydrochloride, to name a few, as well as common water imbibition in the series of weathering products from an original biotite to vermiculite. This process is observed by a systematic increase in the basal spacing from 10 Å. to finally 14 Å. (14.36 Å. in magnesium-saturated vermiculite), and with a distinct influence of the cations adsorbed (cf. C. ¶ 282), as in experiments with Li^+, K^+, or NH_4^+ vermiculites. The configuration

[165] *Nature*, **163**, 1949, 726; *Ibid.*, **166**, 1950, 695; especially *Trans. Intern. Congr. Soil Sci. 4th Congr., Amsterdam*, 1950, Vol. II, 62-67 (with A. A. Milne). For an excellent detailed investigation, see Chapter VII of the monograph "X-Ray Identification and Structures of Clay Minerals,' Mineralogical Society of Great Britain, 1951, pp. 199-223.

of the water layers is particularly affected by the size and polarizing action of the exchangeable cations; a replacement of Mg^{2+} by monovalent and divalent cations causes changes in the *c*-dimension of the unit cell, as observed after leaching with potassium salt solution. The original basal spacing of 14.3 Å. is reduced by the change from the Mg-saturated vermiculite to 11.0 Å. for the K-vermiculite (diffuse line), whereas an ammonium-saturated vermiculite gives a broad line at about 10.6 Å., a barium-saturated material at 12.3 Å. The X-ray diagram of the K-vermiculite is strikingly similar to that of biotite or phlogopite.

87. Walker studied with particular care the possibilities of an accurate distinction of vermiculite from smectites and chlorites[166] by using various treatments either with glycerin or by heating. Thus, montmorillonite swells to a basal spacing of 17.7 Å., and vermiculite and chlorite persist in the spacing at 14 Å. After heating to 500°C. it is shifted to a spacing of 9 to 10 Å. for montmorillonite and vermiculite, whereas that of chlorite persists unchanged. Vermiculite is then positively identified by leaching with an ammonium salt solution which causes a displacement of the basal spacing from 14 to 11 Å., as mentioned above.

88. The direct evidence of the occurrence of vermiculite in weathering soils of mica-rich rocks in north-eastern Scotland was given by G. F. Walker.[167] The prevailing vermiculite type is that derived from biotite. The (060) spacing at more than 1.52 Å. speaks for its trioctahedral character, and the transitions could be well pursued by applying glycerin to the different products of weathering as given in Fig. B. 13 for the changes in shape of the 10 Å. and 14 Å. peaks of the X-ray diagrams. Dioctahedral illite may coexist with the trioctahedral products. The attempts of Walker and A. A. Milne[168] to apply a one-dimensional Fourier synthesis to the vermiculite structure show that the exchange positions are normally occupied by Mg^{2+} ions. The water interlayers consist of double sheets of H_2O molecules, of 5 Å. thickness, with the Mg^{2+} ions in the central positions. As soon as these Mg^{2+} ions are replaced, the water layers also show the changes in spacing as mentioned above.

89. The structure as it was developed by A. McL. Mathieson and G. F. Walker[169] differs from that proposed by Hendricks and Jefferson (1938) in excluding the hypothetical hexagonal water layer structure for vermiculite. The determinations were based on single crystal measurements with the results given in Fig. B. 14, in a projection on (010). Each water molecule is linked to a single oxygen in the silicate surface layer by hydrogen bonding, and weak bondings of this type also act be-

[166] *Nature*, **164**, 1949, 577; *Clay Minerals Bull.*, 3, 1957, 154-163, concerning the great difficulties in discriminating vermiculites and smectites by different physical methods.

[167] *Mineral. Mag.*, **29**, 1950, 72-84.

[168] *Clay Minerals Bull.*, **1**, 1951, 171-174.

[169] *Am. Mineralogist*, **39**, 1954, 231-254.

tween the water sheets. These are held together by the exchangeable cations around which the water molecules form regular hydration shells. It is important that the structure as determined by Walker and Mathieson does not need direct electrostatic interactions between these cations and the silicate surface layers, but that they

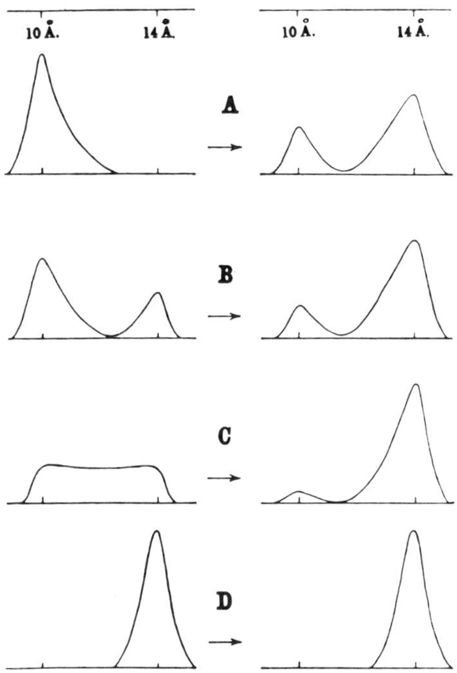

FIG. B. 13. Effect of glycerin treatment on the shapes of the 10 and 14 Å. interference lines at various stages during the change of trioctahedral clay mica into vermiculite (Walker): (*A*) development of the 14 Å. line, and a reduction of that with 10 Å.; (*B*) relative increase in intensity for the 14 Å. line; (*C*) the sharp line 14 Å. resolved from the band 10/14 Å.; (*D*) the fully developed 14 Å. line is not affected by treatment with glycerin.

act in a function of their natural ionic hydration behavior. This explains most satisfactorily the specific effects of Li^+, Mg^{2+}, K^+, or NH_4^+ cations so well described in Walker's investigations. The monoclinic unit cell dimensions are $a_0 = 5.33$ Å., $b_0 = 9.18$ Å., $c_0 = 28.90$ Å., and $\beta = 97°$.

90. Mathieson[170] refined the structure determination of magnesium vermiculite with $c_0 = 14.36$ Å. referring to the electron density distribution in the interlamellar region, which indicates that either the water molecules vibrate asymmetrically or there is a regular displacement of the H_2O positions from the hexagonal arrangement

[170] *Am. Mineralogist*, **43**, 1950, 216-227.

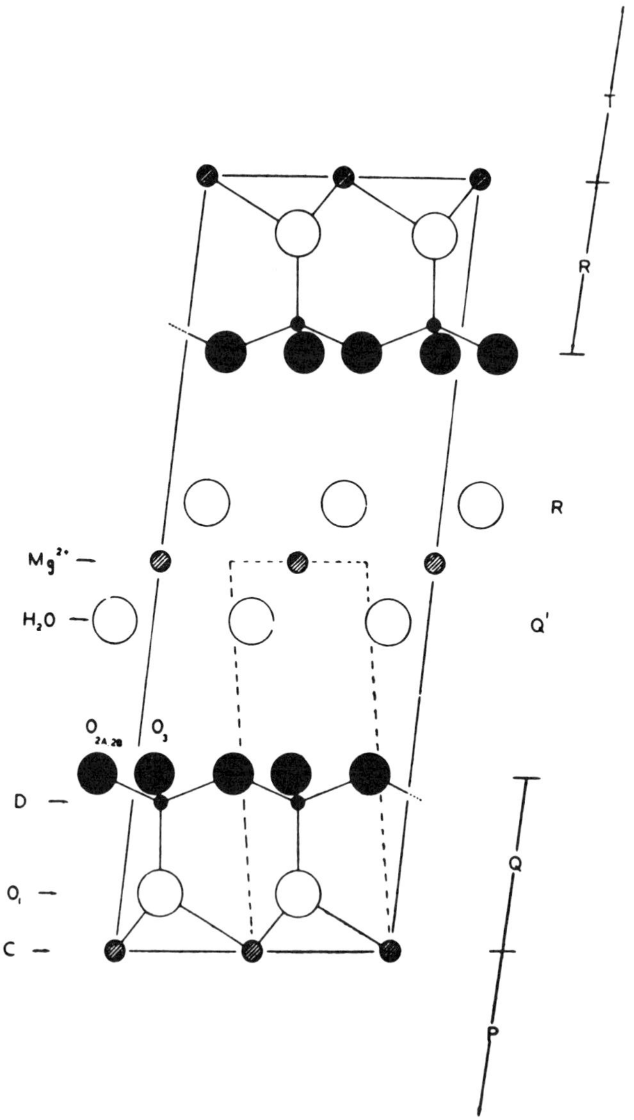

FIG. B. 14. Crystal structure of vermiculite, projected on (010) (Mathieson and Walker).

in a direction just opposite to that proposed in the structure first given by Walker and Mathieson. Probably, the latter interpretation approaches nearer the reality.

91. Extremely instructive also is a direct comparison of the structures of magnesium vermiculite and chlorite (Fig. B. 15), with H_2O–Mg^{2+}–H_2O double sheets in the former and HO–Mg^{2+}–OH double sheets in the latter, but with the same re-

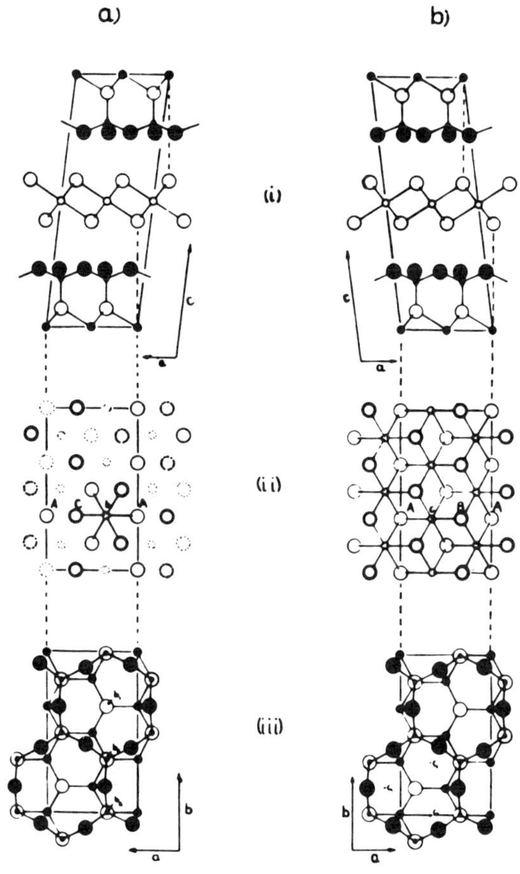

FIG. B. 15. Comparison of the structures of magnesium vermiculite (*a*), and chlorite (*b*) (Mathieson): (i) projections on the (*ac*) plane; (ii) projections of the interlamellar region, normal to the (*ab*) plane; (iii) projection of half the silicate layer normal to the (*ab*) plane.

lationship of the sheets to the adjacent silicate layer surface, the O^{2-} being vertically above the origin Mg atoms in a hexagonal array. The chief difference between both sheets is the manner in which the two silicate–H_2O (or OH) halves of the "sandwich" are held together by the Mg^{2+} ions. The difference in the $d(001)$ spacings is small: c_0 for magnesium vermiculite is = 14.36 Å., for chlorite = 14.2 Å., and the H_2O–

Mg^{2+}–H_2O intersheet distance is in magnesium vermiculite = 2×1.14 Å., HO–Mg^{2+}–OH in chlorite = 2×1.06 Å.

92. In connection with the cation-exchange problems of vermiculite, it is of particular interest that A. Weiss and G. Thamerus[171] were able to replace the K^+ ions of muscovite and biotite by the polyvalent ions Al^{3+} and Si^{4+}. This is not possible in the direct way but must be done exclusively over intermediate exchange products with long-chain *n*-alkyl ammonium ions (cf. C. ¶ 299). The organic mica substitutes react promptly with $Al(OR)_3$ and $Si(OR)_4$, whereas the mica parcels are entirely unchanged. The Al- and Si-substituted micas are not reversible. Since only coordination positions with 6 and 12 O^{2-} ions of the adjacent tetrahedral sheets are available, we see in the Si-mica an analogue of *ortho*-dihydroxyarylene silicic esters which contain the characteristic $[SiO_6]$ coordination group.

93. These observations of Weiss and Thamerus are a consequent supplement to the previous experiments of U. Hofmann, A. Weiss, and A. Mehler[172] on base exchange and innercrystalline swelling phenomena in mica minerals with alkyl ammonium ions, and their replacement by Na^+, Mg^{2+}, Ca^{2+}, and others. The basal spacings of the resulting products are so similar for micas, smectites, and vermiculites that there cannot be any doubt of the complete structural analogy of the silicate layers building them up. This fact is seen as one of the strongest arguments for the correctness of the structure of montmorillonite as proposed by U. Hofmann, K. Endell, and D. Wilm in 1934.

94. The studies of U. Hofmann *et al.* on the cation exchange are particularly instructive if one considers the transitions possible between the smectites and vermiculites.[173] The *(hkl)* interferences in the powder diagrams are always observed if the exchangeable cations are present in an amount higher than 0.55 equivalents with the $[(Si,Al)_4O_{10}]$ units, and vermiculite-type minerals appear as in the characteristic "*batavite*" from the graphite deposits of Kropfmühle, Bavaria. The crystalline order of such vermiculites is higher than in the smectites, but still lower than that in the micas which have one equivalent charge per $[Si_3AlO_{10}]$ unit.

95. R. Roy *et al.*[174] investigated the weathering of micas to form vermiculite, and the hydration behavior of the latter mineral. In this respect, vermiculite shows

[171] *Naturwissenschaften*, **48**, 1961, 70.

[172] *Z. Naturforsch.*, **11** b, 1958, 435-438. On organophilic vermiculite and batavite see especially pp. 431-434.

[173] See, e.g., U. Hofmann, A. Weiss, and G. Koch, *Ber. deut. keram. Ges.*, **32**, 1955, 12-17, especially pp. 14 ff.

[174] *J. Geol.*, **65**, 1957, 603-610 (with L. A. Romo); *Ibid.*, **66**, 1958, 324-326 (with Fr. A. Mumpton). We do not share the opinion of Roy and Romo that vermiculite should be classified as members of the montmorillonite (smectite) family. See also the general remarks on the weather-

above 300°C. a redistribution of the Mg^{2+} ions in the mica parcels of its structure and the tendency to arrange in a more chlorite-like (pseudo-chlorite) structure, which is not swellable in ethylene glycol, and a typical "hybrid" (cf. Fig. B. 16). Vermiculite is therefore never a hydrothermal mineral, not even under mild-metamorphic conditions, but a real weathering phase. Also striking is the relative rarity of modifications of vermiculite low in Fe^{3+} ("batavite" of Kropfmühle is an example), because the leaching of K^+ from the original micas is combined with an oxidation of Fe^{2+} to Fe^{3+} to balance the charges initially in trioctahedral vermiculite.[175] This

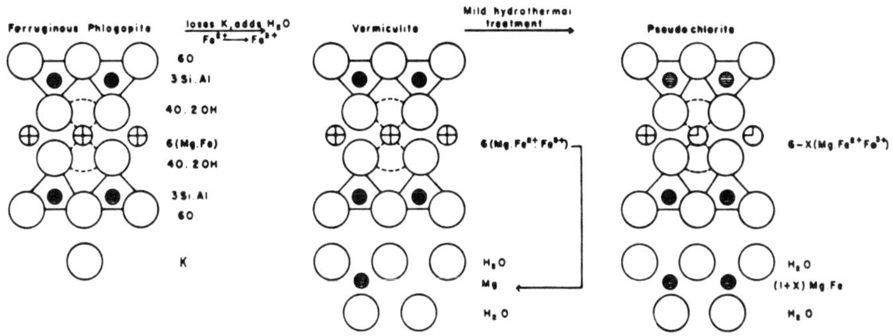

FIG. B. 16. Change of the structural composition involved in the origin of magnesium vermiculite from mica, and in its change to pseudo-chlorite (Roy and Romo).

mechanism also illustrates very well why such mixed-layer structures of chlorite and vermiculite may be so easily found in nature (see below). In every case, the removal of the entire potassium from trioctahedral mica brings about in the new-formed vermiculites very high base exchange capacitities, which are even higher than those commonly observed in smectites. It is on the other hand interesting that in phlogopite crystals low in iron the formation of a 7 Å. chlorite was observed by electron microscopic examination. It is perhaps somewhat surprising that no epitaxis phenomena between both minerals were found.

96. A particular *copper*-containing vermiculite (with 2 to 7 per cent CuO) was described by W. A. Bassett[176] from the copper ore deposits of northern Rhodesia,

ing of micas and chlorites by I. I. Ginzburg, *Trudy Mineral. Petrograf. Inst. im. Levinson-Lessinga, Akad. Nauk S.S.S.R.*, **2**, 1953, 139-160. A trioctahedral smectite very similar to vermiculite, and formed by weathering of biotite was described by D. M. C. MacEwan as "*cardenite,*" *Clay Minerals Bull.*, **2**, 1954, 120-126. The occurrence is in soils of northeastern Scotland, from which also G. F. Walker (1949) derived the vermiculites of "clay biotite."

[175] We emphasize the remark of Roy and Romo that from the V^{3+}-rich *roscoelite* a dioctahedral vermiculite should result.

[176] *Am. Mineralogist*, **43**, 1958, 1112-1133, with extensive dehydration studies and differential-thermal analysis curves. We also refer to the interesting observations of N. V. Shishkin, E. A.

the Cu^{2+} being in the exchange positions of the vermiculite structure. Such a variety is also easily synthesized from magnesium vermiculite by elementary base exchange reactions with 1 *N* $CuCl_2$ solution at 100°C. On the contrary, it is difficult to remove the rather firmly fixed Cu^{2+}, both in the natural as in the synthetic phase. It is remarkable that a copper vermiculite can also be produced directly from biotite by a long-period treatment with $CuCl_2$ solution, in this case as an interstratification of the copper vermiculite with biotite ("*hydrobiotite*" type of J. W. Gruner, 1934). The copper vermiculite is decidedly a low-temperature formation in the supergenic oxidation zone of the Rhodesian copper ores.

97. Somewhat different in their origin are the large vermiculite deposits of Libby, Montana, described by W. A. Bassett,[177] in which the weathering process of a primary augite pyroxenite was preceded by a metasomatic-hydrothermal change to biotite, vermiculite being formed later. Experiments on the base exchange in 0.001 *M* magnesium (and calcium) chloride solution could widely imitate these changes, in which the mixed-layer phase of hydrobiotite (either regularly or irregularly interstratified) is especially evident from differential-thermal curves and X-ray diffraction diagrams. Bassett does not entirely adopt Roy's and Romo's interpretation of the hydrothermal changes of vermiculite into pseudo-chlorite. He supports the suggestion of a hydrolysis reaction under hydrothermal conditions, providing the Mg^{2+} ions with OH ions in hydration envelopes. Nevertheless, the low-temperature formation of the vermiculite deposits is out of the question and chlorite was never observed in these. The solutions which produced the hydrobiotite have probably been even more dilute and have required more time than those altering biotite into vermiculite in the experiments. According to B. Droste and J. C. Tharin,[178] weathering minerals in the Illinoian till clays of northwestern Pennsylvania contain vermiculite as a product of changes from chlorites by decomposition of the brucite layer in their structure, and a reorganization of the hydrated components, together with mixed-layer minerals of montmorillonite plus illite.

98. The water content of the vermiculites as seen from Hendricks and Jefferson's hypothesis of hexagonal networks of water was revised by W. F. Bradley and J. M. Serratosa,[179] to accomodate the *higher* water contents found by weight loss determinations; this is not considered in Walker and Mathieson's structure model. According to the new proposal two octahedra of $[Mg^{2+}-(6H_2O)]$ and 4 H_2O squares

Krogius, and P. A. L'vovich, *Zapiski Vsesoyuz. Mineral. Obshchestva*, **87**, 1958, 682-686, on the analogous base exchange phenomena in the equally supergenic basic iron sulfates of the alunite-jarosite group, especially with the characteristic exchange of potassium by hydroxonium, H_3O^+. See also Fr. Halla, and R. van Tassel, *Naturwissenschaften*, **43**, 1956, 80, on hydroxonium jarosite.

[177] *Am. Mineralogist*, **44**, 1959, 282-299.

[178] *Bull. Geol. Soc. Am.*, **69**, 1958, 61-68.

[179] *Proc. Natl. Conf. on Clays and Clay Minerals*, **7**, 1959, 260-270.

centered about 5.2 Å. along *a* between the octahedra form a three-cell unit of the dimensions 9×15.6 Å. (Fig. B. 17). This arrangement includes twenty hydrogen bonds near 2.75 Å., namely twelve between the silicate and water layers, and eight between the water layers, further 32 hydrogen bonds near 3 Å. within these sheets. Residual 4 H atoms are not active in the bonding system; the environment of each

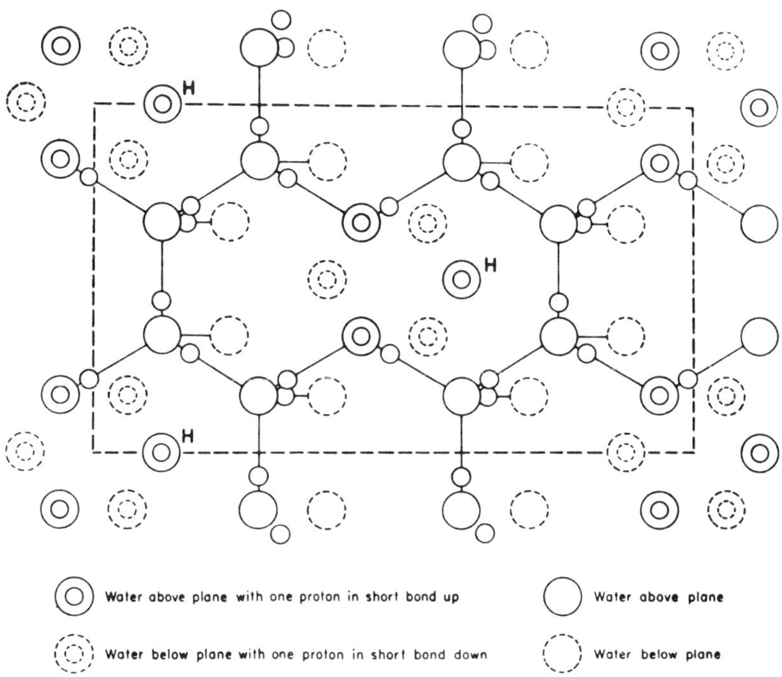

FIG. B. 17. Three-cell unit 9 × 15.6 Å. in the arrangement of interlayer water in vermiculite (Bradley and Serratosa).

H_2O molecule is a distorted tetrahedron. This proposal is in harmony with the infrared absorption spectra which show maxima of extinction at 3600, 3450, and 3350 cm.$^{-1}$ ($\lambda = 2.78$, 2.90, and 2.98 μ) at normal incidence, with increased activity in the highest and lowest energies when a flake is tilted. The hydroxyl axes in the shortest bonds are more inclined to the basal plane than are the intermediate length axes (cf. A. ¶ 46).

99. The particular genetic position of vermiculite in Middle-German serpentinites was considered by S. Matthes.[180] The mineral here often occurs in characteristic

[180] *Fortschr. Mineral.*, **26**, 1950, 88-89. As certain analogues we mention the weathering phenomena in the South-Ukrainian crystalline massive described by E. G. Kukovskiĭ and I. D. Sedletskiĭ, *Doklady Akad. Nauk S.S.S.R.*, **108**, 1956, 719-723.

pseudomorphs after biotite. In these basic rock environments there is also no sign of any postmagmatic-hydrothermal origin of the vermiculites; they are true weathering products with an intermediate mixed-layer hydrobiotite. A subsequent change of vermiculite into chlorite, however, is plausible, with the formation of characteristic pseudomorphs. Vermiculite is entirely absent in the "green schist" facies of metamorphism as defined by P. Eskola (cf. Vol. IV, Section A).

100. The occurrence of vermiculite in connection with *basaltic* effusives is described by Cl. D'Amico and R. Mezzetti[181] from the southern Alps (Trentino). Here the mineral sequence is evidently that from olivine (high in forsterite), first to saponite and then to vermiculite, which in its turn is later changed to bowlingite, iddingsite, and iron oxide minerals. The vermiculite is prevailingly a trioctahedral calcium derivative, slightly swellable with ethylene glycol to a basal spacing of 14.85 Å. This spacing is again reduced to 11.93 Å. by a treatment with K^+ salt solutions, and to 12.61 with Ba^{2+} solutions. Interesting is the increase in iron content of the silicate minerals in this genetic sequence from forsterite to saponite and then to vermiculite, and the oxidation of divalent to trivalent iron in the same time. D'Amico and Mezzetti do not agree with the opinion of Roy and Romo, or that of Bassett, that vermiculite is not stable under hydrothermal conditions. At least for the basaltic vermiculite described, a low-temperature formation of this mineral appears to be certain because of the simultaneous presence of chalcedony and zeolites.

101. The peculiar double-layer mica mineral described as *allevardite* by S. Caillère, A. Mathieu-Sicaud, and S. Hénin[182] is according to G. W. Brindley[183] characterized by interlayer water, with four hydration states, namely with 0, 1, 2, and 3 layers of water molecules between the silicate sheets. The mineral is irreversibly dehydrated at about 450°C. The X-ray diffraction diagram of the material calcined at 550°C. is not muscovite-like as Caillère *et al.* had concluded, and also differs from vermiculite. Complexes with ethylene glycol or other diol alcohols give the basis for a comparison of the structure with vermiculite and with montmorillonite. The separation of $O–H_2O$ is about the same as in vermiculite, but the $H_2O–H_2O$ separation of the layers is considerably less in this mineral due to the presence of cations between the water layers. The $O–H_2O$ and $H_2O–H_2O$ in allevardite are the same if cations are absent, as was shown by a Fourier synthesis. The distribution of the water layers in the four different hydration states (with $d(001) = 14.4$, 16.1, 22.00, and 28.38 Å.) is shown in Fig. B. 18, with the characteristic $O–H_2O$ separation 2.95, 2.92, and 3.07 for the different states, and $H_2O–H_2O = 2.92$ or 2.83 Å.

102. From the Yonago Mine, Nagano Prefecture, Japan, S. Shimoda and T. Su-

[181] *Acta Geol. Alpina,* No. 7, 1959, 121-159, especially pp .145-159 on vermiculite.
[182] *Bull. soc. franç. minéral. et crist.*, **73**, 1950, 193-201.
[183] *Am. Mineralogist,* **41**, 1956, 91-103.

do[184] described another allevardite-like mineral with a $d(001)$ of 26.14 Å. spacing which was interpreted as a complex of a true mica structure (A) with $d(001) = 10$ Å., and a silicate layer structure (B) of 12.48 Å. spacing, alternating with a single water layer, of finally a 15.16 Å. silicate sheet structure (C) alternating with a double layer of H_2O molecules replacing the K^+ cations of clay mica. The 26 Å. spacing is

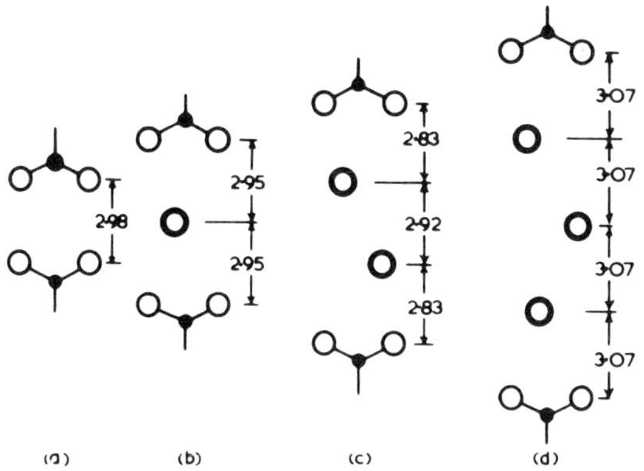

FIG. B. 18. Four hydration states of allevardite (Brindley). The heavy circles represent the water layers.

thus nearest to that attributed to pilings of (A) + (C) and (B) + (C) structures. Even in the latter case small amounts of K^+ are retained. This type of allevardite-like minerals is an alteration product of micas by leaching. Characteristic is that the 26 Å. spacing is usually accompanied by a "tailed" reflection in the small-angle side. The 26 Å. spacing is raised to 29.2 Å. by a continued treatment with water; whereas ethylene glycol in some specimens is not taken up at all, in others the basal spacing is widened to 28 Å. The complex mixed-layer type of such structures is evident.

103. Another analogue of allevardite is described by N. M. Ivkin, N. S. Kitaĭgorodskiĭ, D. D. Kotel'nikov, and Yu. M. Korolev[185] from Dagestan, occurring in hydrothermal carbonate veins in travertine. The electron micrographs show lath-shaped crystallites up to 1 μ in length. In its chemical composition this mineral is similar to the French and the Japanese occurrences. The Dagestan allevardite, however, is swellable in water to double its initial volume, and its structure is *di*octahedral. The basal spacing of 24 Å. is reduced to 19 Å. after calcination at 600°C. to that of mica parcels. With glycerin the spacing is raised to 27.7 Å., combined with a considerable increase in intensity for the interference lines. The structure is characterized by subperiods indicated by doubled ($00l$) reflections.

[184] *Am. Mineralogist*, **45**, 1960, 1069-1077.

MIXED-LAYER CLAY MINERALS

104. We emphasized several times in the previous chapters how often clay minerals occur which do not show the simple characteristics of the kandites, smectites, or vermiculites, but are of a "complex" nature due to *interstratification* of other structural units, either mutually between these three principal families, or even with foreign-layer structure compounds. In chlorites the interstratification of brucite layers between those of pyrophyllite or talc parcels is evidently quite *regular*, and in such case it is not feasible to speak of a mixed-layer type. But as soon as one-dimensional dislocations (in the definition given by H. Jagodzinski) occur, say a stacking of different units in a more or less *irregular* (randomness) sequence, the characterization of the mixed-layer clay minerals is necessary. In the actual state of investigations in this highly attractive but complex field it is not yet possible to develop a strictly rational classification of mixed-layer systematics. It is objectionable to overload the complex nomenclature of the clay minerals by the introduction of new mineral names for every "new structural" type. We share entirely the viewpoints given by Th. Ernst, W. Forkel, and K. von Gehlen[186] that it would be best to develop useful symbols for the composition of mixed-layer minerals of this class, although in the present state of our knowledge it is still difficult to do this in an accurately quantitative manner. In the meantime, however, in most cases it is sufficient to give an approximate ratio of the shares of the interstratified components for a given mixed-layer mineral.

105. Irregular interstratification of layers at random was first emphasized in its general significance by J. W. Gruner (1934) for the structure of vermiculite in "*hydrobiotite*," then by S. B. Hendricks and E. A. Teller[187] for the X-ray interference phenomena in partially ordered layer structures, and then W. F. Bradley[188] applied these principles for clay minerals. A sufficiently precise identification of mixed-layer structures is possible only by highly accurate X-ray methods to distinguish between incidental (physical) mixtures (which show the characteristic reflections of the individual components) and irregular interstratifications. In the case of regular sequences in stacking, a larger unit cell must appear, with a multiple of the individual layers and corresponding (*hkl*) interferences. If the interstratification is

[185] *Zapiski Vsesoyuz. Mineral. Obshchestva,* **88,** 1959, 555-563.

[186] *Ber. deut. keram. Ges.,* **36,** 1959, especially p. 16.

[187] *J. Chem. Phys.,* **10,** 1942, 147-167. See also J. Méring, *Acta Cryst.,* **2,** 1949, 371-377, and G. Brown and D. M. C. MacEwan in Chapter XI of the monograph "X-Ray Identification and Crystal Structure of Clay Minerals," Mineralogical Society of Great Britain, London 1951, pp. 266-284. Furthermore, see J. C. McAtee, *Am. Mineralogist,* **41,** 1956, 627-631, especially on the application of Fourier transforms in random-interstratified montmorillonites with variable amounts of monovalent and divalent cations.

[188] *Am. Mineralogist,* **30,** 1942, 704-713; *Anal. Chem.,* **25,** 1953, 727-730.

at random, it makes some difference whether the layers of the second interstratified type are present only in a low concentration or with more than 10 per cent. In the first case the dominant layers will rule the character of the interference pattern, with slight modifications, nevertheless. In the second case, new diffraction effects will arise, such as a nonintegral series of (00*l*) interferences which are composites of adjacent reflections of the same orders of the different layers, and at an intermediate position between them, or a composite of overlapping reflections of different orders for different layers, and variable intensities. Composite peaks, or more properly bands, will vary with the relative abundance of the two layers. One-dimensional Fourier syntheses based on intensity measurements are here of great importance for the correct interpretation of the complex X-ray patterns, from which the calculation of the phase relationships may then start, as Ch. E. Weaver[189] demonstrated.

106. More recently differential-thermal analysis has become an important tool in the discrimination of possible types of interstratification, especially in combination with determinations of the base exchange properties of the mixed-layer minerals in question. G. Nagelschmidt[190] gave an excellent example for the interstratification of illite and glauconite with alternating layers of 10 and 15 Å. units of the structural periods stacked in the *c*-axis, or for illite cores surrounded by montmorillonite. Another essential aid in identification of mixed-layer structures is the artificial interstratification with ethylene glycol, e.g., in the direction of the *c*-axis which W. F. Bradley[191] described for *rectorite*, with one pyrophyllite and one vermiculite component. There is a most remarkable ambiguity of the origin of structural piling, depending on whether one considers the layers to be geometrical or chemical units. Very characteristic in this respect is also the interstratification of vermiculite and chlorite, with an approximate period of 30 Å. in the *c*-direction, as described by T. Sudo *et al.* (see below) in the so-called "*G*" lattice.

107. Even refined methods of differential-thermal analysis do not in every case give sufficient information on the presence of mixed-layer interstratifications. W. F. Cole[192] refers to illite-montmorillonite intergrowths, for which the sizes of the endothermic peaks in the 500° to 700°C. region do not give any relation to the proportion of the constituents. An examination of the portions from 525° to 625°C., and from 650° to 725°C., however, shows structural differences in these which are not revealed

[189] *Am. Mineralogist*, **41**, 1956, 202-221.

[190] *Mineral. Mag.*, **27**, 1944, 59-61.

[191] *Am. Mineralogist*, **35**, 1944, 591-595.

[192] *Nature*, **175**, 1955, 384. On the temperature stabilities of montmorillonite and vermiculite ethylene glycol complexes, cf. W. F. Bradley, R. A. Rowland, E. J. Weiss, and Ch. E. Weaver, *Natl. Acad. Sci.-Natl. Research Council, Publ.*, **566**, 1958, 348-355 (cf. Vol. V, Section B), using the oscillating heating powder X-ray diffraction method, combined with a one-dimensional Fourier synthesis for the analysis of the interlayer liquid distribution.

by X-ray methods, indicating "anomalous illite" and "anomalous montmorillonite" in the meaning of R. C. Mackenzie, G. F. Walker, and R. Hart[193] and of A. K. Bose and P. Sengupta.[194] Differential-thermal analysis therefore is able to indicate the *direction* in which clay mineral transformations take place in mixed-layer structures, in this case of illite and montmorillonite. An additional auxiliary method is the combination with thermal exposure, and rehydration in steam as recommended by R. D. Hill.[195]

108. Although they do not exclusively apply to mixed-layer clay minerals but generally to all expandable types of these, we here recommend the methods of W. D. Johns, R. E. Grim, and W. F. Bradley[196] as particularly useful for soil investigations, and for study of recent sedimentary rocks, namely a combined differential-thermal analysis and ethylene glycol treatment. We further cite the vapor pressure glycolation of oriented clay mineral samples used by G. Brunton,[197] and the more general studies of W. D. Johns and R. T. Tettenhorst[198] on the behavior of different montmorillonite samples swelling in different polar liquids, especially with water, glycerin, or ethylene glycol (cf. C. ¶ 271) after heating at 200° to 300°C. In this case, very specific influences of the cations adsorbed on montmorillonite sheets sharing the mixed-layer structures became evident. Glycerin, in this case, is more specific than ethylene glycol. A method specifically adopted to the elements present as cations in water interlayers like those of vermiculites and chlorites is boiling with an 0.5 N NaOH solution for $2\frac{1}{2}$ minutes, after a dehydroxylation at 400°C. for 4 hours, as shown by J. B. Dixon and M. L. Jackson[199] for soil analysis.

109. J. S. Hosking[200] in his classification of the clay minerals (cf. B. ¶ 1) is of the opinion that diphormic (1 : 1) layer kandites do not form real mixed-layer structures, whereas triphormic (2 : 1) layer smectites and tetraphormic (2 : 1 : 1) layer chlorites easily do so. Practically all samples with diphormic kandites in alleged mixed-layer combination have proved, in his opinion and after detailed examination, to be only physical mixtures, rather than true mixed-layer types. One of the cases in which principal intergrowths of dickite interstratified with hydromica were consid-

[193] *Mineral. Mag.*, **28**, 1949, 704-713.

[194] *Nature*, **174**, 1954, 40-41.

[195] *Trans. Brit. Ceram. Soc.*, **52**, 1953, 589-613.

[196] *J. Sediment. Petrol.*, December 1954, 242-251.

[197] *Am. Mineralogist*, **40**, 1955, 124-126.

[198] *Am. Mineralogist*, **44**, 1959, 894-896.

[199] *Science*, **129**, 1959, 1616-1617. The indicated method is more convenient than the sodium citrate treatment recommended by T. Tamura, *Soil Sci.*, **9**, 1958, 141-147; see also the discussion of such methods *Am. Mineralogist*, **42**, 1957, 107-110, warning to use great care in assigning the 14 Å. spacing either to vermiculite or to chlorite in soil minerals.

[200] *Second Australian Conference in Soil Science, Melbourne, Victoria, August*, 1957, Vol. I (1), 1957, 26 pp.

ered is the so-called *alushtite* from the Crimea, described by N. V. Logvinenko and V. A. Frank-Kamenetskiĭ.[201] The X-ray diagram is that of dickite but with lines of a mixed-layer type with montmorillonite and hydromica, showing $d(00l) = 14.1$, 17.6, and 28.5 KX, from (001) to (00.16). The electron microscope shows (in double-replica techniques) pseudo-hexagonal dickite crystals and irregular parcels of montmorillonite. This mineral alushtite is evidently a low-temperature hydrothermal product (in quartz veins), and similar to what T. Sudo (see below) described as intergrowths of a kaolinite family structure, with a hypothetical aluminum chlorite in the so-called "*G*" mineral structure (cf. B. ¶ 106).

110. The most frequent mixed-layer clay minerals are those with 2 : 1 layer types which are abundant in sedimentary rocks. Ch. E. Weaver[202] discussed the X-ray diffraction patterns for randomly interstratified clay minerals composed of illite-montmorillonite, chlorite-vermiculite, illite-chlorite-montmorillonite, and regularly interstratified chlorite-vermiculite. Among a rich material of sedimentary rocks from all over the United States, ranging from Cambrian to recent ages, more than 70 per cent of the samples contained some variety of mixed-layer minerals. In marine diagenesis the intergrowths of illite-montmorillonite and chlorite-montmorillonite are more abundant than pure illites and chlorites, whereas weathering reactions produce more of the chlorite-vermiculite mixed-layer structures.

111. Starting with a *regular* interstratification of the type already discussed by Bradley (1950), Weaver differentiates such intergrowths from random structures on the basis that the $(00l)$ reflection series occurs in an integral sequence and that large $d(001)$ values (24 to 30 Å.) are in themselves *not* sufficient proof of regularity. In random mixed-layer clays where the form factors are sufficiently different in addition to an average reflection of the adjacent (001)/(001) pair of interferences, a low-angle reflection will also occur because the contributions from the layers with higher scattering power are not completely cancelled by those of lower power. In a *random* layer material containing say 50 per cent illite and 50 per cent montmorillonite (with 14 Å. normally) layers, a given interval of 120 Å. will contain about 5 illite units of 10 Å. and 5 montmorillonite layers. The stronger scattering will be encountered 5 times per 10 Å., with an ill-defined period of about 24 Å. causing a diffuse scattering maximum. In the regularly mixed layer clay mineral, however, the value 120/5 is consistent as the resulting sharp reflection. Such regular mixed-layer structures are relatively scarce in sedimentary rocks. Figure B. 19 gives the X-ray diagrams of two identical clay minerals of this kind from a Pennsylvanian shale, and a Mississippian limestone residue. They are composed of alternating 14 Å. chlorite and 14.8 Å. vermiculite layers, with a (001) spacing of about 29.2 Å. The

[201] *Doklady Akad. Nauk S.S.S.R.*, **105**, 1955, 554-557; *Ibid.*, **137**, 1961, 1441-1444.
[202] *Am. Mineralogist*, **41**, 1956, 202-221.

agreement of the reflections $(00l)$ up to (00.10) is very satisfactory for the untreated material, but that treated with ethylene glycol showing $d(001) = 31.0$ Å., after heating at 550°C., indicates this spacing reduced to 12.0 Å.

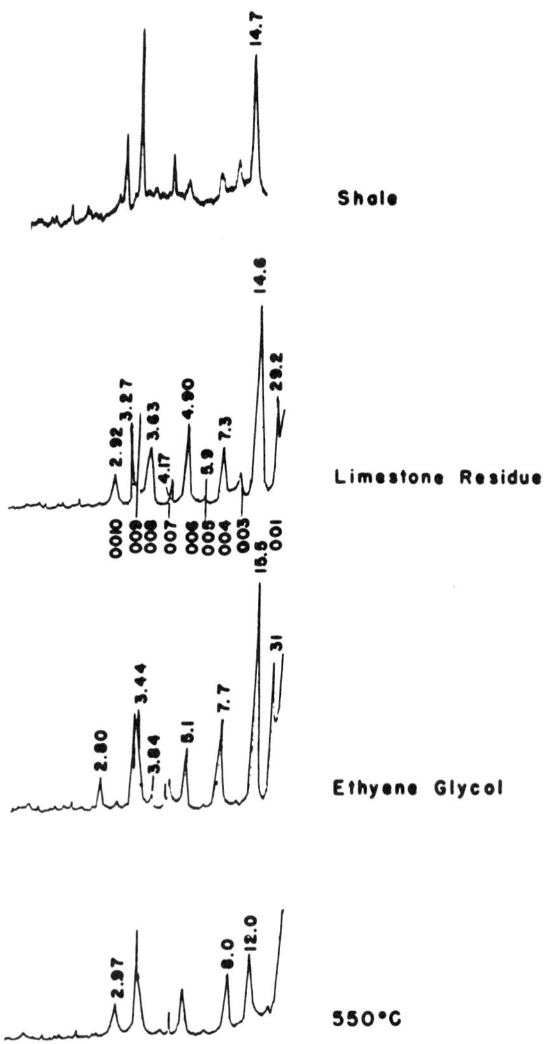

FIG. B. 19. X-ray diffraction diagrams of regular mixed-layer clay minerals, after various treatments (Weaver).

112. Another impressive example is given by J. W. Earley, G. W. Brindley, V. J. McVeagh, and R. C. van den Heuvel[203] for regularly interstratified chlorite-montmorillonite, with $d(001) = 29$ Å., from Ward Co., Texas. The treatment with glycerin raises the spacing to 32 Å. (17.8 Å. for the montmorillonite glycerin complex included). Normal to (001) the electron density distribution curve shows clearly the alternating layers (Fig. B. 20), including the brucite layer in the chlorite share, and the two water layers in the montmorillonite share. The endothermal effects on the differential-thermal curve at 630° and 830°C. are characteristic of these interlayers. Earley *et al.* also discuss the rather difficult problem of the nomenclature of mixed-layer structure clay minerals, in general, and of regularly interstratified types, in particular. They propose to use new mineral names only if the interstratification is regular.

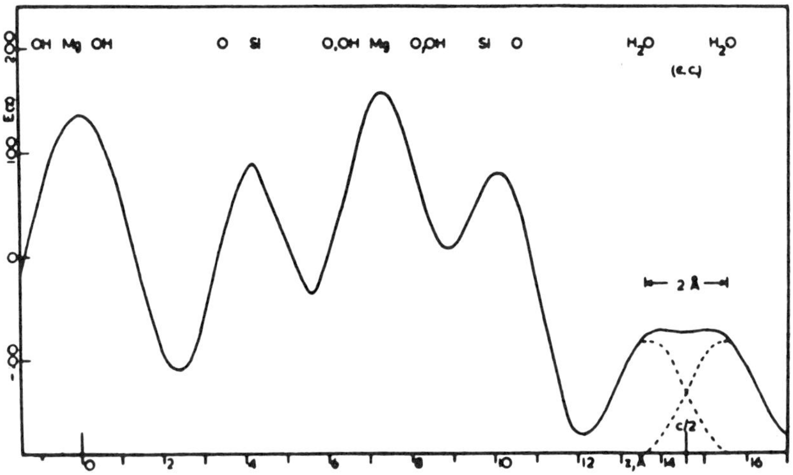

FIG. B. 20. Electron density distribution perpendicular to (001) of regular chlorite-montmorillonite interstratification (Earley, Brindley, McVeagh, and van den Heuvel).

113. In this respect we turn to the suggestion of F. Lippmann[204] to call *corrensite* a "*swellable chlorite*"; this mineral is from the southern German Keuper clay sediments, with $d(060) = 1.54$. Corrensite, therefore, is a *tri*octahedral clay mineral. In air-dry samples the vermiculite layers of corrensite are most regular in their structure if Ca^{2+} is introduced as the exchangeable cation, bringing about strongest 29 Å. reflections. Odd-numbered reflections, like (005) and so on, are then observed,

[203] *Am. Mineralogist*, **41**, 1956, 258-267.

[204] *Heidelberger Beitr. Mineral. u. Petrog.*, **4**, 1954, 130-134; see also D. B. Honeyborne, *Clay Minerals Bull.*, **1**, 1951, 150-155, and F. Lippmann, *Fortschr. Mineral.*, **35**, 1957, 28-30; *Ibid.*, **28**, 1960, 176-182, as a particularly valuable introduction into the actual problems of mixed-layer minerals of the mica-montmorillonite type, using the projection methods of Méring, and his superposition principles of X-ray interferences in regular mixed-layer sequences; see finally *J. Sediment. Petrol.*, **26**, 1956, 125-139. Expandable chlorite was recently described by M. N. A. Peterson, *Am. Mineralogist*, **46**, 1961, 1245-1269, from carbonate rocks of the Cumberland Plateau, Tennessee.

whereas alkali ions as exchangeable cations reduce considerably the intensities in the higher orders. Corrensite is also swellable in glycerine, methyl and ethyl alcohol, acetone, acetonitrile, and water. After heating to 500°C. the 14 Å. line is shifted to 12 Å., or the second order of 24 Å., in the complex of chlorite (14 Å.) and dehydrated vermiculite (12 Å.).

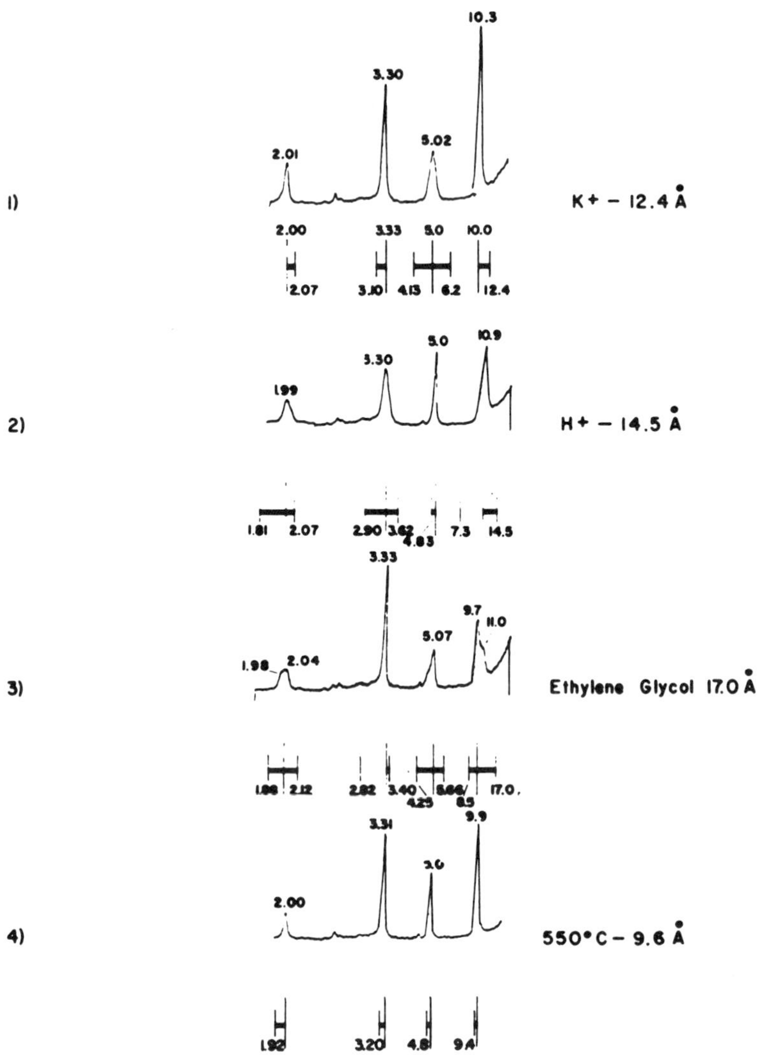

FIG. B. 21. X-ray diffraction diagrams of a randomly interstratified mixed-layer clay mineral, after different treatments (Weaver). The mineral contains 20 per cent expandable and 80 per cent nonexpandable constituents.

114. W. F. Bradley and Ch. E. Weaver[205] described from the Upper-Mississippian limestones of Colorado a thoroughly identical, regularly interstratified chlorite-vermiculite mineral, in an occurrence remarkably free from any accessorial illite or other disturbing clay minerals. The fundamental period is 29 Å. It is genetically interpreted by a replacement of the brucite layers in chlorite by water layers and exchanged bases. In the *nonregular* sequence of randomly interstratified mixed-layer clay minerals the $(00l)$ series is not integral. The $d(00l)$ values *average* from the simultaneous scattering of both layer types, as shown in Fig. B. 21 for an interstratified clay with 20 per cent expandable and 80 per cent non-expandable layers, the dark horizontal lines indicating which adjacent reflections combine to produce the average reflections. The precise position depends on the relative amounts of the two layers, the form factors, and especially in the case of the first maximum the polarization factor. In the latter, it is necessary to apply the formulas and curves given by G. Brown and D. M. C. MacEwan,[206] of which we give only one characteristic

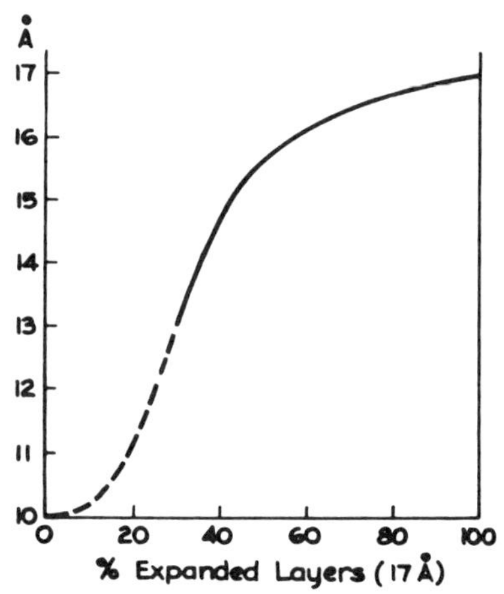

FIG. B. 22. Curve showing the migration of the (001)/(001) peaks of randomly interstratified 10 and 17 Å. layers in mixed-layer clay mineral (Weaver).

curve corresponding to Weaver's paradigm in Fig. B. 22. This author discussed in detail the conditions of some selected cases of X-ray patterns of the original mixed-layer clay minerals, with variable cations, treated with ethylene glycol or thermally.

115. We are not able to give here many details of the multiplicity of phenomena possible in this group of mixed-layer minerals, with variable percentages of expanda-

[205] *Am. Mineralogist*, **41**, 1956, 497-504.

[206] *J. Soil Sci.*, **1**, 1950, 239-253. A very convenient routine procedure for clay mineral analysis was recommended by K. Oinuma, K. Kobayashi, and T. Sudo, *Clay Science* (*Japan*), **1**, 1961, 23-28.

ble layers, or varying with the smallness of the crystallites for illite-chlorite-vermiculite intergrowths occurring in sediments, or of binary, or even ternary, clay mineral composition. The latter are rather common in shales. For the discrimination of such complex types it is indispensable to use systematically X-ray diffraction, differential-thermal analysis, base exchange methods, treatment with ethylene glycol, and so on. Not only for the identification of the clay minerals as such, but also for the correct zoning of sedimentary rocks, are such systematic investigations of great importance.

116. It is most fascinating to study the rich literature on the discovery of mixed-layer clay minerals in different geological environments. We shall start such a discussion from a regional viewpoint by mentioning the results of the extensive investigations of P. Gallitelli *et al.* on the clayish sediments of the Apennine Mountains[207] in which the occurrence of a "*swellable chlorite*" is the most striking phenomenon. The observations of Gallitelli are in perfect analogy to those of Lippmann in the description of corrensite. The association with fireclay type kandites and illite, in addition, is remarkable for the Italian clay sediments in question, called ophiolites. Also interesting in this connection is the experimental change of "swellable chlorite" occurring in the iron ores (minettes) of Lorraine into montmorillonite by an acid treatment, or by a solution of sodium hydrosulfite and sodium tartrate, or by electrodialysis as was demonstrated previously by S. Callière, S. Hénin, and J. Esquevin.[208] Normal chlorites do not show a corresponding change.

117. The genetic conditions of the mixed-layer type of corrensite is also characteristic of diabases changed to clayish products, often observed in Apenninian rocks, and changes of biotite to trioctahedral montmorillonite (cf. G. F. Walker, 1950) in Aberdeenshire, Scotland. The regularity of the alteration is greatest in the chlorite-vermiculite mixed-layer products from diabase, but is gradually reduced in the "red clays" and later decomposition products. In spite of the large abundance of the chlorite-vermiculite mixed-layer minerals in the rock type here discussed, Gallitelli distinguished in one case (from diabase of the Groppo Maggio, Emilia) another type, which does not show the characteristic 29 Å. interference but is highly irregular, and interpreted it as a mixed-layer structure of a *chlorite-saponite* hybrid. The brucite

[207] *Clay Minerals Bull.*, **2**, 1955, 275-280; *Atti soc. toscana sci. nat. Pisa. Mem. Ser. A*, **62**, 1955, 88-124; *Rend. soc. mineral. ital.*, **12**, 1956, 139-144; *Ibid.*, **16**, 1960, 1-16; *Atti accad. nazl. Lincei. Rend. Classe sci. fis. mat. e nat.*, (VIII) **21**, 1956, 146-154; *Ibid.*, **25**, 1958, 13-20; *Atti e mem. accad. sci. lettere ed arti Modena*, [6] **2**, 1960, 172-198; *Congresso Geologico Internacional XXa, Mexico City*, 1956. Sobretiro del Comité Internacional para el Estudio de las Arcillos, 1956-59, 23-26; *Colloq. Internat. Centre Nat. Rech. Sci.* (*Paris*), No. 105, 1961, "Genèse et Synthèse des Argiles," 191-195. On the determination of chlorites in sediments, see also W. F. Bradley, *Natl. Acad. Sci.-Natl. Research Council, Publ.*, **327**, 1956, 324-334.

[208] *Clay Minerals Bull.*, **2**, 1954, 166-170.

layers of chlorite are replaced by water with the cations exchangeable like those in a typical smectite. The mineral is similar to the "pseudo-chlorite" of R. Roy and L. A. Romo[209] (see B. ¶ 95, ¶ 97), or to the mineral from the Gulf of Mexico described by R. E. Grim and W. D. Johns,[210] evidently a product of geological environments with a pH of 7 to 9. A weakly hydrothermal treatment, as a matter of fact, changes the saponite-chlorite hybrid into a product more similar to chlorites.

118. Many analogues are also found between mixed-layer clay minerals described by A. Alietti[211] from the Parma Apennines, and stevensite (cf. B. ¶ 33). The *saponite-talc* alteration layers here afford ten orders of basal reflections, with the 24.8 Å. spacing in the original material changed to 27.1 Å. by treatment with glycerin. A *chlorite-saponite* and chlorite-vermiculite mixed-layer mineral shows analogous changes from 29.4 to 31.6 Å. The genesis of this rich variability of mixed-layer minerals is observed from serpentinites and ophiolites, in good agreement with the hydrothermal experiments of R. Roy and Fr. A. Mumpton (cf. Vol. IV, Section A), the mixed-layer phases being transitions into talc-chlorite rocks.

119. Talc in veins with chalcopyrite, near serpentine rock, is always hydrothermal in its origin. Saponite is also hydrothermal in veins with chalcopyrite, but formed at lower temperatures. Mixed-layer minerals of saponite-talc are low-hydrothermal, but *not* formed by weathering. Chlorite-vermiculite is derived from weathering of basic rocks, or by mild hydrothermal action, and chlorite-saponite mixed layers are probably not weathering formations. Their origin is uncertain, nevertheless.

120. The position of *stevensite* (cf. B. ¶. 33) was not uniformly adopted. G.W. Brindley[212] classified it as a mixed-layer mineral of the smectite group, with a poorly developed sequence of non-integral type in the (00*l*) spacings, the maximum being about 24 to 25 Å. Also after treatment with glycerin, ethylene glycol, or other diols, no improvement in the development of these sequences could be observed; the spacing is raised to about 26 Å., whereas the heat exposure reduces it to 9.6 Å. Brindley interprets stevensite as a mixed-layer mineral consisting of 10 and 14 Å.

[209] *J. Geol.*, **65**, 1957, 603-610. The reasons for failure of attempts to synthesize deliberately mixed-layer clay minerals by hydrothermal equilibrium methods is evidently their metastability; cf. L. A. Romo and R. Roy, *Bull. soc. franç. minéral. et crist.*, **78**, 1955, 433-448.

[210] *Natl. Acad. Sci.-Natl. Research Council, Publ.*, **327**, 1954, 81-103.

[211] *Clay Minerals Bull.*, **3**, 1959, 207-211; *Periodico Mineral. (Rome)*, **28**, 1959, 65-110; *Atti accad. nazl. Lincei. Rend. Classe sci. fis. mat. e nat.*, (VIII) **21**, 1956, 201-207; *Ibid.*, **28**, 1960, 893-898. In their extensive studies on talc, J. A. Pask and M. F. Warner, *J. Am. Ceram. Soc.*, **37**, 1954, 118-128, also speak of multiple transition forms of talc and chlorite as mixed-layer complexes, to the extent that the number of brucite layers in chlorite are decreasing. Normally, regular talc lines do not appear shifted, but some diffuse scattering occurs, explained by S. B. Hendricks by randomly oriented interlayers (1939).

[212] *Am. Mineralogist*, **40**, 1955, 239-247.

units. The latter share is swellable, and may correspond to saponite; the other share, to talc. On the other hand, we have presented the reasons why G. T. Faust and K. J. Murata[213] defended the independent nature of stevensite as a member of the montmorillonite family, closely related to saponite, in spite of its low base exchange capacity and the low swelling degree. The detailed studies of G. T. Faust, J. C. Hathaway, and G. Millot[214] confirm these conclusions. These authors also discuss Alietti's opinion on the saponite-talc mixed-layer mineral which is interpreted as a montmorillonite-type structure with regularly interstratified non-expandable layers in the ratio 1 : 1, as concluded from X-ray diffraction data. Alietti, on the other hand, comes to the conclusion that the difference between the saponite-talc mixed-layer mineral of the Val di Taro occurrence, and that of Brindley's studies from Springfield, New Jersey, is caused by the regularity in the Italian occurrence and the irregular structure in that from New Jersey. The treatment of the mineral by 10 *N* HCl solution does not destroy the mixed-layer crystal structure, but changes the mineral more to a talc-like structure, which is still different from real talc by its characteristic endothermic effect at 865°C., in spite of the effects of the acid treatment. In distinction from Faust's stevensite the saponite-talc mixed-layer mineral is *not* a "defect structure" in the ionic positions of the octahedral layers.

121. Ch. E. Weaver and T. F. Bates[215] described as *metabentonites* (cf. B. ¶ 58), mixed-layer clay minerals from Ordovician and Middle Devonian sediments of central Pennsylvania; these minerals consisted of 22 Å. hydrous mica (montmorillonite) and illite in the ratios 1 : 4 and 1 : 3. X-ray examination and electron microscopy of the Schletta ceramic clays by O. E. Radczewski and H. J. Balden[216] gave the interesting result of mixed-layer structures of prevailing illite and montmorillonite (ratio 2 : 1) in irregular alternation, in agreement with the shiftings observed by glycerin and ethylene glycol treatment. Interesting mixed-layer minerals of montmorillonite and illite are also mentioned by W. F. Cole and A. R. Carthew[217] from Tasmania, corresponding to "*bravaisite*" (W. F. Bradley, 1945), with a random interstratification of 2/5 montmorillonite, 3/5 dioctahedral illite, and others similar to the ordered *rectorite* (Bradley, 1950), with alternation of vermiculite and pyrophyllite, separated by pairs of layers of water molecules, but better interpreted as ordered complex structures of montmorillonite and illite.

[213] *Am. Mineralogist*, **38**, 1953, 973-987.

[214] *Am. Mineralogist*, **44**, 1959, 342-370; especially p. 367 on Alietti's opinions.

[215] *Clay Minerals Bull.*, **1**, 1952, 258-261; *Am. Mineralogist*, **41**, 1956, 359-362, more detailed in *Bull. Geol. Soc. Am.*, **64**, 1953, 921-944.

[216] *Fortschr. Mineral.*, **37**, 1959, 74-78; *Silicates inds.*, **25**, 1960, 61-64.

[217] *Roy. Soc. Tasmania Papers & Proc.*, **87**, August 1953, 12 pp. See also W. F. Cole and M. E. Neilson, *Mining Geol. J. Victoria*, **6**, 1958/59 (3), 3-15, distinguishing well and poorly organized kandite and clay minerals containing triphormic types in mixed-layer alternation with chlorites, vermiculites or smectites as constituents, some of these similar to corrensite.

122. A particularly fascinating field of mixed-layer minerals are the Japanese clays which were especially investigated by T. Sudo *et al.*, and interpreted to be clay minerals derived from volcanic glasses. Chiefly important is the discovery of a mineral of montmorillonite type with a basal spacing of about 30 Å.[218] This mineral type is from the Kurata Mine, Yamaguchi Prefecture, and is accompanied by disordered kaolinite like that in the fireclays of England which were previously discussed by G. W. Brindley and K. Robinson (1948). The smectite mineral shows a reflection at 29.8 KX, and a line 14.7 KX proportional in intensity to the low-angle interference. The (060) spacing is 1.49 KX suggesting that the amount of Al^{3+} replacing Si^{4+} is low. Treatment with ethylene glycol changes the two basal interferences to 15.6 and 32.3 KX, and saturation with water changes them to 16 and 32 to 33 KX. This unusual smectite called preliminarily a "*G*" phase is interpreted as a superlattice compound of a normal montmorillonite structure with two sheets of water molecules interlayered (*M*-lattice), with 15.4 KX cell height, and the *G*-lattice with a 14.0 KX consisting of a hydrargillite layer interleaved between two adjacent silicate layers (with pyrophyllite parcels). For a lattice of the *G* type very special genetic conditions are noted, such as a boundary zone between different kinds of alteration and/or an area affected by successive hydrothermal actions.

123. Also known from Japanese clay occurrences are long-spacing modifications with 26.7 Å. associated with gypsum deposits in which the structural principles of *sepiolite* interfere. Thus, the clay minerals of the Hanaoka Mine described by T. Sudo and H. Hayashi,[219] in spite of many analogies with those of the Kurata Mine are quite different from them, e.g., in the (060) spacing = 1.506 Å., revealing that the number of Al^{3+} atoms in tetrahedral coordination in this case is *not* very low but corresponds to beidellites and certain sericites. We add the observations of T. Sudo and H. Hayashi[220] on a randomly interstratified *kaolinite*-montmorillonite intergrowth occurring in "acid earth" deposits, with a spot-wise distribution of the kaolinite, perhaps better interpreted as a simple epitaxis phenomenon (Fig. B. 23). Such a complex is much disordered, especially if swelling of the montmorillonite takes place in ethylene glycol.

[218] T. Sudo, H. Takahashi, and H. Matsui, *Japan. J. Geol. and Geography, Trans.*, **24**, 1954, 71-85; *Nature*, **173**, 1954, 261; *Clay Minerals Bull.*, **2**, 1954, 193-203.

[219] *Sci. Repts. Tokyo Kyoiku Daigaku*, Section C, No. 25, 1955 (3), 254-294, 281-316. See also Sudo's book "Mineralogical Studies on Clays of Japan," Maruzen, Tokyo, 1959, especially pp. 105-122; *Clay Minerals Bull.*, **2**, 1954, 96-106. See also H. Hayashi, A. Inaba, and T. Sudo, *Clay Science* (*Japan*), **1**, 1961, (3/4) 12-18, on complex clay mineral mixtures occurring in amygdules of basalts; K. Shimosaka and T. Sudo, ibid. 19-29, on clay minerals of Tertiary iron sands; M. Osada and T. Sudo, ibid. 29-40, on chlorite-rich clays in gypsum rocks; T. Sudo, H. Hayashi, and S. Shimoda, *Proc. 9th Nat. Conf. on Clays and Clay Minerals*, **8**, 1962, 378-392, on "intermediate clay minerals."

[220] *Nature*, **178**, 1956, 1115-1116.

124. The 30 Å. chlorite-montmorillonite mixed-layer clay mineral is described by T. Sudo and H. Hayashi from an interesting diaspore-bearing sphalerite-pyrite deposit of the Hanaoka Mine mentioned above,[221] the ore being cemented by the clay minerals. An acid earth clay from Ishikawa Prefecture is rather unusual in its mixed-layer structure described by T. Sudo[222] as an intergrowth also of kaolinite and montmorillonite in random type. Both constituents are *di*octahedral (spacing of (060) lower than 1.50), with very thin crystallites, and highly variable proportions

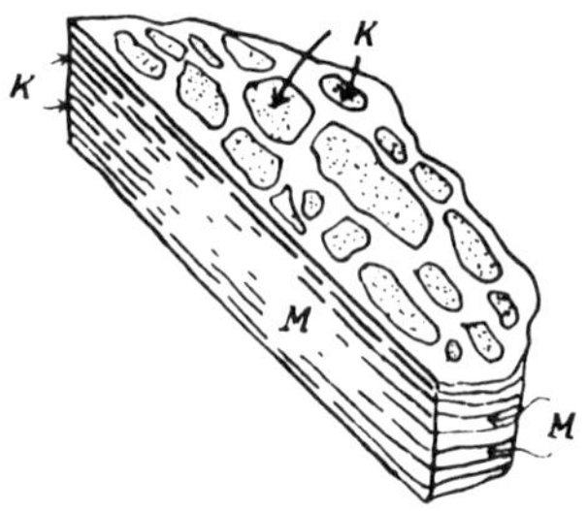

FIG. B. 23. Scheme for a mixed-layer clay mineral consisting of kaolinite (*K*), and montmorillonite (*M*) layers. (Sudo and Hayashi).

of the layers for each constituent (cf. on "alushtite," B. ¶ 109). Accordingly, the treatment with ethylene glycol causes extremely fine folds in a small crystallite by lattice expansion in the montmorillonite layers. The X-ray interferences are then much weakened in intensity by this folding and collapse. From a pyrite deposit of Aromori Prefecture, T. Sudo and H. Kodama[223] also describe a rather regular mixed-layer intergrowth of chlorite-montmorillonite with dioctahedral structure elements (spacing of (060) = 1.506 Å.). This dioctahedral chlorite could be an aluminum chlorite component in characteristic distinction from the commonly known trioctahedral chlorite sharing in mixed-layer structures of clay minerals.

[221] *Mineral. J. (Sapporo)*, **2**, 1957, 187-192.

[222] *Sci. Repts. Tokyo Kyoiku Daigaku*, Section C, No. 43, 1956, 39-55, especially p. 53.

[223] *Z. Krist.*, **109**, 1957, 379-387. In this connection we mention observations of G. Müller, *Neues Jahrb. Mineral., Monatsh.* 1961, 112-120, on a dioctahedral layer silicate of the chlorite group, and a proposal by W. von Engelhardt, G. Müller, and H. Kromer, *Naturwissenschaften*, **49**, 1962, 205-206, to call dioctahedral four-layer silicates with chlorite structure (from soils or decomposed acidic eruptive rocks) "sudoites."

Section C

Silicate Dispersoids

INTRODUCTION: DEFINITIONS AND METHODS

1. It is remarkable that again and again critical reexamination of the fundamental problems of definition of colloids must be made. This problem is particularly complicated by the multiple and sometimes contradictory uses of the word "colloid" for states of matter; it is necessary to determine whether a colloid consists of micro- (or, better, crypto-) crystalline particles, or really amorphous particles, or otherwise, and what are the dimensions which should be designated as typical of colloidal particles. The colloids were first defined by Th. Graham (1861) as "glue-like" materials which cannot migrate through semipermeable membranes. An interesting rediscussion of this basic problem was made by N. Riehl and N. Trubetzkoy[1] who repeated relatively elementary experiments of P. Walden (1892) to study the separation effects of "permeation-capable" (i.e., "crystalloid") and of "permeation-unable" (i.e., "colloidal") substances. The classical dialysis methods with paper, parchment, or cellophane as the diaphragms were supplemented by using semipermeable membranes of high efficiency, such as films of copper ferrocyanide, heavy metal sulfides, iron hydroxide, or nickel silicate hydrate, which can be made according to definite specifications of thickness; these were deposited in cellophane membranes serving as carriers, of about 40 μ in thickness. The results prove that there is *no* sharp boundary between permeation-capable and permeation-unable materials; however, the permeation coefficient, λ, defined by the exponential equation $C = C_0 \cdot e^{-\lambda t}$ for the decrease in concentration of the material dissolved in the semipermeable cell in the course of the dialysis, was measured in relation to that of glucose. This relative permeation $\lambda/\lambda_{gl.}$ was found for acids (also including organic acids) to be particularly high (= 5 to 20). For salts with increasing ionic charges, alkaloids, and others it varies from 5 to 1. For higher molecular unpolar organic materials it gradually ap-

[1] *Z. anorg. u. allgem. Chem.*, **253**, 1945, 54-70.

proaches zero as the molecular weight increases. Molecular weights of some 1000 are the limit of measurable permeations. Semipermeable membranes are superior to common dialysis membranes in making possible separations in the class of the sub-colloidal "crystalloids," as by separations of organic materials from traces of acids (demonstrated in the example of the systems glucose–HCl, H_2SO_4, or amino acids).

2. Other extremely interesting discussions of the problems of colloids and of the "amorphous" states of solid matter were presented by W. A. Weyl and E. Ch. Marboe,[2] especially from the viewpoint of the electron band theory, and of "potential energy profiles" in solids, in general. In this view, it is possible to differentiate between "really amorphous" materials—the state of which is called "chaotic" by Weyl and Marboe—and glasses; furthermore one may distinguish between "porous solids" or "rigid foams," to which are classified the chaotic-type hydrogels of silica and alumina, and more "glassy" gels occurring in unmixing (phase exsolution or separation) phenomena [such as emulsoids or pyronephelites in the original meaning of these terms (1925)], as well as opalescent dispersoids and others. Transitions are typical of chaotic materials with a potential energy profile as shown in Fig. C. 1 (*a*) which,

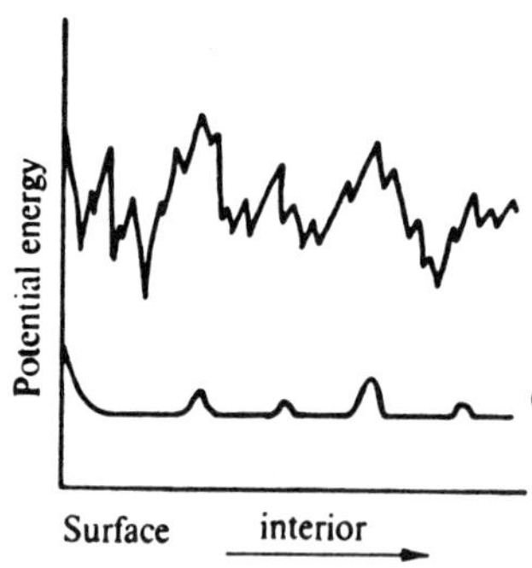

FIG. C. 1. Potential energy profiles of an amorphous solid in amorphous solid in the "chaotic" state (*a*) before and (*b*) after annealing (Weyl and Marboe).

by annealing, are changed to a more or less defective crystal phase of high dispersity, with definite peaks in the bands (*b*). Inversely, amorphization"" is possible by a very intense grinding of crystalline powders (cf. C. ¶ 9), or by "isotropization" to a metamict mineral under the action of intense γ-radiation, or on the surface of solids (chiefly metals, sulfides etc.) by polishing to form so-called Beilby layers (cf. C. ¶ 117 ff.).

3. Concerning the degree of dispersity, or what we call the "colloid dimensions," it would appear to be best to remain on the basis conventionally established by Wo.

[2] *J. Soc. Glass Technol.*, 43, 1959, 191-210.

Ostwald's scale (1927), thus in the limits of 100 and 1 mμ of particle diameters. The eight classes of dispersity types given by the present author in the 1954 edition may also be taken further as an empirical principle, with some additions and revisions which have become necessary due to advanced methods of examination and discrimination.

4. The example given for *solid* dispersed phases in a *solid* dispersion medium by spinels (cf. F. Rinne, 1928) was rediscussed by V. V. Lapin[3] for natural and synthetic spinel crystals from the viewpoint of structural defects (cf. Vol. V, Section A). In a natural picotite crystal this author determined that 5 per cent of the R^{2+} positions are left empty, corresponding to a relative excess in $R_2O_3 = 1 : 1.25$. In so-called "magnalum oxide" of the composition type $(Mg, Fe)_5(Al, Fe)_{18}O_{32}$, the crystalline solubility is that of a supersaturation in alumina over the basic $MgAl_2O_4$ and unmixing. Other spinels show the "classical" spinel molecules associated with TiO, or with cubic $MgO.3ZrO_2$, in variable degrees of dispersity and unmixing, especially observed in slag spinels. Structural details of the unmixing mechanism from alumina-supersaturated spinels were particularly studied by H. Saalfeld and H. Jagodzinski.[4] We mention from these extremely interesting investigations only the appearance of a monoclinic intermediary structure which in its composition is near that of pure alumina (e.g., about $Al_{26}MgO_{40}$). Characteristic of this highly dispersed phase are its complex twinning phenomena and its incompletely occupied lattice with superstructures, related to corundum as a "quasi-cubic analogue" in its oxygen spherical packing, with all signs of a one-dimensional defect structure in Jagodzinski's definition. Also important is the intimate intergrowth of the true spinel with the intermediate structure, in such a degree that there are nearly no separate diffraction phenomena to be discussed. In the unmixing process inner tensions in the lattice, analogous to the "one-phase" unmixing zones in metal alloys (e.g., the Guinier-Preston zones of copper-aluminum alloys), first appear before the intermediary monoclinic phase can be distinguished in the X-ray diffraction diagrams, followed by

[3] *Doklady Akad. Nauk S.S.S.R.*, **104**, 1955, 611-614. On the defects in spinel lattices see also W. Eitel, "Structural Conversions in Crystalline Systems," Special Paper No. 66, Geol. Soc. of Am., New York, 1958, pp. 46-49. The unmixing mechanism of alumina-enriched nickel spinels was recently studied by H. Saalfeld, *Ber. Deut. Keram. Ges.* **39**, 1962, 52-54.

[4] *Z. Krist.*, **109**, 1957, 87-109; 388-409. Quite recently S. Hafner and F. Laves, *Ibid.*, **115**, 1961, 321-330, 331-358, applied infrared absorption methods for a detailed investigation of spinel unmixing and disorder effects. F. Rinne (1928) already observed the intermediary structure but interpreted it as γ-alumina which would be extremely unstable because of the lacking electrostatic neutrality in smallest structural areas. H. Jagodzinski (*in* "Beiträge zur Physik und Chemie des 20. Jahrhunderts" (O. R. Frisch, F. A. Paneth, F. Laves and P. Rosbaud, eds.), Vieweg, Braunschweig, 1959, especially on pp. 206-209) described recently the very particular structural problems of the spinel unmixing, in connection with a critical revision of our general definition of phase boundaries and crystalline solutions.

the crystallization of α-Al_2O_3 (corundum). The high work of nucleation in this last phase is the principal reason of this complex mechanism in higher-dispersed states of new-formations, as seen in characteristic diffuse asterisms (stripes and haloes) of the Laue-diagrams. The end states are microscopically visible corundum crystallizations in excellent orientations of the lamellae parallel to the ((111)) planes of the spinel.

5. A second important group of solid-in-solid dispersed systems are colored minerals, such as blue quartz crystals with titania (rutile) as dispersed material. J. von Vultée and J. Lietz[5] repeated the previous studies of A. Postelmann (1937) on this phenomenon and confirmed in essential points his conclusions from ultramicroscopic observations, concerning rutile needles in a thickness near the limit of optical resolution, namely 100 to 300 mμ and also finer dimensions down to 20 mμ occur. The increase in resolution introduced by the electron microscope is here most fortunate and confirmed the previous estimations as a valuable basis for calculations of the optical properties of such systems. Much more complex are the problems of the "clouded" plagioclase feldspars described by A. Poldervaart and A. K. Gilkey.[6] Such phenomena are not to be confused, however, with common weathering or hydrothermal alteration products (so-called "filled" feldspars with clinozoisite or sericite), nor with trivial inclusions by overgrowths. Strong cloudiness is observed only in the intermediate plagioclase compositions which at lower temperatures are unmixed to extremely fine-dispersed mixtures of albite- and anorthite-enriched plagioclase crystalline solutions (cf. Vol. III, Section B).

6. A great theoretical and also practical importance must be attributed to the *gas*-in-*glass* dispersoid systems in which the gas-filled pores are often extremely small, as in the pumices or ignimbrites of volcanic ejectamenta. The recent rich literature in this field is introduced by a most fascinating mathematical discussion of the mechanism of the volcanic ash formation, as given by J. Verhoogen.[7] We will not consider here the mechanism of the eruptions as such (cf. Vol. IV, Section A), nor the physicochemical conditions of the rising gas-saturated magma melt, whereas the process of the development of bubbles from the clusters of a few hundred water molecules, and their growth in vesiculation, is the important point. They are frozen-in by the suddenly consolidated air-quenched ejectamenta rock material. The original foam structure, which is in the opinion of H. Hentschel[8] an *intra*telluric phenomenon, is connected with the undercooling effect in the melt by several hundred

[5] *Neues Jahrb. Mineral. Monatsh.*, 1956, 49-58.

[6] *Am. Mineralogist*, **39**, 1954, 75-91. See also previous interesting analogues observed by E. Dittler and A. Köhler, *Mineral. u. petrog. Mitt.*, **38**, 1925, especially p. 259; F. Reinhold, *Ibid.*, **29**, 1910, 121.

[7] *Am. J. Sci.*, **249**, 1951, 729-739.

[8] *Fortschr. Mineral.*, **33**, 1955, 141-142. Recently, *Bull. Volcanol.*, **25**, 1963, 291-313.

degrees centigrade in the explosive chilling. The resulting consolidated material as a tuff or an ignimbrite is in Hentschel's opinion, therefore, not welded on its way through the air in the explosion cloud. In this view, the famous incandescent clouds of the Mount Katmai (Valley of Ten Thousand Smokes) are *not* an ignimbrite in its proper definition. The ignimbrites from volcanoes in southern Peru, described by W. F. Jenks and S. S. Goldich,[9] are also not welded. Related conditions have a very great practical importance in the modern industrial production of foam glasses (cf. Vol. II, B. ¶ 73 ff.). M. Hübscher[10] demonstrates, as the best illustration of the structural constitution of such artifacts, the principles developed by E. Manegold[11] for foams in general, characterizing the foam glass as formed in polyhedral, closed-wall cellular systems, whereas analogous structures with disrupted walls and intercommunicating channels would be sponge-like materials. The pore dimensions in these industrial products are larger by far than in foams of colloidal types, but not in those products desirable for building construction purposes (with pores 0.3 to 2.0 mm. in diameter). However, it is not difficult to manufacture really colloidal foams, if needed for uses with such definite physical properties as average heat conductance, mechanical strength, or other special factors.

7. The great importance of solid material powders in grain sizes in, or near to, the limits of colloid dispersity for modern solid state physics and technology, as in powder metallurgy,[12] makes an accurate definition of the *grain size distributions* absolutely indispensable. Based on the purely empirical formulation of the functional relations existing between grain sizes produced in the common grinding process and the weight distribution as expressed in the formula of P. Rosin and E. Rammler (1933), $R(x) = 100 \cdot \exp(-bx^n)$, and an equation given by S. W. Martin[13] concerning the particle number frequency, $N(x) = a \cdot \exp(-bx)$, V. Zehler[14] discussed the mathematical correlations between these parameters and the residue characteristics. A sharp maximum is seen in the weight frequency as a function of the diameters. For different samples the determination of these maxima offers direct relations between the grinding time and the sizes, as applied to the production of highly reactive ferrite powders for electronic ceramics. The tremendous increase in

[9] *Am. Mineralogist*, **40**, 1955, 323.

[10] *Silikat Tech.*, **5**, 1954, 243, 247. Hübscher demonstrates as the most characteristic types of such foams the spherical, polyhedral, and alveolar shapes of the gas-filled cavities.

[11] "Schaum," Chemie und Technik Verlag, Heidelberg, 1953, 520 pp.

[12] On the use of ultra-fine powders in powder metallurgy see, e.g., H. H. Hausner, *Materials & Methods*, **24**, July 1946, 98-102, especially for metal alloy compacting.

[13] Symposium on New Methods for Particle Size Determinations in the Subsieve Range, Am. Soc. Testing Materials, Washington Spring Meeting, 1941, pp. 66-89. See also, as an introduction, the book of H. E. Rose, "Measurement of Particle Size in Very Fine Powders," Chemical Publ., New York, 1954, 127 pp.

[14] *Kolloid-Z.*, **160**, 1958, 16-20.

our knowledge in the mathematical characteristics of the grain size distribution laws in the technology of industrial dusts and related artifacts is well illustrated by the special deductions of E. Feifel.[15] This author gave a detailed description of the calculation and graphical evaluation methods, which are particularly useful for the characterization of the statistical determination of changes in grain size. Good examples are the problems of dust condensation of fly ashes, or the important conditions of granulation of the raw materials for Portland cement kilns[16] in which the presence of definite clay minerals is necessary for a stability of the granules during the calcination process in order to avoid dusting. We will come back to these important points in Vol. V, Section C.

8. The problem of how the *grinding* process[17] itself, or the manner in which mills are acting, influence the grain size distribution of the fine-dispersed products, is a highly practical one. G. F. Hüttig, W. Ebersold, and H. Sales[18] gave a fascinating description of these conditions for exact analysis of grinding operations, with definite "grinding functions." In Hüttig's opinion the grinding process is not only a statistical-mechanical problem, but also a chemical one, especially of bonding strengths, to be attacked by kinetic and thermodynamical methods at the same time. The grinding functions as depending on the grinding time are characterized in diagrams, two of which we show in Fig. C. 2. One is for quartz and another curve for talc because of their fundamentally different type. For quartz the grinding is preferably a crushing; for talc, prevailingly an abrasion process.

9. Grinding is one of the operations which contribute to what we called above an "amorphization" of crystalline powders. It is characteristic of micas (mostly muscovite in industrial milling) that they are easily changed in their crystalline struc-

[15] *Radex Rundschau*, 1952, 235-254; *Ibid.*, 1953 (6) 8-26, with valuable references. See also the discussion remarks of H. zur Strassen, *Ibid.*, 1954, 143-151; *Ibid.*, 1955, 345-348. On crystallographic investigations of dusts and the silicosis problems see H. Jung and W. Drees, *Silikattechnik*, **12**, 1961, 294-300 (cf. II. A. ¶ 335*, Footnote 553).

[16] In this respect we refer to the studies of G. Hill and H. E. Schwiete, *Zement-Kalk-Gips*, **11**, 1958, 181-196, and Vol. V, Section C.

[17] We recommend study of the chapter on "Comminution and Particle Size Distribution" in the book of E. Ryshkevich, "Oxide Ceramics," Academic Press, New York, 1960, 472 pp., especially pp. 12-30, with bibliography of this important technology. In addition, see the important studies of P. P. Budnikov and M. I. Nekrich, *Poroshkovaya Met., Akad. Nauk Ukr. S.S.R.*, **1**, 1961, (6) 17-26; *Silikattechnik*, **12**, 1961, 452-457.

[18] *Radex Rundschau*, 1953, 489-493; *Monatsh. Chem.*, **84**, 1953, 272; *Ibid.*, **85**, 1954, 588-596 (with O. Staufenberger); *Z. Elektrochem.*, **57**, 1953, 534-539; *Proc. Intern. Symposium on Reactivity of Solids. Gothenburg*, 1952 (Pub. 1954), Vol. II, 979-999. Hüttig uses a very specific terminology for the different factors and parameters involved in the grinding process. On corresponding studies of the amorphization of quartz, by X-ray diffraction effects, see G. S. Khodakov and P. A. Rebinder, *Doklady Akad. Nauk S.S.S.R.*, **130**, 1960, 1316-1318, *Kolloid. Zhur.*, **22**, 1960, 365-375.; *Ibid.* **23**, 1961, 482-490; *Silikattechnik*, **13**, 1962, 200-208.

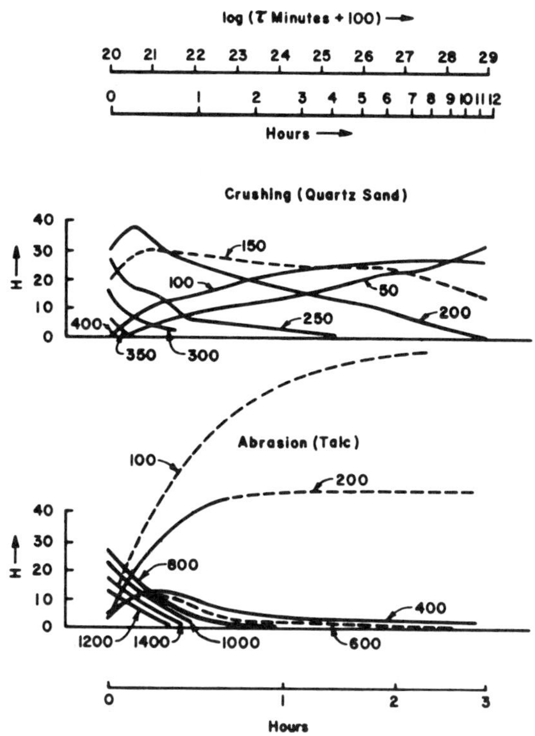

FIG. C. 2. Different grinding characteristics of quartz sand and talc in diagram showing the frequency, H, of different grain size classes (from 50 to 50, and 100 to 100 μ diameters), as a function of the grinding time (Hüttig, Ebersold, and Sales).

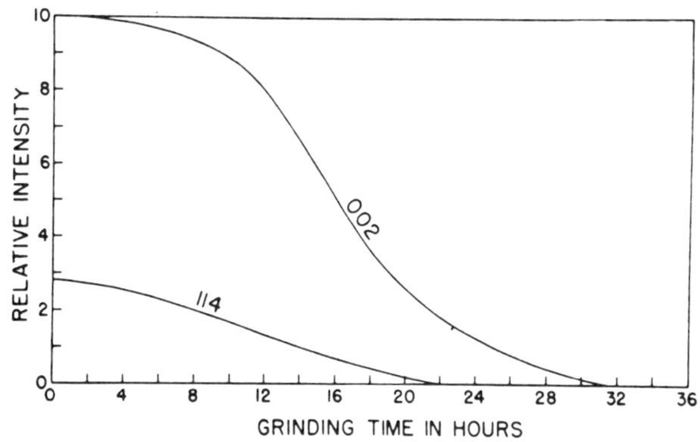

FIG. C. 3. Relative intensity of the (002) and (114) interference line of X-ray diffraction diagrams of natural muscovite ground up to 36 hours (Yoder and Eugster).

ture by intense grinding. R.C. Mackenzie and A. A. Milne[19] observed a considerable increase in the cation exchange properties of finely ground muscovite, and a more or less complete breakdown of the crystal structure as observed by diffuse X-ray diffraction spots. H. S. Yoder and H. P. Eugster[20] extensively discussed these destruction effects as seen in Fig. C. 3 by the steep decrease in intensity of the peaks for (002) and (114) in powder diagrams after 10 to 12 hours of grinding. After 36 hours the pattern is practically that of an amorphous material, and even in the electron diffraction diagrams only sporadic spots indicate residual crystalline areas. Also the differential-thermal curves show, in such a material with completely obliterated structures, no more characteristic peaks, as was demonstrated by R. C. Mackenzie and R. Meldau[21] for muscovite ground either in dry or in wet processes. Dry grinding and also simple filing develop on the mica material very high shearing stresses and local temperature increases (cf. Fig. C. 4). In dry-ground mica, the infrared absorption band at 2.8 μ which is characteristic of ionic hydroxyl groups is gradually

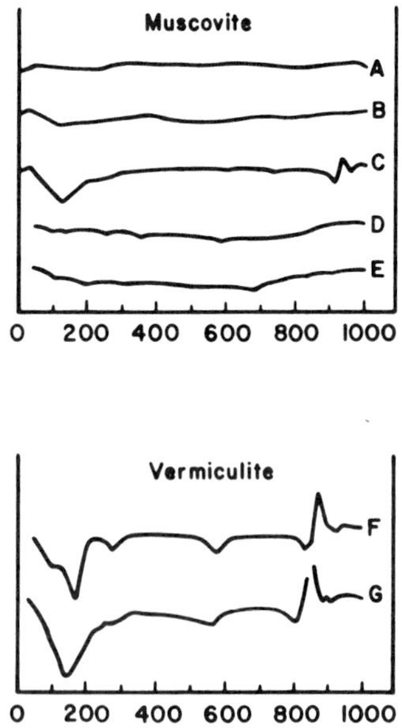

FIG. C. 4. Differential-thermal curves of muscovite and vermiculite powders after different grinding operations (Mackenzie and Meldau): (*A*) material after filing; (*B*) after dry-grinding for 2 hours; (*C*) after dry-grinding for 24 hours; (*D*) wet-ground for 24 hours; (*E*) commercial grinding; (*F*) dry-ground for 24 hours.

[19] *Mineral. Mag.*, **30**, 1953, 178-185.

[20] *Geochim. et Cosmochim. Acta*, **8**, 1955, especially pp. 257 ff.

[21] *Ber. deut. keram. Ges.*, **33**, 1956, 222-229. See also a study of the dehydroxylation peaks on differential-thermal curves of Indian "ruby" and "green" micas, by B. M. Bishui, N. R. Dhar, and S. S. Mandal, *Central Glass & Ceram. Research Inst. Bull.* (*India*), **8**, 1961, 15-22.

obliterated the longer the grinding is extended, and the band at 3.0 μ, characteristic of hydrogen bonded hydroxyl groups, increases in intensity. In the same way, kaolinite clays show quite analogous phenomena which are practically important as a so-called "colloidization" of clays by intense grinding, as described by J. Wiegmann and G. Kranz.[22] The grinding in sintered alumina mortars rapidly obliterates the (001) line and the range of $d = 4.5$ to 3.7 Å. in the X-ray diffraction diagrams, as in Zettlitz kaolin and fireclays. Less sensitive are the (060) lines in this process. Dry-grinding is always more destructive than wet-grinding.

10. The very important changes in *solubility* for particles of finest sizes are particularly evident in the classical example of gypsum (G. A. Hulett, 1904), and in slaking lime hydrate for which R. Hedin recently[23] gave an excellent quantitative investigation. The determination of the solubility by conductometry (cf. Vol. V, Section C), combined with the direct measurement of the particle sizes, resulted in the equation $\log C_{D,T} = \log C_{\infty,T} + 0.13 \cdot 1/D$, in which $C_{D,T}$ is the equilibrium concentration at temperature T, $C_{\infty,T}$ that for large particles (from ground crystalline powders), D the diameter in μ of colloidal or larger sizes (above 0.4 μ). The constant slope of the lines in Fig. C. 5 gives a good illustration of the effects of fine-dispersion of a characteristic, relatively easily soluble powder. The surface tension, σ, of the $Ca(OH)_2$ crystalline particles as a function of temperature is $\sigma = 2.0\ T$. Calcium hydroxide is, at the same time, a good example of the tendency of colloidal particles to grow by recrystallization (in the present case: transcrystallization) from supersaturated solutions, or during precipitation. The great geochemical importance just of this process was discussed by D. P. Grigor'ev,[24] in analogy with the growth of the crystals of metals as a function of the degree of deformation (J. Czochralski's principle, cf. Vol. V, Section A). Such phenomena are particularly important in the formation and aging of sulfide ore deposits or carbonate rocks which show in their general textural habitus their "collomorphic" origin and when they are recrystallized. Grigor'ev also gave a very instructive table demonstrating these qualitative facts in quantitative solubility data as function of the particle radii, following the exponential law $C_r = C_\infty \cdot \exp(2\sigma M/RT\ \varrho \cdot r)$, or Ostwald's and Freundlich's equation $RT/M \cdot \ln(C_r/C_\infty) = 2\sigma/\varrho\ (1/r - 1/r_\infty)$.

11. In modern colloid research the use of *ultrasonic* equipment for fine-dispersion purposes has become general, and a great progress is made by these safe and efficient methods. Especially for the operation of ceramic clays much new apparatus was constructed, of which we here mention the ultrasonic mills described by R. R.

[22] *Silikat Tech.*, **8**, 1957, 520-523.

[23] *Svenska Forskningsinst. Cement Betong vid Kgl. Tek. Högskol. Stockholm, Handl.*, No. 27, 1955, 14 pp. It must be noted that in calculation of the surface tension for $Ca(OH)_2$ directly the concentration values were taken, not the activity coefficients.

[24] *Zapiski Vsesoyuz. Mineral. Obshchestva*, **85**, 1956, 147-170.

West, R. E. Gould, L. B. Coffin, and J. F. Lux.[25] The so-called Kady mill may be understood as a more accomplished modification of the Poulsen colloid mill, with rotation speeds of 16,000 r.p.m., and a cavitation effect developed between specially shaped stators and rotors with slots. Such an instrument is a most efficient help in dispersing slurries to highest uniformity and to excellent electron-microscopic resolution, and is, therefore, much superior to common single-propeller mixing devices. They are equally suitable for a separation of montmorillonite from kaolinite particles in secondary clay agglomerates. Ultrasonic piezo-quartz-controlled dispersion devices (operating with frequencies up to 960 kilocycles) were used by G. Ludwig[26] for the same purpose in the clay research laboratory applying the principles previously given by Agnes Mathieu-Sicaud and G. Levavasseur (1949) and more

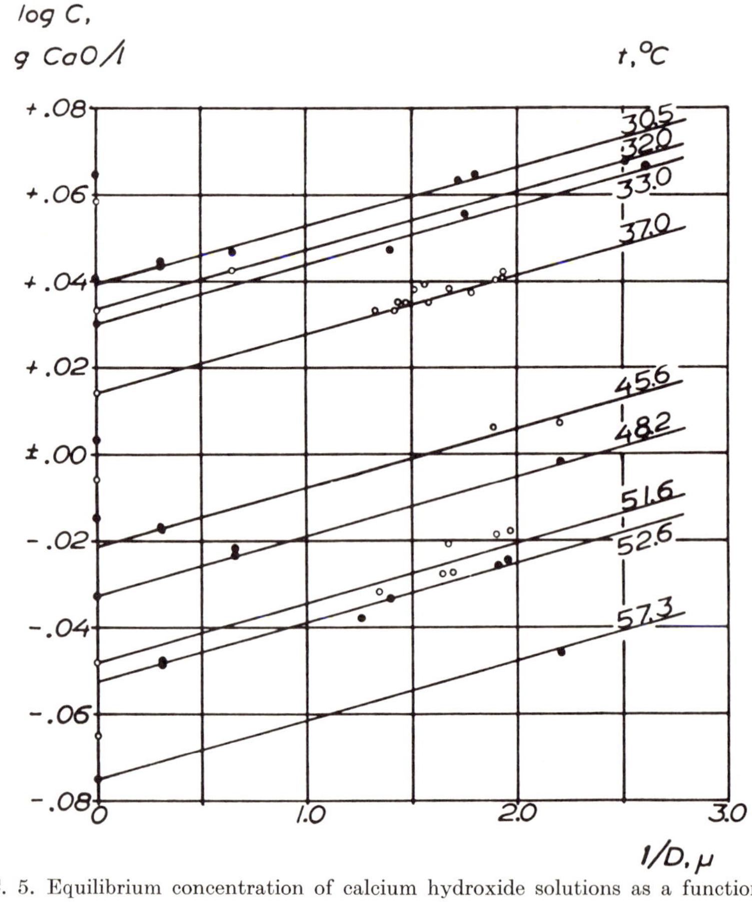

FIG. C. 5. Equilibrium concentration of calcium hydroxide solutions as a function of temperature and particle size (diameter D in μ) (Hedin).

[25] *Am. Ceram. Soc. Bull.*, **39**, 1960, 1-6.

[26] *Fortschr. Mineral.*, **36**, 1958, 78-79.

recently by M. S. Crowley and A. P. Welch.[27] The addition of suspension aids, especially of sodium metaphosphate and of the commercial "Daxad," is much recommended. It is particularly remarkable that the piezo-ultrasonic treatment does not destroy the primary particles, but is restricted in its vibrational energy effect to the subdivision of the secondary agglomerates. Ultrasonic dispersion methods are also of great significance if the dispersed material is to be fixed after dispersion in a material of high viscosity, as in a thermoplastic melt. Industrially this problem was solved very elegantly by E. C. Bernhardt[28] for rutile (titania) particles of colloidal sizes in polyethylene thermoplasts.

12. In the microscopic discrimination of very small mineral particles still visible above the optical resolution limit the use of modern *phase contrast* methods is another very important development. C. W. Correns[29] especially developed and recommended these methods in clay mineralogy and petrology investigations. The principle was outlined as early as 1932 by F. Zernike,[30] and a rapid evolution towards use of suitable instruments started about 1948 when these were made easily accessible as accessories to the common petrographic microscope, especially by constructions of E. Leitz-Wetzlar, C. Zeiss-R.Winkel-Göttingen, Spencer Lens Co., and other well known optical firms. To convert a brightfield microscope into a phase contrast microscope, the condenser must be replaced by a special phase condenser. The usual objective must also be replaced by a phase objective, combined with a diffraction plate with a well-defined phase shifting (by a very thin metal layer). To make the mode of acting of these parts, together with phase deviating effects of the specimen visible, the total amplitude differences must be reduced in the final microscopic image. A sketch of the arrangment of a typical phase contrast microscope (by courtesy of Spencer Lens Co.) is given in Fig. C. 6 according to O. W. Richards.[31] The great advantage of phase contrast microscopy may even be considerably enlarged if these principles are applied to very slight differences in the refractive indices of the mineral specimen and the embedding medium. In this manner, the mentioned

[27] *J. Am. Ceram. Soc.*, **37**, 1954, 433-439; see also T. A. Dickinson, *Ceram. Age*, **65**, 1954 (3) 13-14, 27.

[28] *Ind. Eng. Chem.*, **46**, 1954, 742-746.

[29] *Clay Minerals Bull.*, **2**, 1954, 128-133; "Handbuch der Mikroskopie in der Technik" (Hugo Freund, ed.), Umschau-Verlag, Frankfurt am Main, Vol. IV, "Mikroskopie der Silikate," Pt. 1, 1955, with H. Piller, pp. 722 ff.; see also the latter author, *Heidelberger Beitr. Mineral. u. Petrog.* **3**, 1952, 307-334. See also an elegant method of optical staining of mineral powder samples for diagnostic differentiation, as recommended by O. E. Radczewski, *Ber. deut. keram. Ges.*, **38**, 1961, 389-395.

[30] See especially *Ned. Tijdschr. Natuurk.*, **9**, 1949, 686 ff.

[31] *Science*, **124**, 1956, 810-814, with a valuable bibliography of advancement in phase contrast microscopy as applied in different branches of science; also *Wallerstein Labs. Communs.*, **15**, 1952 (49), 155-162.

advantages can be combined with the well-developed precision methods of immersion in routine petrography, in which the differences of the refractive indices for variable wavelengths, as a function of temperatures in mineral and immersion liquid, are applied for the accurate determination of the refractive indices by striking dispersion effects. This method combined with phase contrasts makes these effects visible in color differences, as an elegant "optical staining" method of great sensitivity, and especially useful in the field of highly dispersed sedimentary minerals.

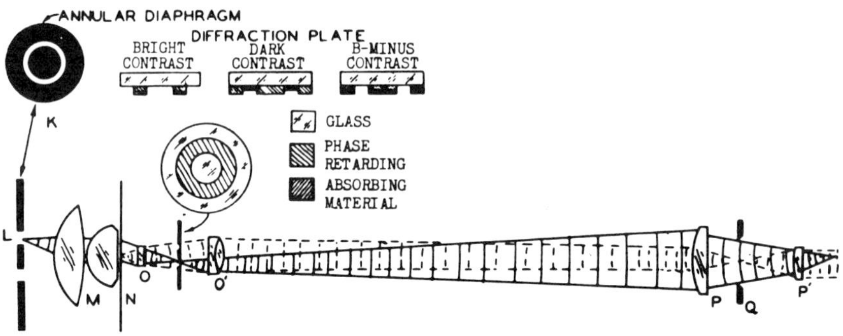

FIG. C. 6. Principle sketch of the Spencer phase contrast microscope (Richards).

13. If one wishes to have a three-dimensional impression of the shapes of fine-dispersed materials, as in silt sediments, the electron microscope offers a series of very valuable shadow-casting methods which may be transferred to light microscopy. A. E. Wright[32] shadowed silt dispersion samples with gold deposited in a cathode-ray vacuum, analogous to the shadow-casting for electron microscopy (C. ¶ 114 ff.). A test material of spherical particles of known dimensions may be added to the sample to determine exactly the shadowing angle. In all the other details, the method of evaluation in the binocular microscope from magnified photographs is identical with that used in electron microscopy. An accurate determination of the sphericity of the particles is made possible with a relatively elementary method.

14. For routine determinations of the sizes of fine-dispersed particles, as in clays, and especially for industrial products like gypsum plaster, Portland cement, and others, the *air elutriation* and *permeation* methods are favored because of their experimental simplicity. Among the latter, the R. L. Blaine method has been standardized[33] and is generally used as a conventional basis for the classification of many

[32] *J. Sediment Petrol.*, **27**, 1957, 306-312.

[33] *ASTM Bull.*, **108**, 1941, 17-20; ASTM Standards, 1949, Pt. 3, 103-105 and C 204-51. See also the air permeation method of F. M. Lea and R. W. Nurse, *J. Soc. Chem. Ind.* (*London*), **58**, 1939, 277-283 T. As an excellent introduction into the theory and practical application of this by no means elementary field of dispersoid physics we recommend the study of the article

of the products mentioned above. A discussion of the different instruments used in this field, and for critical experiments, was given by I. D. McLachlan[34] from the practical viewpoint, and then by F. Gille.[35] Especially for the finest dust of pozzolanic reactivity character (cf. Vol. V, Section C), K. M. Alexander[36] examined the reliability of the Blaine test in the range of 2,000 to 65,000 cm.^{2}/gm. The observed variability in widest limits, if the determinations are made at constant porosity, is caused by variations in the uniformity of the pore texture within the filter bed. If this factor is taken into account, the measurements can be useful up to 40,000 cm.^{2}/gm. for determinations of the specific surface area of diatomaceous earths, opaline rocks, some scoriae, etc. Great difficulties occur if the ground powder shows a tendency to agglomerate. The phenomena of molecular flow, besides viscous flow, must be considered, and the elementary Blaine formula based on the Kozeny (1927) theory for the specific areas:

$$S_w = \frac{14\,x}{d(1-e)} \cdot \sqrt{\frac{e^3 A \bar{P} t}{V n L}},$$

or for the finest powders:

$$S_w = \frac{14\,x}{d(x-e)} \cdot \sqrt{\frac{e^3 A \dot{P} t}{V n L}}$$

must be replaced by the equation given by P. C. Carman and P. le R. Malherbe[37] (originally derived for finest paint pigments):

$$\frac{Q_1 L p_1}{t A \Delta p} = \frac{p F_1}{5 n S_0^2} + \frac{0.96 F_2}{S_0} \cdot \sqrt{\frac{RT}{M}}$$

(for the meaning of the single parameters one must refer to the original literature, e.g., pp. 349 ff. of Alexander's publication [36]).

15. A critical investigation of the theoretical basis of the Blaine method was further given by W. Wieland[38] who applied corrections of the Blaine-Carman equations as given by J. Svensson,[39] based on the introduction of a molecular flow fac-

of H. Walther, *Silikat Tech.*, **4**, 1953, 551-558, giving equally a basic discussion of the permeation methods as practical examples. See also, H. Börner, *Tonind.-Ztg. u. Keram. Rundschau*, **78**, 1954, 204-209; 234-239.

[34] *Australia, Commonwealth Sci. Ind. Research Organization. Div. Bldg. Research, Tech. Paper*, Report G. 3 - 1.1, July, 1949, 14 pp. The bibliography includes especially the older types of laboratory instruments for air flotation, with the theoretical background.

[35] *Schriftenreihe Zementind.*, No. 7, 1951, 14 pp.

[36] *Australian J. Appl. Sci.*, **6**, 1955, 316-326.

[37] *J. Soc. Chem. Ind. (London)*, **69**, 1950, 134-143 (May).

[38] *Zement-Kalk-Gips*, **10**, 1957, 81-89.

[39] *Jernkontorets Ann.*, **133**, 1949, 33-86.

tor which reduces the end results of the specific surface considerably (cf. Fig. C. 7). The specific surface should be a constant, and independent of the porosity, ε, but it shows especially for higher porosity degrees much too low data. Wieland developed

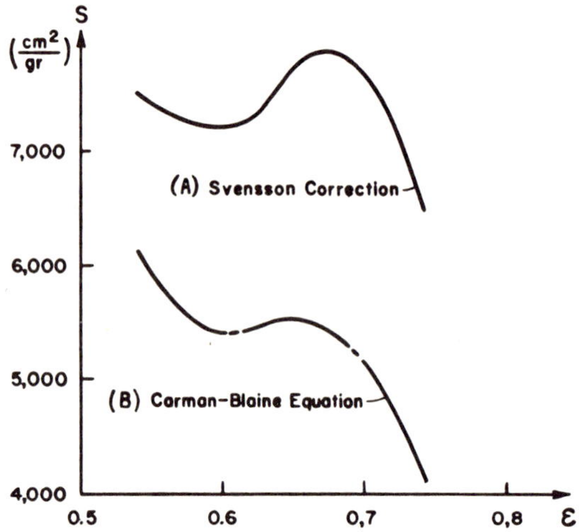

FIG. C. 7. Specific surface as a function of molecular flow through structural pores of powders in the air permeation method (Svensson): (*A*) data calculated from the Blaine-Carman equations; (*B*) after correction of molecular flow (Wieland).

an improved air permeation method and apparatus, as shown in Fig. C. 8. The bed of the powder is stepwise compacted by the piston *S*, and for each height, *h*, of the bed the time of the permeation, *t*, is determined as a function of the porosity, ε. The measured permeation time is then reduced by the relation $t_r = h_0/h \cdot (t - t_0)$, with h_0 = a standard height of the permeation time for the paper filters at the bottom and top of the powder compact. The reduced times, t_r, are plotted in a logarith-

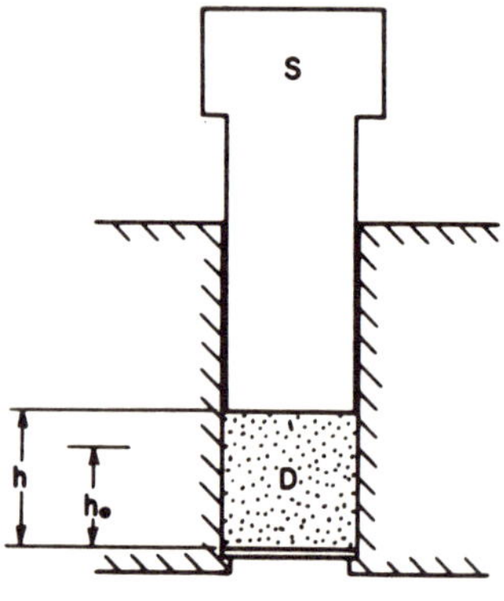

FIG. C. 8. Principle sketch of improved air permeation apparatus for powders (Wieland).

mic diagram as a function of ε (Fig. C. 9) according to the equation $\log t_r = a \cdot \varepsilon + b$, and the specific surface is

$$S_v = C\sqrt{t_r \exp\left[2.303\, a(\varepsilon_0 - \varepsilon)\right]} = C\sqrt{\frac{h_0}{h}(t - t_0)\exp\left[2.303\, a(\varepsilon_0 - \varepsilon)\right]}$$

with ε_0 = a constant porosity for which different powders are to be relatively compared. This end formula is valuable over the wide range from 2,000 to 12,000 cm.²/gm. (whereas the Blaine method has its limit above 3,000 cm.²/gm.), with the only

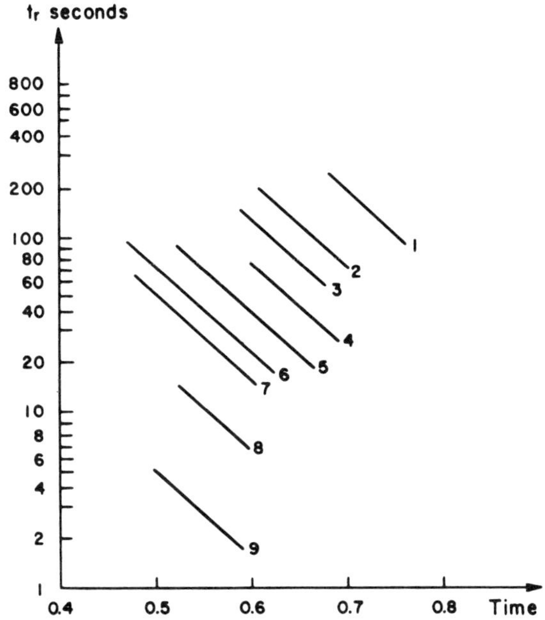

FIG. C. 9. Air permeation times for different powders and cements, as a function of porosity (Wieland).

restriction that it is not more valid for coarser powders with porosities above 0.5, and for finest powders which cannot be compacted to $\varepsilon = 0.55$, as would correspond to the practically optimum conditions of compaction, e.g., for cements. The Wieland formula strictly follows an additivity relation for polymict powders.

16. The principles of pure air *elutriation* were studied from the practical aspects of cement manufacturing by K. Chujo and Sh. Sekino,[40] as based on the exponential

[40] *Semento Gijutsu Nenpo*, 4, 1950, 129-139. For the Wagner turbidimeter standardized method cf. ASTM standard test *SS - C* - 158 *b*; cf. the critical studies of K. Matsuoka and M. Ueda, *Ibid.*, 4, 149-153, for particles of 2.5 μ diameter as the lower limit.

relation between the amounts of particles larger than a certain size x, in μ. For particles up to 1 μ of the density ϱ, with k a constant, and x the Guttmann equivalent diameter in μ, one finds the specific surface

$$S = \frac{6}{\varrho} k \int_{10^{-4}}^{10^{-2}} \frac{\exp(-kx)}{x} dx + \frac{(1 - \exp 10^{-4} k)}{5 \varrho} \cdot 10^5,$$

in cm.^{2}/gm. The results of the calculations show with relation to the Blaine permeation method and also to the Wagner turbidimetry (1933) approximately linear proportionalities. The sedimentation in *stagnant* air suspensions was especially studied more recently by F. S. Eadie and R. E. Payne,[41] who examined particle size distribution by microscopic measurements of fractions deposited and sampled at regular intervals in an apparatus called a "*micromerograph*," with a removable balance pan on an automatically servo-controlled torsion balance. The cumulative weights are determined by signals from a balanced differential transformer, amplified, and demodulated to develop, proportional to the signal, a current that passes through a coil on the balance beam between a pair of permanent magnets. The weight on the pan is compensated by the current passing through the coils and plotted on paper by a recording pen. This ingenious apparatus transferring many of the advantages of the Sven Odén balance (1918/1919) to air sedimentation gives conclusive results for particle sizes down to 3 μ in only 20 minutes, and down to 1 μ in less than 3 hours. The resulting accuracy is claimed to be superior to that of sedimentation in water.

17. The principle of a planar spiral-shaped air current used by K. Wolf (1941) is verified in commercial *spiral classifiers* ("Alpine M-P 800," Alpine Corp., Augsburg, Germany) of which Fig. C. 10 shows a sketch following the description given by E. Felton and R. Watterlond.[42] It consists of two chambers, the feed material entering the first chamber (a) on the inward side of the vanes (b), while air enters between the vanes and the outer edge of the chamber. Adjustment of the air velocity or the injection angle is used to separate the material to defined fractions. Centrifugal force carries the larger particles outward while the smaller particles are carried inward by the air. The coarse particles are directed to an exit by a knife (c) and the fine material, with the air, leaves the classifying chamber through a central opening into the second chamber. Both the coarse and fine fractions are carried away by screw conveyors. The apparatus has proved to be most efficient in the fractionation of talc (steatite) for ceramic production (cf. Vol. V, Section B).

18. Concerning the classical Stokes law of sedimentation rates in suspensions either of uniformly spherical particles (as the strict theory requires) or of angular particles, there is an extensive special literature, of which we here select, as an example,

[41] *Brit. Chem. Eng.*, **1**, 1956, 306-311.

[42] *Am. Ceram. Soc. Bull.*, **39**, 1960 (12), 717-720, especially pp. 717-718.

the very valuable report of H. H. Steinour.[43] Especially for cement pastes two equations developed by this author and by T. C. Powers[44] produce approximately equal results. B. W. Wilson[45] examined by a photographic method the sedimentation of spherical glass particles from 2.5 up to 10 μ in diameter, and found Stokes's law strictly valid if the volume concentration of the suspension was only 0.0001 per cent. Then uniformity of the temperature during the experiment is of a particularly critical

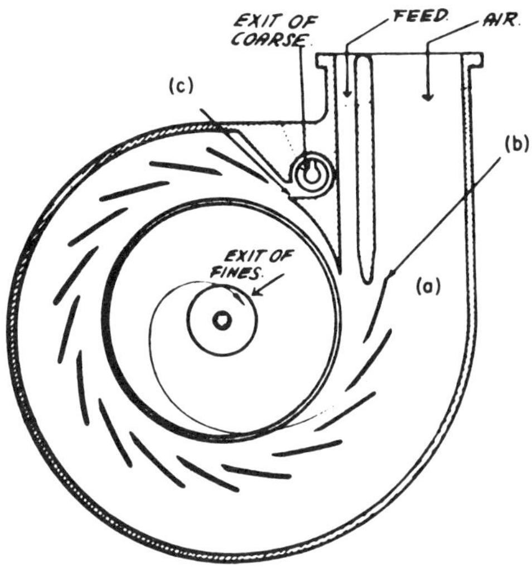

FIG. C. 10. Cross section of the "Alpine M-P 800" spiral classifier (Felton and Watterlond).

importance for exact measurements. With certainty it can be said from Wilson's experiments that no retardation effects occur by electrical double layers on the surface even of such fine particles. For larger sizes (5 to 75 μ), W. Bostock[46] constructed a torsion balance sedimentograph with direct deflection reading, as a simplified Sven Odén balance type. Another very simple spring balance for sedimentation analysis was developed by J. G. Rabatin and R. H. Gale[47] the principle of which is shown in Fig. C. 11, with a shutter mechanism interrupting a parallel light beam to

[43] *Ind. Eng. Chem.*, **36**, 1944, 618-624; 840-847; 901-907; *Research and Develop. Labs. Portland Cement Assoc. Research Dept. Bull.* **3**, *October*, 1944, 52 pp. This article series also includes special effects of deflocculation etc., model systems of emery powders, and the influence of concentration of flocculated suspensions, e.g., of glass, limestone, silica, and burned shales.

[44] *Research and Develop. Labs. Portland Cement Assoc. Research. Dept. Bull.* **2**, 1939, 160 pp., especially on the phenomenon of "*bleeding*" by sedimentation unmixing of cement mortars and concretes (cf. Vol. V, Section C).

[45] *Australian J. Appl. Sci.*, **3**, 1952, 252-256.

[46] *J. Sci. Instr.*, **29**, 1952, 209-211.

[47] *Anal. Chem.*, **28**, 1956, 1314-1316.

a photocell, and with automatic recording of the intensity of the photocurrent. The results are claimed to be of equal reliability as those of the Andreasen pipette, for particles between 2 and 32 μ in diameter as indicated by cumulative weight curves.

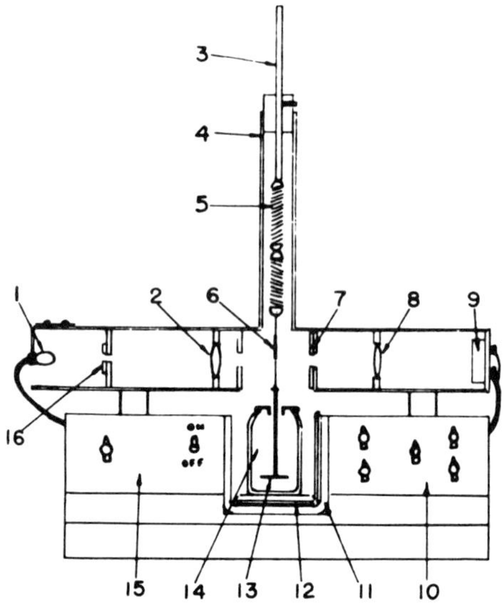

FIG. C. 11. Simplified sedimentation balance (Rabatin and Gale): (13) sedimentation pan in settling vessel (14) in water bath (11), with heating coil (12); (5) sensitive spring; (9) G. E. photovoltaic cell.

19. The old-classical sedimentation tube of the long (Travis) type appeared in a modified form as recommended by D. G. Sammarone and H. E. Saunders[48] especially for ceramic color pigments and milled enamel products of subsieve sizes. In a comparison of the results with a somewhat modified hydrometer method, preference must be given to the latter, especially because of its rapid operation, but the sedimentation tube is valuable for comparison tests of powder fineness. An especially short water elutriator for ceramic purposes was further described by D. R. Wilder and E. S. Fitzsimmons,[49] with a characteristic funnel-shaped lower portion and cylindrical upper ones, with overflow decantation. It is more the type of a classifier in ceramic technology, but evidently equally valuable as the H. Harkort (1927, 1952) classifier with a calibrated spigot for the inflowing water (Fig. C. 12).

[48] *Am. Ceram. Soc. Bull.*, **36**, 1957, 340-346; cf. on P. M. Travis's tube, *ASTM Bull.*, **102**, 1940, 29-32.

[49] *Am. Ceram. Soc. Bull.*, **34**, 1955, 109-114. For very valuable note of fine-ore dressing elutriation and simple equipment for this purpose, see A. M. Gaudin, J. O. Groh, and H. B. Henderson, *Ind. Eng. Chem.*, **22**, 1930, 1363-1366; they used acetone as the suspension medium.

20. The well-known discrepancies becoming evident by a comparison of the particle sizes measured directly in light or electron microscopes, and those calculated on the basis of the Stokes law, create a very important problem, especially for clay mineralogy and petrology. An interesting statistical treatment of this question was made by J. Cartwright[50] who discussed the volume-surface diameter, D_{vs}, as a sta-

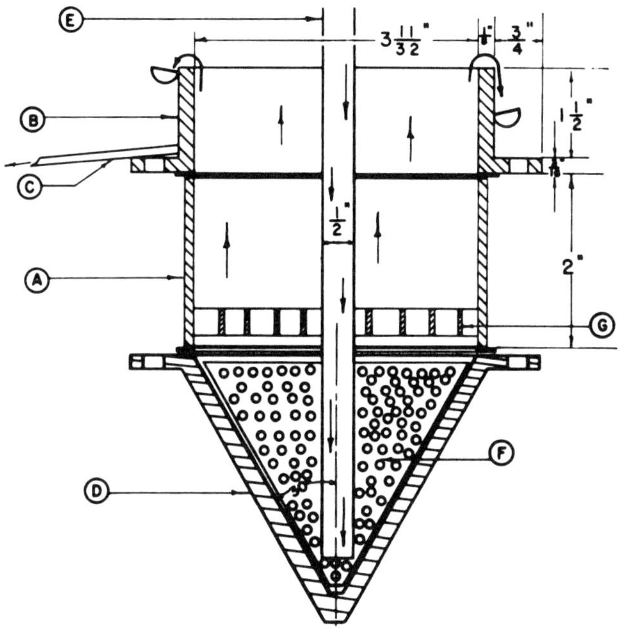

FIG. C. 12. Short-column water elutriator for fractionation by decanting (Wilder and Fitzsimmons): (*D*) metal cone; glass cylinder (*A*) topped by metal cylinder (*B*); (*C*) spiral discharge trough. Water flows through (*E*) into (*D*); (*F*) are impellors (perforated metal sheet); (*G*) grid for removal of turbulence.

tistical value $= \Sigma ND^3/\Sigma ND^2$, where N is the number of particles counted, with the diameter D. The specific surface is then calculated using conventional shape (α_s) and volume factors, α_v, with the result that

$$D_{vs} = K \cdot \frac{a_v}{a_s} \cdot \frac{\sum N D^3}{\sum N D^2}$$

K is for perfect spheres equal to 6 and for clay minerals of different types and occurrences about equal to 8. If one measures the specific surfaces directly by nitrogen adsorption at low temperatures (BET method, cf. Vol. IV, Section A), K is found

[50] *Clay Minerals Bull.*, 3, 1956, 31-35.

equal to 16 to 30, corresponding to a material of high rugosity (surface roughness). For these reasons it is evident that great inconsistencies must be expected between the microscopic data and the results of specific surface calculations. The problem involved here was also estensively treated by A. H. M. Andreasen and B. Nielsen[51] by a comparison of the common sedimentation analysis, the Blaine air permeation method, and the nitrogen adsorption isotherm (BET). The experimental determinations on well-defined porcelain powders show relatively large differences in results. If one takes the pipette analysis as the standard method, recalculates its results as specific surfaces, and these are plotted as abscissae in Fig. C. 13, one finds the best

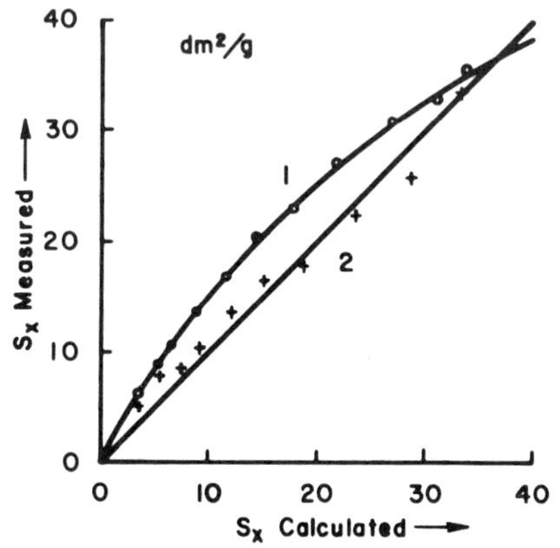

FIG. C. 13. Comparison of specific surfaces as determined by pipette sedimentation analysis (in the abscissae date), with results of the Blaine Test (curve 1, circles), and BET adsorption data (curve 2, crosses) (Andreasen). A reduction factor of 0.45 is applied in the calculation of the data on both curves.

analogy with the BET method (after application of a reduction factor of 0.45 to these). Blaine's method shows values definitely too large for the coarser material. The reduction factor mentioned indicates how the nitrogen adsorption also includes the inner surfaces of cracks and pores. Andreasen comes to the conclusion that characterization of a given ceramic powder product by *one* parameter alone, namely the Blaine specific surface, is rather uncertain.

21. In comparison studies on the suitability of different methods for the determination of specific surface areas of fine powders, further progress was made by D. M.

[51] *Ber. deut. keram. Ges.*, **29**, 1952, 377-379. K. H. Woodside and W. C. Ormsby, *J. Am. Ceram. Soc.*, **43**, 1960, 671-672, recently recommended the adsorption of glycerin on kaolinite for a determination of the specific surface. New results of grain size analysis of kaolins, using the pipette method, see by H. M. Köster, *Ber. Deut. Keram. Ges.* **39**, 1962, 432-438.

Matthew[52] and J. Cartwright, K. Wheatley, and K. S. Sing,[53] who examined different powders with the nitrogen adsorption (BET) test and the Blaine permeation test, modified by P. J. Rigden.[54] In the experimental series of Matthews the results of these two methods were found to differ considerably. The specific surface areas obtained from the BET asdorption isotherm were always the largest. This is explained in the same way as in the experiments of Andreasen and Nielsen, namely, by inner surfaces accessible to the nitrogen but not to the flowing gas in permeation. The examinations of both methods, in comparison with microscopic determinations by Cartwright *et al.* for quartz, tridymite, cristobalite, and fused silica powders, and for glass spheres of 1 to 12 μ in diameter, show a rather constant ratio of surface areas calculated from the microscopic measurements and the BET test, especially if the samples had been etched previously with dilute hydrofluoric acid. The agreement with Andreasen's and Nielsen's observations is evident in the approximate ratio of 2 : 1 for the data of the BET test to the directly measured data, for particles down to 6 μ size. Finer particles, however, show in the air permeation (Blaine-Rigden) test particularly low surface areas. These experimental investigations make evident that, especially for the air permeation, no reliable results can be expected if the porosity of the bed compacts is not also taken into consideration. This is just the point which plays an important role in finest color pigments and Portland cements and caused a fundamental re-examination of the theoretical basis of the Blaine methods by P. C. Carman and P. le R. Malherbe[55] (see above; also Vol. V, Section C). The formula given by these authors includes the effects of molecular and "slip" flow in the pores. The optimum conditions are valid for porosities between 0.4 and 0.5. The general agreement of the results from the improved permeation formula with those of the BET method is, for such a "normal" porosity, within 32 per cent.

22. A very interesting modification of the permeation test was made by using *liquids* as the flowing medium in powder beds of definite porosities; e.g., Fr. Wolf and H. Ritzmann[56] applied such a manometric method, developed by H. J. Kamack,[57]

[52] *J. Appl. Chem. London,* **7**, 1957, 610-613.

[53] *J. Appl. Chem. London,* **8**, 1958, 259-264.

[54] *J. Soc. Chem. Ind. (London),* **62**, 1943, 1-4; *Nature,* **157**, 1946, 268.

[55] *J. Soc. Chem. Ind. (London),* **69**, 1950, 134-143; see also K. M. Alexander, *Australian J. Appl. Sci.,* **6**, 1955, 316-326. For *coarser* powders, G. Ringqvist, *Svenska Forskningsinst. Cement Betong vid Kgl. Tek. Högskol. Stockholm, Handl.,* No. 28, 1955, 26 pp., applied a method of D. Payne, *Nature,* **172**, 1953, 261, based on the determination of the work,

$$\Delta A = \frac{1}{\gamma} \int h \, dv ,$$

done during the withdrawal of a liquor, v, from the capillaries of a moist powder, h being the force of suction (γ the surface tension of the liquid phase).

[56] *Silikat Tech.,* **10**, 1959, 335-344.

for an examination of *dusting* dicalcium silicate in industrial products. The specific surfaces were also determined by the adsorption of stearic acid (cf. C. ¶ 221). Xylene and mercury were used as permeation liquids. Determination by liquid permeation proved to be very suitable for powders between 5 and 0.2 μ in diameter. The Kozeny-Carman equations based on laminar flow were supplemented by a correction factor $1 + Z\lambda S_v[(1 - \varepsilon)/\varepsilon]$, with $Z = 2.5$, λ the free path length of the molecules of the flowing medium, S_v the surface per unit volume. The modified Knudsen coefficient, X, is the ratio of the average free path of air molecules to the radius of the flow channels,

$$X = 0.75 \cdot \left(\frac{1 - \varepsilon}{\varepsilon}\right) / D_k$$

(D_k is the average Kozeny-Carman diameter). The agreement of the calculated specific surfaces of the powders with Blaine's standard data is satisfactory if reduced to the normal porosity $\varepsilon_N = 0.45$. Tabulated data for a comparison of the results of the Kamack method (S_k), those of the stearic acid adsorption (S_s), and the S_v calculated from the grain size distribution, shows in the ratio S_k/S_v a valuable measure for the deviations from ideal (cube or sphere-shaped) particles, found in the range between 1.5 and 3.6. With increasing fineness of the dusting products the heterodisperse structure of the grain size distribution is decreased, but it does not follow the Rosin-Rammler law (see C. ¶ 7), since the particles are not dispersed by rapidly changing forces as they occur in the grinding process, but by uniformly acting inner tensions during the inversion expansion in the dusting Ca_2SiO_4.

23. For the optimum application of the routine *sedimentation pipette* analysis, A. H. M. Andreasen[58] gave a series of most commendable laboratory instructions, especially concerning the stabilization of clay or soil suspensions with sodium pyrophosphate in a 0.001 to 0.003 M. solution, or by normal sodium citrate solution. For suspensions of finest chalk particles in paraffin oil, a slight addition of linseed oil is an important aid; for metal powders, that of cyclohexanol. A very valuable tabulation of the most suitable peptizing agents is given, too. Especially for ceramic clays, A. Ph. Welch and Ch. L. Folkers[59] studied 1 per cent suspensions in various organic liquids. Higher alcohols, such as hexanol, glycols, and glycerin, proved to be excellent suspension media and their alkali salts to be good peptizing agents.

[57] *Anal. Chem.*, **26**, 1954, 1623-1628; this method is originally an *air* permeation test, but is easily modified to liquid media.

[58] *Staub*, 1956 (No. 43), 454-461. On the action of the peptizing agents see also H. Lehmann, *Tonind.-Ztg. u. Keram. Rundschau*, **78**, 1954, 326-328; for an improvement of the Andreasen pipette by the application of a stainless steel pipette tube with fine-adjustment in the vertical direction to a scale, see *Ibid.*, 329-331.

[59] *J. Am. Ceram. Soc.*, **38**, 1955, 454-461. A. J. Launer, *Am. Mineralogist*, **37**, 1952, especially p. 768, used propyl alcohol in his investigation of the infrared absorption spectra of particles of fractionated silicates, smaller than 2 μ.

The minimum viscosity necessary to prevent turbulent flow is for such liquids determined by the formula of K. Kammermeyer and J. L. Binder[60]

$$\eta_{\text{min.}} = \sqrt{0.37\ (r_{\text{max.}})^3\ g\ \varrho_l(\varrho_s - \varrho_l)}$$

(were g is the gravitational constant, ϱ_s the density of the suspended solid, ϱ_l that of the suspension medium) for all of the organic liquids tested (alcohols, ketones, acetates, propionates, cyclic compounds). As deflocculents (cf. C. ¶ 356, ¶ 371) the alkali derivatives investigated are less ionized in the organic media than in water, whereas in water the often used "Daxad" (a polymeric sodium salt of substituted benzoid alkyl sulfonic acids) is most active as deflocculent in an addition of 2 per cent. The alkali alcoholates require a concentration of 5 per cent in the organic suspension media. The action of all of these alcohols and their derivatives is based on van der Waals attraction forces which become smaller with decreasing polarity of the alcohol molecules and increasing size (cf. C. ¶ 306). Brownian movement is reduced in the alcohol suspensions; fulfillment of the Stokes law is improved. Analogous to Andreasen's tabulation, Welch and Folkers also give an instructive compilation of fundamental physical data of the properties of the recommended liquids, including their viscosity, surface tension, and specific heat at 20°C.

24. An original application of the sedimentation grain size analysis for thoria particles was made by G. W. Leddicotte, R. E. Druschel, H. H. Miller, and J. P. McBride.[61] The thoria was first activated by irradiation with neutrons, dispersed at 0.5 per cent concentration in a 0.005 M. solution of sodium pyrophosphate, and settling started past a scintillation counter connected to a count rate meter and a recorder. As the scintillation effects are proportional to the concentration in activated thoria, the counter data can be directly combined with the common calculation based on Stokes law. The strong *orientation* effects which are observed in the sedimentation of pronouncedly fine-tabular particles, as in mica powder, were made useful in the production of so-called "integrated" mica products; H. George and L. Metzger[62] developed an industrial process in which this idea for making foils of pure mica was verified.

25. In the class of the *hydrometer* (areometer) sedimentation measurements the Bouyoucos-Casagrande apparatus (1934) was used by G. Peco[63] for ceramic batches

[60] *Ind. Eng. Chem. Anal. Ed.*, **13**, 1941, 335-337.

[61] ONR Report 2160. Cf. V. Dallred, S. R. Buxton, and J. P. McBride, *J. Phys. Chem.*, **61**, 1957, especially p. 117.

[62] *Rev. gén. élec.*, **59**, 1950, 519-524. The product was called "Samica" mica paper; the mica is in this process first calcined to lose about half its hydroxyl water, then changed to a "pulp" with soda solution. See also R. D. Jackel, *Monthly Research Rept. Office Naval Research*, December, 1951, 6-9.

[63] *Ceramica* (*Milan*), **10**, 1955, (12) 51-56. Similar nomograms see also by R. Gandahl, *Geol. Fören. i Stockholm Förh.*, **74**, 1952, 497-512.

and their granulometric characteristics, with routine and/or precision spindles. Nomograms were given for the evaluation of the four independent variables involved, and for the important corrections for temperatures different from the standard 20°C. For rapid soil fineness analyses L. de Leenheer, M. van Ruymbeke, and L. Maes[64] combined the Köhn pipette (1928) method with the hydrometer, making the advantages of both useful for the accurate determination of the densities of clay-water suspensions in exactly defined levels of the liquid column. For sandy and heavy clay soils a mixture of 0.01 N solution of sodium oxalate with a 0.002 N soda solution is particularly recommended. Any gravimetric determinations or water-bath evaporations of the suspensions are avoided, and the fractions from 0 to 2 μ, 0 to 10 μ, and 0 to 20 μ are directly measured. Only, if more than 2.5 per cent of organic material is present, deviations from the normal sedimentation behavior are observed. In this case, the organic compounds must be destroyed by a treatment with a 30 per cent solution of hydrogen peroxide. The hydrometer spindle method was especially recommended to the ceramists by S. G. Maguire and G. W. Phelps[65] because of its great simplicity and the convenient use of plotting the observation data on commercial probability paper nomograms. The method was especially compared with the results of precision centrifuge measurements of the particle sizes. This examination confirms definitely the practically satisfactory accuracy of the hydrometer method. An example for the plotting mentioned is given in Fig. C. 14.

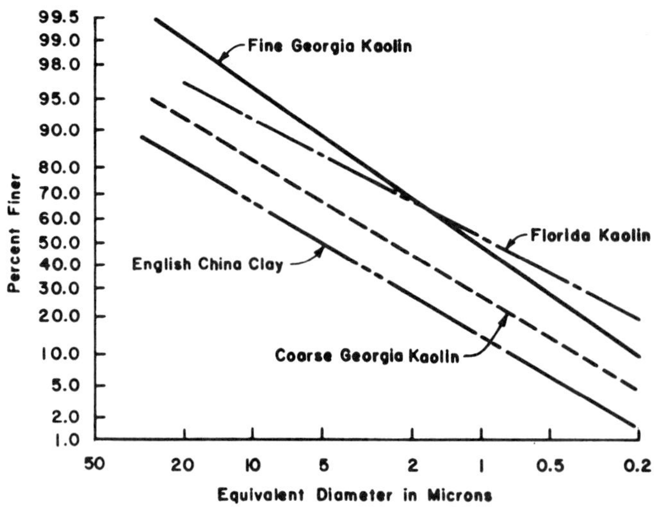

FIG. C. 14. Diagrams showing particle size data (equivalent radii) in logarithmic probability projection for some ceramic kaolins and British china clay (Maguire and Phelps).

[64] *Silicates inds.*, 20, 1955, 237-243; *Bull. soc. belge géol. paléontol. et hydrol.*, 61, 1952, 138-164.
[65] *J. Am. Ceram. Soc.*, 40, 1957, 403-409.

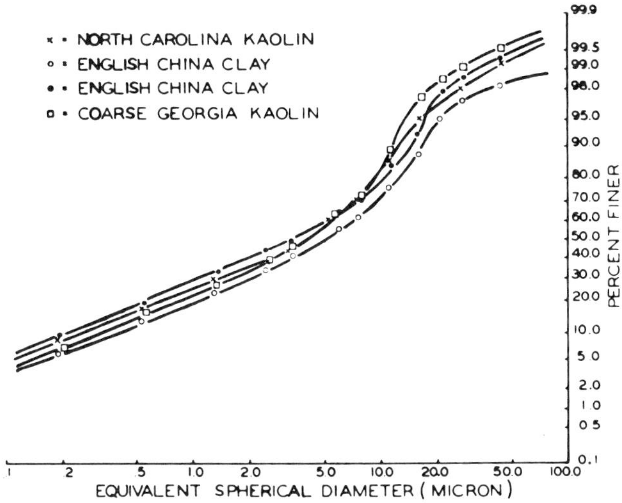

FIG. C. 15. Logarithmic probability diagram of particle size distribution of several kaolins with an inflection near 10 μ of equivalent spherical diameter (Phelps).

Only for particles below 1 μ in size, the hydrometer fails. Further examination of the method as recommended by Phelps,[66] however, showed that the linear logarithmic probability projections are limited to the range of about 0.2 to 10 μ equivalent spherical diameters. Finer or coarser particles show deviations from the straight-linear relations. For whiteware clays, in the range mentioned, the relation is linear, but a rapid change of grain sizes is indicated beyond these limits. Evidently, coexisting distributions of a different nature, superimposed on the clay particle distribution influence the sedimentation process, and one obtains curves such as those seen in Fig. C. 15 with an inflection near 10 μ by "skewed" distributions of foreign minerals, especially those of higher densities (quartz, feldspars, heavy minerals), or of different morphological characteristics (e.g., mica flakes).

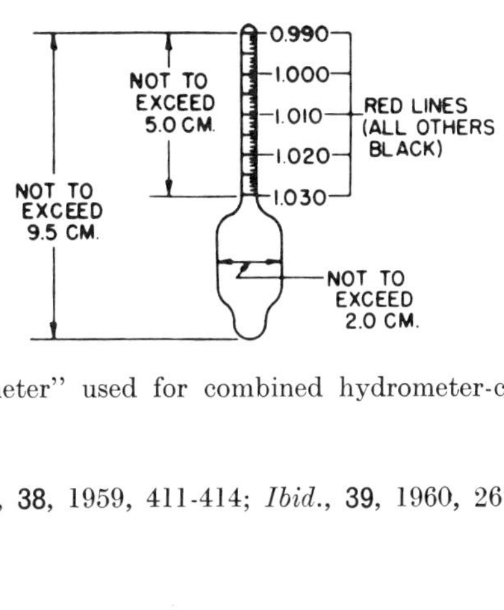

FIG. C. 16. "Soil hydrometer" used for combined hydrometer-centrifuge subsieve particle size measurements (Lester).

[66] *Am. Ceram. Soc. Bull.*, 38, 1959, 411-414; *Ibid.*, 39, 1960, 267-269.

26. A very elaborate hydrometer method for practical purposes was described by R. H. Lester,[67] with valuable nomograms, based on a revision of the basic conditions of hydrometer measurements (see Fig. C. 16) in general. For the improved method, a direct applicability to particles down to 0.1 μ sizes is claimed, with considerable savings in time of operation. In eight hours distributions could be measured equivalent to 700 hours of gravitational settling, if the hydrometer method is combined with centrifuging. A typical size distribution curve for English China clay is given in Fig. C. 17.

27. The *dipping hydrometer* method of S. Berg (1940, 1942) was further developed by this author[68] by a combination of the spindle immersion with a determination of the concentration of the sedimenting suspension, either by pipette methods or by a hydrometer. The range in which the sedimentation can be used in the advanced method is much larger than that of common hydrometry, both for fine and for relatively coarse particles (up to 250 μ in diameter), which often occur in ceramic batches. The sedimentation tube shown in Fig. C. 18a with a short side tube near its bottom for the tapping of samples is connected at definite time intervals with a 10 ml. pipette (Fig. C. 18b). The suspension medium is commonly water (with an addition of sodium pyrophosphate, to a 0.002 M solution). In some cases more viscous liquids such as ethylene glycol would be indispensable for the Andreasen pipette. The dipping hydrometer method is not very suitable for highly viscous liquids. For the larger particles which do not exactly follow the Stokes formula, the equation of H. S. Allen[69] must be applied:

$$v = \frac{1}{2}\left[\frac{\varrho_s - \varrho_l}{\varrho_l}\right]^{2/3} \cdot \frac{(r - \frac{2}{3}\bar{r})}{(\eta/\varrho_l)^{1/3}},$$

in which $\bar{r}$ is the "critical radius" $= \sqrt[3]{9\,\eta^2/2\,\varrho_l(\varrho_s - \varrho_l)}$, the indices s and l referring to the solid and liquid phases of the suspension, respectively.

28. Among the *turbidimetric* (nephelometric) methods for the determination of subsieve grain sizes, the Wagner turbidimeter has attained the ranks of Standard Specifications.[70] A critical comparison of its results with those of the Andreasen

[67] *Am. Ceram. Soc. Bull.*, **37**, 1958, 129-134. The centrifuge was Model SB V 52.574, International Equipment Co., with head No. 277.

[68] *Ber. deut. keram. Ges.*, **33**, 1956, 229-234; *Trans. Intern. Ceram. Congr., 4th Congr., Florence*, 1954, 65-74.

[69] *Phil. Mag.*, [5], **50**, 1900, 323.

[70] ASTM Test C 115-53; see also "The Turbidimetric Particle Size Analysis," Symposium on Particle Size Measurement, 1958, pub. 1960 in *ASTM Spec. Techn. Publ.* No. 234, 1960, 172 ff., by J. L. Musgrave and H. R. Harner, recommended for critical studies, and *Schriftenreihe Zementind.*, No. 16, 1954, pp. 43-55. We mention also in this connection the use of the Coulter counter for particle size analysis for cement powders, etc. recommended by O. E. Brown,

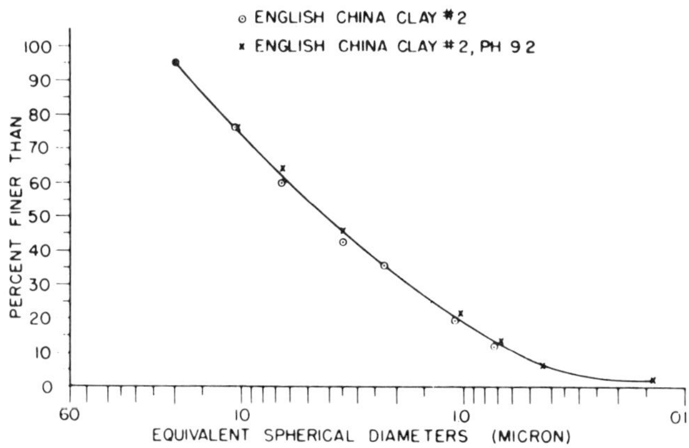

FIG. C. 17. Particle size distribution of English china clay (Lester).

pipette method was made by O. Telle.[71] It shows good agreement in both ways for talc powder and certain organic (penicillin) suspensions. For finest color pigments, however, the photoelectric method generally results in smaller particle sizes than the pipette method. Evidently, the optical properties of the suspended substance affect the results considerably, especially by diffraction and interference phenomena. The

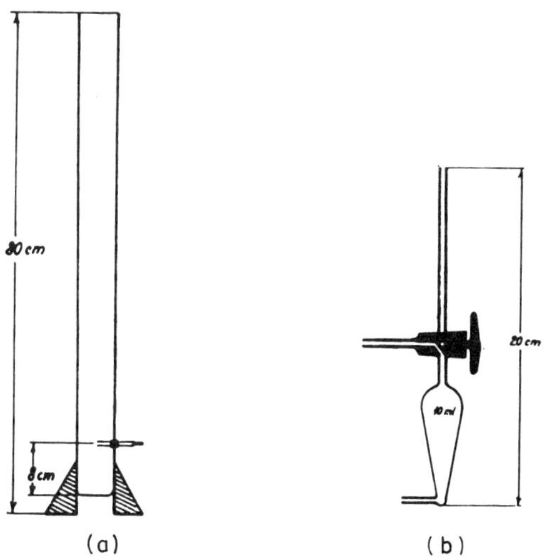

FIG. C. 18. Sedimentation cylinder according to S. Berg (*a*), with removable pipette (*b*).

Am. Inst. Mining Eng., Cement and Gypsum Comm., New York, February, 1960, 14 pp. (Manuscript). The principle of this ingenious apparatus is a counting (after amplification) of electrical pulses produced by particles passing through an aperture of defined dimension (e.g. 140 μ).

[71] *Tonind.-Ztg. u. Keram. Rundschau*, **76**, 1952, 369-373.

light absorption coefficients as a function of the grain sizes below $2r$ of $5\ \mu$ show characteristic anomalies if the aperture angle is smaller than 10° to 15° (in the Musgrave-Harner apparatus it is 15° to 30°). For $2r$ larger than $5\ \mu$ the absorption coefficient can be assumed to be = 1. For very low turbidities, Sh. Tomozane, Sh. Sano, and Sh. Suetoshi[72] constructed a highly sensitive turbidimeter with a vacuum phototube and a photomultiplier, connected with a current-ratio meter. The sensitivity could be raised to such a degree that the light scattering of distilled water could be measured, or that of dilute salt solutions. In this respect the apparatus fulfills the requirements of instruments for determining scattering phenomena in finest colloid dispersoids as discussed in C. ¶ 67 ff. Following the Musgrave-Harner turbidimeter construction (Fig. C. 19), Q. S. Lamar[73] applied the photoelectric meth-

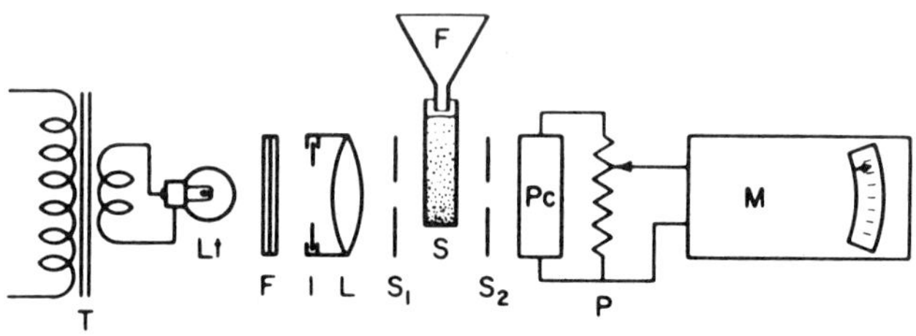

FIG. C 19. Schematic diagram of the Musgrave-Harner turbidimeter (Lamar): (*T*) constant-voltage transformer; (*Lt*) light source; (*F*) heat absorber filter; (*I*) iris diaphragma; (*L*) lens; (S_1, S_2) slits; (*S*) suspension; (*Pc*) photocell; (*P*) variable resistor; (*M*) current ratio meter.

od to talc and clay suspensions and some other minerals or ceramic raw materials, in comparison with the Andreasen pipette analysis. The usefulness of the photoelectric principle in turbidimetry is evident for the wide range from 60 to $0.25\ \mu$ equivalent spherical diameters, and has the great advantage of its rapidity. Lamar's results, however, point to a severe shortcoming of the Wagner method. There are certain materials which apparently cannot be adapted to the turbidimeter analysis, such as carbon black and other dark-colored compounds like ultramarine, Prussian Blue, whereas for kaolins and ballclays, feldspar and quartz powders, limestone, mica, talc, diatomaceous earths, and other bright materials the agreement of the results of turbidimetry and pipette analysis is excellent. The curves also show a good correlation with those from screen analysis (Fig. C. 20).

29. The *centrifuge* is a powerful tool in sedimentation analysis of grain sizes. An approach in this field which has been somewhat forgotten in research practice is

[72] *Asahi Garasu Kenkyu Hokoku*, **3**, 1954, 196-214.

[73] *J. Am. Ceram. Soc.*, **37**, 1954, 386-390; cf. J. Dasner, *U. S. Bur. Mines Inform. Circ.*, No. 7224, November, 1942, 16 pp., and J. M. Dalla Valle, "Micromeritics, The Technology of Fine Particles," 2nd ed., Pitman, New York, 1948, especially pp. 340 f.

that of C. E. Marshall,[74] who recommended the application of a thin layer of a colloid dispersion solution on top of a relatively thick layer of a liquid of higher viscosity. By an adaptation of the layer thicknesses and of the viscosity, η, of the reference liquid, a finely controlled separation of fractions is possible. For clay suspensions a ratio of layer thicknesses of 1 : 4 is perhaps optimum, and η for the reference liquid

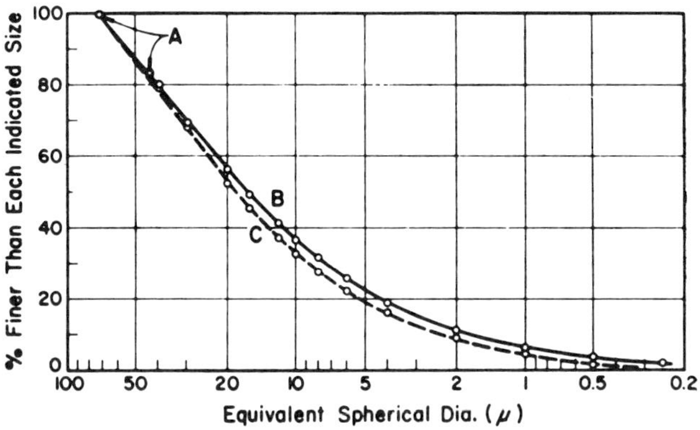

FIG. C. 20. Particle size distribution of a tremolitic talc (Lamar). Results of screen analysis on curve A, of turbidimetric method on curve B, and of Andreasen pipette analysis on curve C.

can be adjusted by solutions of sugar or urea of controlled concentrations. This method is highly useful, especially for particles smaller than 2 μ, down to 20 mμ. The centrifuge type may be chosen for containers of about 50 ml. capacity and of a medium speed type. As one of the many examples of the usefulness of centrifuge methods in ceramic materials of different kinds, we mention the investigations of A. E. Jacobsen and W. F. Sullivan,[75] especially for titania opacifiers and ceramic pigments. The centrifuge method is easily combined with electron microscopy, and the Sven Odén method of tangential intercepts for the sedimentation curves

$$\int_0^{l_m} F(D)\,dD = (1 - p) + t\,\frac{dp}{dt},$$

with p = the total weight fraction sedimented and $F(D)\,dD$ as the weight fraction with diameters D and $D + dD$. For $t(dp/dt)$ is substituted the more easily determined expression $(R - S) \cdot (dp/dS)$ in which R is the distance from the center of rotation to the bottom of the container and S is the corresponding distance to the surface of the dispersion, and $(R - S)$ is the "height of suspension". Nomograms of $100(1 - p)$ plotted versus $(R - S)$ (between 1 and 7 cm.) are most convenient for the graphic

[74] *Proc. Roy. Soc. (London)*, A 126, 1930, 427-439.
[75] *Ind. Eng. Chem. Anal. Ed.*, 18, 1946, 360-364.

determination of the integral mentioned equal to 1 minus the tangential intercept in the graphs. Very favorable experience from the viewpoint of sediment mineralogy and petrology was further discussed by K. Jasmund,[76] who also gives an accurate analysis of the path of the settling particles (Fig. C. 21) in the gravity field of the centrifuge from a hydrodynamic basis. He discusses the amounts of the sedimented material of grain sizes in distinct fractions as a "spectrum" of the quantity distribution

$$W_y = \sum_{D_{\text{min.}}}^{D_{\text{max.}}} C(y, D) = \sum_{D_{\text{min.}}}^{D_{\text{max.}}} \frac{(R_2^2 - [f(y, D)^2]}{(R_2 - R_1)}.$$

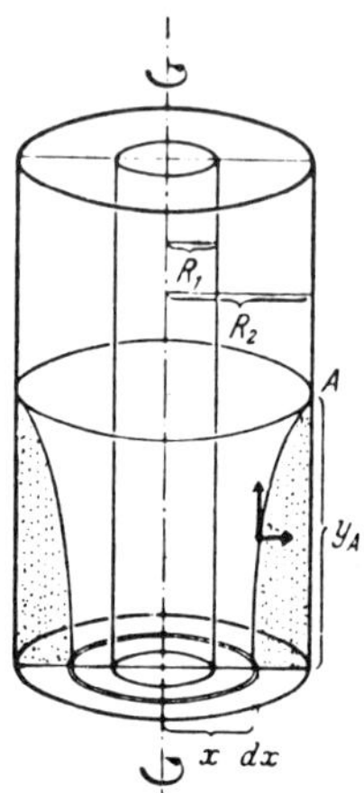

FIG. C. 21. Sedimentation process in rotating column of a suspension in the Sharples super-centrifuge (Jasmund).

For a quartz suspension the distribution curves and the corresponding "spectra" (dotted curves for the calculated spectra; full curves for the experimentally determined spectra) are given in Fig. C. 22. The centrifuge used for the measurements was in this case one of the compressed air-driven "Cepa" (Sharples) super-centrifuge types.

30. We may here only briefly mention the remarkable progress made in modern *ultracentrifuge* construction; molecular weight determinations have been performed in the range from 10^7 down to 10^3, and recently even down to 10^2. J. W. Beams[77] has given a fascinating report on the latter development. He demonstrated the most recent type of ultracentrifuges with magnetic-supported rotors running *in vacuo* and being operated at 20,000 r.p.s. over long time periods.

[76] *Geol. Fören. i Stockholm Förh.*, **77**, 1955, 349-375, with valuable discussion remarks of Å. Grudemo, especially on rapid-working methods for the production of larger amounts, such as montmorillonite fractions from 1 μ down to 0.1 μ.

[77] *Science*, **120**, 1954, 619-625, with literature references.

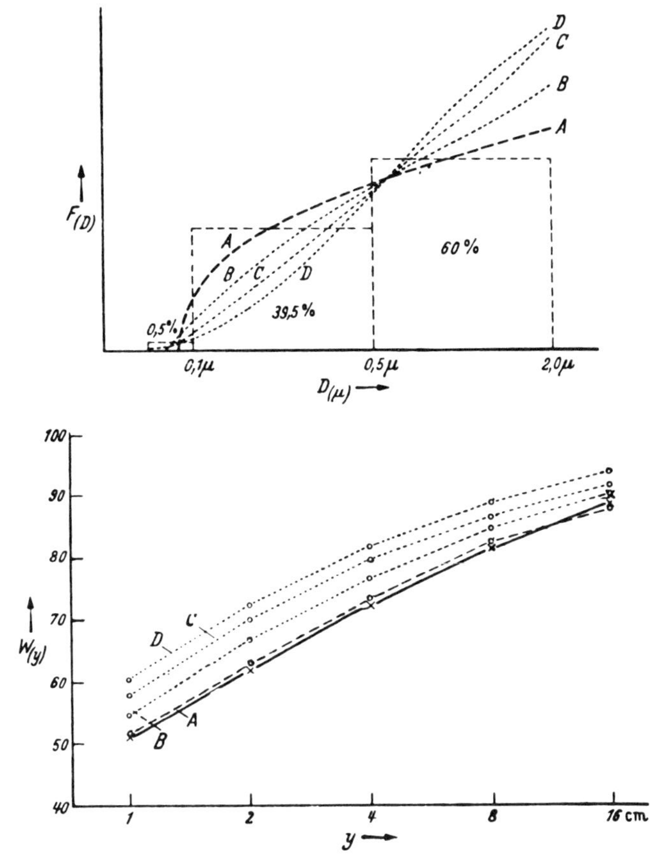

FIG. C. 22. Distribution curves of quartz suspensions, and corresponding quantity "Spectra," for the y-values 1, 2, 4, 8, and 16 cm. (Jasmund). Dotted curves for the calculated data; solid curve for experimentally determined data.

COLLOID SOLUTIONS AND SOLS

31. Semipermeable membrane migration of dissolved material and the distinction between crystalloid (low molecular) and colloid (high-molecular, polymerized) aggregation states have been discussed earlier with respect to the recent studies of N. Riehl and N. Trubetzkoy[78] (cf. C. ¶ 1); studies of the simplest system of silicates, namely silica itself, involve the boundary problem of whether or not a crystalloid solubility of silica in water is possible at all. This is the central question of the existence of the so-called *monomeric* silicic acid in its ortho-form H_4SiO_4, or as the meta-

[78] *Z. anorg. u. allgem. Chem.*, **253**, 1945, 54-70.

silicic acid, H_2SiO_3; this will be discussed from the purely chemical viewpoint in Vol. IV, Section A. There is no doubt that one is justified in speaking of a crystalloid solubility of the monomeric acids, but the problem is complicated by the many transitions to higher polymerizates. Among these we will generally distinguish between the paucimolecular *oligo*silicic acids and the high-polymeric *poly*-acids in order to describe the gradual transition from crystalline solubility to colloidal dispersoid systems and flocculates, or gels. This transition in its multiple phenomena is the field which attracted the particular attention of geochemists and biochemists. The geochemists were interested primarily in the silica cycle in nature; the biochemists were involved in the tremendous task of deciphering problems and causes of specific diseases, of which the so-called *silicosis* is the most fascinating phenomenon.[79] From the geochemical viewpoint, the "budget" and "cycle" of silica was especially investigated by R. Siever,[80] who considered the problems of weathering of silicate rocks and the redissolution of silica as well as the reconstitution of secondary silicate minerals. For these reasons we shall discuss this central investigation more extensively in Vol. IV, Section A, and from the special viewpoint of diagenesis in C. ¶ 162 ff. As for the biochemical side of the problem of the crystalloid solubility of silica and its importance *in vivo*, we must here be very brief; this field will be discussed in connection with the formation of organosilicate compounds chiefly in Vol. II, A. ¶ 306 ff. Silicosis problems are interpreted by one school of biochemical investigators as being a consequence of silica solubility. Fascinating suggestions of adsorption[81] and even of epitaxis (cf. A. ¶ 257 ff.) regularities have been put forth in another direction by H. Seifert[82] in his matrix theory.

32. The true *solubility of quartz* in water was recently investigated by G. S. Khodakov and E. R. Plutsis,[83] using powders having specific surfaces (measured by the

[79] For an introduction into various theories of the causes of silicosis, see W. Stöber and G. Bauer, *Staub*, **19**, 1959, 1-5, which is particularly recommended for study, with an excellent bibliography.

[80] *Am. Mineralogist*, **42**, 1957, 820-842.

[81] We mention here the so-called silica adsorption theory of silicosis as proposed by K. Söffge, *Neues Jahrb. Mineral. Monatsh.*, 1957, 12-22, as a most interesting physicochemical interpretation, based on pH conditions between 5.5 and 8.0, and in physiological solutions between 6.8 and 8.0. The problem of which kinds of dusts in silica and silicate operating industries are dangerous by causing serious silicosis disease, is particularly of interest from the hygienic- statistical viewpoint, chiefly supporting the opinion that dissolution of silica exclusively as quartz is detrimental; cf. K. G. Schmidt, *Ber. deut. keram. Ges.*, **31**, 1954, 355-363. The great reactivity of finest quartz dusts with magnesium hydroxide is often discussed as a model for the analogous solubility reactions in dust lungs; cf. N. Hast, *Arkiv Kemi*, **9**, 1956, 343-360.

[82] *Wiss. Forschungsber. Naturw. Reihe*, **63**, 1954, 199-227; *Ibid.*, **66**, 1958, 3-26, in K. W. Jötten and W. Klosterkötter, *Die Staublungenerkrankungen*, **2**, 1954, and **3**, 9158, with extensive literature references, and particular discussions of the "matrix" theory.

[83] *Doklady Akad. Nauk S.S.S.R.*, **123**, 1958, 725-728.

BET method) of 70,000 to 90,000 cm.²/gm. (grain sizes in the average of 0.3 μ). The solubility was determined by the colorimetric method with the molybdenum reaction, after vigorous shaking in glass cylinders for several hours. The kinetic curves of the solutions in pure water did not give any indication of the presence of colloidal silica, and followed the equation for the concentration changes: $C = C_p[1 - \exp(-k\,\tau)]$ (τ in hours). The plot of $\ln\,[C_p/(C_p - C)]$ versus τ is a linear function with the tangent of the slope angle expressing the rate of solution (parameter k). At 20°C. the solubility reaches 120 mg./liter, which is 20 times higher than the solubility data given by R. K. Iler[84] for coarse-crystalline quartz. As a function of temperature, the solubility rapidly increases with this parameter. The anomalies of the absolute solubility data show evidently how much they depend on the intensity of the grinding process, i.e., on the degree of partial destruction of the crystal structure ("amorphization," cf. C. ¶ 9) as the most important factor. This also explains the reactivity of these finest quartz powders with distinctly soluble hydroxides, especially of calcium or magnesium, at room temperature, as in hydraulic binders (cf. Vol. V, Section C).

33. Whereas one may question the direct possibility of obtaining monomeric H_4SiO_4 or H_2SiO_3 by intense grinding of quartz powders, the preparation of pure "monosilicic acid" was demontrated by G. B. Alexander[85] in a 0.1 molecular solution with pH = 3 by the reaction of a slurry of sodium metasilicate with a cation exchanger resin, at about 0°C. This author emphasizes, however, that the polymerization automatically sets in, even at this low temperature and at optimum pH conditions for stability of the low-molecular silicic acid. Whereas in the electrolyte characteristics of hydrated crystalline sodium metasilicate the presence of silicate anions is certain, as soon as this salt is dissolved the polymerization tendency begins. The silicate ions remain monomeric in the aqueous solutions only in sodium orthosilicate. The classical method of preparation of monosilicic acid by the hydrolysis of the methyl ester of the orthosilicic acid in purest water was followed quite recently by R. Schwarz and W. D. Müller.[86] The solution contained 1 to 15 mg. SiO_2 per

[84] Cf. his book "The Colloid Chemistry of Silica and Silicates," Cornell Univ. Press, Ithaca, New York, 1955, pp. 8 ff., especially on the older data and the application of the Ostwald-Freundlich equation for the solubility as a function of the particle sizes: $\ln(S_r/S_\infty) = 2EV/RTr$ (E = the surface energy in ergs/cm²; V = molar volume in cm³ : r = the radius of the particles concerned).

[85] *J. Am. Chem. Soc.*, **75**, 1953, 2887-2888.

[86] *Z. anorg. u. allgem. Chem.*, **296**, 1958, 273-279. The critical preparation of a pure orthosilicic methyl ester (tetramethyl oxysilane) was described by R. Schwarz and K. G. Knauff, *Ibid.*, **275**, 1954, especially pp. 177 ff. It is particularly difficult if one starts from $SiCl_4$ because of the formation of ester chlorides. A large excess in methyl alcohol must therefore be used. Of great importance is the observation of H. Funk, *Z. Naturforsch.* **11**, b, 1962, 199; *Kolloid-Z.* **184**, 1962, 154, that low-molecular silicic acid is better soluble in water-free solvents than in water, and more stable, e. g., in methanolic HCl. See on the condensation process in such solutions. H. Funk and R. Frydrych. *Naturwissenschaften*, **49**, 1962, 419.

100 ml. The electrolytic conductance of these solutions gave a dissociation constant of 1.24×10^{-10}, but a continuously decreasing tendency in the conductance versus time curves indicates the rapid setting-in of condensation to oligosilicic acids (cf. Vol. IV, Section A).

34. W. Stöber[87] gave a very complete discussion of the solubility problem of "silicic acids," and of the difficulties involved in the study of equilibria in their solutions which make an accurate determination of the physicochemical constants highly laborious, especially for the monomeric acids. We shall discuss these details in Vol. IV, Section A, but emphasize here the great tendency to automatic condensation in the solutions of true monomeric silicic acids. We must caution the reader against the conclusion frequently erroneously formed, that the particularly low rate of dissolution must be connected with a decreased solubility. The dissolution process is, as a matter of fact, irreversible for the different modifications of silica. Much more important, however, is a precise understanding of the automatic condensation process: the monomeric acids form the oligo-acids, probably with a chain-fiber structure, or perhaps also with some branched chain configurations. Although we have not been able to give here the fascinating mathematical formulations which were given by Stöber for the equilibria conditions of the condensation reactions and for the limit concentrations to be expected, we have combined his conclusions with the results of G. B. Alexander, W. M. Heston, and R. K. Iler,[88] and of H. Baumann,[89] in Fig. C. 23 concerning the molybdate-reactive silicic acid. It shows the very characteristic

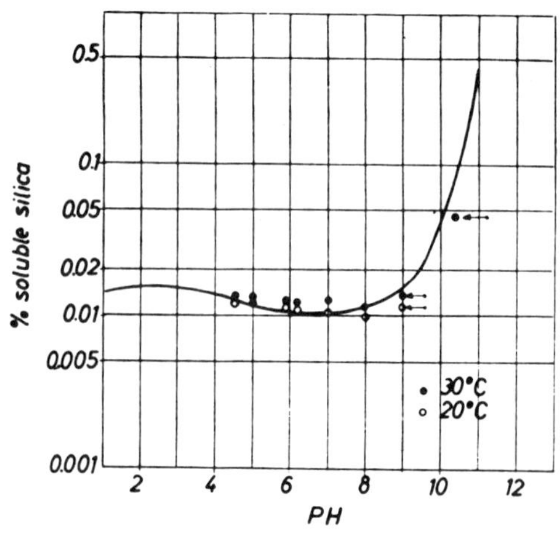

FIG. C. 23. Equilibrium concentration of molybdate-reactive silicic acid as a function of pH. (Alexander, Heston, Iler, and Baumann).

[87] *Kolloid-Z.*, **147**, 1956, 131-141.

[88] *J. Phys. Chem.*, **58**, 1954, 453-455, with very important literature references.

[89] *Beitr. Silikose-Forsch.*, No. 37, 1955, 47-71.

function of pH, and a minimum at pH of about 7.0. As a matter of fact, ammonium molybdate reacts primarily only with the monomeric acids. Such determinations were only possible by avoiding the use of ground quartz and similar coarse forms of silica, but starting from extremely fine-dispersed silica formed by the combustion of silane, or by the hydrolysis of silicon tetrachloride in steam ("Aerosil"). Such materials have very high specific surfaces, of several 100 m^2/gm. But even then, the equilibria are not settled before a period of one to three weeks. Baumann gave the instructive diagram Fig. C. 24 for the relative frequency of silica in different forms

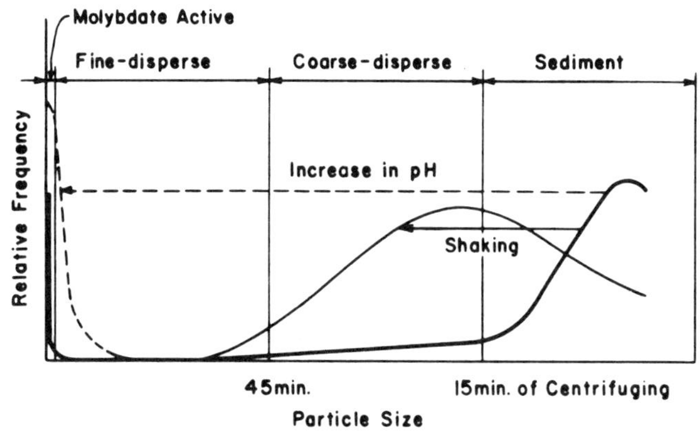

FIG. C. 24. Schematic diagram for distribution of concentration of different "silicic acids," as a function of particle size (Baumann).

in solution as a function of the particle sizes, showing that intensive shaking is able to influence considerably the distribution, shifting it towards higher dispersity degrees. The rate of the hydrolysis of Aerosil as a function of pH is impressively shown in Baumann's curve Fig. C. 25. The basic importance of this group of investigations is evident, especially for an examination of silicosis problems, although exclusively presented from the viewpoint of the silica solubility theory.

35. For characterization of the oligosilicic acids in their solutions H. Baumann[90] gave a very interesting chromatographic reaction, by spraying the chromatograms with a buffered solution (for pH = 1.5) of ammonium molybdate, and curing in moist atmosphere overnight. The yellow silico-molybdate complex is easily identified in ultraviolet light, or changed into molybdenum blue by reduction with hydrazine sulfate, and development with ammonia, with a limit concentration for the determination of 0.2 μg. SiO_2/cm.^{2}. The best conditions for hydrolysis of the silicic acid are pH = 3 because the rate of condensation is here a minimum (with acetate or citrate buffer solution). For the monomeric acid the chromatographic constant is

[90] *Naturwissenschaften*, **43**, 1956, 300-301.

$R_F = 0.75$, and for the more condensed oligo-acids it may be up to 1.0. After 5 days the oligo-acid $H_2Si_2O_5$ has its optimum stability at pH = 2. In weakly acid, neutral, or weakly alkaline solutions the condensation reactions are so rapid that the lower polymerizates are omitted, and higher acids appear with maximum stabilities at pH = 4 (with veronal acetate buffer). In their studies on the organic alkyl esters of

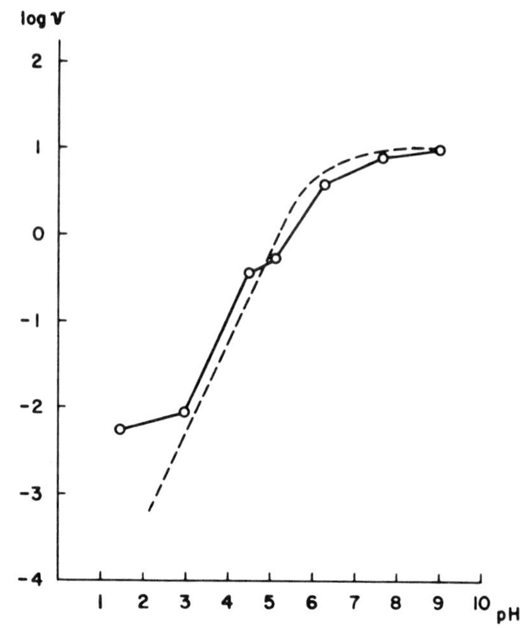

FIG. C. 25. Rate of hydrolysis of Aerosil as a function of pH (Baumann).

silicic acid, R. Schwarz and K. G. Knauff[91] also particularly discuss formation of alkoxysilanes and oligo-esters, starting from hydrolysis of tetramethoxysilane. A di-, tri-, and tetra-ester could be isolated with distinct arrest points on the distillation curve at 42°, 87°, and about 110°C., respectively. These oligoesters are much inclined to "dismutate" to the orthomonoester plus polyesters, but by small additions of phosphoric acid this undesirable disproportionation can be suppressed. The oligoesters are all of the chain type; no cyclic esters were observed. The determination of the molecular weights by the cryoscopic method for the hydrolysis products gave a distint picture of the condensation as a function of time, from the orthoesters to the oligoesters and finally to polyesters (cf. Fig. C. 26). The orthosilicic acid is *not* volatile with water vapors. The hydrolysis of oligo-acids forming orthomonosilicic

[91] *Z. anorg. u. allgem. Chem.*, **275**, 1954, 176-192. Particularly remarkable is the often explosive start of the dismutation of the oligoesters, connected with a strong vesiculation in a material of rapidly increased viscosity, forming pumice-like products in the residues of distillation.

acid is a most unexpected reaction type, characterized by the disruption of Si—O—Si bridge bonds, such as

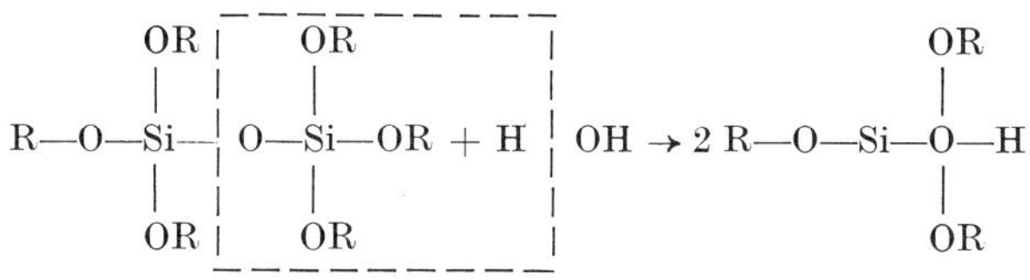

unexpected insofar as the same bridging is very stable in silicones which can be disrupted only with strong acids at elevated temperatures. Alkoxy-groups evidently loosen the Si—O—Si bonds.

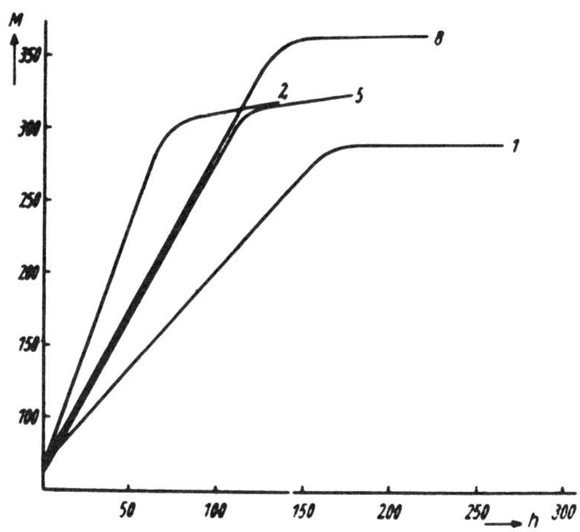

FIG. C. 26. Condensation of monomeric silicic acid (1, 2, and 5), and of disilicic acid (8), as a function of time (Schwarz and Knauff).

36. Concerning the constitution of the hydrolysis products of tetraethyl oxysilane, O. Glemser and G. Rieck[92] studied the infrared absorption spectra of monomeric and oligosilicic acids (Fig. C. 27). They make it evident that the majority of the water is present in the form of hydroxyl groups. A strong hydrogen bridge bonding is indicated by the shoulder at 3.45 μ, corresponding to the distance O—H···O of 2.6 Å. The 2.75 μ band indicates the undisturbed hydroxyl groups on the surface. Glemser and Rieck characterize the silicic acids as "condensed hydroxides," as distinct from non-stoichiometric hydroxides, such as those of aluminum and chromium in which the hydroxyl groups can be removed from the oxide core and can be restored, without any change in the crystalline structure.

[92] *Z. anorg. u. allgem. Chem.*, **297**, 1958, especially pp. 177-181.

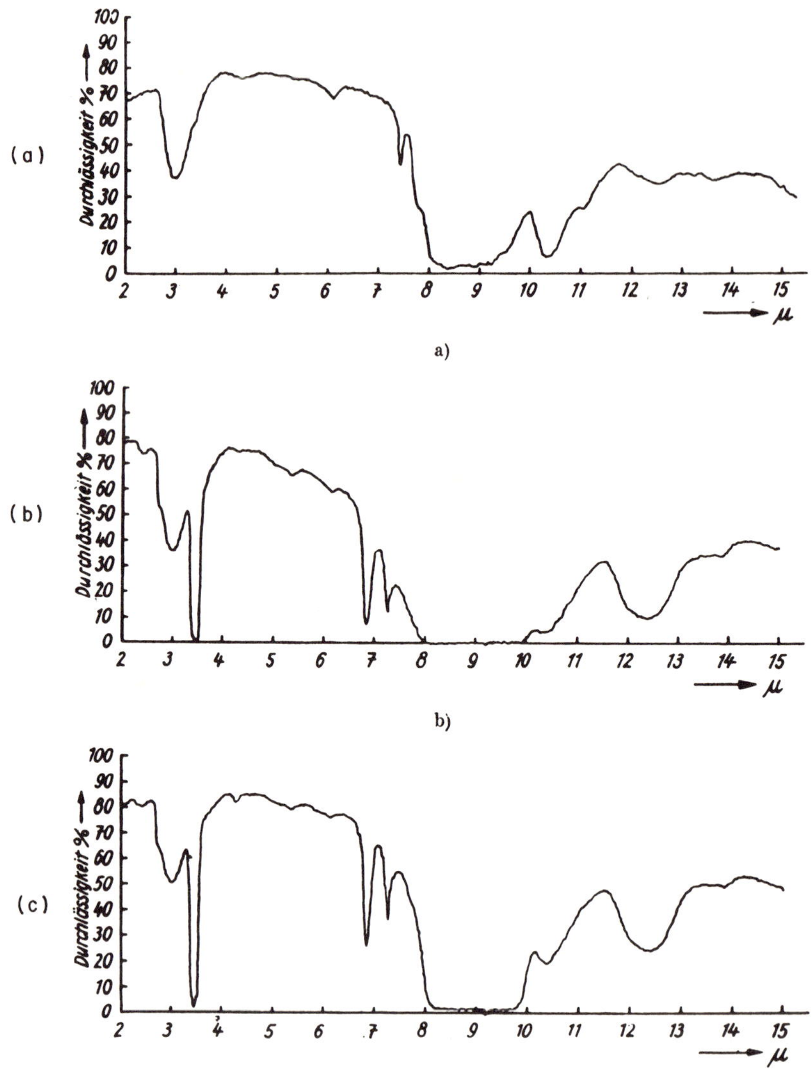

FIG. C. 27. Infrared absorption spectra of silicon hydroxides: (*a*) with 12.7 per cent water; (*b*) with 12 per cent water, heated to 280°C.; (*c*) heated to 290°C. (Glemser and Rieck).

37. A very important phenomenon is the response of silica to its true molecular solubility in dilute mineral acids. Th. H. Elmer and M. E. Nordberg[93] determined the solubility of silica from porous Vycor-type glass, and of a commercial silica gel, in dilute nitric acids solutions at 95°C., and established typical curves of the dissolution rates and of the solubilities as a function of the concentration of the acid

[93] *J. Am. Ceram. Soc.*, **41**, 1958, 517-520.

(Fig. C. 28). The solubility at constant temperature decreased with increasing concentrations of the nitric acid solution, and for a constant acid concentration the same was observed with increasing temperature. The initial dissolution rate has a maximum at about 0.8 N of the acid. The solubility at constant temperature is of the form $\log C = -a \cdot N - b$ (N the normality of the acid solution, a and b empirical constants). This general behavior is important for industrial leaching processes, as in the production of glass fibers high in silica. The principles here involved may for a temperature of 36°C. also have much importance for the biochemical role of silica in body fluids, if the solubility theory of the origin of silicosis is accepted (see above).

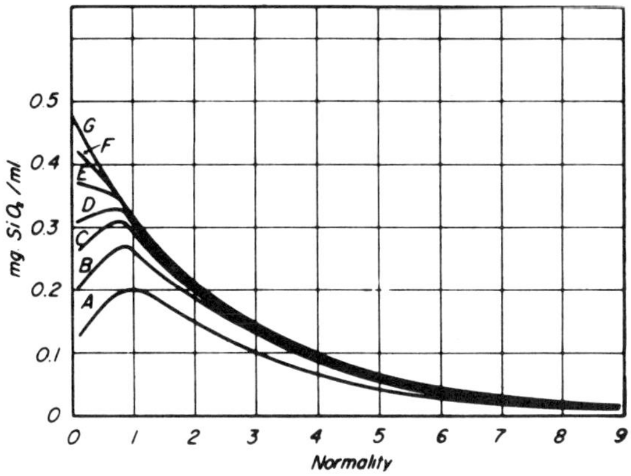

FIG. C. 28. Amount of silica dissolved from porous glass plate in dilute nitric acid at 95°C: (A) for 10 hours; (B) for 20 hours; (C) for 30 hours; (D) for 40 hours; (E) for 60 hours; (F) for 100 hours; (G) equilibrium curve (Elmer and Nordberg).

38. Of no less significance is the often discussed problem of the participation of foreign anions, first of all of Cl^- and F^- in the solubility phenomena of silica. In 1931, G. Jander and H. Heukeshoven,[94] in a series of studies on amphoteric oxyhydrates of acidic character, discussed the possibility of complex chlorido-silicic acid anions influencing the solubility mechanisms of silica in true molecular solutions. H. Sadek[95] measured with galvanic cells containing an AgCl electrode the values of pCl in mixtures of sodium silicate, hydrochloric acid, and silver chloride, and found a distinct increase in this parameter over that calculated under the assumption of all chlorine present in the form of free Cl^- anions. The straight line projection in a diagram of pCl as a function of ($1 + \log SiO_2$) is interpreted by the conclusion that some compounds were formed, especially chlorosilicate anions, in the ratio 1 HCl : 1 SiO_2. Also G. B. Alexander, W. M. Heston, and R. K. Iler[96] discussed the possibility

[94] *Z. anorg. u. allgem. Chem.*, **201**, 1931, 361-382, especially pp. 374 ff.
[95] *J. Indian Chem. Soc.*, **29**, 1952, 507-510.
[96] *J. Phys. Chem.*, **58**, 1954, especially p. 455.

of acids other than HF promoting the solubility of silica, with reference to the authors mentioned above. Complex anion formation may be very analogous to that occurring with HF-containing silica solutions in which the $(SiF_6)^{2-}$ anions are formed, analogous to the configuration $(Si(OH)_6)^{2-}$, recently also discussed by W. A. Weyl and E. A. Weyl and E. A. Hauser[97] in the light of the screening theory. These ideas have a particular interest especially for the possibility of illustrating "catalytic" effects of fluorine in the hydrolysis of ethyl orthosilicic acid ester, and the condensation-polymerization of aqueous silicic acid in general (as seen from Fig. C. 29), for the chain-condensation process and the final syneresis to a silica hydrogel.

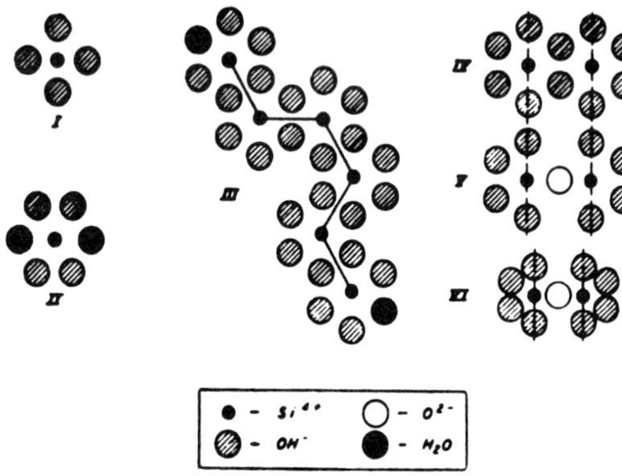

FIG. C. 29. Schematic transitions from the molecule $Si(OH)_4$ (I) to the same plus 2 molecules of water (II); forming a chain arrangement (III); neighboring chain groups (IV); condensation (V); and syneresis (VI) (Weyl and Hauser).

39. We mentioned before the polymerization of monosilicic acid to oligo-acids as observed with cryoscopic measurements, which were extended by R. K. Iler[98] to molecular weight determinations. In the initial states of silicic acid formed by the decomposition of a sodium silicate solution (3.3 Na_2O,1.0 SiO_2) by mineral acids, the average molecular weight was 200, corresponding to a polymerization degree of 3.25 SiO_2 units per molecule. The rate of polymerization is proportional to the square of the concentration in silica, the average molecular weight increasing in proportion to the square root of time. The polymerization to a molecular weight of 6000 (in a 2 to 6 percentage concentration) reaches that point at which gelation occurs, and a three-dimensional framework is formed, the mechanical rigidity of which is greater the greater the amount of silica contained in the structure. A very characteristic maximum is observed in the curve shown in Fig. C. 30 representing the time required for silicic acids to polymerize to an average number of 28 silicon atoms per molecule

[97] *Kolloid-Z.*, **124**, 1957, 72-76.

[98] *J. Phys. Chem.*, **57**, 1953, 604-607.

in a 1 *M* silica solution at 25°C. as a function of pH. H. Baumann[99] used for determination of the polymerization degrees over the range of pH from 0.5 to 9.0 at 30°C. the molybdate reactivity of the resulting silicic acids (Figs. C. 31 and 32). Both polymerization and depolymerization, are strongly influenced by the pH and the in-

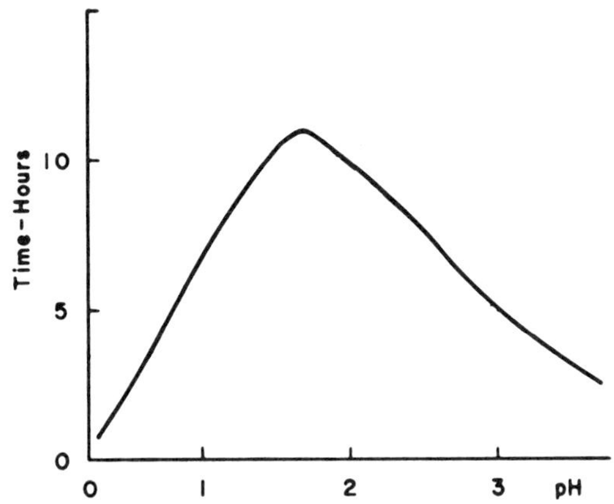

FIG. C. 30. Time required for the polymerization of silicic acid to an average number of 28 Si atoms per molecule of the polymerizate, as a function of pH; 1 *M* concentration in SiO_2; temperature 25°C. (Iler).

itial concentration, the rates of polymerization being very low at pH = 2 to 4 (see above). If pH is raised to 4.0, the rate increases proportional to the OH^- ion concentration. The equilibrium concentration in silica reaches 100 to 150 μg./cm.^{3}, and is nearly constant over the whole range of pH investigated. Characteristic is the behavior of the fluorine anions present; these ions undergo a true chemical reaction and form the complex $(SiF_6)^{2-}$ anions which considerably influence the course of polymerization (reducing it) and depolymerization (increasing it). On the other hand, chlorine and bromine anions accelerate both processes as a consequence of some catalytic effects. Different cations present do not influence the degree of polymerization.

40. Coming back to the problem of *dialysis* of silicic acids in their different states of condensation and polymerization, we find in the studies of L. Holzapfel[100] an example of the connection of dialysis with the solubility theory of silicosis. The surface layer of high solubility on quartz which was postulated in this theory was localized by this author to the edges and corners of quartz fragments, or to layer units disrupted from the crystal surface. If organic material is present in the intense-

[99] *Kolloid-Z.*, **162**, 1959, 28-35.

[100] *Z. anorg. u. allgem. Chem.*, **273**, 1953, 186-192.

Curve		Initial Concentration (γSiO_2/ccm)
PH 5.8	1	1550 γ/cc
	2	1050 "
	3	765 "

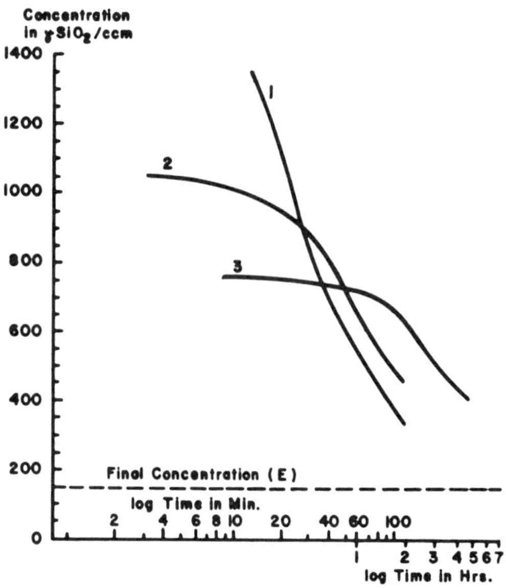

FIG. 31 *a*

Curve		Initial Concentration (γSiO_2/ccm)
PH 7.2	1	2020 γ/cc
	2	1010 "
	3	645 "
	4	490 "

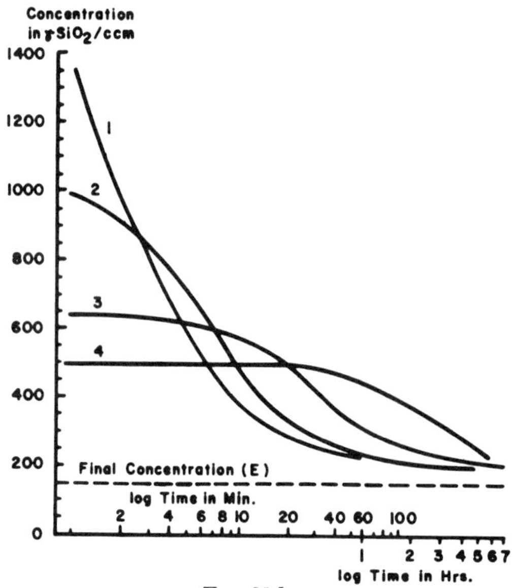

FIG. 31 *b*

Curve		Initial Concentration (γSiO_2/ccm)
PH 9.0	1	1012 γ/cc
	2	600 "
	3	435 "

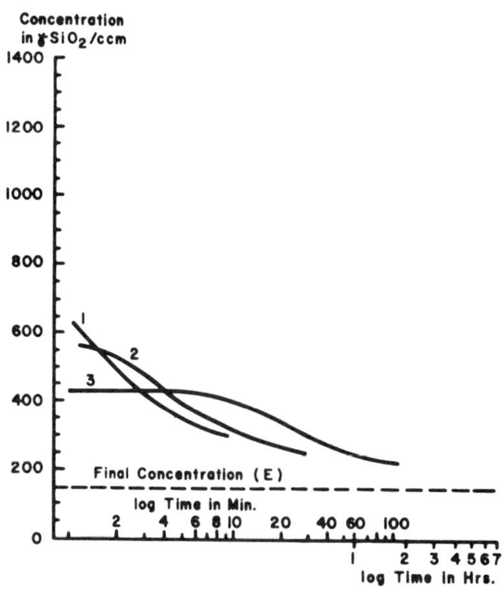

FIG. 31 *c*

FIG. C. 31. Concentration in molybdate-reactive silicic acid during polymerization at 30°C., as a function of time, and in the range from pH = 5.8 to 9.0 (Baumann).

ly shaken suspensions of quartz powder, amino acids (e.g., glycine, L-leucine, tyrosine, cystine) or urea markedly change the dialysis of the silica, as interpreted by the formation of organosilicic complexes, in ionized states. Highly specific phenomena occur if oleinic acid is added to amino acid–quartz suspensions; this completely counteracts and finally suppresses the effects of the amino acids and reacts directly with the quartz particles. The full importance of these observations lies in the field of the organosilicic complex formation (Vol. II, A. ¶ 306 ff.). For such and related questions, careful

Curve	1	2	3	4	5	6	7	8
PH	0.45	1.20	4.00	5.35	5.80	7.20	8.30	9.00

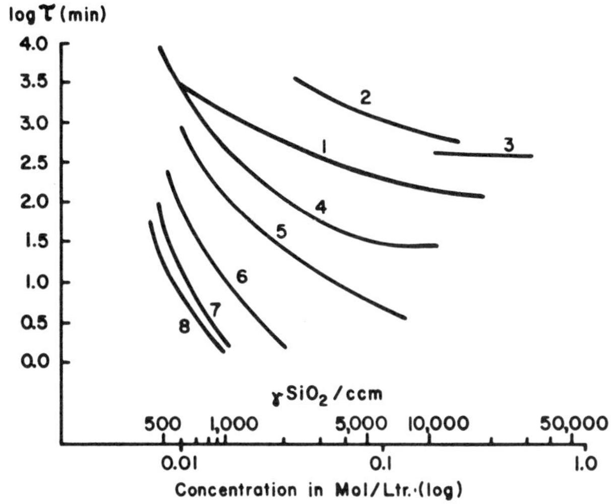

FIG. C. 32. Half-value time of polymerization rate as a function of the initial concentration in silica and of pH (Baumann).

definition of the grain size in the ultrafine range is indispensable, as was demonstrated by G. B. Alexander and R. K. Iler[101] for the commercial "Ludox" colloid silica (which is alkali-stabilized), with nonporous particles of 16 to 60 mμ in diameter as determined by the nitrogen adsorption, turbidimetric, and electron microscopic direct methods. L. Rothardt[102] applied for such problems an impulse echo method of ultrasonic absorption (frequency of 8 Mc.) which responds to particles of one-tenth of the wavelength of the ultrasonic vibrations (practical upper limit of particles with 10 μ in diameter) for the primary as well as for the secondary (aggregate) particles. Optical (light scattering) methods for the study of the polymerization in mixtures of sodium metasilicate and ammonium acetate solutions were

[101] *J. Phys. Chem.*, **57**, 1953, 932-934.

[102] *Naturwissenschaften*, **45**, 1958, 486-487.

applied by S. A. Greenberg and D. Sinclair,[103] chiefly by turbidimetry, dissymmetry, and depolarization measurements, with the remarkable result that the silica particles behave like isotropic *spheres*. This conclusion is the more remarkable as in previous authors' opinion the formation of *fibrils* or fibers is the characteristic form of the silica from hydrolyzed silicate solutions. The spherical particles of the higher condensation degrees also show a high degree of *solvation* as concluded from the diameter versus time curves (see Fig. C. 33) when extrapolated to a zero time intercept on the ordinate axis, at about 200 Å. *Dried* silica gels and aerogels show spherical particles of 30 to 60 Å. diameter. Polymerization and aggregation simultaneously occur and finally aggregation alone predominates to form the gel.

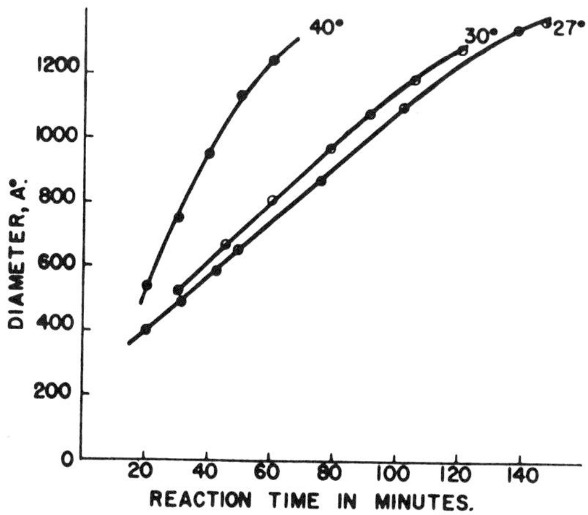

FIG. C. 33. Rate of change of diameters of silica sol particles formed from equivolume mixtures of 1 per cent sodium metasilicate and 3 per cent ammonium acetate solution, at various temperatures (Greenberg and Sinclair).

41. The polymerization reaction is interpreted by Greenberg and Sinclair not by the elementary condensation of uncharged silicic acid molecules, but by a real reaction of ionized acids with nonionized acids, in the following way:

$$\mathrm{HO{-}\overset{\displaystyle OH}{\underset{\displaystyle OH}{\overset{|}{\underset{|}{Si}}}}{-}O^{-}{-}} + \mathrm{HO{-}\overset{\displaystyle OH}{\underset{\displaystyle OH}{\overset{|}{\underset{|}{Si}}}}{-}OH} \rightarrow \mathrm{HO{-}\overset{\displaystyle OH}{\underset{\displaystyle OH}{\overset{|}{\underset{|}{Si}}}}{-}O{-}\overset{\displaystyle OH}{\underset{\displaystyle OH}{\overset{|}{\underset{|}{Si}}}}{-}OH} + \mathrm{OH^{-}}$$

[103] *J. Phys. Chem.*, **59**, 1955, 435-450. See also K. Goto, *Ibid.*, **60**, 1956, 1007-1008, who emphasizes the pseudo-equilibria and states of the polymerization reactions and the acceleration of this process at higher *pH* conditions when the colloid particles are highly charged and depolymerize rapidly.

This reaction was first proposed by K. D. Ashley and W. B. Innes,[104] and is analogous to the condensation of trimethylsilanol in basic solutions which was studied by W. T. Grubb.[105] As the principal reason for the correctness of this assumption, Greenberg and Sinclair emphasize: (1) that no reaction takes place at high pH when the solutions mainly contain silicic acid ions repelling one another, and polysilicic acids would be expected to depolymerize immediately; (2) the reaction is slow, if taking place at all, at low pH where the concentration in nonionized silicic acid is high; and (3) it proceeds at the maximum rate in the range of pH of 8 to 9, when the concentration in silicic acid and the singly ionized forms are both reasonably high. The ratio $(H_3SiO_4)^-/(H_4SiO_4)$ is at pH = 9 about 1.6/10, and at pH = 8 it is 1.6/100. We further emphasize the extensive studies of the polymerization phenomena in dilute aqueous solutions of silica as determined by M. F. Bechtold[106] in connection with spontaneous gelation, the greatest rate of which is observed by titration for pH = 5.05 (shifted to 5.85 by the polymerization), and an inflection point at pH = 4.65 (shifted correspondingly to 5.30). The turbidity measurements (determined with a Zeiss-Pulfrich mechanical photometer, with a turbidity head) show

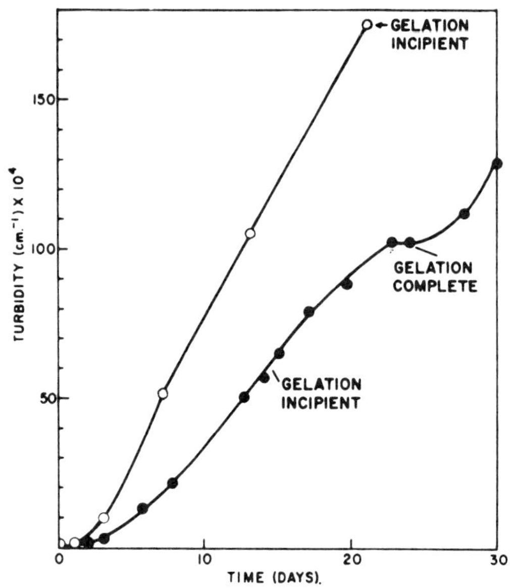

FIG. C. 34. Turbidity in aqueous silicic acid solutions versus polymerization time at room temperature (Bechtold): solid circles refer to a concentration of 1.83 per cent SiO_2; open circles to 1.78 per cent silica; ratio SiO_2/Na_2O = 4450, and pH = 4.5.

[104] *Ind. Eng. Chem.*, 44, 1952, 2857-2863.

[105] *J. Am. Chem. Soc.*, 76, 1954, 3408-3414. Because of the marked increase in basicity during the aging of silica gel catalysts, this important conclusion was generalized.

[106] *J. Phys. Chem.*, 59, 1955, 532-541.

the very characteristic initial and final gelation effects according to Fig. C. 34. The preparation of the silica sols was in this case made by a hydrogen ion exchange resin ("Nalcite" MX, a condensation product of formaldehyde with α, β, γ-phenylsulfonic acids), with sodium metasilicate solution.

42. According to R. K. Iler and R. L. Dalton[107] the *viscosity* of silica hydrosols is a sensitive indicator of the degree of hydration of its (spherical) particles—of the water bound on their surface. The calculations were based on an equation of M. Mooney[108]: the relative viscosity is $\eta_r = \exp[2.5\, c(1 - k \cdot c)]$ (k corresponding to a "self-crowding" factor), and the parameter $\ln \eta_r = 2.5 \cdot c/(1 - 1.43\, c)$, in which c is the volume fraction of the dispersed phases. It is a function of S, which is the weight per cent concentration of the spherical silica particles in the suspension, and of P, the weight percentage of the silica in solution:

$$S = \frac{P}{0.00566\, P + c\,(1 - 0.00566\, P)}$$

The particles may be written in the hydrated form $(SiO_2)_x(H_2O)_y(H_2O)_{n(2y)}$ in the dispersed phase. They indicate how n layers of water are fixed on the surface of the chemically bonded silanol complex $(SiO_2)_x(H_2O)_y$, and always contain two silanol groups equivalent to one molecule of water; thus is

$$\frac{S}{100} = \frac{60\, x}{60\, x + y(2\, n + 1) \cdot 18}\,.$$

From the average diameter, d, of the particles in millimicrons, one derives

$$x = \frac{22}{6}\, \pi\, d^3 \qquad \text{and} \qquad y = \frac{\pi}{2}\,(2.80\, d)^2$$

and further

$$n = \frac{1}{2}\left[\frac{8480\,(100 - S)}{S\,A} - 1\right]$$

A = the surface area of the particles in cm.^{2}/gm., determined by a titration method of G. W. Sear,[109] and S as derived from Mooney's equation given above. Iler and Dalton found for a series of different silica sols, $n = 0.6$ to 1.1, in average $= 0.9$, showing that a monomolecular layer of water is immobilized on the hydroxylated surface of the silica particles, with particle diameters between 3.3 and 7.5 mμ.

[107] *J. Phys. Chem.*, **69**, 1956, 955-957.

[108] *J. Colloid Sci.*, **6**, 1951, 162-170.

[109] *Anal. Chem.*, **28**, 1956, 1981-1983; the titration of silica hydrosols in a 20 per cent NaCl solution is made with NaOH solution, and the increase of pH observed from 4.0 to 9.0 as a linear function of the specific surface of the colloidal silica particles.

43. A most elaborate calculation of the equilibria of silica polycondensation was recently given by W. Stöber,[110] starting from the molybdate-reactive silicic acids as the limit condensates which reach a constant concentration of reversible establishment for relatively low amounts dissolved. If there is a true thermodynamic equilibrium possible between the single condensation steps, a formula can be developed, for pH below 12, for the total solubility of silicic acid, which is in reasonable agreement with the experimental data. The rather complicated equation makes evident that the influence of pH is restricted to the low-molecular range, with the upper limit of molecular dispersity. In acidic or neutral media the influence of pH is very low for the total solubility. Another equation was derived which is valid for pH below 7. In alkaline media with pH above 9, a strong increase in monosilicic acid is noted which changes the concentrations with increasing pH by more than one order of magnitude. However, both equations given by Stöber are inadequate for calculation of the total solubility because the maximum condensation step cannot be experimentally fixed with sufficient accuracy. An estimate and comparison with the data given by H. Baumann[111] do show that a satisfactory approximation is possible in the calculation. The first term in the equation for pH below 7 shows for molybdate-reactive silicic acids very characteristic limit concentrations, and the percentage share of the low-molecular acids which is, e.g., for the saturation value $g_y = 78.000$ μg. SiO_2/ml. about 74 per cent, for $g_y = 4.950$ μg./ml. about 4.7 per cent, but for $g_y = 0.089$ μg./ml. below 0.001 per cent. This percentage is the only portion of silicic acid which underlies the influence of the pH changes.

44. Because of the fundamental importance of the gel characteristics of polymeric silica (and even more, of mixed gels of silica and alumina) for the modern hydrocarbon cracking catalysis processes of the petroleum industry, the reactions of gelation and their properties have been extensively studied. Although many of the facts observed here illustrate details of the constitution of such materials, and should perhaps be discussed in connection with the investigations demonstrated above, we may here only refer briefly to one or more representative results, and refer *in toto* to the discussion of mixed gels in a later section (C. ¶ 162-¶ 175) as an important supplement. We mention here only the studies of C. J. Plank[112] on the extremely rapid gelation of certain hydrosols of this kind and the sensitivity of the gelation time to changes in pH demonstrating that *ionic* reactions direct this process.

[110] *Kolloid-Z.*, **151**, 1957, 42-47.
[111] *Beitr. Silikose-Forsch.*, No. 37, 1955, 45-71.
[112] *J. Colloid Sci.*, **2**, 1947, 399-412; 413-427.

ELECTROSTATIC CHARGES OF COLLOIDAL SUSPENSIONS AND FLOCCULATION

45. From the practical viewpoint of electroinsulator manufacturing, the *electrophoretic* deposition of ceramic bodies has a continuously increasing significance, not restricted only to common porcelain batches, but even more specifically for oxide ceramics because of the high degree of primary compaction of the coatings deposited on conducting surfaces, as on the tungsten heating elements of electronic tubes. The electrophoretic deposition of alumina from an organic suspension medium (mostly methyl alcohol) in electronic tube manufacturing is one of the examples given by R. Bahn[113] in which this method was particularly successful in solving a technological problem. The electrokinetic conditions for the stability of oxide suspensions by the formation of electrostatic double-layers are based on the investigations of A. Cohen and O. Schafmeister,[114] and of H. Freundlich and H. P. Zeh.[115] The addition of aluminum or chromium citrate solution is recommended for a correction of the electrostatic potential of the particles near its inversion. The suspension is an organic liquid (mostly alcohols) which is particularly sensitive to traces of water, and these are detrimental for the formation of a homogeneous and dense deposition of oxides by an electrolytic decomposition effect. Because of this sensitivity, one may replace the alcohols by mixtures of acetone and amyl acetate as the suspension medium, with a slight addition of collodion to avoid the formation of framework structures in the suspension (concerning thixotropy, see C. ¶ 333 ff.). Colloid-disperse oxides are often highly hygroscopic. The electrophoresis is conducted either cathodic or anodic in deposition. The distribution of the electrostatic field lines strongly affects the anaphoretic depositions, but affects the cataphoretic phenomenon very little, in the relative thickness of the oxide deposit. Also for mixed oxides, such as alumina and chromium oxide, or alumina with gallium oxide, and others, the electrophoretic method is superior to all the other ceramic deposition principles, but the optimum conditions must be very carefully specified.

46. The *flocculation* of colloid suspensions is particularly important in the behavior of ceramic clays. R. K. Schofield and H. R. Samson[116] studied in this respect the behavior of kaolinite which in a highly purified state, but containing sodium, exhibits rapid flocculation, in contrast with illite and montmorillonite clays. A stabilization of kaolinite suspensions (deflocculation) is possible. In addition to the classical methods of applying solutions of sodium hydroxide, sodium pyrophosphate, oxalate, or alginate, deflocculation may be achieved by introducing negatively

[113] *Silikat Tech.*, **9**, 1958, 299-303.
[114] *Z. physik. Chem. (Leipzig)*, **125**, 1927, 401-419 on the ζ potential.
[115] *Z. physik. Chem. (Leipzig)*, **114**, 1924, 65-68.
[116] *Discussions Faraday Soc.*, **18**, 1954, 135-145.

charged tabular particles of another clay mineral to overlie the positive charges on the edges of the larger kaolinite crystals. An "edge-to-face" flocculation by hydrochloric acid is different from the "salt" flocculation by sodium chloride. In the first case, the apparent viscosity is high and the flow birefringence practically zero; in the second case, the apparent viscosity is low and there is some flow birefringence observed. The electrostatic attraction of the positively charged edges with the negatively charged faces of the crystals is the reason for flocculation. Surfaces of small particles are never homogeneous, and the mutual attraction of the positively, or negatively, charged regions coexisting on the same kaolinite particle thus brings about the flocculation. Calcium hydroxide is less efficient than sodium hydroxide as a deflocculent since the calcium cations are more strongly bound to the surface of the kaolinite crystals. The interparticle forces are reduced, and the attraction between edges and faces increased (cf. C. ¶ 336).

47. The *sedimentation volume* of bentonitic clays was studied by M. S. Fedotova and K. A. Pospelova[117] after flocculation of suspensions, with 3 to 5 per cent solid material by aluminum chloride (concentration of its solution varied between 0.0003 and 0.10 per cent), in respect to the permeability of a bentonite-treated sand bed. The rate of permeation through such a bed is expressed in an equation of the form $\log v = a + \gamma \cdot \log t$, in which γ is measured as the tangent function of the slope angle for the parallel lines corresponding to the equation, b being the volume of the filtrate. The best impermeability effect is observed if the bentonite is first completely impregnated into the sand bed, and then flocculated as a dense deposit on the grains, but it is difficult to have a deeper flocculation layer than 5 cm. If the constant a (determined for a filtration time of 10 minutes) is plotted versus the concentration of the different salt solutions (chlorides of sodium, potassium, calcium, magnesium, and barium), the efficiencies for the flocculation by NaCl, $MgCl_2$, and $AlCl_3$ are in the ratio 1 : 4 : 113. The effect of aluminum chloride can be more or less controlled by the reverse deflocculation effect of soda solution.

48. L. Wiklander and Elisabet Koutler-Andersson[118] investigated the "floc volume" of Swedish clays and of related minerals in fractionated grain sizes. The influence of the concentration of the suspensions on the floc volume is shown in Fig. C. 35, for nontronite particles smaller than 0.14 μ. The floc volume for bentonite is rather constant in the pH range from 7.4 to 9.4 but at pH = 10 it shows a considerable decrease. Nontronite and montmorillonite give about the same volumes, but illite shows rather inconsistent data of the floc volume indicating the structural heterogeneity of this mineral. Mixtures of bentonite and of vermiculite with nontronite show additive properties, but in mixtures of vermiculite (25 per cent) and mont-

[117] *Doklady Akad. Nauk S.S.S.R.*, **94**, 1954, 1129-1132.
[118] *Acta Agr. Scand.*, **2**, 1955, 215-223.

morillonite (75 per cent) a considerable depression occurs, evidently caused by disturbance in the montmorillonite structure by the much larger flakes of vermiculite. In soil clays, the relations are sometimes very complex, and the clay minerals do not altogether exist as discrete particles, but are more or less interlayered. In general, the floc volume increases with decreasing particle sizes. A great floc volume always indicates a considerable content in montmorillonite group minerals, with a subordinate content in kaolinite. Very low floc volumes, on the other hand, always indicate kaolinitic minerals as dominant.

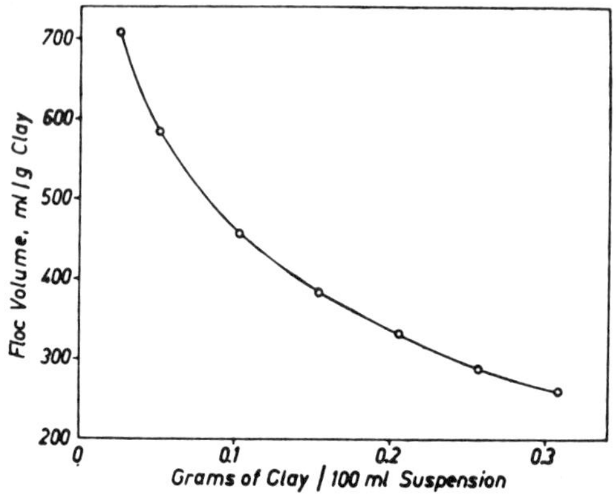

FIG. C. 35. Effect of concentration of a nontronite suspension with particles smaller than 0.14 μ in sedimentation over 24 hours, on its floc volume (Wiklander and Koutler-Andersson).

49. K. Pate, W. Noble, and F. H. Clews[119] discuss the sedimentation volume of clays as a particularly characteristic property, of a rank equal to that of plasticity and drying shrinkage for ceramic raw materials, especially in relation to their contents in particles smaller than 2 μ in size (cf. C. ¶ 257 ff.). The sedimentation volume is only slightly dependent on the temperature of sedimentation, on acidity of the aqueous medium or on mechanical vibration effects, but very sensitive to changes in concentration of the suspensions, as Wiklander and Koutler-Andersson described. In continuation of his previous (1943) investigations, A. von Buzágh and F. Szántó[120] studied the sedimentation volume of quartz suspensions in polar media (water, ethyl alcohol, ether, acetone), and found it also independent of temperature. In nonpolar suspension media, however (in benzene or carbon tetrachloride), the floc volume decreases with rising temperature, with a slight hysteresis effect. Traces of water in a nonpolar medium increase the sedimentation volume immediately. Very interesting is the behavior of propanol which contains polar and nonpolar aliphatic groups,

[119] *Trans. Brit. Ceram. Soc.*, **55**, 1956, 14-156.
[120] *Kolloid Z.*, **141**, 1955, 1-4.

because of the directed adsorption, with the aliphatic portion oriented away from the quartz crystal surface.

50. In this, and related respects, especially the particles of clay minerals behave as *colloid electrolytes*, in the light of the Debye-Hückel theory.[121] They act as definite ionized compounds dispersed to colloidal particle sizes, but with a true equilibrium existing between the surrounding medium and the particle composition, the electrostatic charges of which are due to the ionization of the compound itself. This is distinguished from the second type of colloidal "solids" in which the bulk of the particle is made up of insoluble material with no equilibrium, and electrostatic charges are due chiefly to the preferential adsorption of *foreign* ions. C. E. Marshall[122] classified the montmorillonites as true colloid electrolytes of the first type. The suspension of this clay mineral is like a dissociated electrolyte, the ions of which fall within the colloidal range of dimensions.

51. The dissolution of a silica hydrogel by sodium hydroxide commonly called the *peptization*, is seemingly the reverse of the process of polymerization which was extensively discussed above. Ch. B. Hurd and St. H. Laning[123] studied this apparently elementary process by observing in the microscope the changes of thin silica films during the attack by a 1 M sodium hydroxide solution.

52. *Electrodialysis*, as a method of purifying ceramic raw materials, especially of clays and kaolins, plays a continuously increasing role in modern quality porcelain manufacturing, since the basic experiments of Graf B. von Schwerin (1918). We mention in this connection the studies of S. K. Chatterjee[124] with high-voltage (440 volts) dialyzing equipment, especially to reduce the total alkalies from feldspar containing raw materials, although a complete destruction of the feldspars is not possible in this way. A more basic discussion of the *flotation* phenomenon in the light of the theory of the electrokinetic potential, ζ, was given by V. Borisov.[125] Electrostatic double-layers surrounding the suspended particles determine, together with the effects of flotation agents added, the "electrokinetic index," or the differences in potentials (measured in millivolts) of the mineral surface in the flotation solutions

[121] Cf. G. S. Hartley, *Trans. Faraday Soc.*, **31**, 1935, 31-50, 68.

[122] *Soil Sci. Soc. Am. Proc.*, **1**, 1936, 23-31; the definitions given here are those recommended by Margaret D. Foster, *Am. Mineralogist*, **38**, 1953, especially p. 1005.

[123] *J. Phys. Chem.*, **58**, 1954, 914-915; see also Ch. B. Hurd, M. D. Smith, F. Witzel, and A. C. Glamm, *Ibid.*, **57**, 1953, 678-680, on polymerization equilibria studied by dialysis experiments.

[124] *J. Indian Chem. Soc.*, **29**, 1952, 549-553; see also S. Stacha, *Zpravy Českoslov. keram. a sklarské spolecnosti*, **24**, 1948, 60-66.

[125] *Doklady Akad. Nauk S.S.S.R.*, **99**, 1954, 427-430. Fibers of montmorillonite on apatite ores (from Kola) are highly troublesome in flotation practice. They can be explained by electrophoresis and coprecipitation as was demonstrated by O. B. Dubkin, *Zapiski Vsesoyuz. Mineral-Obshchestva*, **89**, 1960, 572-576.

containing the electrolyte alone, and the collector plus the electrolyte. This index is governed by the ion that is selectively adsorbed in the double-layer.

53. Borisov discusses the flotation of non-ore minearls, like apatite, calcite, dolomite, and barite, in solutions containing sodium oleate. The state of hydration of the mineral is lowest, and the flotation activity is highest, near the isoelectric point. For the optimum flotation efficiency of the collector agent (here sodium oleate) one must add the corresponding electrolyte in such an amount that the potential ζ is reduced to zero (e.g., sodium sulfate for barite, making a specific separation of barite from calcite possible). Minerals with a negative ζ potential do not show any anionic adsorption as collectors. Examples are quartz, nepheline, and aegirite, in soda solutions with 150 mg./liter sodium oleate. The surface of these minerals becomes more hydrophilic with increasing soda concentration. Sodium oleate is not adsorbed and does not influence ζ, the electrokinetic index is constantly $= 0$, and no flotation effect is possible. Sodium pyrophosphate abruptly increases ζ for apatite, calcite and dolomite, and the flotation is suppressed by an increasing hydrophilic behavior.

54. A very important application of electrodialysis combined with electrolytic effects, is that in *soil consolidation* to make clays water-stable by electrical conditioning. N. I. Maksimova, N. N. Serb-Serbina, N. I. Titkov, and A. S. Korzhuev[126] treated a bentonitic soil (with a natural base exchange capacity of 60.7 meq./100 gm.) in distilled water, with graphite electrodes, and in a 1 per cent solution of calcium chloride with iron electrodes, or a 1 per cent solution of aluminum chloride with aluminum electrodes, under a direct current potential of 30 to 50 volts, with a current density of 5 to 50 mA./cm.^{2}. The consistency of the products, P_m, as the load of plastic deformation, was tested with a conical plastometer, under variable moisture conditions (W per cent). The consolidation effect by the electrolysis is most evident in the chloride solutions with soluble electrodes, and the secondary precipitation of iron and/or aluminum hydroxide in the soil sediment on the cathode electrode, with the new-crystallization of calcite. This precipitation effect is by far more efficient in consolidating than a previous admixture of the hydroxides, or of limestone powder to the soil. At the same time the soda-soluble silica is drastically reduced, the base exchange capacity also considerably diminished.

55. E. V. Ballou[127] studied the electroosmotic flow of water in homoionic kaolinite clays as a function of the adsorbed counter-ion and the water content, but could not find a simple interpretation of the phenomena in terms of the ion hydration. The relation of the electroosmotic flow to the hydraulic permeability at variable wa-

[126] *Doklady Akad. Nauk S.S.S.R.*, **132**, 1960, 413-416.

[127] *J. Colloid Sci.*, **10**, 1955, 450-460; on the theoretical details here discussed see S. Glasstone, "Textbook of Physical Chemistry," Van Nostrand, Princeton, New Jersey, 1946, p. 1223.

ter contents, however, is in agreement with the electrokinetic theory of Helmholtz and von Smoluchowski, but not with the equations of G. Schmid[128] for fine-porous systems. The apparatus used by Ballou is shown in Fig. C. 36. The sample is packed between porous discs, E, E', of alundum (high-sintered alumina) in a Lucite tube, D, with the ends ground flat. The voltage is applied to platinum gauze screens, F, F', in the end compartments, and measured between platinum gauze screens, G, G', cemented onto the alumina discs, to determine the potential drop. The water flow was measured by weighing the overflowing liquid in the tubes H, H'.

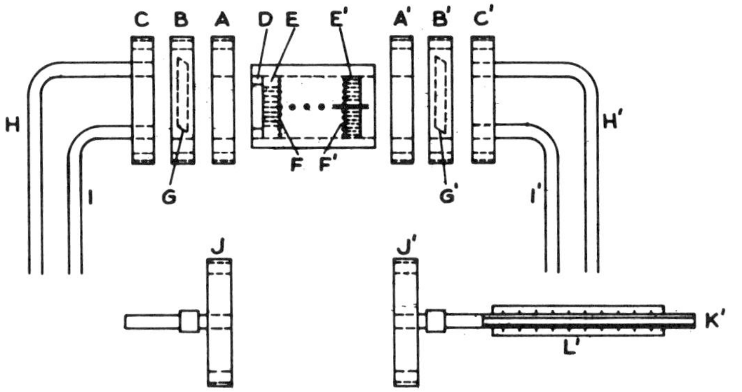

FIG. C. 36. Apparatus for measurement of electroosmotic flow rate and hydraulic permeability of kaolinite (Ballou): A, B, C, A', B', C' are Lucite blocks; D, Lucite ring; E, E', alundum discs; F, F', G, G', platinum gauze screens and electrodes; H, H', overflow tubes; I, I', drainage tubes; J, J', Lucite end blocks for permeability tests; K', "Trubore" capillary tube; L', scale.

56. A similar study of counter-electroosmosis in its action on the permeability of kaolin was observed by A. S. Michaels and C. S. Lin.[129] These authors proved that the counter-electroosmosis exerts an additional resistance to the flow of water in compacted kaolinite clay beds. Very important is the effect of fixed sodium ions on H-kaolinite which brings about a pronounced reduction in permeability. This effects is irreversible, and interpreted as caused by increased dispersity of the clay in the bed. For common flow calculations based on the elementary aspects of viscous flow laws, the effect of counter-electroosmosis is less important. Decisive, on the other hand, for the aggregation of the clay which rules the permeability, are the pH conditions and the ionic characteristics of the permeating solutions (Fig. C. 37).

57. M. Bender[130] determined in particular the ζ potentials on glass walls against a solution by determining the difference of the intensity of electroosmotic currents

[128] *Z. Elektrochem.*, **55**, 1951, 229-237.
[129] *Ind. Eng. Chem.*, **47**, 1955, 1249-1253.
[130] *J. Colloid Sci.*, **9**, 1954, 400-408.

in a rectangular closed glass cell, and from the hydrodynamic conditions caused in the same cell when an exterior electric potential is applied to it. The experimental results show positive changes in the hydrodynamic behavior, in a parabolic curve for the cell height and the particle velocity, if the pH in the medium in the cell is equal to, or larger than 7.0, approaching a straight line when the charge on the glass

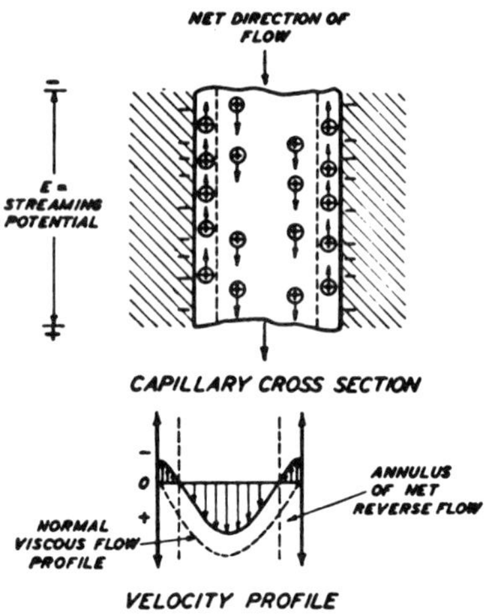

FIG. C. 37. Schematic diagram showing the counter-electroosmosis effect in flow through a fine capillary by "downstream" transport of ionic charges in the diffuse part of the double-layer (Michaels and Lin).

surface decreases and the electroosmotic currents are minimized. Electroosmotic flow as ζ potentials are given for the plane glass surface in contact with the electrolyte solution. The change is negative in 0.0001 M soda solution and decreases with decreasing alkalinity, and positive in an acidic medium (e.g., 0.002 M sulfuric acid solution).

58. From the geochemical viewpoint the action of silica hydrogels as *protective colloids* is of the highest importance, as was particularly emphasized in the recent investigation of the classical pyrite-barite deposit of Meggen in Westphalia, with its very characteristic association of these two minerals in spheroidal forms. The description of characteristic colloidal structures as given by E. Nickel[131] in the "petrified muds" or "fossil slurries" of this occurrence, demonstrates that the hydrophobic systems in the iron sulfide were under the action of protecting films of silica

[131] *Chem. Erde*, **18**, 1956, 99-136, especially pp. 123-129. On the spheroids of iron sulfide here discussed, see the previous studies of B. Doss (1912) and H. Jegurow (1897), cited in C. Doelter's "Handbuch der Mineralchemie," Steinkopff, Dresden and Leipzig, 1926, Vol. IV, Pt. 1, pp. 583 ff.

gel, which brought about the spheroidal shapes of the concretions ("coagels") of only a few microns in diameter. Silicic acid is one of the most common and most effective protective colloids in nature. Also in inorganic chemistry the role of a silica hydrosol or gel is most evident, as in the stabilization effects of tin(II) hydroxide which is observed as a compound of bright-yellow color in widely varied sizes of colloidal particles. The initially precipitated tin(II) hydroxide, as induced by an intense stirring of a mix of solutions of sodium metasilicate, sodium hydroxide, and tin(II) chloride, is white but it shows a sudden change to yellow with a characteristic X-ray diffraction diagram during treatment in the autoclave. The diagram is interpreted by H. Weber[132] in connection with color pigment characteristics in comparison with cadmium sulfide (the latter is superior in color saturation effects).

THE OPTICAL PROPERTIES OF COLLOIDAL SUSPENSIONS

59. The great interest developed in the last decade on the constitution of *opalescent* borosilicate glasses, especially in those characterized as unmixing dispersoids for manufacturing the Vycor glass type (cf. Vol. III, Section B; Vol. V, Section B) is reflected in a series of highly fascinating investigations by Russian authors.[133] Such porous glasses were especially described by S. P. Zhdanov[134] in their dispersoid characteristics and adsorption properties (cf. Vol. II, A. ¶ 404), and by M. M. Gurevich[135] in their spectral dispersion as a function of the light wavelengths. The measurements were made with a sensitive photoelectric method, in a hemispherical diffraction range (similar to the well-known Ulbricht sphere), for the basic glass with 7 per cent Na_2O, 23 per cent B_2O_3, and 70 per cent SiO_2, in different states of opalescence. The optical density

$$DD_{\text{diffr.}} = m \cdot \log \lambda - \log (k \cdot x)$$

(x being the thickness of the glass sample) is shown in Fig. C. 38, and the diffraction transmittance, τ, $= 10^{-kx/\lambda^m}$. As a function of the particle radius (r) the light flow, F, is determined by the Rayleigh equation

$$F = U \frac{(r)^6}{\lambda^4} \left(\frac{n + n_0}{n^2 + 2\, n_0^2} \right) (n - n_0)^2$$

[132] *Z. anorg. u. allgem. Chem.*, **301**, 1959, 109-112.

[133] See especially in the *Stroenie Stekla, Inst. Khim. Silikatov Akad. Nauk S.S.S.R., Trudy Soveshchaniya, Leningrad*, 1953 (Publ. 1955). See also D. I. Levin, S. P. Zhdanov and E. A. Poraĭ-Koshits, *Izvest. Akad. Nauk S.S.S.R., Otdel. Khim. Nauk* 1955, 197-207, 395-402.

[134] *Doklady Akad. Nauk S.S.S.R.*, **92**, 1953, 597-600; *Stroenie Stekla, Inst. Khim. Silikatov Akad. Nauk S.S.S.R., Trudy Soveshchaniya, Leningrad*, 1953 (Pub. 1955), 162-175.

[135] *Stroenie Stekla, Inst. Khim. Silikatov Akad. Nauk S.S.S.R., Trudy Soveshchaniya, Leningrad*, 1953 (Pub. 1955), 202-206. See also *Zhur. Tekh. Fiz.*, **23**, 1953, 986-993.

(n being the refractive index of the particle material, n_0 that of the medium). By introduction of the Cauchy dispersion formula,

$$(n - n_0) = B/\lambda^2 - B_0/\lambda^2$$

and

$$F = U'(r)^6/\lambda^8$$

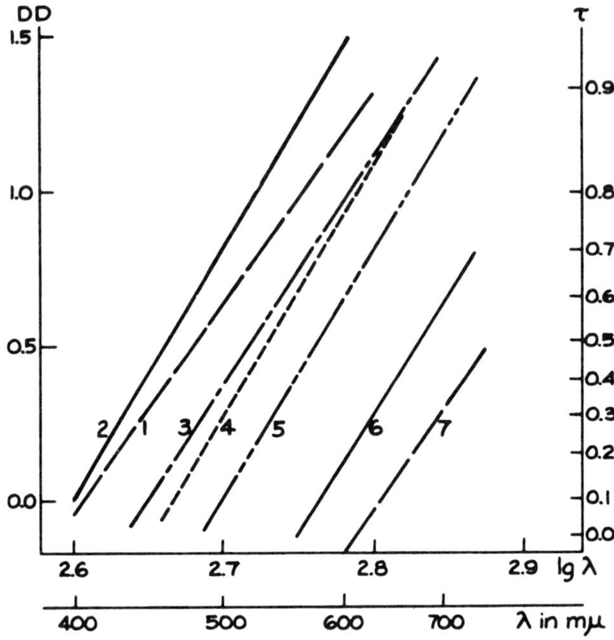

FIG. C. 38. Functional relation between the optical density of diffraction, *DD*, and the transmittance, τ, with the wavelength in the diffraction phenomenon of opalescent borosilicate glasses (Gurvich).

is shown the tremendous influence of the particle size and of the wavelengths on optical diffraction phenomena. D. I. Levin[136] demonstrated the very strong influence of chemical composition (alkali or lead contents) on the Rayleigh diffraction effects, and also that of depolarization of diffracted light, by such changes in composition. In addition, the development of the opalescence phenomenon in sodium borosilicate glasses by thermal treatment is particularly striking. In Fig. C. 39 is shown the rapid increase in the relative intensity (I that at the experimental temperature, I_0 that at room temperature) in the range from 550° to 700°C., followed by a steep slope.

[136] *Stroenie Stekla, Inst. Khim. Silikatov Akad. Nauk S.S.S.R., Trudy Soveshchaniya, Leningrad,* 1953 (Pub. 1955), 198-200. On Rayleigh diffraction in glasses, see A. Winter, *Verres et Réfractaires,* 16, 1962, 273-281.

The dispersion effect is here quite anomalous, since in place of the factor λ^{-4} in the Rayleigh equation now appears a factor λ^{-7} to explain the effects observed. The structural heterogeneity of the opalescent glasses is most evident in this anomaly in the understanding of the crystallite theory (cf. Vol. II, A. ¶ 203 ff.), by particles of the order of magnitude of 10^2 Å. We will come back to these problems in connection with Debye's theory of light scattering in liquids and glasses. The crystallites developed up to 700°C. are then liquefied, corresponding to the steep slope of the curve in Fig. C. 39. The low depolarization coefficient, ϱ, of the non-opalescent boro-

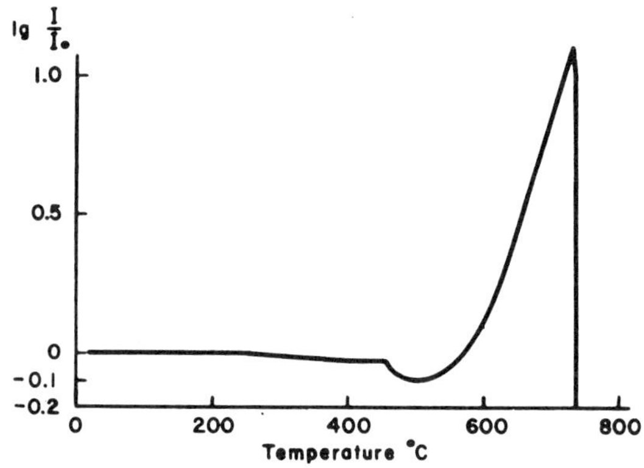

Fig. C. 39. Change in intensity of scattered light from a sodium borosilicate glass as a function of temperature (Levin).

silicate glass in the range up to 500°C. is increased rapidly to a maximum at 500°C. but then it remains constant in the temperature range of the opalescence. The coefficient is at a minimum when the intensity ratio I/I_e (I_e being the intensity of diffraction of ethyl ether at room temperature as a basis of reference) is equal to $10^{2.4}$. It should be noted, as a comparison, that $I_0/I_e = 18 = 10^{1.26}$.

60. E. A. Poraĭ-Koshits, D. I. Levin, and N. S. Andreev[137] extended the previous studies on the effects of thermal treatment of sodium borosilicate glasses by comparing the diameters of the pores of the opalescent glass after leaching the alkalies by acids; they plotted the results as a function of the temperature in the range of opalescence. The gradual increase in intensity of the scattered light at constant temperature is a consequence of the development of chemical heterogeneities. For high intensities, the permeability coefficients were determined for the optimum wavelength of 650 mμ, with the photoelectric spectrophotometer. The leached glasses, in addition, were subject to low-angle X-ray diffraction (down to 10′ as angles of incidence)

[137] *Izvest. Akad. Nauk S.S.S.R., Otdel. Khim. Nauk* 1955, No. 1, 31-39; No. 3, 395-402; 1956, No. 3, 287-293.

in a vacuum camera, with strictly monochromatic Cu Kα radiation. In agreement with Gurevich's data, the coefficient m in the exponential formula for the light scattering intensity was above 4, mostly between 6 and 8. The pore diameter calculated was 50 to 250 Å., increasing the longer the glass samples were thermally exposed and the higher the temperatures (experimental temperatures between 600° and 650°C.). The decreasing intensity of the X-ray scattering in the low angle range is explained by the lower scattering power of structures with larger pores.

61. It is highly instructive to compare the light scattering phenomena in opalescent glasses with the analogous optical phenomena in classical *unmixing dispersoids* of the type of system KCl–NaCl, on which an investigation by E. Scheil and H. Stadelmaier[138] gave important experimental details in the study of synthetic water-clear monocrystals. The measurement of the diffraction-transmittance was made with a photoelectric turbidimeter. The relative intensity $I/I_0 = \exp(-kd)$ (d is the thickness) made it evident that there is no additivity of the extinctions [the extinction defined as the negative logarithms of (I/I_0)] for the crystalline solutions as a function of their composition (see Fig. C. 40). The spectral dispersion

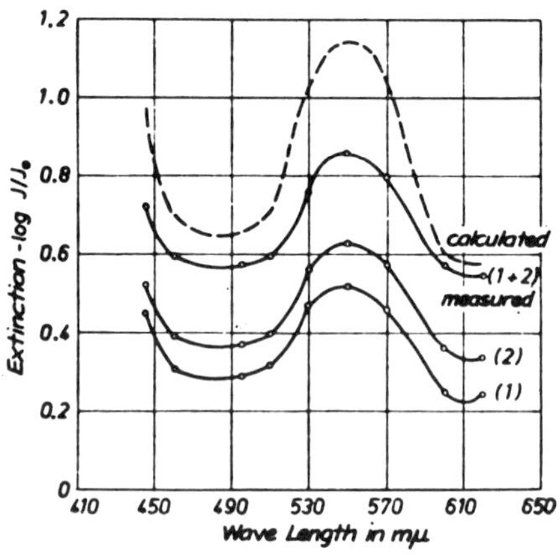

FIG. C. 40. Extinction coefficient as a function of wavelengths in a crystalline solution of 65 mol. per cent NaCl, and 35 per cent KCl, at 305°C. (Scheil and Stadelmaier).

indicates at the same time the changes in extinction coefficients as a function of wavelengths and of the time of thermal annealing, for a constant composition (Fig. C. 41). A peculiarity of these curves is the relatively higher transmittance in the

[138] *Z. Metallk.*, **43**, 1952, 227-235. Very similar problems are studied [as in the precipitation of barium sulfate or in the cleaning of river water with alum by C. K. Sloan, *J. Phys. Chem.*, **59**, 1955, 834-840] by using the light scattering method over an angle range of about 70° to 0°; this is especially helpful in following aging phenomena of the precipitates.

blue which cannot be explained from the Rayleigh equation, and a maximum in the yellow-green. The size and structure of the diffracting particles evidently is relatively unchanged, in spite of such different degrees of turbidity. The extinction is proportional to—and therefore a direct measure of—the amounts of unmixed material, and valuable conclusions on the kinetics of the unmixing process can be drawn from the optical data.

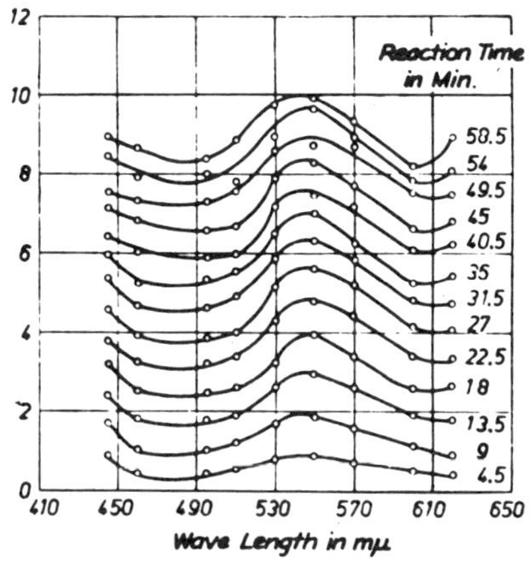

FIG. C. 41. Change of extinction constants, k, as a function of the wavelengths for different periods of thermal annealing at 305°C. (Scheil and Stadelmaier).

62. Among the different instruments for routine quantitative measurements of light scattering phenomena, those constructed by L. Prod'homme[139] were designed for study of the diffusivity in optical glasses, based on the principle of a double-beam (beam separator) photometer with a Lummer cube, for relative diffusivity constants between 10^{-7} and 10^{-3}. Best silica glass shows (variable with different wavelengths) $R = 1$ to 4×10^{-6}, commercial Crown Glass 4 to 20×10^{-6}, Flint Glass $= 10$ to 50×10^{-6}, and synthetic plastics (Plexiglas) $= 1$ to 5×10^{-4}. The existence of relatively high molecular diffusivities in glasses might be explained by anisotropic molecular aggregations (Krishnan effect), or by incidental structural heterogeneities (Raman, 1927). Prod'homme[140] determined the diffusivity $R_{20°} = I_{90°}/I$ (for purest benzene as a basis of reference) $= 10^{-5}$ (for 546.1 mμ), and tabulated the quotients $R_{gl.}/n_{gl.}$ as a function of the refractive index n. This ratio for silica glass is equal to 0.17×10^{-5}; for an extra-heavy Flint Glass, it is equal to 2×10^{-5}, in a simple linear relatio of $I/I_{benz.}$ versus $(n^2 - 1)^2$ (Fig. C. 42). The anomaly that sodium borosilicate glass as well as optical glass of the borosilicate Crown class show deviations from this rule is in agreement with D. I. Levin's results (see above), that the intensity of the

[139] *Rev. opt.*, **32**, 1953, 615-618.

[140] *Compt. rend.*, **245**, 1957, 300-302.

diffracted light can be considerably changed (nearly doubled) if the relative content in R_2O is decreased by 50 per cent. These facts speak strongly in favor of the presence of chemical heterogeneities in the glass constitution.

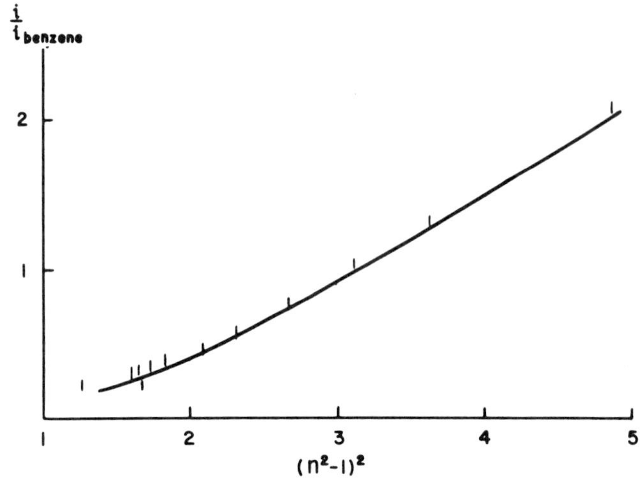

FIG. C. 42. Diffusivity of different optical glasses, plotted versus parameter $(n^2 - 1)^2$ (Prod'-homme).

63. Great progress has been made in modern instrumentation in the field of light scattering in the low-angle range (down to 0.05°) and up to 140°, as seen in the high-resolution scattering and recording photometer developed by W. H. Aughey and

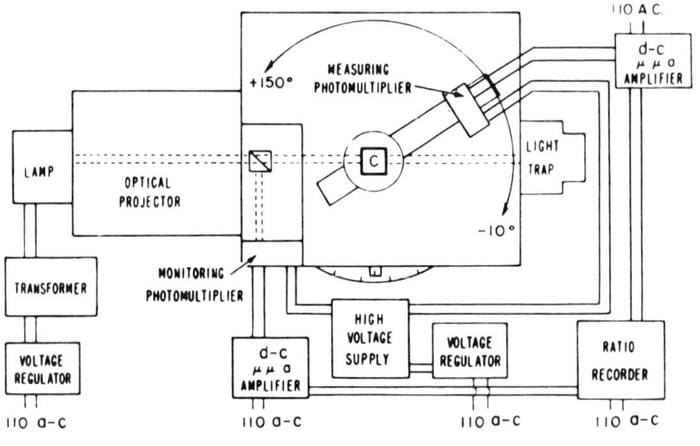

FIG. C. 43. Schematic sketch of the absolute light scattering photometer of McIntyre and Doderer.

F. J. Baum[141] for particles in the size of large molecules, up to colloid particles near 1 μ, and even larger heterogeneities with a radius range up to 100 μ for the lowest angles and high resolution. The instrument must therefore also be furnished with the possibility of variations in the intensity of the light in wide limits (ranging in the instrument mentioned here by a factor of 10^8). Especially for solutions of high polymers another instrument of these general characteristics was described by D. McIntyre and G. C. Doderer,[142] with automatic recording and determination of the ratio of the intensities of the scattered and the incident light. Whereas Fig. C. 43 gives a general impression of the construction of this latter instrument, Fig. C. 44

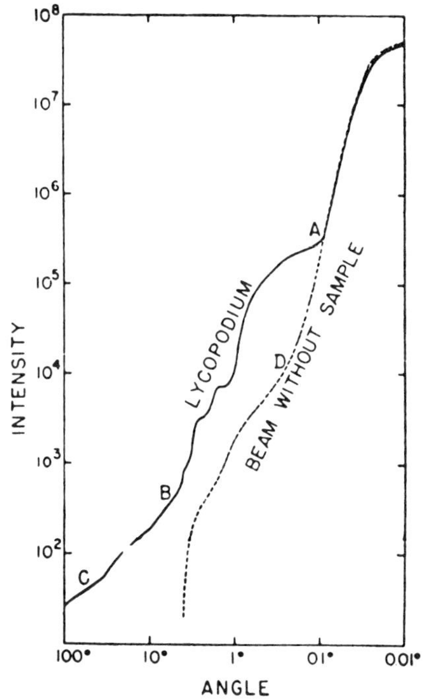

FIG. C. 44. Light scattering curve of lycopodium powder in low-angle range (Aughey and Baum).

gives a light scattering curve of Lycopodium powder as recorded in Aughey and Baum's instrument, demonstrating how the course of the curve from A to B can be used to calculate the diameter of the particles.

64. The theoretical background for the light scattering method applied to silicate solutions or in colloidal silicate suspensions is given by the fundamental investi-

[141] *J. Opt. Soc. Am.*, **44**, 1954, 833-837.

[142] *J. Research Natl. Bur. Standards*, **62**, 1959, 153-159.

gations of P. Debye, first with R. V. Nauman,[143] and then with E. W. Anacker.[144] In the first series of experiments, carefully purified (filtered through sintered stainless steel and/or platinum filters) sodium silicate solutions (with the ratio SiO_2/Na_2O ranging from 0.5 to 3.75) were measured for their light diffusivity by determining the turbidity at an angle of 90° to the incident beam, then the angular scattering pattern, and the viscosity, as functions of the concentration. The turbidity is expressed here by

$$\tau = \frac{32\,\pi^3}{\lambda^4}\left(\frac{\mu_0^2(\mu - \mu_0)^2}{n}\right)$$

in which $n\ (= N \cdot c)$ is the number of particles per cm.³, μ_0 is the refractive index of the solvent (H_2O), and μ is that of the solution. It is most important to know that up to $SiO_2/Na_2O = 2.0$ the solutions of sodium silicates do not give any indication of an aggregation into colloidal particles, the solutions being similar to those of sugar. Above 2.0 the turbidities are higher than in sugar solutions, with perhaps some aggregation, but in every case very large particles are entirely absent. For $SiO_2/Na_2O = 3.75$ the solutions show increasing molecular weights up to 5,000 or 10,000. The viscosity curves show a similar relation, and at high concentrations the viscosity rapidly increases because the molecules are very near to one another, although even then no flocculation of large colloidal particles occurs.

65. Debye's studies with Anacker concern the *shapes* of micelles, as in solutions of detergents like *n*-tetradecyl and hexadecyl trimethylammonium bromide, in solutions of potassium bromide of variable concentration. Here the relationship existing between the intensity of the scattered light, the angle of observation, Θ, and a length dimension characteristic of the given micelle model, L, is written as a function of a parameter

$$X = \frac{2\,\pi\,L}{\lambda} \cdot \sin\frac{\Theta}{2}$$

By plotting I as a function of Θ for various chosen L, and of $I_0/I_{(180-\Theta)}$ (called the dissymetry coefficient, or for short "dissymetry") for any given pair of angles symmetrical about 90° as a function of L, one can determine that value of L which characterizes the shape of the micelles. For *n*-tetradecyl trimethylammonium bromide (in 0.083 M. KBr solution) the dissymmetry is constant (nearly = 1.02) corresponding to particles no larger than 10 per cent of the wavelength of the exciting radiation, or no larger than 33 mμ. For *n*-hexadecyl trimethylammonium bromide, however, in a 0.178 and 0.233 M. KBr solution, there is an increasing dissymmetry with increasing concentration in the detergent and increasing molecular weights. The calcu-

[143] *J. Phys. Chem.*, **51**, 1947, 18-32; *Ibid.*, **55**, 1951, 1-9.

[144] *J. Phys. Chem.*, **55**, 1951, 644-655.

lation shows that in this case the presence of spherical or disc-like particles is excluded. Rodlike shapes fit best to the observed scattering data, and a model is given for the micelles of such elongated particles of circular cross-section, with polar heads determining the detergent properties lying on the periphery, the hydrocarbon tails filling the interior. The ends of the rods may act as the polar heads do, but the model is distinctly different from an ordinary cylindrical model in that the hydrocarbon tails would be perpendicular to the symmetry axis and not parallel to it. The intensity of the scattered light from such a rod (of negligible diameter) would be

$$I = \frac{1}{X} \int_0^{2X} \frac{\sin u}{u}\, du - \left(\frac{\sin X}{X}\right)^2$$

with

$$X = \frac{2\pi}{\lambda} L_{\text{rod}} \sin \frac{\Theta}{2}$$

whereas for the spherical shapes the intensity would be

$$I = \left[\frac{3}{X^2} (\sin X - X \cos X)\right]$$

X in relation to L_{sphere} (the diameter of the sphere). The coincidence of the theoretical and experimental parameters I versus log $[(2\pi/\lambda) L_{\text{rod}} \sin(\Theta/2)]$ is practically ideal for the rod shape model.

66. S. A. Greenberg and D. Sinclair[145] applied Debye's theories of light scattering to the polymerization process of silicic acid suspensions using a photometer constructed by B. A. Brice, M. Halwer, and R. J. Speiser[146] for the determination of molecular weights of polymers. The suspensions were prepared by mixing highly purified sodium metasilicate and ammonium acetate solutions. The measurements of the polymerization by the determination of turbidity and dissymmetry were also combined with measurements of the degree of depolarization, using horizontally (h), vertically (v), or unpolarized (u) incident light with the depolarization parameters:

$$\varrho_v = H_v/V_v \qquad \varrho_h = V_h/H_h$$

$$\varrho_u = (H_v + H_h)/(V_v + V_h) = H_u/V_u$$

(H and V referring to the intensities of the horizontal and vertical components). Examples of the usefulness of the optical methods may be seen in Fig. C. 45, a diagram of the turbidity, $\tau_{\text{corr.}}$, increasing as a function of the reaction time, for different concentrations of the sodium silicate solution in equivolume mixtures with a

[145] *J. Phys. Chem.*, **59**, 1955, 435-440.
[146] *J. Opt. Soc. Am.*, **40**, 1950, 768-778.

3 per cent ammonium acetate solution; and in Fig. C. 46, the dissymmetry $(I_0/I_{(180-\Theta)})$ versus Θ, for two solutions showing very good agreement of the experimental data with Debye's theory of light scattering for spherical particles, and also in agreement with the depolarization measurements for particles up to 150 μ in diameter. The

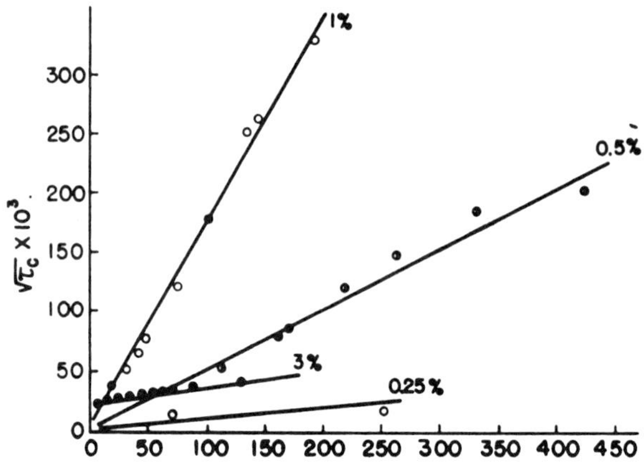

FIG. C. 45. Turbidity of silica solutions as a function of reaction time, for different concentrations of sodium metasilicate solutions mixed with equivolume 3 per cent ammonium acetate solution (Greenberg and Sinclair).

particle diameter as a function of the reaction time is seen from Fig. C. 33 for different temperatures, from a 1 per cent sodium silicate solution mixed in equivolume ratio with a 3 per cent ammonium acetate solution. The significance of these results in leading toward an understanding of the polymerization process is quite evident (cf. C. ¶ 31 ff.).

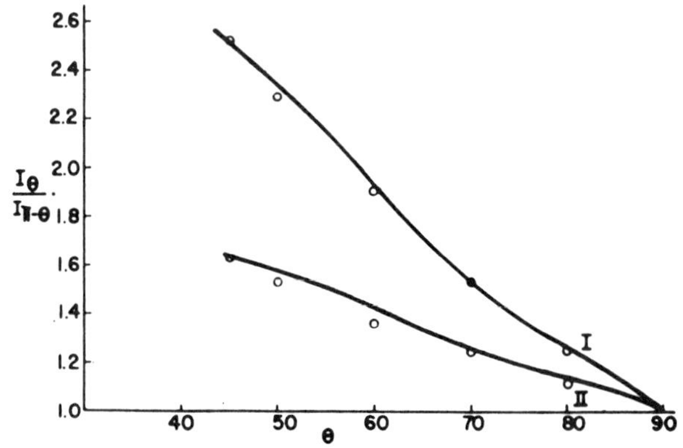

FIG. C. 46. Dissymmetry and diffraction angle of two silica solutions showing coincidence of the theoretical curve for spherical particles (solid curves), and experimental data (circles) (Greenberg and Sinclair).

67. In this respect the light-scattering method is equally useful in silicate research as it is for the investigation of organic polymerization problems, such as those studied by A. Oth and J. Bisshops[147] for cellulose derivatives, or for polyvinyl and polyacrylnitrile polymerizates. Especially for the polysilicic acids, R. K. Iler[148] further demonstrated how the initial number-average molecular weights (200), determined from viscosity measurements, agree with those (325) determined by the light scattering method in Debye's and Nauman's esperiments. The gel point is reached at molecular weights of about 6,000. Evidently, the polysilicic acid particles are the units which finally combine to build up the three-dimensional framework of the hydrogel. This framework becomes infinite at the gel point, even before all of the units have definitely joined the structure, thus explaining the relatively low molecular weight observed at the gel point (cf. C. ¶ 39). "Ludox" silica sols were also studied by G. B. Alexander and R. K. Iler[149] in a comparison of the results from Debye's light scattering method with electron microscopy and the BET nitrogen adsorption method. The light scattering data are generally somewhat larger than the average weight particle diameters estimated from electron micrographs, and the BET nitrogen data are the smallest. However, all these experimental diameters are larger than those calculated from the theoretical relation of particle growth in the volume X containing G grams of SiO_2 per liter withdrawn at time t, X being $= 3 \times 2.303 \cdot \log_{10}(D/D_0)$ (D being the diameter at time t, D_0 that at $t = 0$).

68. The light scattering conditions of the three-dimensional framework gels were studied by J. Yudelson and B. Rabinovitch[150] using a photometer of the Brice-Halwer type (see above), of high sensitivity, and for systems of polyglyceryl sebacate and polyglyceryl hydrophtalate as polymerizates. It is very instructive that such gels, although they are rigid in the bulk, still behave in light scattering as liquids, following the equations given by M. von Smoluchowski (1908) referring to the thermal compressibility and the optical dielectric constant of the medium with anisotropic scattering units. This may be compared with the results of P. Debye and A. M.

[147] *Congr. chim. ind., Brussels*, 1954, No. 3; *Ind. chim. belge*, **20**, Special Issue, 1955, 426-429. The measurements were made with the photometer-turbidimeter of A. Oth and V. Desreux, *Bull. soc. chim. Belges*, **63**, 1951, 261-264, which was calibrated with "Ludox" silica sol and polystyrene, methyl-ethyl ketone, and toluol. A similar highly sensitive photometer is that constructed by Sh. Tomozane, Sh. Sano, and Sh. Suetoshi, *Asahi Garasu Kenkyu Hokoku*, **3**, 1954, 196-214, for liquids. The Rayleigh constant for benzene was established by Oth and Desreux as a basis reference $= 48.4 \times 10^{-6}$ at $\lambda = 435.8$ mμ (cf. *J. Polymer Sci.*, **10**, 1953, 551-557). For very important comparison of the molecular weights of polystyrene as a standard basis, determined by viscosimetry, light scattering, and other methods, see comments by H. P. Frank and H. Mark, *Ibid.*, **17**, 1955, 1-20.

[148] *J. Phys. Chem.*, **57**, 1953, 604-607.

[149] *J. Phys. Chem.*, **57**, 1953, 932-934.

[150] *J. Phys. Chem.*, **60**, 1956, 1512-1518.

Bueche[151] with solids like Lucite, polystyrene, and glass, with large dissymmetries and a residual intensity at large scattering angles attributed to small heterogeneities. Yudelson and Rabinovitch are of the opinion that this residual intensity is the scattering of the pure solid, measured in polyglyceryl hydrophthalate, whereas in polyglyceryl sebacate the scattering behavior more resembles that described by Debye and Bueche. In spite of the tremendous complications involved by foreign contaminations on the radiation envelopes for condensed systems in the determination of the molecular properties of solids by the light scattering method, the conclusion on the analogy of the optically perfectly pure amorphous solids with liquids in the scattering behavior is of great principal importance.

69. The *color* of metals dispersed in colloidal suspensions was reexamined under the aspects of G. Mie's theory of diffraction polarization (1908) by J. Turkevich, G. Garton, and P. C. Stevenson,[152] combined with direct measurements of the particle sizes in the electron microscope. An excellent agreement could be established between the experimental data and those particle sizes calculated from the absorption curves of the suspensions. This establishment of the full correctness of Mie's theory for spherical particles, e.g., in gold sols, is a most valuable basis for other problems of colloid-colored suspensions occurring in silicate systems, such as in glass of the gold ruby type. For typically *anisotropic* particles, say of the tabular crystal habitus of clay minerals, the majority of the methods of determination of electro-optical birefringence (Kerr effect) proved to be very applicable, as demonstrated by A. Kahn and D. R. Lewis[153] for sodium montmorillonite in aqueous suspensions of a concentration of 0.05 per cent of the solid. The rate of decay on removal of the electrostatic field according to H. Benoit,[154] is determined by the exponential equation

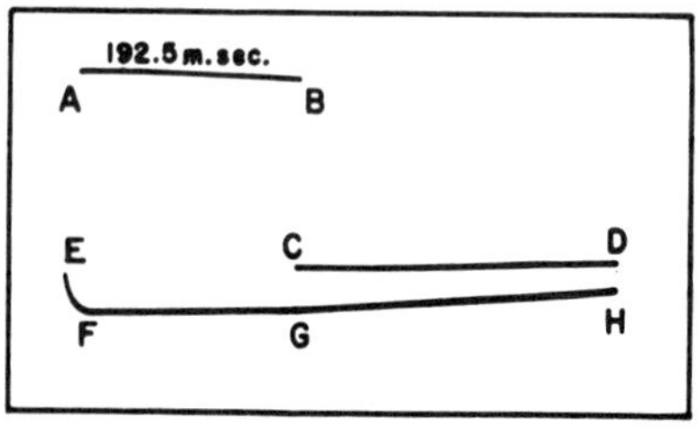

FIG. C. 47. Electrooptical birefringence for a sodium montmorillonite suspension with coarse particles (above 140 mμ) (Kahn and Lewis). ABCD voltage pulse, EFGH birefringence curve.

[151] *J. Appl. Phys.*, **20**, 1949, 518-525; *J. Chem. Phys.*, **18**, 1950, 1423-1425.

[152] *J. Colloid Sci.*, Suppl. 1, 1954, 26-35.

[153] *J. Phys. Chem.*, **58**, 1954, 801-804.

[154] *Ann. phys.*, [12] **6**, 1951, 561-609; see also C. T. O'Konski and B. H. Zimm, *Science*, **111**, 1950, 113-116.

$\Delta n/\Delta n_0 = \exp(-6\,Dt)$, as a function of time, t, and the rotational diffusion constant, D, if n_0 means the initial birefringence at time t_0 and n that at the time t. The parameter D can be correlated with the viscosity of the suspension, η, and the semimajor axis, a, of the oblate-spheroid-shaped particle (the semiminor axis being very much smaller than a), in a formula previously developed by J. Perrin:

$$a^3 = \frac{3}{32} \cdot \frac{k\dot{T}}{\eta\, D}.$$

A typical decay curve is shown in Fig. C. 47, as recorded by an electronic dual beam cathode ray oscillograph. Pulses are in the range of 2 to 200 msec. For monodisperse fractions, varying in the equivalent spherical radii from 140 down to about 10 mμ, the D values increase with finer particle sizes (Fig. C. 48) from 0.23 to 29 sec.$^{-1}$,

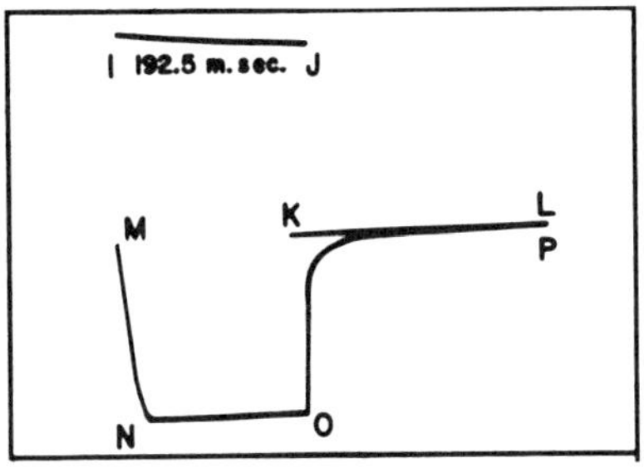

FIG. C. 48. Electrooptical birefringence effects for a sodium montmorillonite suspension with finest particles (7-23 mμ in diameter) (Kahn and Lewis). IJKL voltage curve, MNOP birefringence curve.

and a from 1230 to 250 mμ. Not only the decay curve, but also the electro-optical birefringence "rise curve," $S - T$ in Fig. C. 49, gives valuable information on the electric properties of the montmorillonite particles, although Benoit's equations on combined permanent and induced dipoles in the particles did not fit.

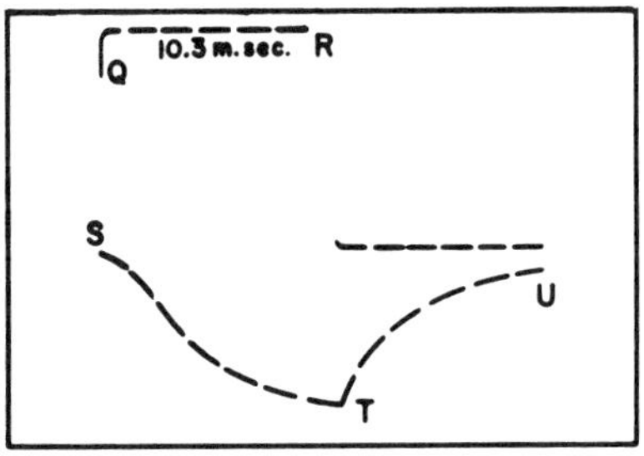

FIG. C. 49. Electrooptical birefringence effects for finest sodium montmorillonite particles, same as in Fig. C. 48, but with expanded time scale (Kahn and Lewis). Distance QR corresponds to 10.3 milliseconds.

70. The anisotropy of the molecular units in polymerized phosphates is indicated by changes in the electric conductance whether the solution is stirred or stagnant. U. Schindewolf[155] could demonstrate that flowing solutions of polyphosphates (e.g., with 1000 units in the chains) show a distinct orientation effect in tubes if the liquid is propelled by nitrogen. If the electrostatic field lines are perpendicular to the tube axis, the conductance is smaller than in stagnant solution; if they are paralleel to the axis, the conductance is higher. The difference $\Delta\varkappa = \varkappa_{\text{stagn.}} - \varkappa_{\text{flowing}}$ is increased with the flow speed to a definite limit. With increasing concentration in the polyphosphate and a constant flow speed, $\Delta\varkappa$ is decreased. The difference $\Delta\varkappa$ depends on the length of the polymerized chain molecules; the shorter the chain, the lower $\Delta\varkappa$.

71. The same phenomenon appears in a solution of sodium thymonucleinate or of soap micelles. Analogous anisotropies in ceramic clays cause anomalies in extrusion structures or pressed bodies and, therefore, excessive shrinkage phenomena by particle alignment; the great importance of this was shown for halloysite by Br. F. Hampel and I. V. Cutler[156] (cf. Vol. V, Section B). Excellent models of such alignments may be observed in gels with oriented micelles of (ionotropic) cellulose glycolate, with included crystal nuclei of lead oxalate originated by an ion exchange of diffusing lead cations, as described by H. Thiele and H. Krönke.[157] Many biological phenomena, such as the oriented growth of hydroxyl apatite in fibrils, may be analogous in their origin. In boron trioxide glass fibers, very interesting crystallite orientation effects were observed by B. K. Banerjee[158] with colloidal inclusions of gold, silver, or platinum halides. In this case, however, the X-ray diffraction diagrams showed very distinct fiber textures in aging specimens exclusively of the boron trioxide fibers, whereas in borax or silica glass no such orientation effects are developed. Pure boron trioxide glass fibers do not indicate any orientation type. The crystallites are caused by a kind of "devitrification" of B_2O_3, and not an orientation of the dispersed noble metal particles which, however, act as "intruder nucleation catalysts." The crystallites of the devitrified material are not perfectly oriented, they show a scattering in a cone around the fiber axis.

[155] *Naturwissenschaften*, **40**, 1953, 435; *Z. physik. Chem.* (*Leipzig*), **1**, 1954, 129-133, 134-141; *Z. Elektrochem.*, **58**, 1954, 697-702; *Intern. Union Pure and Appl. Chem., Sect. for Inorg. Chem., Colloq. Münster, Westphalia* 1954, 271-277 (with K. F. Bonhoeffer). K. Heckmann, *Naturwissenschaften*, **40**, 1953, 478; *Z. physik. Chem.* (*Leipzig*), **9**, 1956, 318-337; *Ibid.*, **20**, 1959, 42-52; *J. Colloid Sci.*, **13**, 1958, 266-272 (with K. G. Götz) studied the anisotropy of electrical conductance of soap solutions. On the theory of these phenomena see also M. Eigen and G. Schwarz, *Ibid.*, **12**, 1957, 181-194, and *Z. physik. Chem.* (*Leipzig*), **19**, 1959, 286-314.

[156] *J. Am. Ceram. Soc.*, **36**, 1953, 30-34.

[157] *Naturwissenschaften*, **42**, 1955, 389.

[158] *J. Am. Ceram. Soc.*, **36**, 1953, 294-298.

72. Interesting problems of *colored minerals* have often arisen from studies on blue rock salt, which is the most striking occurrence, or from studies on intensely colored fluorite. We here recall the previous studies of M. Savostyanova,[159] who made a remarkable attempt to solve the problem of blue rock salt from Mie's theory, applied to the system Na–NaCl. Spectrophotometric measurements of the absorption curves of natural blue salt, and of synthetic crystals colored by an additive method in liquid sodium, or of a salt obtained from yellow crystals by the action of heat and pressure under the influence of light, are interpreted as a confirmation of F. Cornu's early hypothesis of the colloid-dispersed state of sodium metal particles as the coloring pigment. A reexamination of this hypothesis by F. V. Chukhrov,[160] however, could not agree with the colloidal nature of the color centers, but gave evidence that radioactive K^{40} causes the color changes. V. M. Vinokurov[161] investigated deep-blue rock salt crystals from the carnallite horizons of the Solikamsk salt deposits, measuring with a highly sensitive pH-meter the changes in pH of the solutions of pure NaCl and of the blue crystals in purest water. Whereas the pure NaCl showed pH values between 5.15 and 5.40, the solution of the blue crystals reached a pH between 5.35 and 5.91, corresponding to a concentration in neutral sodium, Na^0, of 0.7 to 5.8×10^{16} atoms/gm. The magnetic susceptibility, χ, measured by a highly sensitive method of J. Joussot-Dubien, gave a change of from -0.510×10^{-6} for colorless salt crystals to -0.463×10^{-6}, or a paramagnetic susceptibility share of the sodium atoms of $\Delta_\chi = 0.047 \times 10^{-6}$, and for one sodium atom $= 2.13 \times 10^{-27}$. The concentration in Na^0 would then be $= 22.1 \times 10^{18}$ atoms/gm. The discrepancy from the value of the order of magnitude 10^{16} speaks for the wide limits of errors of both series of measurements.

73. In spite of the rich experience on the radioactive origin of the highly variable colors in fluorite, such as that of the classical occurrence of Wölsendorf, Upper Bavaria,[162] the nature of the coloring pigments is again and again questioned. A much detailed investigation by P. D. Allen[163] gave (besides valuable data on the spectral-analytical determination of accessorial elements, among which strontium is particularly important) precision data of the refractive index at 20°C., varying by 0.00111. Fluorites with a principal light absorption at long waves are

[159] *Z. Physik*, **64**, 1930, 262-278.

[160] In "Kolloidy v zemnoĭ Kore" (Colloids in the Earth's Crust), Acad. of Sci. U.S.S.R., Moscow, 1953.

[161] *Zapiski Vsesoyuz. Mineral. Obshchestva*, **87**, 1958, 504-507. The method of J. Joussot-Dubien, B. Lemanceau, and A. Pacault, *J. Chim. Phys.*, **54**, 1956, 198-205, is based on the changes in frequency of an oscillator circuit as a measure of susceptibility.

[162] We mention here only the studies of Luise Goebel, *Sitzber. Akad. Wiss., Wien, Math. naturw. Kl.*, I, **39**, 1930, 373-391, with excellent micrographs of radioactive haloes, and of ultramicroscopic (dark-field) evidence given of colloid-suspended particles.

[163] *Am. Mineralogist*, **37**, 1952, 910-930.

characterized by abnormally low refractive indexes, and vice versa. Those have higher refractive indexes which show the chief absorption in the short-wave range. The anomalously low refractive indexes are explained by deficiencies in fluorine, which are electrostatically compensated by neutral colloidal calcium, the anomalously high indexes by excess calcium ions in structural holes. Purest synthetic calcium fluoride has the cube edge unit $a_0 = 5.4629_5 \pm 0.0001_0$ Å. but no changes of a_0 are found in the colored crystals. The maximum of absorption in the short wave-range is produced by the color centers. With increasing sizes of the light-scattering centers the colors change from green to blue to violet. The chief light-scattering centers are the noncharged calcium atoms which in the yellow-brown colored fluorites are coagulated to colloidal particles with a change of the color to blue. For further details, the reader is referred to the extensive bibliography given by Allen on the characteristic luminescence phenomena of colored fluorite, in accord with R. W. Pohl's theory (1936).[164]

74. The color centers caused by defects in the structures are not resolved by ultramicroscopic methods, and therefore it is not justifiable to draw, as often attempted, a direct parallel between the typically colloidal-colored suspensions of metals in organic solvents and the so-called "metal clouds" or "pyrosols" in fused metal salts. All the systems of the type metal-salt melt have proved to be true *molecular* solutions as was confirmed by the recent investigations of M. A. Bredig,[165] for the system Na–NaCl, or by G. G. Urazov and A. S. Karnaukhov[166] for the systems Pb–$PbCl_2$ and Cd–$CdCl_2$, in which either the thermal analysis gave evidence of true solubility, or of the formation of a subhaloide, such as CdCl, which is stable, however, only in narrow temperature limits, wherein two immiscible liquid phases occur. Bredig found a rapid increase in the mutual metal-salt solubility with increasing temperature in the alkali metal-alkali halide systems, and complete miscibility of the melts

[164] We must also refer to many special investigations on the structural defects in colored alkali and alkaline earth halide crystals, which are compilated and discussed in the reports of G. Isetti, *Periodico Mineral.* (*Rome*), **26**, 1957, 227-241; *Ibid.*, **28**, 1959, 19-29, and of S. M. Stishov, *Zapiski Vsesoyuz. Mineral. Obshchestva*, **89**, 1960, 166-179. This author discusses, e.g., the colors of spodumene (cf. Esther W. Claffy, *Am. Mineralogist*, **38**, 1953, 919-931) in connection with the replacement of cations with different electrostatic charges, in this case of $Mn^{2+} \rightleftharpoons Li^+$. E. D. Belyankina, *Voprosy Petrog. i Mineral., Akad., Nauk S.S.S.R.*, **2**, 1953, 77-87, found by spectroscopic determinations not only rubidium and cesium enriched in amazonite, but also lead, and Stishov who speaks of 0.05 to 0.09 per cent PbO as possible accessories in this mineral, emphasizes the exchange of $Pb^{2+} \rightleftharpoons K^+$ as the probable source of the defect coloring of amazonite, not taking into account the rubidium contents. For an easily readable introduction to the rich experimental and theoretical material of R. W. Pohl we recommend the study of his book "Einführung in die Optik," Springer, Berlin, 1944, 2nd/3rd ed., especially pp. 190 ff.

[165] *J. Am. Chem. Soc.*, **77**, 1955, 307-312, for sodium-sodium halides, and for the corresponding systems of potassium and cesium: *Ibid.*, 1454-1458.

[166] *Doklady Akad. Nauk S.S.S.R.*, **96**, 1954, 535-538.

to be expected above 1000° to 1100°C. in contrast with the systems of alkaline earth metals and their salt melts which do not show such a high degree of solubility. In near relation to this group of particular dispersion phenomena is the "contamination" of alkali halide crystals with platinum, if grown from a melt in platinum crucibles. A yellow fluorescence of the sodium chloride crystals was in this case described by W. J. Fredericks and A. B. Scott[167] with appreciable absorption at 204 mμ similar to that of the $(PtCl_6)^{2-}$ complex ion in aqueous solution. The highest contamination found was 0.8 p.p.m. by weight.

75. The most intensely investigated color phenomenon in the silica system is that of *smoky quartz* on which considerable data on optical measurements is available in the literature, especially of the visible spectrum and in the ultraviolet. We mention among the more recent studies those of N. E. Vedeneeva and L. G. Chentsova,[168] especially on thermoluminescence and bleaching relations, but also on changes in the electrostatic field. The authors established the identity of the color and of the luminescence centers. The spectrographic examination of the contaminations revealed a great many possibilities for the chemical nature of these centers, in the alkalies, magnesium, calcium, aluminum, manganese, iron, copper, titanium atoms present in distinct amounts. The maximum absorption at $\lambda = 440$ to 460 mμ (for the extraordinary wave) is particularly stable in the natural crystals of smoky quartz, less in the crystals colored artificially by X-ray irradiation of thermally bleached samples. Chentsova and Vedeneeva distinguished three kinds of color centers: (1) those bleached by increasing the temperature up to 250°C.; (2) those which disappear at 251°C. constant; and (3) those which are stable to higher temperatures. The isotherms for the centers (2) show an exponential type, characterizing centers of two different life times, independent of whether the crystals were naturally colored or artificially stained. They are, however, different in the relative concencentrations in the centers. The origin of the color centers is assumed to be the removal of electrons from the crystal structure of quartz, fixed in levels of variable depths, as determined by the presence and the nature of foreign elements. These are able to migrate into the free interstices of the framework structure of low-quartz, with an open diameter of 2 Å. The color centers are therefore analogous to the F-centers in irradiated (subtractively colored) alkali halide crystals. Lithium and sodium present in the channels are not centers for the coloring by the irradiation, and also do not activate luminescence.

76. Freshly irradiated quartz is highly thermoluminescent. The presence of aluminum atoms as the principal color centers was established by measurements of

[167] Paper presented to the Meeting of the Am. Chem. Soc., Division Phys. and Inorg. Chem., San Francisco, 1958, April 13/18, No. 63 Q.

[168] *Doklady Akad. Nauk S.S.S.R.*, **87**, 1952, 197-199; *Trudy Inst. Krist. Akad. Nauk S.S.S.R.*, 7, 1952, 159-190; 191-202; *Ibid.*, **11**, 1955, 195-199.

the paramagnetic resonance effects after irradiation with neutrons, by J. H. E. Griffiths, J. Owen, and I. M. Ward.[169] Especially Mary C. M. O'Brien[170] gave evidence that the hyperfine structure of the paramagnetic resonance spectrum shows that each color center is associated with an aluminum atom in the site of a silicon atom in the tetrahedral framework, with the unpaired electron in a localized orbital on one of the nearest oxygen atoms (Fig. C. 50). This basic model was fully confirmed

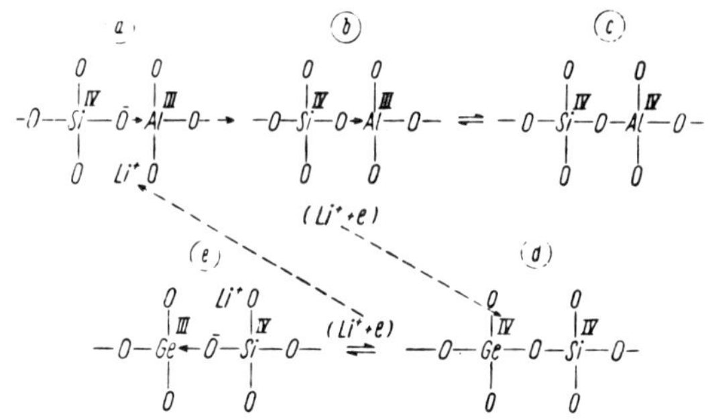

FIG. C. 50. Schematic representation of impurity centers in quartz (O'Brien): (*a*) substitutional aluminum accompanied by interstitial alkali before irradiation; (*b*) aluminum color center formed, electron on aluminum; (*c*) electron is on oxgyen (most of time). Aluminum has charge similar to a substitutional silicon; (*d*) substitutional germanium before irradiation; (*e*) germanium color center formed after acceptance of lithium, and electron during formation of (*b*) and (*c*). After thermal treatment or light-bleaching, the lithium and electron return to aluminum and regenerate (*a*) with bleaching of the color centers which change back to (*d*) (Cohen).

by the extensive investigations of A. J. Cohen,[171] especially the anisotropy of the color center effects. The basic fact is that the anisotropic color center at 460 mμ is assigned to aluminum, but always accompanied by the presence of sodium and/or lithium to compensate for the electrostatic neutrality. This fact is most evident from the quantitative determination by spectral analysis of the contaminating elements in smoky quartz, in agreement with the theory of "stuffed" structural isomorphism of quartz with eucryptite, emphasized by M. J. Buerger[172] through the structural analogy of Si[SiO_4] in quartz and LiAl[SiO_4] in eucryptite. One is thus able to explain the introduction of the elements lithium and sodium coupled to the aluminum in quartz. The aluminum here *substitutes* for the silicon and thus develops a real color center. The lithium or sodium ions find their places in the channels open for them in the structure of the quartz. The anisotropy of the center is evidently the position

[169] *Nature*, **173**, 1954, 439-440.

[170] *Proc. Roy. Soc.* (*London*), A **231**, 1955, 401-404.

[171] *J. Chem. Phys.*, **25**, 1956, 908-914; *Ibid.*, **28**, 1958, 401-405; *Am. Mineralogist*, **43**, 1958, 58-68.

[172] *Am. Mineralogist*, **39**, 1954, 603 ff.

of the absorption bands at 460 mμ for ε, and at 485 mμ for ω. The dichroism, π, as the ratio of the optical densities for the ε/ω rays is stronger in the natural quartz than in those bleached and irradiated with X-rays, or in synthetic quartz (cf. Fig. C. 51).

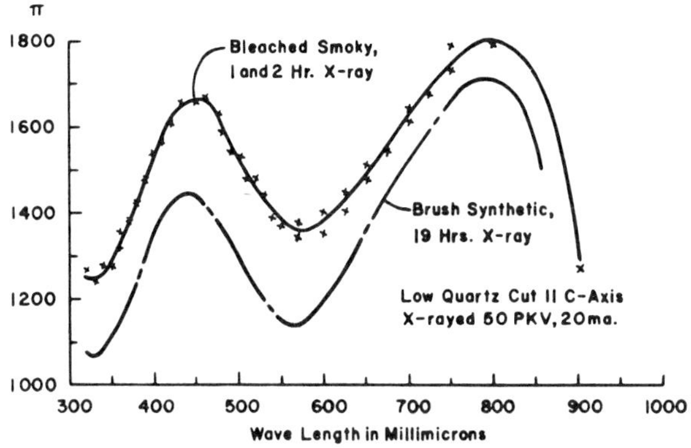

FIG. C. 51. Comparison of the anisotropy, π, of natural quartz, following various times of X-ray irradiation, and relative anisotropy of a natural and a synthetic quartz (Cohen).

77. There is another characteristic color center at 275 mμ in synthetic quartz which is connected to *germanium* substituting for silicon in the framework structure of low-quartz. Germanium-doped synthetic crystals show it just as do the natural germanium-containing crystals. The anisotropy of these centers is also evident, with the maximum absorption for the electric vector oriented perpendicular to the c-axis, whereas parallel c the effect is poorly resolved in two bands. This germanium center is bleached more rapidly than the other centers present in smoky quartz crystals in which it is commonly associated with that at 460 mμ for aluminum centers.

78. For the germanium centers the model (Fig. C. 52) previously derived by O' Brien and then by J. H. Anderson and J. A. Weil[173] from the paramagnetic resonance spectrum, with hyperfine splitting observed in oscilloscope recordings, is valid. Germanium substitutes for silicon atoms located on a twofold symmetry axis, surrounded by a (slightly distorted) tetrahedron of oxygen atoms. The alkali atoms which are indispensable in their association with aluminum are located interstitially in the c-axis channels, at positions indicated by α and β in Fig. C. 52. The lithium and sodium atoms have isotopes (Li^7 and Na^{23}) with spins 3/2 to account for the paramagnetic resonance effects, whereas the hyperfine structures from interaction with germanium (Ge^{73}) nuclei (with 9/2 spins) are so few that their effect cannot easily be detected experimentally.

[173] *J. Chem. Phys.*, **31**, 1959, 427-434.

79. As one of the most important results of the investigations of Cohen (with E. S. Hodge),[174] the very distinct specificity of some of the "impurities" in the crystal structure of natural or artificially stained smoky quartz, during their growth must be emphasized. Entirely nonspecific is the replacement, however, of silicon by germanium in a zonal growth of the crystals. Germanium is uniformly distributed over the whole crystal substance. When germanium is present, the aluminum appears in the zone of the *c*-axis, [0001], as well as a substitute for silicon (see above), but

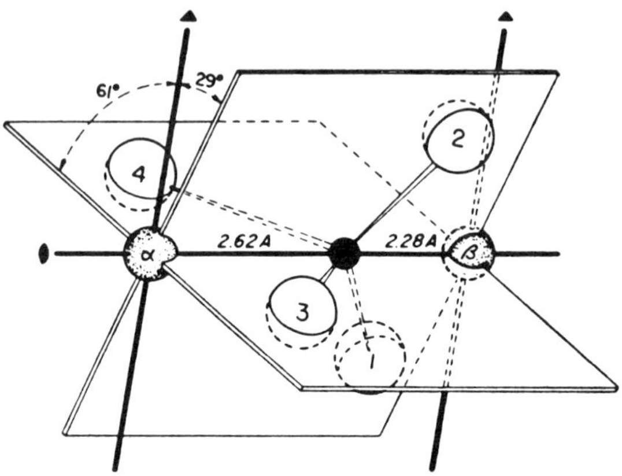

FIG. C. 52. Open circles indicate positions of oxygen anions surrounding one silicon cation (solid circle) in β-quartz (Anderson and Weil). Trigonal and digonal symmetry axes are marked. Shaded circles indicate possible sites for interstitial alkali cations.

(as was more recently found) aluminum also occurs in interstitial positions, together with lithium and sodium, always following the aluminum on the channels of the structure. The substituted occurrence of aluminum is coupled with that of germanium in distinct growing zones, whereas the interstitial aluminum (plus the ionic charge balance-giving alkali atoms) are incorporated in other zones of the growing crystals. The major portion of the aluminum is substitutional, and able to develop color centers by irradiation in different concentrations and different growth zones. If no germanium is present, aluminum is interstitially deposited in (0001) but becomes substitutional when germanium is present. Rhombohedral growth is particularly essential for interstitial aluminum, as seen from Fig. C. 53, which shows the zonal distribution of colored and noncolored zones in synthetic quartz crystals. The interstitial aluminum can be "exsolved" to colloidal pyronephelite particles visible in the Tyndall light phenomenon and in the dark-field microscope by thermal treat-

[174] *Phys. and Chem. Solids*, **7**, 1958, 361-362; *Ibid.*, **13**, 1960, 321-325. See also the instructive observations of J. A. Bastin and E. W. J. Mitchell, *Am. Mineralogist*, **46**, 1961, 1227-1244, on light scattering in heat-treated synthetic quartz.

ment above 573°C. and slow cooling to avoid intense cracking of the crystals. The exsolvate is an aluminosilicate of lithium (and/or sodium), most probably of eucryptite type. Color centers are exclusively related to the substitutional aluminum, and not to the interstitial type. Color distributions in different cross section areas and growth zones are, therefore, characteristically developed (as seen in Fig. C. 53), and can be systematically studied in synthetic quartz crystals with controlled contents of aluminum and germanium.

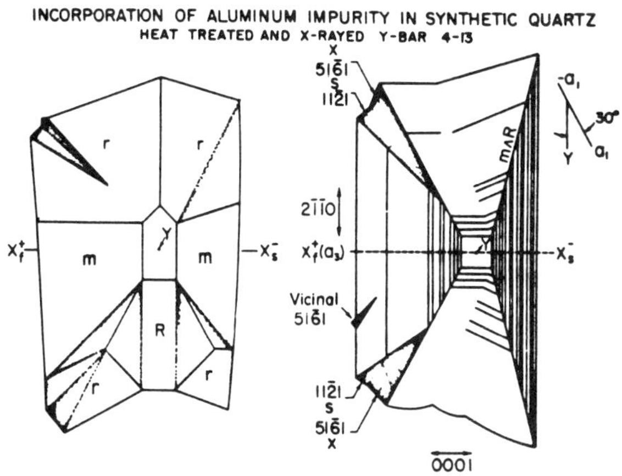

Designation	Face	Index	
m	Hexagonal Prism	$(10\bar{1}0)$	
r	"Positive" Rhombohedron	$(10\bar{1}1)$	mr = 38° 13'
z or R	"Negative" Rhombohedron	$(01\bar{1}1)$	mR = 66°52'
s	Trigonal Bipyramid	$(11\bar{2}1)$	ms = 37°58'
x	Trigonal Trapezohedron	$(51\bar{6}1)$	mx = 12° 1'

Substitutional Al impurity incorporation x>s>r

FIG. C. 53. Sketch of substitutional and interstitial distribution of aluminum in Y-bar specimen of synthetic quartz, cross-section, as elucidated by thermal treatment and X-ray irradiation (Cohen).

80. The situation is much more complex in the problems of the color centers in *amethyst.* Rich material on the optical absorption spectra of natural blue, and of rose quartz has been collected by J. von Vultée and J. Lietz,[175] especially concerning the presence of titanium as a coloring pigment in the form of ultramicronic crystallites of rutile included (cf. C. ¶ 5). In rose quartz also a dilute coloring by mangani-compounds is most probable, causing the absorption maximum at about 500 mμ.

[175] *Neues Jahrb. Mineral. Monatsh.*, 1956, 49-58. The optical phenomenon of pleochroism in amethyst is described in *Fortschr. Mineral.*, **35**, 1957, 36 ff., and *Neues Jahrb. Mineral. Monatsh.*, 1956, 217-232; *Naturwissenschaften*, **43**, 1956, 443-444 (with W. Münchberg). Optically, amethyst is not different from quartz, the red-blue "pleochroism" is merely a physiological effect. V. A. Frank-Kamenetskiĭ, *Silikat Tech.*, **12**, 1961, 306-308, speaks of "microisomorphism" of the "contaminating" elements in quartz; cf. A. ¶ 230*, Footnote[364].

J. Lietz and W. Münchberg[176] further studied the thermal destruction of the amethyst color. Radioactive γ-irradiation restores the violet color in some of the discolored crystals, whereas in others it only develops a smoky-gray hue. Initially colorless zones are also not changed by intense irradiation. In addition to the color centers which correspond to maxima at 350 and 530 mμ, in some crystals others also appear which are typical of the smoky quartz. J. Lietz and M. P. Hänisch[177] discovered recently an interesting development of an intense brown color of smoky quartz type, or amethyst coloring, by electrolysis, especially the migration of deeply colored zones from the anode in an electrostatic field (200 to 1200 volts) above 760°C. Starting from the anode, this colored "front" is built up advancing parallel to the *c* axis, leaving behind it a pale-yellow colored zone. The potential color centers responding to electrolysis (abbreviated as ε-centers) differ in their much higher thermal stability (up to 1000°C.) from those centers developed by X-ray and γ-irradiation (called γ-centers), but both centers deliver the same absorption spectrum. Electrolysis also restores the color of amethyst bleached at 400° to 500°C. For the origin of the ε-centers a probably spontaneous removal of electrons is characteristic, and the crystals are stained by ionic conduction reactions, starting from the anode.

81. A. J. Cohen[178] confirmed in general the observations of J. Lietz and his coworkers, adding an adsorption band at 950 mμ, and in some amethyst specimens also those at 225 and 266 mμ. The 340 mμ band is anisotropic, with 360 mμ for the electrical vector vibrating parallel *c*. Amethysts from Brazil often show, after bleaching at elevated temperatures, a green color which is due to a separate mineral (perhaps chlorite). The assignment of specific impurities by means of spectral analysis to the different absorption bands is difficult. The 540 mμ band is perhaps most analogous to the F-centers in alkali halides; that at 360 mμ may be caused by the substitution of electropositive atoms for silicon in the structure, with a concomitant introduction of additional impurities, namely of atoms in interstitial positions (nickel may act in this way, with different valencies). These would also explain the F-like centers and coupled intensity ratios, electroneutrality being preserved by an appropriate number of oxygen-vacancies. Halide anion impurities may account for the bands at 225 and 266 mμ, associated with cation-vacancies. The fine-lamellar *d–l* twinning which is so characteristic of amethyst, is genetically connected with the presence of a contamination associated with one or more color centers, and may assist spiral growth

[176] *Neues Jahrb. Mineral. Monatsh.*, 1957, 25-33.

[177] *Fortschr. Mineral.*, **37**, 1959, 72; *Naturwissenschaften*, **46**, 1959, 67-68. For further interesting experiments on the migration of cations and the formation of color centers at the electrolysis of quartz plates cut perpendicular to the *c*-axis, see H. H. Pfenninger and F. Laves, *Ibid.*, **47**, 1960, 276. See also the detailed Inaug.-Dissert. Univ. of Zürich, 1961, 130 pp., by H. H. Pfenninger. A detailed report of J. Lietz and M.R. Hänisch on their experiments is given in *Cursillos y Confer. Inst. Lucas Mallada, Madrid*, **7**, 1960, 83-90.

[178] *Am. Mineralogist*, **41**, 1956, 874-891.

in nucleation to relieve inner strains. Precision crystal parameter measurements for detecting such effects with many natural and synthetic quartz crystals (including smoky quartz and amethyst) show systematic variations, not only in the ratio c_0/a_0, but even more pronounced in the "increment ratio"

$$\left(\frac{\Delta a}{a} \Big/ \frac{\Delta c}{c}\right),$$

as a function of c_0/a_0, depending on the ratio of interstitial to substitutional replacement of silicon. The fact that color centers characteristic of substitutional aluminum are not present in amethyst is most likely due to the fact that the aluminum impurity is mainly interstitial and not in the proper environment for the formation of color centers. Interstitial impurities expand the c-axis direction markedly, whereas substitutional replacement causes expansion of both axes. The increment ratio therefore is a good indicator for the degree of interstitial and substitutional replacement of silicon by aluminum.

82. The color development in *glasses*, as far as it is not of the ionic type (to be discussed in Vol. II, B. ¶ 44 ff.), is in many respects analogous to the defect crystal structure phenomena discussed here. We therefore find a near relation between disperse colors and nucleation effects of crystallization for which a special example is the influence of noble metal particles on the texture development in boron trioxide glass fibers (see C. ¶ 71). The laws of the growth of gold particles in sols by the reduction of a gold chloride solution were established in the classical experiments of R. Zsigmondy and P. A. Thiessen (1925), and reexamined by J. Turkevich, P. C. Stevenson, and J. Hillier[179] for the sizes between 20 and 150 mμ. The particle size distribution, however, does not follow, as was previously assumed, a Gauss error curve, but is determined by the kinetics of the reduction process, namely the nucleation stage and the growth stage. For the first step a nucleation theory can be based on the reduction reaction over an intermediate complex compound between the gold ion and the organic agent (sodium citrate or sodium acetonedicarboxylate solution); for the second step a typical exponential growth relation

$$D = D_0 \sqrt[3]{(W_{\mathrm{Cl}} + W_{\mathrm{Au}}) / W_{\mathrm{Au}}}$$

(W_{Au} the weight portion of the gold nuclei, W_{Cl} the weight portion of gold in $AuCl_3$) is observed as in hydroxylamine hydrochloride–$AuCl_3$ solution (Fig. C. 54). This relation follows a curve with an inflection point for the systems with sodium citrate as the reducing complex, but in the sodium acetonedicarboxylate solution there is no inflection point, although in the later stages the course is practically the same as

[179] *J. Phys. Chem.*, **57**, 1953, 670-672. See also R. Zsigmondy and P. A. Thiessen, "Das kolloidale Gold," Akademische Verlagsges., Leipzig, 1925.

for the citrate system. The nucleation process for the total number of particles with diameters greater than D follows the law

$$N(t) = \int_{D(t)}^{\infty} n(D) \, . \, dD$$

and the definite integral corresponds to the shaded area under the distribution curve, as indicated in Fig. C. 54. The lower integration limit is then given by the law of growth noted above, and $\log D(t) = \log D_0 + c \cdot t$.

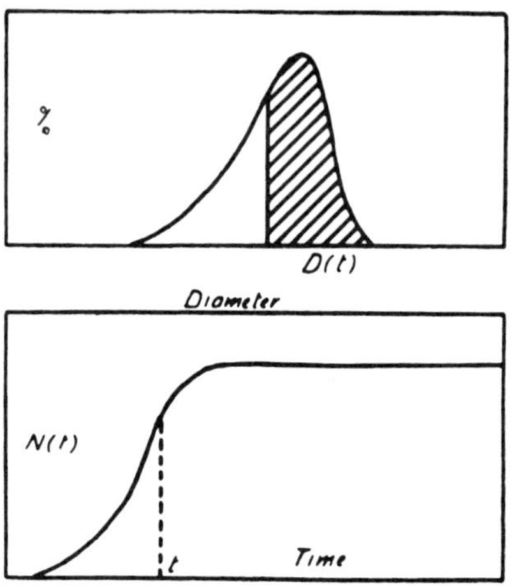

FIG. C. 54. Relation between particle size distribution (above) and nucleation curve (below) in formation of colloidal gold (Turkevich, Stevenson, and Hillier).

83. As colloidal gold sols are the best examples of nucleation and particle growth of dispersed metals in a suspension medium, we may extend our discussion on the importance of nucleation mechanisms for supersaturation and phase changes, such as phase unmixing phenomena or the precipitation from mixed electrolyte solutions, in which the time of nucleation and the particle growth play the decisive role. This is particularly important in metallurgical problems and in silicate technology, many special examples of which we will discuss later in this text. V. K. La Mer[180] has given an excellent introduction to the general treatment of nucleation in phase transitions which we can only outline broadly here. His discussion is based on previous work, especially on that of M. Volmer[181] dealing with phase boundaries of metastable states of matter. On a strictly thermodynamic and mechanical-statistic basis it is possible

[180] *Ind. Eng. Chem.*, **44**, 1952, 1270-1277; *Ceram. Age*, **61**, 1953 (3), 23-30; see also G. M. Pound, *Ind. Eng. Chem.*, **44**, 1952, 1278-1283.

[181] See his book "Kinetik der Phasenbildung," Dresden and Leipzig, Steinkopff, 1939; Ann Arbor, Michigan, Edwards, 1945, 220 pp.

to determine the process of formation of what La Mer calls the "embryos" and nuclei. The example of well-defined suspensions of sulfur, from the molecular-dissolved state after nucleation to distinct sulfur hydrosols, is a second model for the analogous reactions and nucleations in silicate systems, chiefly the type of metal-silicate systems. The so-called ruby glasses are the classical representative of dispersed colloidal-colored pigments (Siedentopf and Zsigmondy, 1905). The newer types of photosensitive glass were discovered by S. D. Stookey (1949). Finally, colloid-dispersed metal particles have been used in nucleation of devitrification centers in the production of the so-called "Pyroceram" and "Vitrokeram" (cf. Vol. III, A. ¶ 39 f.).[182]

84. Among the many types of colloidal phenomena to be discussed here we shall first mention the relation between the formation of color pigments and the property of microporosity of glasses. M. A. Bezborodov[183] described a copper-ruby staining on simple contact of a copper metal fragment with silica glass heated in a graphite crucible to 1800° or 1900°C. The color was concentrated in a surface layer, whereas the deeper portions of the glass remained colorless. The particles have sizes of 10 to 20 mμ (determined by electron microscopy), they are therefore of a size comparable to the larger units of the polyhedral framework in the structure of silica glass (e.g., 540 units of [SiO_4] groups), at a viscosity of the order of magnitude of 10^{13} at 1100°C., and 2×10^2 at 2000°C. as determined previously by M. P. Volarovich and A. A. Leont'eva (1940). The smaller copper particles are, therefore, trapped in the structural cages of the silica glass after diffusion into the glass in a molecular dispersion. The density of the glass (2.20) corresponds to a number of empty cavities in the glass, which is 17 per cent higher than in the structure of quartz. Bezborodov generalizes that the staining of glasses by the dispersion of metal particles is a consequence of their microporosity, which is not only a property of silica glass, as in the Vycor type leached from borosilicate glass, but also occurs in silicates, borates, and phosphates. It is not more logical to restrict the staining process to base exchange reactions; for example, in the classical exchange of silver for sodium ions, entirely alkali-free glasses also show the same silver-staining phenomenon.

85. The problems involved with the metals existing in the atomic state in glasses were especially discussed by W. A. Weyl[184] who, however, emphasizes that with

[182] We mention here the important studies of W. Hinz and P.-O. Kunth, *Silikat Tech.*, **10**, 1959, 119-122; *Ibid.*, **11**, 1960, 506-511; also, see B. Löcsei, *Ibid.*, **10**, 1959, 589-595, with discussion remarks of W. Hinz and Fr. Wihsmann, *Ibid.* p. 596. See also the fascinating reports of J. Robredo, *Verres et Réfractaires*, **16**, 1962, 225-230, 282-289.

[183] *Doklady Akad. Nauk S.S.S.R.*, **95**, 1954, 349-351. See the previous opinions of S. D. Stookey, *J. Am. Ceram. Soc.*, **32**, 1949, 246-249, who postulated the solubility of noble metals in glasses in the ionic form, before their reduction precipitates the free metal, as the Zsigmondy gold sol condensation reaction also demonstrates.

[184] *J. Phys. Chem.*, **57**, 1953, 753-757.

respect to the noble metals, glasses can respond to low concentrations (below 0.01 per cent) in the same general fashion as do crystals. Glasses may contain metals frozen-in from the vapor state by rapid cooling. Chilled glasses of this kind are colorless, and also do not show any fluorescence when exposed to ultraviolet light. If they are, however, annealed, the metal particles grow by condensation to colloidal sizes, and then show fluorescence. Particularly well studied as a model is the behavior of silver, dissolved as the chloride in sodium chloride, and then reduced by hydrogen gas at only 150°C. to a strongly fluorescent system. The original silver ions are changed to the atoms by trapping an electron, the proton entering the electron cloud of neighboring chlorine anions. The same happens with a silver glass in which silver replaces the sodium cations and is reduced to silver atoms causing a strong fluorescence in ultraviolet light ($\lambda = 3650$ Å. as an optimum). The formation of coherent silver films (mirrors) on the surface of silver-containing glasses was observed by G. E. Rindone.[185] This belongs to the group of phenomena in metal-silicate systems, in this latter case with alkali-free alumino- borosilicate glasses and in the presence of water vapor. Although these films, which show an electrical conductivity, may be formed at room temperature in the dark, elevated temperatures and light accelerate the process. In such experiments the occurrence of an aggregation state of the metal atoms, intermediate between an atomic and a metallic character, is of interest.

86. The reduction of lead dissolved in glasses in the ionic state by hydrogen was studied by V. I. Shelyubskiĭ,[186] who observed excellently developed cubic crystals of the metal in the black surface film on such a reduced glass, with forms ((100)) ((111)) ((110)), of 10 to 100 mμ in diameter, growing to larger sizes. The connection between optical and electrical conductivity properties in such glasses is illustrated in Fig. C. 55. The same problem was investigated by H. H. Funk[187] in lead glasses reduced on their surface with hydrogen at 300° to 500°C. The glasses were of silicate composition, to which systematically increased additions of boron trioxide were introduced. The corresponding reduction of bismuth borate glasses brings about products with very peculiar electrical resistance properties being anomalously low

[185] *J. Soc. Glass Technol.*, **37**, 1953, 124-128. It is impossible to give here a complete discussion of the abundance of experimental facts which support the theoretical ideas of W. A. Weyl and G. E. Rindone. We may only mention here the phase-separation phenomenon under the action of platinum particles observed in the electron microscope by G. E. Rindone and R. J. Ryder, *Glass Ind.*, **38**, January 1957, 31-51, in sodium metaphosphate melts before crystallization occurs. The concentration of the colloidal platinum may be as low as 0.002 to 0.006 per cent (cf. Vol. III, Section A). Glasses with still lower platinum contents are gray colored but not devitrified.

[186] *Stroenie Stekla, Inst. Khim. Silikatov Akad. Nauk S.S.S.R., Trudy Soveshchaniya, Leningrad,* 1953 (Pub. 1955), 219-223.

[187] *Glastech. Ber.*, **31**, 1958, 269-272.

in certain compositions (cf. Vol. II, A. ¶ 130). X-ray diagrams show that the color pigments are really lead and bismuth metal particles. The coarser the structure of the surface layers, the higher the conductivity. No amorphous semiconductor effects are detected, also no Hall effect, nor any changes in the resistance by a magnetic field.

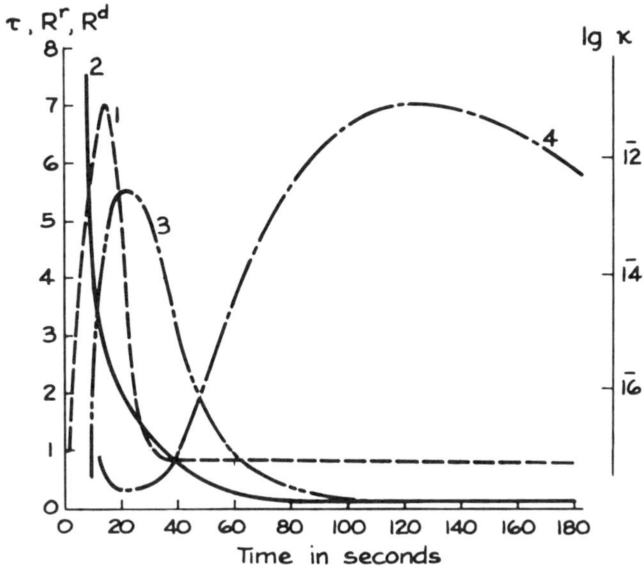

FIG. C. 55. Change of electrical surface conductance, ϰ, and of optical characteristics of reduced lead films from lead silicate glass (with 30 per cent PbO), as a function of time (Shelyubskiĭ). Curve (1) refers to log ϰ; (2) to light transmittance × 10; (C) to mirror reflectivity, $R^r \times 0.05$; (4) to diffuse reflectivity, $R^d \times 100$.

87. G. E. Rindone and J. L. Rhoads,[188] and R. J. Ryder and G. E. Rindone[189] studied the optical properties of platinum particles suspended in alkali-lead silicate, borate, and phosphate glasses into which the noble metal was introduced as $PtCl_4$. We have already mentioned that stable gray-colored platinum glasses are observed in sodium-lead silicate glass compositions, which show a distinct Tyndall scattering phenomenon of high sensitivity, since even 1 p.p.m. can easily be detected in this way. In phosphate glasses, ionic colors (yellow-orange prevail over gray, with a stabilization effect of lead oxide for the ionic solubility of platinum, if the oxygen potential is kept high. This stabilization is also observed for borate and phosphate glass compositions. The optical transmittance of the gray glasses is rather constant

[188] *J. Am. Ceram. Soc.*, **39**, 1956, 173-180.
[189] *J. Am. Ceram. Soc.*, **41**, 1958, 415-422.

in the visible spectrum range, but rapidly cut off at 420 mμ to 350 mμ for the yellow glasses, with strong absorption bands in the ultraviolet and the infrared above 1.5 μ. Particularly strong is the stabilization effect of oxidizing agents, such as cerium dioxide and sodium nitrate, in platinum-containing glasses, which tend to retain the platinum in the ionic state. Rindone and Rhoads observed quite analogous phenomena if palladium or rhodium are introduced in silicate, borate, and phosphate glasses, showing yellow-orange colors if the metal is dissolved as ions and gray colors if it is in the metallic-colloidal state, with corresponding typical absorption bands in the visible spectrum. For rhodium, brown colors are typical if it is present in the trivalent state; for palladium, yellow colors are typical if it is present in the divalent state. Concerning the role of the alkalies influencing the solubility of platinum in glasses, it is characteristic that potassium glasses can retain five times more platinum than lithium glasses, and the sodium glasses can retain twice as much as the lithium glasses, before showing the color. In mixed alkali silicate glasses, the combination of potassium and lithium resembles sodium glass alone in the minimum platinum concentration that gives a gray color. Sodium and lithium, or potassium combined, do not change the minimum platinum concentration from that observed in lithium silicate and potassium silicate glasses.

88. Red-colored silver glasses with an adsorption band at 501.7 to 502.5 mμ in borate and phosphate compositions, analogous to certain organosols of silver, were studied by Bh. K. Banerjee.[190] In calcium-sodium or lead silicate glasses such a red silver color is only possible in the presence of some chloride anion, as E. Forst and N. J. Kreidl[191] emphasized. Banerjee identified the silver metal particles by the strong X-ray interference line at 4.059 Å. (coincident with the strongest line of cristobalite, but diffuse). W. A. Weyl[192] attributes the red color to the irregular shape of the silver particles grown in a melt with alkali halide present. In copper ruby this coincidence is not so evident. In the electron diffraction patterns of several industrial glasses containing devitrification centers of cristobalite and tridymite as given by L. A. Afanas'ev,[193] the presence of the characteristic copper lines could not be established. In their place, those of copper alloys with tin appear to be formed by

[190] *J. Am. Ceram. Soc.*, **37**, 1954, 288-290. H. Tober, *Glastech. Ber.*, **34**, 1961, 456-460, studied the red-staining of glasses by AgCl, Ag_2SO_4, etc. under similar conditions. On the application of Mie's theory of colloid silver and the colors evolved in its dispersoids, see E. Wiegel, *Z. Physik*, **136**, 1954, 642-653.

[191] *J. Am. Ceram. Soc.*, **25**, 1942, 278-280.

[192] "Coloured Glass," Monograph of the Society of Glass Technology, Sheffield, England, 1951, p. 378.

[193] *Stroenie Stekla, Inst. Khim. Silikatov Akad. Nauk S.S.S.R., Trudy Soveshchaniya, Leningrad*, 1953 (Pub. 1955), 224-226. The presence of tin for making copper ruby glass is *not* absolutely necessary, as was shown recently by A. Ram, S. N. Prasad, V. K. Vaish, S. S. Verma, and V. G. Upadhyaya, *Central Glass & Ceram. Research Inst. Bull. (India)*, **6**, 1959, 34-35.

the addition of tin dioxide to the copper ruby glass batch compositions as a stabilizer of the color. The intermetallic compounds CuSn and $Cu_{20}Sn_6$ were thus confirmed, in addition to some lines of cuprous oxide. The identification of copper metal and of copper oxides was previously made by E. Zschimmer and L. Riedel (1929). It is still not quite clear how K. Beyersdorfer[194] identified the crystallites observed in copper ruby glass as the metal by electron microscopic observation (sizes 10 to 30 mμ, in single crystals up to 200 mμ). The identification was probably made owing to their morphological resemblance to the coarse light-microscope-visible particles in copper aventurine glass. S. F. Brown and F. H. Norton,[195] in an extensive investigation on copper-red glazes, also identified the pigments with copper metal by model systems in a water-soluble sodium silicate melt. From such a melt a fine-dispersed residue is obtained by leaching which gives the X-ray diagram of the pure metal. In every case the multiple play of oxidation and reduction cycles (especially if carbon monoxide is present in the kiln atmosphere) is essential for production of the best commercial copper pigments, with nuclei formed between 800° and 1250°C. and grown to colloidal particles at about 650°C. (see Fig. C. 56). If the growth ("striking") reaches sizes that are too large, the color is changed to muddy-red and even black, with spherical particles of macroscopic sizes, up to 0.5 mm. In place of tin, Brown and Norton used silicon carbide powder as an addition to the glaze mixes. The complex origin of copper ruby glass was studied in detail by A. Ram and his school; they came to the conclusive result that Cu^+ ions have no coloring effect, Cu^{2+} ions color *deep* blue, whereas cuprous oxide is the characteristic pigment of the ruby glass. Its disintegration according to the scheme $Cu_2O \rightarrow Cu + CuO$ at higher temperatures explains the X-ray results of Zschimmer and Riedel (1929).

[194] *Silikat Tech.*, **3**, 1952, 119-125. Very coarse crystallizations of a similar type may also occur in certain glasses and glazes with copper-silver pigments, as described by O. S. Levi, *Am. Ceram. Soc. Bull.*, **34**, 1955, 119-121, or in the famous "oxen-blood" glazes of Chinese ancient porcelain manufacturing, as observed by J. M. Matthews, *Emporia State Research Studies*, **6**, 1957, (1), 31 pp. Some of these compositions are highly complex, as mixtures of copper, silver, zinc, and tin salts used in these. The color palette of such ceramic products is very rich in beautiful samples. Matthews again emphasizes how indispensable tin must be, as a correlation to the amount of copper, in the glaze batches of this kind. We further remark the recommendation of G. Yamaguchi, *Yogyo Kyokaishi*, **61**, 1953, 549-553; *Glass Ind.*, 1955, (2), 83 ff., 106, to use a spinel-like compound containing aluminum in a "lower valency" (Al^+ or Al^{2+} assumed) state, as a reduction agent in copper-red glass manufacturing.

[195] *J. Am. Ceram. Soc.*, **42**, 1959, 499-503. For the important recent papers of A. Ram, and S. N. Prasad and K. K. Vaish, see *Central Glass & Ceram. Research Inst. Bull.* (*India*), **5**, 1958, 4-10; *Ibid.*, **6**, 1959, 35-36; 86-88; *Ibid.*, **7**, 1960, 49-53, 110-116, and *Proc. Natl. Inst. Sci. India. Pt. A*, **26**, Suppl. I, 1960/61, 12-25. W. E. S. Turner and H. P. Rooksby, *Glastech. Ber.*, **32**, K, 1959, 17-28, demonstrated the presence of Cu_2O in ancient copper-red glasses. Colloidal suspensions of cuprous oxide, for a comparison of their color with that of copper ruby glass, were recently studied by A. Ram, S. N. Prasad, A. Sarker, and K. P. Srivastava, *Central Glass & Ceram. Research Inst., Bull.* (*India*), **8**, 1961, 99-108.

89. The great influence which *irradiation effects* may have on color in glasses, analogous to color phenomena in defect structures of alkali halide crystals, was made evident in a series of highly important investigations by I. Kuan Han Sun and N. J. Kreidl;[196] these were discussed from the modern aspects of electron exchange reactions on energy levels in solids. Various types and energy contents of radiation may be applied on glass, either neutrons of charged particles with a mass from zero (in the photon) to that of atoms (in the fission fragments from uranium), with energies ranging from that of slow neutrons to 10^{16} e.v. in cosmic radiation. They bring about a wide spectrum of different responses of glasses to radiation impact. A large number of free-moving electrons cause electron deficiencies in particular locations and concentrations of these free electrons in electropositive sites, connected with oxidation and reduction reactions, or in negative ion traps with a characteristic color center developed. The excited electronic energy may also bring about luminescence phenomena. The displacement of atomic nuclei and the formation of holes (vacancies) is another consequence. Finally, atomic nuclei may be transformed into other

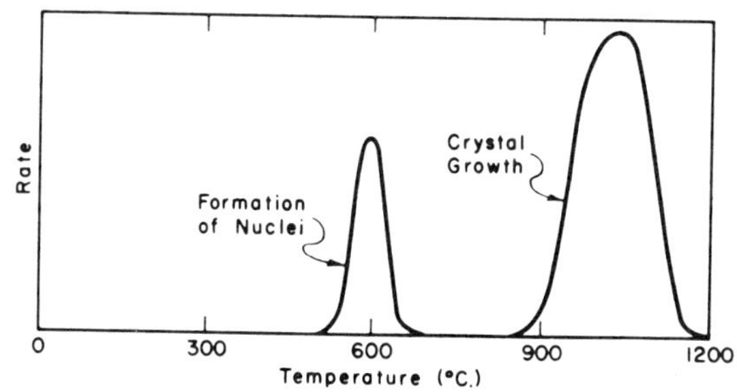

FIG. C. 56 Conditions of controlled nucleation and crystallization of copper ("striking") in copper-red glazes (Brown and Norton).

nuclei, e.g., B^{10} into Li^7, Si^{30} into Si^{31} into P^{31}, and so on. Chiefly the range of radiation energies below 10^{16} Röngtgen units is investigated, because above this value the glass becomes black and nontransparent. The characteristic responses observed are the so-called solarization, the production of colors by oxidation-reduction processes, luminescence, color center formation, and photosensitization by a reduction of ions to metal, which is coagulated by heat treatment to colloidal colored particles (these are the effects which interest us particularly at present); such responses may be made useful in the development of dosimeters. Very useful in this respect is a cerium-vanadium silicate glass combination, recommended by L. O. Upton (1940), or a silver-containing potassium-barium-aluminum phosphate glass,[197] recommended by W. A. Weyl *et al.*

[196] *Glass Ind.*, **33**, 1952, 511-514, 546; 589-594, 614; 651-653, 674.

[197] Cf. J. H. Schulman, W. A. Weyl, R. J. Ginther and I. W. Evans, *J. Electrochem. Soc.*, **95**, 1949,

90. The *photosensitive glass* first described by S. A. Stookey (1949) under the action of ultraviolet irradiation requires the presence of sensitizers, such as tetravalent cerium. This element is also known as a powerful agent preventing discoloration of glass (cf. Vol. II, B. ¶ 54 ff.). Stookey[198] further developed his important invention by finding what he called "chemical machining" processes based on nucleation in crystallizing lithium silicate glass (with some potassium and aluminum) containing traces of silver and tetravalent cerium. The silver metal particles are evolved by developing of the positive "image" of the normal photosensitive reaction. The positive image is built up in a three-dimensional photographic-visible form by the final heat treatment. If the particles of this image now nucleate the crystallization of the lithium silicate, it is possible to use the very great difference of solubility of the crystallized and of the glassy silicate in dilute hydrofluoric acid in order to etch away the irradiated portions. It is thus possible to obtain very precise matrices for photoengraving processes, for miniaturization of electrical components in electronic instruments, special filters, and similar equipment. The crystallites in the nucleated lithium silicate surrounding the positive image reach about 4 μ in size. They are (expressed in terms of volume or mass of material removed per unit time) soluble about 10^5 times faster in dilute hydrofluoric acid solutions than the remaining glass in the nonexposed parts, and are, therefore, rapidly etched away. Other processes were found by Stookey[199] to give to the photosensitive glass particularly valuable mechanical properties, and to transfer a silver image from a photographic emulsion onto photosensitive glass ("photostaining").

91. The nucleation and growth of the metal particles (especially gold) in Stookey's photosensitive glasses was studied by R. D. Maurer[200] with light absorption and scattering methods. It is evident from these studies that the nucleation starts from a supersaturated solution, which was in these experiments an antimony-containing glass composition. No spontaneous nucleation is possible in this case and photoelectrons are required to induce the process. The number of growth sites increases with the exposure time, whereas the rate of the growth is constant for all the variable exposures. If a spherical shape of the growing particles can be assumed, absolute values can be calculated, using a theory developed by E. Zener[201] for the initial change of particle size as a function of time, and for the number of nuclei, as a function of exposure intensity. Justification for the assumption of spherical particles is seen

70-79; J. H. Schulman, R. J. Ginther, C. C. Klick, and E. W. Claffy, *Am. Ceram. Soc. Bull.*, **30**, 1951, 97; *J. Appl. Phys.*, **22**, 1951, 1479-1487; R. S. Alger and R. A. Levy, *Phys. Rev.*, **81**, 1951, 657.

[198] *Ind. Eng. Chem.*, **45**, 1953, 115-118.

[199] *Ind. Eng. Chem.*, **46**, 1954, 174-176.

[200] *J. Appl. Phys.*, **29**, 1958, 1-8. See also the article by S. D. Stookey and R. D. Maurer on "Catalyzed Crystallization of Glass-Theory and Practice," in Progress in Ceramic Science, Vol. 2, 1961, 77-101, Pergamon Press, New York.

[201] *J. Appl. Phys.*, **20**, 1949, 950-953.

in the agreement of measured depolarization data from the particles of the silver image in the glass, namely data below 0.005 for the R. Gans depolarization coefficient (1912). This is sufficiently near zero to permit extrapolation from the ratio b/a of the ellipsoidal shape of the particles, related to the depolarization coefficient, from 0.9 to 1.0 for ideal spheres. Taking these additional remarks into consideration, the findings of Maurer are in principle thoroughly in agreement with those previously discussed in this volume concerning the nucleation and growth of particles, represented by equations and absorption curves for Zsigmondy-Thiessen-type gold hydrosols (cf. the discussion of the work of J. Turkevich, G. Garton, and P. C. Stevenson in C. ¶ 82). The gold first migrates into the glass as dissolved ions. This diffusion current of gold ions must take place at the same time with the diffusion of the precipitating ions, such as those of antimony, Sb^{3+}. The nuclei then grow to colloidal particles until a fixed amount of gold is precipitated, as determined by the equilibrium of the reaction $2\ Au^+ + Sb^{3+} \rightarrow Au^0 + Sb^{5+}$. The maximum size will be in the order of magnitude of a few hundred Å., like that observed in the hydrosols.

92. A very detailed investigation of the optical properties of photosensitive glasses, with silver, cerium, Ce^{4+}, and antimony, Sb^{3+}, as the photochemically active ingredients was recently made by K. Barth[202] especially on the spectral degree of transmittance, τ = the ratio of the transmitted light amount, Φ_τ, to the impacting current, Φ, at constant wavelengths, and the so-called blackening, $S = \log (\Phi/\Phi_\tau)$, determined over the range of $\lambda = 250$ to 700 mμ. The chemical composition of the experimental glass plays a definite role: the color intensity is increased if the contents in silver and sodium are raised in the glass (for silver there is an upper limit of concentration of 0.06 moles); it is increased further if aluminum is replaced by barium. The addition of tetravalent cerium has an optimum concentration at 0.01 moles CeO_2. Deficient or excess trivalent cerium is detrimental to the color evolution. A solarization effect running side-by-side with the development of the latent image by irradiation is caused by a contamination of the glass by traces of divalent iron, and by the reaction $Fe^{2+} + h\nu \rightarrow Fe^{3+} + e^-$. This solarization electron regenerates trivalent cerium to its reducing action of the photosensitive interchange with the silver ions, but the solarization ions do not affect the latent image, since the electrons are retained in definite positions in the glass structures as long as the temperature is below that required for the "striking," i.e., for the start of the development of the latent to the visible image. In the striking process no shifting occurs of the characteristic absorption maxima to longer wavelengths, as postulated by Mie's theory, if the growth of the neutral silver particles to colloidal ones takes place by mere apposition. The increasing color intensity during striking, therefore, is a consequence of the continuously increasing number of particles, not of their growth beyond a (rather low) limit. Characteristic is the width of absorption maxima in

[202] *Silikat Tech.*, 11, 1960, 101-105.

the transmission degree (see Fig. C. 57), distinct from the rather narrow bands corresponding to Mie's theory. This width indicates a certain variability in particle sizes. If one calculates the curve of transmittance from an assumed distribution of 85 per cent of particles up to 10 mμ in diameter, 8 per cent of those up to 20 mμ, and 7 per cent of those up to 60 mμ, the good agreement of the experimentally determined and the calculated transmittance curves in Fig. C. 58 is convincing enough to indicate that this assumption cannot be far away from reality.

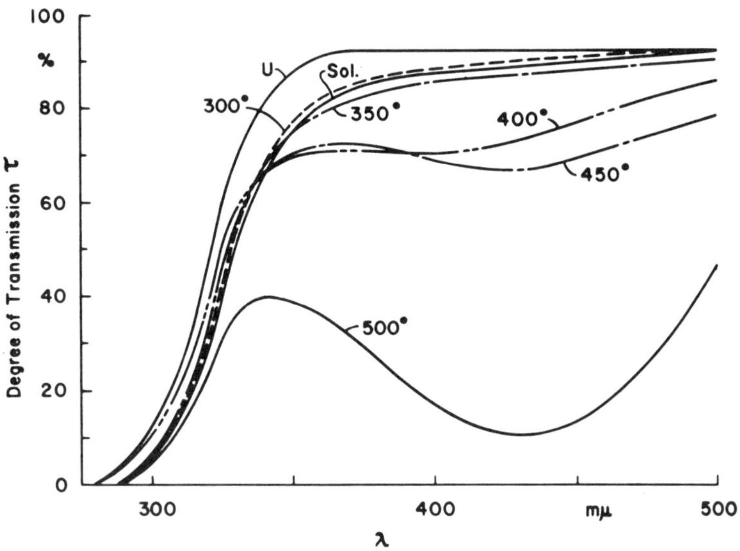

FIG. C. 57. Transmission degree of photosensitive silver glass at different temperatures of thermal development of the latent image (Barth). Curve *U* corresponds to non irradiated glass; *Sol.* to solarization effect. Irradiation time was 60 minutes; development time, 30 minutes.

93. Color phenomena in glasses containing *sulfides*, *selenides*, or *tellurides*, or the free divalent anions, or complex sulfoferrite-type anions, have considerable practical interest. This difficult field, which finds its industrial significance especially in the glasses of amber-colored bottle glass or in selenium-ruby glass and related products, has been studied extensively, but there are still many fundamental questions open to discussion. Valuable contributions to knowledge of amber glasses were made by W. L. Spix and F. R. Bacon,[203] and by F. R. Bacon and C. J. Billian.[204] The absorption curves of such glasses show a very characteristic dependence on their previous thermal treatment ("thermal history"). If they are soaked over a certain period between 625° and 510°C. and then air-quenched, they show a transmittance curve

[203] *J. Am. Ceram. Soc.*, **36**, 1953, 377-383.
[204] *J. Am. Ceram. Soc.*, **37**, 1954, 60-66.

characteristic of this temperature range, and with deeper soaking temperatures the transmittance is decreased in a corresponding manner. There is evidently a state of equilibrium characteristic of the soaking temperature, reached at 625°C. in about 10 minutes and at 510°C. in not less than about 15 days. or even more. The absorption

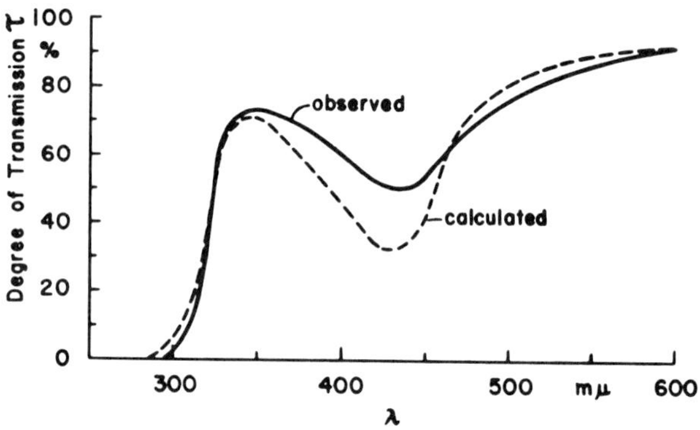

FIG. C. 58. Transmission degree of a photosensitive silver glass, for an assumed distribution of particle sizes in the visible image, in comparison with that calculated from Mie's theory for 6×10^{-6} mm.^{3} Ag/mm.^{3} (Barth).

coefficient can also vary greatly as a function of the soaking temperature for a constant wavelength, as seen from Fig. C. 59, in which the negative logarithm of the transmittance is introduced as "absorbance." The greatest change in absorbance is in the ultraviolet and the visible violet of the spectrum, the greatest relative change occurring for $\lambda = 400$ and 650 mμ. The sulfoferrite complex anion theory developed by A. Dietzel and E. Neumann (1940) for "carbon-yellow" glasses gives the best

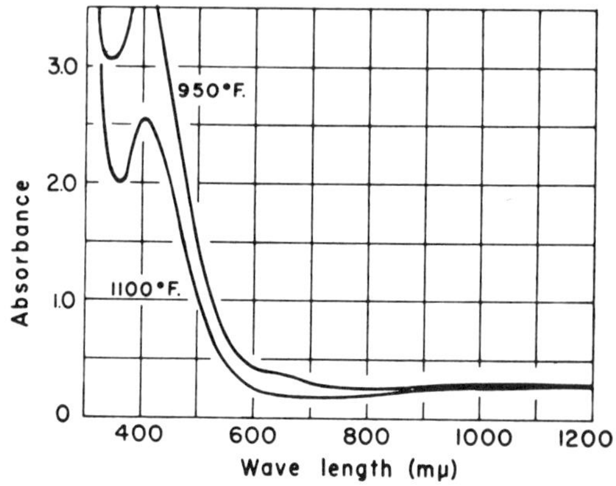

FIG. C. 59. Curves of the absorbancy of an amber-colored bottle glass, as a function of temperature of thermal treatment and of wavelengths (Spix and Bacon).

explanation of these coloration effects. Bacon and Spix, however, did not observe any colloidal precipitation of particles of iron sulfide, FeS, which would have caused a large increase in absorbance over the whole visible spectrum, with a prominent graying and darkening effect. Bacon and Billian especially studied the transmittance of iron sulfide amber glasses at 550 mμ; it is decreased with increasing divalent iron content or increasing sulfur content. This transmittance, however, is very sensitive to other factors, such as the presence of fluorine anions, or to thermal history effects which influence very much the physiological color attributes of given amber glasses, witness the conventional "yellowing factor" or the "redness ratio" of the C.I.E. (Commission Internationale d'Éclairage) system.

94. In the *selenium ruby* glasses the presence of crystalline solutions of cadmium selenide in cadmium sulfide particles near the boundary between colloidal and microscopic sizes, as the coloring pigment, is made certain by X-ray diffraction precision measurements, specifically those of T. Yamauchi and Y. Endo.[205] The more cadmium selenide present in the composition of the crystalline particles, the deeper the red color of the glass. *Antimony ruby* glasses are based on the intense color effects of antimony trisulfide as such, or of an "antimony cinnabar" which was investigated by E. Dönges and R. Fricke,[206] with particular emphasis given to a comparison with the mineral kermesite (pyrostibite), of the composition Sb_2S_2O, and to the photochemical changes of the crimson-red trisulfide in ultraviolet or sun light. The antimony cinnabar is a mixture of the trisulfide and trioxide, with some sulfur, but not an oxysulfide of antimony. Its high degree of dispersion is expressed in the very high specific surface area of this artificial product. The conditions for manufacturing an antimony ruby glass were studied by T. M. Tomilina.[207] She came to the remarkable conclusion that the sodium dithiometaantimonate, $NaSbS_2O$, is an intermediate reaction product in corresponding batches, later changed to the cherry-red compound $Sb_2O_3.Sb_2S_3$, and dispersed in colloidal state in the glass. Pure trisulfide colors brown-red. Its formation is not desirable when it is formed under an excess of sulfur. On the other hand, too little sulfur in the batch brings about not less undesirable formation of the trioxide which is no color pigment at all. Additions of fine-dispersed carbon are required for stabilizing the thioantimonate complex, protecting it from "burning away." In its color effect, antimony ruby is intermediate between copper and selenium ruby, with an optical density of 0.14 to 0.15 at $\lambda = 720$ mμ, and above 3.0 at 574 mμ. Additions of potassium to the glass composition reduce the optical

[205] *Yogyo Kyokai Shi,* **60,** 1952, 417-421, especially on p. 420 a valuable table of X-ray data for different cadmium sulfide-selenide pigments. For recent studies on the colors produced in glass by CdSe and CdS, see F. G. West-Oram, *Central Glass & Ceram, Research Inst., Bull. (India),* **8,** 1961, 131-134.

[206] *Z. anorg. u. allgem. Chem.,* **253,** 1945, 2-8.

[207] *Belorus. Politekh. Inst. im. I. V. Stalina, Sbornik Nauch. Rabot,* No. 47, 1955, 47-53.

density in the red even to 0.1. Lithium increases it in the red, decreases it in the green, and zinc oxide decolorizes the antimony ruby glass. Both lithium and zinc therefore are undesirable batch ingredients. Antimony trisulfide, as such, is a glass-forming compound, and because of its complex composition in the system K_2O–Sb_2O_3–Sb_2S_3 it is a most valuable basic component for infrared-transmitting glasses, as B. W. King and G. D. Kelly[208] have demonstrated, particularly transmittent between 3.5 and 6.0 μ.

95. Concerning glass coloring by *tellurium*, we mention the fascinating studies of Y. Hasegawa and S. Kawakubo,[209] who worked with alkali calcium silicate glasses in the presence of strongly reducing agents like aluminum and zinc metal powder, tartar, tartaric acid, and Seignette salt. These authors are of the opinion that the intense colors observed are caused by alkali tellurides or polytellurides, in analogy to the well-known alkali sulfides and polysulfides. Beautiful red-violet colors are produced in potassium silicate glasses, with aluminum as the reducing agent. Sodium silicate or borosilicate glasses evolve brownish colors, with some difficulty. The absorption spectra show maxima at 480 to 490 mμ, identical with those observed in a concentrated solution of potassium hydroxide for tellurium (as potassium polytelluride).

INVESTIGATION OF COLLOIDS BY X-RAY AND CORPUSCULAR RADIATION EFFECTS

96. The great importance of X-ray diffraction methods for the determination of colloid particle sizes by the measurement of the broadening effects of the interference lines in powder diagrams was recently illustrated for ceramic purposes in the investigation by W. F. Sullivan and S. F. Cole[210] dealing with titania as used in enameling techniques. We shall have to do here with finest particles of the anatase modification of titania, which on heating at 700° to 950°C. are inverted into rutile (cf. Vol. II, B. ¶ 268 ff.). Precision X-ray diagrams showed that before this inversion the crystal structure of the anatase is slightly widened, and that also the first nuclei of the rutile show an expanded unit cell. The primary particles visible in the electron microscope, of only 10 mμ in size, are agglomerated to secondary particles of 50 mμ, and even up to 400 mμ. The thermal growth of the crystallites from a water and sulfate anion containing raw material of titania could be studied by Sullivan and Cole using the Scherrer equation for the crystal dimensions perpendicular to

[208] *J. Am. Ceram. Soc.*, **41**, 1958, 367-371.

[209] *Glastech. Ber.*, **30**, 1957, 332-335; see also Y. Hasegawa, M. Yanagi and S. Kawakubo, *Yogyo Kyokai Shi*, **65**, 1957, 179-186.

[210] *J. Am. Ceram. Soc.*, **42**, 1959, 127-133.

a given reflecting plane (*hkl*)

$$D_{\text{hkl}} = \frac{0.9\,\lambda}{\cos\vartheta \cdot \beta_{1/2}}$$

($\beta_{1/2}$ is the half-maximum line breadth), applying a correction $B^2 = \beta^2_{1/2} + b^2$ (B is the experimentally measured breadth, b is the breadth of a line of the same material with crystal sizes above 100 mμ). The average dimension, $\overline{D}$, is equal to $(D_{200} \cdot D_{004} \cdot D_{101})^{1/3}$ taken from the line diagram. As a function of the temperature of calcination, D is given in Fig. C. 60. $\overline{D}$ is a basis of comparison used to eliminate the major anisotropy effects of the morphological development of the primary crystals.

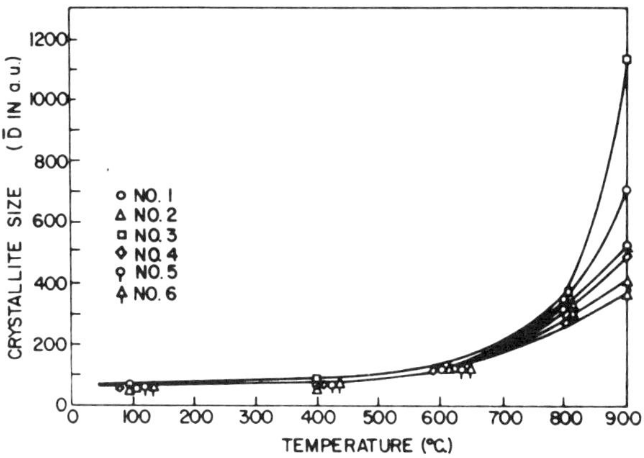

FIG. C. 60. Change in the average crystallite dimension, $\overline{D}$, with the temperature of calcination of hydrous titanium dioxide (Sullivan and Cole).

97. Because of the great significance of grain sizes of reactive particles in carbonate-silicate batches, as in the raw mixes of the Portland cement industry or related sintering products, the studies of R. A. W. Haul and F. R. L. Schöning[211] on the thermal decomposition of calcium, magnesium carbonate, and dolomite should be emphasized. These authors used the methods of broadened X-ray interference lines for determination of particle sizes. In the first stages of the dissociation these are 6 to 13 mμ in diameter, rapidly increasing with raising calcination temperature, especially above 900°C. with primary particles larger than 100 mμ, the upper limit of the applicability of the method. By Geiger counter measurements of the interfer-

[211] R. A. W. Haul with H. Stein, J. D. Louw, H. Wilsdorf, H. Heystek, and J. Markus, *Nature*, **167**, 1951, 241-242; 727; 945-946; *Acta Cryst.*, **5**, 1952, 250-255; *Am. Mineralogist*, **37**, 1952, 166-179; *Naturwissenschaften*, **38**, 1951, 283; *Proc. Phys. Soc. London*, **65**, B, 1952, 528-535. Regarding the investigations of R. A. W. Haul and F. R. L. Schöning discussed here, see *Z. anorg. u. allgem. Chem.*, **269**, 1952, 120-134.

ence intensities, it was possible to measure: (1) the integral intensity; (2) the intensity at the maximum; (3) and from the ratio of (1) to (2), the integral breadth. The "line profiles," $B(x)$, for the broadened lines and $b(x)$ that of a sharp standard line, are expressed as a function of the form

$$y = c/(1 + k^2x^2)^2$$

to solve the general integral equation for $B(x)$. The calculation gives, for the important line (220) and its broadening, the relation of $B(x)$ and of the crystallite size calculated from it, shown in Fig. C. 61 as a function of the temperature of calcination,

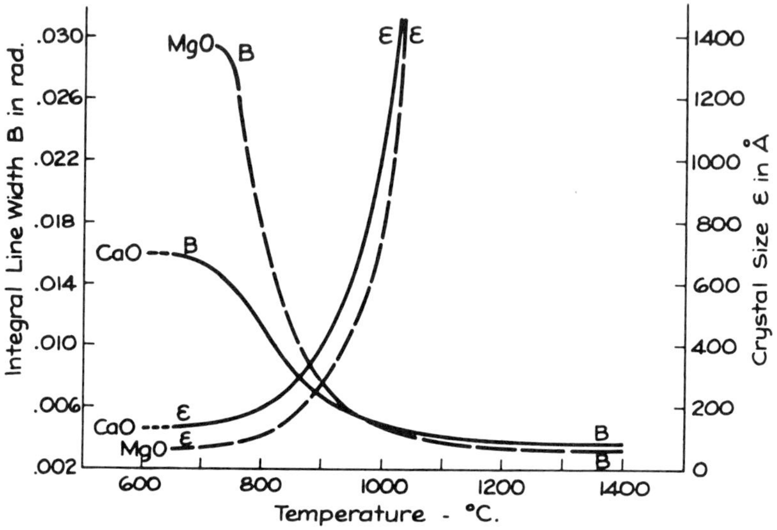

FIG. C. 61. Functional relation of the integral breadth of the interference line (220), B, and the crystallite size, ε, with calcination temperature of dolomite (Haul and Schöning).

for dolomite, and in Fig. C. 62 their growth as a function of time. No lattice distortions were observed in their effects on the integral intensities, nor were strains which would have increased the broadening of the lines with increasing diffraction angles Θ. The particles are probably iso-dimensional, in analogy with the cube-shaped magnesia particles visible in the electron microscope. For calcite, however, there are indications of slight anisotropies with flat-rhombohedral forms in the calcination products at 650° to 750°C. The agreement of the sizes calculated from the X-ray interferences, and the observation in the electron microscope is very good, whereas the data from the BET nitrogen adsorption measurements of the specific surfaces are only satisfactory for the smallest particles up to 5 mμ. Those above 10 mμ show in the nitrogen adsorption data only 50 per cent of the morphological crystallite surface; particles of sizes up to 38 mμ show only 2 per cent of the geometric surface. Both methods therefore must consider fundamentally different structural elements in the determinations. Specific factors of crystal growth and nucleation are valid as by the partial dissociation of dolomite to magnesia and calcite, caused by orientation effects of the new-formed calcite crystallites relative to the initial dolomite (cf. Vol. V, Section A).

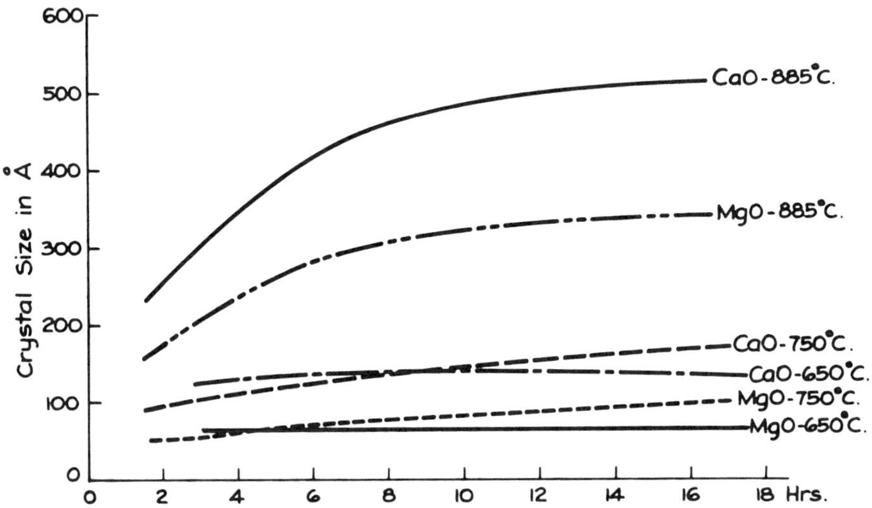

FIG. C. 62. Crystallite growth with time for magnesia and calcium oxide formed by the calcination of dolomite (Haul and Schöning).

98. Another example of the determination of crystallite sizes in calcination products which is important in silicate technology is the formation of γ-alumina. R. Fricke and G. Weitbrecht[212] studied this oxide as it is formed by dehydration of boehmite, γ-AlO(OH), after calcination at 500° to 900°C. for two hours, with X-ray diffraction measurements. A series of corrections was applied, among which we mention as the most important the correction for a finite diaphragm diameter, because the equations of M. von Laue (1926) and R. Brill (1928) are in this case fully valid for particles with low X-ray absorption. For γ-alumina it was made evident also that the observed broadenings of the interferences are true particle size effects and are not caused by variations in the lattice plane distances by strains, in spite of the trivial lattice defects always present in such calcination products in "*statu nascendi e solido.*" The electron microscopy of the γ-alumina showed secondary particles of a spongy nature. The data from X-ray interferences are lower for the primary particles than those determined in the electron microscope, and characteristic is a relatively slow growth in size with increasing temperature. The electron microscope

[212] *Z. anorg. u. allgem. Chem.*, **253**, 1945, 9-29, 30-36.

is not able to detect very small particles of γ-alumina because of its low absorption coefficient, and the lack of contrasts to the surroundings. Fricke and Weitbrecht further investigated the role of small amounts of silica present in the γ-alumina. This impurity, if present in considerable amounts (2 to 10 per cent) chiefly impedes the growth of the primary particles whereas, surprisingly enough, the growth is accelerated if there is a silica content of 0.5 per cent (curve II in Fig. C. 63). This favorable phenomenon was interpreted by a Hedvall effect, either by the inversion of the silica of quartz type into cristobalite near 900°C., or by the solid state reaction of silica with alumina to form reaction rims with mullite (cf. Vol. V, Section A).

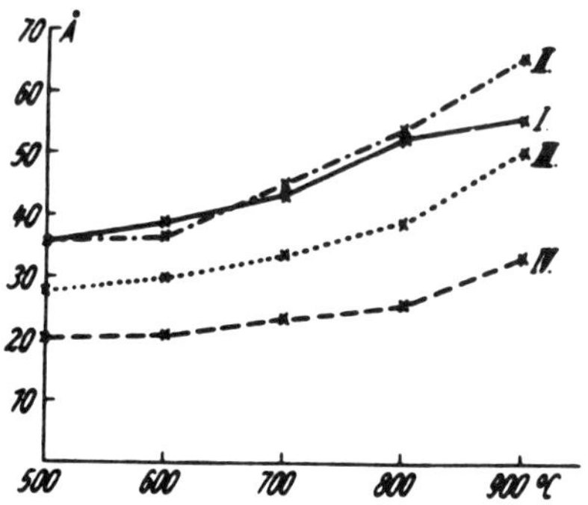

FIG. C. 63. Growth of γ-alumina particles as a function of temperature in presence of silica (Fricke and Weitbrecht). Curve (I) corresponds to pure γ-alumina; (II) to a sample with 0.5 per cent silica; (III) to one with 2.7 per cent; (IV) with 10.55 per cent silica.

99. G. W. Brindley[213] discussed the possibilities of using the X-ray method of broadened interference lines for the determination of grain sizes in dispersed clay minerals, taking into account structural defects in these, and strains distorting their lattice, which often appear combined and make the calculations highly difficult. In comparison with the basis of the classical theory of X-ray diffraction conditions for three-dimensional crystal structures, the clay minerals often show two-dimensional conditions of diffraction. In many cases, therefore, the electron diffraction may be superior to X-ray methods in clay mineral determination. The order of magnitude of halloysite and montmorillonite near 10^1 to 10^2 mμ as determined by the X-ray method may be taken as an empirical average, but determination of the thickness of the tabular crystals, as in kaolinite and dickite, is highly difficult if electron microscopic-stereoscopic and shadow-casting methods cannot be applied. On the particular problems introduced by the frequent occurrence of mixed-layer structures, see B. ¶ 104 ff.

[213] *Discussions Faraday Soc.*, No. 11, 1951, 75-82.

100. Similar problems in the clay mineral group were studied by A. Clauss, H. P. Boehm, and U. Hofmann[214] with highly surface-active and catalytically valuable carbon soots. For these a comparison of the X-ray diffraction data with the results of adsorption isotherms (cf. C. ¶ 200) is highly instructive. The interferences (200) on the diffraction diagrams are pure cross lattice effects, or (20) is the indexed interpretation of this characteristic line, with a very sharp declining intensity toward higher Bragg angle values. The maximum is also slightly shifted to higher Θ angles, simulating a shorter distance C—C than is real. In addition, the interference line (201) often overlaps the maximum for (20). In contrast with (20), the interference line (02) is nearly symmetrical for graphitized soots. The electron microscope showed for such soots mostly spheroidal particles. The great disorder in the soot particles concerns the directions of *a* and *b* in the crystal structure, whereas the carbon planes of the graphite structure remain parallel. With increasing degrees of graphitization the interference lines become sharper, and even (*hkl*) interferences appear, indicating the gradual development of a three-dimensional order, but still distortions occur even in the best-graphitized soots. The comparsion of the sizes, calculated from the X-ray diffraction data and by electron microscopic measurements, differ by one or two orders of magnitude. Crystal growth is practically negligible at temperatures below 1300°C.

101. R. Fricke and E. Eberspächer[215] recognized the great importance of the application of *low-angle* X-ray diffraction for samples of the γ-alumina type, prepared from boehmite and bayerite by dehydration. They were concerned with the "liquid-like" arrangement of the single monocrystals in aggregation forms, intergrown to secondary units, as a "dense-packed colloid system," in the meaning that O. Kratky[216] had defined in his theory of "exterior interferences." Fricke and Weitbrecht used the low-angle diffraction camera developed by O. Kratky, A. Sekora, and R. Treer,[217] with a pentaerythrite monochromator crystal and a silver pinhole diaphragm, for a variable divergency of the primary beam. For a reduction of the exposure time the

[214] *Z. anorg. u. allgem. Chem.*, **290**, 1957, 35-51.

[215] *Z. anorg. u. allgem. Chem.*, **265**, 1951, 21-40. A great help for understanding the general meaning of low-angle diffractions is seen in optical models of diffraction microscopy which the reader will find in the fascinating article of R. Hosemann, *Naturwissenschaften*, **46**, 1959, 545-552. A further introduction into the significance of low-angle diffraction methods for colloid physics and for high-polymers, including a detailed discussion of the theoretical background, camera construction, etc. was given by R. Hosemann in *Ergeb. exakt. Naturw.*, **24**, 1951, 142-221. The importance of this method is demonstrated for the study of alumina and mixed alumina-silica hydrogels used as cracking catalysts (cf. C. ¶ 208-215). Cf. also P. W. Schmidt, J. Krogh-Moe, and H. D. Bale, *J. Phys. Chem.*, **60**, 1956, 1580-1582.

[216] *Monatsh. Chem.*, **71**, 1947, 325-340.

[217] *Z. Elektrochem.*, **48**, 1942, 587-601. An excellent low-angle diffraction camera constructed by H. Jagodzinski and K. Wohlleben, *Ibid.*, **64**, 1960, 212-216, was used for the structure analysis of curled chrysotile crystals (cf. A. ¶ 159*, Footnote[252]).

photographic film was warmed up to a temperature of 30°C. The evaluation of the diffraction diagrams gave characteristic maxima of the occurrence frequency for a secondary average distance of structural planes near 15 mμ for γ-alumina from boehmite, whereas for samples from bayerite no maximum of this kind was observed. C. G. Shull and L. C. Roess[218] gave a complete discussion of the theory and application of low-angle X-ray scattering for the finely dispersed particles of other oxides, and especially for the amorphous silica gels, with intensity curves derived for different distribution types, such as for the Gauss, Maxwell, and "rectangular" size distributions, to achieve a comparison of the theoretical and experimental scattering curves. The agreement of the results with determinations of specific surface measurements is very satisfactory.

102. Taking clay particles suspended in water, with pronounced swelling characteristics, W. J. West[219] applied the low-angle diffraction method (with a Du Mond apparatus) to study different actions of sodium pyrophosphate on the suspensions, as used in drilling muds, influencing the viscosity and gel strengths (cf. C. ¶ 332). The decrease in particle sizes in such pyrophosphate-containing suspensions is particularly important. A much more detailed study of clay minerals (with Kratky's camera with slit diaphragms) down to Bragg angles Θ of 13′ was made by U. Hof-

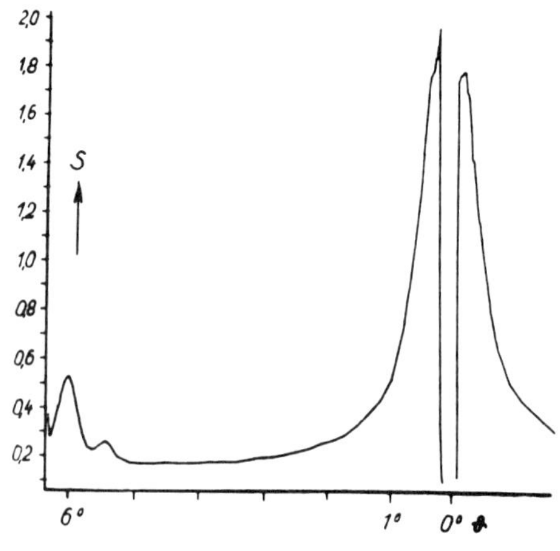

FIG. C. 64. Photometric curve of low-angle diffraction (equator) diagram of a chrysotile fiber (from Arizona) (Hofmann and Nemetschek).

[218] *J. Appl. Phys.*, **18**, 1947, 295-307. On the theory for spheroidal particles, discs, and rods, see *Ibid.* 308-313, for both a Maxwell and a rectangular mass distribution.

[219] *J. Colloid Sci.*, **7**, 1952, 295-305; cf. also R. J. Arnott, Inaug.-Dissert., Columbia Univ., New York, 1954, 63 pp.; *Am. Mineralogist*, **40**, 1955, 309. This investigation illustrates the difficulties in applying the low-angle methods to illitic and montmorillonitic clays, as well as to halloysite and attapulgite, because of their flaky or fibrous-tubular crystal habitus.

mann and Th. Nemetschek,[220] especially to examine the innercrystalline swelling phenomenon of montmorillonite, vermiculite, and batavite. One remarkable result proves that there is no larger period perpendicular to the layers than the average distance of the layers themselves, namely about 15 Å. for montmorillonite, 14.3 Å. for batavite, 12 to 14 Å. for vermiculites, dependent on the exchangeable cations present. A particularly strong low-angle dispersion is characteristic of illite from Sárospatak, attapulgite (cf. the photometric curves in Fig. C. 64), and chrysotile asbestos (Fig. C. 65) with fiber texture diagrams along the equator direction. The

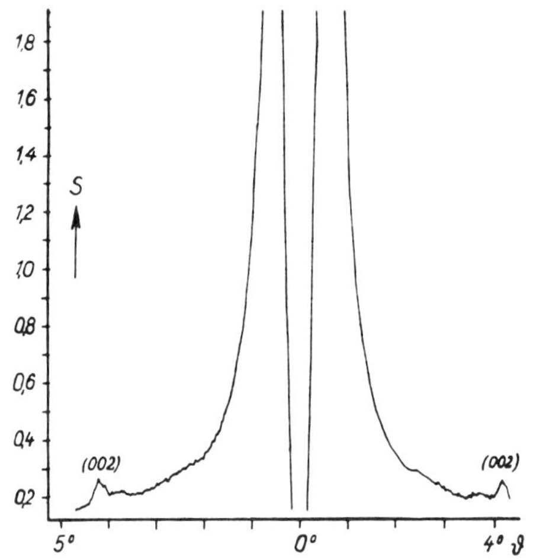

FIG. C. 65. Photometric curve of low-angle X-ray diffraction (equator) diagram of attapulgite, fraction below 1 μ (Hofmann and Nemetschek).

fibrillar texture with units of 10 to 20 mμ thick single fibers, makes understandable the very high specific surface activity and the excellent catalytic activity of attapulgite. The same may be said of illite with very thin flakelets of about 10 mμ thickness (cf. the diffraction curve in Fig. C. 66).

103. Another important field of application of low-angle X-ray diffraction is the study of *porous glasses* as described by E. A. Poraĭ-Koshits and V. N. Filipovich.[221] The radii of heterogeneities indicated by the diffraction effect can be referred equally either to the dispersed material or to the presence of empty pores. The mathematical discussion of the diffraction phenomenon is based on the so-called Babinet principle, but it is not immediately possible to interpret observed intensity functions of the diffraction alone to decide the pore-or-particle problem, or which kind of hetero-

[220] *Kolloid-Z.*, **130**, 1953, 83-89.
[221] *Izvest. Akad. Nauk S.S.S.R., Otdel. Khim. Nauk*, 1955, 21-30.

geneities is characterized by the radii. The glasses studied by Poraĭ-Koshits and Filipovich are the typical microporous, potassium hydroxide-leached sodium borosilicate glasses of the Vycor type from which the silica deposits filling the pores were removed by a second leaching with dilute hydrofluoric acid solution. By this differential-etching method it was then possible to change pore sizes systematically and to observe the resulting effects. The shifting of the diffraction maxima from 29.5′ to 19′, thus to smaller Bragg angles after leaching, with increasing pores, distinctly indicates that the diffraction originates from the pores and not from particles of the glass. In the measure of the progressive etching the pores are widened and the wall

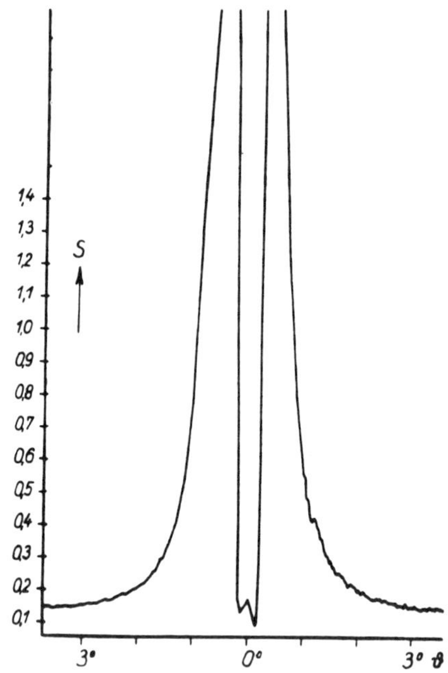

FIG. C. 66. Photometric curve of the low-angle diffraction (equator) diagram of illite from Sárospatak, fraction below 1 μ (Hofmann and Nemetschek).

thicknesses are reduced in the cavities of the glass, comparable to the conditions studied by C. B. Walker, J. Blin, and A. Guinier[222] for silver-aluminum alloys, at 525° and 600°C. A secondary diffraction effect is caused by the particle sizes of the etched glass itself. N. S. Andreev and E. A. Poraĭ-Koshits[223] worked with the vacuum camera of O. Kratky (Fig. C. 67) for the determination of chemical heterogeneities in sodium borosilicate glass (as a 2 cm. thick, well-polished plate annealed at 500°,

[222] *Compt. rend.*, **235**, 1952, 254-255.

[223] *Doklady Akad. Nauk S.S.S.R.*, **118**, 1958, 735-737; *Nature*, **182**, 1958, 375 f.; *J. Soc. Glass Technol.*, **43**, 1959, 235-261.

then heated to 600° to 650°C., and air-quenched). The diffraction curves show characteristic slopes from the maximum intensity for increasing distances from the central beam, indicating the rather uniform size of the primary structural heterogeneities and of the pores after leaching. The general type of the curves is the same for the original, and for the leached glasses. Pores in the original glass are 10.5 to 16.3 mμ in diameter, those in the leached glass 18 mμ, if treated at 600°C., but 38 mμ after soaking for one hour, and 61 mμ, if treated for six hours, at 650°C.

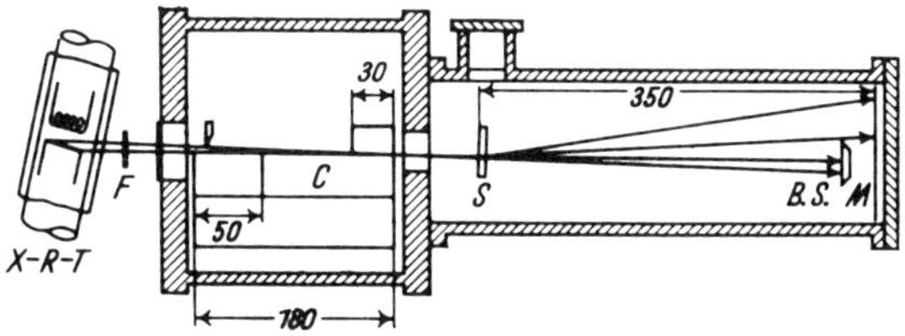

FIG. C. 67. Vacuum camera for small-angle X-ray diffraction, with Kratky-collimator (Poraĭ-Koshits): (*XRT*) X-ray tube; (*F*) filter; (*C*) collimator; (*S*) sample; (*BS*) impact of primary beam on photographic film (*M*).

104. E. A. Poraĭ-Koshitz[224] emphasizes that the general theory of microheterogeneities in the glass constitution is the principal point in which the classical framework theory of Zachariasen failed to give a satisfactory explanation for the physical and chemical properties of glass, especially in connection with the opalescence phenomena of borosilicate glasses. In the low-angle diffraction of X-rays it is very important in the evaluation of the diffraction diagrams to eliminate grain effects of the exterior surface of the glass samples used in the experiments before leaching. Therefore, glass powders are highly unsuitable for this purpose; not until well-polished samples were used was the complete elimination of those disturbing effects possible. A general impression of resulting photometric curves is given in Fig. C. 68. Their evaluation demonstrated that close relations exist between the structure of the orig-

[224] *Glastech. Ber.*, **32**, 1959, 450-459. The criticism of the results of previous experiments of L. C. Hoffman, W. O. Statton, and L. Cartz, *Nature*, **176**, 1955, 561-562, who worked with glass powders, is also directed partly against the results of P. Debye and H. Brumberger, *J. Phys. Chem.*, **61**, 1957, 1623-1624, who used sinter-pressed Pyrex glass discs, and thus obtained the misleading surface diffraction effects mentioned above as influencing the measurements. Calculation of the intensity of low-angle diffraction is given as previously by P. Debye and A. M. Bueche, *J. Appl. Phys.*, **20**, 1949, 518-525, for light scattering (cf. C. ¶ 68). The correlation distances characteristic of Vycor glass are given by Debye and Brumberger in the order of magnitude of 17.5 and 39 Å. for the first, and 80 to 90 Å. for the second type of these distances; in Pyrex glass ultrafilter pressed bodies they are 30 to 60 Å. and 130 Å., respectively, much smaller than for the holes of the sintered discs (750 to 1200 Å.). This is the reason why Debye and Brumberger assumed strain effects (differences in electron density) as the origin of the scattering observed.

inal glass and of the porous glass obtained after the leaching process; also closely related is the optical opalescence penomenon (cf. C. ¶ 126). When the original glass loses its opalescence at higher temperatures, a strong change in the silica framework occurs by the replacement of the larger channels by finer ones. The glass becomes porous, not in a dense packing, as indicated in the slope from 6.5′ to 15′ on the curves.

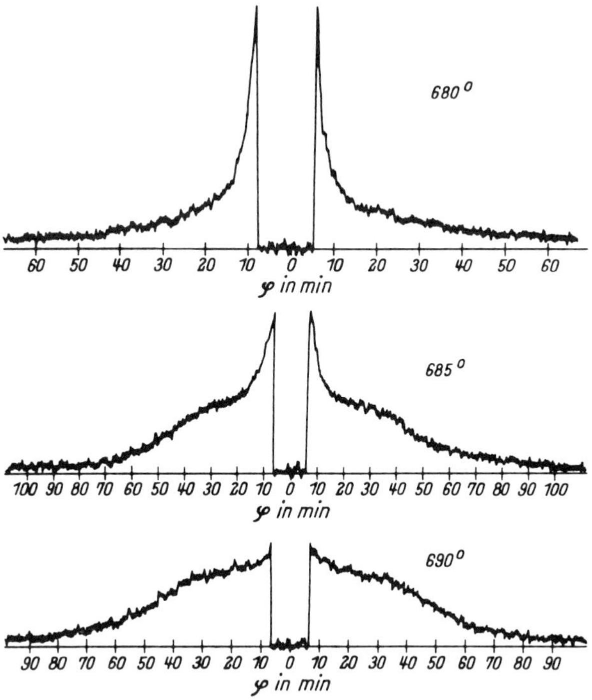

FIG. C. 68. Microphotometric curves for a porous sodium borosilicate glass, after annealing at different temperatures (Poraĭ-Koshits).

The smoother slope above $\Theta = 90'$ shows the finer glass structure and dense packing of the finer diffraction centers. Andreev's experiments give direct evidence of microheterogeneity of sodium borosilicate glasses, in the series of intensity versus diffraction angle diagrams for samples having different thermal treatment, shown in Fig. C. 69; these demonstrate the complete continuity of the changes from the original to the porous glasses. The shifting to smaller angles of diffraction with increasing temperatures of annealing is also evident, indicating the increase in heterogeneity spaces in the original and in the leached glasses. This also means that the distances, shapes, and distributions of the particles and pores are the same in both groups of glasses (on the general significance of these conclusions for the constitution of silicate glasses, cf. Vol. II, A. ¶ 207).

105. Previous expectations regarding the use of *neutron diffraction* for the study of high-dispersed systems have been confirmed. An especially important advantage of this method is seen in the investigation of "amorphous" systems, as in liquids.

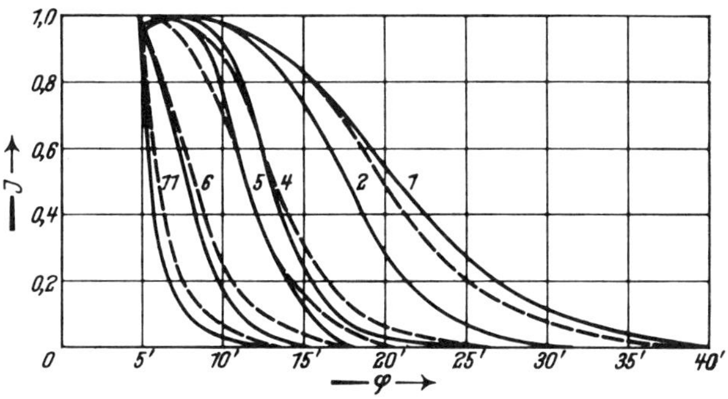

FIG. C. 69. Intensity curves for low-angle diffraction of original glasses (full curves), and of porous glasses (dotted curves). Thermal treatment of glasses: (1) at 500°C. for 5 hours; (2) at 600°C. for 1 hour; (4) at 600°C. for 4 hours; (5) at 500°C. for 8 hours; (11) at 650°C. for 6 hours.

As an instructive example of this kind we mention here the investigations of O. Chamberlain[225] on the diffraction of neutrons emitted from a nuclear reactor, by molten sulfur, lead, and bismuth. The diffractometer shown in Fig. C. 70, is a modification of that of C. G. Shull and E. O. Wollan (1948). In Fig. C. 71 are seen typical diffraction diagrams as an inversed function of the wavelengths. The nearest distances separating neighboring pairs of atoms in sulfur were determined to be 2.1 Å.; in lead twelve nearest neighboring atoms are at a distance of 3.4 Å.; in bismuth eight next neighbors are 4.4 Å. apart. The chief advantage of neutron diffraction over that of X-rays is that the positions of light elements in compounds can be much more favorable in their scattering amplitudes (corresponding to the form factors in X-rays) and be of the same higher order as for X-rays with heavy elements. Unfortunately, the resolution power of the neutron spectrometer is much smaller than that for X-rays, restricting the optimum usefulness of neutron diffraction to structures of high symmetry, of small unit cell dimensions, or for the study of order-disorder relations. For glasses, however, the conditions in neutron diffraction are very favorable, as A. H. Weber[226] and especially W. O. Milligan[227] demonstrated. For sil-

[225] *Phys. Rev.*, **77**, 1950, 305-313. See also G. E. R. Schulze, *Fortschr. Mineral.* **35**, 1958, 135-150; especially for magnetite, ferrites of spinel structures, etc., cf. G. E. Bacon and F. F. Roberts, *Acta Cryst.*, **6**, 1953, 57-62. On cementite, Fe_3C, cf. D. Meinhardt, *Arch. Eisenhüttenw.*, **30**, 1959, 51-55, as an example of the position of light carbon atoms in a structure with heavy metal (iron) atoms.

[226] *Nucleonics*, **7**, December, 1950, 31-39.

[227] *Phys. Rev.*, **83**, 1951, 226-227.

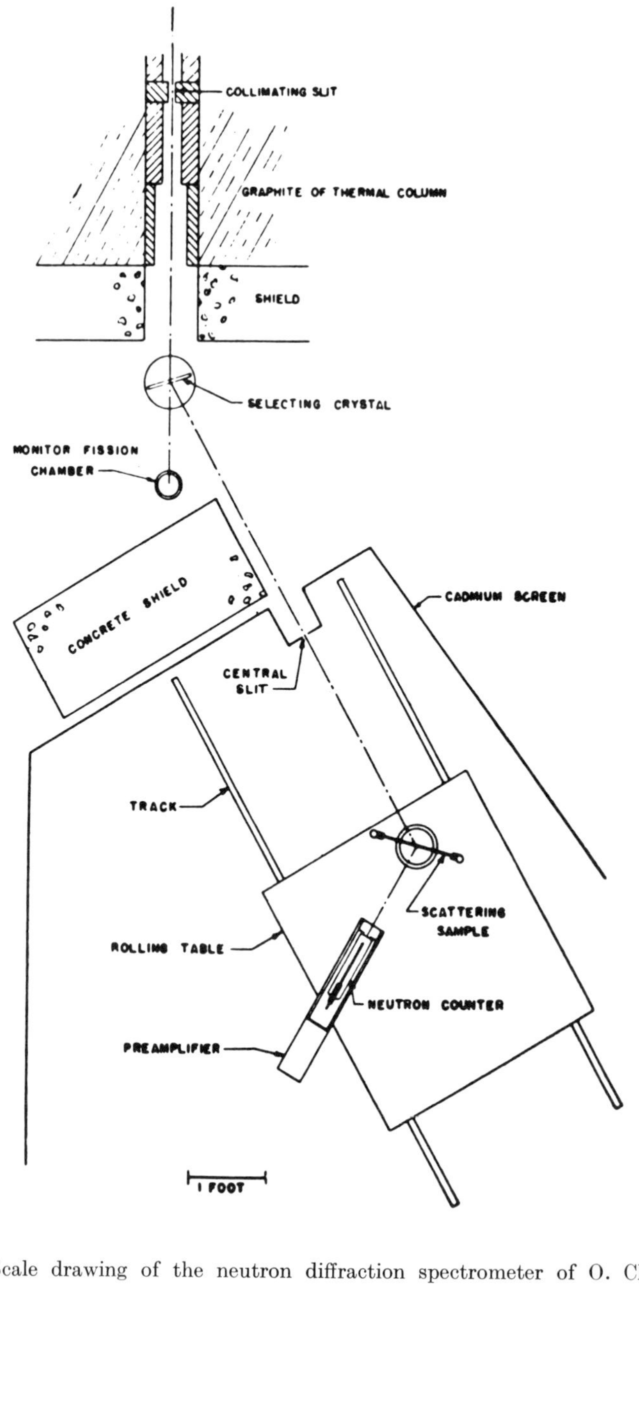

FIG. C. 70. Scale drawing of the neutron diffraction spectrometer of O. Chamberlain.

ica glass, which shows in B. E. Warren's (1934) X-ray diagrams only one single broad peak and indistinct minor maxima, Milligan gave five well-developed peaks at distances of 1.58, 2.56, 3.02, 4.0 and 5.1 Å., and more diffuse peaks at 5.9, 6.6, 7.4, 9.0 and 9.7 Å. Weber also gave a diagram for a soda-lime silicate glass, with seven peaks at 1.09, 1.21, 1.45, 1.82, 2.29, 2.72 and 3.42 Å. Plexiglas (an organic polymeric plastic) shows an approximately constant scattering as a function of sin Θ/λ.

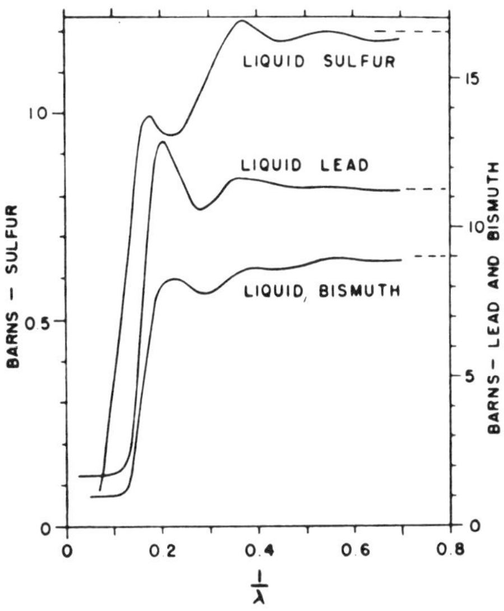

FIG. C. 71. Total scattering cross-sections as inverse functions of wavelengths (Chamberlain), for sulfur, lead, and bismuth.

106. For the special problem of the position of hydrogen atoms and of protons in crystalline structures, the neutron diffraction method is probably of high usefulness, not only in organic compounds, but when combined with nuclear magnetic resonance spectra. In comparison with results of X-ray diffraction, D. E. Elleman and D. Williams[228] could locate the protons in the important layer structure of brucite, $Mg(OH)_2$, in parallel planes with hexagonal symmetry, with mutual distances of adjacent protons of 3.12 ± 0.01 Å. in this plane, but only 1.93 ± 0.93 Å. between one given proton and the next one in the adjacent plane. The distance of the proton planes is 0.69 ± 0.02 Å.

107. We will discuss electron diffraction phenomena in the following chapter on the electron microscope, since all of the modern larger instruments are furnished

[228] *Natl. Bur. Standards (U. S.) Circ.* **527**, 1954, 345-354. For determination of the hydrogen positions in $Ca(OH)_2$ by magnetic nuclear resonance method see D. M. Henderson and H.S. Gutowski, *Am. Mineralogist*, **47**, 1962, 1231-1251 (cf. I. A. ¶ 6*, Footnote[10]).

with diffraction accessories. We should emphasize here, however, the importance of field-emission (point projection) microscopy of electrons and ions as one of the most instructive and promising modern methods for the study of adsorption layers and for making visible material of molecular size. E. W. Müller[229] has given a most fascinating report on the investigations available until 1953, and E. Schmid and F. Stangler[230] a report for the study of condensation films on crystallographically oriented metal crystal faces. Among the molecules made visible we note naphthalene, anthracene, phtalocyanine, and a series of other compounds of medium molecular weights.

ELECTRON MICROSCOPY OF DISPERSED SYSTEMS

108. The rapid evolution of the electron microscope to an indispensable instrument of the modern silicate research laboratory, and the very large literature in this important field, can here only be reviewed briefly. We mention as particularly valuable the bibliography (up to 1950) in the monograph of Cl. Marton, S. Sass, M. Swerdlow, A. van Bronkhorst, and H. Meryman.[231] M. Swerdlow[232] described in a special report the great advance made in the years up to 1954, and B. von Borries[233] gave a very interesting discussion of the newest results and of the limits of electron microscopy, as seen up to 1956, especially of the progress in the preparation of samples. In the series of modern electrostatic electron microscopes that of the VEB C. Zeiss, Jena, called Elmi D.2 is probably the most efficient available instrument, as described by H. Müller.[234] In comparison with the accomplished construction of the Siemens "Elmi"-microscope for transmittent electron radiation, the advance in mirror electron microscopes for reflected beams is still rather slow. G. Bartz, G. Weissenberg, and D. Wiskott[235] described an attempt made in this direction by the E. Leitz Optical Works, Wetzlar, presenting possibilities of developing a secondary emission microscope method. The principal difficulty in this field is seen in deviations of the electron paths on a perfectly even plane sample, by the grain boundaries of metal

[229] *Ergeb. exakt. Naturw.*, **27**, 1953, 290-360; see also *Natl. Bur. Standards (U. S.) Circ.* **527**, 1954, 345-354.

[230] *Radex Rundschau*, 1956, 173-188.

[231] *Natl. Bur. Standards (U. S.) Circ.* **502**, 1950, 87 pp.

[232] *Anal. Chem.*, **26**, 34-42, with many additional references.

[233] *Radex Rundschau*, 1956, 200-225; see also E. Brüche, *Naturwissenschaften*, **44**, 1957, 601-610. As an excellent introduction, with most valuable descriptions of sampling techniques and preparation in electron microscopy, we recommend the study of the recent book of L. Reimer, "Elektronenmikroskopische Untersuchungs-und Präparationsmethoden," Springer, Berlin, 1959, 300 pp. See also the extensive report of H. König, *Ergeb. exakt. Naturw.*, **27**, 1953, 188-247.

[234] *Jenaer Rundschau*, **4**, 1959, 147-153.

[235] *Radex Rundschau*, 1956, 163-172.

crystallites, and the build-up of boundary potentials. It is not possible to etch metals in conformity with the different contact potentials and different energies of electron emission.

109. Another highly promising attempt to make immediately visible surface structures of opaque objects in the electron microscope is that by G. Möllenstedt and M. Keller,[236] namely by a secondary electron emission from ion-bombarded metallic surfaces under the action of positive rays of 25 kv. potential. The resolution limit of such an electron microscope is 50 mμ, not as high as in the common transmission instruments but satisfactory for many important metallographic purposes. It may be possible to improve the resolution considerably by very small apertures of the Brüche-Johannson-type electron immersion objective. The secondary electron emission is a specific property of the ion-irradiated material. It is therefore easy to differentiate, by variable intensities of emitted radiation, the constituents of a complex sample, as could be demonstrated for graphite, cementite (Fe_3C), iron sulfide, and α-iron in cast iron structures, or for different orientations of the crystals of an austenitic steel. The contrasts are particularly favorable if the sample is heated during exposure to 150°C.

110. In the elementary transmittent-beam electron microscopy considerable progress has been made by the study of the structure of glass fibers. By an ingenious simple fiber spinning process, P. Kuhn and G. Schimmel[237] were able to make useful the ligh resolution power (1 mμ) of modern microscopes to detect finest heterogeneities in silica glass. A final decision on the existence and nature of A. Smekal's surface flaws ("*Kerbstellen*"), however, (cf. Vol. II, A. ¶ 300) was not possible since strong changes occur in the fibers during irradiation in the electron beam. It is possible to "spin" fibers of 1.5 mμ in thickness from 2 μ fibers under the caloric effect of electron bonbardment. The finest endings of such extremely thin fibers may consist of a few single [SiO_4] groups developing an extremely high surface energy. Peculiar spherical structures near the endings cannot yet be identified. They may be influenced by carbon layers derived from the vacuum of the electron microscope. Darker wavy structural details are observed in the endings of alkali-free calcium- and aluminum-rich refractory glasses, with chain-like configuration and a surprising regularity. A borosilicate glass (with 11 per cent Na_2O) showed

[236] *Radex Rundschau*, 1956, 153-162. With particular pleasure we note, during the final revision of the manuscript of our text, that on the basis of the investigations of G. Möllenstedt *et al.*, the problems of the development of a highly useful emission electron microscope have been solved. H. Düker, *Radex Rundschau*, 1960, 406-420, reports on the construction of such an instrument by which important details of changes on the surface of iron by oxidation could be observed; cf. *Z. Metallk.*, **51**, 1960, 314-321, 377-381; *Schweiz. Arch. angew. Wiss. u. Tech.*, **26**, 1960, 199-201; *Proc. European Reg. Conf. on Electron Microscopy, Delft*, 1960, Vol. I, 456-459.

[237] *Glastech. Ber.*, **30**, 1957, 463-470.

boiling by alkali evaporation, from the interior of the fibers, to a foamy aspect. Selenium ruby blass also shows the boiling very strikingly, with a segregation of metallic selenium on the surface. Lead glasses show an increasingly intensive irradiation, with phase unmixing and the agglomeration of dark material, which slowly evaporates and leaves a skeleton of foamy glass. The freshly spun fibers are especially sensitive to such phase separations, but aged glasses are rather highly stabilized.

111. The observation of structural heterogeneities in glasses of different chemical compositions was previously described by A. F. Prebus and J. W. Michener[238] in transmittance electron micrographs copied in high contrasts. The sizes of the variously shaped structural elements (either granular or chain-like arrangements) is given between 2 and 20 mμ. The significance of these findings in correlation with the heterogeneity described by E. A. Poraĭ-Koshits (see C. ¶ 103) and other Russian authors is most evident. The order of magnitude of 2 to 20 mμ coincides with that for the ordered three-dimensional areas assumed in the crystallite theory of glass constitution (cf. Vol. II, A. ¶ 206 ff.). A granular structure in glasses and a typical phase separation are also observed in the micrographs given by F. Oberlies,[239] especially for sodium borosilicate glass (of Vycor type), but also for common soda-lime silicate glass on fresh fractures, by replica techniques (see below). This direct evidence confirms the previous assumptions of A. Dietzel[240] on latent unmixing tendencies to form alkali-enriched and alkali-impoverished portions in the glass structure. The order of magnitude of the unmixing was observed by Oberlies as 10 to 20 mμ, identical with those concluded by Prebus and Michener. In silica glass the heterogeneities shown by electron microscopy are of different types, namely not yet liquefied residual quartz fragments or first nuclei of devitrification products, such as cristobalite or tridymite.

112. Whereas the compact structure and rigidity of glasses requires fracturing to create surfaces suitable for electron microscope methods, the specific foliate structure of micas, chlorites, glauconite, and others makes the sampling difficult because of the smoothness of the material. G. Pfefferkorn, H. Themann, and H. Urban[241] found the use of an ultramicrotome (with a glass blade as knife) most useful

[238] *Ind. Eng. Chem.*, **46**, 1954, 147-153.

[239] *Naturwissenschaften*, **43**, 1956, 224; *Ibid.*, **44**, 1957, 488-489.

[240] *Glastech. Ber.*, **22**, 1949, especially p. 213.

[241] *Naturwissenschaften*, **43**, 1956, 535; *Fortschr. Mineral.*, **35**, 1957, 55-56. See further G. Pfefferkorn, S. Kienow, W. Ohnemüller, and H. Urban, *Naturwissenschaften*, **47**, 1960, 322-323, on ultramicrotome sections through mullite crystals (tubular, with fillings of iron oxides), and of common roof tiles. On the ultramicrotome of B. von Borries see *Radex Rundschau*, 1956, especially pp. 216 ff. which was used also for the study of coals and of silicotic lung alveolae. See F.-J. Eckardt, *Fortschr. Mineral.*, **38**, 1960, 126 f.; especially *Z. Krist.*, **116**, 1961, 36-47; he used the ultramicrotome for sections of 300 Å. thickness through clay sediments. See further

in preparing thinnest sections (of 0.1 to 1.0 mm.^{2} in size) of foliate minerals. Impressive electron micrographs of the mostly dendritic growth aggregates, namely of glauconite grains of radiant crystal "felts" with a very high pore volume (above 50 per cent), result from this method. Kaolinite grains, if sedimented only from a height of a few mm., are at random in their flakelets, whereas pronounced parallel textures are observed if the sedimentation height was 100 mm. A very important success of ultramicrotome techniques using a *diamond* knife is the definite evidence given to the tubular structure of chrysotile fibers as shown by M. Maser, R. V. Rice, and H. P. Klug.[242] The micrographs show the circular cross sections, and end-on views of concentric tubes.

113. For the sampling of the finest dispersed solids, dusting directly on the collodion supporting film is the best method, as was shown by W. Behne and W. Müller.[243] If the particles of the dispersoid were pronouncedly anisotropic in shape, as in the tubular laths of halloysite, these authors applied an evaporation method for a water suspension in the electrostatic condensor (potential field of 80 kv., from a Kleinst-van de Graaf generator); alternatively, they sprayed the suspension onto a metal diaphragm above the drop. The advantage of this method is that the crystals are oriented in all directions in the space, and one can observe cross sections of typical annular shape in halloysite. For very similar purposes, especially for the study of halloysite or of natural and synthetic chrysotile asbestos, garnierite, or others, W. Noll, H. Kircher, and W. Sybertz[244] suspended the mineral powder in vaseline by an a.c. vibration mixer and applied it in a thin smear onto the collodion supporting membrane. The vaseline was then removed by leaching in petrol ether or xylol, until the dispersed material was entirely dry and purified.

114. For morphological study of surface structures, electron microscopy offers optimum conditions for application of the *replica technique* of H. Mahl (1941). Its subsequent development came to fruition through the work of a great many authors. We must here omit details of the many proposals made for the use of various organoplastics, which involved the fundamental idea of preparing either primary replicas directly drawn from the surface of the sample and observed in the trans-

O. E. Radczewski and J. Schädel, *Physik. Verhandl.*, **12**, 1961, 153, especially on the distortion of kaolinite crystals by a diamond kuife cutting process, and a subsequent dehydration in the vacuum of the electron microscope. The $d(001)$ parameter appears anomalously shortened. See also *Ber. Deut. Keram. Ges.* **39**, 1962, 48-51. Powder samples and ultramicrotome thin sections of goethite ooids were studied by O. E. Radczewski and P. Schneiderhöhn, *Beitr. Mineral. Petrogr.* **8**, 1962, 349-353, in the electron microscope.

[242] *Am. Mineralogist,* **45**, 1960, 680-688. The Servall Porter-Blum ultramicrotome was used with a diamond knife, the specimen being embedded in Araldite, an epoxy resin superior in adhesion to polymethacrylate resins.

[243] *Naturwissenschaften*, **41**, 1954, 138.

[244] *Kolloid-Z.* **157**, 1958, 1-11.

mittent electron beam, or reinforcing the (negative) image of the primary replica by another of particularly desirable tenacity and leaching away the primary replica and the sample residues adherent to it. Since the contrasts in the directly drawn replicas are not very high, it is most important that the (negative or positive) image be reinforced by shadow-casting with metal vapors (or related substances) deposited on the surface in a cathode vacuum chamber. This shadowing is chiefly done with platinum, or with carbon or silicon monoxide, with many special variations of experimental procedure. The reader will do best to study the special descriptions in the literature cited below, here only discussed in relation to their important results.

115. We must emphasize the great progress made in our knowledge on the constitution of clays and other mineral aggregations by replica techniques, first of all concerning the intricate morphological development of the clay minerals. Impurities by associated minerals of a different chemical character are in most cases easily detected. It was sometimes also surprising for investigators to see that halloysite and fibrous magnesium silicate hydrates are much more frequent in clay sediments than was previously assumed.[245] Truly excellent, instructive micrographs of kaolinite crystals in beautiful single forms, or in worm-shaped aggregates, with very smooth basal surfaces, were later shown by J. H. Koenig and S. C. Lyons, and by these authors with J. J. Comer[246] who especially worked out successfully the replica and shadowing methods for ceramic clay raw products. Comer developed shadow-casting with carbon to a high level. Carbon is particularly valuable because of the extreme fineness of its particles and the structureless coating on the shadowed portions of the image, as seen in Comer's studies with J. W. Turley[247] on the surface of bulk clays. Best results were obtained with a preshadowing layer of platinum, backed by a layer of evaporated carbon or by "Parlodion" (see Fig. C. 72), prior to the removal of the clay with hydrofluoric acid solution. Even the finest details are shown by highly porous halloysite, with its tubular structure, and also by attapulgite fibers. The great advantages of stereoscopic electron micrographs for an examination

[245] Cf. E. G. J. Wiegmann, *Silikat Tech.*, **4**, 1953, 546-547.

[246] *Ceram. Age*, July, 1955, 8-14; *Ceram. Ind.*, October, 1956, 125-127, 148, 150. On the smoothness of surface of kaolinite flakes, see H. W. Stetson, *J. Am. Ceram. Soc.*, **38**, 1955, 462. The apparent roughness described by W. D. Kingery and J. Francl, *Ibid.*, **37**, 1954, especially p. 602, is caused by losses of image-forming electrons by a diffraction process, and replica images are freefrom such disturbances. Very remarkable electron micrographs of clay minerals, of scorodite, gearksutite, and a series of other minerals in ore deposits were given by G. S. Gritsaenko, A. I. Gorshkov, and K. E. Frolova, *Zapiski Vsesoyuz. Mineral. Obshchestva*, **87**, 1958, 269-276; *Ibid.*, **90**, 1961, 266-269.

[247] *J. Appl. Phys.*, **26**, 1955, 346-350; see also "Symposium on Microscopy," Special Tech. Publ., ASTM, No. 257, 1959, 94-120. The replica and shadow-casting techniques are especially described in the second section of the book of L. Reimer, "Elektronenmikroskopische Untersuchungs-und Präparationsmethoden," Springer, Berlin, 1959; the carbon replica is discussed on pp. 150 ff.

of stripping surface layers of clay particles is also quite evident from the paper of Comer and Turley. A series of beautiful electron micrographs by J. J. Comer with G. W. Brindley[248] and Th. F. Bates[249] may here be recommended for detailed study.

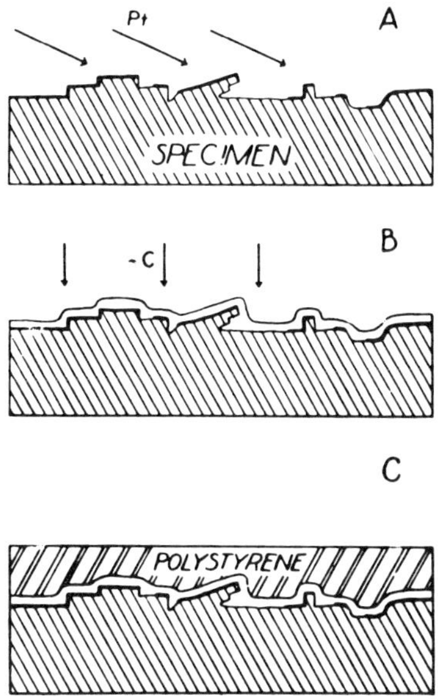

FIG. C. 72. Stages in preparation of preshadowed carbon replicas (Comer).

These advanced replica techniques are very useful for studying changes occurring in clay minerals, as well as for study of mullite formation in kaolinite and metakaolinite after firing to high temperatures (cf. Vol. V, Section B), or in the recrystallization phenomena of oxide ceramics.[250]

[248] *Natl. Acad. Sci.-Natl. Research Council, Publ.*, **456**, 1956, 61-66; *Ibid.*, **395**, 1955, 26-30, with L. B. Sand.

[249] *Natl. Acad. Sci.-Natl. Research Council Publ.*, **395**, 1955, 1-9. *Mineral Inds. Expt. Station, College of Mineral Industries, Pennsylvania State Univ., Circ.*, **51**, September, 1958, 61 pp., containing an atlas of electron micrographs of many fine-dispersed minerals and non-clay minerals. K. Saji, *Zement-Kalk-Gips*, **12**, 1959, 418-423, could by carbon-replica-shadowing methods even make visible details in the surface of hardening Portland cement pastes (cf. Vol. V, Section C).

[250] See H. Pfisterer, *Ber. deut. keram. Ges.*, **30**, 1953, 216-220; *Ibid.*, **31**, 1954, 72-77; R. Dallendörfer, *Silikat Tech.*, **9**, 1958, 510-513. These authors worked with Triafol BN (cellulose acetatebutyrate, without addition of softener), and chose silicon monoxide as the shadow-casting material in the study of different porcelain structures. A combined carbon replica-platinum-palladium shadowing method for the study of glass surfaces was developed by Sh. Hashi, *Asahi Garasu Kenkyu Hokoku*, **8**, 1958, 96-100.

116. The use of silica replicas was early shown by F. A. Hamm and J. J. Comer,[251] and then by Comer and Turley in their studies mentioned above, for clay minerals, combined with platinum shadowing. A similar technique is mentioned by V. I. Shelyubskiĭ[252] for the study of glasses, namely the lead-containing glasses described in C. ¶ 86, or porous leached borosilicate glasses.

117. In a series of most interesting studies, E. Brüche and H. Poppa[253] gave a complete electron-micrographic analysis of the technologically basic process of glass polishing. Often rather contradictory hypotheses and theories had been developed in the literature. Exclusively chemical viewpoints were assumed, on one hand, referring to intense reactions combined with the abrasion of the grains of the polishing agent. On the other hand, purely physical aspects were assumed in the polishing process, namely a finest grinding process of the glass surface, building up a so-called Beilby layer, i. e. a completely amorphous aggregation of colloidal particles of the polished material intimately mixed together with the lubricant agent, as commonly added. In analogy to the process on polished metals, Beilby layers were also discussed by Brüche and Poppa[254] together with a critical comparison with opinions of A. Kaller[255] (cf. Vol. II, A. ¶ 413 ff.). Without discussing here the details of the distinction between optical polishing and industrial plate glass polishing processes, the electron microscope, combined with refined replica methods for a finest development of relief structures, cannot determine chemical reactions occurring in the polishing mix on the glass surface. However, the very instructive diagram Fig. C. 73 may give an impression of Brüche's and Poppa's explanation of how the sharp profiles of the "surface mountains" are gradually removed, in certain analogies to the polishing process of metals (e.g., by the anodic-electrolytic etching polishing of gold, or silver).

118. *Plastic* deformation in glass polishing is not the crucial process, and the typical "smearing" of Beilby layers only occurs with iron oxide layers of a pasty consistency. Normally, grooves on the surface are not filled up with loose polishing material. The surface layer also does not consist of pure glass material, but of a

[251] *Anal. Chem.*, **20**, 1948, 861-870.

[252] *Stroenie Stekla, Inst. Khim. Silikatov Akad. Nauk S.S.S.R., Trudy Soveshchaniya, Leningrad*, 1953 (Pub. 1955), 219-223.

[253] *Silikat Tech.*, **6**, 1955, 378-384; *Ibid.*, **8**, 1957, 516-519; *J. Soc. Glass Technol.*, **40**, 1956, 513-519; *Glastech. Ber.*, **28**, 1955, 232-242; on the so-called double-replica technique, combined with carbon-shadowing, see especially *Ibid.*, pp. 233 ff.; *Ibid.*, **29**, 1956, 183-192; *Ibid.*, **30**, 1957, 163-175, 387-393; *Ibid.*, **31**, 1958, 341-348.

[254] *Z. angew. Phys.*, **8**, 1956, 486-492.

[255] *Naturwissenschaften*, **43**, 1956, 158; *Silikat Tech.*, **7**, 1956, 380-390, 399; *Jenaer Jahrb.*, 1957/58, 145-167. The chemical theory of polishing was especially defended by I. V. Grebenshchikov (1937), and N. M. Kachalov (1946).

gel-like glass modified by the polishing agent, in a very thin, smooth layer. The previous theory of F. W. Preston,[256] stating that the reappearance of overpolished scratches on highly polished glass after etching is caused by the decomposition of the amorphous glass material smeared into the grooves, cannot be correct. The pressure under which a scratching diamond pointer was conducted over the glass surface

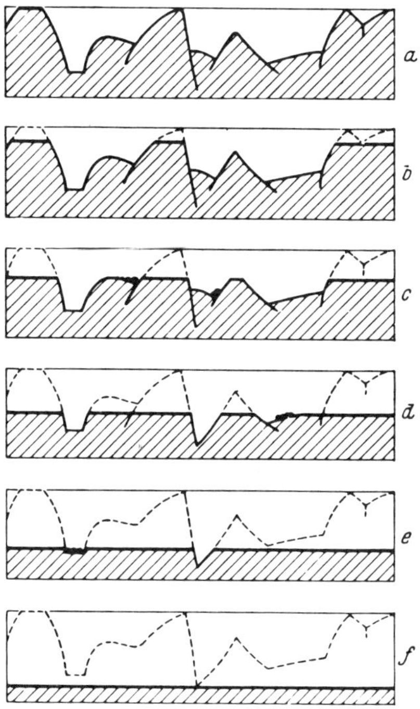

FIG. C. 73. Diagram illustrating progress of industrial polishing of plate glass with felt and rouge (Brüche and Poppa).

cuts into the glass about 30 μ and brings about zones of strains which make the glass sensitive to etching. This explanation is in agreement with the thoroughly brittle behavior of glass in the dry state, whereas in water adsorption layers on glass a more hydrogel-like behavior is thinkable (A. Smekal's "microplasticity").[257]

[256] *Trans. Opt. Soc. London*, 23, 1922, 141-161.

[257] *Glastech. Ber.*, 28, 1955, 81. Important recent contributions to the investigation of the glass polishing process were made by V. A. Dubrovskiĭ and T. S. Dubrovskaya, *Steklo i Keram.*, 16, 1959 (7), 30-35, by applying the radioactive isotope tracers Ca^{45} and Fe^{59} in the cation exchange of calcium from the glass with iron and zinc in polishing accelerators ($FeSO_4$, $FeCl_3$, $ZnSO_4$) added to the polishing rouge. A cation exchange of this kind controls accelerating or retarding phenomena. Also interesting is the nature of the protection film on the glass formed. It is a silica hydrogel only if the polishing liquid is acidic; under neutral conditions insoluble calcium

119. Fresh-broken glass surfaces show in the electron microscope image smooth-walled grooves of scratches, identical if produced in the open atmosphere, or under water, alcohol, benzene, trichloroethylene, or sodium-dried petroleum, all which act as lubricants. The great rigidity of dry open glass surfaces and high exterior friction cause the splintery shapes of the cracks, if no lubricant is present. In summary, Brüche and Poppa's studies on glass polishing demonstrate most impressively the high degree of conclusiveness and reliability of the electron-microscopic method in a systematic experimental application to a silicate problem of such a complex character. It is interesting in this connection that W. Rebentisch and F. Dingelacker[258] used the carbon-replica technique also for a study of the size, crystal habitus, and aggregation of polishing rouge in its very distinct and specific influence on the suitability for optical polishing of glass. Pseudomorphs of iron oxide after crystals of iron oxalate and characteristic recrystallization phenomena were detected in the electron micrographs given by those authors. γ-Fe_2O_3 is formed from iron formate. It is stable up to 700°C., with a strong color change from brown to red above this temperature (to α-Fe_2O_3).

120. The carbon-shadowing method is also especially recommended by Irmtraud Pfeiffer[259] for the study of the finest details (with a resolution of 6 mμ) in the surface of metal alloys, using her direct-shadowing method. The carbon is deposited from an intense electric arc (with 20 to 30 amp. burning one to two seconds only) in a vacuum of 10^{-5} mm.Hg, the carbon being applied in a thickness of about 30 mμ. Very important is also the experience of R. V. Rice and A. J. Cohen[260] in the study of surface "sculpture" of twinned quartz crystals by a successive examination of specific areas by optical microscopy and then by electron microscopy. These authors used a kind of focusing action of a meniscus on the thin carbon replica floating on a liquid, the replica positioned on a grid (100, 400, or 1000 mesh copper nets) as the liquid is drawn down through a glass ball-and-socket joint (cf. Fig. C. 74). Supporting films are in this method dispensable, and a maximum resolution is verified in the exceptionally sturdy carbon replicas which are able to withstand the repeated transfers into 48 per cent hydrofluoric acid to dissolve quartz particles adherent to them.

121. To what a high degree of accomplishment and instructiveness the modern replica techniques, combined with shadow-casting methods, using carbon, silica, silicon monoxide, or even silicon carbide, have been developed, is demonstrated by

silicate hydrogels occur, which in basic solution phases also contain sodium (see also recently *Structure of Glass., Proc. 3rd All-Union Conf. on Glassy State, Leningrad*, 1960, (English Translation), 382-384).

[258] *Glastech. Ber.*, **32**, 1959, 321-327.

[259] *Naturwissenschaften*, **42**, 1955, 508-509.

[260] *Am. Mineralogist*, **43**, 1958, 25-33.

F. Grasenick[261] in high quality electron micrographs, not only of metal alloys but also of metallurgical products, including refractories and their reaction products, in stereoscopic (anaglyph) diagrams. If the sample contains ferromagnetic components, it is also possible to use electron spin exchange with the electron beam of the

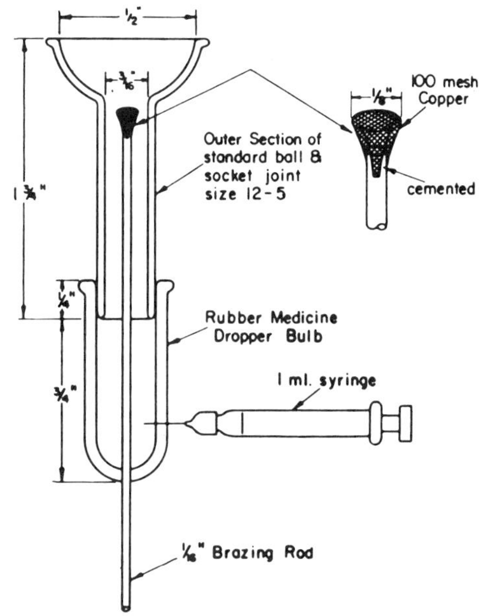

FIG. C. 74. Schematic diagram of replica-orienting apparatus (Rice and Cohen).

microscope and the resulting deviations for diagnostic purposes. Sh. Yamaguchi[262] speaks of an electron polarization microscopy in this sense. He has demonstrated the results in the extraordinarily strong contrasts obtained in the electron micrograph of a magnetite film formed by the oxidation of iron, a contrast which it is not possible to observe with common techniques, only with alumina replicas (Mahl, 1950).

122. The investigator must be well acquainted with the multiple phenomena of structural details in samples when studied under the electron microscope, modified under the *action of heat* evolved in the high vacuum of the instrument by electron bombardment. Particularly typical phenomena of this kind include the considerable changes in finest fibers of crysotile asbestos, as demonstrated by W.

[261] *Radex Rundschau*, 1956, 226-246; 1957, 843-856. We mention here also the fascinating results of U. Hofmann, H.-P. Boehm, and W. Gromes, *Z. Anorg. Allg. Chem.* **308**, 1961, 143-154, on measurements of the size and thickness of clay mineral crystals in the electron microscope, combined with determinations of anion exchange (cf. C. ¶ 271f.).

[262] *Naturwissenschaften*, **44**, 1957, 555.

Noll and H. Kircher,[263] and by B. Nagy and Th. F. Bates.[264] The originally tubular morphology of the chrysotile is thus changed to fibers looking like flat laths with serrated and pitted surfaces, and in higher degrees of decomposition the edges of the fibers appear rounded ("fused"). Another very troublesome phenomenon is the contamination which was observed, e.g., by J. J. Comer,[265] if brass or copper specimen holders are used from which the electron beams evolve vapors of the metal, contaminating the sample with copper, zinc, zinc oxide, or even lead. Also a copper grid to support the specimen can cause deposits of the metal on the sample.

123. Particularly troublesome and often misleading in morphological studies by electron microscopy are the moiré-like *interference phenomena* caused by superimposed, or slightly disoriented crystallites. Very early these phenomena were recognized in their significance as causes of errors in the interpretation of the images. H. König[266] discussed well-selected examples in his monograph report mentioned repeatedly. Th. Nemetschek, W. Grassmann, and U. Hofmann[267] compared the interference phenomena in the fibrils of collagen with those in hornblende asbestos which show a great sensitivity in the interference patterns in the intensity and position of small changes in the angle between the electron condensor planes. Corresponding phenomena were described by G. W. Brindley, J. J. Comer, R. Uyeda, and J. Zussman[268] as interference fringes on the tabular crystals of Manchurian "Yu Yen stone" (dense antigorite), with distinct correlations of the fringe separation and the lattice parameter a_0 (= 38.5 Å.). As an example the equally spaced fringe separation was given as 100 ± 10 Å. (connections with the superstructure of antigorite are discussed in A. ¶ ¶ 156–163). They have a great similarity with the electron interferences (fringes) observed by O. Rang and H. Poppa[269] in superimposed gold crystals, anal-

[263] *Naturwissenschaften*, **39**, 1952, 188.

[264] *Am. Mineralogist*, **37**, 1952, 1055-1058.

[265] *J. Appl. Phys.*, **28**, 1957, 820. Another kind of contamination by carbon deposition from decomposed organic material was studied by H. G. Heide, *Naturwissenschaften*, **47**, 1960, 313-317. Not only traces of oxygen cause this troublesome reaction, it occurs even in nitrogen or argon, and is evidently caused by ionization effects exerted by the electron beam. Heide's investigations give, on the other hand, full evidence that in special gas cells it is possible to observe samples under pressures up to 760 mm. Hg in the electron microscope and even to avoid desiccation by volatilization of water.

[266] *Ergeb. exakt. Naturw.*, **27**, 1953, especially pp. 232-241.

[267] *Z. Naturforsch.*, **10**, *b*, 1955, 61-68.

[268] *Acta Cryst.*, **11**, 1958, 99-102. G. Kunze (cf. A. ¶ 160) gave in his comprehensive monograph on the structure of antigorite, *Fortschr. Mineral.*, **39**, 1961, 206-324, a detailed derivation of the moiré patterns in electron micrographs and their evaluation (see *Ibid.*, pp. 297-315).

[269] *Naturwissenschaften*, **45**, 1958, 239; see also J. Hillier, *Natl. Bur. Standards (U. S.) Circ.* **527**, 1954, 413-416, for plate-like crystals, slightly curved or in multilayers; see also E. Brüche and J. Demny, *Naturwissenschaften*, **46**, 1959, 225, for the detection of moiré patterns which are not in crystal pairs, but in single crystals of gold, with spiral growth (cf. Vol. III, Section A),

ogous to interference figures observed by J. W. Menter[270] in copper and platinum phthalocyanines. A complete geometric theory for the derivation of moiré patterns and of related interference phenomena in electron microscopy for superimposed positions was recently proposed by O. Rang.[271] It completely confirms the observation that double crystals give reflections like from a single crystal only where the appropriate net planes may be regarded as the continuation of the planes of one crystal to those of the other crystal. The theory provides a firm basis, not only for reliable conclusions concerning structural flaws in the crystals, but also to measure the coherence distances of the electrons. Purely geometrically one may distinguish between parallel moirés and distortion moirés, both often combined.

124. The particular usefulness of ion bombardment for an "*etching*" of crystal surfaces for electron microscopic-diagnostic purposes was demonstrated by T. K. Bierlein, H. W. Newkirk, and B. Mastel[272] in their investigation of highly refractory porcelains, oxides, and especially cermets. Such an etching is analogous to the ionic bombardment etching of metallographic studies, producing a very clean surface free from stains and corrosion products. The ion bombardment-etched surface can also be studied especially well by light-optic, such as phase contrast microscopy. Fine details of the crystalline structure in porcelains and cermets are made visible. Concerning the study of etching figures used for crystal orientation in the electron microscopic samples, G. Pfefferkorn[273] earlier developed a replica technique which gave surprising details in the geometric symmetry conditions of crystal growth, expecially if the etching is made in a supersonic frequency field to promote a uniform dissolution process by the etching solvent.

125. The great advantages of electron microscopy replica methods in ceramic investigations are so obvious that it is justifiable to select here some of their characteristic applications for special ceramic problems which at the same time demonstrate special preparation methods. Fr. Oberlies and N. Köppen[274] studied the very interesting details of ancient (especially Greek) ceramic techniques on the surface of terra sigillata vases, and the alternate oxidizing and reducing conduction of the furnace atmosphere to produce the well-known deep-black and red deposits in the ceramic structure. The electron microscope shows the single steps of the mechanical orientation of the platy clay crystals in the sintered (not fused-glazed) surface layer of such artifacts. Replica and platinum shadowing are the standard methods for

in two or more layers. Also iron nitride crystals show interference patterns caused by crystal flaws, as described by W. Pitsch, *Arch. Eisenhüttenw.*, **29**, 1958, 125-128.

[270] *Proc. Roy. Soc.* (*London*), A **236**, 1956, 119-135.

[271] *Z. Krist.*, **114**, 1960, 98-119.

[272] *J. Am. Ceram. Soc.*, **41**, 1958, 196-200.

[273] *Fortschr. Mineral.*, **29**/**30**, 1954, 34-38.

[274] *Ber. deut. keram. Ges.*, **30**, 1953, 102-110; Ibid., **31**, 1954, 287-301.

preparing the samples; they are also useful for the study of the highly fragile weathered surface layers of ancient or modern glasses as was shown by Fr. Oberlies.[275] Characteristic of such iridescent ("blind") glasses are weathering crystals of calcite in typical twins. Such electron microscopic studies are so sensitive that they show on window or plate glass after only three days a "swelling layer" in which sodium silicate and carbonate hydrate crystals are identified, starting from single points on the surface.

126. We repeatedly mentioned the excellent conclusiveness of electron micrographs of *porous* or leached sodium borosilicate *glasses* of the Vycor type. In the majority of cases they are treated by replica techniques with metal or carbon shadowing, as in the photographs given by W. Skatulla, W. Vogel, and H. Wessel,[276] futher by Kl. Kühne and W. Skatulla.[277] A combined evaporated carbon film and silicon monoxide-platinum replica technique was controlled in its fine-granular structure by the comparison with the very smooth surface of molybdenum trioxide crystals sublimated on the samples. Not speaking here of the structural heterogeneity of the borosilicate glasses from the viewpoint of glass constitution (cf. Vol. II, A. ¶ 239; Vol. III, Section B), the electron micrographs show beautifully the drop-like unmixing structures of the B_2O_3-enriched and impoverished areas (phase separation). Also in fluoride-containing glasses analogous structural marks are observed (cf. Vol. II, A. ¶ 384). Stereophotograms of the borosilicate glasses show the finely dispersed silica in the interstices of the spongy structure with its pores and secondary elements of the unmixed phases in Pyrex and other glasses, often combined with a distinct cloudiness and opalescence, which is in high degree dependent on the thermal history of the samples.

127. Identical observations were further demonstrated by M. Watanabe, H. Noake, and T. Aiba,[278] employing a special two-step replica technique. Even the very complex constitution of the ground-coat/iron interface in steel sheet enamels could be studied by electron microscopy, as I. Buck[279] showed, especially for the adhesion layer in the glassy interface. Buck used a water-soluble polyvinyl alcohol ("Elvanol") for the initial replica and Formvar (polyvinyl formal) for the second (positive) replica of high sturdiness and flexibility. Needle-shaped crystals in the interface

[275] *Glastech. Ber.*, **29**, 1956, 109-120.

[276] *Silikat Tech.*, **9**, 1958, 51-62, 323.

[277] *Silikat Tech.*, **10**, 1959, 105-119; see also selected electron micrographs prepared by W. Skatulla in *Jenaer Rundschau*, **4**, 1959, 79-83, shadowed by wolfram trioxide vapors. Systematic electron-microscopic studies of binary, ternary, and polynary glasses were recently presented by F. K. Aleĭnikov, P. V. Dundzis, R. B. Paulavichius, and V. A. Slizhis, *Lietuvos TSR Mokslu Akad. Darbai, Ser. B.* **2** (29), 1962, 95-108.

[278] *J. Am. Ceram. Soc.*, **42**, 1959, 593-599.

[279] *Am. Ceram. Soc. Bull.*, **32**, 1953, 114-117.

could be identified as tridymite by electron diffraction, besides other undetermined crystal phases. It is highly interesting that Sh. Yamaguchi and T. Nakayama[280] found titanium films deposited by evaporation in a high vacuum on glass surfaces to be an excellent structure-free replica material and particularly suitable for preparing optical gratings for determination of the magnification in electron microscopy. The sculpture of smooth subjects in loose powder form, sprayed on a liquid replica resin film, such as Formvar, is recommended as a "wrapping" method by H. Sitte,[281] combined with metal shadowing. This method was especially developed by H. König with most promising results (see above). The "wrapping" techniques and carbon replicas were also applied by G. Schimmel[282] in the electron microscopic investigation of extremely soft materials, namely calcium hydroxide suspensions (pulps) in slaking lime and of calcium carbonate in hydraulic mixes. The drying of such materials like slaked lime requires special operations, such as the application of low-temperature evaporation of the water in a high vacuum.

128. Especially for refractory silicate bodies with mullite and corundum, L. Degueldre[283] used a combined collodion-silicon monoxide replica method developed by E. Kellenberger,[284] as for polished sections of Pythagoras porcelain, etched by a 2 to 4 per cent hydrofluoric acid solution. The crystallographic development of both mullite and corundum is excellently illustrated in the micrographs, with characteristic single mullite needles torn from the polished surface and collected in the first negative collodion replica. The sharpest details are observed in repeated collodion transfers, especially in the fifth film drawn. Also P. P. Budnikov and V. S. Fadeeva[285] showed previously very good crystallizations in glazed mullite bodies, with about 8 per cent of fluxes, using a collodion-gelatin two-step replica technique, and shadowing with chromium. The older techniques of preparing alumina films for supporting electron microscopic samples is still important if the material is to be subjected to a thermal treatment in the microscope, and to be observed in its structural changes during the exposure. Those changes which the alumina films undergo were extensively studied by G. Hass,[286] by combined electron microscopy, electron diffraction, and optical (Drude) polarization methods, especially in growth of the oxide layers on aluminum metal, in thicknesses of 15 to 20 Å. initially, in

[280] *J. Appl. Phys.*, **24**, 1953, 658. W. S. Smith and W. G. Kirchgessner, *Ibid.*, **24**, 1953, 662, recommend Formvar and chromium shadowing, followed by evaporated silica for the studies of glass surfaces in the electron microscope.

[281] *Naturwissenschaften*, **42**, 1955, 366-367.

[282] *Zement-Kalk-Gips*, **10**, 1957, 134-138; *Ibid.*, **11**, 1958, 46-49. See also H. von Zander, *Ibid.*, **11**, 1958, 41-45.

[283] *Bull. soc. franç. céram.*, 1955, No. 28, 29-35.

[284] *Experientia*, **4**, 1948, 449-450.

[285] *Ogneupory*, **17**, 1952, No. 5, 228-230.

[286] *Z. anorg. u. allgem. Chem.*, **254**, 1947, 96-106.

glass-clear behavior and smoothness. The material is first practically amorphous, but gradually changed into γ-alumina (with $a_0 = 7.90$ Å.) in the temperature range from 600° to 680°C. Analogous studies of G. Hass[287] on silicon deposited in a high vacuum evaporation chamber (on rock salt or alumina) show in the electron diffraction diagram three diffuse rings; it develops full crystallinity above 600°C. (with $a_0 = 5.42 \pm 0.01$ Å.). The oxidation of silicon in the open air is very sluggish, even at 700°C.

129. The use of modern electron microscopes also for *electron diffraction* operations is so general nowadays and has reached such high instrumental precision that the full structure determination of crystal phases has become possible (cf. A. ¶ 149; B. ¶ 7, etc.). In addition, diffraction patterns can be obtained from very small areas of the sample surface (less than 1 μ^2) with special probing diaphragms. O. E. Radczewski[288] used this very simple method systematically in preparing diffraction diagrams for identification of clay minerals in sediments, ceramic raw products, and other materials such as nontronite from different soils. The diffraction diagrams obtained can be evaluated in the same way as are common X-ray Laue diagrams, by gnomonic projection and derivation of the reciprocal lattice by well-known graphic methods. O. E. Radczewski and H. Goossens[289] further discussed the possibility of obtaining oriented "conoscopic" diagrams of single crystals, e.g., of kaolinite, and of dark-field images with Bragg reflections, of (330) or in another case of (060) identified from defocussed diffraction patterns, and determined in the latter case by the zone [010], thus indicating that the crystal was elongated in the direction of the *b* axis. In the kaolinitic-illitic clay from Schletta (derived from a pitchstone), O. E. Radczewski and H. J. Balden[290] detected by analogous X-ray and electron diffraction methods, the presence of a lath-shaped mixed-layer mineral (one third montmorillonite, two thirds illite) (cf. B. ¶ 127). Beyond that, the electron diffraction makes a clear-cut distinction of kaolinite and illite possible, even in the case of their identical morphological development.

130. Electron diffraction diagrams of textured clay mineral aggregates, especially

[287] *Z. anorg. u. allgem. Chem.*, **257**, 1948, 166-172.

[288] *Fortschr. Mineral.*, **31**, 1953, 59-60; *Optik*, **10**, 1953, 163-169; *Sprechsaal*, **93**, 1960, 155-159. The important problems of the correlation of electron diffraction diagrams with common Debye-Scherrer photograms were facilitated by H. Kudielka and H. Rohde, *Arch. Eisenhüttenw.*, **31**, 1960, 503-507, by the use of a suitable rotation apparatus to produce rotation-symmetric diagrams, for a direct comparison of both methods.

[289] *Fortschr. Mineral.*, **34**, 1956, 61; *Optik*, **13**, 1956, 307-313. See also very recently G. W. Brindley and C. de Kimpe, *Am. Mineralogist*, **46**, 1961, 1005-1016, on the identification of clay minerals in single crystals from the electron diffraction diagrams by accurate measurements especially of the b_0 parameter.

[290] *Fortschr. Mineral.*, **37**, 1959, 74-78; *Silicates inds.*, **25**, 1960, 61-64; *Staub*, **22**, 1962, 312-323; Science of Ceramics, Confer. Oxford, June 1961, Proc., vol. I. 85-105.

with their basal reflections, were used by R. A. Mitra and M.V.R.K. Rao[291] for halloysite and montmorillonite, deposited as very fine films from an aqueous suspension on highly polished stainless steel discs, reflection patterns of which (with three faint haloes) may be taken into account. The "orientation reflection" method is particularly valuable for rapid identification of clay minerals in soil colloids. For celadonite, B. B. Zvyagin and R. A. Shakhova[292] observed unexpectedly characteristic interference spots from $(0kl)$ and $(hk0)$ planes, with the plane (a^*b^*) of the reciprocal lattice projected into the electron diffraction diagram. Pairwise reflections appear as $(0k0)$–$(0k1)$, $(hl0)$–$(hk1)$, further on the textures parallel c^* axis in the basal series: $(02l)$–$(\pm 11l)$ $(\pm 20l)$ $(\pm 13l)$, and so on. These textures are explained by the typical ribbon-shaped habit of celadonite crystals.

131. Elaborate electron diffraction studies combined with excellent electron micrographs were developed by J. Zussman, G. W. Brindley, and J. J. Comer,[293] on the serpentine minerals (on the structural details cf. A. ¶ 156–¶ 157), in their different varieties as silky chrysotile (with tubular elements), of clino- and ortho-chrysotile, similar to the X-ray fiber diagrams. Essentially platy are the so-called lizardites and bastites, further the antigorites and picrolites show a typical superlattice structure and corresponding diffraction diagrams. Since the serpentine minerals are highly sensitive to the heating effect in high vacuum from the electron beam impact, as seen from weakened and broadened diffraction spots, careful and rapid photographic operations are required. It is most desirable in such cases (as is possible in modern electron microscopes) that the electron-optical system can be changed from microscopy to diffraction without disturbing the specimen. A universal-stage-like adapter of the kind proposed by J. Hillier, R. F. Baker, and V. K. Zworykin (1942) is most profitable for controlled tilting of the specimen holder possible, especially for the determination of the platy and lath-shaped particles.

132. Besides the electron diffraction methods of collimating the fine electron beam to impinge on areas as small as 1 μ^2, the idea of developing a point-by-point determination of the chemical characteristics of the impacted material by emission of X-rays in a "microanalyzer" with a curved quartz crystal and a Geiger counter detector was discussed by R. Castaing and A. Guinier.[294] Although only metallic

[291] *Naturwissenschaften*, **42**, 1955, 119-120.

[292] *Kristallografiya*, **2**, 1957, 181-183.

[293] *Am. Mineralogist*, **42**, 1957, 133-153. As a remarkable result of electron diffraction analysis we also mention the studies of G. W. Brindley and J. J. Comer, *Mineral. Mag.*, **32**, 1960, 363-365, on the somewhat enigmatic mineral scarbroite which showed an excellent tabular orthorhombic crystallinity, and is in every case not pseudohexagonal as W. J. Duffin and J. Goodyear, *Nature*, **180**, 1957, 977; *Mineral. Mag.*, **32**, 1960, 353-362, had concluded from indexing X-ray powder diagrams.

[294] *Recherche aéronaut.*, 1951, No. 23, 41-50; *Anal. Chem.*, **25**, 1953, 724-726. See also R. Castaing, *Natl. Bur. Standards (U. S.) Circ.* **527**, 1954, 305-308. 309-313.

systems hare been examined up to the present, much attention should also be paid to existing possibilities in silicate investigations because of the high accuracy of such emission-spectroscopic methods. Further, samples down to 2 μ in diameter can be readily determined in their chemical nature. It may be recalled that J. Hillier and R. F. Baker (1944) recommended a similar idea, based on the determination of the nonelastic impact of electrons on the sample and corresponding energy losses.

SILICA AND SILICATE HYDROGELS

133. We will no longer restrict the discussion of hydrogels to silica hydrogels, but treat equally of these and of silicate gels, with which they have many of their physicochemical characteristics in common. The basic modern concept of the genesis of hydrogels from hydrosols as a polymerization—or more explicitly a polycondensation equilibrium—was given by W. Stöber.[295] (The reader is advised to review the work which was discussed in C. ¶ 34 ff., especially the experiments of H. Baumann.[296]) Typically lyophobic suspensions, such as silica hydrosols, agglomerate finally in the gelation process to what E. Manegold[297] called "coagels" which play a basic role in geochemistry, especially as protecting colloids, as in certain ore-barite deposits for which E. Nickel[298] gave a particularly instructive example in the large deposits of pyrite with barite near Meggen, Westphalia.

134. The multiple parallels existing between silica gels with mixed silica and alumina gels have, however, a much more general importance in the chemical industry. It is chiefly the great impulse given by modern investigations on catalysts and catalyst "carriers," which has deepened our knowledge of the characteristics of these hydrogels. From the viewpoint that structural requirements are essential for the activity of silica (or related, e.g., silica-alumina) gels as catalysts, the studies of C. J. Plank and L. C. Drake[299] may be seen in the foreground. The gels are notably characterized by their extremely rapid gelation under certain pH conditions, and the gelation-time reaction indicates their ionic nature. The intermicellar bonds are hydrogen bonds the rupture of which in acid media brings about a splitting to form micellar particles of a higher density and a higher specific surface. In the mixed silica-alumina gels, the hydrogen bonds joining the silica and alumina groups are the

[295] *Kolloid-Z.*, **151**, 1957, 42-47.

[296] *Beitr. Silikose-Forsch.*, No. 37, 1955, 45-77.

[297] *Kolloid-Z.*, **96**, 1941, 186-210.

[298] *Chem. Erde*, 1956, 99-136, especially p. 124.

[299] *J. Colloid. Sci.*, **2**, 1947, 399-412; 413-427. Plank and Drake chiefly emphasize the differences existing between silica and mixed silica-alumina hydrogels, which are of fundamental importance for their catalytic specificity. In this connection it is also interesting that these authors prepared mixed silica-zirconia hydrogels with "zeolitic" properties.

most important for what we call the *acidity* of the gels (cf. C. ¶ 211 ff.); this rules their activities as cracking catalysts, and shall therefore be treated extensively in our considerations. Basic media promote the rupture of the hydrogen bonds mentioned above and bring about smaller particles of higher specific surface, with increased micellar electrostatic charges but with reduced density. All calculations in the evaluation of experimental data of density measurements (by mercury displacement methods) and of specific surface determinations (by the BET nitrogen adsorption method) are based on the assumption of *spherical* particles in the gel micellae. The higher the contents of the sols in solids [Plank and Drake introduce a parameter *pc* defined as the total weight of the oxides (silica and/or alumina) divided by the volume of the sol], the lower were the densities in the resulting gels. At the same time, the specific surface was reduced in silica gels, but increased in mixed silica-alumina gels. A very important factor also is the *aging* which moderates in its general effects

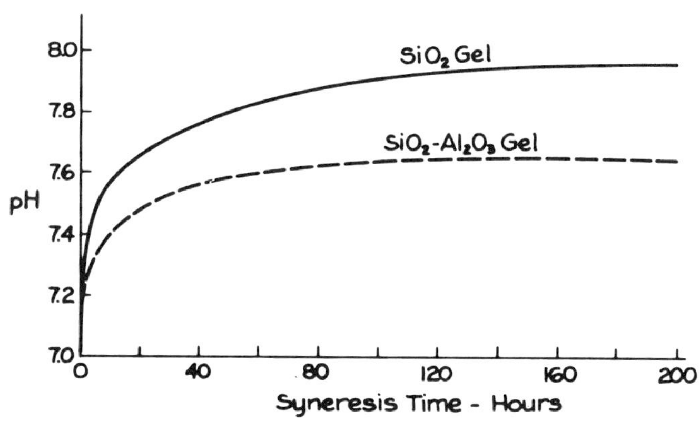

FIG. C. 75. Change of pH with time of syneresis of silica and silica-alumina hydrogels (Plank and Drake).

any changes which would be produced by subsequent base exchange treatment on the gels. In the course of the syneresis of the gels, a characteristic change in pH as a function of time occurs, as seen from Fig. C. 75. The volume of the water squeezed out by syneresis is also correlated to the time factor of this process in a characteristic curve shown in Fig. C. 76.

135. The squeezing out of water from hydrogels under high pressure is a phenomenon quite generally observed, as in clay muds, described by P. A. Kryukov and N. A. Komarova,[300] who refer to the effects of natural rock pressure conditions in

[300] *Doklady Akad. Nauk S.S.S.R.*, **99**, 1954, 617-619; cf. more recently M. Tschapek, *Naturwissenschaften*, **46**, 1959, 489, for kaolinite and montmorillonite. The water squeezed out can be replaced by aniline or butyl alcohol.

mines or to tectonic displacements. The experimental reproduction and measurement of such effects with silica hydrogel, bentonite, Askangel, and others shows a linear relation of residual (equilibrium) water contents as a function of the logarithm of pressure. Organic gels are anomalous, having hyperbolic functions of the water contents versus pressure. Askangel (a hydrogen montmorillonite of high purity) has an extremely steep dehydration curve with increasing pressure, but no break points

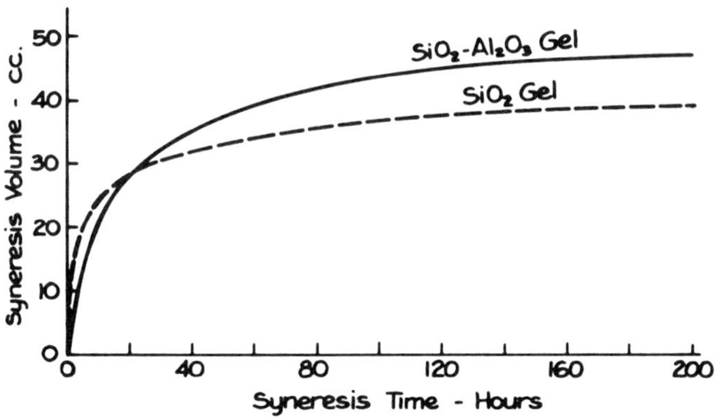

FIG. C. 76. Syneresis volume as a function of syneresis time (Plank and Drake).

which might correspond to "molecular moisture capacities" (cf. C. ¶ 363). E. Forslind[301] fundamentally discussed the conditions of particular properties of the water molecule and its association affecting syneresis of hydrogels in the gelation process. He started from the electronic structure of the water molecule and hydrophilic groups on the surface of the solid phase in a given colloidal system. The more loosely the micellae of the solid are held together by weak, nondirected forces, the larger is the part of the water in determining the equilibrium configuration of the interface. By these interactions, well developed "water lattice" structures are stabilized, comparable to the crystalline structures with water interlayers of halloysite, or vermiculite, etc. (cf. B. ¶ 57, ¶ 88).

136. The mechanism of gelation appears then as a rheological phenomenon in the sol/gel transition. Especially for silica gels, Sh. Kurosaki[302] studied the *dielectric properties* of water sorbed on the gels, indicated by changes of the apparent dielectric constant, ε'', and of the specific polarization of the water molecules, as a function of the concentration. The apparent loss factor, as a function of the water content, could be measured over a wide frequency range, from 2×10^3 to 10^6 cycles/sec.

[301] *Svenska Forskningsinst. Cement Betong vid Kgl. Tek. Högskol. Stockholm, Handl.*, No. 18, 1953, 50-63.

[302] *J. Phys. Chem.*, **58**, 1954, 320-324.

(Figs. C. 77 and 78). One may distinguish three different states existing for the water in the gel: (1) up to 15 mg. H_2O/gm. gel the specific polarization is very low, the water is firmly bound to the gel surface, and the heat of adsorption is high; (2) from 15 to 94 mg./gm. the state of the water in the gel may be compared with that of water dissolved in organic solvents; and (3) above 94 mg./gm., corresponding to three-molecular, and thicker, layers of water, the water is "capillary-condensed,"

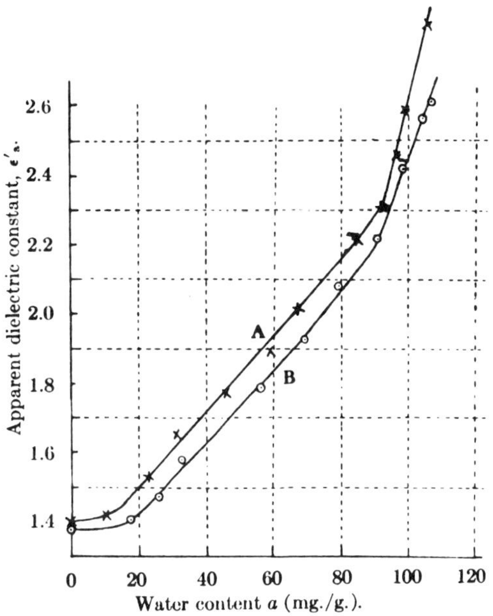

FIG. C. 77. Apparent dielectric constant of silica hydrogels as a function of sorbed amount of water, at 25°C. (Kurosaki).

with a maximum loss factor for $\nu = 10^4$/sec. A semiquantitative explanation of the apparent dielectric constant in its maximum can be made by the R. W. Sillars'[303] theory of binary aggregates. The capillary-condensed water has a stronger hydrogen-bonding (comparable to that in ice) than liquid water shows.

137. G. Kämpf and H. W. Kohlschütter[304] found for a series of silica gels quite analogous results for the apparent dielectric constant of the adsorbed water and also the characteristic maxima of this constant ε'' for discrete frequencies, by a more detailed analysis of extensive experimental data. These authors were able to distin-

[303] *J. Inst. Elec. Engrs.* (*London*), **80**, 1937, 378-394.

[304] *Z, anorg. u. allgem. Chem.*, **294**, 1958, 10-21. For new precision equilibrium data (isotherms and isopiestic curves) for silica gels and water vapor see St. S. Hubard, *Ind. Eng. Chem.*, **46**, 1954, 356-358.

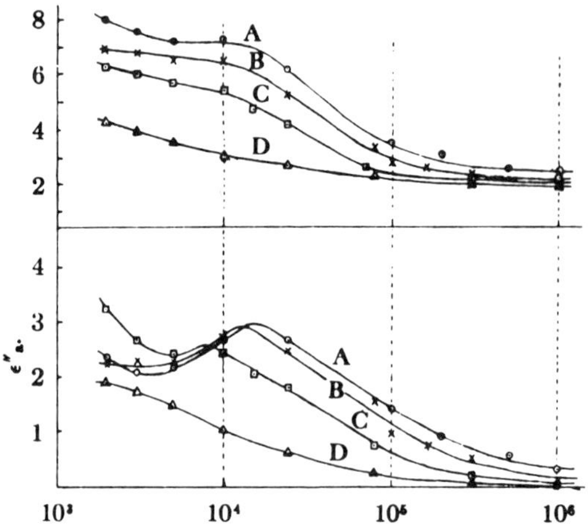

FIG. C. 78. Apparent dielectric constant (upper part of figure), and loss factor (lower part), as functions of frequency, for silica hydrogels (Kurosaki). Water content in mg./gm. gel: A — 143; B — 107; C — 93; D — 85.

guish between that share of the total dielectric properties of the gel system which is caused by the atomic and electronic polarization of the water molecules themselves and the other share contributed by the orientation polarization of the water molecules. The latter share is caused by variations in the mobility of the water molecules, or by their more or less firm bonding to the surface of the gel framework. With increasing water charges (in mm.³/gm. gel, written x/m in Fig. C. 79) onto the

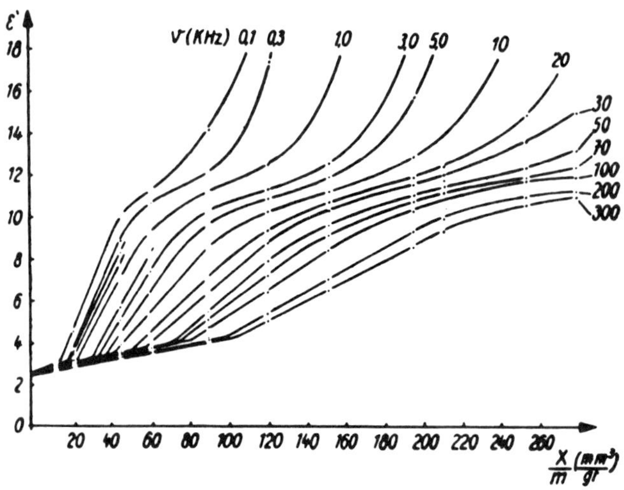

FIG. C. 79. Apparent dielectric constants of silica hydrogels as a function of their water contents (x/m), for variable frequencies (Kämpf and Kohlschütter).

gel, the dielectric response shows the phenomena of anomalous dispersion, which is particularly conclusive for the recognition of orientation effects of the water. In general there are thoroughly continuous transitions in the mobility of the water molecules and, therefore, also in the strengths of the binding. This continuity is the same phenomenon which is observed in the continuity of the differential heat of adsorption with increasing x/m. For low values of x/m it was not yet possible to distinguish between chemically bonded water in SiOH groups and physically adsorbed water molecules. It is also remarkable that with increasing temperature the maxima for the apparent dielectric constant as a function of the frequencies are shifted according to the linear relation $\ln \varepsilon''_{max.} = \text{const.} \cdot 1/T$, corresponding to a true relaxation mechanism in P. Debye's theory. The activation energies of the place exchange of the adsorbed water molecules are increased with decreasing x/m, as from 5,080 cal./mole for 132 mm.^{3}/gm., to 7,200 cal./mole for 45.4 mm.^{3}/gm. They are, therefore, in the order of magnitude of proton mobility in ice and liquid water, as determined by A. Gierer and K. Wirtz, and also by N. Riehl.[305]

138. In an attempt to characterize the molecular water or hydroxyl contents of a fine-granular and porous silica hydrogel by a chemical reaction, H. W. Kohlschütter, P. Best, and G. Wirzing[306] slowly reacted the gel with liquid trimethyl silicon monochloride, $(CH_3)_3SiCl$. The capillary water of the gel is volatilized as the compound $(CH_3)_3SiOSi(CH_3)_3$ (boiling point, 100°C.) the principal reaction product, but the chemically bonded water, in hydroxyl groups on the gel surface brings about a weight increase of the solid after the hydrolysis and especially an increase in the ignition residue. The stoichiometry of the reaction[307]

$$(CH_3)_3SiCl + HOSi\,(O/_2)_3 \rightarrow (CH_3)_3Si\text{—}O\text{—}Si(O/_2)_3 + HCl$$

as the basis for the gravimetric analysis, speaks in favor of the reality of the hydroxyl groups on the surface of the gels. By calcination they give off water and at room temperature they are reformed in moist surroundings. The reaction of silica hydrogels with trimethyl monochlorosilane is, therefore, a sensitive and reliable indicator of the hydroxyl groups on surface-active gels, independent from capillary water, which is removed by the reaction with the chloride. Kohlschütter and Kämpf[308] further observed that the gels, if not calcined above 1000°C. to distinctly sintered products, show relatively more active places (per m.^{2} of the remaining surface), which are remodeled to SiOH groups at room temperature, than the original hy-

[305] *Naturwissenschaften*, **43**, 1956, 145-152, especially pp. 149-150; see also A. Gierer and K. Wirtz, *J. Chem. Phys.*, **17**, 1949, 745-746; *Ann. Physik*, [6] **6**, 1949, 257-304.

[306] *Z. anorg. u. allgem. Chem.*, **285**, 1956, 236-245; cf. W. Noll, K. Damm, and R. Fauss, *Kolloid-Z.* **169**, 1960, 18-28.

[307] The symbol $(O/_2)_3$ Si means the binding of a Si atom in the SiO_2 framework.

[308] *Z. anorg. u. allgem. Chem.*, **292**, 1957, especially pp. 304 ff.

drogel with the larger surface development. This result is in contrast to the results of J. E. Lukesh,[309] who found by precision X-ray diffraction measurements for a leached Vycor glass, after impregnation with water, no physical or chemical evidence for the presence of "anything else but liquid water" in the pores of the porous silica body. The direct evidence for the presence of liquid water is given by the practically exact coincidence of the H_2O diffraction bands for silica glass plus water, and liquid water alone.

139. Evidently, the porosity of Vycor glass is coarse in comparison with that of most of the silica gels. Much care was given in numerous investigations to the determination of the pore volume of active silica gels and related systems in which we may include, e.g., also the hydration products of hydraulic binders, with their quite particular and highly important porosity properties.[310] Especially for silica gels used for drying purposes, A. V. Kiselev[311] discussed the question of whether the pores of these adsorption-active materials are channels of cylindrical, or bottle-shaped form, either free on both ends or closed at one end, wedge-shaped, and so forth. It is not sufficient to apply the well-known Thompson formula, as given by J. S. Andersson (1914) for the calculation of a fictive pore radius, R, from the capillary condensation limit indicated in gels on the adsorption curve, namely by the equation

$$R = \frac{2\,\alpha\, s_0}{\varrho\, p_s \ln(\varrho/\varrho_s)}$$

(with ϱ = density, α = surface tension of the liquid phase forming the meniscus, s_0 the vapor density on the surface, p the pressure above the meniscus, p_s the saturation pressure), because one will in most cases find results contradictory with those derived from other methods. Kiselev starts from the idea of spherical particles in the gel structure, building up more or less densely packed arrangements of an experimentally determined total density. This packing of homogeneous particle sizes is possible with different degrees of contacting, with coordination numbers usually between 4 and 12, in the same way as such packings were discussed in crystal

[309] *J. Phys. Chem.*, **59**, 1955, 669-670.

[310] We may here refer only very generally to the discussions in Vol. V, Section C, especially the fundamental investigations of T. C. Powers and T. L. Brownyard, *Research and Develop. Labs. Portland Cement Assoc., Research Dept. Bull.*, No. 22, 1948; St. Brunauer, *Ibid.*, No. 80, 1957, 428-468, especially pp. 443-458, gave a recent report on the results. See also "The Science of Engineering Methods," by Wiley, New York, 1957, pp. 428-468.

[311] *Doklady Akad. Nauk S.S.S.R.*, **98**, 1954, 431-434. The ideas of Kiselev here discussed are in a high degree analogous to the ideal structure of "cement gels" as given by Powers and Brownyard (see above). An order of magnitude of the pores up to 10^{-7} cm. was estimated by B. Tavasci, *Cemento*, **48**, 1951 (5), pp. 3-8, as the most probable one, from optical data, and in agreement with the calculations of Andersson.

structures or for glasses. For different silica gels characterized by their specific surfaces measured by the BET nitrogen adsorption method, the sizes of the spherical elements, D, and also those of the pores, d, can be calculated, as well as the meniscus diameters for capillary condensation in Thompson's equation. For silica gels with a specific surface between 300 and 800 $m.^2/gm.$ one finds contact coordination numbers, n, between 3.2 and 7.0, in a pressed gel with 300 $m.^2/gm$ $n = 5.5$. In the same way as n is decreased, the ratio d/D is increased, d being the pore diameter of the "bottlenecks" between the spheres. Fine-porous gels are, therefore, dense packings of primary silica particles, coarse-porous gels, loose packings with interlocking chains of such intergrown particles.

140. An interesting attempt to use the *freezing point depression* of water condensed in the capillaries of silica gels, based on dilatometric measurements, was made by B. Rai Puri, Raj Sharma, and M. L. Lakhanpal.[312] This method is particularly remarkable because of its inherent analogy with the industrially important problem of the frost actions on Portland cement pastes and concretes, also studied by T. C. Powers and T. L. Brownyard (cf. Vol. V, Section C). The apparatus used by the Indian authors is shown in Fig. C. 80, as applied to silica gels, bentonite, clay loam, and a synthetic barium aluminosilicate gel (with 0.5 M BaO; 1 M Al_2O_3; 1 M SiO_2; aqu.) for variable vapor pressures. Down to the capillary pores of about 37 Å. size, the agreement of the observed and the theoretical data is satisfactory; not, however, below 28 Å. In still finer-porous systems, the density and surface tension properties are probably changed from those known in the bulk form as it is tacitly assumed in the Thompson equation, and even the heat of vaporization of a capillary condensate may be different from that of the same liquid in the bulk, so that the Thompson equation is no more applicable. Another very remarkable use of this equation concerns the occurrence of surprisingly porous, although homogeneous-appearing magnetite crystals from metasomatic ore deposits in the Ural Mountains, changed to greenalite in their central parts. L. N. Ovchinnikov and A. S. Shur[313] measured the depression of the vapor pressure of benzene (with a quartz spring balance) in its adsorption equilibria on these magnetite crystals, and calculated the pore volume, w, (in $cm.^3/gr \times 10^4$) as a function of the pore radius (in $m\mu$). The results are: 0.55 per cent porosity for the magnetite of the exterior portions and 0.92 per cent in the greenalite core portion, with pores mostly between 10 to 50 $m\mu$ in diameter, but also down to 10 $m\mu$ and up to 100 $m\mu$.

[312] *J. Phys. Chem.*, **58**, 1954, 289-292. For, similar experiments for carbon tetrachloride and benzene, see I. Higuchi and Y. Iwagami, *Ibid.*, **56**, 1952, 921-927. On critics of the Thompson equation see also C. Pierce and R. N. Smith, *Ibid.*, **57**, 1953, 64-68.

[313] *Doklady Akad. Nauk S.S.S.R.*, **82**, 1952, 977-979. For the very interesting genetic conditions of metasomatic formations of such highly porous minerals, see L. N. Ovchinnikov, *Trudy Gorno-Geol. Inst. Akad. Nauk S.S.S.R. Ural. Filial*, 1955 (26), 30-49, especially pp. 37-48. The microporosity is also observed as in rock-forming garnets.

141. The *gas permeability* methods for the determination of the porosity of fine-dispersed solids and coarse-porous gel-like products, as in catalyst carriers and ceramics, are based on the formulae of M. A. Knudsen (1909), establishing the diameters of capillary channels in a solid, with flow of gas through the material, and the kinetic theory of gases. If the diameter of the channels is large, in comparison with the free path length of the gas molecules, the gas flow is viscous, but for very small

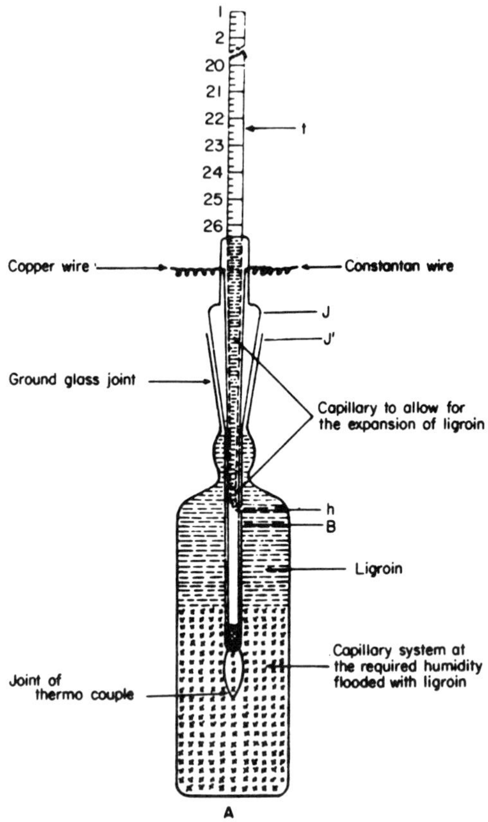

FIG. C. 80. Apparatus for the determination of the freezing point depression of capillary-condensed water on gels (Rai Puri, Raj Sharma, and Lakhanapal).

pores and channels the molecular flow of light gases must be considered. We will come back to these problems especially in connection with the porosity of ceramic bodies (cf. Vol. V, Section B), but may here mention that the porosity of different cokes (in the meaning of the Knudsen theory a macroporosity of an order of magnitude of several microns) was determined in this way by S. Ergun and J. Owen,[314]

[314] *Anal. Chem.*, **25**, 1953, 1222-1226.

as an example that porosities of a similar kind are the paths for material diffusing in framework and channel structures of rock-forming silicates. This is a viewpoint of great general interest for the genetic explanation of so-called "clouded" feldspars (preferably intermediate plagioclases, cf. C. ¶ 5) as described by A. Poldervaart and A. K. Gilkey.[315] There are many transitions, from such still rather macroporous reaction paths in framework structure minerals, to the very characteristic pseudomorphs of silica gel and iron hydrogels after pyroxenes (hedenbergite), and olivines, as described by M. Federico and M. Fornaseri.[316] Particularly remarkable for such pseudomorphs is the preservation of the chains of pyroxene type in the aggregates which show an excellent Wiener form birefringence, and a complete parallel intergrowth of the goethite with [010] parallel to (110) of the hedenbergite, and of [100] of the goethite with [001] of the pyroxene embedded into the anisotropic silica hydrogel.

142. It is the special merit of electron microscopy that the channels are made perceptible in structures like that of chrysotile, fibrous serpentine asbestos, and also in garnierite, halloysite, and the like, which are of the order of magnitude of the capillary adsorption systems in silicate minerals, as W. Noll, H. Kircher, and W. Sybertz[317] could impressively demonstrate. These authors made extensive measurements of the capillary water contents in the hollow fibers of chrysotile, with the characteristic steep branch of the adsorption curve (isotherm) at (p/p_s) values above 0.6 (cf. Fig. C. 81) for asbestos of the specific surface of about 100 m.²/gm. The length

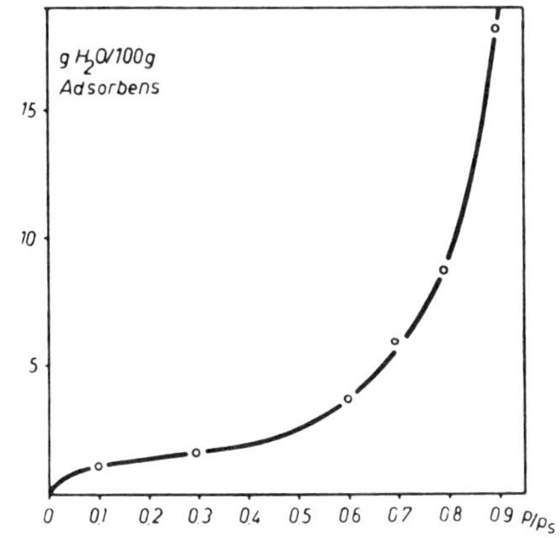

FIG. C. 81. Typical adsorption curve of water on synthetic chrysotile, at 20°C. (Noll, Kircher, and Sybertz).

[315] *Am. Mineralogist*, **39**, 1954, 75-91.

[316] *Periodico Mineral. (Rome)*, **22**, 1953, 107-134; compare the older observations of K. Spangenberg, *Heidelberger Beitr. Mineral. u. Petrog.*, **1**, 1949, 535-547, on pseudomorphs of silica gel and goethite after olivine.

[317] *Kolloid-Z.*, **157**, 1958-1-11; see also *Fortschr. Mineral.*, **29/30**, 1950/51, 391-392, on the

of the "free" channels is not necessarily the same as the length of the fibers as seen in the electron micrograph. On the contrary, lateral displacements and distortions of the fibrils cut off the free channels considerably, as schematically shown in Fig. C. 82. A somewhat bothersome explanation proposed by E. J. W. Whittaker[318] for the reduction of the "free" channels in the fibers, namely that these channels may have been clogged by fragments of the curved layer parcels, is thus dispensable.

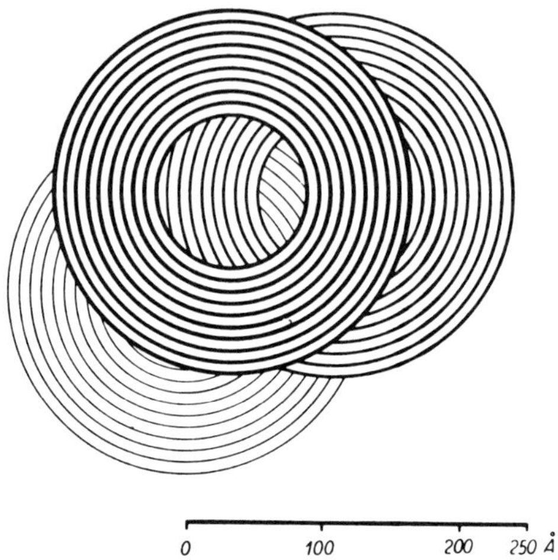

FIG. C. 82. Displacement of tubular fibrils of chysotile asbestos, clogging the tubular capillary systems (Noll, Kircher, and Sybertz).

143. Another kind of channel influencing the capillary transmittancy of hydrogels was studied by R. Fricke and K. Jockers[319] in the behavior of *radioactive emanation gas* (Thoron from ThX, with a half-life time of 3.8 days) permeating a gel-like material, such as γ-alumina or active zinc oxide. An evident effect of this kind is observed by the change of diffusion of the emanation gas through empty and through water-filled capillaries. There is another influence of the diameter of the channel systems which can be observed if the gel is mechanically compressed. Aging effects are observed as spontaneous contractions of the channel diameters. The emanation capacity, EC, itself is in a very elementary relation with the specific surface, S, of the gel, given by the equation $S = \text{const.} \cdot EC$. The surface was calculated as

hollow channels in chrysotile, whereas hornblende asbestos shows massive fibers. Synthetic cobaltum chrysotile has a considerably higher specific surface than magnesium and nickel chrysotile (165 to 190 m.^{2}/gm. in comparison with 110 to 125 m^2/gm. in the latter); cf. *Naturwissenschaften*, **45**, 1958, 489; *Beitr. Mineral. u. Petrog.*, **7**, 1960, 232-241, especially pp. 235-237.

[318]. *Acta Cryst.*, **10**, 1957, 149-156.

[319] *Z. anorg. u. allgem. Chem.*, **259**, 1949, 299-302; *Ibid.*, **265**, 1951, 49-55.

usual from the monomolecular layer adsorption of water. Since capillary condensation affects the emanation capacity, it is not possible to calculate S accurately from EC measurements in moist air.

144. Coming back again to the capillary system in chrysotile asbestos, G. J. Young and F. H. Healey[320] found great differences in the adsorption type of gases and vapors of nonpolar argon, carbon monoxide, acetylene, butane, and polar water, ammonia, di- and tri-methyl amine, insofar as only water and ammonia vapors showed an anomalously high adsorption. If the asbestos, however, was heated to increasing temperatures (activated) before the adsorption, the surface areas of the system, calculated in m.^{2}/gm. become gradually similar for water and nitrogen, and finally at 425°C. they are the same (Fig. C. 83). It is evident that at 425°C.

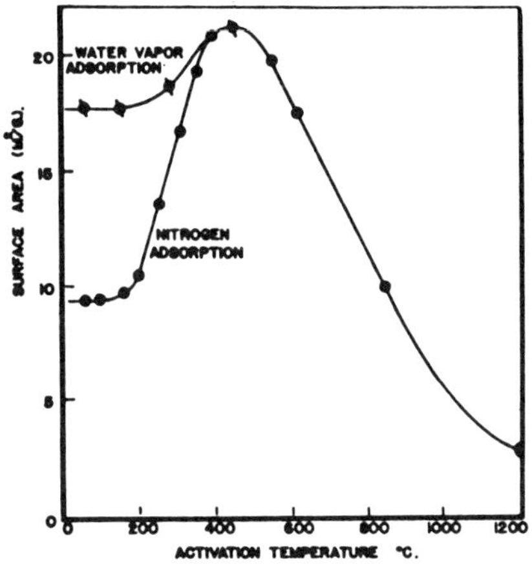

FIG. C. 83. Adsorption on chrysotile asbestos at various temperatures of activation (Young and Healey).

there is a change in the surface structure of the asbestos, and above 450°C. the surface is definitely altered until at much higher temperatures the internal crystalline structure breaks down. The anomalous specific adsorption for water and ammonia, however is explained by a kind of "blocking effect." In other words, firmly bound water in the adsorption-active capillaries permits the entrance and permeation of water and ammonia, but not that of the nonpolar gases and vapors. This behavior is to a high degree specific for the capillary system of chrysotile. In a somewhat different manner it is also observed in the tabular antigorite, but not in tremolite and anthophyllite which do not have capillary fibrils but massive fiber structures.

145. The filling of the capillary system with adsorbed water is particularly strik-

[320] *J. Phys. Chem.*, **58**, 1954, 881-884.

ing if the heats of immersional wetting are determined and the "absolute method" of W. D. Harkins[321] is applied to measurements previously given by A. C. Zettelmoyer, G. J. Young, J. J. Chessick, and F. H. Healey (cf. C. ¶ 318).[322] After coverage of the capillaries with a monolayer of water, the heat of wetting curve falls to the heat of wetting for a bulk water surface (= 118 ergs/cm.^{2}). The surface area is calculated as a function of the relative pressure of the adsorbed water film (Fig. C. 84).

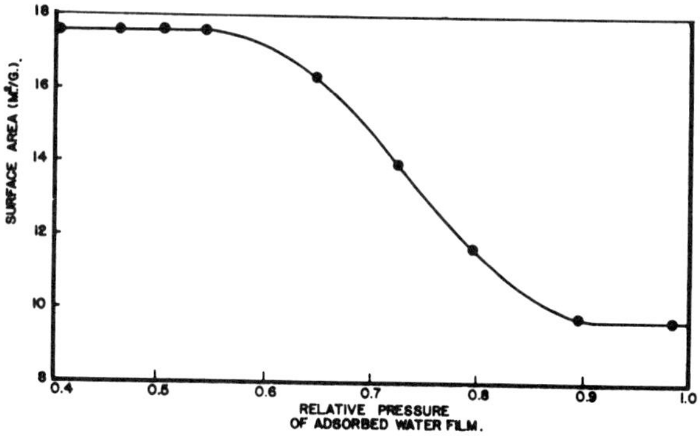

FIG. C. 84. Surface area of chrysotile asbestos calculated by W. D. Harkins' "absolute method," as a function of relative pressure, p/p_s (Young and Healey).

The slope is characteristic if the coverage is $p/p_s = 0.4$ to 0.6, the surface area calculated from the heat of immersion is that from the water vapor adsorption, or the external plus the internal area not occupied by water plugs. From $p/p_s = 0.6$ to 0.85, the capillaries are filling up, and the surface area calculated is a function of the size distribution for the capillary diameters, until at 0.9 the capillaries are completely filled, and the area available is then determined only by the external surface as given by the nitrogen adsorption. This is perhaps the best possible illustration for the real existence of the capillary system in chrysotile asbestos.

146. Also the heat of immersion of an activated asbestos in water is much higher (850 ergs/cm.^{2}) than for nonactivated asbestos (420 ergs/cm.^{2}); that for wetting the capillary surfaces and their filling up (1425 ergs/cm.^{2}) is also much higher than the heat of wetting values known for common solid surfaces. It is, therefore, not surprising that the water plugs retain so firmly other vapors than water and ammonia, and are also not completely removed by evacuation alone at 10^{-5} mm., if not heated (activated) by raising the temperature with the upper limit of 425°C. They are easily re-formed at lower temperatures in moist surroundings. Healey and Young[323]

[321] Cf. the book "The Physical Chemistry of Surface Films," Reinhold, New York, 1952.
[322] *J. Phys. Chem.*, **57**, 1953, 649-652.
[323] *J. Phys. Chem.*, **58**, 1954, 885-886.

concluded further that the surface of nonactivated asbestos is homogeneous to water but heterogeneous to argon. They give distribution curves for the adsorption site energies of activated and nonactivated surfaces from the isosteric heats for argon adsorption. The energy of the surface of the capillary structure made available to adsorption by the activation is lower in respect to nonpolar gases than that for the external surface of the fibers. A distribution function $f(e)$ of the site energy, e, can be given, which for argon is a relative basis for discussing the potential fields influencing the argon atoms by adsorption on non-activated, and activated surfaces (Fig. C. 85).

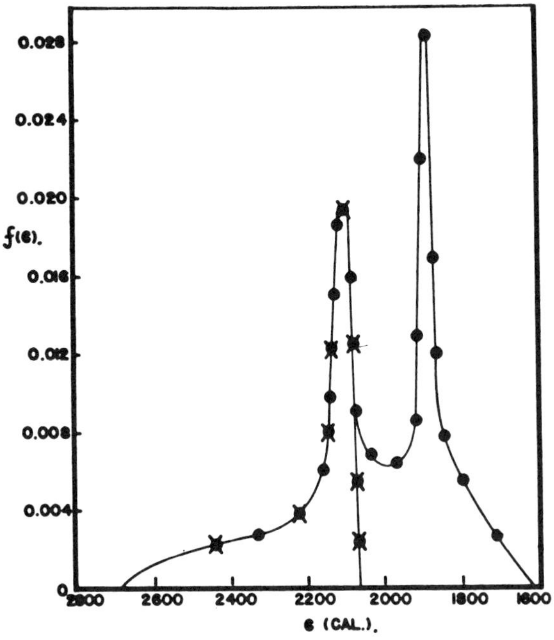

FIG. C. 85. Site energy distribution for adsorption of argon on chrysotile asbestos, in the non-activated and the activated state (Young and Healey); ✖ unactivated, ● activated at 425°C.

147. The method of adsorption of nonpolar vapors for the determination of the external area of given silicate structures is by no means restricted to those with capillary fibrils. It is equally feasible for the tabular clay minerals, as M. J. Brown and S. J. Gregg[324] demonstrated for kaolinite, and its adsorption for carbon tetrachloride as a function of different thermal treatments (activations). Similar effects of water driven out by heating are observed, and internal layer coverage is re-formed in a moist surrounding, if the temperature of activation does not reach the irretriev-

[324] *Clay Minerals Bull.*, 1, 1947-1952, 228-235.

able limit of 400° to 500°C. For normal silica gels, H. W. Kohlschütter and G. Kämpf[325] gave extensive measurements of the adsorption (and desorption) isotherms in different stages of sintering by thermal treatment, further changes in the dielectric constant, the dielectric losses, and as an indication for the hydroxyl contents, the reaction with trimethyl monochlorosilane at room temperature. The differences of the ordinates in the adsorption-desorption curves in the hysteresis areas (for constant p/p_s) indicate the amount of water which is firmly bound to the surface of the gel, in agreement with the previous results of Kohlschütter, Best, and Wirzing mentioned above.[326] The differential heat of adsorption is a monotonous function of the water load of the framework for all the sintering steps experimentally studied.

148. In a rather similar way F. Wolf and H. Beyer[327] applied a so-called *gas-chromatographic* method for the study of silica hydrogels in varying thermal treatments, and their changes in pore size, inner surface, structural water content, based on Kohlschütter's ideas on the surface-existent silanol groups, Si—OH. The pore size distribution follows a Gauss error integral function, with a slight shifting of the maxima to smaller sizes (from 70 to about 65 Å.) above 700°C. Gas chromatography based on the use of well-defined silica gels involves the stronger polarization effects of the surface-existent siloxan groups, Si—O—Si, in reference to the Si—OH groups, for saturated hydrocarbons. Therefore, the more the surface-existent hydroxyl groups are reduced by calcination to form Si—O—Si groups, the more the adsorption for saturated hydrocarbons is increased—that for olefins, however, decreased. If one defines by U the parameter in gas chromatography[328] as the volume of an eluation gas required to drive a given gas component through the column filled with one gram of the sorbent, one finds according to Fig. C. 86 for n-butane and β-butylene characteristically different U-curves as a function of the calcination temperature of the gel. The initial increase in U on the curve for butane up to 500°C. which complies with the change of the silanol to siloxan groups, is compensated above 500°C. by the reduction of the surface evolution with increasing temperature.

149. The tremendous influence of the heat treatment of silica gels on their final physical properties is equally important for the theoretical as for the practical-industrial viewpoints, especially for catalysis and gas separation by adsorption (gas chromatography as mentioned above). We mention in the latter aspects the

[325] *Z. anorg. u. allgem. Chem.*, **292**, 1957, 298-310.

[326] *Z. anorg. u. allgem. Chem.*, **285**, 1956, 236-245. See more recently W. Noll, K. Damm, and R. Fauss, *Kolloid-Z.*, **169**, 1960, 18-28.

[327] *Z. anorg. u. allgem. Chem.*, **300**, 1959, 33-40.

[328] Cf. N. M. Turkel'taub, *Zhur. Fiz. Khim.*, **31**, 1957, 2102-2108; *Zavodskaya Lab.*, **23**, 1957, 1023-1024; with A. I. Kolyubyakina and M. S. Selenkina, *Zhur. Anal. Khim.*, **12**, 1957, 302-312. See further J. Janak, *Chem. listy*, **47**, 1953, 817-827; *Collection Czechoslov. Chem. Communs.*, **18**, 1953, 798-802; *Ibid.*, **19**, 1954, 684-699.

recent discussion of M. Malanchuk and E. B. Stuart,[329] who showed the characteristic curve of the changes in specific surface of a typical industrial silica gel with increasing temperature (Fig. C. 87) of "retorting". In this operation the time of thermal exposure is a very important factor because of the strong insulating effects of the porous structure on the heat transfer. The phenomena of crystallization of the silica gels, analogous to the natural aging of gels to hyalite, then to opal, chert, chalcedony, and so on, is extensively studied by X-ray diffraction methods, especially by O. W. Flörke[330] with refined differential-thermal analysis and dilatometric methods. Al-

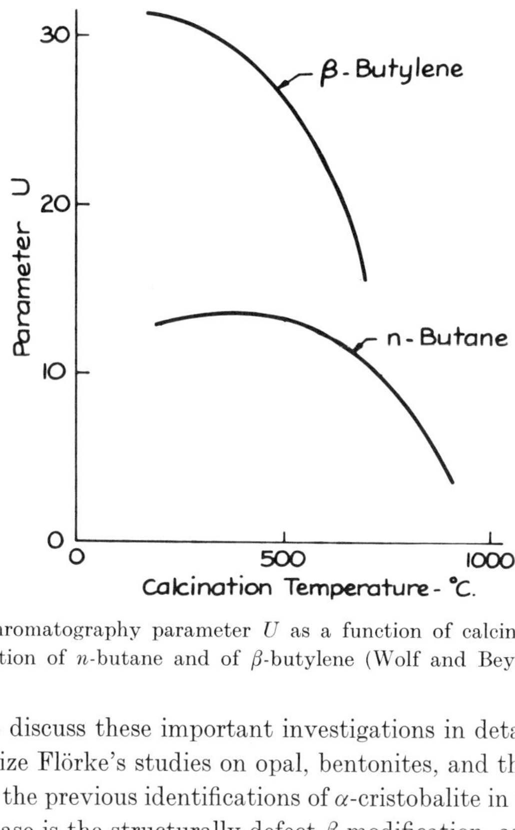

FIG. C. 86. Gas chromatography parameter U as a function of calcination temperature of silica gel, for adsorption of n-butane and of β-butylene (Wolf and Beyer).

though we have to discuss these important investigations in detail in Vol. III, Section B, we emphasize Flörke's studies on opal, bentonites, and the like, in which he demonstrated that the previous identifications of α-cristobalite in these are incorrect. The cristobalite phase is the structurally defect β-modification, as seen from the absence of some characteristic interference lines in the X-ray diagrams and the shift-

[329] *Ind. Eng. Chem.*, **50**, 1958, 1207-1210. The calcination time must be more than two hours as a minimum; N. V. Kiselev (see C. ¶ 219) chose as a rule 12 hours.

[330] *Neues Jahrb. Mineral. Monatsh.*, 1955, 217-224; *Veröffentl. Max-Planck-Inst. Silikatforsch.*, **15**, 1955, 11-17; *Silicates inds.*, **26**, 1961, 424-426. See also combined X-ray and electron diffraction, differential-thermal analysis and infrared absorption spectra examinations of opals and lussatite by V. Z. Zabelin, *Doklady Akad. Nauk S.S.S.R.*, **139**, 1961, 184-187.

ings of some other maxima. The crystallization at low temperatures is the reason for the high defectiveness of the cristobalite structures frequently observed.

150. Flörke follows a recommendation of F. Laves (1939) in calling the cristobalite-containing silica gel products *lussatites*, and those with quartz, *chalcedonies*. The wood opal from Giessen (Hesse) and the fire opal from Zimapan (Mexico) are thus typical lussatites. The inversion effects of the intensely defective cristobalite in bentonites occur at an anomalously low temperature (60° to 70°C.), and are extended

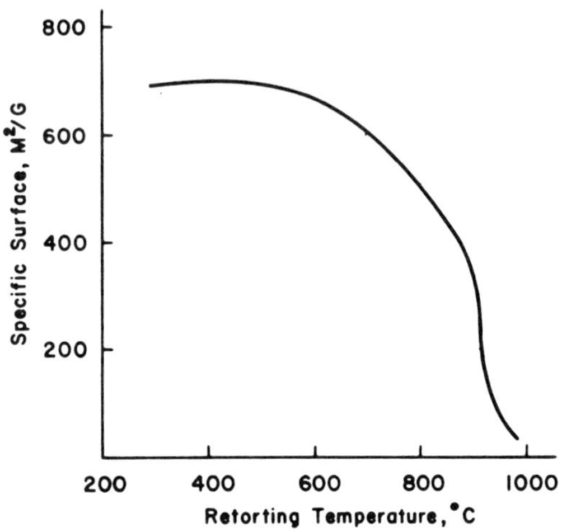

FIG. C. 87. Decrease of specific surface of silica gel with increasing calcination ("retorting") temperature (Malanchuk and Stuart).

over a temperature range of 40° to 60°C. A more refined investigation of natural opals by O. W. Flörke[331] confirms how rare in nature really amorphous silica gels are (as in hyalite, some gem opals, and organogenic diatomites and "sinters"), most of the opals being either defective cristobalite (lussatite) or quartz (chalcedony), in hornstones, lydites, silicified sandstones. Opal is mostly found in very young formations, the transition into quartz is the rule, with paramorphic textures of quartz

[331] *Fortschr. Mineral.*, **37**, 1959, 73-74. The so-called melanophlogite from Sicily is here interpreted as a pseudomorph of quartz after fluorite or some other undetermined mineral. Br. J. Shinnar and D. E. Appleman, *Geol. Soc. Am., Spec. Papers* N. 73, 1963, 245, however, insist on the opinion that melanophlogite is a distinct phase, different from other modifications of silica (cf. III. B. ¶ 2), of cubic symmetry, density 2.05, and with a unit cell of $a_0 = 13.402 \pm 0.004$ Å. On related textures see O. Braitsch, *Heidelberger Beitr. Mineral. u. Petrog.*, **5**, 1957, 331-372. On the geochemical role of the solubility of silica in the formation of cherts, etc., see also R. Siever, *Am. Mineralogist*, **42**, 1957, especially pp. 822 ff., 838 f.

after cristobalite. A. Maas[332] was able to make visible the crystallization of silica by replica techniques in the electron microscopic image of wood opal.

151. The differential-thermal curves and X-ray interferences are generally those of several modifications of crystalline silica phases at the same time (cristobalite, tridymite, and quartz), with a strong broadening of the interference lines, and an intense background effect. It was also attempted by etching techniques to differentiate the silica modifications. The presence of quartz in chalcedonies was thus again confirmed, whereas in opals no direct identification was possible because of the extremely small dimensions of the particles embedded in an entirely structureless matrix. Similar determinations of the silica crystallizations in very young bentonites of hydrothermal origin, were described by R. Pellizzer and G. Guidetti,[333] with the clear establishment of cristobalite, in analogy with the hydrothermal synthesis of this mineral by J. Wyart.[334] The cristobalite occurs in the Capalbio clays described by Pellizzer and Fornaseri in globular aggregates as shown in electron micrographs, with excellent electron diffraction diagrams.

152. The cristallinity of silica gels after calcination at moderately high temperatures was also revealed on curves of true heat capacity versus temperature, as demonstrated by Ts. Takamura[335] for the range between 370° and 620°K. The measurements of the true specific heat were made with a vacuum radiation (adiabatic) calorimeter. Undoubtedly, full cristobalite X-ray diagrams are observed in samples of silica gels calcined at 1370°K. for 5 hours, with the corresponding heat capacity data, whereas those samples which were calcined at 630°, 873°, and 1073°K. did not give definite patterns of cristobalite in X-ray diffraction, even though the heat capacity data distinctly indicated the presence of cristobalite and/or tridymite, with the inversion effects characteristic of these SiO_2 phases, and typical annealing effects discussed extensively in Vol. III, Section B. The relations of the adsorption isotherms of benzene (at 0°C.) to the crystalline state of the calcined gels were in no case distinct. It is concluded that the surface properties of the silica gel are independent of the crystallinity state. In every case, there is no "surface melting" below 1070°K. The silica gel used had a high surface area, and it was observed that there is an anomalous increment of heat capacity due to the high dispersity and to crystallinity defects, even in the high temperature range, as it is typical also of finest-

[332] *Electron Microscopy*: Proc. Stockholm Conference Sept. 1956, Academic Press, New York, 1957, 339-341.

[333] *Rend. soc. mineral. ital.*, **15**, 1959, 181-190.

[334] *Bull. soc. franç. minéral.*, **66**, 1943, 479-489. Compare also the cristobalite containing clay in fuller's earth from Georgia, described by G. W. Brindley, *Clay Minerals Bull.*, **3**, 1956, 167-169 (cf. C. ¶ 216).

[335] *Kolloid-Z.*, **157**, 1958, 11-16.

dispersed titania gels, according to J. A. Morrison, D. Patterson, and J. S. Dugdale.[336]

153. Recent descriptions with silica phase determinations in tripolites and diatomaceous earths were given by G. I. Bushinskiĭ, and V. A. Frank-Kamenetskiĭ,[337] as raw materials for pozzolanic binders (cf. Vol. V, Section C) of high chemical reactivity. Also these tripoli earths show (besides quartz as a contamination) the typical cristobalite diffraction diagrams, whereas most diatomaceous earths are X-ray amorphous. The size of the cristobalite crystallites in tripolite is about 5×10^{-6} cm. In the Middle Volga Basin such tripolites and diatomaceous earths are of Palaeozoic age. The crystallization of cristobalite and/or tridymite and quartz, therefore, is much more advanced than in less aged natural sedimentary silica deposits, as described by V. V. Vlasov, U. G. Distanov, Yu. I. Sozin, and V. S. Verkhunov.[338] The maximum amount of cristobalite (25 per cent) was found in a tripolite from the Ulyanov District, and also marls from the same region contain 10 to 20 per cent. Cristobalite, on the other hand, is entirely absent in the diatomaceous earths which are still completely X-ray amorphous.

154. The gradual transition of sedimentary colloid gels by aging and crystallization to mineral aggregates and rocks is the important geochemical process called *diagenesis*. The different grades in the evolution of sandstones were recently discussed by E. C. Dapples[339] in their physicochemical and geological conditions. The interesting occurrence of quartz in soft coal seams which must have originated at temperatures not much higher than normal is a remarkable phenomenon of similar changes from silica precipitates under the conditions of lowest-grade metamorphism and diagenesis, as described by J. H. Hellmers and J. Ottemann.[340] H. H. Murray and C. P. Gravenor[341] especially investigated the presence of colloidal silica in sediments of different continental and marine types, as river and sea water deposits, and under different climates. Not only the pure silica precipitates, but chiefly the *coprecipitates* with the hydrogels of aluminum and trivalent iron, are important as the diagenetic roots of the formation of clays and slates, or even of other silicates of low-temper-

[336] *Trans. Faraday Soc.*, **52**, 1956, 761-771; *Proc. Roy. Soc.* (*London*) A **224**, 1954, 228-235.

[337] *Doklady Akad. Nauk S.S.S.R.*, **96**, 1954, 817-820.

[338] *Doklady Akad. Nauk S.S.S.R.*, **128**, 1959, 1254-1257. A very interesting characterization of the silica in diatomeae shells, their spongy structure in analogy to silica gels, etc. see by J.-G. Helmcke, *Naturwissenschaften*, **41**, 1954, 254-255. Even electron diffraction patterns do not show any crystallinity of the silica material.

[339] *Bull. Geol. Soc. Am.*, **67**, 1956, 1684.

[340] *Silikat Tech.*, **3**, 1952, 51-56.

[341] *Science*, **118**, 1953, 25-28. On the important question of marine silica deposition see especially the opinions of K. Rankama and Th. G. Sahama, "Geochemistry," Univ. of Chicago Press, Chicago, 1950, pp. 555 f.

ature genesis. In the colloidal fractions of many typical sediments of this kind, Murray and Gravenor made evident the presence of colloidal silica and quartz. The delta sediments are usually higher in colloidal silica than the average shales, deposited by inorganic-chemical precipitation from solutions, or by flocculation from sols in contact with electrolyte solutions, mostly as a coprecipitate with iron and aluminum hydroxides. F. Schwarz[342] stressed the importance of such coprecipitation reactions for the diagenetic consolidation, which he compares with hardening of hydraulic binders in its fundamental relation to the role of water in corresponding reactions. The typical association of cherty nodules with iron-manganese hydroxide ore deposits, is described from the Shizouka Prefecture (Japan) by T. Sudo,[343] with characteristic banded structures and rounded-oolithic surfaces, of extremely fine-grained quartz, delessite, and hematite, often in peculiar vermicular shapes.

155. A very detailed study of the lydites and other types of cherty slates of the Kulm and Silurian formations was given by Hildegard Hoss.[344] They show in the X-ray diagrams all of the quartz interference lines, even with normal intensities, but in spite of that, no $\alpha \rightleftharpoons \beta$ inversion effects are observed on the differential-thermal curves. This same phenomenon is typical of novaculite as O. F. Tuttle and M. L. Keith (1952) had already emphasized. It is, therefore, a characteristic of the Palaeozoic lydites that they contain quartz in a highly structural-defect form, although no cristobalite is identified in them, as in Miocenic hornstones, described by N. L. Taliaferro (1935). The transition of opaline silica gel into cristobalite and then into microcrystalline quartz (chalcedony or quartzine) is the typical diagenetic sequence made evident. We can not discuss here the origin of silica in connection with the Kulm diabase volcanisms. Hoss assumes that silica organisms (chiefly radiolariae) essentially contributed to the primary deposition of the opaline colloidal silica. The fact that even the diagenesis of geologically ancient silica rocks (as the lydites are) did not change them completely to normal quartz, but that these contain the quartz in a highly structural-defect form, is most remarkable and makes doubly impressive how slowly such diagenetic transitions may develop.

156. V. D. Lomtadze[345] investigated the diagenetic conditions of *clayish* sediments, especially in their consolidation to slates and argillitic schists, with a corresponding change in density, water content, porosity, and other physical properties of practical geological-engineering significance. Van der Waals forces are discussed as the reason for the binding of the originally loose colloidal aggregates into consolidated shales

[342] *Silikat Tech.*, **8**, 1957, 418-422.

[343] *Japan. J. Geol. and. Geography, Trans.*, **23**, 1953, 109-116.

[344] *Heidelberger Beitr. Mineral. u. Petrog.*, **6**, 1957, 59-88; cf. N. L. Taliaferro, *Am. J. Sci.*, **230**, 1935, 450-474.

[345] *Doklady Akad. Nauk S.S.S.R.*, **102**, 1955, 819-822; *Zapiski Leningrad. Gorn. Inst.*, **33**, 1958, 102-108.

(with the participation of biochemical factors); a skeleton of solid particles is cemented together by a coagulate of silica-aluminum-iron hydroxide gels. The mechanical characterization of such argillaceous products, their deformation and plasticity as well as the elastic relaxation behavior, are understood through this transition state between fluids and rigid substances. Like the latter they yield by creeping phenomena in the consolidated state of crystalline rocks. Gradual changes in the nature of the clayish sediments under the action of diagenesis factors were also studied by I. D. Zkhus,[346] based on the very old sediments of the Ural, with kaolinite as the last member of a "weathering series," formed in an acidic medium. In a slightly alkaline medium the kaolinite is changed into beidellite or montmorillonite, as I. I. Ginzburg and I. A. Rukavishnikova[347] observed. The same transition was described by Zkhus from the Carbonian marine clays of the Moscow Basin, with beautifully developed pseudohexagonal pseudomorphs after kaolinite. Such hexagonal forms are entirely unusual for common montmorillonite, and it is not easy to understand this diagenetic change since montmorillonite is a three-layer, kaolinite a two-layer clay mineral. The fact of the transition of kaolinite into montmorillonite, however, cannot be denied since in the bottom sediments of Lake Baĭkal the same changes were observed.[348]

157. *Glauconite* is commonly assumed to be the type mineral formed under halmyrolytic-diagenetic conditions, but it is recently discussed by M. G. Dyadchenko and A. Ya. Khatuntseva[349] in a typically continental formation, namely in alluvial and eluvial Quaternary sediments of River Irsha, near the eruptive complex of Korosten, Ukraine. This continental glauconite is lower in silica, alumina, and iron oxides, but higher in potash than the marine glauconites, in characteristic association with weathering minerals, and with marcasite and siderite, both indicating reducing conditions during its crystallization. Colloidal grodnolite, an amorphous calcium phosphate mineral, is also typical of this very particular association in which biochemical factors may have played some additional role. In general, glauconite is a very early diagenetic mineral, originating from mixed gels, as Ida Valeton[350] em-

[346] *Doklady Akad. Nauk S.S.S.R.*, **115**, 1957, 376-379. See also *Ibid.*, **94**, 1954, 797-799.

[347] "Minerals of the ancient Weathering Crust of the Ural Mountains," Izdatel'stvo Akademii Nauk S.S.S.R., Moscow, 1951, 714 pp.

[348] Also G. Kulbicki, *Clay Minerals Bull.*, **2**, 1954, 183-188, discussed the observation of montmorillonite formed from a two-layer clay mineral, namely from halloysite, under the action of solutions containing dissolved silica in slightly alkaline pH conditions. He contests the reality of the formation of montmorillonite from kaolinite.

[349] *Zapiski Vsesoyuz. Mineral. Obshchestva*, **85**, 1956, 49-57.

[350] *Fortschr. Mineral.*, **37**, 1959, 93; *Mitt. geol. Staatsinst. Hamburg*, **23**, 1958, 88-131. We mention here also the valuable report of H. Braun on the genesis of the North German marine-sedimentary iron ore deposits, in *Z. Erzbergbau Metallhüttenw.* **14**, 1961, 466-464 (with contributions by C. W. Correns. H. Borchert et al.

phasized. She found the glauconite in Oligocenic clays, with characteristic transitions from the 1 M type of the micas to mixed layer types with chlorite, combined with systematic decreases in trivalent iron and potassium, as from a glauconite, with full potassium positions in the mica-like interlayer, to a chlorite.

158. The structural factors ruling the diagenetic changes were discussed from the crystallochemical viewpoint by R. E. Grim and W. F. Bradley,[351] especially emphasizing the great contrasts existing between the kaolinites, and the mica-type clay minerals. If the diagenesis is characterized by important changes in the clay mineral assemblages taking place at the same time with changes in the geological environments, then changes of montmorillonite into chlorite are better understood than, say that of kaolinite into montmorillonite (see above). There is also a certain "memory" factor, as when glauconite forms coatings on non-clay minerals in Tertiary sands (in the coastal area of the Gulf of Mexico), indicating a series of steps from an apparently amorphous colloid to glauconite. The "flint-clays" of Carbonian age are another example, with extremely fine-grained but very pure kaolinite, often formed in situ in "worms". Grim and Bradley emphasize that the rich variability of diagenetic phenomena and of their products may also be influenced basically by different rates of primary sedimentation, the rapidly deposited products trapping large amounts of foreign materials in their more porous space, and developing a sediment of great instability, as in glacial outwash, falls of volcanic ashes, and such events. It may be expected that the crystallochemical study of mica-type minerals as chlorites, and their mixed-layer structures, may be instructive for a more detailed understanding of "degrading" and "aggrading" processes in diagenesis.

159. The problem of whether it is correct to speak of silicic acids or of silica hydrates was answered in favor of the latter by P. A. Thiessen and A. Körner (1929, 1930). Interesting recent studies of this question, however, make a revision of this idea indispensable. The compound $H_2Si_2O_5$, commonly called "disilicic acid," is a typical two-dimensional "*lepidoid*" as it was defined by H. Kautsky and R. Irnich.[352] The basic copper silicate

$$\boxed{(Si_6O_6)O_{6/2}}\ (OCuOH)_6$$

is easily prepared by the reaction of the disilicic acid, written in Kautsky's terminology:

$$\boxed{(Si_6O_6)O_{6/2}}\ (OH)_6\,,$$

[351] *Geol. Rundschau*, **43**, 1955, 469-474.

[352] *Z. anorg. u. allgem. Chem.*, **295**, 1958, 193-205. On the surface evolution of lepidoids see H. Kautsky and H. Pfleger, ibidem 206-217 (cf. Vol. IV, Section B). An excellent recent instruction for the preparation of lepidoid silicic acid was presented by W. Hinz, H. Ruttloff, and A. Täufel, *Silikattechnik*, **13**, 1962 378-381.

with tetrammine copper salt solutions. For the "net" silicic acid as $H_2Si_2O_5$ was called for convenience, the two-dimensional structural units are loosely superimposed on one another, building up a three-dimensional structure. The green copper silicate loses one molecule of water at 300° to 400°C. and forms the brown compound

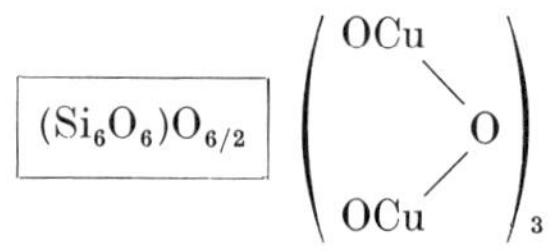

This dehydration is reversible, and in moist air the green silicate is easily re-formed. A break-down of the structure, however, takes place at 600°C. Kautsky's formulation of these compounds is in agreement with the derivation of the nets

$$\boxed{(Si_6O_6)O_{6/2}}$$

from siloxene, $Si_6H_6O_3$, (cf. Vol. IV, Section B) with the undulated unit

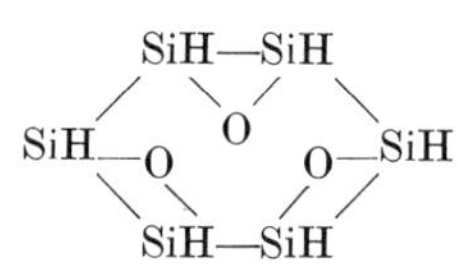

which are interlinked by Si—O—Si (siloxan) bonds with six surrounding oxygen atoms to build up the two-dimensional layers with annular units. In the disilicic acid, the Si_6O_6 rings are interlinked by Si—O—Si bridges; each silicon atom is connected to one hydroxyl group.

160. Concerning the acidic nature of the so-called silicic acids, R. Schwarz and K. G. Knauff[353] very positively defended their existence as *true acids*, which can be prepared by the hydrolysis of their alkyl esters. For this reaction special experimental details were developed. As another suitable method, Schwarz[354] describes the hydrolysis of pure sublimated crystalline silicon disulfide in an isothermal process. He especially emphasizes the strictly topochemical character of the reaction which can be observed in the electron microscope as the change of the long-fibrous sulfide crystals into amorphous products but without any change of the exterior shapes. The reaction products are interpreted first as orthosilicic acid,

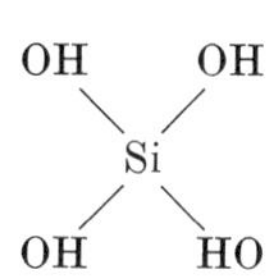

[353] *Z. anorg. u. allgem. Chem.*, **275**, 1954, 176-192.
[354] *Z. anorg. u. allgem. Chem.*, **276**, 1954, 33-46.

which immediately loses water to form the chain-like acid

$H_{10}Si_4O_{13}$ = 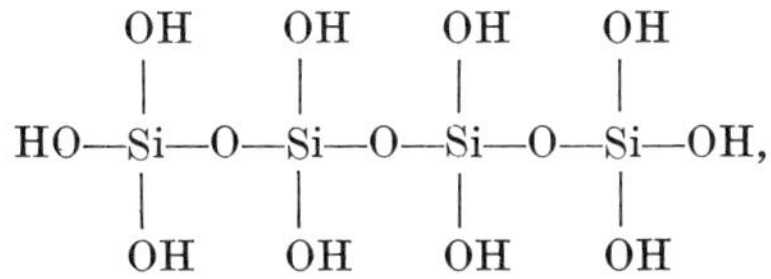

followed by a condensation to an *annular* type of the tetrasilicic acid $H_8Si_4O_{12}$, or a polymeric metasilicic acid with the constitution formula:

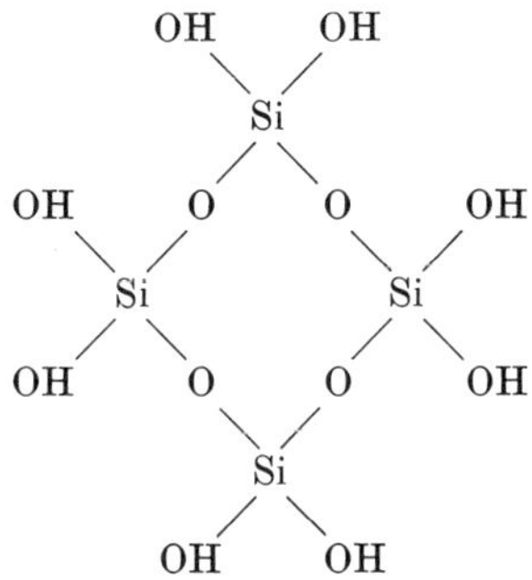

This acid is relatively stable and causes the sharp break at a water content of 23 per cent on the 30°C. isotherm. The formation of a ring structure is also known from the isopropyl and isobutyl esters of the tetrasilicic acid, prepared by R. Schwarz and A. Kessler,[355] and corresponding compounds formed by the reaction of brenzcatechol and salicylic acid with silicon tetrachloride, studied by Schwarz and W. Kuchen.[356]

161. A very mild dehydration using dioxane as the eluation agent at 20°C. brings about the immediate condensation of the annular tetrasilicic acid to $H_{14}Si_8O_{23}$, with two rings of the same type. With slowly elevated temperatures analogous higher condensation products are verified. These compounds do not react with sulfuryl chloride, giving evidence to prove the absence of free water. With an ether solution of diazomethane (G. Berger's reaction, 1941) qualitative evidence, however, is given that the Si—OH groups are of acidic character (to this side of the problems involved here we will come back in C. ¶ 211 ff.). The results of the investigations of R. Schwarz and of his collaborators are in remarkable contrast with the previous opinions and conclusions of P. A. Thiessen and A. Körner (1929, 1930), who denied every acidic character of the condensation products of orthosilicic acid. Schwarz emphasizes that $Si(OH)_4$ is unknown in the solid state, and in every case so unstable that its

[355] *Z. anorg. u. allgem. Chem.*, **263**, 1950, 15-30.
[356] *Z. anorg. u. allgem. Chem.*, **266**, 1951, 185-192; *Chem. Ber.*, **85**, 1952, 624-630.

polymerization to "silica hydrates" in a series of dehydration products alleged by Thiessen and Körner is most unlikely. As a matter of fact, H. B. Weiser, W. O. Milligan, and W. J. Coppoe[357] could not reproduce the results of Thiessen and Körner in a series of different "individual silica hydrates." Only the so-called disilicic acid, $2SiO_2 \cdot H_2O$, has a certain stability if prepared from gels produced from alkali silicate solution and dilute mineral acids. Only the ortho-acid $Si(OH)_4$ is volatile with steam. The same may be the case with $Si_2O(OH)_6$, as E. L. Brady[358] concluded (cf. Vol. IV, Section A).

REACTIONS IN MIXED GELS OF SILICA AND ALUMINA; ORIGIN OF CLAY SEDIMENTS

162. It is most evident that a modern challenge to the problems involved in the diagenetic change of mixed silica-alumina hydrogels to clay minerals must be based on crystallochemical principles. Ch. Kiefer[359] in his general discussion of the classification and stability of layer silicates (phyllosilicates) in the clay mineral group used that approach. On the other hand, many successful attempts were made for a better understanding, from the crystallochemical standpoint, of energy changes in the diagenetic transitions from the hydrogel mixtures, precipitated under given pH conditions in weathering systems, to form first cryptocrystalline hydroxides (hydrargillite, boehmite), which then react with silica hydrogel to form *allophane* as the first step of silicate formation. This procedure was also studied on the basis of extensive observations of natural occurrences, as by G. D. Sherman[360] dealing with "latesol" soils of Mauna Kea, on the island of Hawaii, with the formation of hydrargillite-allophane crusts on andesitic volcanic ashes, and by T. Tamura and M. L. Jackson[361] in a discussion of the full way from the hydrogels of alumina, iron oxide, and silica, to the crystalline kaolinite, montmorillonite, nontronite, and so on. Allophane, as the typical intermediate formation is in this series by no means

[357] *J. Phys. Chem.*, **43**, 1939, 1109-1120.

[358] *J. Phys. Chem.*, **57**, 1953, 706-710.

[359] *Rev. matériaux construct. et trav. publ.*, **422** to **432**, November/September 1950/51; Thesis, Faculty of Sciences, University of Paris, 1951, No. 288, 55 pp. H. Mase, *Bull. Chem. Soc. Jap.* **34**, 1961, 214-225, established important classification principles of silicate structures and their reactivity to dilute acids. In general, neso-, soro-, ino-, and phyllosilicates with bonding energies below 30,500 kcal./mole, and tectosilicates with energies below 33,500 kcal./mole, are decomposed by acids, forming free silicic acid.

[360] *Science*, **125**, 1957, 1243-1244. See also the fascinating article on rock weathering and clay formation by Th. F. Bates, *Mineral Inds. Exp. Sta. Circular*, **29**, 1960 (8), pp. 1, 4-6, with a full discussion of Sherman's observations. Halloysite and allophane are the typical weathering silicates in these formations; there is also a strong desilication from halloysite to hydrargillite.

[361] *Science*, **117**, 1953, 381-383.

any longer a mixed gel. Its infrared absorption spectrum studied by P. F. Kerr and H. H. Adler[362] shows distinctly individual features, reminding us somewhat of the aluminum silicates (andalusite or kyanite), with bands different from those of the hydrogels. In every case, allophane is also not completely X-ray-amorphous, showing four distinct interference bands and having a decidedly higher order than most glasses, as was observed by W. A. White,[363] with allophane from Lawrence Co., Indiana.

163. For the industrially important Middle-European kaolin-clay deposits, H. Urban[364] claimed an analogous origin directly from weathering solutions of the Lausitz granite-granodiorite massive, in stratigraphic horizons with tropic lateritization, and subsequent diagenesis *in situ*. This is in contrast with the older geological theory of the prevailing role of humic acid suspensions in "Tertiary swamp waters" decomposing the granite to kaolinitic residues. The presence of boehmite in those kaolin clays is still an indication of the origin from mixed gels in the sequence given above, and in agreement whith C. W. Correns's theory of kaolinization (cf. Vol. IV, Section A). It gives a full confirmation of the ideas of H. Harrassowitz (1926) on the lateritization process.

164. Experimental investigations to imitate to a certain degree the formation of argillaceous minerals at low temperatures were successfully made by Stéphane Hénin and O. Robichet,[365] starting from mixtures of silicate and aluminate solutions, buf-

[362] *Am. Petrol. Inst. Project 49, Clay Mineral Standards,* Report No. 8, 1950, especially p. 23 and Plate 19.

[363] *Am. Mineralogist,* **38**, 1953, 634-642.

[364] *Silikat Tech.,* **4**, 1953, 299-300; *Fortschr. Mineral.,* **32**, 1954, 68-69; this article also discussed certain difficulties in the nomenclature of clay sediments and kaolins which are of practical importance. We further emphasize the studies of R. Neumann, *Silikat Tech.,* **5**, 1954, 252-259, concerning the ceramic kaolins of the Halle region, derived from porphyries, also through a lateritization phase of weathering. The soft coal overlying cannot have been responsible for the kaolinization (cf. H. Stremme's theory, 1908); they only contributed to local infiltrations, indicated by marcasite-pyrite concretions. It is interesting to compare with the European clays those occurring under totally different (recent-tropical) conditions in Australia, as described by J. S. Hosking, M. E. Neilson, and A. R. Carthew, *Australian J. Agr. Research,* **8**, 1957, 45-74, and compare Hosking, *J. Council Sci. Ind. Research,* **21**, 1948, 21-50 (cf. Vol. IV, Section B). Hosking is also of the opinion that the influence of humus substances on clay deposition was much overestimated. On the other hand, B. E. Balme and J. D. Brooks, *Australian J. Sci.,* **16**, 1953, 65, described peculiar kaolinite petrefacts with preserved plant tissues from a coal seam, with transitions to vitrinite coal material. There is no doubt about the origin of this kaolinite *in situ.*

[365] *Compt. rend.,* **236**, 1953, 517-519; *Clay Minerals Bull.,* **2**, 1954, 110-115. Both authors are doubtful whether simple aging of coprecipitates may produce clay minerals. See also previous studies in *Ibid.,* **1**, 1947, 24-25. On experimental methods to prepare mixed silica-alumina hydrogels of high reactivity from orthosilicic acid esters and aluminum isopropoxide or other metal-organic compounds, see R. Roy, *J. Am. Ceram. Soc.,* **38**, 1956, 146.

fered with sodium or ammonium acetate at boiling temperatures, or of a solution of magnesium and aluminum salts (concentrations about 40 mg./liter). The slow sedimentation observed over several weeks developed from aluminate and silicate solutions, first as a precipitation of boehmite, at pH = 6.0 to 6.5, then the crystallization of kaolinite. If magnesium chloride (or acetate) is present at pH = 8.0 to 8.5, a silicate-aluminate solution in the ratio 4 : 1 brings about magnesium montmorillonite, at 7.5 to 8.0 antigorite, all of the crystalline phases being identified by their X-ray diffraction diagrams. More recent experiments of St. Hénin[366] included the role of di- and trivalent iron, or nickel, in the presence of silica, at different pH conditions, analogous to the formation of magnesium montmorillonite, if pH is above 7.0. The silica dissolved by the hydrolysis of glass container walls, is sufficient to bring about the precipitation of antigorite in magnesium (and/or nickel) salt solutions. If potassium ions are present, the typical illite was found, but it was not possible to synthesize, in this procedure, a pure aluminum montmorillonite without the presence of magnesium.

165. A series of apparently reversed reactions occur in aridic climates, such as the desilification of halloysite to hydrargillite, as observed by Th. F. Bates,[367] and of kaolinites of the Congo changed under the action of pure water or magnesium salt solutions, as described by M. C. Gastuche,[368] and in the experiments of A. Oberlin and C. Tchoubar.[369] Gastuche not only studied the electron micrographs of corroded kaolinite crystals, and the development of characteristic fractures parallel to the hexagonal outlines, but demonstrated the decomposition of the kaolinite in experiments, as by treating it with oxalic acid solutions. The kaolinite crystals then show a tendency to curl up. The formation of boehmite is striking and analogous to the reaction products observed in Hawaii. Oberlin and Tchoubar studied kaolinite clays from the Sahara, also with curled and folded edges, and their characteristic fragmentation to lathlike shapes. They gave evidence also of the formation of metahalloysite, with its typical two-dimensional grating patterns (*hk* interferences) in the electron diffraction diagrams. The mechanisms of these changes are not yet understood, especially not the specific action of magnesium sulfate solution promoting the splitting and fracturing of the kaolinite crystals (cf. Vol. IV, Section A).

[366] *Natl. Acad. Sci.-Natl. Research Council, Publ.*, **456**, 1956, 54-60.

[367] *Mineral Inds. Expt. Sta. Circular*, **29**, 1960 (8), p. 5-6. It is called by Bates the dominant process of weathering in Hawaii. On the halloysite and hydrargillite formation in these rocks see also Th. F. Bates, *Proc. Nat. Conf. Clays and Clay Minerals*, **9**, 1962, 315-328.

[368] *Silicates inds.*, **24**, 1959, 237-244, 313-320. An interesting attempt of hydrothermal aging of a mixture of hydrargillite with "Ludox" silica sol to synthesize at low temperatures a typical kandite structure was made by G. W. Brindley and C. de Kimpe, *Nature*, **190**, 1961, 254; *Am. Mineralogist*, **46**, 1961, 1370-1381 (with M. C. Gastuche), cf. IV. A. ¶ 466. Foonote 748.

[369] *Silicates inds.*, **24**, 1959, 193-202.

166. An interesting synthesis of an *iron allophane* entirely analogous to common allophane was described by K. M. Feodotev and V. G. Starostina,[370] based on the observation of minerals containing iron halloysite and iron allophane on the contacts of Carbonian limestones to Jurassic clays in the Moscow Basin. The clay minerals are associated with hyalite, aluminum and iron hydroxides, jarosite, and others. By the reaction of alkali silicate solution with a mix of aluminum and iron (III) chloride in equivalent ratios $R_2O_3 : SiO_2 = 1 : 2$ to 5, and $Al_2O_3 : Fe_2O_3 = 0.10$ to 1.07 at low temperatures, stable allophanoids were obtained which, however, easily split off iron oxide at slightly elevated temperatures.

167. The geological literature is rich in discussions on the role of *organic substances* in the low-temperature genesis of clay sedimentation. Often the question was examined whether or not there are complexes formed between colloidal or crystalline aluminosilicates of the clay type with "humic acids." A very complete compilation of the literature up to 1954 was given by E. Gruner,[371] as seen from the viewpoint of the ceramist. Starting from the lignin of the organisms, intermediate so-called "humo-lignite" complexes have been observed forming a series of transition compounds of true humus materials and finally into humus coals. All these substances are highly complex mixes of different chemical compounds, dependent in their ratios on the age and many physicochemical factors of the environment. In every case, they are basically different from the aliphatic acids derived from cellulose, and from the bituminous substances derived from waxes, resins, and others. Methoxyl groups are contained in the intermediate humo-lignite bodies, which are split off in the transition to the humic acids, and to alkali-soluble humus derivatives. Gruner is of the opinion that much of the chemical reactivity of the humus materials with the colloids and cryptocrystalline aluminosilicates is not so important in clay diagenesis than is their action as protecting colloids in the transport of the colloidal weathering products.

168. In this sense, they may be called adsorption compounds, or even organosilicate complexes, especially in the (still rather hypothetical) gray-colored humus horizons. One may not omit, in the chemical changes of the humus material itself, the role of bacteria which are so important in the ceramic process of fermentation or "souring" (cf. C. ¶ 233). P. Coraducci[372] also discussed the humo-silicates of the gray horizons, in their particular influence on the ceramic properties of clays (plasticity, viscosity of their suspensions, deflocculation characteristics, etc.) (cf. C.

[370] *Voprosy Petrog. i Mineral., Akad. Nauk S.S.S.R.*, **2**, 1953, 220-227.

[371] *Ber. deut. keram. Ges.*, **32**, 1955, 169-172; 199-203. See also M. C. Gastuche and J. Uytterhoeven, *Silicates inds.*, **26**, 1961, 341-351. An interesting study of the humic acids extracted from clays was made by K. Ojiri, *Asahi Garasu Kenkyu Hokoku*, **9**, 1959, 128-136.

[372] *Ceramica (Milan)*, **12**, 1957 (4) 67-71.

¶ 358). It is in this connection interesting that W. A. Garrett and G. F. Walker[373] studied the possibility of intercalation complexes of halloysite (from Djebel-Debar, Algeria) with salts, having cations of low hydration energy (potassium), whereas for the same clay mineral there is no inner base exchange capacity. The intercalation complexes chiefly depend on the anion and its size. A complex of halloysite with the potassium salt of trimethylacetic acid, $(CH_3)_3C.COOK$, is rather stable, but by washing with saturated NaCl solution it is split, although no sodium chloride complex is formed in the place of the previous compound.

169. We repeatedly mentioned above the particular phenomena that hydrargillite or related hydroxide minerals are formed in weathering sediments in tropical and aridic climates by a final desilication process, as in the weathering sediments of the Hawaian Islands. This phenomenon generally known in connection with the formation of *lateritic* soils was recently discussed by V. T. Allen[374] from Venezuela, with a distinction of siliceous (argillaceous) laterites (with kaolinite as the characteristic mineral), ferrugineous laterites (with goethite, limonite, hematite), and aluminous laterites (with hydrargillite, boehmite) to which also belong the typical bauxitic laterites (with kaolinite as an accessorial mineral). In the examples described by Allen the argillaceous laterites often preserve the structure of the primary rocks from which they were derived and contain characteristic residual minerals which were chemically resistant enough to the weathering agents. The clear distinctions of the different types of laterites recommended by Allen, according to their characteristic mineral contents, may generally avoid the misleading use of the word laterite as a mere field term, so often remarked in the economic-geological literature. In Allen's definition the well-known alumina-enriched slates (Schiefertone) of Neurode in Silesia, are intermediate between clay sediments and argillaceous (dickite containing) laterites, as O. E. Radczewski[375] showed by electron microscopy and by the identification of dickite (pholerite) associated with diaspore, both minerals being characteristic of hydrothermal aftereffects in this special sedimentary rock.

170. The Neurode aluminous slates have great similarity with the important alumina-enriched Missouri fireclays also of typical sedimentary origin, which were especially investigated by W. D. Keller, J. F. Westcott, an A. O. Bledsoe.[376] Diaspore "burleys" (oolites) and boehmite are found in these, with the highest degree of silica leaching in the Southern portion of the large basin, whereas the kaolinitic-illitic facies (the so-called semiplastic or semiflint) are developed more in the marginal Northern portions of the slowly sinking depositional region. In the definition of C. R.

[373] *Clay Minerals Bull.*, 4, 1959, 75-80; cf. A. Rivière, *Ibid.*, 1, 1950, 121-124.
[374] *Bull. Geol. Soc. Am.*, **68**, 1957, 1693.
[375] *Ber. deut. keram. Ges.*, **28**, 1951, 119-138; *Fortschr. Mineral.*, **28**, 1949, 68-69.
[376] *Natl. Acad. Sci.-Natl. Research Council, Publ.* **327**, 1954, 7-46.

van Hise (1904), both processes of destruction and construction (katamorphism and anamorphism) appear to have operated in the formation of the clay minerals. The pH conditions of the sedimentation were probably nearly neutral, or slightly acidic, in the oxidation potential parameter (*Eh*) negative (reducing) as indicated by the organic residues and by pyrite. The absence of hydrargillite in the alumina-enriched clays is highly remarkable. The appearance of boehmite, and even of diaspore, in its place, is interpreted by Keller *et al.* as an indication of waterlogged and reducing conditions, whereas hydrargillite (and laterites) indicate leaching by freshened, circulating waters with a positive *Eh* potential (oxidizing). The diaspore is also very remarkable having "inherited" the structural characteristics of primary kaolinite from which it was derived. Its formation, however, is not easily compatible with some of the conclusions made by G. Ervin and E. F. Osborn (1951) from their studies in the system Al_2O_3–H_2O, since no geological evidence provides a source of the high temperatures postulated (above 275°C.) for the stability of diaspore. No hydrothermal paragenesis was ever observed, nor was dickite which is so characteristically developed in the Neurode slates.

171. Everything speaks in favor of a low-temperature deposition of the diaspore, and of the "burley" oolites in diagenetic cycles, probably as a metastable crystallization product with its structural inheritance from the layer structure of primary kaolinite. W. D. Keller[377] extensively discussed the influences of the so-called "microclimates," namely locally restricted climatic conditions governing the evolution of the clay mineral changes, as indicated in the Missouri fireclays, in terms of the concentration of active ions, the oxidation potential, and the temperature in the natural reaction system, either in ground water zones, paludal, lacustrine, or marine conditions. It is the complex action of those different factors like *pH*, *Eh*, concentration, and so on, which direct the evolution of the sediments, as W. C. Krumbein and R. M. Garrels[378] demonstrated in a fundamentally important study. Keller, on such a wide basis, discussed the characteristic genetic environment factors for the kaolinite group, for montmorillonite, illite, and bauxitic laterite formations which illustrate the geological-chemical history of typical clay and aluminum ore deposits described in a rich geological literature.

172. In distinction from the Missouri diaspore fireclays, the bauxites of Southern Europe are enriched with boehmite as the principal aluminum mineral. Diaspore is observed in Southern Europe only in connection with Palaeozoic rocks, as J. Bridge[379] emphasized. A high hydrargillite content, and some boehmite, is typical of

[377] *Bull. Am. Assoc. Petrol. Geologists*, **40**, 1956, 2689-2710.

[378] *J. Geol.*, **60**, 1952, 1-33.

[379] *Am. Inst. Mining Met. Eng. Tech. Publ.*, "Comments in Problems of Clay and Laterite Genesis, a Symposium," Am. Inst. Mining and Metallurg. Engineers, Symposium, 1952, pp. 212-213.

the bauxites of Victoria, Australia, as described by J. McAndrew.[380] They are derived from basalts, in parallel with Upper-Silesian hydrargillite bauxites, described by K. Spangenberg[381] as associated with metahalloysite occurring on a deeply corroded Karst horizon, comparable with the "sinkholes" surface typical of the Missouri strata. No boehmite was found in this remarkable occurrence of Silesia. The rather frequent nests of red bauxites in the Northern Urals are interesting because of their contents in dickite below the contact of an ore body with limestone, as described by E. S. Gutkin,[382] further in the bauxite deposits of Cheremukhovo. The genesis of the dickite is interpreted as derived from allophane-halloysite weathering products, under typically low-temperature (*not* hydrothermal) conditions. Hydrargillite slates, in the soft-coal districts of Northern Bohemia were examined by V. Tománek[383] because of their exceedingly high content in titania (6.64 per cent) and their good ceramic properties. French ceramic clays high in alumina of similar character were described by Louise Halm.[384] They are equivalent to the Missouri fireclays and contain free diaspore. Some of them also show boehmite and hydrargillite, even with the same oolitic textures observed in the Missouri clays.

173. An X-ray investigation of the bauxitic aluminum ores of British Guiana was made by G. W. Brindley and W. H. Sutton[385] in which hydrargillite is the principal alumina mineral associated with kaolinite in varying states of crystallinity. Also these ores are remarkably high in titania (2 per cent), in the form of anatase, but low in boehmite. The frequently low degree of crystalline order in the kaolinite is the reason why it was previously often identified as halloysite, namely by the continuous scattering in the range between $2\Theta = 20°$ and $27°$ in the powder diagrams (Cu Kα radiation). A high titania content is also typical of many of the Middle Asiatic bauxites, as that of Kaïrak (Uzbekistan) described by S. K. Gipp,[386] with anatase and leucoxene (3.5 to even 5.5 per cent) dispersed all over the bauxite, or in concretionary form, evidently coagulated from titania hydrogel. Diaspore is here the principal aluminum ore mineral, associated with chlorite, but often the diaspore is replaced by a secondary kaolinite, starting from the core in concentric shells. Metahalloysite is observed in the upper weathering zones. The geochemical association of aluminum with titanium is most characteristic, although the latter element has

[380] *Australian J. Sci.*, **16**, 1953, 64.

[381] *Neues Jahrb. Mineral. Monatsh.*, 1949, A, 152-156; the occurrence was previously described by J. Kuhl, *Arch. minéral. soc. sci. Varsovie*, **9**, 1933, 105-115, as "bauxitic argillites," but not recognized in their full significance.

[382] *Zapiski Vsesouyz. Mineral. Obshchestva*, **87**, 1958, 614-617.

[383] *Ber. deut. keram. Ges.*, **34**, 1957, 326-328.

[384] *Bull. soc. franç. céram.*, 1951 (12), 31-39; *Am. Ceram. Soc. Bull.*, **31**, 1952, 79-84.

[385] *Econ. Geol.*, **52**, 1957, 391-399.

[386] *Doklady Akad. Nauk S.S.S.R.*, **126**, 1959, 859-862.

a lower mobility. S. I. Beneslavskiĭ[387] especially discussed the influence of the mineral composition of Russian bauxites (chiefly with hydrargillite, but also with diaspore, and as a particularity, with corundum) on their leachability in the industrial alkali extraction process. In spite of their high bulk alumina content, there is no quantitative output because the corundum share is not decomposed by the alkali treatment at 105°C. and is a loss in the operation by incomplete dissolution, even if the temperature is raised to 190° and 225°C. in autoclaving. There are, furthermore, in southern Ukraine and Tatary (Barzas) bauxites which show an incomplete alkali solubility although no corundum could be found, but an alumina-containing goethite of a surprisingly high stability, such as in the residues from bauxites of the Yeniseĭ District, with approximate composition $Al_2O_3.2Fe_2O_3.5H_2O$, also rather high in titania. Another very remarkable phenomenon observed in bauxites of northwest Kazakhstan, in a typical Karst environment, is the replacement of quartz by hydrargillite in bean-ore-like aggregates, as described by N. A. Lisitsyna,[388] often goethite associated with the hydrargillite, in all transitions from the beginning to the complete replacement. There are even regular pseudomorphs of hydrargillite after quartz in silicified limestones and dolomites.

174. The particular occurrence of the so-called terra rossa in Miocenic clayish-calcareous rocks was studied by L. Edlmann[389] as a typical leaching residue product of tropic weathering (laterites), with ratios $SiO_2 : Al_2O_3 = 1 : 1.1$ to 1.2 in gray sandy "ash soil" pockets, of the same origin as terra rossa. It becomes red like the latter, after calcination and is high in iron oxide with some manganese. Terra rossa formations are highly desilified, originally siallitic, clay sediments. The extracted silica is often redeposited in cherty nodules in the surrounding limestones, also with new-crystallized illite, as a secondary reaction product of the colloidal silica with alumina of the laterites. Recent processes of this type were described by J. A. Ferguson[390] from Queensland, Australia, as soil types, in dark red to black colors, derived from weathered basalts. The detailed mineral analysis showed montmorillonite as the first clay mineral formed after an intense release of calcium and magnesium from the basalt under the recent weathering conditions. Later the montmorillonite is decomposed by the loss of alkalies and of iron, and changed to kaolinitic or halloysitic products, which in their turn are changed to hydrargillite and iron hydroxides which cause the intense coloring of the "kraznosems" (red soils), chocolate soils,

[387] *Doklady Akad. Nauk S.S.S.R.*, **95**, 1954, 1077-1080; *Ibid.*, **113**, 1957, 1130-1132.

[388] *Doklady Akad. Nauk S.S.S.R.*, **109**, 1956, 1023-1025. The corrosion of quartz in subtropic climatic conditions is a rather general phenomenon in Mesozoic and Cenozoic weathering crusts of Kazakhstan, as described by N. A. Lisitsyna, *Doklady Akad. Nauk S.S.S.R.*, **114**, 1957, 862-864.

[389] *Ricerca sci.*, **22**, 1952, 1749-1754; *Ibid.*, **24** 1954, 326-332.

[390] *Australian J. Agr. Research*, **5**, 1954, 98-108.

red loams, prairie soils, and others, of the laterite type. The changes are in the sequence:

Basalt	→ *Montmorillonite* and *Carbonates*	→ *Kaolinite* and/or *Halloysite*	→ *Hydrargillite*
(largely plagioclase and Fe–Mg silicates)	(in black clay soils)	(in red soils)	

A resynthesis of the montmorillonite is highly unlikely and was never observed.

175. R. M. Garrels[391] made a most remarkable attempt to give an exact thermodynamic basis to the general geological reactions governing the changes on the earth's surface. Although we have to discuss this most interesting investigation in other connections with the hydration of aluminosilicates (cf. Vol. IV, Section A), we may here emphasize that the reversibility of the reaction

$$\text{Aluminum Hydroxide Minerals} + \text{Silica Solutions} \rightleftharpoons \text{Kaolinite} + \text{Water}$$

is fully evident. The reaction goes to the right side of the equation, as in the resilication of Arkansas bauxites with waters of overlying swamps,[392] to the left side when the concentration in silica dissolved in circulating waters is constantly kept low by removal. The monohydrates of alumina, AlO(OH), boehmite and diaspore, seem to be more easily kaolinized than hydrargillite. The equilibrium constant for the reactions mentioned is $10^{7.54}$, the free energy $\Delta F^0_R = -1.364 \times \log 10^{7.54} = -10{,}300$ cal.

DIFFUSION AND RHYTHMIC PRECIPITATION IN GELS

176. Since the classical experiments of R. E. Liesegang (1913), many contributions to the description and genesis of banded agates and of their peculiar rhythmic structures have been made in recent years. W. Fischer[393] described agates deposited in the typical amygdaloid cavities (gas bubbles) in consolidated effusive rocks, and assumed that circulating solutions rich in silica derived from the hydrothermal alteration of the lavas brought about the material filling the cavities. The material of the agates is in general not so homogeneous as was previously often assumed. Fischer distinguishes not only the deposition of quartzine, chalcedony, opal, but also of zeolites and calcite in the banded agate structures by rhythmic diffusion, under variable conditions of temperature, concentration, and velocity of the circulating solutions.

[391] *Am. Mineralogist,* **42,** 1957, 780-791, especially pp. 784 ff.

[392] Cf. M. I. Goldman, *Econ. Geol.,* **50,** 1955, 586-610.

[393] *Neues Jahrb. Mineral. Abhandl.,* **86,** 1954, 367-392. We also mention here the amygdules in andesite described by E. Ingerson, *Am. Mineralogist,* **38,** 1953, 1057-1064, with excellently banded agate fillings. Ingerson assumes rather low temperatures in their formation.

Entirely in agreement with Liesegang's early generalization that rhythmic precipitations are not only restricted on the agates as the most striking occurrences, but also a general geological phenomenon, V. F. Petrun[394] discussed colored quartz veins in salbands of Devonian clayish sandstones of the Ulu-Tau Range (Kazakhstan), with fine intergrowths of rutile, goethite, and iron-manganese ores in rhythms of the pigment segregations. Three-dimensional rhythms in a colloidal matrix were especially studied by J. D. Carl and G. C. Amstutz[395] from laboratory model experiments by rhythmic diffusion and precipitation of silver chromate. Particular emphasis was given to formation of secondary rings (often five or six) intercalated between the stronger primary rings. The secondary rings are whitish, and evidently chemically different from the intensely red silver chromate rings, perhaps originating from contaminations of the gelatin. Interesting details of natural rocks could be reproduced in the three-dimensional model, as in a volcanic rock, the stopping of rhythmic weathering rings of limonite at mineral grains, or the waving around them. The Liesegang phenomenon offers a perfect explanation also of the exfoliation of igneous or sedimentary rocks, or of spheroidal weathering, making the previous theories of "weathering pressures" unnecessary.

177. Many attempts have been made to explain the Liesegang phenomena and to give a full theory of diffusion in gels; A. Packter[396] studied the geometric progression in the spacing between the rings of silver chromate, silver arsenate, copper pyridyl rhodanide in gelatin, and of lead iodide in agar. The constant k in the geometric progression equation is $k - 1 = A + B/C_0$ (with C_0 the initial concentration of the diffusing solution, A an inverse function of the peptization ratio for the sol, B dependent on the minimum cation concentration at which precipitation occurs). The diffusion of silver nitrate, copper sulfate, or lead nitrate into the gels is not affected by the presence of anions which might cause the precipitation. Strong protecting effects of the sols are observed if peptizing agents are added, such as organic acids, with a very distinct and specific influence of their constitution (as carboxylic, amino acids, hydroxyacids, chelating, and nonchelating acids). For precipitations of silver chromate, the spacing of the rings is a function of the concentration in silver ions, with different gelatin contents, or with additions of traces of chlorine anions, or detergents like sodium dodecyl sulfate, hexadecyltrimethyl ammonium acetate, or formaldehyde or other aldehydes. The spacing between the rings is the closer as the

[394] *Zapiski Vsesoyuz. Mineral. Obshchestva*, **83**, 1954, 226-233; see also the previous discussions of O. Mügge, *Neues Jahrb. Mineral. Geol. Beilage-Bd.*, **58**, 1928, 303-348.

[395] *Bull. Geol. Soc. Am.*, **69**, 1958, 1467-1469. A. Kieslinger, in the book "Zerstörungen an Steinbauten, ihre Ursachen, und ihre Abwehr," Leipzig and Vienna, F. Deuticke, 1932 p. 57, presumed early that Liesegang diffusion phenomena are responsible for rock exfoliation in weathering.

[396] *J. Colloid Sci.*, **10**, 1955, 46-62 (with R. Matalon); *Ibid.*, **11**, 1956, 96-106; *Kolloid-Z.*, **142**, 1955, 109-117.

gel becomes more fluid. Chlorine ions or anionic detergents in acidic gelatin, or cationic detergents in alkaline gelatin, decrease the spacing as a function of the relative concentration. Aldehydes increase the spacing, especially formaldehyde. Also a reprecipitation of silver chromate after the dissolution of the rings in ammonia, and diffusion of acetic acid into the clear gelatin sol, follows the equation given above, with the constant A increasing in the sequence $Cl^- < HAsO_4^{2-} < CrO_4^{2-} < C_2O_4^{2-}$. Organic anions (those of amino acids) have a stabilizing effect in the presence of ammonia, with wider spacings of the rings. Excess in silver ions above 20 per cent brings about a complete flocculation. Packter comes to the conclusion that a coagulation theory illustrates the rhythmic precipitation phenomena better than the Ostwald supersaturation theory which, nevertheless, was used by St. Prager[397] in calculations for a periodic precipitation mechanism.

178. A totally different new aspect, however, is given for the Liesegang phenomenon by C. J. van Oss and P. Hirsch-Ayalon,[398] based on the so-called Hirsch effect which is observed when two solutions, say containing silver and chromate ions, diffuse toward each other through a gel (or membrane) endowing the latter with a perfect impermeability to the ions which participate in the reaction, but leaving it permeable to other ions and to the solvent. The model experiment is the precipitation of barium sulfate in a cellophane membrane from 0.1 N solutions of barium hydroxide and sulfuric acid, with a membrane potential $E = 2.3026 \cdot RT/F\ (pH_{\mathrm{I}} - pH_{\mathrm{II}})$. For the pair Ba^{2+}/SO_4^{2-} is $E = 670$ mv., for Na^+/SO_4^{2-} or $Na^+/Cl^- = 65$ mv. (in this case no more semipermeability exists). The multiple rhythm of precipitation is explained in the following way: the precipitate layer inside the membrane remains impermeable to the forming ions only as long as small amounts of these are present in solution on either side of the membrane; as soon as one of these ions is lacking on one side, the precipitate layer can be crossed, after a shorter or longer time period. The important point of complete depletion of one of the forming ions between precipitation bands is not only made evident by the colorless zone already mentioned earlier by Liesegang, but also in recent experiments of E. S. Hedges (1932) and H. N. Holmes(1926).

179. Similar diffusion phenomena into gels and sols with *anisometric* particles were described by H. Thiele and H. Krönke[399] in fibrillar gels of cellulose glycolate and

[397] *J. Chem. Phys.*, **25**, 1956, 279-283.

[398] *Science*, **129**, 1959, 1365-1366, with an excellent bibliography. On the Hirsch-Ayalon effect, see *Rec. trav. chim.*, **75**, 1956, 1065-1079; *J. Polymer Sci.*, **23**, 1957, 697-704; *Chem. Weekblad*, **53**, 1956, 557-559.

[399] *Naturwissenschaften*, 42, 1955, 388. Analogous to flow birefringence caused by the anisometric particles of clay suspensions, R. Pellizzer, *Rend. soc. mineral. ital.* **16**, 1960, 347-358, observed highly characteristic orientation effects in strong magnetic fields (with interesting bibliography).

the precipitation of lead oxalate, locally determined by the micelles. In this manner one may speak of an anisotropic-micellar crystallization of such precipitates parallel and/or perpendicular to the fibrils. Although here the deposition is not typically rhythmic as in Liesegang's precipitations, the mechanism of the anisometric orientation is of great importance in biochemical processes, as in the deposition of hydroxyl apatite parallel to the elastic fibrils in tissues. The precipitations observed by Thiele and Krönke are partially reversible if the precipitate is able to undergo ion exchange reactions. By the sorption of anisometric gels for dye-stufls, iodine, and the like, a typical orientation pleochroism can be developed. For silicate fibrillar systems, as in asbestos, analogous reactions may be sometimes of particular interest.

180. Rhythmic-periodic structures in technological silicate artifacts are repeatedly striking. W. Geilmann[400] described such phenomena in Egyptian and medieval soda-lime silicate glasses which were exposed to surface weathering for a long time in soil (on the chemical durability of such glasses cf. Vol. II, B. ¶ 180). Also this process is determined by topochemical factors such as gel reactions. In the leached surface layer of the glasses, often beautifully developed rhythmic (agate-like) banding appear, or hemispherical forms of progressive intrusion of the weathering products, plug or spindle-shaped corrosion forms which coalesce more or less, often with surprising analogies to structures as described in the micrographs of Liesegang in his studies on artificial rhythmic precipitations and natural banded forms. That diffusions in silicate melts may develop quite similar patterns was made evident by H. Jebsen-Marwedel[401] from industrial glasses, corresponding to previously described patterns in ceramic bodies and glazes (cf. K. Endell, 1913; J. A. Pask and C. W. Parmelee, 1943; P. Beyersdorfer, 1943, in copper aventurine glasses). Such "pyrogenetic" diffusions are observed as devitrification rings, in rhythmic-spherulitic forms with the crystallization of acicular $Na_2O.3CaO.6SiO_2$ (devitrite), also in a mixture with cristobalite. Although the analogy with the typical rhythmic precipitation forms in gels is only rather approximate and external, Liesegang himself emphasized rhythms observed in magmatic systems, as in sphere diorites, or diffusion banding in contact-metamorphic rocks, and so forth.

FUNDAMENTAL PHENOMENA OF ADSORPTION ON SILICA AND SILICATE GELS

181. We will discuss equally pure and mixed silica gels, and even colloid-disperse silicate minerals, in this introductory section. Initially we consider the clay minerals because they have, in common with true colloid gels, high development of specific

[400] *Glastech. Ber.*, **29**, 1956, 145-168.

[401] *Glastech. Ber.*, **31**, 1958, 311-315; *Verres et réfractaires*, **12**, 1958, 324-329.

surface and porosity which are the basic properties in adsorbent activities. Of great importance from geochemical and economic-geological viewpoints, is the fact that adsorption phenomena bring about enrichments in widely distributed elements, in some cases even in deposits of useful ore minerals. Omitting here any discussion of soil formation phenomena, we may emphasize the particular importance which cation and anion adsorption on clay minerals have attained in recent times for the disposal of radioactive fission products resulting from the operation of nuclear reactors, as discussed by W. L'a. R. Rice.[402] If undesirable fission wastes are adsorbed on clay minerals, they may then be fixed by thermal treatment to form sintered bodies which can be appropriately sited. The fixation of iron and lanthanum on clays was studied under these aspects as model systems. A concentration of 5,000 p.p.m. of clay was found most effective in removing lanthanum from solutions, and the adsorption amount of this element increases with increasing pH of the solution. Iron is even better adsorbed than lanthanum by clays because of its greater ease of forming complexes in solution.

182. *Fixation* of cations or anions on a given colloidal silicate, chiefly on clay minerals, is a process particularly important in agricultural chemistry. S. Goldsztaub, S. Hénin, and R. Wey[403] gave an excellent example in the fixation of *phosphate* anions on kaolinite and montmorillonite at different pH conditions by determining the excess $(H_2PO_4)^-$ ions after 30 minutes contact of the mineral with a solution of known concentration. The optimum fixation was observed for pH between 3 and 6. For pH = 3.3 to 3.8, montmorillonite adsorbed 18 meq. of $(H_2PO_4)^-/100$ gm., kaolinite less of it. Evidently, the aluminum positions on the edges of the clay mineral particles are the active centers of the specific adsorption. This phenomenon is thoroughly analogous to the specific sorption of *fluorine* anions by kaolinite and montmorillonite, and by aluminum hydroxide gel, as observed by H. R. Sampson,[404] which takes place with the formation of a relatively stable complex. "Desorption" of phosphate anions from the complex with clay minerals as the basis of sorption, always means a chemical replacement, as by the anions of oxalic, citric, or hydrofluoric acids, or by dissolution also of aluminum to which the anions were sorbed, and a more stable complex with aluminum formed.

183. The role of fluorine ions in such processes is understood as an exchange of hydroxyl and fluorine ions, as A. Weiss, A. Mehler, G. Koch, and U. Hofmann[405]

[402] *Phys. Rev.*, **99**, 1955, 340. The use of bentonite and vermiculite for decontamination of waste waters from nuclear pile plants is demonstrated also by W. Hoffmann and U. Hofmann, *Nukleonik*, **3**, 1961, 195-200.

[403] *Clay Minerals Bull.*, **2**, 1954, 162-166.

[404] *Clay Minerals Bull.*, **1**, 1952, 266-272.

[405] *Z. anorg. u. allgem. Chem.*, **284**, 1956, 247-271. The presence of calcite, dolomite, limonite, and so on as contaminations considerably disturbs the quantitative exchange amounts. Even

demonstrated. These are chiefly localized in the margins of the octahedral layers of the clay mineral structure. There is also a distinct influence of charges of the cations near the margins. In nontronite the influence of these cations can be studied by changing the octahedral central cations, namely of iron, by reduction and reoxidation cycling. The remarkable fact was established that in nontronite all of the divalent iron ions are localized on the margins of the crystal. This fact explains the surprisingly easy reduction and oxidation of the iron in nontronite, and its color changes from green to brown if oxidized in alkali hydroxide solution, or to green again if reduced by sulfur dioxide. Kaolinite is the typical two-layer clay mineral with free hydroxyl groups on one basis plane of the monocrystal. Because of these special positions, the exchange $OH^- \rightleftharpoons F^-$ depends in its amount on the thickness of the crystal units in the clay suspension, a factor which does not play any important role in the same exchange reaction with three-layer minerals like montmorillonite or vermiculite. Halloysite behaves in this respect like kaolinite, although a basis plane is not the site of the hydroxyl groups but the inner walls of the tubular structure. Halloysite is specific in that the reaction is not stopped with the exchange introduction of fluorine, but the structure breaks down to a mixture of an amorphous silica hydrogel and ammonium cryolite. Ammonium fluoride is introduced into the halloysite as is water.[406] For this reason the decomposition in halloysite is much more rapid than in kaolinite (Fig. C. 88). A schema of the reaction is given in Fig. C. 89. Magnesium-rich clay minerals (saponite, hectorite, vermiculite) are also relatively easily decomposed by ammonium fluoride solution, with precipitation of magnesium fluoride. The anion exchange takes place in three steps: (*a*) exchange $OH^- \rightleftharpoons F^-$ on the prism faces of the structure; (*b*) on the basis planes; (*c*) removal of hydroxyl groups from the crystal structure and ensuing destruction.

184. One of the most typical geochemical adsorption phenomena is the specific fixation of *vanadium* on clay minerals. This phenomenon is of great importance in ceramics because vanadium-containing clays bring about troublesome discolorations in porcelain glazes, and in fuels a vanadium containing clay mineral causes an intense corrosion of refractories. A. J. Hammer[407] investigated the conditions of the

bromine ions are exchangeable for hydroxyl groups in nontronite. Many of the details given on the *local* effects of adsorption on the margins, and on the basis faces recall the experiments of P. A. Thiessen, *Ibid.*, **253**, 1945, 161-169, on the adsorption of gold particles on kaolinite crystal fragments under the influence of different charge signs. See also the interesting experiments of G. W. Brindley, R.H . S. Robertson and R. C. Mackenzie, *Am. Mineralogist*, **39**, 1954, especially pp. 135 f. on gold sol adsorption on kaolinite from Pugu, Tanganyika.

406 Ammonium fluoride is rather similar to ice in its molecular properties and even forms crystalline solutions (up to 10 per cent NH_4F) with ice, as R. Brill and S. Zaromb observed (cf. *Intern. Congr. Cryst., Paris*, 1954).

407 *Ceram. Age*, **65**, 1955 (4), 47-50; (5), 7-10; (6), 19-20; (8), 16-17; (9), 18-19, 32; (10), 28-29; (11), 24-27, with bibliography.

vanadium contents of clays in a series of publications. Omitting here the mineralogy of vanadium, especially in the form of sulfides, an isomorphous substitution of aluminum by vanadium takes place, as in roscoelite, a vanadium mica. Beyond that, there is a specific adsorption-enrichment of vanadium on clays, or on the hydrogels of iron and aluminum hydroxide, with an optimum at pH = 3.5. In fuel oils and in

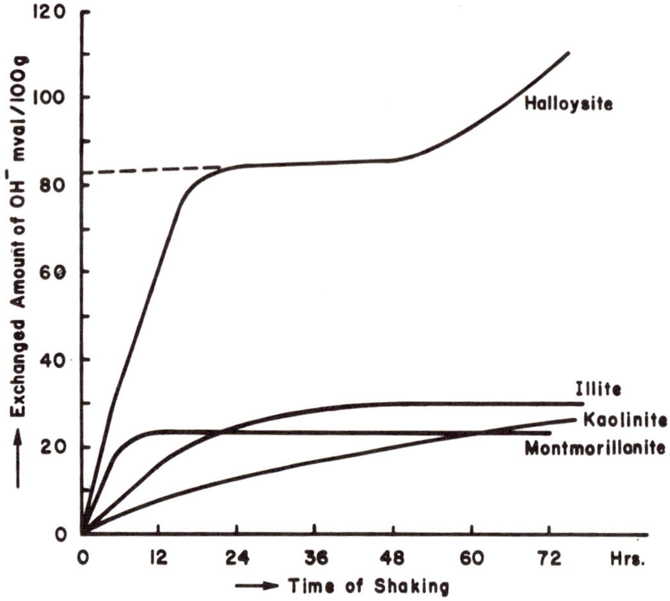

FIG. C. 88. Amounts of fluorine anions adsorption exchanged for hydroxyl anions on the surface of clay minerals (Weiss, Mehler, Koch, and Hofmann), as a function of duration of shaking with ammonium fluoride solution (pH initially 7.1: concentration of the solution in F^- 5000 meq./100 gm.

the carbonaceous material of sediments (derived from azotobacteria in soils) there are organic complex compounds with vanadium, of great stability. This latter type is responsible for most of the corrosion effects on the refractory linings of combustion chambers, described by W. A. Lambertson (1949) (cf. Vol. II, C. ¶ 84). Similar in

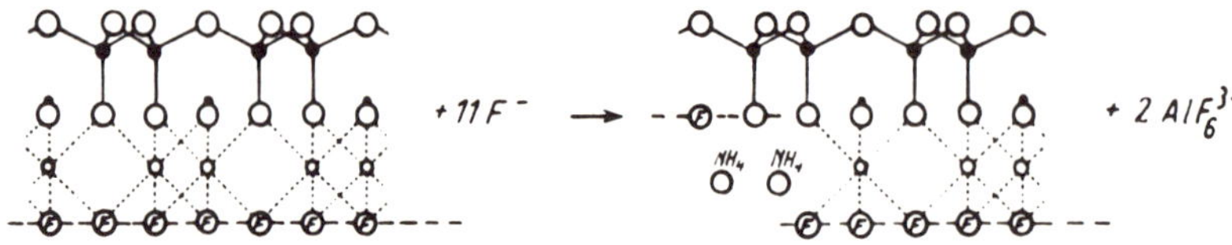

FIG. C. 89. Schema of destruction of halloysite or kaolinite beyond the hydroxyl-fluorine anion exchange (Weiss, Mehler, Koch, and Hofmann).

its specific fixation by adsorption on clay minerals is *molybdenum* as I. Goldberg[408] observed, in the form of $(MoO_4)^{2-}$ anions.

185. According to P. T. Davies and T. R. Scott[409] one may conclude tnat *uranium*-containing solutions, such as those formed by acid leaching of low-grade uranium ores (in which uranium is hexavalent) show a strong tendency to adsorb the element on suspended particles of clay minerals (halloysite, montmorillonite, vermiculite). Up to 0.02 meq. of uranium are fixed from sulfate solutions by adsorption. The mechanism, however, is quite different from the common base exchange capacity rules; the amount of uranium taken up is far below that which would be possible in the case of normal base exchange. The planful adsorption of *gold* and other precious metals on montmorillonitic clays is discussed by C. L. Mantell[410] in mining practice, especially for exploiting the "lost or locked-in" gold in barren land sediments, like those of Western Australia. For prospecting of rare elements adsorbed, as on limonite gels of the oxidation zone of ore deposits, electrodialysis is recommended as an important and helpful method by A. A. Ivanov and Z. G. Sheĭna.[411]

186. It is in the practically important process of *flotation* that the electrokinetic conditions accompanying the adsorption phenomena must be carefully considered. This process is not only of considerable significance in the separation of metal ores from the so-called gangue (much of silicate material is included in this general term), but in enrichment of the high-grade silicate materials themselves. In this regard E. Stacha[412] gave valuable information on the electroosmotic flotation of kaolins, and R. B. Booth[413] reported on "nonmetallic ores," including the flotation of kyanite. The theoretical background of the problems here involved, seen from the viewpoint of the electrokinetic theory of cationic adsorption, was discussed by V. M. Borisov.[414] He defined in the "electrokinetic index" the difference of the electrostatic potentials of the mineral surface in the flotation solutions containing the electrolyte alone, and those containing the "collector" plus the electrolyte. This index is governed by that ion which is selectively adsorbed on the double-layer around the suspended particles. Cationic adsorption is indicated by positive values of the ζ potential; anionic adsorption by negative values. Sodium oleate, a common collector reagent, is in optimum efficiency if ζ is reduced to 0 by the systematic addition of an electrolyte (e.g., sodium sulfate). As floated minerals in a soda solution, examples of nepheline

[408] Thesis in Soil Science, Univ. of California, Berkeley, Calif., 1950.

[409] *Nature*, **178**, 1956, 1195.

[410] *Chem. Eng. Progr. Symposium Ser.*, **50**, 1955 (15), 21-24.

[411] *Zapiski Vsesoyuz. Mineral. Obshchestva*, **85**, 1956, 413-416.

[412] *Zpravy Českoslov. keram. a sklarské spolecnosti*, **24**, 1948, 60-66.

[413] *Ind. Eng. Chem.*, **47**, 1955, especially pp. 555-558, with extensive references.

[414] *Doklady Akad. Nauk S.S.S.R.*, **99**, 1954, 427-430.

and aegirite are given by Borisov. H. Kirchberg and W. Schubert[415] described the considerable success of planfully directed flotation in the enrichment of the feldspar of so-called pegmatite sands to high concentrates, using lauryl pyridinium sulfate as the flotation addition (collector).

187. A detailed structural analysis of the adsorption reactions on the surface of crystalline quartz was further given by H. Seifert, R. Buhl, and K.-F. Seifert,[416] who specifically demonstrated how important it is not to neglect the *anisotropic* orientation effects accompanying the adsorption phenomena, and showed schematic diagrams of the spherical packings on the faces (0001), $(10\bar{1}0)$ of *d*- and *l*-quartz, further of $(11\bar{2}0)$, $(10\bar{1}1)$, and $(21\bar{3}1)$ of *l*-quartz alone. Structural defects in these ideal packings may even increase the adsorptivity of polar substances, chiefly through the bridging bonds with hydroxyl or amino groups in these. Although the principal importance of Seifert's theoretical discussions is in the field of epitaxy (cf. A. ¶ 257 ff.), it is sufficient to emphasize that they also serve to exclude the previous flotation theories of A. F. Taggart, T. C. Taylor, and A. F. Knoll,[417] especially concerning agents like the lauryl derivatives. For such studies of the adsorption on crystal faces of larger size than in clay mineral particles, as of oleic acid on fluorite cleavage faces, the study of the infrared spectra was found highly instructive in an investigation by R. O. French, M. E. Wadsworth, M. A. Cook, and I. B. Cutler.[418] We cannot discuss here the details of the important behavior of the *cis* and *trans* peaks in the spectrum of oleic acid and in oleates, and their changes in adsorption, but we will come back to this interesting method in C. ¶ 306, discussing similar investigations of the catalysis of silica gels.

188. In analytical determinations of alkalies in silicate minerals or in industrial glasses, losses by the adsorption in the glazes of porcelain receptacles and dishes are often observed. W. Geilmann and A. Gänssle[419] studied this important problem by the introduction of radioactive tracer isotopes, such as K^{42}, Rb^{86}, and Cs^{134}, with sensitivities of weight displacements of 0.1 μg. Considerable alkali losses were thus equally established in the Lawrence-Smith method and decomposition of the silicate specimens by hydrofluoric acid in mixtures with sulfuric or perchloric acid, also of sodium in the zinc–uranyl acetate complex, or of potassium in the tetraphenyl-B complex. Especially at elevated temperatures, common feldspar glazes of porcelain beakers and like products show a specific adsorption of alkalies as shown by the tracer radioactivity indications. An analogous geochemical phenomenon is the lo-

[415] *Silikat Tech.*, **2**, 1951, 196-199.

[416] *Kolloid-Z.*, **141**, 1955, 146-159.

[417] *Am. Inst. Mining Met. Eng. Tech. Publ.*, **312**, 1930, 3-33; *Trans. Am. Inst. Mining Met. Engrs. Iron Steel Div.*, **87**, 1930, 217-248.

[418] *J. Phys. Chem.*, **58**, 1954, 805-811.

[419] *Glastech. Ber.*, **27**, 1954, 80-88.

calization of radioactive elements in different minerals of pelagic clays, studied by G. Arrhenius and E. D. Goldberg,[420] with a specific enrichment in authigenic phillipsite, acting as a scavenger of radioactive elements in marine deposits. Radioautographs of α-particles emitted from such phillipsite crystals embedded in nuclear track emulsion are highly conclusive. Another most characteristic chemisorption is that of calcium on silica gels, a process which plays a fundamental role in the theory of the hydration products of Portland cements and other hydraulic binders, and especially in pozzolanas (cf. Vol. V, Section C). K. G. Krasil'nikov[421] extensively discussed this important phenomenon from the viewpoint of general adsorption, as in the chemisorption of lime on tripolite and diatomaceous earths, determined by common analytical methods. This problem is particularly important in the formation of definite calcium silicate hydrates of which Krasil'nikov mentions those with the ratios $CaO : SiO_2 = 0.9$, 1.5, and 2.0, all in highly dispersed states, and discusses their crystallization phenomena.

189. The formation of natural deposits of *bentonites* is perhaps one of the most conclusive examples of the manner in which adsorption (especially of alkalies) and diagenesis are interrelated in the sedimentation process. Among the rich literature in this field we especially emphasize the extensive investigations of T. Sudo[422] on the alteration of volcanic glasses in Japan, with all intermediate transitions from the original glass ejectamenta to typical montmorillonite-bentonites, allophanes, halloysite, kaolinite, and even mica clays (with sericite, celadonite). The occurrence also of clay minerals with mixed-layer structures may be particularly emphasized in these products (cf. B. ¶ 122 ff.), which often have a characteristic cristobalite content. Much detailed studies were further made by W. D. Keller, G. A. Kiersch, and P. Howell[423] on the argillation of diabase tuffs to bentonite in Arizona, in the complex process of hydration and hydrolysis as the first steps toward the final changes into bentonite, and even pure montmorillonite. The net result of argillation by weathering is to concentrate the hydrogen ions from the hydrolysis into the clay minerals which remain temporarily on the land, and to concentrate the hydroxyl in the oceans (with $pH = 7.0$). The typical alkali-bentonites are characterized by the addition of metal cations to the hydrolysis products in an environment of relatively stagnant waters (as in lake basins). Bleaching earths of commercial importance were deposited under such conditions in the Navajo country, Arizona, as shown by W. D. Keller

[420] *Tellus*, **7**, 1955, 226-231; cf. E. D. Goldberg, *J. Geol.*, **62**, 1954, 249-265.

[421] *Trudy Soveshchaniya Khim. Tsementa, Nauch.-Tekh. Obshchestvo Prom. Stroitel. Materialov*, 1956, 351-380.

[422] *Clay Minerals Bull.*, **2**, 1954, 96-106; *Sci. Repts. Tokyo Kyoiku Daigaku* **5**, 1956 (43) 39-55. See also the important book of T. Sudo, "Mineralogical Study on Clays of Japan." Maruzen, Tokyo, 1959, 328 pp., especially pp. 194-225, including the "acid clays."

[423] *Natl. Acad. Sci-Natl. Research Council, Publ.*, **395**, 1955, 413-420.

and G. A. Kiersch[424] who studied the course of the alteration of vitric (latitic) volcanic ash tuffs to bentonite in all details of transition by field observation, microscopic, and X-ray examination. The first step of this transition was a hydration, followed by the addition of magnesium, and the removal of silica, and alkalies, but aluminum remained static and was flocculated in the presence of calcium and magnesium. The regional character of the deposits in a shelf area, with typical basin (or channel) structures is evident in the local stratigraphy. E-an Zen[425] investigated in a similar way marine coastal shallow bottom sediments from Peru and Chile, with a high content of volcanic glass in multiple varying states of "devitrification," in other words, diagenetic-metamorphic conditions of a very early grade. Whereas montmorillonite is relatively rare in observed sediments, a non-swelling mica mineral, related to muscovite, is particularly common, nearly ubiquitous. In the second place one also finds kaolinite. Illite, randomly interlayered with montmorillonite is the product of alteration of vitric (fresh-vesicular rhyolitic glass) volcanic ashes occurring near Denver, Colorado, as described by J. Schlocker and R. van Horn.[426]

ADSORPTION OF ORGANIC DYESTUFFS ON SILICA AND SILICATE GELS

190. A very interesting example of a highly surface-active silica gel is the "silica wool" (Kieselwatte) prepared by H. W. Kohlschütter, H. Mattner, and K. Beyersdorfer.[427] Treatment of organic cotton wool fibers by impregnation with silicon tetrachloride is followed by hydrolysis of the latter in water vapor. Thin layers of silica gel are deposited on the wool fibers which, after combustion of the cellulose material in oxygen at 850°C., yield a regular silica pseudomorph of great fineness in the electron microscopic image. It is easy to stain this amorphous silica film pseudomorph with organic dyestuffs, such as methylene blue in aqueous solution. In comparison with common gels, the staining of the silica wool is a rather rapid process although the final amount of the adsorbed dyestuff is much less than in normal gels of the same thermal treatment. The silica wool, in its electron microscopic aspect is more a lamellar film than a fibrillar aggregation. The adsorption of methylene blue follows the elementary Freundlich adsorption isotherm of the type $\log a = \log k + 1/n \cdot \log c$.

[424] *Econ. Geol.*, **50**, 1955, 469-494. By acid treatment the well-known activated "Filtrol" is commercially produced.

[425] *Am. Mineralogist*, **42**, 1957, 889-903; *Am. J. Sci.*, **257**, 1959, 29-43, especially on the relations between the clay minerals and carbonates in marine sedimentation. The ratio $Na_2O : SiO_2 = 1 : 3.75$ is still in the range of non-colloidal suspensions; cf. Vol. IV, Section A, according to P. Debye and R. V. Nauman, *J. Phys. Chem.*, **55**, 1951, 1-9.

[426] *J. Sediment. Petrol.*, **28**, 1958, 31-35.

[427] *Z. anorg. u. allgem. Chem.*, **262**, 1950, 122-128.

191. For sodium silicates of medium-high silica content (type $Na_2O : SiO_2 = 1 : 3.75$) in aqueous solution, B. Carroll and E. Freeman[428] showed that adsorption of crystal violet or methylene blue as indicators is a good means to study, by changes of the optical absorption spectra, the effects of aging and ion exchange in colloid solutions. Aging is characterized by a loss in the adsorption affinity for the dyestuff. If salts are added, the absorption bands in the spectrum are shifted backward toward that of the free dyestuff in water, thus indicating displacement of the dyestuff by the salt, an observation of great importance for the explanation of corresponding shiftings in the absorption spectra of the dyestuff adsorption products on clays (see below). The nature of the anion in the added salt is not particularly important, whereas the cations determine the binding affinity of the salts on the silica amicrons in the solution, with a pronounced "displacement series" $Ca^{2+} > Ba^{2+} > Sr^{2+} > K^+ > Na^+ > Li^+$. This effect is so sensitive that less than 1 p.p.m. of a divalent cation could be observed in its result. The adsorption isotherm for crystal violet gives the maximum number of dyestuff molecules bound per formula weight of the sodium silicate. Assuming n binding sites per group $Na_2O.SiO_{3.75}$, an equation given by G. Scatchard (1949) can be applied:

$$(r/A) = nk - rk$$

with $r =$ the ratio of moles of the bonded dyestuff to the total moles $Na_2O.SiO_{3.75}$, A the concentration of the unbound dyestuff, k the "intrinsic binding constant." A linear graph is given for r/A versus r, with the intercept n on the abscissae axis $=$ 0.96. The free energy, $\Delta F^o = -RT \cdot \ln(nk)$ is $= -6{,}960$ cals. The isoelectric point of the sol is seen from Fig. C. 90 near pH $= 2.9$, in the graph showing the optical density measured in the spectrophotometric curve for the absorption maximum of a 1.0×10^{-5} M solution of methylene blue, versus pH, in a $7.72 \cdot 10^{-5}$ M sodium silicate solution.

192. The diagnostic differentiation of *clay minerals* by adsorption tests of organic dyestuffs (*staining* reactions) was extensively studied in several papers of the last decade. B. S. Emodi[429] recommended the determination of montmorillonite by dyestuff reactions as attributed to base exchange, if the dyestuff itself forms cations to undergo such an exchange on the surface of the clay mineral particle. The structural positions of the adsorbed dyestuff ions are determined between opposite silicate layers. The interlayer water of montmorillonite is expelled and the basal spacing adjusted to the size of the dyestuff molecule. R. C. Mielenz and M. E. King[430]

[428] *J. Phys. Chem.*, **53**, 1954, 335-338.

[429] *Clay Minerals Bull.*, **1**, 1949, 76-79. See also V. S. Ramachandran, K. P. Kacker, and N. K. Patwardhan, *Am. Mineralogist*, **47**, 1962, 165-169.

[430] *Am. Soc. Testing Materials, Proc.*, **51**, 1951, 1213-1233.

attempted to standardize staining tests for a rapid identification of clay minerals, as in soils and sedimentary rocks. They recommend the use of Safranine Y or Malachite Green in nitrobenzene solution to examine the pH in adsorbed water films, and an aqueous solution of benzidine for oxidation-reduction reactions as they are typical of aromatic amines changed to specifically colored forms. Induced pleochroism of the crystals of kaolinite and anauxite on one hand, and of nacrite and dickite on the other hand, can thus be distinguished. Benzidine color change reactions are equally discussed for a distinction of montmorillonite, palygorskite, kaolinite

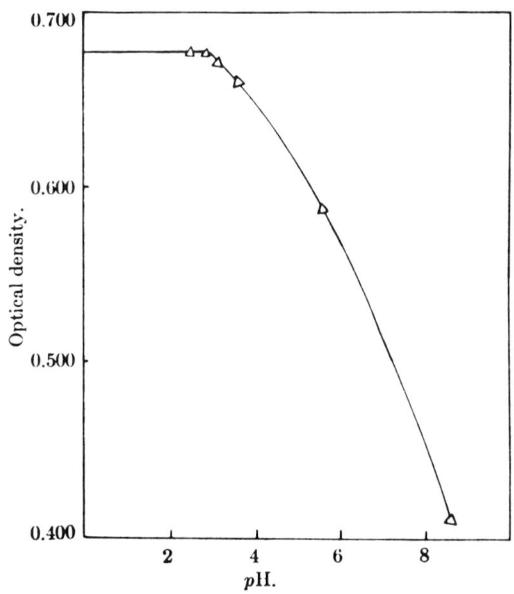

FIG. C. 90. Optical density as a function of pH in dilute solutions of methylene blue and sodium silicate (ratio $Na_2O : SiO_2 = 3.75$) for the wavelength $\lambda = 668$ mμ (Carroll and Freeman).

and halloysite. The complementary use of both, acid-base reactions, *and* oxidation-reduction reactions is what Mielenz and King find most conclusive in a systematic differential diagnosis of clay minerals in staining. The use of dyestuffs with strong specific fluorescence colors, such as Acridine Yellow, Acridine Orange, berberine sulfate, Auramine, Rhodamine B, Thioflavine, Titan Yellow, eosine, fluoresceine, were examined by R. Fahn and M. Gennrich[431] as a suitable method for differential diagnosis of clay minerals. Only nitrogen-containing dyestuffs, however, were found very useful, whereas eosine and fluoresceine are not adsorbed in sufficient amounts, and Titan Yellow shows steric hindrance effects. Kaolinite always had the lowest adsorption; illite, calcium bentonite, and sodium bentonite had increasingly better

[431] *Tonind.-Ztg. u. Keram. Rundschau,* **79,** 1955, 233-236.

responses, roughly symbatic with increasing base exchange capacity and Enslin test values. These fluorescent dyestuffs also show displacement effects of interlayer water and corresponding basal spacing changes as a function of the size of the dyestuff molecule, in a typical base-exchange reaction.

193. N. E. Vedeneeva[432] attains particular merit in having most extensively studied the problems of staining reactions for the identification of clay minerals, first investigating the topochemical conditions of the adsorption of benzidine on the surface of these crystalline materials. She emphasized the indispensable adequateness (epitaxy) of the benzidine molecule to the lattice dimensions and of the electrostatic binding conditions. Bathochromic effects are shown by methylene blue and Chrysoidine on montmorillonite and hydromica if ammonium or potassium salts are added. Malachite Green (with methyl groups) and Bright Green (with ethyl groups) also behave differently because of different adequacy of steric effects. Vedeneeva[433] further recognized the importance of thiazine (or better, phenothiazine) dyestuffs even in the absorption spectra of the stained clay minerals. A specific effect of methylene blue on montmorillonite is recognized, which is shared neither by benzidine nor by thionine, a steric effect corresponding to the specificity of distances in the grouping of the dimethylamine groups in methylene blue (12.5 Å. as indicated in the formula):

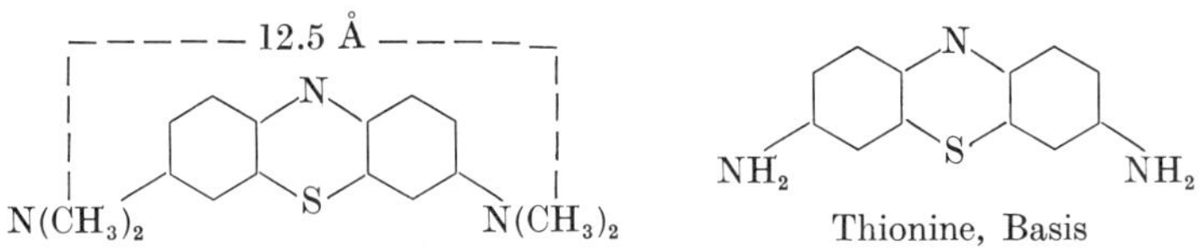

Methylene Blue

Thionine, Basis (Thionine is the Chloride)

The electrostatic field effects of cations present in the staining solution is expressed by the potential $e \cdot Z/r$ which is lower for K^+, Rb^+, Cs^+, NH_4^+, and H_3O^+, higher for Na^+, Mg^{2+}, and Ca^{2+}. Distinct dislocations and intensity changes of characteristic absorption bands at $\lambda = 580$ and 670 mμ are also caused by elevated temperatures (e.g., 100°C.) when another new band with $\lambda = 650$ mμ appears. The band at 580 mμ is practically suppressed in potassium- and hydroxonium-containing (electrodialyzed) montmorillonite suspensions with methylene blue. Montmorillonite stained with thionine does not show these specific changes since in this dyestuff only simple aminogroups, NH_2, are present. The epitaxy effect mentioned above is understood by the regular orientation of the laminar dyestuff molecules on the (001) plane of

[432] *Trudy Soveshchaniya Vsesoyuznogo Khromatografii, Akad. Nauk S.S.S.R.*, 1950 (Pub. 1952), 144-146.

[433] *Doklady Akad. Nauk S.S.S.R.*, **98**, 1954, 585-588. On the structure of methylene blue see G. S. Zhdanov, Z. V. Zvonkova, and L. G. Vorontsova, *Kristallografiya*, **1**, 1956, 61-65, with the cation $(C_{16}H_{18}N_3S)^+$ in planar configuration.

the clay mineral. This orientation phenomenon is also different if the dyestuff appears monomeric or dimeric. The band at 580 mμ observed with methylene blue is assumed to indicate the dimeric state of the dyestuff.

194. For the practical use, Vedeneeva and M. A. Rateev[434] worked out spectrophotometric methods for the determination of montmorillonite and monothermite (from Chasov Yar) in methylene blue (alone, or with KCl), and benzidine (alone, or with ammonia). Differentiation of montmorillonite and monothermite is only possible by X-ray diffraction analysis. Beidellite sometimes shows anomalous adsorption color reactions with methylene blue in so far as potassium chloride addition does not change the color from blue-violet to green, as is normal for montmorillonites, but changes it even more to violet. The absorption maxima at 580 mμ are shifted to 560 mμ. Evidently, calcium and magnesium ions affect the reaction with methylene blue and KCl. In such cases, the calcium- and magnesium-enriched beidellites have a better crystallographic development than the normal montmorillonites show. The "double-cation fixation" effect means perhaps a stabilization of the methylene blue complex against the change by the potassium salt.

195. The "doublet cation fixation" was further investigated by Vedeneeva[435] in a detailed study of the behavior of the 580 mμ absorption band which is common to all of the clay minerals because of the identical adsorption on (001) with the active groups $N(CH_3)_2$ in methylene blue, or NH_2 in benzidine (cf. Figs. C. 91 and 92). A strong shifting towards 670 mμ is observed in montmorillonite adsorbates if the suspension is dried. In kaolinite stained with methylene blue both extinction maxima persist side-by-side but in hydromicas the $\lambda = 670$ mμ band is stronger only if the intensity of the staining is weak, the 580 mμ band stronger in the samples of highest staining intensity. The 670 mμ maximum is assumed to correspond to a firm binding of the dyestuff ions on the clay mineral. The transition of the doublet into this firm ionic binding is easiest in montmorillonite because of the high polarizability of the cations in methylene blue by the anions on the surface of the clay mineral. Also benzidine-stained montmorillonite and beidellite shows the doublet adsorption of the amino groups on (001) in a series of complex bands of $\lambda = 590$ to 660 mμ. Monothermite from Chasov Yar, however, is basically different in acidic or neutral suspensions, and no transition from the doublet type into the firm ionic fixation occurs. By drying the monothermite stained in acidic suspension a color change

[434] *Doklady Akad. Nauk S.S.S.R.*, **100**, 1955, 771-773; see also M. A. Rateev, *Ibid.*, **96**, 1954, 813-816, on beidellites of the Maĭkop sediment horizons, and the change of their color reactions with methylene blue in the absorption maxima at 656 to 670 mμ (from blue-violet to green or yellowish-green) in a few minutes. With benzidine the reaction is very sluggish. H.-D. Bartholomä and H. E. Schwiete, *Ziegelindustrie*, **13**, 1960, 87-105, 421-427, studied the influence of grinding intensity on the sorption of methylene blue on the surface of clay minerals.

[435] *Doklady Akad. Nauk S.S.S.R.*, **105**, 1955, 1248-1251.

from blue to nearly colorless is also observed. Weakest is the binding on kaolinite. The doublet fixation phenomenon shows in many respects analogies with the crystallochemical relations during the crystallization of inorganic salts (experiments with barium and lead nitrate were made by Vedeneeva and S. N. Slavnova,[436] in the presence of methylene blue, with characteristic changes in the spectrophotometric curve).

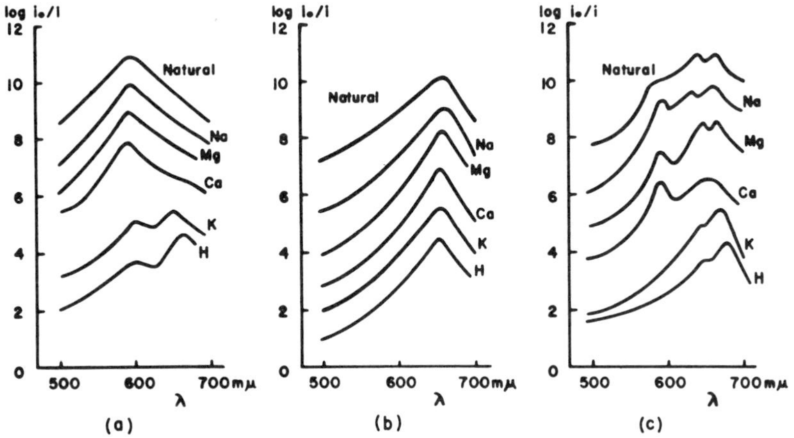

FIG. C. 91 Spectrophotometric curves of absorption by methylene blue, adsorbed on a montmorillonite clay No. 1 (Vedeneeva): (*a*) suspension in initial state; (*b*) after addition of KCl; (*c*) after heating up to 100°C. The cations taken up by base exchange are noted on the curves.

196. The influence of varying pH conditions on the staining of montmorillonitic clays by methylene blue was especially studied by B. M. Mikhaĭlov[437] with spectrophotometry between $\lambda = 500$ and 700 mμ, in the range of pH = 2.60 to 9.25, both in distilled water and in a dilute solution of potassium chloride. Whereas alkaline solutions do not change the character of the absorption curves, considerable changes are observed in acidic solutions, visible in a color change from violet to blue, corresponding to an intensive increase of the absorption maximum at 660 mμ which is, according to Vedeneeva, characteristic of the hydroxonium ions adsorbed. The KCl containing suspensions are green for pH = 6.6, and become more yellowish green with higher acidities, with an extinction maximum near 742 mμ. No striking changes are observed with kaolinite, beidellite, and hydromica; their natural occurrences are usually in media with pH of 6 to 8.5, in kaolins, rather exceptionally also of 5.2 to 5.7.

[436] *Trudy Inst. Krist. Akad. Nauk S.S.S.R.*, **7**, 1952, 135-158; cf. E. N. Slavnova, *Doklady Akad. Nauk S.S.S.R.*, **107**, 1956, 694-696.
[437] *Doklady Akad. Nauk. S.S.S.R.*, **105**, 1955, 818-819.

197. Vedeneeva and especially M. F. Vikulova[438] developed the methylene blue staining method to a standard diagnostic laboratory testing for every kind of clay mineral in soils, ceramic raw materials, and so on. In general, montmorillonite responds better to staining than kaolinite. The latter is better determined by the differential-thermal analysis method. Electron microscopic results are rather uncertain if the minerals are not developed in their characteristic shapes. X-ray and electron diffraction, however, are equally reliable for qualitative and even quantitative de-

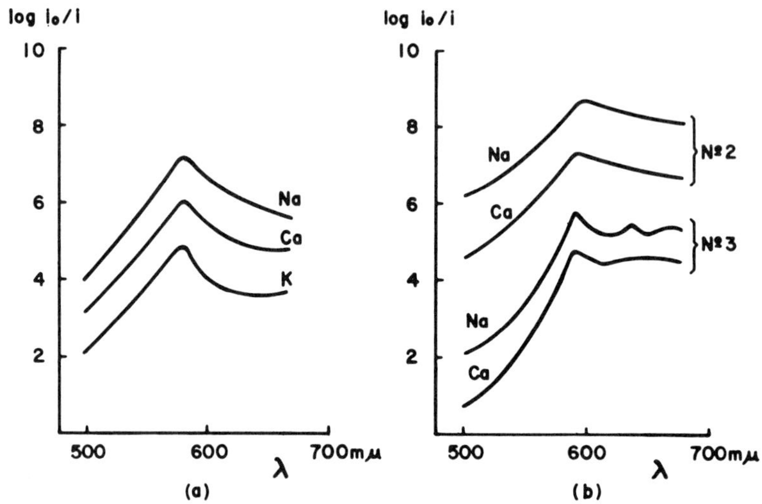

FIG. C. 92. Spectrophotometric curves of absorption by benzidine, adsorbed on montmorillonite No. 1 (Vedeneeva). (*a*) Corresponding curves of absorption by methylene blue, adsorbed on two other montmorillonite clays, (*b*), Nos. 2 and 3). The cations taken up by base exchange are noted on the curves.

[438] In the Symposium Report on "General Characteristics of Clays," *Trudy Vsesoyuz. Nauch. Issledovatel. Geol. Inst.*, 1957, Acad. Sci. U.S.S.R., 361 pp., with numerous contributions on the analytical and petrographic methods, differential-thermal analysis, and so forth. Particularly important are the reports on pp. 7-90, and 229-269 on staining methods. See also M. F. Vikulova. *Doklady Akad. Nauk S.S.S.R.*, **130**, 1960, 173-175. The practical use of staining methods is confirmed by M. M. Shukevich, *Doklady Akad. Nauk S.S.S.R.*, **94**, 1954, 327-328, for clay sediments of Kazakhstan. Anomalous staining behavior of beidellite with methylene blue was further discussed by N. E. Vedeneeva and M. A. Rateev, *Ibid.*, **100**, 1955, 559-562. Most remarkable are the more recent observations of W. E. Worrall, *Trans. Brit. Ceram. Soc.*, **57**, 1958, 210-217 that besides the cation exchange character of the primary adsorption of methylene blue on clays there is, beyond a certain limit value, a common physical adsorption obeying the Freundlich isotherm rule. It is not possible, however, to base on that limit a method for the determination of the cation exchange capacity (cf. C. ¶ 280 ff.). A valuable critical discussion of the suitability of different determination methods for clay minerals, in comparison with the staining method, was recently presented by M. F. Vikulova in the Symposium "Fizicheskie Metody Issledovanii Osadochnykh Porod i Mineralov," *Akad. Nauk S.S.S.R.* 1962, 14-36.

terminations. Amorphous material may occur in kaolinite clays up to 30 per cent, in montmorillonitic clays up to 60 per cent. Vikulova especially gave valuable data on the comparison between results of the different methods of identification in polymict clay mineral systems, for which a combination of X-ray or electron diffraction with the staining method may be most conclusive. X-ray diffraction is most suitable for the detection of small amounts of one or the other definite minerals.

198. The epitactic intergrowths of the phenothiazine nucleus in methylene blue on (001) of montmorillonite, and the influence of the electrostatic potential $e \cdot Z/r$ of cations is analogous to interesting morphological changes observed by A. Mathieu-Sicaud, J. Méring, and I. Perrin-Bonnet[439] in the electron microscopic behavior of hectorite, namely the formation of pseudocrystalline scales of the mineral, with defect marginal zones. Vedeneeva and L. I. Tsukerman[440] connect these changes with those of the light absorption maxima in suspensions of montmorillonite with methylene blue in the presence of different cations. As a matter of fact, violet-stained montmorillonite particles form pseudocrystalline aggregates of peculiar spine-shaped forms, coiled to elongated rolls because the initial scales stick together on the peripheries. This coiling to spines is not affected by sodium, calcium, or magnesium ions, but decidedly impeded by potassium ions. The blue-green stained films decay in potassium-containing solutions to lumps without doublet absorption spectra. The coiling is best studied in transverse reflected light from a parabolic mirror on the stage of a light microscope. The same phenomenon is observed with benzidine-stained samples. The films are disaggregated by heating but can be "reorganized" on cooling if potassium ions are absent.

ADSORPTION OF SILANOL GROUPS ON THE SURFACE OF SILICATE GELS; ACIDITY PROBLEMS

199. The continuously increasing importance of the water adsorption on silica gels and on highly dispersed silicate systems, such as of the clay group minerals, especially their behavior as catalysts and as catalyst carriers for the hydrocarbon industry, make a fundamental study of problems of the chemical nature of the surface reactions indispensable. Here, we have to distinguish between the binding characteristics in the inner of the "silica skeleton" of simple silica gels with siloxane groups

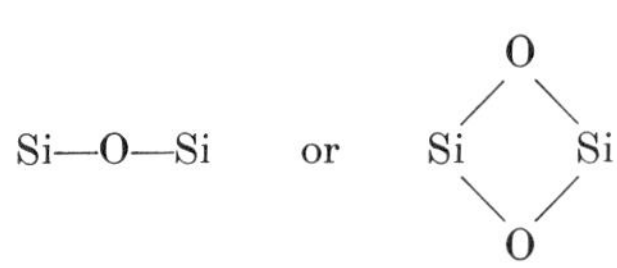

[439] *Bull. soc. franç. minéral. et crist.*, **74**, 1951, 439-456.
[440] *Kristallografiya*, **2**, 1957, 72-77.

and the silanol groups on the surface

$$Si—O—H .$$

The latter bonds show specific hydrophilic reactions and the important ability to undergo hydrogen bindings with polar groups, as with the amino groups of dyestuffs which we have discussed in the previous paragraphs. Changes of the hydrophilic character of the silanol groups on silica gels by an exchange of the hydroxyl groups for fluorine anions to hydrophobic characteristics were studied by I. E. Neĭmark, R. Yu. Sheĭnfaĭn, and L. G. Svintsova.[441] By this exchange the surface becomes pronouncedly hydrophobic because the fluorine anions are not able to undergo hydrogen bindings. The experiments confirm this expectation completely, for example, if a silica gel is impregnated with silicon tetrafluoride dissolved in alcoholic solution. The hydrophobicity is combined with an apparent change in porosity. The fluorinated gels show a completely changed behavior in the adsorption characteristics with nonpolar molecules (benzene) or polar alcohols to which their surface becomes inert.

200. On the other hand, there is an important problem of how water is adsorbed on the surface of crystalline silica, especially on quartz. S. P. Zhdanov[442] devoted a series of investigations to this question. Purest rock crystal was powdered by a solenoid-operated agate mortar sealed in a high-vacuum cylinder with a liquid nitrogen trap to exclude every contamination by foreign vapors. The powder was then transferred, still in the vacuum, to the adsorption container, the protecting envelope broken, and the powder exposed to water at 18°C. The specific surface of the powder was separately determined by the adsorption of argon at — 195.6°C. and found to be 16 m.^{2}/gm. The first adsorption cycle of water on the quartz surface is somewhat unusual, represented by a stepwise curve indicating the complex physical and chemical reactions occurring on the crystal surface and especially the irreversible character of the hydration. Repeated adsorption $\rightleftharpoons$ desorption cycles show then a hysteresis phenomenon. The same operation for a silica gel (and also for porous glass of the Vycor type), dehydrated previously *in vacuo* at 600°C., gave adsorption-desorption curves which lie *below* those for quartz hydrated at 100°C. For vacuum-treated silica gel, only 15 per cent of the hydroxyl groups remain of those which are present in the state of highest hydration.

201. Particularly instructive are the distinctions between a freshly formed quartz surface (obtained by grinding *in vacuo*), and a surface dehydrated by calcination *in vacuo*. In the first case a much more adsorption-active surface is formed. Evi-

[441] *Doklady Akad. Nauk S.S.S.R.*, **108**, 1956, 871-874.

[442] *Doklady Akad. Nauk S.S.S.R.*, **115**, 1957, 938-941. On the important studies of I. W. Obreimow on the surface energy of mica determined by grinding *in vacuo* see *Proc.. Roy. Soc.* (*London*). A **127**, 1930, 290, and the very valuable discussion in the book of K. L. Wolf, "Physik und Chemie der Grenzflächen," Springer, Berlin, 1957, Vol. I, pp. 15, 122; 1959, Vol. II, p. 299.

dently, the structures of the two types of surfaces are different. The grinding process in vacuo brings about by the disrupture of Si—O bonds in the siloxane skeleton free electrostatic charges of the type >Si⁺—O⁻, similar to those observed after grinding mica in vacuo. In the dehydration process by calcination *in vacuo*, on the other hand, the hydroxyl-laden surface of quartz brings about only reactions between neighboring hydroxyl groups and the removal of water, with a compensation of the valences, to form closed groups like

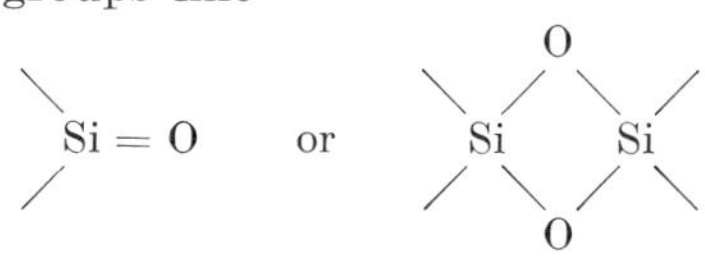

without remaining electrical charges. In quartz broken *in vacuo*, however, the remanent electrostatic charges are the reason for the specific adsorption of polar water molecules on the freshly hydrophilic surface. The rapid increase in the amounts of the adsorbed water, as a function of the relative pressure p/p_s, indicates the chemical sorption effect of a true hydration.

202. In silica hydrogels the difference between physically adsorbed, and chemisorbed water is most evident by the ease with which the first type of "moisture" is removed by heating in dry air at 120°C., and the firm binding of the hydroxyl groups in the silanol group Si—OH of the surface. J. H. de Boer, M. E. A. Hermans, and J. M. Vleeskens[443] observed the first small losses of bound water at 150°C., and a continuous dehydroxylation above 380°C. In the same way that hydroxyl groups are slowly disrupted and neighboring silanol groups undergo chemical reactions with removal of water and the formation of condensed siloxane bindings, the ability of the gel to adsorb polar molecules, taking lauric acid as a typical test material, is reduced. The previously strongly hydrophilic surface character becomes more and more hydrophobic. The process of dehydroxylation is, however, reversible in the temperature range mentioned, and liquid water at 90°C. is not only adsorbed as a polar molecule but the silanol groups are restored in a time period of several hours. The pores of the silica gels studied by de Boer *et al.* were of the order of magnitude of 20 Å. as determined by mercury displacement after replacing the water first by ethyl alcohol, and this then by ether, and rapid heating. The dehydration and rehydration of a gel, as with 6.2 OH groups per 100 Å^2 of surface, is declined after

[443] *Koninkl. Ned. Akad. Wetenschap. Proc.*, **60**, B, 1957, 45-53; 54-59; 234-244; *Ibid.*, **61**, B, 1958, 2-11: 88-93. C. Okkerse and J. H. de Boer, *Silicates Ind.* **27**, 1962, 195-202, studied extensively the reactivity of amorphous silica in aqueous solutions. Analogous phenomena play an important role in the ageing processes of silica-alumina cracking catalysts on steam-treating; cf. W. G. Schlaffer, C. Z. Morgan, and J. N. Wilson, *J. Phys. Chem.* **61**, 1957, 714-722, and the electron-microscopic observations of C. R. Adams and H. H. Voge, *Ibid.* 722-727 (cf. ¶ 211*, Footnote 460). Further, on the hydroxyl content in Aerosil silica gel, J. J. Fripiat and J. Uytterhoeven, *ibid.* **66**, 1962, 800-805.

repeated dehydration and rehydration to a limiting value of 4.6 ± 0.2 OH groups/ 100 Å^2. This decrease in the course of an "accomodation" is explained by a smoothening effect, the disappearance of initial irregularities in the surface. A particular anomaly is the unstable binding of more than one hydroxyl to one silicon atom of the surface layer.

203. The stable formation of Si—OH with one hydroxyl per Si atoms is reached after the accomodation which is of a fundamentally different character than a reduction of the surface area, as by a growth of smaller particles to larger ones, not excluding, however, that primary particles grow together on definite points. The mechanism of the condensation of the hydroxyl groups in the silanol-laden surface can occur in different manners, of which the most common is the condensation to Si—O—Si bonds, and the elimination of water from parallel polyhedral faces. There may also be reactions of hydroxyl groups on edges or corners of one polyhedron reacting with hydroxyl groups on edges or faces of neighbors, and these condensation reactions would bring about irreversible dehydroxylation. Reversible would be condensations to Si—O—Si bonds from Si—OH of adjacent silicon atoms on the same surface, or on surfaces containing silicon atoms with more than one hydroxyl group, as mentioned above.

204. Interesting chemical reactions with the surface hydroxyl groups of amorphous silica gels ("Aerosil," a product of Fa. Degussa) were described by H.-P. Boehm and M. Schneider,[444] citing a *chlorination* of the surface by a treatment with thionyl chloride, the same reaction which H. Deuel and G. Huber[445] used for their studies of the chlorination of clay minerals (cf. C. ¶ 268). In this respect the hydroxyl groups behave like the carboxyl groups of organic acids, which can be changed by thionyl chloride to the acyl chlorides with the group —COCl. The possibility of the introduction of methoxyl groups, —OCH_3, is offered in using G. Berger's reaction (1941) with diazomethane (in ether solution), or an esterification by boiling with alcohols, as R. K. Iler[446] demonstrated, or with trimethyl chlorosilane a change to a complex with —O—$Si(CH_3)_3$ groups, as H. W. Kohlschütter, P. Best, and G. Wirzing[447] described (cf. C. ¶ 138). Beyond that, Boehm and Schneider were able to react the chlorinated (distinctly hydrophobic) silica gel with lithium-phenyl to form a product with phenyl groups fixed on the silicon atoms (indicated by a distinct, although weak peak in the infrared spectrum). The silanol groups of the surface behave as

[444] *Z. anorg. u. allgem. Chem.*, **301**, 1959, 326-335. See also the important studies of K. H. Ebert, *Monatsh. Chem.*, **88**, 1957, 275-287, and recently H.-P. Boehm and G. Kämpf, *Z. phys. Chem. N. F.* **23**, 1960, 257-266, on water vapor adsorption on silica surfaces.

[445] *Helv. Chim. Acta*, **34**, 1951, 1697-1701.

[446] U. S. Patent No. 2,657,149 (1953).

[447] *Z. anorg. u. allgem. Chem.*, **285**, 1956, 236-245.

acidic groups, and a slight but real acidity was observed also by S. A. Greenberg[448] in a reaction of silica gel with calcium acetate or nitrate, lowering the pH slightly to acidity.

205. The reaction of silica gel with calcium hydroxide is thoroughly that of a neutralization, and plays as such an important role in the hydration of Portland cement, and especially of pozzolana cements (cf. Vol. V, Section C). In contradistinction to Greenberg's results, however, Boehm and Schneider postulate an equimolecular (not equivalent) reaction of silica gel with calcium ions, more specifically a reaction of the type

$$\geq Si{-}OH + Ca(OH)_2 \rightarrow H_2O + \geq Si{-}O^-.Ca^+.OH^-$$

as seen from the electric conductance measurements evaluated to the concentrations (cf. Fig. C. 93) (about double of the number of reacting OH^- groups determined by the acyl chloride formation in the $SOCl_2$ reaction). An explanation of this interesting effect may be revealed in the relatively large distance of the negative charges of the hydroxyl groups which are in Aerosil about 32 Å^2 per charge. A similar phe-

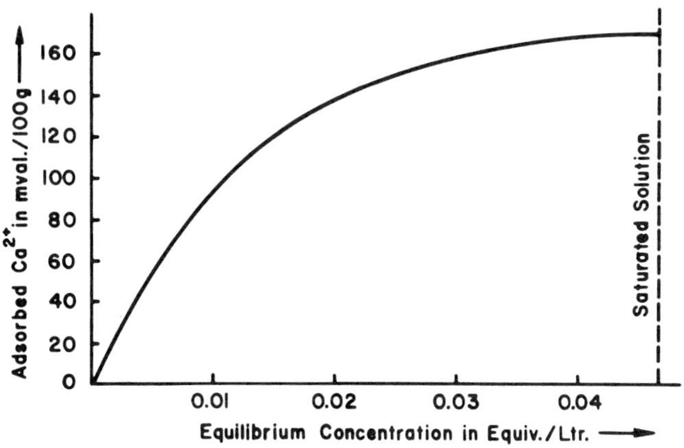

FIG. C. 93. Reaction of calcium hydroxide solution with silica gel (Aerosil) (Boehm and Schneider).

[448] *J. Phys. Chem.*, **60**, 1956, 325-330; H. Kunowski and U. Hofmann, *Angew. Chem.*, **67**, 1955, 289-300. U. Hofmann, K. Endell, and D. Wilm (1934) previously postulated the acidic nature of hydroxyl groups on the margins of clay mineral particles. See also R. K. Iler, "The Colloid Chemistry of Silica and Silicates," Cornell Univ. Press, Ithaca, New York, 1955, pp. 99 ff. and 234 ff. See also the ideas of P. C. Carman, *Trans. Faraday Soc.*, **36**, 1940, 964-973.

[449] *Kolloid-Z.*, **158**, 1958, 22-28. From $AlCl_3$ solutions, Al^{3+} ions are rather firmly bound on silica gel surfaces, as made evident by H.-P. Boehm and M. Schneider, *Z. anorg, allg. Chem.* **316**, 1962, 128-133.

nomenon was found by A. Weiss[449] in the reaction of calcium ions with kaolinite for similar reasons (distance 23 Å^2 per charge).

206. The considerable change in the adsorption characteristics of methylated and chlorinated silica gels for polar *and* nonpolar vapors was made evident by K. D. Shcherbakova and K. I. Slovetskaya.[450] Starting from a silica hydrogel (with a specific surface of 325 m.^{2}/gm.) the reactions with diazomethane and thionyl chloride, and a second reaction of the chlorinated gel with methyl alcohol, yielded the "modified" gels adsorption behavior, which showed with methyl alcohol and benzene vapors a structure not much changed from that of the original gel, but with replacement of the hydroxyl groups by methoxyl groups and by chlorine. The constant C, however, in the equation for the calculation of the specific surface by the nitrogen adsorption method (cf. Vol. IV, Section A), is considerably changed by the substitution, and the quantitative amounts of the vapors adsorbed (in μmoles/gm.) are equally reduced. Also W. Stöber[451] studied a similar phenomenon in the chemical reaction of dimethyldichlorosilane, and of trimethylchlorosilane on silica gels, determining the amounts of combustion products of the adsorbed organic complex in oxygen after evacuation. The saturation of the surface (originally containing 6.2 μmoles/m.^{2}) is reached at 2.5 μmoles/m.^{2} of the silane. This considerably lower value of surface occupation is a steric effect. Also Stöber observed the characteristic reduction of the adsorption capacity for ammonia and water, relative to the original silanol surface (cf. Figs. C. 94 and 95). The silane coverage can be hydrolyzed in about 18 hours and the silanol surface restored. The adsorption reactions are, in principle, the same on silica gel and quartz crystals.

207. We will come back to these important investigations later (C. ¶ 327 ff.). It may be sufficient here to emphasize that, in perfect agreement with these previous observations, A. V. Kiselev, N. V. Kovaleva, A. Ya. Korolev, and K. D. Shcherbakova[452] were able to modify silica gels to hydrophobic materials (industrially useful as fillers in high-polymers, or as thickeners in lubricants) by a treatment with trimethylchlorosilane. The adsorbed amounts of water are about one order of magnitude smaller than in the original silanol surface. The firm bonding of the $—O—Si(CH_3)_3$ groups on the surface is demonstrated by the durability in contact with liquid water for several months. Even near $p/p_s = 1$ no densely packed monolayer of water is formed, the capillary condensation is completely suppressed and no wetting is observed. In the opposite direction, a suspension of soot, which is highly hydrophobic and has a very low adsorptivity for polar molecules (water, ammonia, methyl amine, etc.), can be made hydrophilic by treatment with sodium chlorite solution at 75°C.

[450] *Doklady Akad. Nauk S.S.S.R.*, **111**, 1956, 855-859.
[451] *Kolloid-Z.*, **149**, 1956, 39-46.
[452] *Doklady Akad. Nauk S.S.S.R.*, **124**, 1959, 617-620.

(pH = 3.75). Carboxylic and phenolic surface groups analogous to the silanol groups on silica gels are formed. The modified soot becomes a colloid-electrolyte and has a tremendously increased adsorption for water. At high p/p_s it may show capillary condensation phenomena.

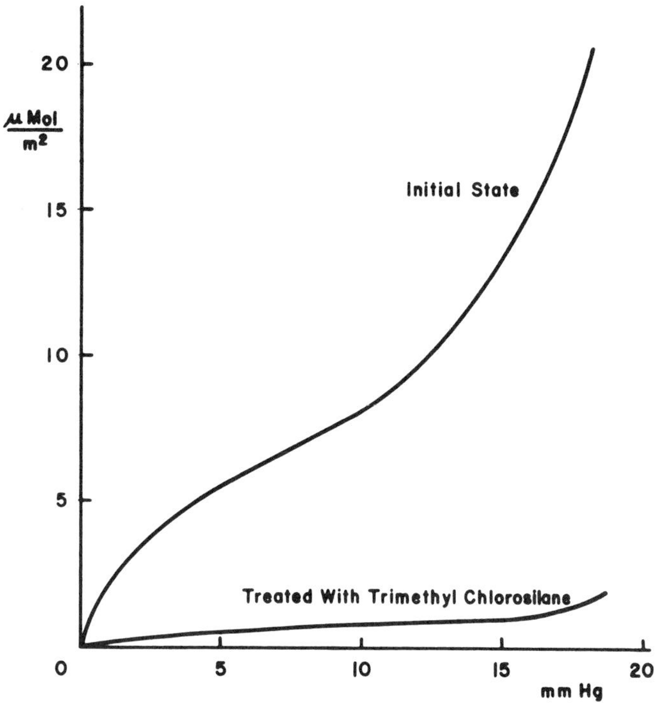

FIG. C. 94. Reversible adsorption of water (at 25°C.) on original silica gel, and the same gel after treatment with trimethylmonochlorosilane (Stöber).

208. A. V. Kiselev[453] extensively discussed the problem of the acidic-basic mechanisms of the adsorption on normal silica gels, emphasizing the chemical interaction and activity of the adsorption process in bringing about complexes of various stability, as a function of specific conditions in the interactions between adsorbed molecule and the adsorbent. It is therefore the *acceptor-donor* relations which rule the acid-base characteristics, especially if hydrogen bonding occurs, as in π complexes indicated by infrared spectroscopy,[454] or the adsorption of trimethyl amine on silica gel[455]

[453] *Doklady Akad. Nauk S.S.R.*, **106**, 1956, 1046-1049.

[454] On this important analogue in the interaction of HCl with aromatic hydrocarbons, see H. C. Brown and J. D. Brady, *J. Am. Chem. Soc.*, **74**, 1952, 3570-3582.

[455] Cf. A. N. Terenin, and N. G. Yaroslavskiĭ, *Doklady Akad. Nauk S.S.S.R.*, **66**, 1949, 885-888, using infrared spectroscopy.

and on porous glass. In all these cases, the adsorption is a step forward to true chemical reactions finally forming silica ethers. In the special case of the interaction of silica gel with aromatic compounds there is an interaction with the π electrons in the aromatic core similar to the interaction of benzene with hydrogen chloride:

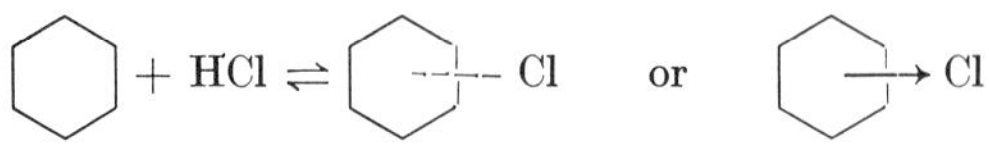

and a π complex is formed. With phenol the same interaction is supplemented by the reaction of the silanol group with the oxygen atom in C_6H_5OH:

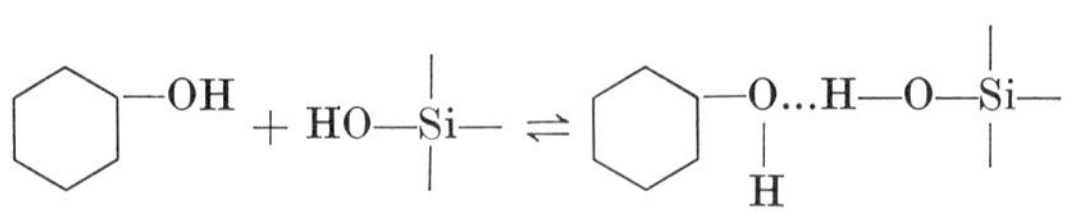

and analogous reactions with triethylamine, methyl alcohol, and the latter may be symbolized in a similar manner in the π bondings by ···. The selectivity of such chemical adsorption effects is important for the removal of olefins and aromatics from mixtures with saturated hydrocarbons, of naphthalene from toluene, or finally

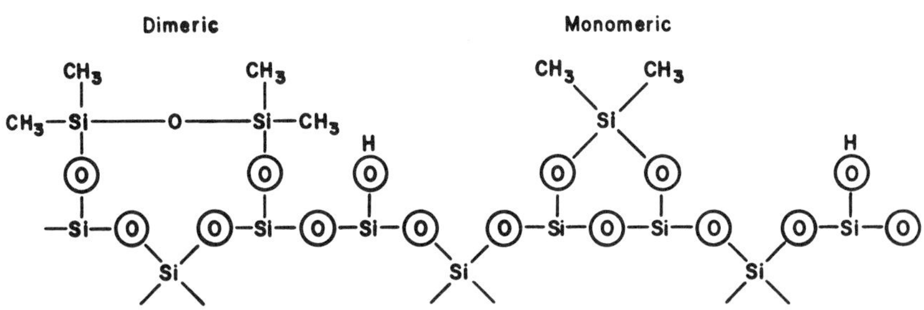

FIG. C. 95. Possible bonding of the condensation product of a silica gel surface with dimethyldichlorosilane (Stöber).

for chromatographic reactions in which silica gels are used as sorbents, or as electron acceptors and proton donors. For phenol, the donor-acceptor scheme is even double, as well as for the aromatic core, referring to the oxygen atom of the hydroxyl:

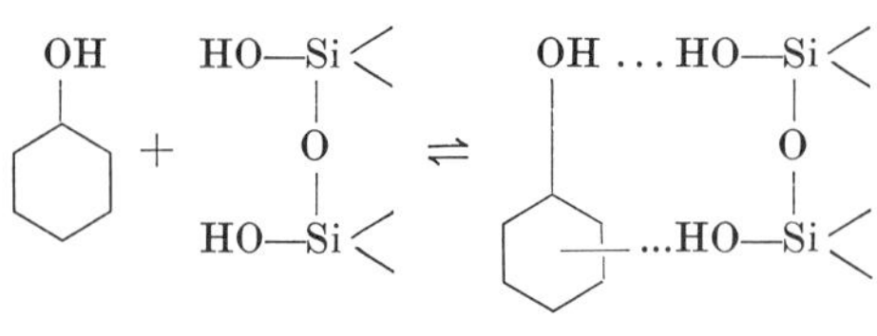

The full significance of these ideas for the acidity-basicity problem will be discussed later in connection with the characterization of mixed alumina-silica gels as cracking catalysts (C. ¶ 211 ff.).

209. Ultraviolet spectroscopy as a means for the study of adsorption phenomena was applied by E. I. Kotov,[456] such as for an investigation of the benzidine reaction on silica gel, on alumina-silica gels, and bentonite. Characteristic of the sorbents is a maximum absorption at about 300 mμ. Especially for alumina-silica gels and bentonite additional bands occur at 450, 760, and 850 mμ (which belong to the benzidine ion), and in bentonite another at 600 mμ (attributed probably to water). The conclusions of Kotov[457] from his study of analogous ultraviolet absorption spectra of adsorbates of aromatic amines (aniline and 1- and 2-naphthylamine) on the same gels are important. He found specific centers on the surface, acting as strong proton donors and electron acceptors forming the cations $C_6H_5NH_3^+$ and $C_{10}H_7NH_3^+$, thus creating a direct analogue to the complex formed between aniline and aluminum chloride of the type $C_6H_5NH_2AlCl_3$ in which $AlCl_3$ is the "Lewis acid," aniline the "base".[458] The importance of this conclusion for the theory of cracking catalysts is fully obvious. The spectrophotometric examination of alumina-silica gels of different compositions, when used as catalysts of this kind, and of their alkali-"poisoned" derivatives in the ultraviolet and visible range, gives evidence of the active centers with electron acceptor properties. They ionize the aromatic amines (aniline, dimethyl-*p*-phenylenediamine) to form addition complexes, as Kotov and A. N. Terenin[459] demonstrated in detail (characteristic absorption curves in Figs. C. 96–98).

210. Characteristic shiftings occur in the band $\lambda = 380$ mμ for aniline in silica gels to 260 mμ in active alumina-silica gels (with a specific surface of 300 to 500 m.^{2}/gm.). The latter band corresponds exactly to that observed in HCl or H_2SO_4 solutions of aniline with the cation $C_6H_5NH_3^+$. For dimethyl-*p*-phenylenediamine the corresponding (visible) absorption bands appear at 570 to 550 mμ. Very characteristic are the "poisoning" effects by the presence of sodium and lithium ions in the alumina-silica gels which suppress the shiftings so typical of the complex formation, which is in pure alumina-silica gels sharply shifted to shorter wavelengths (e.g., for aniline). Although the acidity, i.e., the electron acceptor qualities of the adsorption centers, is preserved and these centers remain available for foreign molecules, the activity as cracking catalysts is seriously reduced or lost. The conditions of this specific activity, therefore, are not necessarily combined with the presence of such centers; this opinion is in contrast with previous theoretical assumptions on cracking catalysts effects. The proton donor centers are evidently extinguished by the poisoning and are identified by their reaction with the adsorbed molecules.

211. There is a very extensive literature on the problem of the acidic nature

[456] *Optika i Spektroskopiya*, **1**, 1956, 500-506.

[457] *Optika i Spektroskopiya*, **3**, 1957, 115-122. On the electron bands in the spectra of solutions in general, see G. Kortüm, *Z. physik. Chem.* (*Leipzig*), **42**, B, 1939, 39-66.

[458] Cf. V. A. Khadeev and A. N. Terenin, *Izvest. Akad. Nauk S.S.S.R.*, 1940 (1), 59-69.

[459] *Doklady Akad. Nauk S.S.S.R.*, **124**, 1959, 865-868.

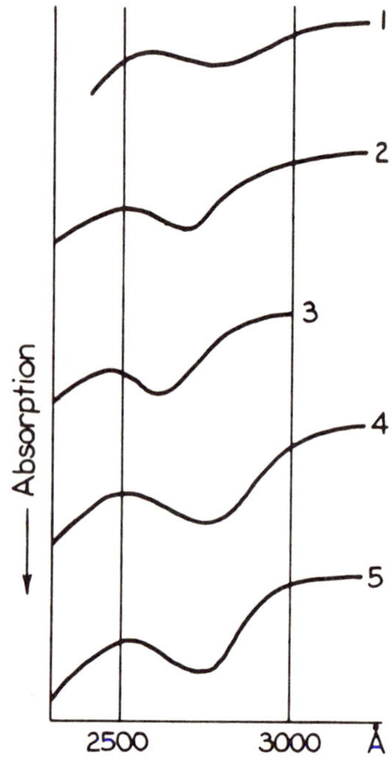

FIG. C. 96. Ultraviolet absorption spectra of aniline adsorbed on different silica and alumina-silica gels (Kotov and Terenin): (1) on silica gel; (2) on alumina gel; (3) on alumina-silica gel (*A*) 50 : 50; (4) on alumina-silica gel (laboratory-prepared) 50 : 50 (gel *D*); (5) on gel *D*, poisoned by lithium.

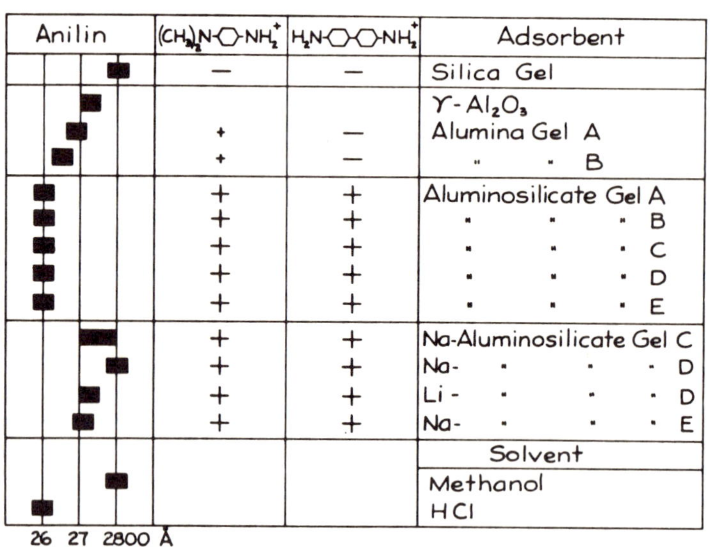

FIG. C. 97. Comparison of ultraviolet absorption bands characteristic of aniline and dimethyl-*p*-phenylenediamine, adsorbed on different gels (Kotov and Terenin).

of the active centers in silica-alumina gel catalysts for hydrocarbon cracking; an excellent report on this subject was presented by K. V. Topchieva, G. M. Panchenkov, M. A. Kaliko, A. V. Agafonov, L. I. Pigusova, N. M. Kamakin and Ya. V. Mirsky.[460] Many previous authors have examined the specific problem of whether the protonic (Brönsted) type or the aprotonic (Lewis) type of the acidic reactions on active centers would be the essential factor in explaining the catalytic effects, both qualitatively and quantitatively.[461] As the final result of the comparison of

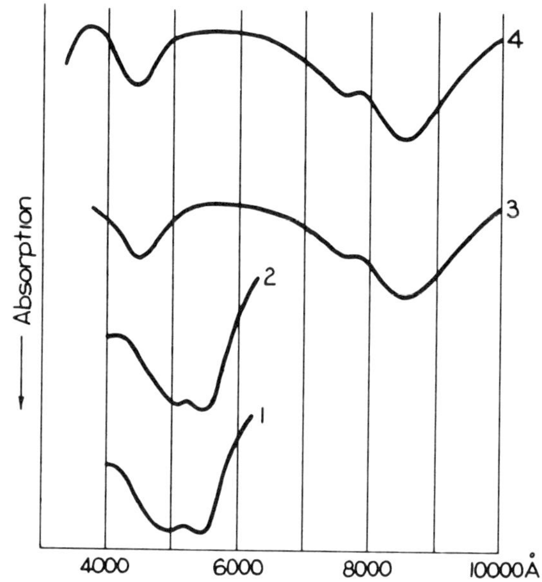

FIG. C. 98. Absorption spectra, in the visible range, of dimethyl-*p*-phenylenediamine and benzidine, adsorbed on different gels (Kotov and Terenin): (1 and 3) adsorption on alumina-silica gel *D*; (2) adsorption on lithium-poisoned, gel *D*; (4) adsorption on sodium-poisoned gel *D*.

[460] *Proceedings of the Fifth World Petroleum Congress, New York,* 1959 (World Petroleum Publications, New York), Section III Paper 10, 16 pp. On the industrial synthesis of high-grade silica-alumina petroleum catalysts, see W. H. Shearon and W. R. Fullem, *Ind. Eng. Chem.*, **51**, 1959, 720-726. The necessity of excessive care in drying the freshly prepared gels is emphasized by K. V. Topchieva, and even ice sublimation drying *in vacuo* is considered as a suitable very mild method of drying; cf. W. G. Schlaffer, C. Z. Morgan, and J. N. Wilson, *J. Phys. Chem.*, **61**, 1957, 719-722, on the aging of the catalysts. On the structure of "FCC" (fluid catalytic cracking catalysts) see also K. Tsukioka, Sh. Uemura, and G. Sato, *Asahi Garasu Kenkyu Hokoku*, **12**, 1962, 125-141, 143-156.

[461] We mention of this wide literature only the following most representative publications: Chs. L. Thomas, *Ind. Eng. Chem.*, **41**, 1949, 2564-2573; M. W. Tamele, *Discussions Faraday Soc.*, 1950, No. 8, 270-279; T. H. Milliken, G. A. Mills, and A. G. Oblad, *Ibid.*, 1950 (8) 279-290; A.S.P. Ballod and K. V. Topchieva, *Uspekhi Khim.*, **20**, 1951, 161-175; *Doklady Akad. Nauk S.S.S.R.*, **75**, 1950, 247-250; K. G. Miesserov, *Uspekhi Khim.*, **22**, 1953, 279-290; *Doklady Akad.*

this rich material, Topchieva states that it is even yet not clear enough whether the concept of acidic properties of the catalyst surfaces has any justification at all because of the lack of independent evaluation methods in the comparatively low concentrations of those centers on the catalyst surface. The donor-acceptor mechanism of the interactions with the reagents makes it impossible to discriminate between the hydrated and nonhydrated forms of the acids. Therefore, it is not possible to apply exclusively chemical (e.g., titration) methods for the characterization of silica-alumina catalysts, but chemical and physical methods and measurements must be combined with reaction-kinetic aspects.

212. Physicochemical methods considered particularly suitable are spectroscopy and isotope tracer methods; also used are deuterium-hydrogen exchange, taking into account the extent of the deuterization reaction product, and "poisoning" of the active centers by alkalies (especially in sodium and lithium acetate solutions) labeled with radioactive isotopes. In this way, determination of the concentration in protonic centers was attempted by K. V. Topchieva and I. F. Moskovskaya (1955) and by G. M. Panchenkov.[462] The non-protonic centres were then determined from successive high-temperature dehydrations of the catalyst and by titration in anhydrous solvents, by the method of Y. Trambouze, L. de Mourgues, and M. Perrin.[463] Cation exchange from salt solutions selectively destroys only the active centers of the silica-alumina gel complexes, not those of the free oxide gels, as J. D. Danforth[464] emphasized. The introduction of the alkali ions brings about a rapid decrease in the velocity constant, α, of the cracking of cumene (isopropyl benzene, $C_6H_5.CH(CH_3)_2$) at 475°C. (Fig. C. 99), and by extrapolation of the linear branch with a steep slope to the activity of zero with progressive poisoning by lithium and sodium ions. The "catalytic acidity" is given by the intersection of the extrapolation with the abscissa axis. The calculated catalytically active surface is only about 1 per cent of the total surface in maximum.

213. The maximum protonic acidity determined by the "catalytic acidity" is

Nauk S.S.S.R., **84**, 1952, 1009-1011; J. D. Danforth, *J. Phys. Chem.*, **59**, 1955, 564-566; *Advances in Catalysis*, **9**, 1957, 558-568; C. Walling, *J. Am. Chem. Soc.*, **72**, 1950, 1164-1168; H. A. Benesi, *Ibid.*, **78**, 1956, 5490-5494; *J. Phys. Chem.*, **61**, 1957, 970-973; O. Johnson, *Ibid.*, **59**, 1955, 827-831; R. L. Richardson and S. W. Benson, *Ibid.*, **61**, 1957, 405-411; K. V. Topchieva, *Uchenye Zapiski, Moskov. Gosudarst. Univ. im. M. V. Lomonosova*, No. 164, 1953, 5-12; Y. J. Trambouze, M. Perrin, and L. De Mourgues, *Advances in Catalysis*, **9**, 1957, 544-550; K. V. Topchieva and I. F. Moskovskaya, *Doklady Akad. Nauk S.S.S.R.*, **101**, 1955, 517-520.

[462] "Trudy Vsesoyuzn. Soveshchanii po khimicheskogo Ochishchenii Mineral'nogo Masla." Izdatel'stvo Akademii Nauk Azerbaĭdzhanskoĭ S.S.R., Baku, 1953, pp. 72-79.

[463] *J. chim. phys.*, **51**, 1954, 723-727. The titration is made in benzene with $AlCl_3$ as the "acid" and dioxan (or ethyl acetate) as the "base". Endpoint of the titration is determined by noting the temperature rise as the base was added; cf. T. Trambouze, *Compt. rend*, **233**, 1951, 648-649.

[464] *J. Phys. Chem.*, **58**, 1954, 1030-1033.

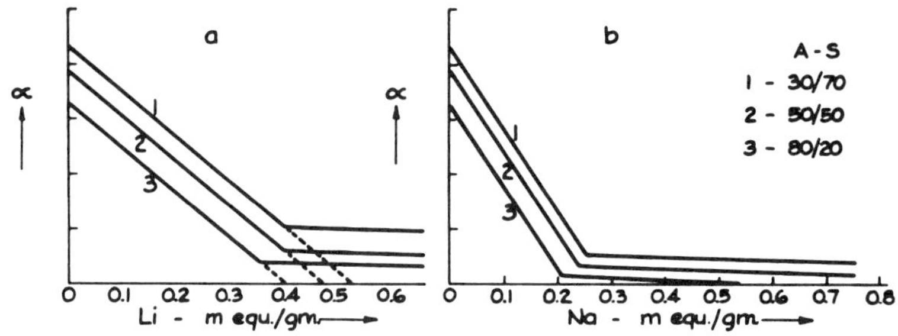

FIG. C. 99. Relation of the rate constant, *a*, of cracking of cumene at 475°C. to amounts of introduced poisoning lithium (*a*) and sodium (*b*), cations in different alumina-silica gel catalysts (Topchieva and Moskovskaya).

found for a ratio $Al_2O_3/SiO_2 = 30/70$ in the gel catalyst (Fig. C. 100). The relationship of the simple activity to the value of protonic acidity is lincar, and one may be tempted to conclude that there is a direct connection between the cracking activity of the gels and the protonic activity. However, the determination of the non-protonic acidity in poisoned alumina-silica gels gives high values for this property. Whereas, pure silica gels have no cracking-catalytic activities at all, and also no non-protonic acidity, pure alumina gels have a nonprotonic acidity comparable with that of alumina-

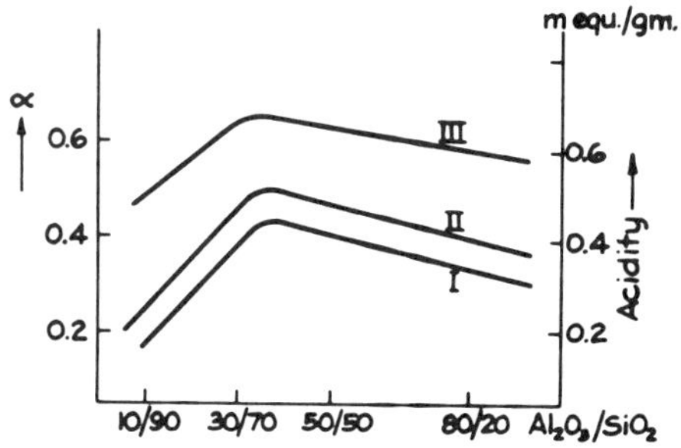

FIG. C. 100. Relation of the rate constant, *a* (*I*), to protonic acidity (*II*), and the aprotonic acidity (*III*) of various alumina-silica gel catalysts, characterized by the ratio Al_2O_3/SiO_2 (Topchieva and Moskovskaya).

silica gels (namely 0.43 meq./gm.) as seen in Fig. C. 101. One must, therefore, conclude that in alumina-silica gel catalysts both protonic and non-protonic centers exist side-by-side. The non-protonic acidity does not participate in the cracking reaction by itself, without a proton, that is in the dehydrated samples. With increasing temperature of calcination the number of non-protonic centers is increased, the catalytic activity decreased. These facts, speaking of the protonic centers being indispensable for catalytic effect, are in agreement with the results of the ultraviolet spectroscopic investigations of E. I. Kotov and A. N. Terenin[465] already discussed above. The specific catalytic activity of alumina-silica gel catalysts, as distinct from the alumina and silica gels themselves, was most evident in those results.

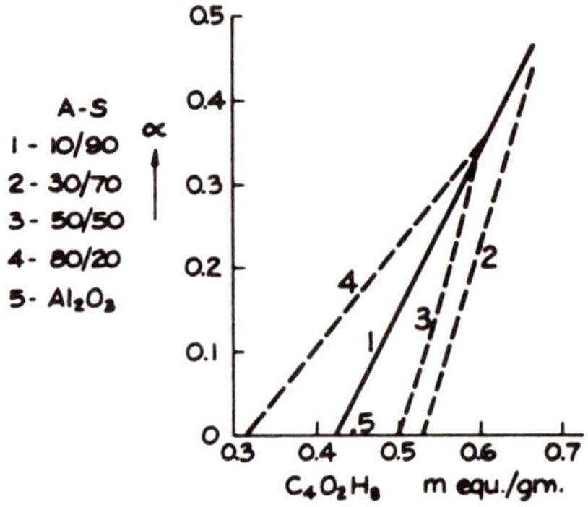

FIG. C. 101. Relation of the rate constant, a, to amount of aprotonic acidity (determined by meq./gm. of dioxane used in titration) for the original state of alumina-silica gel catalysts of different composition (solid line) and for the same catalysts poisoned (dotted lines) (Topchieva and Moskovskaya).

214. Nevertheless, there is strong evidence *contradicting* these apparently plausible conclusions. There are inconsistencies in the rates of hydrocarbon conversions, hydrogen exchange between deuterated hydrocarbon (e.g., isobutane) and the catalyst as observed by S. G. Hindin, G. A. Mills, and A. G. Oblad[466] and quite especially by G. M. Panchenkov, Z. V. Gryaznova, V. M. Emil'yanova, and L. G. Ganichenko.[467] Furthermore, there are tremendous differences between the rates of isomerization of cyclohexene at 350°C. on deuterated silica-alumina gel catalysts, which indicate an entirely different kinetic mechanism of the surface reactions than those in connection with protonic or nonprotonic centers. In other words, there is *no* simple relation between catalytic conversion and hydrogen exchange. Cumene, cracked by deuterated catalysts, shows an exchange decreasing to zero and the percentage of conversion is not affected at all if the space velocity of the cumene is changed by a factor of two (Fig. C. 102). Topchieva comes to the conclusion that there cannot be any direct relation between conversion of hydrocarbons by silica-alumina catalysts and the substitution of hydrogen by deuterium in the catalyst. The rates of

[465] *Doklady Akad. Nauk S.S.S.R.*, **124**, 1959, 866-868; *Optika i Spektroskopiya*, **3**, 1957, 115-122.

[466] *J. Am. Chem. Soc.*, **73**, 1951, 278-281.

[467] "Problemy Kinetiki i Katalisa." Akademiya Nauk S.S.S.R. Moscow 1956; Sovyeshchanii Instituta Fiz.-Khimii, **9**, 1957, 145-151. On the sorption and diffusion of simple paraffins in silica-alumina cracking catalysts see also the recent publication of R. M. Barrer and T. Gabor, *Proc. Roy. Soc. (London)*, **A. 256**, 1960, 267-290.

cumene cracking, and cyclohexene isomerization and hydrogen exchange, however, depend on the *water content* of the catalyst. Very high activities are observed when this is about 0.8 per cent. A mass-spectroscopic study of cumene and benzene, obtained in cumene cracking, revealed the presence of products with any possible degree of deuteration. Protons evidently play a minor role in the process of hydrocarbon conversion. Apparently the main contributing factor in catalysis are the electronophilic aluminum atoms which could be considered as Lewis-type acids.

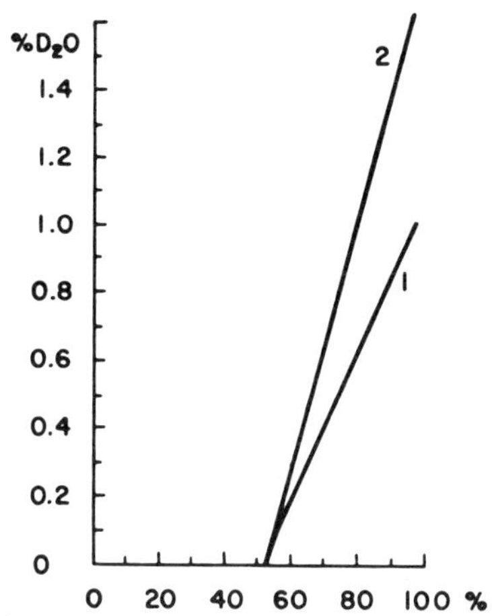

FIG. C. 102. Amount of hydrogen atoms exchanged for deuterium in monomeric (1) and polymeric (2) fractions, versus isomerization degree, in per cent (Panchenkov, Gryaznova, Emil'yanova, and Ganichenko).

215. It is not possible here to develop the ideas of Topchieva *et al.* on the origin of the active centers in alumina-silica catalysts. It may be mentioned only how the radioactive tracers Cs^{134} and Na^{22} are adsorbed on cracking catalysts before carrying out a catalytic reaction. They then occupy both active and inactive centers, as seen by Yu. A. Bitepazh.[468] If a curve (Fig. C. 103) is plotted for the catalytic reaction with a stepwise substitution of ammonium by (labeled) calcium and sodium ions in a catalyst preliminarily covered with those ions up to the point of complete deactivation, the exchange-desorption curve (*A-2*) does *not* coincide with the adsorption curve (*A-1*). The greater is this hysteresis, the higher the temperature of the calcination of the catalyst covered with calcium and sodium. In the preparation of the cracking gel catalyst, therefore, a finer structure is formed inside the porous framework. It may be of the molecular sieve type (cf. Vol. IV, Section B). Because of this superfine structure, cations of smaller radii should be adsorbed in greater

[468] *Zhur. Obshcheĭ Khim.*, 17, 1945, 199-207.

amounts than larger cations, in conformity with observations of J. D. Danforth.[469] The thermal stability of catalyst activity differs greatly with silica gels impregnated with various aluminum salts (see Fig. C. 104, aluminum sulfate in curve 1, oxychloride in curve 2), containing the same amount of aluminum. These facts are qualitatively consistent with the assumed concept of the relative stability of single centers

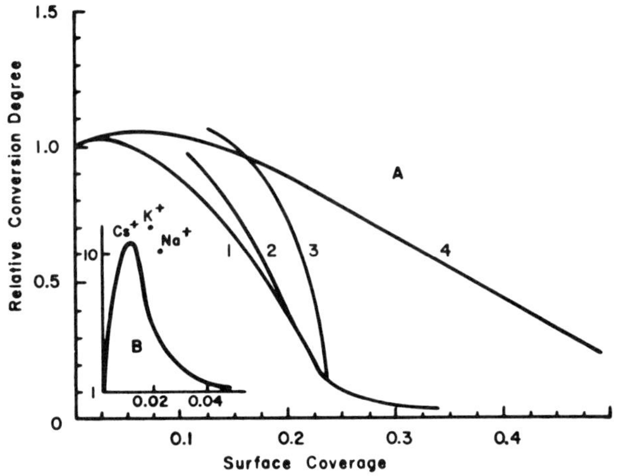

FIG. C. 103. Conversion of cumene versus surface coverage of silica-alumina catalyst: (*A-1*) adsorption with cesium; (*A-2* and *3*) the same after calcination at 400° and 750°C., respectively; (*A-4*) adsorption with sodium; (*B*) catalytic effect in surface covering of low-activity silica-alumina gel catalyst, with different metal cations (Bitepazh).

of aluminum hydroxide particles in the corpuscular structure of the catalyst, which is understood by Topchieva *et al.* in the model of an active center as shown in Fig. C. 105. From this model and its consequences, the Russian authors come to the radical opinion that the adsorption phenomena of the bases on alumina-silica gel catalysts fundamentally contradict the concept of the acidic nature of the surface

[469] *J. Phys. Chem.*, **58**, 1954, 1030-1033. For the use of the reader specially interested in cracking catalyst problems, in order to round up this most fascinating field, we add the following publications: C. J. Plank and L. C. Drake, *J. Colloid Sci.*, **2**, 1947, 399-412 (on the differences between silica, and silica-alumina gels); A. V. Frost, *Vestnik Moskov. Univ. Ser. Fiz. Mat. i Estestven. Nauk*, 1946, 111-117 (on the kinetics of catalytic cracking); K. V. Topchieva and G. M. Panchenkov, *Doklady Akad. Nauk S.S.S.R.*, **74**, 1950, 1109-1112; G. K. Miesserov, *Ibid.*, **91**, 1953, 553-556 (on poisoning by pyridine bases); K. V. Topchieva, *Zhur. Fiz. Khim.*, **27**, 1953, 411-419 (on adsorption isotherms of vapors on alumina-silica catalysts); K. V. Topchieva, *Uchenye Zapiski, Moskov. Gosudarst. Univ. im. M. V. Lomonosova*, 1955, No. 174, 75-100 (relations between catalytic activity and chemical surface properties). See also the excellent report given by Z. V. Gryaznova and G. M. Panchenkov in *Uspekhi Khim.*, **28**, 1959, 1523-1543, on the total problems of the structure and catalytic activity of alumina-silica hydrogels, with a very valuable bibliography.

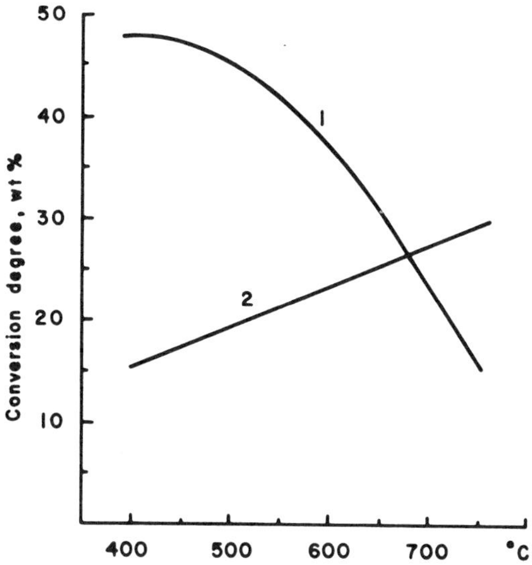

FIG. C. 104. Catalytic activity variation of silica gel impregnated with aluminum sulfate (1), and with aluminum oxychloride (2), versus calcination temperature (Topchieva).

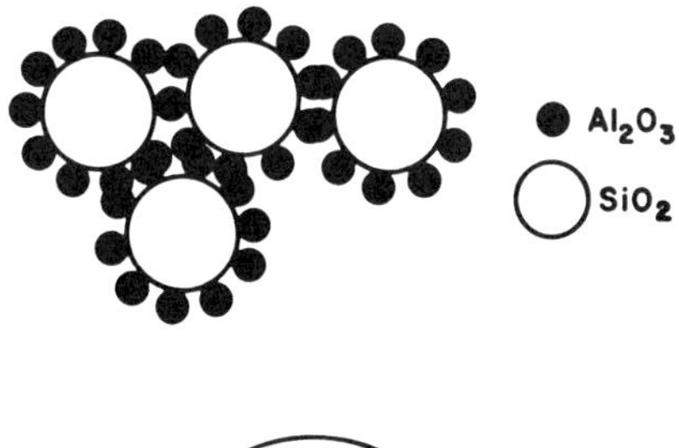

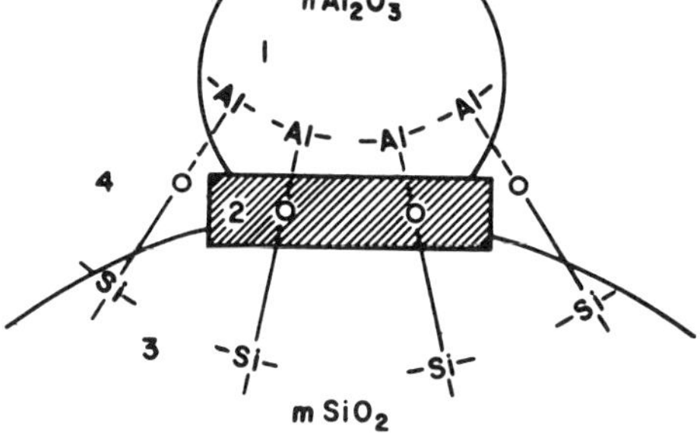

FIG. C. 105. Model of an active center in silica-alumina gel catalyst structure: (1) an alumina particle; (2) inactive silica-alumina particles; (3) a silica particle; (4) position of unsaturated bond, single active center (Topchieva).

which can *not* be considered as a Lewis type (without water), or a Brönsted (with water) acid alone. In their opinion the presence of hydrogels of basic *and* acidic nature is essential, the formation of particles of individual hydroxides in the process of preparation of the catalysts, plus an optimum degree of dispersion.

SURFACE ACTIVITY OF CLAY MINERALS AND OF RELATED MATERIALS

216. The classic *fuller's earths* as surface-active bleaching clays chiefly consist of montmorillonite. G. W. Brindley[470] investigated a montmorillonite and cristobalite clay of this kind, the latter mineral being in a highly disordered state with irregular stacking of the tetrahedral $[Si_2O_5]_\infty$ layers. In the electron micrograph the montmorillonite appears in characteristic lath-like forms; the cristobalite, in globular aggregates. For industrial purposes, a highly purified and enriched montmorillonite called "*Askangel*" is widely used in Russia. A. E. Avsarkisova and M. E. Shishniashvili[471] described the enrichment of this mineral occurring in deposits in Georgia. These products are highly useful as oil-well drilling mud and lubricant emulsions. They are purified from the crude mineral by centrifuging, coagulation of the finest particles by barium chloride solution, and subsequent peptization of the pasty products by sodium sulfate solution at pH = 6 (to avoid formation of barium carbonate). The final pH is 8.25, controlled by additions of sodium hydroxide. The material is highly thixotropic and does not show any tendency of syneresis, even over many months. It is analogous to the American "*Aquagel*," useful as active fillers in rubber and other products.[472] For such purposes, the exact characterization of the particle size diameters plays an important role. The process of activation of montmorillonite by *acid treatment* was extensively studied by Y. Otsubo and Ch. Kato,[473] especially

[470] *Clay Minerals Bull.*, **3**, 1957 (18) 167-169. V. Stubičan, S. Teža and M. Wrischer, *Kolloid-Z.*, **151**, 1957, 33-41, observed besides cristobalite, highly dispersed amorphous silica; if this is removed by reaction with dilute sodium hydroxide, the adsorption activity of the bentonite is decreased after acid treatment (activation).

[471] *Doklady Akad. Nauk S.S.S.R.*, **118**, 1958, 560-562. Analogous highly soap-like montmorillonites are known from Siebenbürgen as "Gaura earths," and even commercially interesting; cf. R. Grengg, *Fortschr. Mineral.*, **33**, 1955, 126-128.

[472] On these industrially important problems, especially on a comparison of so-called "white soots," namely bentonite, montmorillonite, and the like, as fillers in rubber, see A. Clauss, H.-P. Boehm, and U. Hofmann, *Z. anorg. u. allgem. Chem.*, **290**, 1957, 35-51; see also H. Kunowski and U. Hofmann, *Angew. Chem.*, **67**, 1955, 289-300.

[473] *Nippon Kagaku Zasshi*, **74** A, 1953, 311-314; 417-420. See recently V. S. Komarov, N. F. Ermolenko, and V. I. Varlamov, *Doklady Akad. Nauk S.S.S.R.*, **139**, 1961, 665-668, who recommend precipitation of the dissolved alumina and silica after acid treatment by a second treatment with ammonia solution to stabilize the activated material mechanically. Catalysts on the

for the purpose of decolorizing kerosene, but also for cracking heavy mineral oils, polymerization of butylene, and similar uses. Products enriched in silica by acid treatment show a much higher activity than the original montmorillonite.

217. A full equivalent to bleaching earths of the bentonite-montmorillonite type, according to G. P. C. Chambers,[474] are the *sepiolites* because of their very fine-fibrous, fluffy consistence, for refining mineral lubricant oils (but not for vegetable oils containing xanthophyll (lutein) or carotene and also for clearing sugar brines, and the like. In enameling batches, sepiolite (as that from Vallecas, Spain) may entirely replace the montmorillonite, or as a filter aid (like diatomaceous earths). It is more resistent to salt flocculation even than attapulgite. The specific surface may reach 330 m.^{2}/gm., and capillary condensation occurs with organic vapors to a three-molecular layer thickness. The selectivity of the sorption of normal paraffins on sepiolite and attapulgite was especially recognized by R. M. Barrer, N. Mackenzie, and D. M. MacLeod[475] in their studies on molecular sieves (Vol. IV, Section B) with detailed kinetic and thermodynamical characteristics (by the ΔH, ΔG, ΔS parameters) of the vapor sorption phenomena. Rather specific is the behavior of sepiolite outgassed at 20° to 140°C. for the isomers of pentane. *n*-pentane and isopentane are equally strongly adsorbed; neopentane less intensely. For attapulgite the selectivity changes with the temperature of pretreatment. Above 88°C. there is a rapid decrease in intensity of sorption, but it can be restored by a treatment with saturated water vapor at room temperature. The order of selectivity in sepiolite (*n*-pentane and isopentane $\gg$ neopentane) cannot be simply related to distances of the molecular centers of gravity from the sorbing surface. The structural similarity between attapulgite and sepiolite may mean that an explanation in terms of surface corrugation must be attempted for sepiolite. F. L. Pundsack[476] extended the studies to *chrysotile* asbestos, especially to investigations on the stability of its suspensions and variations of surface area in nitrogen adsorption behavior. Surprising analogies to magnesium hydroxide are evident in the colloid-chemical characteristics of such suspensions. The aqueous chrysotile suspensions develop magnesium and hydroxyl ions dissociating from the surface, making the solution phase distinctly alkaline (Pundsack and G. Reimschussel). Pundsack comes to some conclusions which are contradictory to the tubular structure of chrysotile fibers although this fact is well established (cf. A. ¶ 158 ff.). On the better explanation found by W. Noll, H. Kircher, and W. Sybertz[477] concerning the curvature of the chrysotile fibers about the *a*-axis, cf. A. ¶ 164.

basis of magnesium-bentonites from Spanish Morocco were used by V. Aleixandre and T. Fernandez, *Silicates inds.*, **25**, 1960, 243-248, in the polymerization catalysis of butadiene.

[474] *Silicates inds.*, **24**, 1959, 181-189.

[475] *J. Phys. Chem.*, **58**, 1954, 560-568; 568-572.

[476] *J. Phys. Chem.*, **59**, 1955, 892-895; *Ibid.*, **60**, 1956, 361-364; 1218-1222.

[477] *Kolloid-Z.*, **157**, 1958, especially p. 9.

218. Adsorption activity and bleaching capacity of activated clay and related minerals are not necessarily symbatic characteristics. H. Balduin and P. Wieden[478] used a calorimetric method of determining the heat of wetting (accuracy $\pm$ 0.5 cal.), and a colorimetric bleaching test[479] recommended by G. Hamilton and P. Wieden for the examination of industrial bleaching earths, and found no symbatic relation between the data of both tests. Paraffins and cyclic hydrocarbons have very low heats of wetting, aromatic hydrocarbons somewhat higher, aromatic compounds with functional groups still higher. An unusually high caloric effect is observed with pinene (I), evidently due to a chemical reaction (polymerization, oxidation). The extent of dehydration of the bleaching earths is particularly important for the heat effects. There are strong differences between the effects with crude earths, and the same in the acid-activated state. Tetralin (1,2,3,4-tetrahydronaphthalene) (II) is catalytically oxidized to a hydroperoxide (III), and the active oxygen concentration changing in tetralin solution (in an indifferent solvent) can be quantitatively observed. A "peroxide value" can be determined from the transition of the peroxide into α-tetralon ketone (IV) which is changed to α-tetralol (V) by titration with sodium thiosulfate solution. A general parallelism exists between the observed caloric effects of the peroxide decomposition and the bleaching test data. Kaolinite

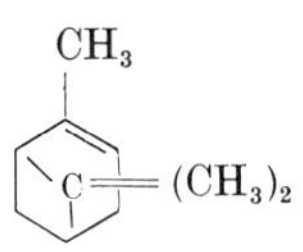

(I)

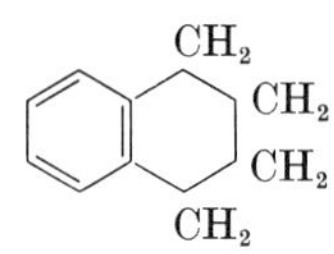

(II)

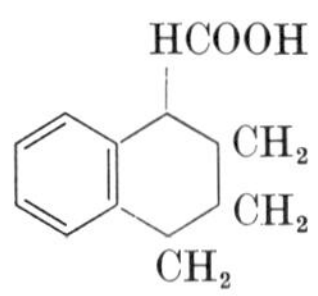

(III)

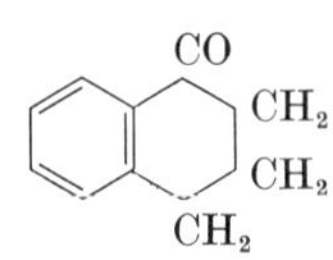

(IV)

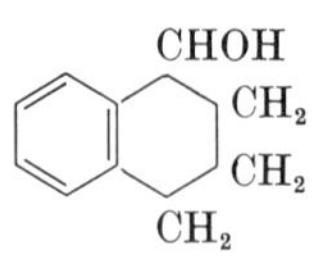

(V)

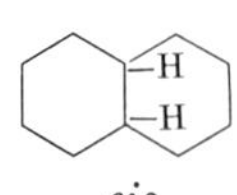

cis

(VI a)

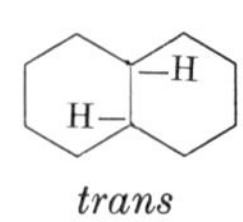

trans

(VI b)

[478] *Ber. deut. keram. Ges.*, **29**, 1952, 344-350; cf. also *Kolloid-Z.*, **125**, 1952, 173-174.

[479] *Erdöl*, 1950, 113.

is practically inert, but the montmorillonite content of the earths can be related rather well to the caloric data. Also the solvents used in the peroxide reaction show different results for xylene, cyclohexane, and *cis*- and *trans*- Decalin (decahydronaphthalene) (VI a, b); constant and reproducible values, however, are obtained in alcohol solutions. It is interesting to compare the activation of montmorillonitic clays by acid treatment with cracking catalytic activity: whereas, with increasing intensity of the acid treatment the active surface continuously is increased, catalytic activity shows a distinct maximum. Evidently, removal of aluminum ions from the octahedral positions causes the decrease in catalytic activity, in concordance with the opinions of Ch. L. Thomas,[480] and provides a theoretical explanation of the reactions in the Tetralin hydroperoxide decomposition mentioned above.

219. The special practical needs of the characterization of activated bleaching earths and related materials made the use of rapid and sufficiently accurate methods of particle size determinations in the laboratory most indispensable, especially use of the fundamental adsorption methods of the BET (Brunauer, Emmett, and Teller) type (cf. Vol. IV, Section A), mostly by the adsorption of nitrogen (or rare gases) at very low temperatures. The nitrogen adsorption was more recently studied extensively by A. P. Karnaukov, A. V. Kiselev, and E. V. Khrapova[481] in its application for quartz powders, and in comparison with these for well-defined silica gels, over a wide range of p/p_s (corresponding to pressures down to 10^{-5} mm. Hg). There are distinctly increased adsorption potentials in the narrower pores of the gels. This fact, characteristic of the adsorption of hydrocarbons, however, is not specific to those compounds, but is also observed in inert gases like nitrogen. It indicates that even in the absence of every polarity of the adsorbed molecules the classic electrostatic potentials are influenced by narrow pores. If the cavity surfaces are covered with hydroxyl ions, as is the case in silica gels, their electrostatic field effects decrease the polarization of the adsorbed molecules, and the induced potential is reduced. On the other hand, there is a continuous transition from the adsorption qualities of quartz to coarser gels. M. G. Kaganer[482] continued studies on the adsorption of nitrogen on silica, silica-alumina gels, and other compounds, for developing a practical adsorption method based on the linear equation for the adsorbed amount, a, $\log a = C - D \cdot [\log (p/p_s)]^2$, in which C equals $\log (w_0/v)$ with v the specific volume of the liquid, w_0 a constant for the sorbent, and $D = 0.43\ BT^2/\beta^2$ (here β is the affinity coefficient). Kaganer's simplified method needs only a few determinations at $p/p_s = 10^{-4}$ to 10^{-2} to characterize the sorbent material in wide limits of the specific surface

[480] *Ind. Eng. Chem.*, **41**, 1949, 2564-2573; with J. Hickey and Gl. Stecker, *Ibid.*, **42**, 1950, 866-871.

[481] *Doklady Akad. Nauk S.S.S.R.*, **94**, 1954, 915-918.

[482] *Doklady Akad. Nauk S.S.S.R.*, **116**, 1957, 251-254; *Ibid.*, **122**, 1958, 416-419.

(about 10 m.^{2}/gm. for nonporous barium sulfate, 1600 m.^{2}/gm. for coconut charcoal), and in tolerably low pressure conditions, avoiding experimentally difficult measurements at extremely low pressures.

220. Similar problems like those of bleaching earth activity are also concerned in the chemical activity phenomena of cements and pozzolanas (cf. Vol. V, Section C) for which simplified adsorption methods were recommended as making it possible to avoid the exact BET determinations. K. G. Krasil'nikov[483] extensively discussed this important problem from the general viewpoint of the kinetics of base adsorption on an acidic sorbent. R. L. Blaine and H. J. Valis[484] discussed results of BET determinations with pastes of hydrated Portland cements, in comparison with the simple adsorption of water at room temperature, and L. F. Gleysteen and G. L. Kalousek[485] developed a full experimental technique for the determination of surfaces of concrete products based on the same, simplified principle. Systematic differences of the results in comparison with the BET data must be taken into account, chiefly dependent on the previous thermal treatment or steam curing conditions (e.g., by autoclave exposure) (cf. Vol. V, Section C). Gleysteen and Kalousek's method is, therefore, generally restricted to purely comparative purposes. For catalysts of high specific area, water adsorption was always much lower than that determined by nitrogen adsorption, whereas for cement pastes the reverse relation was observed.

221. Starting from the needs of practical surface area determinations by simplified adsorption methods which should be useful in the ceramic plant laboratory, Cl. Orr and P. F. Bankston (1952) developed a method characterized by the sorption of stearic acid on a highly dispersed sample. A critical examination of this method by K. S. Birrell[486] demonstrated considerable defects in fatty acid adsorption on kaolins and subsoil materials from New Zealand, in comparison with much higher results of low-temperature gas adsorption. Even alumina in allophane is apparently in a form which is not capable of adsorbing the fatty acid completely, and only some natural alumina samples in which the surface is coated with an active form of alumina gave approximately accurate results. E. Deeg and O. Huber,[487] in a fundamental investigation on the water content of clay bodies and their dielectric properties, mention the great discrepancies of the results for the calculated thickness of adsorbed water layers on clays. The BET method gave a specific area of 2×10^5 cm.^{2}/gm., as a monomolecular layer of 5 per cent of adsorbed water, and about 4

[483] *Trudy Soveshchaniya Khim. Tsementa, Nauch.-Tekh. Obshchestvo Prom. Stroitel. Materialov,* 1956, 351-380.

[484] *J. Research Natl. Bur. Standards,* **42,** 1949, 257-267; see especially also T. C. Powers and T. L. Brownyard, *Proc. Am. Concrete Inst.,* **43,** 1946, 469-504.

[485] *Proc. Am. Concrete Inst.,* **51,** 1954, 234 ff.; *J. Am. Concrete Inst.,* **26,** 1955, 437-446.

[486] *New Zealand J. Sci. Technol.,* **38,** B, 1957, 588-594.

[487] *Ber. deut. keram. Ges.,* **32,** 1955, especially p. 269.

times greater than the surface data postulated by the stearic acid method, because of the large size of the molecule of the fatty acid. It is in every case advisable to use Orr and Bankston's method with great care for absolute surface area determinations and the calculation of the thickness of adsorption layers. The nitrogen adsorption data, however, make possible a direct conclusion as to the amounts of water adsorption.

222. The classic methylene blue adsorption method for determination of sorption layers was first applied by F. Paneth and A. Radu.[488] A. Clauss, H.-P. Boehm, and U. Hofmann[489] made a most interesting comparative study of the suitability of adsorption methods for determination of the active surface of soots (for rubber filling), including the BET nitrogen method, and adsorption of phenol, or of methylene blue. An excellent agreement was found for the data (3.2 m.^{2}/gm.) by nitrogen and phenol (from a 1 per cent aqueous solution) over the entire range of particle sizes studied. Methylene blue (from a 0.075 per cent solution) shows some particularities in so far as the space occupied by the molecule is different on common soots (2.45 m.^{2}/mg.) and on graphitized soots (1.46 m.^{2}/mg.). This observation is highly interesting since J. M. Los and C. K. Tompkins[490] determined the area of 2.44 m.^{2}/mg. for adsorption of methylene blue on a positively charged mercury surface. The graphitization indicates a considerable change in the rugosity (degree of roughness) of the particles by this thermal treatment.

223. A very careful determination of the surface area of typical bleaching bentonites, before and after, activation by acid treatment, was made by J. de Dios Lopez-Gonzales and V. R. Deitz[491] using the BET nitrogen adsorption, and calculating surfaces as well by the BET, the Hüttig, and the "summation" equations (cf. Vol. IV, Section A). The activation process significantly increases the external, or nonswelling, surface of the montmorillonite crystal particles. The ratio of surface areas of activated to original material is about 4 : 1. A simplified view of the reaction during activation may be given by Fig. C. 106 which shows an average packing of ten elementary structural layer units in a montmorillonite particle in which four of these layers are assumed to be decomposed by the acid, leaving silica sheets. In the re-

[488] *Chem. Ber.*, **57**, 1924, 1221-1225.

[489] *Z. anorg. u. allgem. Chem.*, **290**, 1957, 35-51. On the practical use of the phenol adsorption method for a simplified determination of the surface areas not only of soots and similar organophilic materials, but also for clays, see H. Kunowski and U. Hofmann, *Angew. Chem.*, **67**, 1955, especially pp. 290 ff. (with CCl_4 as solvent), and A. Weiss, A. Mehler, G. Koch, and U. Hofmann, *Z. anorg. u. allgem. Chem.*, **284**, 1956, especially pp. 265 ff. See further H.-P. Boehm and W. Gromes, *Angew. Chem.*, **71**, 1959, 65-69. K. H. Woodside and W. C. Ormsby, *J. Am. Ceram. Soc.*, **43**, 1960, 671-672, demonstrated a method of the surface-area determination of kaolinite using glycerin adsorption.

[490] *J. Chem. Phys.*, **24**, 1956, 630.

[491] *J. Research Natl. Bur. Standards*, **48**, 1952, 325-333.

sidual structure of the bleaching earth product, however, the unchanged fragments *A*, *B*, *C* do not necessarily have to be oriented parallel. The nitrogen can enter during the adsorption process into the interspaces *a* and *b*, in this case with a threefold increase in active surface (from originally about 80 to 250 m.^{2}/gm.), or four times if one takes into account the 20 per cent weight loss by the acid reaction, thus a real surface of 320 m.^{2}/gm. As a matter of fact, the process of acid activation must be

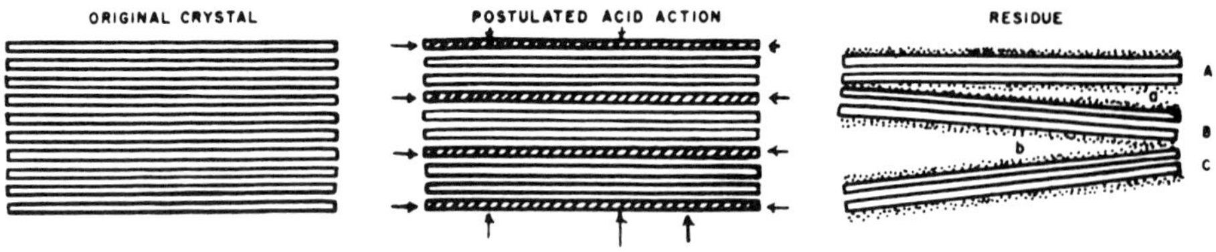

FIG. C. 106. Scheme of mechanism of acid attack on a bentonite (montmorillonite) particle during activation (Lopez-Gonzalez and Deitz).

much more complex than here outlined, the maximum in surface area being observed when all the adsorbed water is removed at 250° to 350°C. The more irregular structure of the activated material, in comparison with the original bentonite, is the reason for its lower thermal stability, since the surface area of activated bentonite is appreciably reduced at temperatures at which the original material does not show any perceptible change. These facts are practically important for the systematic determination of an optimum temperature at which the activated bentonite is to be used with a maximum of useful area, especially if it is used for catalytic purposes.

224. Many conditions for good suitability of bleaching earths in the chemical industry are also valid for other natural and synthetic silicates, among which we may mention some silicates of heavy metals recommended as catalysts, either in their original or in a dehydrated (or dehydroxylated) state, or, after complete removal of the water, undergoing reduction by hydrogen (or carbon monoxide). In this latter case, the silicates become "carriers" for the metal catalysts formed by the reduction. In this respect some new types of layer silicates with $[Si_2O_5]$ groups, prepared by G. Rienäcker and G. Techel,[492] are particularly interesting. Starting from the crystalline silver silicates, Ag_4SiO_4, Ag_2SiO_3, and $Ag_2Si_2O_5$, first studied by E. Thilo and F. Wodtcke[493] (cf. Vol. III, Section B), Rienäcker and Techel observed the highest

[492] *Z. anorg. u. allgem. Chem.*, **303**, 1960, 247-262.

[493] *Z. anorg. u. allgem. Chem.*, **295**, 1958, 247-261. On the acid $H_2Si_2O_5$ see H. Kautsky and R. Irnich, *Ibid.*, **295**, 1958, 193-205. For measurements of the catalytic activity of pure nickel powder for a comparison, see G. Rienäcker and J. Völter, *Ibid.*, **296**, 1958, 210-219. A recent investigation of the reduction of the Ag_2SiO_3 polysilicate by hydrogen was presented by Fr. Wodtcke and Fr. Liebau, *Z. Elektrochem.*, **66**, 1962, 690-697.

surface activity in the phyllosilicate $Ag_2Si_2O_5$, after reduction to fine-dispersed silver on a carrier framework of residual $[Si_2O_5]_\infty$ layers. Analogous highly surface-active nickel, cobalt, and copper silicate carriers were prepared from phyllosilicates, synthesized by a reaction of free acid $H_2Si_2O_5$ with the metal salt ammoniates $[Ni(HN_3)_6]SO_4$, $[Co(NH_3)_6]Cl_2$, and $[Cu(NH_3)_4]SO_4$. The reduction by hydrogen takes place at 375° to 500°C.; the resulting products have specific surfaces of 180 to 250 m.^{2}/gm., and are silica-carriers of metal catalysts for the decomposition of formic acid to carbon dioxide and hydrogen as the test reaction. The silver mica prepared by Thilo and Wodtcke, $AgAl_2[AlSi_3O_{10}](OH)_2$, cannot be reduced so promptly. The nickel silica carrier catalyst is superior in its activity to pure nickel powders.

225. The principle of reduction of a heavy metal silicate to prepare particularly efficient metal catalysts was previously used by G. C. A. Schuit and J. J. B. van Eijk van Voorthuijsen,[494] who started from a nickel antigorite, $Ni_3[Si_2O_5](OH)_4$, or from a nickel montmorillonite of the type formula $Ni_3[Si_2O_5](OH)_2$, reduced at 425°C. and yielding a catalyst of a particularly high thermal stability (up to 1000°C.). The topochemical viewpoint is here very evident in the phyllosilicate catalysts with layer structures protecting the metal formed by the reduction even at relatively high temperatures against collective crystallization.

226. The adsorption of benzene, methane, water, nitrogen, and oxygen on montmorillonites with different cations, especially with the large cations NH_4^+ and Cs^+, was investigated by R. M. Barrer and J. S. S. Reay.[495] They studied the interlayer spacings of these derivative, intermediate between the natural clay mineral and the alkyl ammonium derivatives (cf. Vol. IV, Section B). The *steric* conditions for the alkali ions in the montmorillonite structure are illustrated by Fig. C. 107 for the cavity between two adjacent particle surfaces with $[Si_2O_5]_\infty$ tetrahedral layers in contact with each other, and Fig. C. 108 to show (in a somewhat simplified manner) the freedom of superposition of adjacent sheets for potassium and cesium montmorillonite. The special conditions for cesium and ammonium as alkali ions in montmorillonite make the use of these derivatives for adsorption particularly complex because of the distinct *fixation* tendencies (cf. C. ¶ 290). If in place of cations, organic groups are introduced into the constitution of the montmorillonite, such as the phenyl group, (cf. H. Deuel, cf. Vol. II, A. ¶ 326),[496] the sorption of benzene and water gives

[494] U. S. Patent No. 2,657,875 (November, 1953); Dutch Patent No. 678, 436. See the report of G. C. A. Schuit and L. L. van Reijen, *Advances in Catalysis*, **10**, 1958, especially pp. 242-250, 316 ff.; also J. J. B. van Eijk van Voorthuijsen and P. Franzen, *Rec. trav. chim.*, **69**, 1950, 666-667; *Ibid.*, **70**, 1951, 793-812; *Trans. Intern. Congr. Soil Sci., 4th Congress, Amsterdam*, 1950, Vol. 3, 34-37. Vol. 4, 64. And G. C. A. Schuit and N. H. de Boer, *Nature*, **168**, 1951, 1040-1041; *Rec. trav. chim.*, **70**, 1951, 1067-1084.

[495] *J. Chem. Soc.*, 1958, 3824-3830.

[496] *Ber. deut. keram. Ges.*, **31**, 1954, 1-7.

no evidence that the benzene penetrated between the lamellae, or that any change occurred in the basal spacing.[497] The phenyl montmorillonite, therefore, differs notably from the natural clay mineral in having a little accessible surface for the nonpolar benzene molecule, but none for the polar water molecule. The phenyl groups may be attached covalently at one or more points, in some cases, to adjacent sheets, thus cross-linking them and preventing swelling in the presence of benzene vapor.

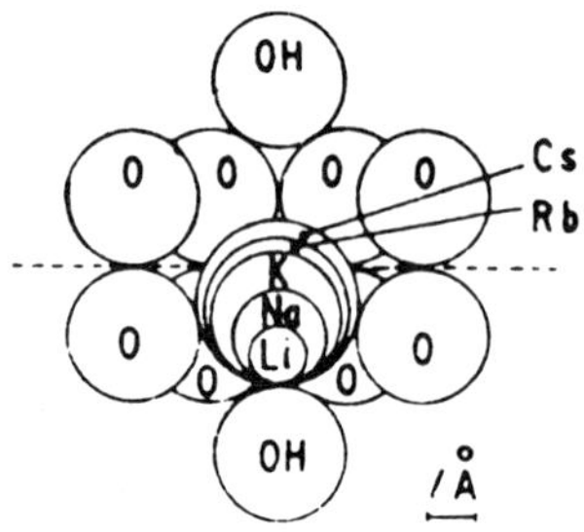

FIG. C. 107. Cross-section of the cavity formed between two montmorillonite layers in contact with each other, showing the size of alkali cations in relation to the cavity size (Barrer and Reay).

227. Finally, we mention the interesting ideas of I. D. Zkhus[498] on the catalytic role of montmorillonite in conversion of organic material in marine sediments into hydrocarbons, specifically in the formation of mineral oil deposits. In the place of montmorillonite, Zkhus constantly identified hydromicas in the petroleum-bearing Jurassic sediments of the Northern Caucasus oil and natural gas fields. The hydromicas evidently were all derived from primary montmorillonite, as seen from details

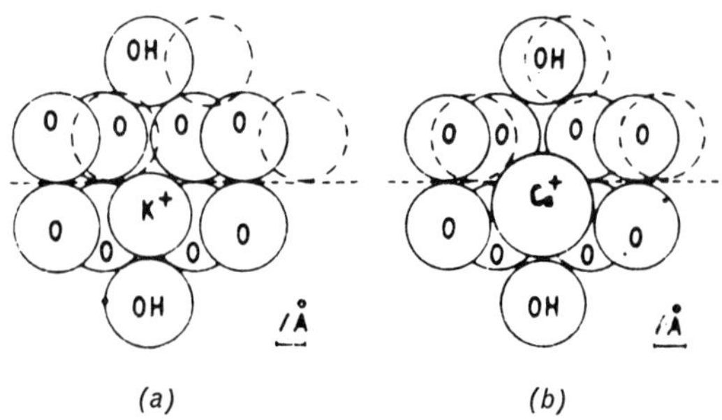

FIG. C. 108. Freedom of superposition of adjacent sheets in (*a*) potassium and (*b*) cesium montmorillonite, without separation of the sheets (Barrer and Reay).

[497] Cf. R. M. Barrer and J. S. S. Reay, *Clay Minerals Bull.*, **3**, 1958, 214-220.

[498] *Doklady Akad. Nauk S.S.S.R.*, **123** ,1958, 353-356. The previous discussions of A. V. Frost, *Uspekhi Khim.*, **14**, 1945, 501-509, may also be emphasized in this connection.

of the electron microscopic examination results and dyestuff adsorption tests. It is assumed by Zkhus that in the primary marine sediments the organic materials was in a strongly reducing surrounding, the montmorillonite changed by chemisorption of potassium into hydromicas. The energy set free in this reaction may be the root of the energy-consuming formation of hydrocarbons from organic material by hydrogenation, both reaction types being coupled in parallel. The intensity of the process may have varied in wide limits. There are such formations with only an incomplete change of montmorillonite into hydromica. Bacteria, on the other hand, may have contributed much, in their energy cycles, to the hydrogenation.

CLAY-WATER SYSTEMS

Plasticity, Workability, and Shrinkage

228. Our actual aspects of the physicochemical nature of the plasticity phenomena in their complex multiplicity were extensively and critically discussed in a report by E. Gruner.[499] He comes to the conclusion that prevailingly chemical forces, combined with crystallographic and physical factors, make up the characteristics of plasticity in systems of solid/liquid phases, as in typical clay slurries. The hydration of the crystal surface of clay mineral particles, supplemented by ionic adsorption, and taking into account the two-dimensional character of these clay minerals, is apparently the main reason why plastic properties are observed in oxide and silicate-water systems. Gruner therefore distinguishes three types of plastic behavior:

1. In truly plastic systems it is a consequence of surface reactions between solid and liquid phases. In its minimum degree it is a chemical property of the system concerned, but can in wide limits be increased by physical factors in the state of the solid phase, namely by grain size effects (if the macrocrystals are dispersed to colloid-dimensioned particles), or by special crystal habitus properties of microcrystals (as it is the case in crystalline clay minerals), but variable in wide limits. Silicates (or sometimes other inorganic compounds) with typical layer structures show a particularly well developed true plasticity as a consequence of their chemosorption of a liquid phase.

2. Plasticity is a consequence of processes, or of the action of additions, which either generate or reinforce surface-active reactions. Both types of "auxiliary" processes can be produced by chemical activation of the crystallite surface (e.g.

[499] *Ber. deut. keram. Ges.*, **31**, 1954, 135-142. One may study this general introduction into the plasticity problems in comparison with the article of U. Hofmann, *Keram. Zeitschr.*, **14**, 1962 (1), 14-19, entirely written under the aspects of modern structural chemistry.

by the formation of an oxychloride),[500] or by the adsorption of polar hydrophilic colloids, mostly of organic origin (such as humus materials, see below). The principal condition which must be fulfilled for the action of auxiliary processes is that the adsorbed materials increase the hydrophilic character of the surfaces.

3. The phenomenon that solid particles of a given size are embedded in highly viscous or adhesive liquids or in gels, Gruner calls a "fictive" plasticity, thus making possible the permanent (nonelastic) deformation, which is the general indication of plastic properties.

229. Gruner's definitions, therefore, do not include all the plastic phenomena observed in monocrystals, or in polycrystalline monophase systems. However, the general theory of plasticity of solid-liquid systems includes not only the common ceramic slurries of clays and kaolins but also those of the important oxide ceramics and other inorganic systems in which the solid phase is able to undergo by chemosorption a formation of a liquid phase hull.

230. H. von Platen and H. G. F. Winkler[501] especially emphasized the close relation of clay mineral-water systems with the phenomenon of thixotropy (cf. C. ¶ 333 ff.). They made extensive measurements of the influence of grain size of accurately defined fractions from the data characterizing both properties, the plasticity being determined by the classic Atterberg method. The specific surface of the samples studied was carefully measured by the BET method, and the plasticity studied also in its relation to the electrolyte content of the liquid phase. Whereas the thixotropy is exclusively determined by the specific surface (for all minerals investigated in a standard 3.6 normal solution of sodium chloride), the plasticity shows a pronounced dependency, beyond that, on the *shape* of the mineral particles. Different kaolins, (meta-)halloysites, illites, and montmorillonites were examined. Their aggregation-free systems, such as those of the kaolin-water group, show a linear relation to the specific surface (cf. Fig. C. 109), and have the same plasticity data for the same specific surfaces. In distilled water, different clay minerals also have different Atterberg plasticity numbers. For a constant specific surface the plasticity of kaolins is lower than that of meta-halloysite, this is lower than that of illite, and that of montmorillonite has the highest value (in the ratio of about $1 : > 1 : 3$ to $4 : 10$). Quartz and feldspar powder fractions show in the systems with water no plasticity, down to the highest dispersion degrees of grain sizes. This result is the more

[500] Cf. the aluminum and zirconium oxychloride formation assumed by O. Ruff and W. Goebel, *Z. anorg. u. allgem. Chem.*, **133**, 1924, 220-229, in the activation of alumina and zirconia slips by hydrochloric acid. See recently P. J. Anderson and P. Murray, *J. Am. Ceram. Soc.*, **42**, 1959, 70-74, with a complete discussion of Ruff's theory of plasticity in oxide/water suspensions and of the rheological behavior of the slips.

[501] *Kolloid-Z.*, **158**, 1958, 3-22.

remarkable as quartz and feldspar show a certain plasticity in apolar liquids. In the 3.6 N NaCl solution the plasticity depends only on the specific surface, and on the shape of the clay mineral particles, but not on the composition of the mineral. It is most evident from von Platen and Winkler's results that the plasticity and thixotropy are properties which have a deep-going inner relation (cf. C. ¶ 342).

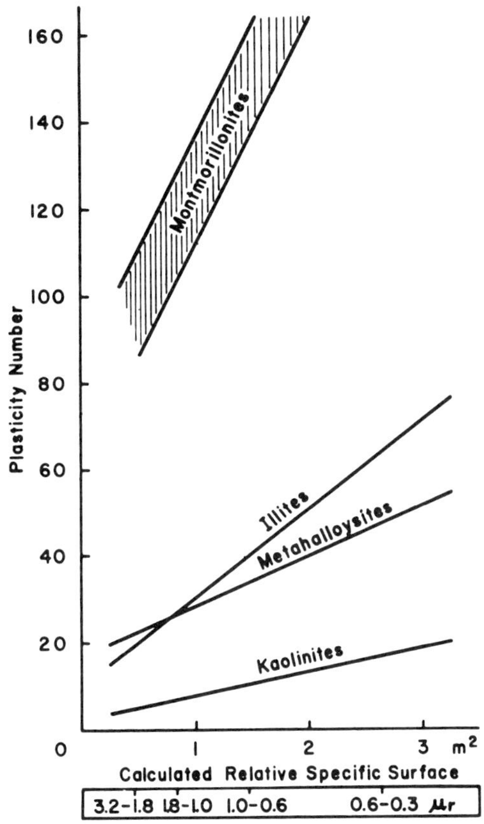

FIG. C. 109. Plasticity data of aggregate-free minerals of different type, as a function of calculated relative specific surface (von Platen and Winkler). The fraction limits are given in μ equivalent radii.

In every case, the characteristics of a given suspension medium exerts the same effects on plasticity *and* thixotropy. If one of the given materials has a high plasticity, it will of necessity also develop a high thixotropy. The specific influence of particle shape on plasticity is not in contradiction with this statement but is explained by the particle displacement mechanism under shear, typical of every plastic deformation.

231. The practical importance of Winkler's investigations is very evident in the manifold phenomena of ceramic production from plastic clays and kaolins. Among the practical, well established results from his theories and experiments, we mention here the stability problem of clays for the production of roof tiles to changes by frost.[502] For this property the degree of firing of the finished tiles is not so essential as the particle size distribution in the plastic clay of the batch. The clays of frost-sensitive tiles have commonly a rather equal distribution of grain sizes, chiefly between 0.2 and 40 μ, whereas frost-resistant clays show pronounced maxima for 20 to 38 μ sizes and a systematically decreasing amount of smaller particles.

232. It is further most interesting to learn from the studies of W. A. White[503] on the allophanes of Lawrence County, Indiana, that these minerals are not plastic at all, not rolling to a thread of $\frac{1}{8}$ in. diameter in the Atterberg test. Since allophanes and halloysites often occur together, it is understandable that sometimes in the ceramic literature allophane is mentioned as a plastic clay material, especially for bonding. The Lawrence County allophane has a poor "green" compression strength which is, e.g., important if it were to be used for bonding as a foundry material. For this purpose highly plastic bonding clays of the bentonite group are chiefly preferred in steel casting operations. We emphasize in this connection the extensive investigations of H. E. Schwiete[504] on foundry clays (cf. also C. ¶ 365). The use of highly plastic additions to improve the plasticity of foundry sands (to "synthesize" such sands) is common in steel plants. In the same way, it is possible to a certain degree to improve by additions of alkali oxalates, of tannin, or sodium acetate, the plasticity of low grade kaolins, as A. Zwetsch[505] demonstrated for West-German

[502] *Ziegelindustrie*, **10**, 1957, 673-678. On the influence of submicroscopic porosity on resistance of bricks to frost, see O. Carlsson, *Trans. Chalmers Univ. Technol. Gothenburg*, **212**, 1959, 13 pp. The minimum of grain sizes near 2 to 20 μ which was postulated by K. Bergmann, *Ziegelindustrie*, **9**, 1956, 253-262, is of secondary importance. It is in every case not possible to produce a frost-resistant tile from frost-sensitive clays, whereas it is quite possible to make frost-sensitive tiles from frost-resistant clays, by an inadequate firing schedule, thus illustrating the importance of controlling the firing conditions to an optimum effect. It is highly instructive to compare Winkler's conclusions on the frost resistivity of plastic clays with the studies of T. C. Powers and R. A. Helmuth, *Highway Research Board, Proc.*, **32**, 1953, 285-297; *Research and Develop. Labs. Portland Cement Assoc. Research Dept. Bull.*, No. 46, September, 1953; No. 71, October, 1956, on the changes by freezing in hardened Portland cement pastes and concretes. Based on careful experiments on the freezing behavior of monoionic clays, G. Rosenthal presented recently a deeper-going interpretation of frost resistivity of clays, in its close correlation to plasticity; see *Ber. deut. keram. Ges.*, **39**, 1962, 304-309; "Science of Ceramics," Academic Press, New York 1962, Vol. I. pp. 139-151.

[503] *Am. Mineralogist*, **38**, 1953, especially pp. 640-642.

[504] *Giesserei*, **44**, 1957 (7), 165-174; in this direction we also mention the studies of H. Gries, *Ibid.*, **39**, 1952 (6), 121-127, on the technology of foundry sands which contain in the place of suitable bonding clay hydrated cement, in steel-casting process (cf. C. ¶ 248).

[505] Especially *Tonind.-Ztg. u. Keram. Rundschau*, **77**, 1953, 88-92; 130-135; 171-174.

kaolin occurrences. However, in these "plastification" endeavors, not so much is the true plasticity improved, as in the high degree of Zettlitz kaolin, but the slurry casting properties are decidedly improved, as is bending strength. Apparent plasticity, in other words, the "feel" or "touch" of mica-containing English china clays used for casting sanitary ware, which are excellent for this purpose (cf. C. ¶ 364), can be "simulated" by Georgia kaolin in mixture with pigment grade mica powder, pyrophyllite, sericite, or talc, as G. W. Phelps[506] has demonstrated.

233. An increase of real plasticity, however, is possible under the still somewhat enigmatic effects of bacterial *fermentation* or "souring" of clays and kaolins. K. Hideo[507] studied this highly interesting problem from the viewpoint of bacteriology in pure culture experiments, controlling the plasticity changes by the J. Roller deformation test in measuring the yield point in gm./cm.^{2} and the tan Θ (Θ is the angle of the linear portion of the deformation curve). The results are highly instructive and recommend further studies in this direction in so far as high-frequency irradiation also promotes bacteria growth and improves the quality of the clay products by increased plasticity. O. Kallauner[508] in his studies on fermentation problems emphasizes the formation of organic acids evolved by the metabolism of fungi, algae, or bacteria in the fermenting clays at 25° to 30°C. as the optimum temperature. Plasticity and binding strength show characteristic hyperbolic functions of the time, similar to adsorption isotherms, suggesting the combination of a Langmuir type adsorption equation with the Gibbs-Szykowski formula for lowering of concentration in a solution by adsorption on an active surface. The resulting hyperbolic kinetic law of mass activity could also be applied to the kinetics of clay fermentation, yielding data for a reasonable time limit for this process in practice. S. Schelinski[509] observed in the fermentation of refractory clays for glass pots a beneficial effect of slight additions of magnesium chloride to increase the workability, evidently due to formation of a reinforced hydration hull around the clay particles due to the adsorption of the magnesium ion and base exchange reactions. The pH of the mix however, is not changed during the fermentation process. Schelinski does not ascribe an intense action to the metabolism of living bacteria. Nevertheless, specific organic agents of the culture medium may really improve their growth.

[506] *Am. Ceram. Soc. Bull.*, **38**, 1959, 411-414.

[507] *J. Kyoto Dyosi Daigaku*, 1952, A. No. 1, 1-4. A recent electron-microscopic study of the fermentation process was made by Fr. Oberlies and G. Pohlmann, *Trans. 6th Intern. Ceram. Congr., Wiesbaden*, 1960, pp. 149-168; *Veröffentl. Max-Planck-Inst. Silikatforsch.*, **20**, 1960, 223-225. The end effect of the fermentation is an increase in colloid material acting as a mixture of organic mucines and decomposition products of clay, kaolin or feldspar.

[508] *Silikat Tech.* **5**, 1954, 216-218.

[509] *Silikat Tech.*, **6**, 1955, 251-254. We remind the opinions of E. Gruner on the general importance of organic materials in clayish sediments: *Ber. deut. keram. Ges.*, **32**, 1955, 169-172, 199-203, discussed in C. ¶ 291, ¶ 329, and related studies of P. Coraducci, *Ceramica (Milan)*, **12**, 1957 (6), 67-71.

234. Some characteristic details in the plastic behavior of clay-water systems were simulated in a dynamic model developed by E. Deeg,[510] based on the strong permanent dipole moment of the water molecule which is imitated by permanent magnetic forces. The models of scaly clay particles are simulated by a ferromagnetic ceramic material (Ferrodure) shaped to tabular fragments. They represent the surface-active particles on which magnetized iron particles are oriented in characteristic patterns, which may well give an impression of the distribution of "active centers" on the sheets, edges, and corners, of the "intermolecular forces," of the "water hull," and so on. Deformation of the model gives a direct illustration of the elastic portion of the affine changes in shape and of relaxation phenomenon, followed by the "plastic" or permanent deformation. Leaning additions to plastic clays can be simulated in the model by weakly magnetic particles added. Particularly well demonstrated are "thixotropy" effects and the process of slip casting. Optical model experiments on details of the plastic extrusion process of clays were made by J. H. Weymouth and W. O. Williamson[511] in pyroxylen-backed films in which the basal faces of the planar parallel-oriented kaolin particles are "traced" by embedded crystals of rutile and tourmaline. Not only the common plastic flow through cylindrical nozzle tubes but also rotational strains could thus be studied, both being of fundamental importance in clay-pressing techniques. The formation of shear joints in the plastic clay makes evident how the clay mineral crystallites, separated by water hulls, are arranged to a lattice in which many of them fail to achieve positions of a minimum potential energy. Similar orientation effects are important for the formation of thermotactic structures (cf. Vol. III, Section A), not only in clay ceramics but also in oxide-ceramic bodies, such as alumina as studied by H. Britsch.[512] In this case the tabular form of the corundum crystallites is the reason for texture effects observed in the finished products. In plastic-shaping operations, therefore, the axial pressure should be replaced whenever possible by uniform hydrostatic pressing methods.

235. Concerning the practical determination of clay body plasticities, the Brabender "plastograph" is again recommended by Cl. L. Marshall[513] as the most convenient standard instrument. He defined a "plastograph factor," which characterizes both the consistency of the sample material and the range of water content during the maximum consistency. This factor is expressed in the area *ABC* (in Fig. C. 110) of the plastograms obtained in the instrument, and represents directly a

[510] *Ber. deut. keram. Ges.*, **32**, 1955, 323-329.

[511] *Am. J. Sci.*, **251**, 1953, 89-108. The practical problems of laminary orientation of clay bodies by extrusion, seen from the viewpoint of rheological theory (cf. C. ¶ 351) were discussed more recently by H. Detaille, *Silicates inds.*, **25**, 1959, 370-375.

[512] *Ber. deut. keram. Ges.*, **36**, 1959, especially pp. 142-144.

[513] *Am. Ceram. Soc. Bull.*, **34**, 1955, 54-56.

relative plasticity value distinguishing the different clay minerals as seen in Fig. C. 111. The effects of different ion exchange reactions for a given clay mineral and of different liquids (e.g., water-alcohol mistures) are very well demonstrated in this

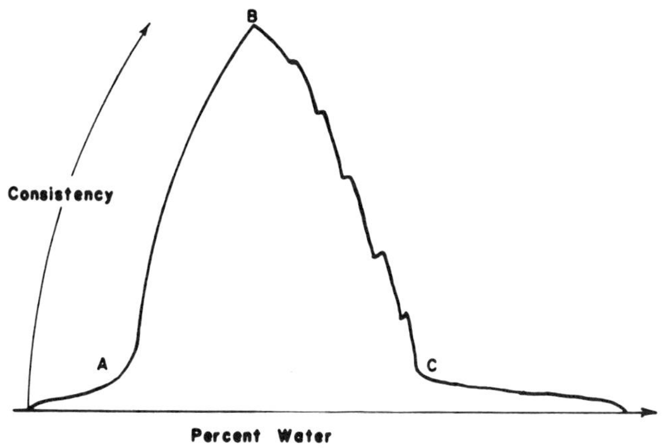

FIG. C. 110. Typical Brabender plastograph diagram with "plastograph factor" in area *ABC* (Marshall).

way. R. R. West and W. G. Lawrence[514] modified somewhat the Brabender plastograph, making it highly sensitive for the determination of small amounts of montmorillonite in clays and establishing its effect on workability. This study is partic-

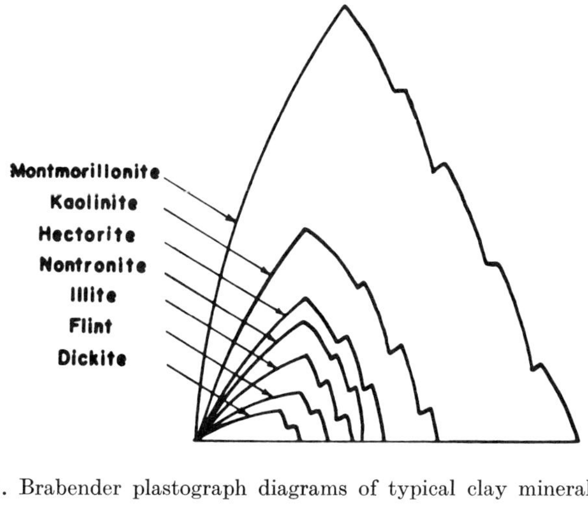

FIG. C. 111. Brabender plastograph diagrams of typical clay minerals (Marshall).

[514] *Am. Ceram. Soc. Bull.*, 38, 1959, 135-138.

ularly important for the characterization of English ballclays,[515] and of their beneficial effects in ceramic bodies, especially by stabilizing them to the action of electrolytes (particularly sulfates) in the mixing water for which the small amount of montmorillonite acts as an "ion scavenger."

236. A recent detailed investigation of the determination of the plasticity by E. Scharrer and U. Hofmann[516] puts in the foreground the tensile strength of the moist clay or kaolin in the form of an extruded rod, as a function of the water content, and the measurement of the deformation pressure in comparison with the classical methods of Atterberg and Pfefferkorn. The apparatus used by Scharrer and Hofmann is shown in Fig. C. 112, for a cross-section of 1 cm.^{2} in the clay rod. The deformation

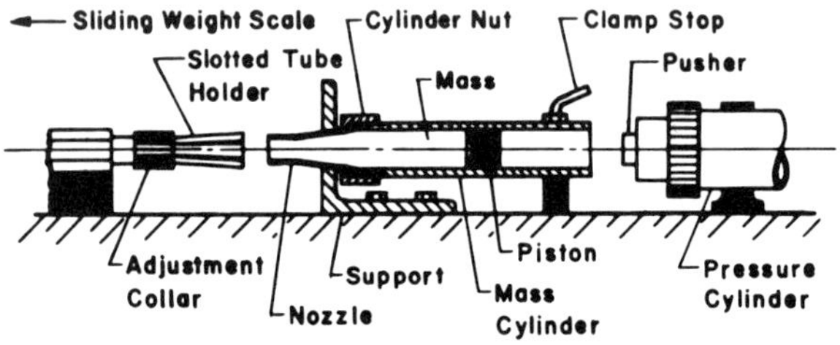

FIG. C. 112. Apparatus for measurement of tensile strength and deformation pressure of moist clay rods (Scharrer and Hofmann).

pressure is defined as that minimum pressure (measured by a manometer) which is sufficient to extrude the clay rod (of 5 cm. length) from the nozzle with a maximum rate of 1 cm./sec. The disruption of the rod is performed by a sliding weight scale acting over a steel rope to the frame holder of the clay rod. This holder is slotted at its end for safe conduction of the soft clay body. The tension stress is very gradually increased by the slow rotation of a hand-wheel at the sliding weight scale. The resulting tensile strength data were determined as the average of six to eight single experiments. The results show the same type of maxima in the diagrams for the strengths as a function of the deformation pressures and also of the water contents, as known from extensive previous studies of G. Keppeler and H. Gotthardt (1930). The maximum ratio of both parameters is indicated separately in Fig. C. 113 as an example for Zettlitz kaolin. A vertical arrow indicates that water content which corresponds to the

[515] Excellent characteristics of English ballclays in their geological-stratigraphic occurrence, their mineralogical composition, and physical-technological properties were given by M. Inzegneri, *Ceramica (Milan)*, **11**, 1956 (1), 37-41, and particularly in an extensive monograph by D. A. Holdridge, *Trans. Brit. Ceram. Soc.*, **55**, 1956, 369-440, with a description of their plastic behavior on pp. 405 ff. See also on the mineralogical composition of ballclays O. E. Radczewski, *Sprechsaal*, **95**, 1962, 285-288 (chiefly electron microscopy and infrared absorption spectra).

[516] *Ber. deut. keram. Ges.*, **35**, 1958, 278-285. See previously M. Linseis, *Ibid.*, **29**, 1952, 35-37.

plasticity number according to Pfefferkorn. An inclined arrow further indicates the maximum ratio of tensile strength to deformation pressure. This ratio is lowest in coarse-granular kaolins, and highest in very fine dispersed kaolins. For common clays and halloysite this ratio is rather irregular and lower.[517] It is interesting to observe that the curve of the compression strength as a function of the water content

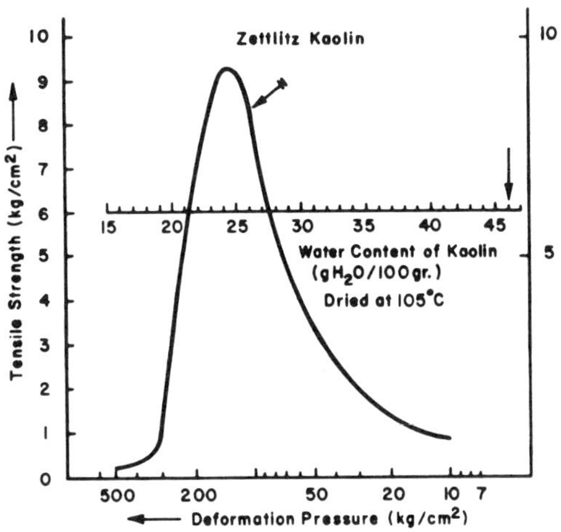

FIG. C. 113. Typical diagram of tensile strength and deformation pressure of Zettlitz kaolin, as a function of water content (Scharrer and Hofmann). Vertical arrow indicates the water content which corresponds to the plasticity number according to Pfefferkorn; inclined arrow to coordinates for the point on the curve for which the tensile strength/deformation pressure quotient is maximum.

shows for a plastic foundry sand (with 5 per cent bentonite) a similar characteristic maximum,[518] and the two points marked by arrows have the same meaning as mentioned above for the Zettlitz kaolin (see Fig. C. 114).

237. For the bonding properties of clay minerals, the general aspects on the mechanism of *adhesion* are of fundamental significance; they were reported by R. Houwink[519] to be a function of molecular forces exerted on solid surfaces. Houwink especially treats of adhesive liquids and organic polymers, for which sometimes (as in sodium silicate solutions and brominated butyl rubber) an intermediate composition shows the strongest adhesion effects. Montmorillonite acts as the best known

[517] Cf. Th. Haase, *Silikat Tech.*, 3, 1952, 265-267; *Ber. deut. keram. Ges.*, 34, 1957, 27-33.

[518] Cf. G. W. Morey and W. C. Taylor, cited by K. Endell and U. Hofmann, *Angew. Chem.*, 60 A, 1948, 237-242.

[519] *Kolloid-Z.*, 151, 1957, 143-147, J. Nuyens, *Silicates inds.*, 26, 1961, 279-283, recently gave an application of the generalized Coulomb theory for the discussion of capillarity and adhesion for clay-water systems, from the viewpoints of soil mechanics.

"mineral glue" among the clay minerals because of its innercrystalline swelling. In exceedingly smooth thin films it is an excellent adhesion material on quartz grain surfaces and is used in most of the foundry sands mentioned above.[520] U. Hofmann, R. Fahn, and A. Weiss[521] studied the adhesion of montmorillonite especially in connection with its thixotropy and its high sedimentation volume (cf.

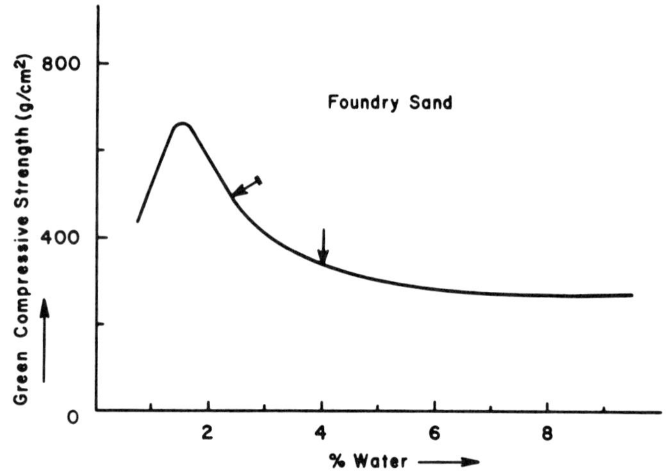

FIG. C. 114. Compressive strength as a function of water content for a foundry sand with 5 per cent bentonite (Morey and Taylor).

C. ¶ 334, ¶ 346), not only in water but also in different polar and nonpolar liquids. The mutual distances and number of contact points of the particles in the "house of cards" structure of the thixotropic system are definite functions of the wetting capacities. The importance of these aspects will be discussed in details later. Adhesion problems become particularly critical if highly nonplastic materials are used in foundry sands which have in their suspensions only very poor thixotropic properties. These factors are very important in modern steel casting with olivine sands (cf. Vol. V, Section B).[522]

238. The *drying shrinkage* behavior of ceramic clays depends very much on the grain size fraction distribution as was first demonstrated by C. G. Harman and F. Fraulini (1940) and recently in the experiments of P. W. Berg and B. Nielson[523]

[520] An introductory discussion of these factors see by U. Hofmann, *Angew. Chem.*, **68**, 1956, 53-61, especially pp. 60 f., and the previous studies of K. Endell and U. Hofmann, *Ibid.*, **60**, A, 1948, 237-242; *Monatsh. Chem.*, **78**, 1948, 258-272.

[521] *Kolloid-Z.*, **151**, 1957, 97-115.

[522] See also Fr. Hofmann, *Giesserei*, **46**, 1959, 49-54.

[523] *Ber. deut. keram. Ges.*, **29**, 1952, 407-410. For recent investigation on the drying process

for ballclays. Particularly high shrinkages occur for Utah halloysite clays, according to B. F. Hampel and I. B. Cutler,[524] which in the extrusion processing (cf. Fig. C. 115) show a pronounced particle alignment and after firing considerable differential shrinkage. The linear drying shrinkage of bars prepared, as from Fox clay, Utah, is perpendicular to the direction of flow of the extruded material much greater than

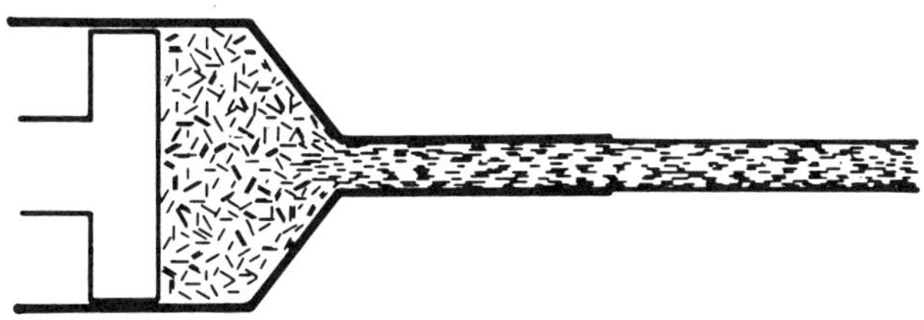

FIG. C. 115. Mechanism of alignment of halloysite particles during extrusion of rods (Hampel and Cutler).

that in the flow direction. Doubtless the well-known lath-like form and the tubular structure of the crystal particles bring about these effects. The collapse of the tubes finally causes anomalies of shrinkage also in the fired bodies. The *drying strength* of clays was studied by W. H. Sutton and F. R. Matson[525] for Georgia kaolin, Kentucky ballclay, and Missouri fireclays bodies in the temperature range of 110° to 800°C. as a function of aging, electrodialysis, and ion exchange phenomena. A very detailed investigation by U. Hofmann, W. Czerch, and E. Scharrer,[526] concerned with the reasons for the bending strength after drying for a series of different clays and kaolins, showed that this property is evidently influenced by the shape of the clay mineral crystallites, and the content of the slip-cast or extrusion-pressed rod samples in solid particles. The influence of cation base exchange is demonstrated by the fact that sodium ions favor development of pronounced textures in the clay bodies in a much higher degree than do calcium ions. After drying shrinkage the sodium clay bodies are, therefore, of a denser consistency (cf. Fig. C. 116). A specific increase in bending strength after drying is observed in potassium clay bodies. This phenomenon is in analogy with the observation of A. Weiss, G. Koch, and U. Hofmann[527] that

of clay-water systems, especially of the contraction phenomena, see T. Nakayama, *Asahi Garasu Kenkyu Hokoku*, **10**, 1960, 1-16. The orientation of kaolinite crystals was made evident by X-ray analysis.

[524] *J. Am. Ceram. Soc.*, **36**, 1953, 30-34. Halloysite-"kaolins" are particularly important for the ceramic industry in South America, based on Brazilian and Venezuela clay deposits; cf. Yvonne Stourdzé-Visconti, B. N. F. Nicot, and E. Goulart de Andrade, *Am. Mineralogist*, **41**, 1956, 67-75; *Bol. Inst. nacl. tecnol.* (*Rio de Janeiro*), **5**, 1954, No. 11, 22-34; T. Eder, J. Magasrevy, T. Temt, and P. Wieden, *Ber. deut. keram. Ges.*, **35**, 1958, 285-294.

[525] *J. Am. Ceram. Soc.*, **39**, 1956, 25-30.

[526] *Ber. deut. keram. Ges.*, **35**, 1958, 219-225.

[527] *Z. Naturforsch.*, **6**, b, 1951, 405-409; *Fer. deut. keram. Ges.*, **32**, 1955, 12-17. On the constitu-

potassium ions impede an innercrystalline swelling (such as that of vermiculite or batavite), whereas sodium and calcium ions favor it. The potassium ions, therefore, cause a higher cohesion of the crystallites on the contact points.

239. Concerning the nature of the water hull on clay particles, or the thickness of the water layer on different clay mineral particles, as a function of the cation exchange phenomena, we mention previous important data given by W. H. East[528]

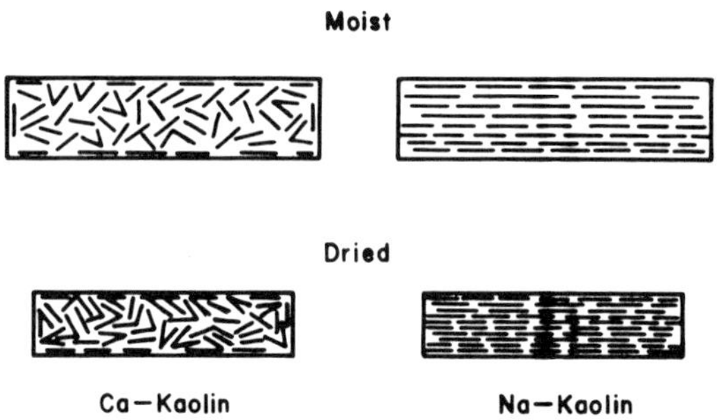

FIG. C. 116. Texture of crystallites of clay minerals in moist and dried extruded rods (degree of orientation much exaggerated) (Hoffmann, Czerch, and Scharrer).

for kaolinite, in comparison with hydrargillite and quartz, and combined with calorimetric measurements of heats of wetting (cf. C. ¶ 319 ff.). Dividing the volume of water loss during shrinkage by the total specimen area, the film thickness was found to be a definite function of the moisture content according to Fig. C. 117.

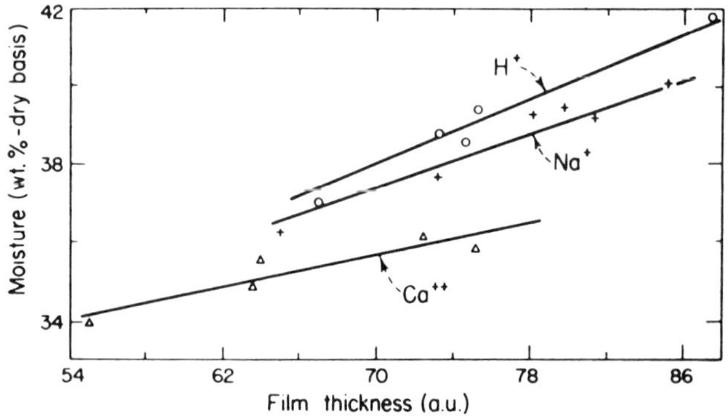

FIG. C. 117. Effect of exchangeable ions and moisture content on thickness of water films on kaolinite (grain size fraction 0.4 to 0.8 μ) (East).

tion, especially the space filling, in dry-pressed ceramic bodies, see recent studies of U. Hofmann, E. Scharrer, W. Czerch, K. Frühauf, and W. Burck, *Ber. deut. keram. Ges.*, **39**, 1962, 125-130.

[528] *J. Am. Ceram. Soc.*, **33**, 1950, 211-218.

For a constant moisture content of 40 per cent on a dry weight basis, it was found that for H^+-kaolinite the film thickness is 80 Å., for Na^+-kaolinite 86 Å., and for Ca^{2+}-kaolinite 106 Å., respectively. The pore volume calculated for kaolinite aggregates of the assumed structure shown in Fig. C. 118 (as the difference of the rec-

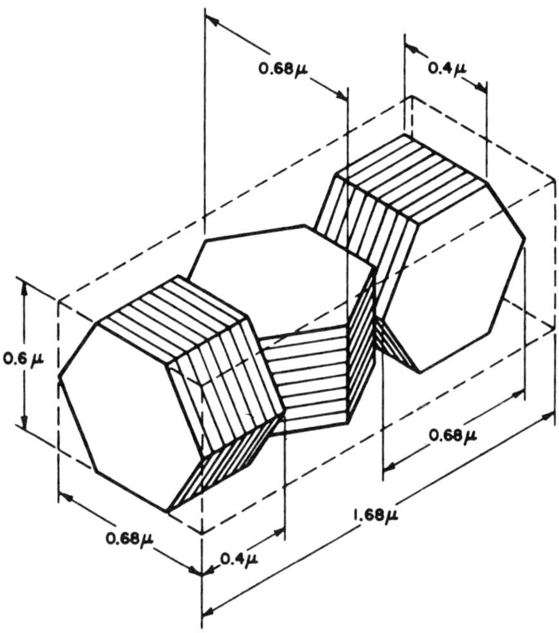

FIG. C. 118. Assumed dimensions of an aggregate of plastic kaolinite particles for calculation of the pore volume (East).

tangular volume outlined by the largest dimensions, and the total particles volume) is, for all of the kaolins studied by East, about 46 per cent. For a comparison, the water film on a sodium-charged quartz particle aggregate is only 22 Å. thick for 40 per cent moisture content, whereas for sodium-hydrargillite a layer of 60 Å. thickness was determined, in this case, under a constant pressure of 1.8 kg./cm^2 (kaolinites show under the same conditions a water film thickness of 68 Å.).

240. A most extensive and fundamental investigation on the drying process of clay bodies in connection with their plastic properties was made by W. D. Kingery and J. Francl,[529] working with a specially developed and accurate apparatus. The concept of plasticity is here based on the definition given by E. O. Wilson (1933) that the product of yield value and maximum deformation is taken as a measure of "*workability*." For this reason the apparatus was constructed to measure, with

[529] *J. Am. Ceram. Soc.*, 37, 1954, 596-602.

a horizontally movable carriage the shear exerted on prismatic clay samples of $1 \times 1 \times 6$ cm. dimensions which are sheared in two 1 cm. cubes. The conventional stress-strain diagram is shown in Fig. C. 119. The results of this important investigation clearly illustrate the tremendous effects of surface tension and of its changes

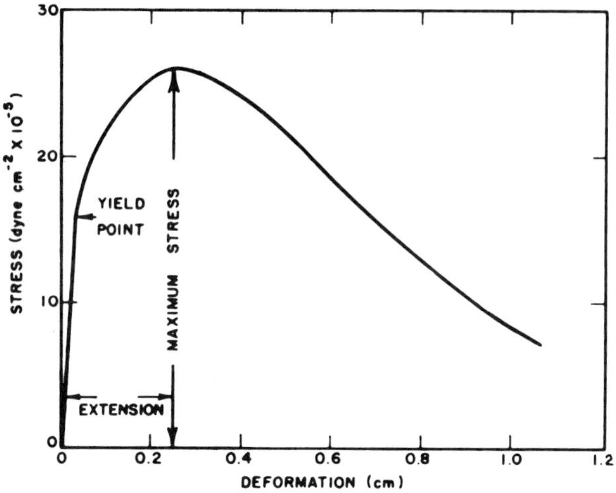

FIG. C. 119. Stress-strain diagram for shearing of clay bodies (Kingery and Francl).

in the liquid phase of the clay aggregates. Kingery and Francl particularly established that workability and yield value are increased with increasing surface tension of the aqueous solutions of different surface-active additives (of cationic and anionic type characterized in Table I) (cf. Fig. C. 120). Workability is a linear function of the surface tension according to the following equation:

TABLE I

AQUEOUS SOLUTIONS USED FOR CLAY MIXTURES, WITH pH = 6 TO 8

	Additives	Concentration in weight per cent	Type of surfactant	Surface tension at 25°C. in dynes/cm.
A	None, pure water	—	—	71.0
B	Nopco 1187-X	0.7	Anionic	60.3
C	Avitone A	0.7	Anionic	43.5
D	Merpentine	0.7	Nonionic	33.3
E	Santomerse	1.0	Anionic	32.8
F	Product "QB"	0.35	Cationic	30.5
G	F-126	0.4	Anionic	16.0

workability = 13.3 × surface tension (cf. Fig. C. 121). The drying shrinkage for a given water content increases linearly as surface tension is increased, whereas the firing shrinkage decreases (Fig. C. 122). Total shrinkage is rather constant. The role of drying for various additives (in the constant rate period) is decreased as the surface tension increases (Fig. C. 123). The progressive changes in the water film with decreasing water contents are schematically represented in Fig. C. 124. The results of

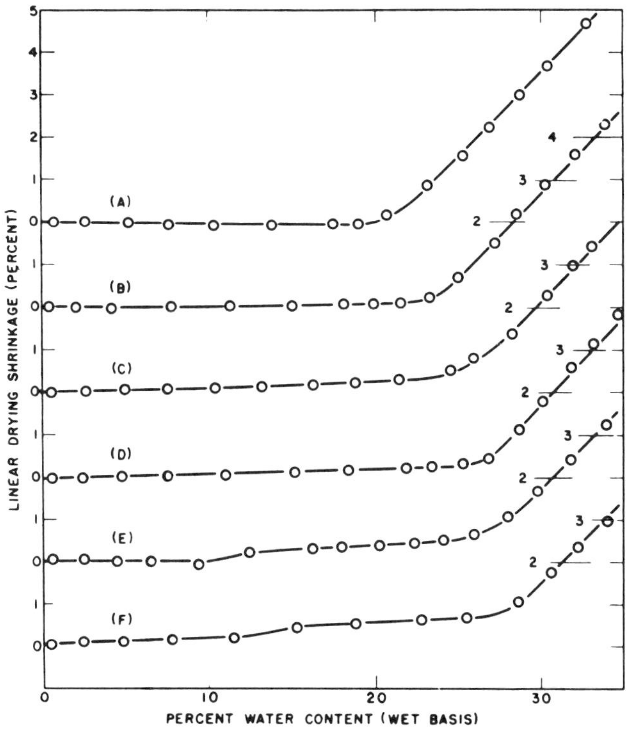

FIG. C. 120. Linear drying shrinkage as a function of the water content for various additives, decreasing in surface tension from A to F (cf. Table I) (Kingery and Francl).

Kingery and Francl's investigations are interpreted by the membrane film theory of plasticity previously developed by F. H. Norton (1948) for clays and different other materials; M. McQuarry,[530] however, made critical remarks concerning the correct meaning of "maximum deformation" and the interpretation of the workability in the experiments here discussed.

241. Most recently, W. C. Ormsby[531] criticized the theory of "stretched membranes"

[530] *J. Am. Ceram. Soc.*, **38**, 1955, 341.

[531] *Am. Ceram. Soc. Bull.*, **39**, 1960, 408-412; cf. W. C. Ormsby, R. M. Witucki, and W. A. Weyl, *Natl. Acad. Sci.-Natl. Research Council, Publ.*, **456**, 1956, 251-272.

of Norton, and especially the conclusions of Kingery and Francl from their measurements of the workability based on the determination of yield points in general, and of the modification of workability by surfactants in particular. Ormsby extended the incomplete series of types of wetting agents used in the experiments of Kingery and Francl by more systematic selections of surface-active substances, namely a larger number of anionic, cationic, and nonionic agents. The conclusion that surface tension would be the essential factor in modifying workability properties is refuted from these additional data given by Ormsby. The fact that nonionic and cationic surfactants produced flocculated systems whereas anionic agents produced deflocculated systems is in Ormsby's opinion a clear indication that the electrokinetic

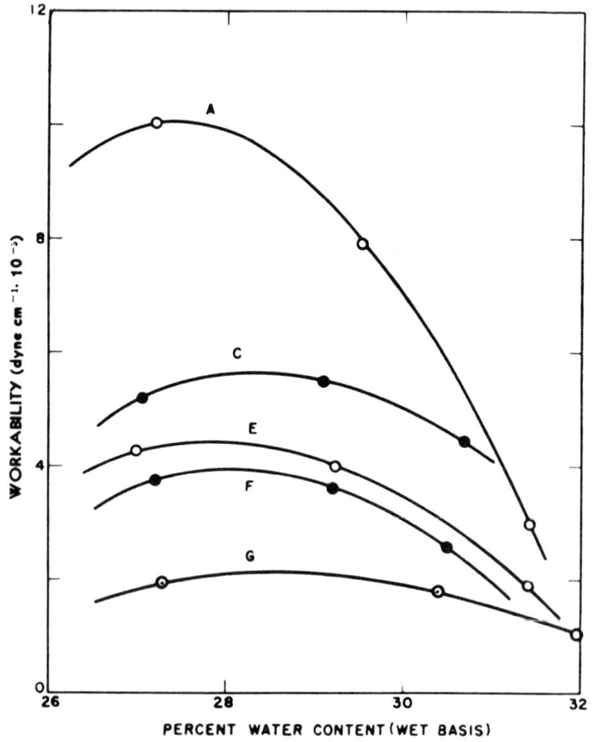

FIG. C. 121. Effect of the water content on the workability of clay, with various additives decreasing in surface tension from *A* to *G* (Kingery and Francl).

factors exerting attractive and repulsive forces on the particles of the clay-water systems are decisive alone, and not surface tension effects. Such ionic interactions can be simply demonstrated, as in the determination of the filtration behavior of kaolin with different wetting agents, as shown in Fig. C. 125. Opposite effects are also observed for the sedimentation volume and the deformation properties of such suspensions, as seen from Fig. C. 126 (cf. C. ¶ 310).

242. The imbibition capacity for water in clayish sedimentary rocks was determined by J. Konta[532] in samples the total and effective porosities of which were measured by standard ceramic methods. Four of the rocks were kaolinitic, three

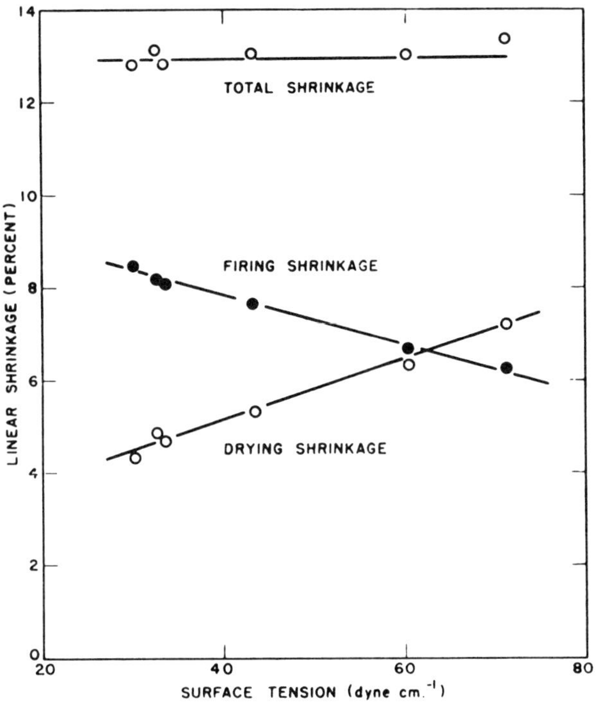

FIG. C. 122. Effect of surface tension on drying and firing shrinkage; total shrinkage is the sum of both (Kingery and Francl.)

prevailingly illitic (one of them ferrugineous), and four montmorillonitic. For the imbibition experiments water and ethylene glycol were used as polar liquids. They were applied in uniform drops on the surface of polished rock samples. Montmorillonitic clays are wetted with the widest extension of the sucked drops, and swollen; kaolinites showed the smallest wetting areas. The viscosity of the imbibed liquids plays an important role in the rates of the sucking-up process, and the specific surface of the crystalline material of the clays acts equally. In kaolinites, therefore, the

[532] *Silicates inds.*, **24**, 1959, 133-136; more extensively in *Am. Mineralogist*, **46**, 1961, 289-303. On the suction method see J. Konta, *Chem. Erde*, **18**, 1956, 179-193.

liquids are sucked up more rapidly than in illites, and especially in montmorillonites of high dispersity.

243. Another most remarkable attempt to measure the thickness of the water film (the adsorption layers) on moist plastic clay particles by *dielectric* property changes with ultrashort electromagnetic oscillations ($\nu = 10^8$ to 10^9/sec.) was made by E. Deeg and O. Huber.[533] Since a direct determination of the water content by the common methods of dielectric constant measurements is not suitable, the indirect "exluation" method with dioxan as the solvent in the so-called dielcometer (or dekameter)[534] was also attempted for this purpose. H. Lehmann, H. Kolkmeier,

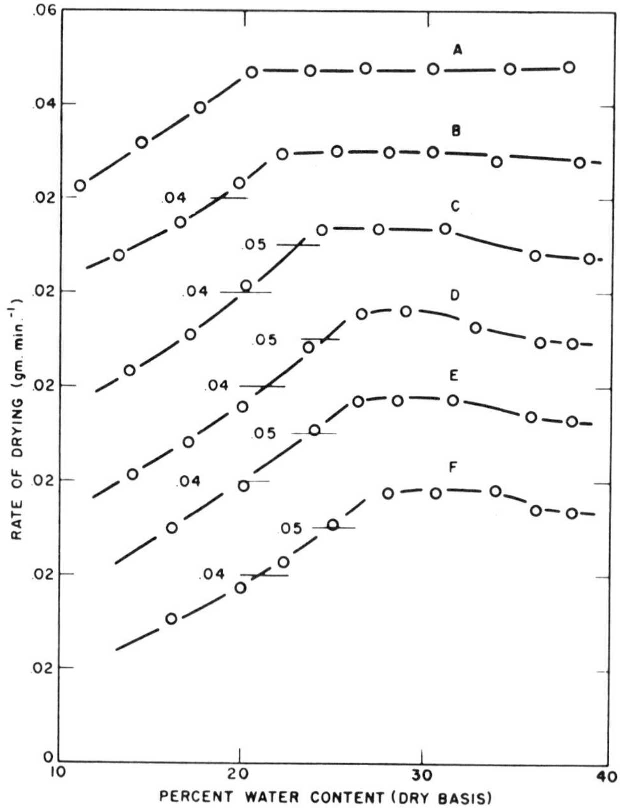

FIG. C. 123. Rate of drying for various additives, decreasing in surface tension from *A* to *F* (Kingery and Francl).

[533] *Naturwissenschaften*, 42, 1955, 507; extensively in *Ber. deut. keram. Ges.*, 32, 1955, 261-272.

[534] The dielcometer was developed by L. Ebert, cf. *Angew. Chem.*, 47, 1934, 305-315. The instrument is based on the immersion method of the sample in a capacity cell; see L. Ebert, R. Eisenschitz, and H. von Hartel, *Z. physik. Chem.* (*Leipzig*), 1 B, 1928, 94-114.

and M. Marenbach[535] determined the specific polarizability of the clay-water samples as a function of the bound water content. Deeg and Huber found an ideal mixing of the hydrated clay particles and free water indicated for water contents above the technological "mixing water" contents for standard consistency, by a linear function (Fig. C. 127). Deviations from the theoretical relation indicate interactions between the water molecules and the clay particles, and a rather discontinuous transition from the normal to a more irregular course of the polarizability curve, corresponding

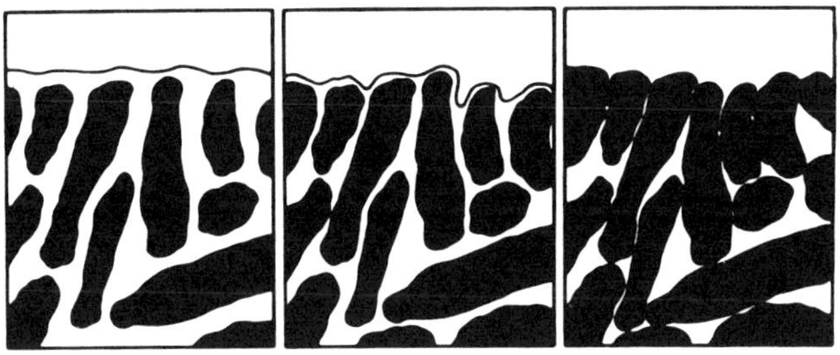

FIG. C. 124. Schematic representation of progressive changes in the water film as water content is decreased (Kingery and Francl).

to the technological "mixing water" content as the critical point, which also is important in the definition of the plasticity. The experimental determination of the dielectric constants in the clay suspensions was based on the principles given by K. S. Knol and M. J. O. Strutt,[536] for frequencies in the 10^8 to 10^9/sec. range. Plottep

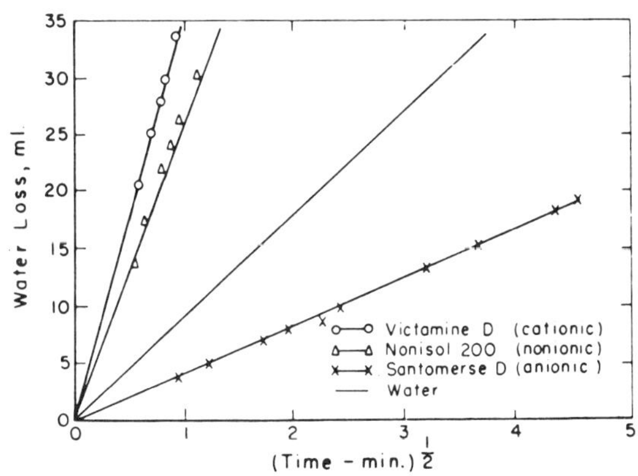

FIG. C. 125. Filtration behavior of kaolin (suspension with 25 per cent solid), as influenced by additions of different wetting agents (Ormsby, Witucki, and Weyl).

[535] *Tonind.-Ztg. u. Keram. Rundschau*, **78**, 1954, 336-338; critical measurements by E. Deeg, *Ibid.*, **78**, 1954, 358-359.

[536] *Ned. Tijdschr. Natuurk.*, **9**, 1942, 577.

as the polarizability was the nonimaginary part of the factor

$$P^* = \frac{(\varepsilon + i\,\eta) - 1}{(\varepsilon + i\,\eta) + 2} \cdot \frac{1}{\varrho}$$

[for a material with the dielectric loss $\varepsilon^* = \varepsilon\,(1 + i\eta) = \varepsilon\,(1 - i\tan\delta)$]; the specific susceptibility, as the parameter

$$\chi = \frac{\varepsilon - 1}{\varrho}$$

the specific polarizability

$$P = \frac{\varepsilon - 1}{\varepsilon + 2} \cdot \frac{1}{\varrho}$$

for $\eta = 0$ (ϱ is the density of the sample material).

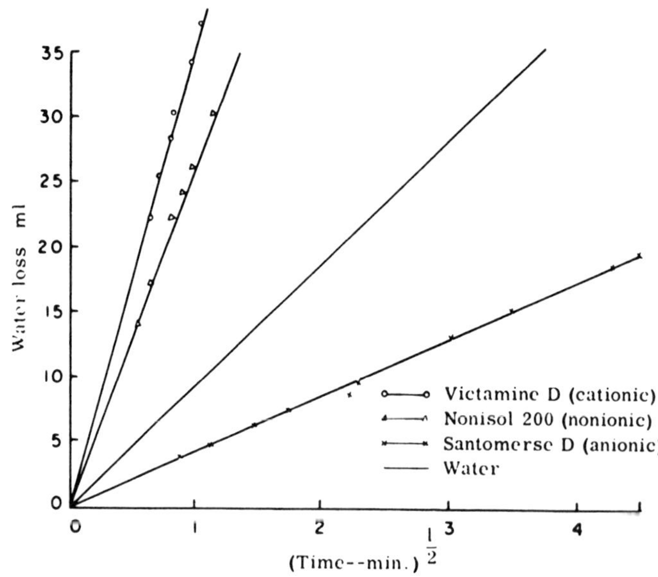

FIG. C. 126. Summary graph of deformation, sedimentation volume, and filtration behavior of wetting agents on kaolin suspensions (Ormsby, Witucki, and Weyl).

244. A typical example of the experimental results is given in Fig. C. 128, showing the specific polarizability of the two clays mentioned above and of Moosburg bentonite. The hatched range on the abscissa axis marks the plasticity range in the definition of R. Rieke (1915) (difference of the rolling limit and standard mixing water contents). The curves show three distinct, physically different ranges of water concentrations in the clay-water mixes. (1) At lowest water contents, the firmly bound water is fixed on the clay particles. For this state the polarizability has a constant value (as 0.4 for Klingenberg clay); the water cannot contribute appreciably to the total polarizability of the system, and the para-electric behavior of water is nearly eliminated; (2) from 7 to 30 per cent water the curve shows most impressively the mutual interactions of the water molecules with the clay particle surface. It is the range also of plastic shaping, extrusion pressing, and so on, in ceramic practice. At 15 per cent there is a remarkable break point ("knee") in the polarizability

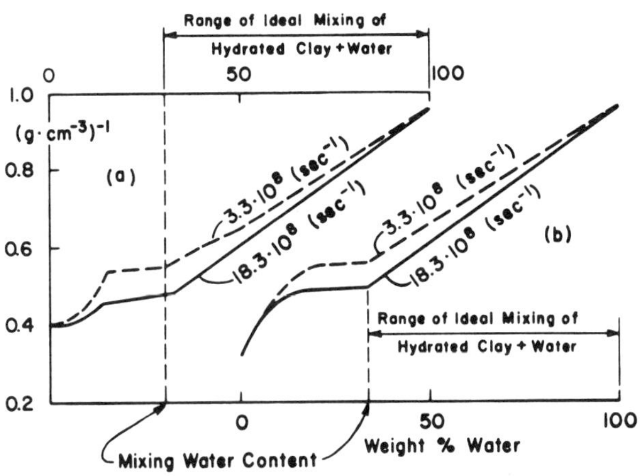

FIG. C. 127. Functional relation of specific polarizability to water content: (*a*) of Klingenberg clay; (*b*) of "Kolloid Kaolin C," for two different frequencies (Deeg and Huber).

curve which does not correspond to any analogous phenomenon in the vapor pressure curve given for comparison in Fig. C. 129. Near the 15 per cent point, a remarkable change in the color of the clay-water mix is observed: the originally white clay becomes bluish-gray in the 15 per cent range because the reflexion of yellow and red light waves is reduced in comparison with the blue part of the spectrum. This range (2) is that of the typical water adsorption on clays, in which the building up of a monomolecular layer of water molecules takes place, followed by additional layers. The break point corresponds at 15 per cent to a four-molecule water layer complex, whereas at 30 per cent up to ten layers are built up. The molecules in the 2 to 3 molecular layers are in a state of anomalously increased mobility, whereas above four molecular layers, pure dipole forces prevail. The range (3) of the fully ideal mixing of the molecules of free water with the hydrated clay particles, as mentioned above, is characterized by simple additivity of the linear function. For montmorillonitic bentonite a particular phenomenon is observed in a high dielectric loss characteristic of the 40 per cent water mix, as determined with a frequency of 3.1×10^8/sec. (cf. Fig. C. 130).

245. The *aeration* process in clay-water suspensions, which in certain respects is analogous to the air-entraining phenomenon in Portland cement pastes and concretes and is important for the frost-resistivity of these systems (cf. Vol. V, Section C),[537] was studied by I. A. Al'perovich and P. P. Budnikov.[538] *De-aeration* destroys the interlinking of air bubbles in the framework of clay particles and causes a "flocculation structure." These aspects are in complete agreement with P. A. Rebinder's theory of flotation processes as adhesion of gases on hydrophobic solids.[539] Experiments with montmorillonites, highly de-aerated *in vacuo*, showed in the grain size

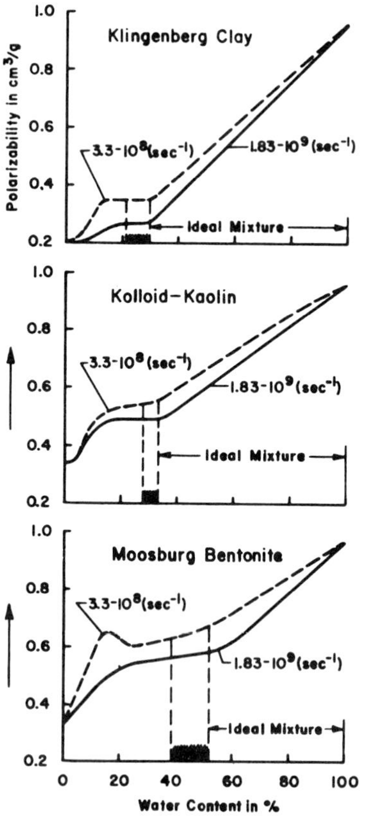

FIG. C. 128. Specific polarizability of three clay samples as a function of water content, with range of plasticity (in Rieke's definition), marked by hatches (Deeg and Huber).

[537] See the fundamental investigations of T. C. Powers *et al.*, *Research and Develop. Labs. Portland Cement Assoc. Research Dept. Bull.*, **49** and **49** A, 1951, 20 and 17 pp.; *Ibid.*, **71**, 1956, 47 pp., especially pp. 37 ff.; *Ibid.*, **90**, 1958, 28 pp. On a calorimeter- strain apparatus for the study of freezing and thawing, see G. Verbeck and P. Klieger, *Ibid.*, **95**, 1958, 22 pp.

[538] *Doklady Akad. Nauk S.S.S.R.*, **116**, 1957, 463-465.

[539] "Rol' Gazov i Reagentov v Protsessakh Flotatsii," Izdat. Akad. Nauk S.S.S.R., Moscow-Leningrad, 1950.

analysis an increase in amounts of the finest particle fractions and a reduction of them in the range between 10 and 50 μ. The aggregation of primary particles, e.g., in a soil sediment, is not only determined by electrolytic coagulation effects but

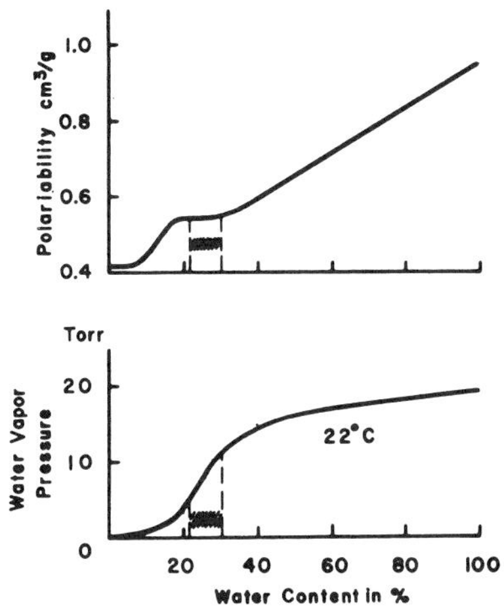

FIG. C. 129. Changes in specific electric polarizability and in water vapor partial pressure, as a function of water content for a clay-water system (Deeg and Huber).

also by the adhesion of microbubbles of air to hydrophobic clay particles. If the concentration in adsorbed cations is high, the amount of adhesion gases is inconsiderable and electrolytic coagulation suppresses the flocculation; under inverse conditions, the aggregation flocculation prevails.

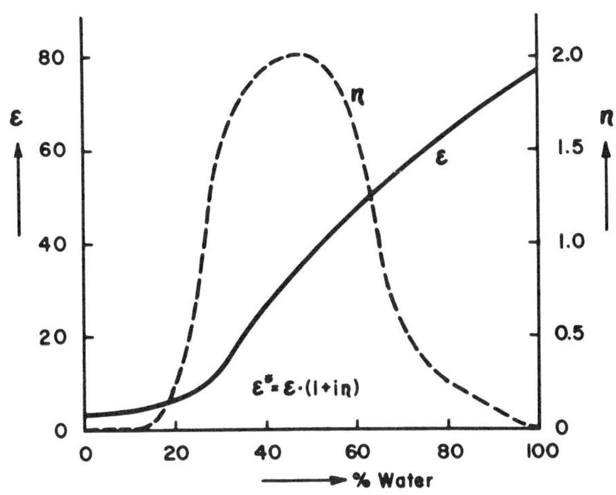

FIG. C. 130. Dielectric constant, ε, and dielectric loss ($\eta = -\tan \delta$), in the complex polarizability of Moosburg bentonite, as a function of the water content (Deeg and Huber).

246. The effects of *freezing* in bentonitic clays and clay-loam (regarding silica gel and a synthetic barium aluminosilicate, cf. C. ¶ 140) were studied from the viewpoint of the theory of capillary condensation by B. Rai Puri, L. R. Sharma, and M. L. Lakhanpal[540] by dilatometric measurements of the volume effect of the freezing reaction. The apparatus used is shown in Fig. C. 80 (C. ¶ 140). The freezing point is observed very sharply as a discontinuity with rising temperature after having the sample frozen under controlled relative water vapor pressures, by noting the level of the ligroin meniscus in the capillary tube, *t*. The relative vapor pressure (p/p_s) is determined by the equation

$$\log (p/p_s) = \frac{2\,\gamma' \cdot M}{r\,\varrho\,RT'}$$

in which γ' is the interfacial tension between the ligroin and water at temperature T', as determined separately. The freezing point depression as a function of (p/p_s) is given by

$$-\log (p/p_s) = \frac{\Delta T \cdot \Delta H}{RTT_0}$$

(T_0 the normal freezing point of the liquid at p_s, from which is calculated from the precision data of the National Bureau of Standards[541]:

$$-\log (p/p_s) = 0.9686 \cdot 10^{-2} \cdot \Delta T - 0.56 \cdot 10^{-6}(\Delta T)^2 + 0.72 \cdot 10^{-8}(\Delta T)^3$$

for the best possible determination of relative pressures. The agreement of measured and calculated values for ΔT for bentonite and loam clay is rather good for capillaries between 410 and 37 Å. in diameter, and is even better than that which was observed by I. Higuchi and Y. Iwagami[542] for the adsorption of benzene and carbon tetrachloride on silica gels. For the finest capillaries, however, the density and surface tension properties of the condensates are probably different from those measured for the same compounds in bulk form (cf. C. ¶ 140, Footnote 312).

247. M. B. M'Ewen[543] applied the potential theory for determination of the *rigidity* of a clay-water framework in end-to-end aggregates, in positions of minimum energy for the interparticle distance and the thickness of Freundlich-Gouy double-layers involved on the particles, as a function of the concentration in electrolytes. The most stable configuration of such clay-water aggregations is that of small thin

[540] *J. Phys. Chem.*, **58**, 1954, 289-292. Other studies on changes in bentonitic gels by freezing and lyophilization see J. L. Ahlrichs and J. L. White, *Science*, **136**, 1962, 1116-1118.

[541] Cf. N. S. Osborne, H. S. Stimson, and D. C. Ginnings, *J. Research Natl. Bur. Standards*, **23**, 1939, 197-260; 643-646.

[542] *J. Phys. Chem.*, **56**, 1952, 921-927.

[543] *Clay Minerals Bull.*, **1**, 1951, 177-178.

plates, as in a "gelated" montmorillonite system. M. B. M'Ewen and M. I. Pratt and with D. L. Mould[544] studied further such rigid frameworks in suspensions of Wyoming bentonite. The linear aggregation with the particles oriented edge-to-edge in the form of flat ribbon is the basic characteristic of the gelation process. If sodium hydroxide is present, a cross-linking of these ribbons occurs through an edge-to-edge association, but no larger sheets or discs were observed, also no corrugated ribbons which would have been formed by an edge-to-surface interaction, or the formation of micelles, caused by surface association of the plates, when pH is varied in the range from 6.9 to 13. M'Ewen and Pratt, on the other hand, propose a three-dimensional structure if ribbons cross-linking at their ends are associated (Fig. C. 131). This structure model may correspond with what U. Hofmann called

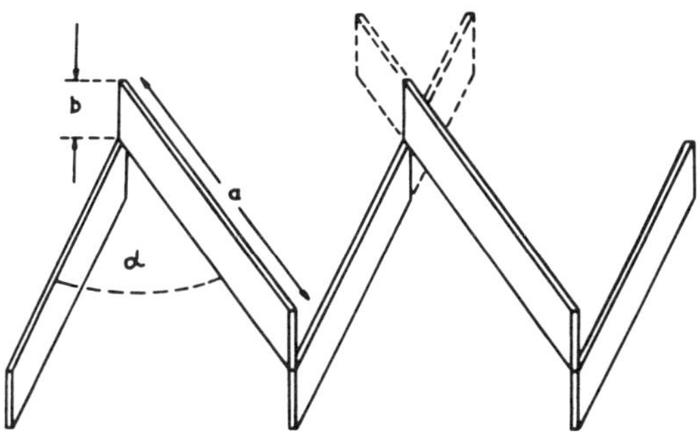

FIG. C. 131. Proposed model of structures in sols of Wyoming bentonite (M'Ewen and Pratt).

the "house of cards" structure (cf. C. ¶ 133), and was assumed to exist in gelated thixotropic montmorillonite emulsions from which the water was slowly removed by ice-sublimation *in vacuo*.[545] Suspensions in pure water (of the so-called *P* type) are of low rigidity, have almost an ideal elastic behavior, and show birefringence at rest. In the presence of alkali (*S* type suspensions) a pronounced rigidity is developed and the deformation behavior is of the viscoelastic nature, no birefringence being observed at rest. These properties are in agreement with the theoretical potential energy curves for the flat-platy particles, as mentioned above. The primary potential minimum corresponding to a low rigidity provided by the theory is that shown by the *P* type suspensions; it is relatively deep, but flat. The combination of attraction and repulsion forces between the projecting points at broken edges of the plates play the decisive role in the formation and the shape of the primary potential minimum (cf. also C. ¶ 336).

[544] *Trans. Faraday Soc.*, **53**, 1957, 535-547; 548-564.

[545] *Angew. Chem.*, **55**, 1942, 283-289, especially pp. 288 f.; see also A. Weiss, R. Fahn, and U. Hofmann, *Naturwissenschaften*, **39**, 1952, 351-352; *Ber. deut. keram. Ges.*, **30**, 1952, pp. 24-25.

248. What the ceramist calls the "green compression strength" of clay bodies was studied by W. F. Bradley[546] for natural beds of bentonite at field moisture conditions. The connection of this parameter with the water dipole layers and their thickness is in a remarkable conformity with what E. Deeg and O. Huber (cf. C. ¶ 221) determined from dielectric measurements, and with the previous studies of D. M. C. MacEvan[547] on the number of such liquid layers. A stable complex for bentonite in the presence of excess water would have four layers, and Bradley found for the natural moisture conditions of montmorillonite beds a thickness range from 3 to 3.5 layers, in excellent agreement. One may establish an empirical equation of the green strength (measured in p.s.i. units) in this state: $= 100 \times \varphi - 50$, in which φ is the bulk volume of the rock which is actually occupied by solids. The strengths of industrially suitable synthetic foundry sands (cf. C. ¶ 232) are of the order of 30 to 40 p.s.i., with φ about 0.9. The green strength of the natural beds is further approximately equal to the load of overburdens in the field. The importance of these considerations for soil investigations is evident, but also for the characteristics of crude kaolin deposits of which R. Neumann[548] gave a very instructive description for the Middle-German (Halle-Meissen District) occurrences, especially in relation to the rocks from which they were derived. This author also stressed the suitability of accurate water suction capacity measurements wih the apparatus of O. Enslin (1933) which was particularly recommended by K. Endell (1938) for soil studies.

249. More recently, H. Lehmann and G. Ackmann[549] used the Enslin apparatus for the measurement of optimum water imbibition and of drying time, with which the Enslin data are in an approximately linear relation. In spite of that, O. Eckart[550] could not observe consistent results if the Enslin device was applied to the study of moulding sand binders (chiefly bentonites) which are better characterized only by their green shearing strengths and gas permeability. Also O. E. Radczewski,[551] in a critical study on different ceramic testing methods, did not find the Enslin device very suitable for examination of montmorillonitic clays, although it shows satisfactory results for fine quartz powders, tile, fireclays, and kaolins. Radczewski speaks especially against the common use of Enslin data as characteristic of "swelling capacities" since they are exclusively functions of water suction capacity. For use of the Enslin device with highly plastic bentonites, the long time periods required for the final establishment of the imbibition equilibrium are particularly unsuitable (4 to 5 days), whereas for kaolins or fireclays this state is reached in a few hours. The

[546] *Natl. Acad. Sci.-Natl. Research Council, Publ.*, **456**, 1956, 41-44.

[547] Cf. "X-Ray Identification and Crystal Structures of Clay Minerals," Mineralogical Society of Great Britain, London, 1951, pp. 86-137.

[548] *Ber. deut. keram. Ges.*; **31**, 1954, 156-167.

[549] *Tonind.-Ztg. u. Keram. Rundschau*, **78**, 1954, 335-336.

[550] *Giesserei*, **40**, 1953 (9), 229-230.

[551] *Ber. deut. keram. Ges.*, **34**, 1957, 297-302, especially pp. 300-302.

conclusions of Radczewski are in general confirmed by a more detailed study of the Enslin test for plastic kaolins and clays in fractions below 2 μ in diameter which was performed by R. Neumann.[552] From extensive numerical data this author presents an important graph illustrating the symbatic relations existing between the Atterberg rolling limit, w_a, the plasticity number $Pl = w_f - w_a$, the flow limit, w_f, and the Enslin data (as the maximum water suction capacity), shown in Figs. C. 132 and 133. The latter illustrates the proportionality between the Enslin value and w_f and Pl.

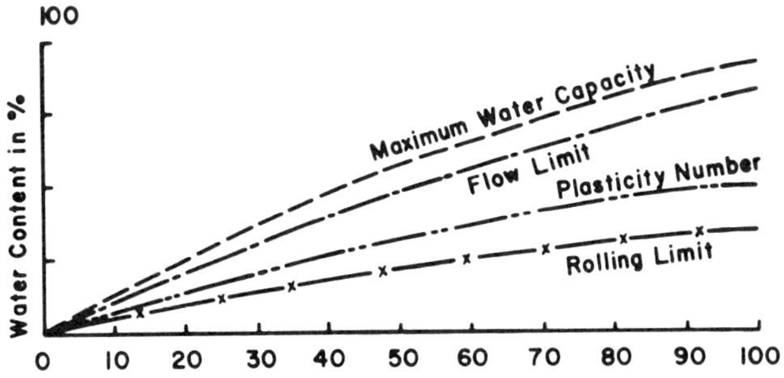

FIG. C. 132. Increase in water contents for the rolling limit (w_a), the flow limit (w_f), the plasticity number ($Pl = w_f - w_a$), and the Enslin value (maximum water capacity $w_{max.}$), with increasing amounts of grain size fraction below 2 μ (Neumann).

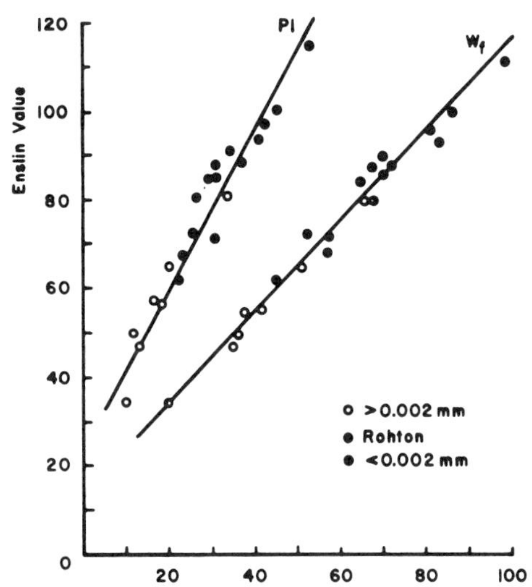

FIG. C. 133. Enslin value (maximum water capacity, $w_{max.}$), as a function of water contents at the flow limit (w_f), and of the plasticity number (Pl), for crude clay, and the fractions above and below 2 μ (Neumann).

[552] *Ber. deut. keram. Ges.*, 34, 1957, 274-281.

Electrostatic Theory of Clay-Water Systems and Base Exchange Phenomena

FUNDAMENTALS; THE DOUBLE-LAYER THEORY

250. There is great probability that the exchange of alkali ions for protons on the surface of layer-structure silicates and of related crystal phases takes place in reality over the hydroxonium, H_3O^+, ions. H. S. Yoder and H. P. Eugster[553] emphasized that especially in micas and also in montmorillonites or in related layer compounds, the exchange of potassium by hydroxonium ions would be common, as was concluded from the interpretation of M. D. Forster's analytical studies on montmorillonites.[554] Recent data from the minerals of the alunite-jarosite group, which have many particularities in common with those of the mica and hydromica family, show the exchange of potassium and ammonium cations for hydroxonium ions as most plausible in the experiments of N. V. Shishkin, E. A. Krogius, and P. A. L'vovich.[555]

251. Theories of the development of electrostatic charges in kaolinite-water systems were re-examined by W. G. Lawrence[556] from a crystallochemical viewpoint, with a calculation of the *bond energies* of crystal sites on fractured discontinuities, especially to $Al^{1.5+}$, Si^{2+}, O^-, and $OH^{0.5-}$ on the halves (Fig. C. 134). The hydration of the cations must be taken into account in such considerations (cf. Table II), as $Na(H_2O)_5^+$ with an effective cation radius of 3.3 Å. according to R. A.

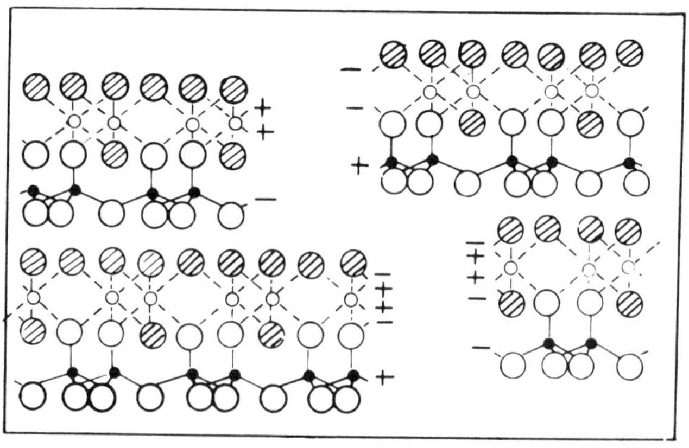

FIG. C. 134. Kaolinite structure fracturated parallel to the *c*-axis (Lawrence).

553 *Geochim. et Cosmochim. Acta*, **8**, 1955, 225-280, especially p. 254.

554 *Am. Mineralogist*, **36**, 1951, 717-730.

555 *Zapiski Vsesoyuz. Mineral. Obshchestva*, **87**, 1958, 682-686.

556 *J. Am. Ceram. Soc.*, **41**, 1958, 136-140; cf. E. Fusarini, *Ceramica* (*Milan*), **13**, 1958 (12), 61-65. On the behavior of water in materials of molecular dimensions see also the fascinating discussion presented by H. Lafuma, "Bulles et Gouttes" Symposium Paris June 1962, Question VI. A, (reprint 4 pp.).

TABLE II

BOND ENERGIES OF VARIOUS CATIONS AND ANIONS WITH CHARGED LATTICE SITES [a]

Cation	Anion	Cation hydration	Hydrated cation radius (Å.)	Bond energy in ergs/bond × 10^{-12}		
				for a —1 Site	for a —½ Site	Corrected for $\varepsilon=3$
Li^+	O^-	7	3.7	4.6	2.3	0.26
Na^+	O^-	5	3.3	4.8	2.4	0.27
K^+	O^-	4	3.1	5.1	2.5	0.28
NH_4^+	O^-	4	3.0	5.3	2.7	0.31
Mg^{2+}	O^-	12	4.4	8.0	4.0	0.45
Ca^{2+}	O^-	10	4.2	8.3	4.2	0.47
Al^{3+}	O^-	6	1.85	21.6	10.8	1.2
H_3O^+	O^-	(Hydrogen bond + O^-—O^- contribution)				22.0
$Al^{1.5+}$	$(OH)^-$			18.0		18.0
Si^{2+}	$(OH)^-$			27.6		27.6
H_2O	O^-	8,000 cal./mole		0.5		0.5
H_2O	H_2O	Ice, 4,500 cal/mole		0.3		0.3

[a] kT for room temperature = 0.04.

Robinson and R. H. Stokes,[557] or for $Li(H_2O)_7^+$ with 3.7 Å., $K(H_2O)_4^+$ with 3.1 Å., $Mg(H_2O)_{12}^{2+}$ with 4.4 Å., $Ca(H_2O)_{10}^{2+}$ with 4.2 Å., $Al(H_2O)_6^{3+}$) with 1.85 Å., and so on. The series of increasing bond energies between cation and negative unsaturated crystal sites is $Li^+ < Na^+ < K^+ < NH_4^+ < Mg^{2+} < Ca^{2+} < Al^{3+} < H_3O^+$, corresponding to cation replacement on kaolinite, and explaining the effects of concentration of electrolytes on replacement and gelation, as A. L. Johnson and F. H. Norton (1941) had previously proposed, although on the basis of a different mechanism. The lowest bond energies of the alkali ions bring about the highest development of charges, and these ions are also most affected by thermal energy changes. Such effects of thermal energy are most simply approximated by an Arrhenius equation for the number of vacant sites, N, as a function of the total number of sites, N_0), and the bond energy, E, by $N = N_0 \cdot \exp(-E/kT)$. For the adsorbed hydrated Na^+ ion, $E/kT = 6.7$ and 0.15 per cent of the sites will be vacant at any given time. For Ca^{2+} ions adsorbed on negative sites, there will be a much lower probability of

[557] "Electrolyte Solutions," Academic Press, New York, 1955, 512 pp. See also the previous tabulation given according to R. E. Grim, by E. Gruner and R. E. Vogel, *Kolloid-Z.*, **118**, 1950, especially on p. 105.

removal; Al^{3+} ions, once adsorbed to satisfy unsaturated anion sites on the crystal lattice, will be removed only with great difficulties.

252. The anomalously high bond energy of the hydroxonium ion, H_3O^+, explains why its bonds will not be broken by thermal energy and no electrostatic charge is developed by it. W. G. Lawrence[558] extended his studies on the connection of ion adsorption on clay particles to plasticity phenomena and rigidity of the water films. By decreasing the charge on the particles as an exchange of monovalent by di- and trivalent cations, plasticity is increased. Additions which would reduce the surface tension of the water must also be avoided, and rigidity of the adsorption water structure is increased (cf. Kingery and Francl, see above). Additions of montmorillonite, especially in its calcium variety, also increase plasticity, not only by the extremely fine grain size of this mineral, but also by the charge effects. Raising temperatures causes a reduction of the rigidity of water films, with lower yield values and a decrease in workability. Also the lowering of the dielectric constant from 80.4 at room temperature, to 55.9 at 100°C. has a plasticity-reducing effect. The consequences of Lawrence's deductions further cover many details in the theory of flocculation and deflocculation in clay-water systems (cf. C. ¶ 354 ff.).

253. The electric double layer on a negatively charged plane surface, as of a kaolinite particle, is calculated by the Gouy-Chapman equation

$$\frac{d^2\psi}{dx^2} = \frac{4\pi}{D} \sum_i z_i\, n_{i0} \cdot e \cdot \exp\,(z_i\, e\, \psi/kT)$$

with ψ = the electrostatic potential at the distance x from the surface of the particle, e the electronic charge, n_{i0} the number of ions of the species i, with valency z_i per unit volume, and D the dielectric constant of the solution. It illustrates how the electrostatic potential gradient changes at any point in the double layer, and the local volume density of the charges, Γ. After reduction to molar concentrations, one may write for the density of charges the equation

$$\Gamma^2 = \frac{4}{\beta} \sum (c_{is} - c_{i0})$$

with c_{is} the concentration on the surface of the particles, c_{i0} that in the ambient solution, and $\beta = 8\, e^2 N/1000 \cdot D \cdot kT$. A. W. Taylor[559] derived from these general equations the concentrations in ions on the surface of clay particles in various electrolyte solutions. Where the surface density of charge is high, the conditions near the surface are independent of the exterior electrolyte concentration. One may apply the

[558] *J. Am. Ceram., Soc.*, **41**, 1958, 147-150.

[559] *J. Am. Ceram. Soc.*, **42**, 1959, 182-184. For an interesting discussion of double layers, see W. A. Weyl, *J. Colloid Sci.*, **6**, 1951, 389-405. (Cf. Vol. II, A. ¶ 392 ff.).

results of the calculation to the concentration of clays in dilute acid solutions, and, from a discussion of effects of neutral salts added, to the acidity of the surface. Taylor's conclusions are also important in explaining the decay of acid-washed clays from hydrogen to aluminum systems. It is practically impossible to inhibit this decay by washing-out the excess acid with distilled water, as it is, on the other hand, impossible to remove aluminum completely by washing with water.

254. The diffuse double-layer is especially important in its extension to the discussion of the *zeta potential,*

$$\zeta = \frac{n \cdot e \cdot \delta}{Dr(r + \delta)}$$

(n is the number of uniformly distributed inner ions and equivalent external ions, r the radius of the firmly fixed ionic hull, δ the distance between the exterior and interior layers of the particle complex assumed to be a spherical condensor) for stability and rheological properties of ceramic clay-water slurries (slips) (cf. C. ¶ 361 ff.). Especially for systems of "nonplastic" oxides and water, P. J. Anderson and P. Murray[560] discussed the characteristic minimum occurring in apparent viscosity (η^*) versus pH curves in terms of the ζ potential of the dispersed oxides. As a matter of fact, the minimum of the apparent viscosity (η^*) corresponds rather exactly to the maximum in ζ, both parameters plotted as functions of pH, such as for thorium or uranium dioxide. This maximum further corresponds to a 50 per cent dissociation of the basic sulfate groups. The effect of the ζ potential on η^* is opposite to that of the common electroviscous effect, which is generally discussed for dilute dispersions of hydrophobic systems. For concentrated slips the electroviscous effect is less important than the effect of a yield stress on apparent viscosity. In clay-water systems with well-established stress-strain diagrams it was found that electrolyte additions, which lead to an increased yield stress, decrease the actual viscosity under flow.

255. For kaolinite suspensions, N. Street[561] measured the ζ potential as a function of additions of a polyanion (introduced as the sodium salt of hydrolyzed polyacrylonitrile). At low concentrations of this anion (about up to 0.3 parts/1000 parts of clay) in H-clay the yield stress is increased, evidently caused by interparticle linkage with the polyanion. A sodium clay, however, shows a reduced yield stress at all concentrations of the same polyanion, due to a neutralization effect of the positive edge charges by adsorption of this anion. The best experimental method for the determination of the ζ potential is the conductometric and potentiometric titration of the kaolinite suspensions, and measurement of the electrophoretic mobility under

[560] *J. Am. Ceram. Soc.*, **42**, 1959, 70-74.

[561] *J. Colloid Sci.*, **12**, 1957, 1-9; *Australian J. Chem.*, **9**, 1956, 450-466 (with A. S. Buchanan); 467-477.

neutralization with the hydroxide solutions of lithium, sodium, potassium, calcium, or barium (method of D. Stigter and K. J. Mysels[562]). The base exchange capacity of kaolinite in neutral solutions thus appears to be combined with an isomorphous replacement effect in the lattice. In alkaline solutions, or after the adsorption of polyvalent anions, the originally positively charged edge faces may become negative and participate in the exchange reaction.

256. Concrete ideas on the configuration of the cations in single interlayers of water molecules in montmorillonite were demonstrated by R. C. Mackenzie[563]: the cations are supposed to occupy flattened spheres of vibration limited by the interlayer distance and distributed in the center of the intersheet space. Complications arise if magnesium ions replace aluminum in the lattice, which in this case is held more loosely than if aluminum ions replace silicon, and an increase in the hydration degree will ensue. The exchangeable cations primarily govern the water sorption. Further consideration of the energy of hydration is in analogy with the discussions of the bond series in W. G. Lawrence's studies (see C. ¶ 251).

257. An important characteristic of clay-water systems is the *sedimentation volume*, determined by elementary methods, as by the method of G. C. Kandilarow (1940). This phenomenon is in connection with the *swelling* properties of different cation-occupied clay minerals, especially of the montmorillonite group, a good example of colloid electrolytes for which G. S. Hartley[564] applied the Debye-Hückel theory to explain the greater swelling of sodium montmorillonite as compared with calcium montmorillonite, namely, by a stronger dissociation of the sodium complex by which a greater number of structural units are left with a negative charge.[565] Variable flocculation of bentonite suspensions with a maximum sedimentation volume of the deposits is an important technological problem in making sands impenetrable for water. M. S. Fedotova and K. A. Pospelova[566] studied this question in highly stable 3 to 5 per cent solid material containing suspensions of a sodium bentonite which was flocculated by a 0.0003 to 0.05 per cent solution of aluminum chloride. The rate of filtration through the voluminous sand-bentonite deposit is of the elementary form $\log v = a + \gamma \cdot \log t$ (γ measured by the tangent function of the slope angles of the linear sedimentation curves, v the volume of the filtrate). The times of flocculation can be controlled by an addition of sodium carbonate which more or less compensates for the flocculating effect of the aluminum chloride. The flocculation of strongly swelling montmorillonite with a high sedimentation volume was also

[562] *J. Phys. Chem.*, **59**, 1955, 45-51, for the micellae of sodium dodecyl sulfate.

[563] *Clay Minerals Bull.*, **1**, 1950, 115-119.

[564] *Trans. Faraday Soc.*, **31**, 1935, 31-50.

[565] Cf. Margaret D. Foster, Am. Mineralogist **38**, 1953, p. 1005.

[566] *Doklady Akad. Nauk S.S.S.R.*, **94**, 1954, 1129-1132.

studied by U. Hofmann,[567] and especially a strikingly sharp boundary of the sediment to the supernatant solution observed. In this sedimentation volume is formed what Hofmann calls a framework of a "house of cards" structure (see Fig. C. 135) which can be isolated by ice-sublimation drying in vacuo (cf. C. ¶ 333).

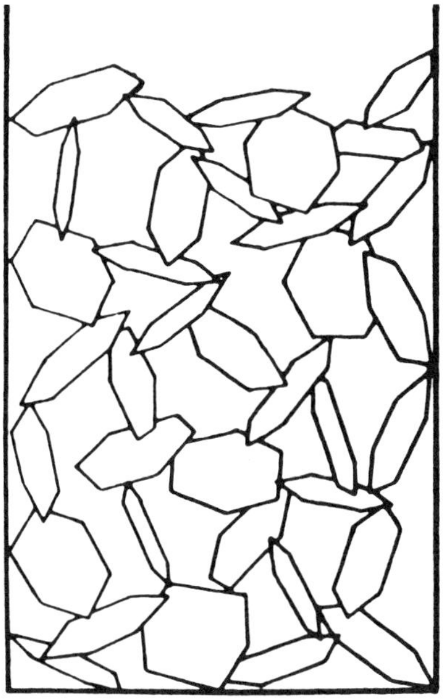

FIG. C. 135. "House of cards" structure of a montmorillonite with a high sedimentation volume (Hofmann).

258. A particular investigation on the sedimentation volume of different clays was made by K. Pate, W. Noble, and F. H. Clews[568] from the practical ceramic-technological viewpoint, in connection with working moisture content in the "leather-hard" state, or the drying shrinkage, "green" curvature, the fineness, and the settling tendency in slips. A very characteristic change in the rates of settling (observed by the shifting of the sharp demarcation line with time) was confirmed according to older data (given by H. Coe and G. H. Clevenger in 1916) in concentrated clay suspensions which settle for a definite period with a uniform rate after which the

[567] *Angew. Chem.*, **68**, 1956, 53-61, especially pp. 59-60. Sedimentation volume in connection with grain size distribution, rheological properties of kaolin suspensions was studied recently by Kl. Frühauf, W. Burck, and U. Hofmann, *Kolloid-Z.* **180**, 1962, 150-160 (particularly pp. 152 ff.), as related to the activities of exchangeable cations. The activities of adsorbed cations Ca > Ba > La correspond to a Hofmeister series. Such slurries are of great importance for their practical use in bore hole muds and "synthetic" foundry sands (cf. U. Hofmann and W. Burck, *Angew. Chem.*, **73**, 1961, 342, and C. ¶ 332).

[568] *Trans. Brit. Ceram. Soc.*, **55**, 1956, 140-156.

settlement proceeds discontinuously and more slowly. The typical shape of the settling curves (Fig. C. 136) with sharp break points noted thereon, is another criterion for the characterization of ceramic clays. There is an approximate proportionality between the sedimentation volumes in the break points and the final volume.

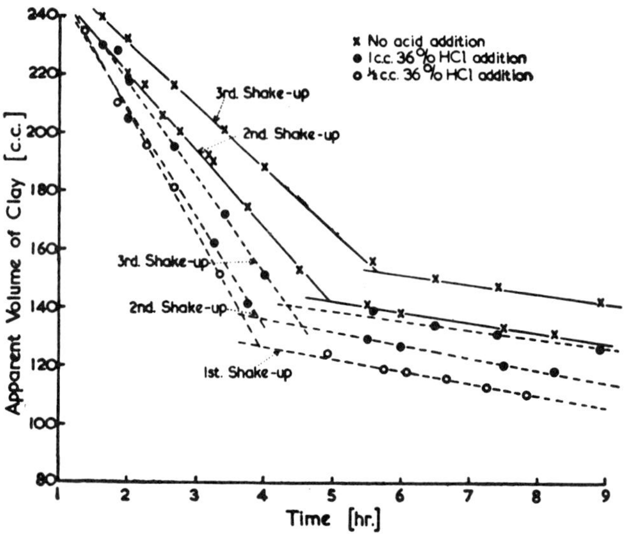

FIG. C. 136. Rates of settling of a given clay, with characteristic break points (Pate, Noble, and Clews).

259. The sedimentation volume of sericite as a "clay mineral" was determined by B. Tamamushi and T. Shirai,[569] from deposits in Japan, with its ζ potential and mobility which is lower than that of montmorillonite. The sedimentation volume in different 1 *M* electrolyte solutions is greater by the ratio of 2 : 1 than those for kaolinite, for all the solutions examined, although the base exchange properties (the Hofmeister cation exchange series) of both minerals are the same. The sedimentation volume of sericite in water is intermediate between that of kaolins and bentonites, but is about double that of bentonite in carbon tetrachloride suspension. In the same medium kaolinite shows a sedimentation volume intermediate between that of bentonite and sericite.

260. The sedimentation volume is also correlated in an interesting way with the heats of wetting and the emulsification of inorganic solids. L. N. Mukerjee and S. N. Srivastava[570] studied this question for different combinations of solids (oxides, hydroxides, sulfates, phosphates) with liquid media (water and kerosene). Bentonite

[569] *Sci. Papers Coll. Gen. Educ. Univ. Tokyo,* **4**, 1954 (1), 35-46.
[570] *Kolloid-Z.* **162**, 1959, 10-14.

is an exception to a rather general rule that the higher the wettability, as measured in higher heats of wetting, the lower the sedimentation volumes. Generally, the amount of sedimentation was lower in water than in kerosene suspension, because the affinity of all the investigated solids is greater for water than for kerosene. The emulsification is also enhanced by low interfacial (surface) tensions. From a similar more practical viewpoint, H. Kuno[571] studied the sedimentation behavior of crushed glass powders (of 12 to 30 μ particle size) in polar and nonpolar liquid media. The sedimentation volume for both changes almost linearly with the density of the liquid, whereas the viscosity does not affect it. Very high viscosities apparently increase the sedimentation volumes of solids.

261. The rate of settling as a function of the polarity of the liquid media was observed by R. Wolff.[572] He found it smaller in polar than in nonpolar liquids. The volume versus time graphs show approximately hyperbolic curves, the eccentricities of which are higher for polar liquids. The time in which half of the sedimentation height is attained is shorter in nonpolar media; increasing temperature also decreases the time. The sedimentation tendency of Portland cement suspensions, with additions of trass, limestone powder, and others, is an important technological factor which was discussed by A. Steopoe.[573] He dealt with the multiple effects of additions, such as plastifiers, surfactants, and electrolytes (cf. Vol. V, Section C). Whereas normal Portland cement has a cellular skeleton structure of the sediments, the addition of agents with electrostatic charges opposite to that of the cement particles will affect the type of this texture. Hydraulic additions and those of basic granulated blast furnace slag will widen the structure. Quartz sand or limestone powder will shrink it, as much as do acidic and crystalline slags.

THE ION CONCENTRATION IN CLAY-WATER SYSTEMS; TOTAL ADSORPTION CAPACITY

262. Because of the paramount importance of cation exchange reactions for clay-water systems, I. B. Cutler and M. A. Cook[574] developed a thorough theory of these phenomena in aqueous media, based specifically on the previous fundamental deductions of C. E. Marshall.[575] We discussed in C. ¶ 211-¶ 215 the coprecipitated hydrogels of alumina and silica in relation to their properties of more or less weak acidity; however, the corresponding conditions in clay minerals with adsorbed hydrogen ions are not entirely explained by this concept. T. H. Milliken, G. A. Mills, and A.

[571] *Kolloid-Z.*, **166**, 1959, 54-55.

[572] *Kolloid-Z.*, **150**, 1957, 71-80.

[573] *Zement-Kalk-Gips*, 12, 1959, 67-69; cf. *Rev. matériaux construct. et trav. publ.*, **508**, 1958, 1-9, especially on electrostatic charges of cement particles relative to additions.

[574] *J. Am. Ceram. Soc.*, **36**, 1953, 165-170.

[575] In his book "Colloid Chemistry of the Silicate Minerals," Academic Press, New York, 1949, 195 pp.

G. Oblad[576] developed a theory of shift-of-coordinations for cracking catalysts, which should be compared with the weak acidity concept, namely ideas of broken-bonds, as the basis of adsorption site exchange reactions. For a cation and anion exchange, M. A. Cook, I. B. Cutler, G. R. Hill, M. E. Wadsworth, and A. G. Oblad[577] previously discussed the general case of a reaction of the type $S_6 + OH^- \rightleftharpoons S_4^- + n\,H_2O$, with n the number of water molecules split out by the coordination shift $S_6 \rightarrow S_4$, say, for a six-coordinated neutral aluminum atom, and a four-coordinated Al^- site, activated for the adsorption of a cation by preliminary OH^- adsorption. If the concentration of water and its activity, $A_{H_2O}n$ is constant, the mass action equilibium constant is

$$K_1 = \frac{A_{S_4^-}}{A_{S_6} \cdot A_{OH^-}}$$

Because of the dissociation of the adsorption complex with the M^+ described by $M^+S_4^- \rightleftharpoons S_4^- + M^+$, free S_4^- sites exist, particularly at high pH, in the negatively charged clay particles of a suspension, and an equilibrium constant

$$K_3 = \frac{A_{S_4^-} \cdot A_{M^+}}{A_{M^+S_4^-}}$$

will be valid. With activity coefficients and concentrations the constants will be written

$$K_1 = \frac{\gamma_1[S_4^-]}{\gamma_2[S_6] \cdot A_{OH^-}}$$

and

$$K_3 = \frac{\gamma_1[S_4^-] \cdot A_{M^+}}{\gamma_3[M^+ S_4^-]}$$

If S_0 is the total number of potential and active sites we have

$$[S_0] = [M^+ S_4^-] + [S_6] + [S_4^-]$$

for the site balance. We derive from the combined equations

$$K_1' = \frac{K_1 \cdot \gamma_2}{\gamma_1} \quad \text{and} \quad K_3' = \frac{K_3 \cdot \gamma_3}{\gamma_1}$$

and

$$K_3'(A_{OH^-} + 1/K_1') = A_{M^+} A_{OH^-} \left(\frac{1 - \Theta}{\Theta}\right)$$

with $\Theta = [M^+ S_4^-]/[S_0]$. In a perfectly analogous way, we may derive for the weak

[576] *Discussions Faraday Soc.*, **8**, 1950, 279-290.
[577] *J. Phys. Chem.*, **57**, 1953, 1-6.

acid dissociation $H^+S_4^- \rightleftharpoons H^+ + S_4^-$;

$$K_2 = \frac{A_{H^+}\gamma_1[S_4^-]}{\gamma_4[H^+S_4^-]}$$

and finally

$$K_3'\left(A_{OH^-} + \frac{K_w}{K_3'}\right) = A_{M^+}A_{HO^-}\left(\frac{1-\Theta}{\Theta}\right)$$

with $K_w = A_{OH^-}A_{H^+}$ and $K_2' = K_2\gamma_4/\gamma_1$. Combined both ways we obtain

$$K_3'(A_{OH^-} + K_0) = A_{M^+}A_{OH^-}\left(\frac{1-\Theta}{\Theta}\right)$$

with K_0 either equal to $1/K_1'$, or to K_w/K_2', for both mechanisms equally.

263. The experimental way to examine applicability either of the weak acid or of the shift-of-coordination concept is titration of the clay suspensions with sodium or potassium hydroxide, or with ammonium hydroxide solution. Cutler and

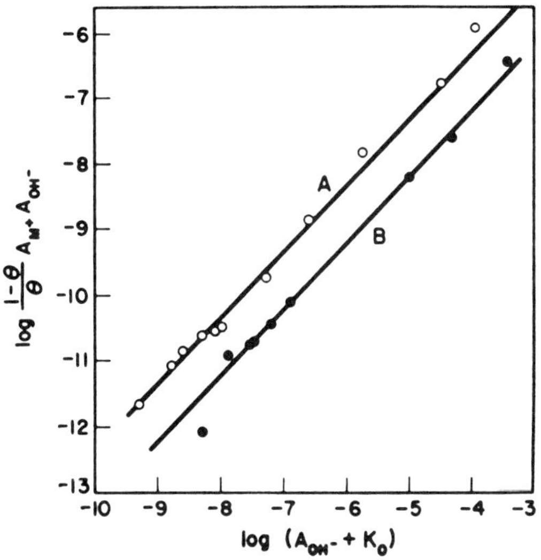

FIG. C. 137. Typical titration curve of montmorillonite with sodium hydroxide, expressed in coordinates of the equilibrium equation for ion exchange (Cutler and Cook). Curve A is for a concentration of 2.8 per cent; curve B for a 1.0 per cent concentration in montmorillonite.

Cook determined from Marshall's previous data the linear titration characteristics of kaolinite, montmorillonite, beidellite, and illite, of which we give here in Fig. C. 137 only one example, with controlled pH, concentration of the clay mineral, and activity of the cation common to the titrating base. They used a clay membrane electrode. A_{H^+} and A_{OH^-} are obtained from pH, A_{M^+} from the membrane electrode measurements. $[M^+S^-]$ is the difference of $[M^+]$ derived from the titration with the aid of activity coefficients,[578] and the original amount of the base added. For S_0 which is not a sharply defined constant, one may approximate $[M^+S_4^-]$ for high pH, by trial and error. K_3' can be derived from graphs as the slope of the straight line; K_0 from the intercept of the line in Fig. C. 137. It is, however, difficult to distinguish between K_w/K_2' and $1/K_1'$ or, in other words to decide between the weak acid and the shift-of-coordination concepts. From the equilibrium constants the changes in free energy, ΔF, are calculated in the well known way from $\Delta F = -RT \ln K$. They are tabulated in Table III for the fundamental reactions. The ΔF values are low, since the alkali cations are rather loosely held. K_3' does not vary with salt con-

TABLE III

APPROXIMATE CHANGES IN FREE ENERGY FOR CATION EXCHANGE REACTIONS

Reaction	Equilibrium Constant	Average ΔF in cal./mole
$S_6 + OH^- \rightleftharpoons S_4$	K_1'	$\cong -13{,}000$
$H^+S_4^- \rightleftharpoons S_4^- + H^+$	K_2'	$\cong -7{,}500$
$M^+S_4^- \rightleftharpoons S_4^- + M^+$	K_3'	$= -4{,}700$ [a]

[a] Approximately the same for Na^+, K^+, and NH_4^+ cations.

centration, but with that in the clay. Its suspension is a colloid electrolyte of higher effective ionic strength than monovalent salt solutions up to a 0.01 *M* concentration. If the colloid electrolyte has the (very small) molecular concentration, *m*, the chemical potential

$$\mu = \frac{1}{2} \sum_i m_i Z_i^2$$

can be high for a high valence *Z*. For $m = 10^{-3}$ to 10^{-6} one may calculate $Z = 10$ to 100, and 10^3 to 10^4 sites per particle would exist of which one tenth or one hundredth are dissociated, leaving an effective charge of the particle of 10 to 100. The ionic strength of the clay colloid electrolyte is also not constant over the entire range of the titration curves.

264. Besides the experimental data collected by C. E. Marshall, as mentioned above, there is a rich literature on *titration methods* and curves for clay suspensions, especially for ceramic clay raw materials and slips. We mention here the older investigations

[578] Cf. H. S. Harned and B. B. Owen, "Physical Chemistry of Electrolyte Solutions," 2nd ed., Reinhold, New York, 1950, 675 pp.

of E. C. Henry and N. W. Taylor[579] in which the neutralization point of the "clay acid" by sodium hydroxide coincides with the minimum content of this base, which is sufficient to produce the minimum of viscosity in the slips. For montmorillonite, J. D'Ans and E. Ulrich[580] titrated the hydrogen-saturated form, with a continuous control of the pH in a glass electrode (upper limit pH = 9). The titration is very slow in the neutral to alkaline range, with end values reached at times in several days. D'Ans and Ulrich speak of montmorillonite, because of its pronounced two-step reaction with neutral salts (KCl), as a dibasic acid, the first H^+ ions of which are highly acidic, the second ions belonging to a weak acid type. A conductometric

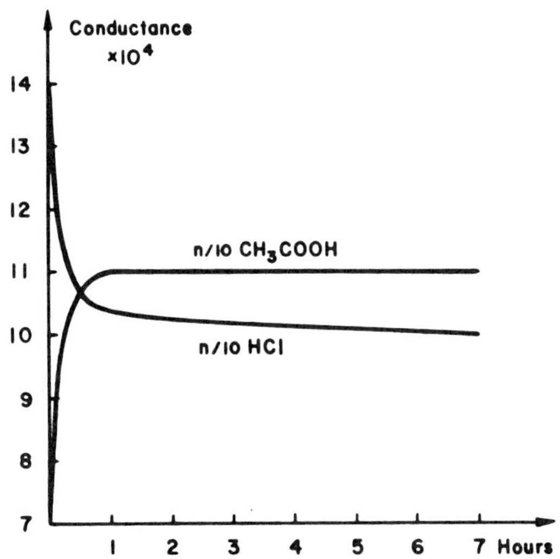

FIG. C. 138. Change of conductance of bentonite suspensions in hydrochloric and acetic acid solutions, as a function of time (Fischbeck and Messmer).

titration method was used for similar purposes by K. Fischbeck and E. Messmer,[581] who emphasized the reversibility of the neutralization reactions in the wide pH range of 2 to 13. Especially the time characteristics of the establishment of equilibria is better observed by conductometry than by common indicators, with typical curves demonstrating the differences in the reactions with dilute hydrochloric and acetic acid (Fig. C. 138) and with phosphoric acid (Fig. C. 139). The effects here ob-

[579] *J. Am. Ceram. Soc.*, **21**, 1938, 165-175, a supplement in **22**, 1939, 207-210. Titration of hydrogen-saturated montmorillonite with NaOH solution informs on edge and interlayer ion exchanges, as indicated by specific pH conditions of neutralization; cf. E. C. Jonas, *Science*, **140**, 1963, 75-76 (with important literature references).

[580] *Naturwissenschaften*, **36**, 1949, 344-345.

[581] *Z. anorg. u. allgem. Chem.*, **274**, 1953, 1-10.

served are explained by differences in the dissociation type of the acids, and of the solubilities of the salts formed by the reaction. The break points in the titration curves are not always sharp, especially so if the clays are not homogeneous in their composition. The conductometric determinations, however, are valuable in the (first) break points of the titration curves for a rapid and relatively accurate measurement of the S-value of the ionic base exchange, in good agreement with the data

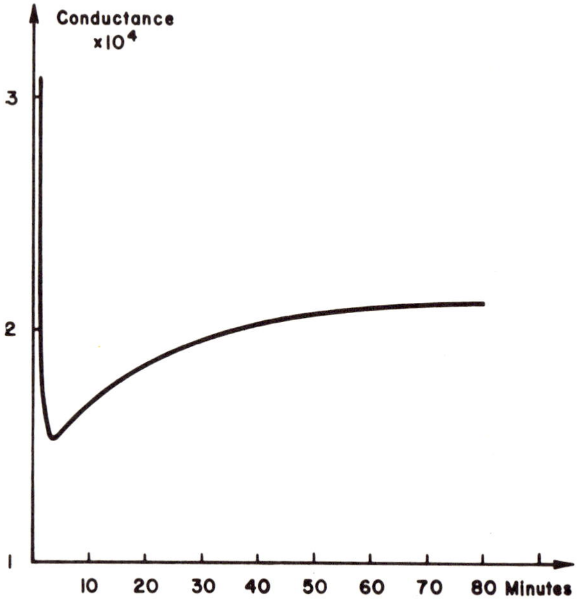

FIG. C. 139. Change of conductance of bentonite suspension in phosphoric acid solution, as a function of time (Fischbeck and Messmer).

of the standard ammonium acetate method (cf. C. ¶ 275). Another investigation of acid-base titration curves of clay suspensions by A. Dietzel and O. Schmitt[582] emphasizes time delays in the establishment of final values, with slow changes to the final pH. These authors describe, besides the smooth titration curves, as with kaolinite as a monobasic acid according to Marshall, also other curves with a "hump" at pH = 4 to 7, characteristic of the dibasic acid as Fischbeck and Messmer concluded previously (Fig. C. 140). Other clays may show such a hump in the more acid range near 2 to 3, as in montmorillonite-illite clays. It is, however, very difficult to draw conclusions about the mineral composition of given clays from the type of titration curves alone. Only this is certain, that clays with smooth-type curves, with one inflection point, may contain kaolinite in which exclusively the hydroxyl groups on

[582] *Ber. deut. keram. Ges.*, 31, 1954, 77-79.

its surface react. One may only conclude from titration and pH values as to whether a given clay contains exclusively kaolinite or (also) montmorillonite- or illite-type minerals. A clear decision is possible only by X-ray diagram data.

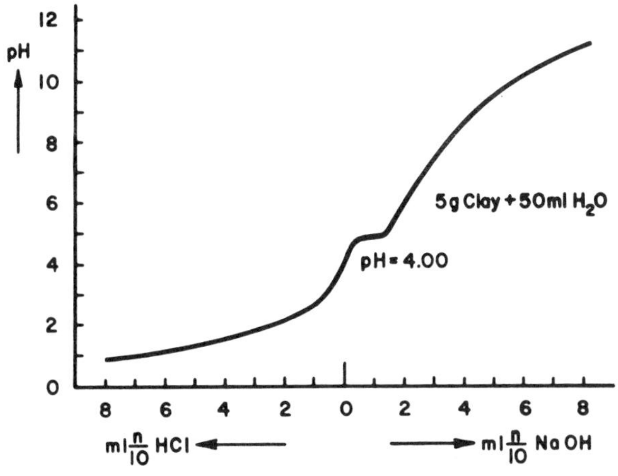

FIG. C. 140. Titration curve of a clay with a "hump" near pH = 5 (Dietzel and Schmitt).

265. For acidic montmorillonite suspensions, B. K. Warkentin[583] also used the conductometric titration method. The neutralization branch of the curve is linear for a monovalent base, curved with an increasing negative slope if a divalent base is used in titration. The results agree completely with what the double-layer theory of Freundlich-Gouy requires. The determination of base exchange capacity yields the same data if mono- or divalent bases were used. R. P. Mitra and H. Singh[584] emphasize the pronounced buffer action of a hydrogen-montmorillonite in the titration curve observed at pH = 5.5 to 6.5. If the H-montmorillonite is treated with a $1N$ hydrochloric acid solution in washing the clay, a material of stronger acidic character, with pH = 3, is observed. Aging, however, gradually restores the normal behavior. The aging effect is only observed in a medium of strong dissociating power, say in water, whereas in methyl alcohol no aging is observed, and the strong-acid state is indefinitely persistent, even after curing for 30 days in methyl alcohol. The aging effect immediately reappears if the clay is transferred into water (see Fig. C. 141). These facts are not compatible with a previous theory of W. H. Slabaugh[585] that the effect of aging is the result of highly reactive hydrogen ions grad-

[583] *Kolloid-Z.*, **153**, 1957, 44-46.

[584] *Naturwissenschaften*, **46**, 1959, 319-320. On the entire problems of the acidity of alumino-silicates, see the extensive report of Z. V. Gryaznova and G. M. Panchenkov, *Uspekhi Khim.*, **28**, 1959, 1523-1543, in connection with their catalytic activities.

[585] *J. Am. Chem. Soc.*, **74**, 1952, 4462-4466. Interesting is the calorimeter used by W. H. Sla-

ually diffusing into the laminar structure of the phyllosilicate. The real seat of the reaction responsible for the aging effect is the liquid medium. The strong acid state is caused by exchangeable H^+ ions on the planar surfaces, not on the edges of the crystallites where Al^{3+} ions are free which only bring about the intermediate acidity or pH = 5. In water, the H^+ and Al^{3+} ions on both kinds of sites dissociate giving electric double-layers with these counter ions. By Brownian motion the two kinds of double layers overlap and make the diffusion exchange of the sites possible, with the end effect of the buffered pH = 4 to 6 state (aged or acid-washed).

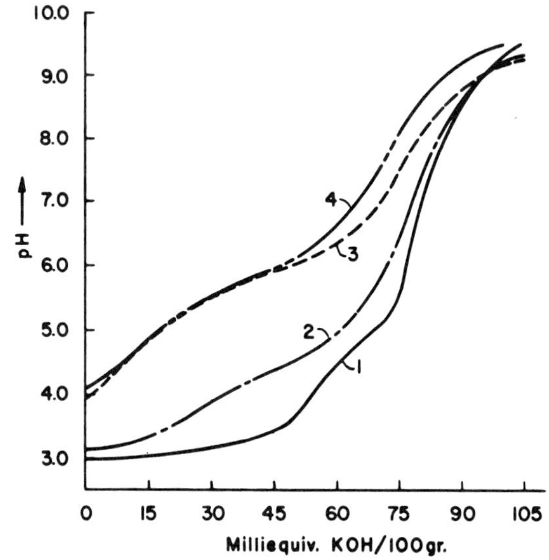

FIG. C. 141. Titration curve of Wyoming bentonite (abscissa meq. KOH) (Mitra and Singh): (1) for bentonite washed with 1 *N* hydrochloric acid solution, and kept in methyl alcohol for 30 days; (2) the same, kept in water for 60 days; (3) the same, kept in water for 90 days; (4) washed in 0.02 *N* hydrochloric acid solution.

266. A remarkable counterpart to montmorillonite is chrysotile which, in its conductometric titration curves, was discussed by F. L. Pundsack and G. Reimschussel.[586] Different chrysotile samples titrated with a 0.515 *N* hydrochloric acid solution at 100°C. exhibit several inflections. The first increments of the acid cause a sharp drop in pH from 10 to about 6.8. It is logical to expect that the hydroxyl groups on the surface of the fibrils are the most accessible to acid attack, reacting to about 4.5 per cent of total fiber basicity, whereas the hydroxyl groups in the structural units $Mg_6(OH)_8[Si_4O_{10}]$ comprise two thirds of the total basicity. This implies that 7

baugh which is a modification of that developed by W. P. White in his classical thermochemical investigations (1916).

[586] *J. Phys. Chem.*, **60**, 1955, 1218-1222.

per cent of the total hydroxyl groups in the structure occur on the surface of the fibrils. A second break on the titration curves corresponds to 67 per cent of complete reaction and complete acid recation with all of the remaining structural hydroxyl groups. This second inflection point occurs at a position corresponding to eight of the total twelve equivalents of base in chrysotile.

267. The investigation of W. H. Slabaugh mentioned above in connection with the aging effects is a good example of the applicability of calorimetric-thermochemical methods in the study of the neutralization of H-bentonites. By a kind of stepwise titration the heat of neutralization was determined, with the result of aging effects also in this method, as well for electrodialyzed bentonite as for a bentonite thoroughly exchanged with a hydrogen resin (Amberlite IR 120 column) (Fig. C. 142). The

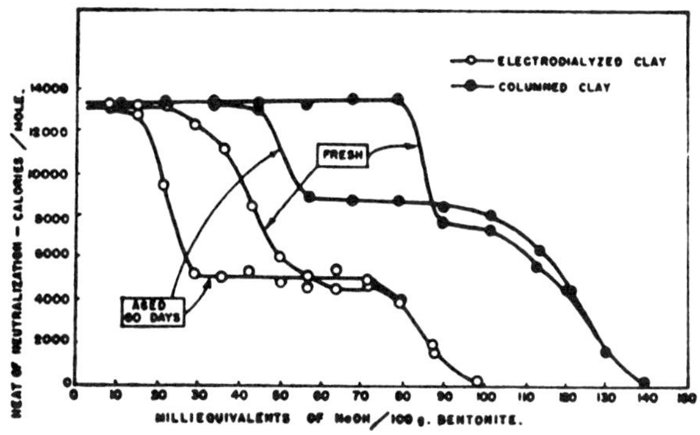

FIG. C. 142. Heat of neutralization and aging effects of hydrogen bentonites, both electrodialyzed, and hydrogen-exchanged with Amberlite IR 120 column (Slabaugh).

so-called thermometric titration method of Y. Trambouze[587] may also be mentioned here, although it was discussed in C. ¶ 212 in connection with the acidity problems of alumina-silica hydrogels (as cracking catalysts). Trambouze used, in his experimental series, aluminum chloride as a Lewis acid, dioxane or ethyl acetate as a Lewis base. The "neutralization" is observed by a distinct break on the thermometer (or, rather, thermocouple)-controlled temperature versus dioxane concentration curves.

268. The fascinating idea of G. Berger (1941) to use the reaction of diazomethane (in absolute ether solution) with hydroxyl groups on clay minerals for a definite characterization of their acidity properties has more recently been refined into

[587] *Compt. rend.*, **233**, 1953, 648-649; *J. chim. phys.*, **51**, 1954, 723-728 (with L. de Mourgues and M. Perrin) and related publications of K. G. Miesserov, *Doklady Akad. Nauk S.S.S.R.*, **84**, 1952, 1009-1011; *Uspekhi Khim.*, **22**, 1953, 279-290; K. V. Topchieva, *Uchenye Zapiski, Moskov. Gosudarst. Univ. im. M. V. Lomonosova*, No. 174, 1955, 75-100; with I. F. Moskovskaya, *Doklady Akad. Nauk S.S.S.R.*, **101**, 1955, 517-520; *Ibid.*, **123**, 1958, 891-894.

a more critical approach. J. L. Martin Vivaldi and St. B. Hendricks[588] studied this reaction with different well-defined clay minerals and found that about half of the exchangeable hydrogen ions in kaolinite and illite are rapidly esterified to methoxyl groups, —OCH_3, but the other half of them react only sluggishly. These are evidently less accessible ions although they are very active in producing hydrocarbon polymers. In H-montmorillonite dried at 300°C. only 10 per cent of the exchangeable hydrogen ions react rapidly, indicating the exclusive methoxylation of external surfaces. A continued exposure to diazomethane causes another 10 per cent of the interlaminar

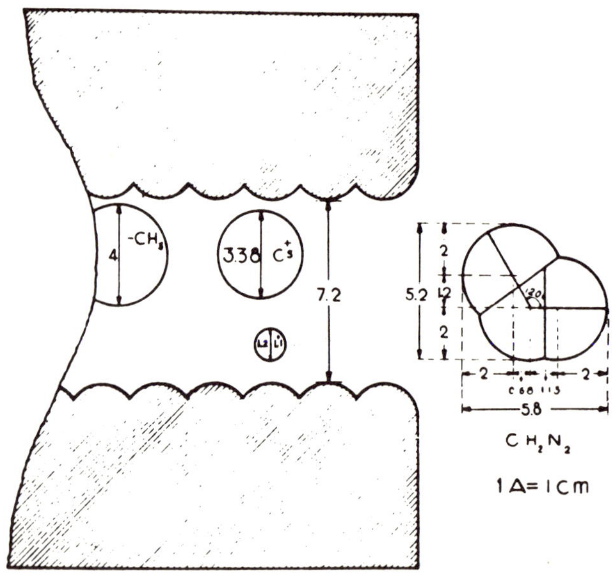

FIG. C. 143. Schematic representation of steric conditions of the methoxylation of the inner surface of kaolinite crystallites, by reaction with diazomethane (Martin Vivaldi and Hendricks).

exchange ions to be methoxylated. In sodium and calcium montmorillonite prepared at pH = 7, the diazomethane reaction shows that 30 meq./100 gm. hydrogen ions are capable of reacting with a separation of the structural layers and without any polymer formation. The steric conditions of the introduction of the methyl groups into kaolinite are shown in Fig. C. 143. R. Schwarz and H. W. Hennicke[589] used

[588] *Anales edafol. y fisiol. vegetal (Madrid)*, **11**, 1952, 601-629. G. Berger, *Compt. rend. de la conférence de pédologie méditerranéenne, Alger-Montpellier*, 1947, pp. 119-122) used the methoxylation reaction by diazomethane for the identification of montmorillonite in soils, with an output of 94.0 mq./100 mg, in comparison with standard base exchange capacity determined by direct methods to be 100 meq./100 gm. for the same montmorillonite. Berger assumes hydroxyl groups of weak-acidic properties in montmorillonite.

[589] *Z. anorg. u. allgem. Chem.*, **283**, 1956, 347-350.

the diazomethane method for the determination of the constitution of crystalline disilicic acid, $(H_2Si_2O_5)_x$ (C. ¶ 159). In this compound only a few of the total ≡Si—OH groups are accessible to the methoxylation reaction (1.7 to 4.9 per cent were determined). Especially after heating at 800°C. there is no reaction at all with thionyl chloride, which had been recommended by H. Deuel and G. Huber[590] for the chlorination and formation of direct ≡Si—Cl bonds in the layer structures, although Deuel observed it for montmorillonite and bentonite. The contradictions between the indications given by Schwarz and Deuel and their collaborators is apparently irreconcilable. Also G. Brown, R. Greene-Kelly, and K. Norrish[591] refute Deuel's conclusions as incompatible with their experimental results. On the other hand, we may see further reactions with organic derivatives of montmorillonite in C. ¶ 304 ff.

269. The base exchange of hydrogen-laden silica (gel or quartz), mica, kaolinite, and montmorillonite in fractions smaller than 2 μ was determined by A. K. Ganguly[592] with the potassium chloride absorption and titration method of the freed H^+ ions. In general, the exchange spots on the surface of the layer structure minerals are not of uniform character. There are definite levels of energy of interaction, and the bonding energies are ruled by the nature and location of the atoms inside the lattice. Ganguly found a decreasing order of the acid function for silicon, aluminum, and magnesium in their bonding to the H^+ ions in the hydroxyl groups. A. K. Ganguly and S. K. Mukherjee[593] further studied the addition of colloid salts to hydrogen clays on the basis of bentonite, kaolinite, illite, and mica, to develop heteroionic systems holding two cations, with the ratio of the ions $M_1 : M_2$ in the range from 9 : 1 to 7 : 3,5 : 5,3 : 7,1 : 9. Such binary heteroionic clay minerals hold the first cations more firmly bound. The differentiation between firmly and loosely bound ions becomes particularly evident when monovalent cations are present in amounts higher than 30 per cent, the divalent cations in 50 per cent of exchangeable bases. Symmetry values for Na^+, K^+, NH_4^+, and for Mg^{2+}, Ca^{2+}, Ba^{2+} colloid salts were determined on exchange with the chlorides of these metals. The symmetry values increase with reducing pH for all cations. The lyotropic effect persists to pH = 3.0 when the values are corrected for H^+ ions. Over the range from 0.4 to 4 times the symmetry concentrations, the exchange isotherms show characteristic inflections. The inherent equality of the bond energies of the cations makes no mathematical formulation for the exchange process fully valid. In the kaolinite structure, Ganguly and Mukherjee distinguish two exchangeable positions, in bentonite and micas, three of these.

[590] *Helv. Chim. Acta*, **34**, 1951, 1687-1701.

[591] *Clay Minerals Bull.*, **1**, 1952, 214-222.

[592] *J. Phys. Chem.*, **55**, 1951, 1417-1428.

[593] *J. Phys. Chem.*, **55**, 1951, 1429-1446.

270. Especially for montmorillonite, J. W. Earley, B. B. Osthaus, and U. H. Milne[594] investigated the exchange properties of different occurrences in the United States and in Canada, with their contaminations by cristobalite, hydrous mica, and even lepidocrocite, their electron microscopy, and numerical exchange capacities, varying in the limits between 66 and 128 meq./100 gm., as a function of chemical compositions with widely varying isomorphous replacements. There is a very distinct correlation between the crystallographic constants and the substitution of aluminum by magnesium and iron in the octahedral layers, and of silicon by aluminum in the tetrahedral sheets. In some of the montmorillonites the base exchange capacities are rather constant; in others there is a marked increase for the fractions below 0.3 μ, due to the impurities in the coarser fractions. Structural factors which limit the maximum size of the crystals are equally characteristic of size development and base exchange capacity. Extensive studies of A. Weiss[595] also concern the montmorillonites and vermiculites for which he finds an interesting correlation of both factors mentioned before and innercrystalline swelling, whereas in micas the thickness of the crystals also influences the quantitative base exchange.

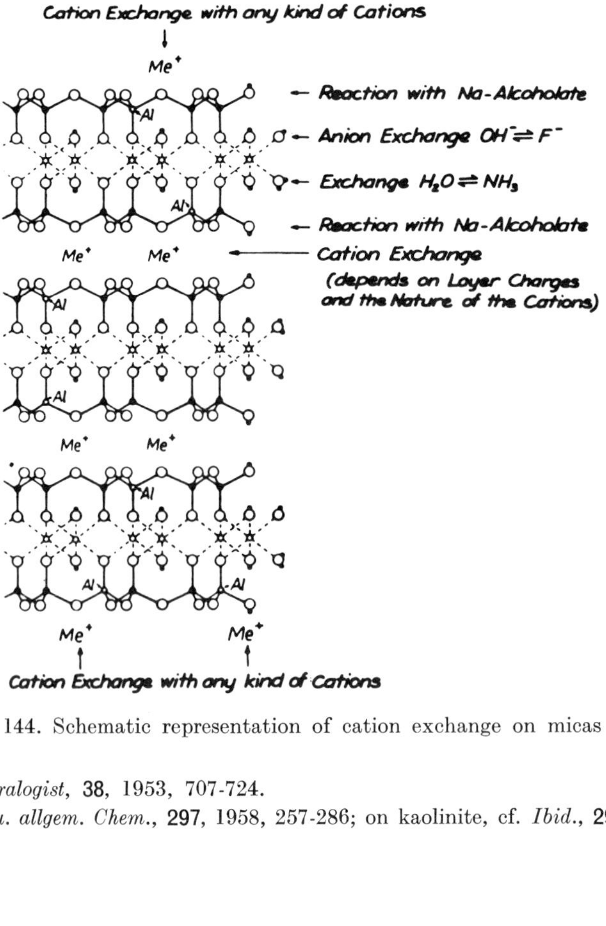

FIG. C. 144. Schematic representation of cation exchange on micas (Weiss).

[594] *Am. Mineralogist*, **38**, 1953, 707-724.

[595] *Z. anorg. u. allgem. Chem.*, **297**, 1958, 257-286; on kaolinite, cf. *Ibid.*, **299**, 1959, 92-120.

271. The inner basic surfaces are made accessible for cation exchange reactions by an exchange of potassium ions by larger alkyl ammonium cations in swelling. The crystal edges are the preferred positions for hydroxyl exchangeable for fluorine anions, or for water molecules exchangeable by ammonia, and amines, forming ammino-compounds. Probably, the ≡Si—OH groups are able to react with alkali alcoholates (fr. Figs. C. 144 and 145). If these groups are blocked by a modification

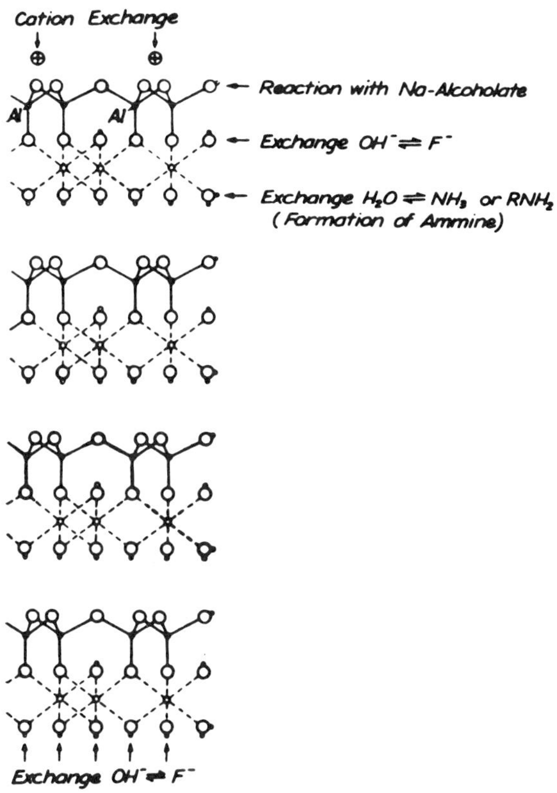

FIG. C. 145. Schematic representation of cation exchange on kaolinite (Weiss).

into groups ≡Si—O—SiR_3, nevertheless the quantitative cation exchange capacity is *not* affected, nor is the exchange of hydroxyl by flyorine anions. Only for the micas (as very thin films) in which the potassium ions are replaced by other cations on the basis face, an $OH^- \rightleftharpoons F^-$ replacement was not observed. For muscovite, biotite, paragonite, and lepidolite, the introduction of lithium, sodium, magnesium, and calcium ions causes a swelling of the products in water, ethylene glycol, or glycerin. In general, the cation exchange capacity of freshly ground micas is only poorly reproducible, but such powders show some hydroxyl exchange by F^- in a neutral solution of ammonium fluoride. The siloxan derivates of micas prepared by a reaction with trialkyl chlorosilane show an unchanged cation exchange capacity, and the swelling is the same as before, although the suspensions show a different thixotropy.

272. Illite behaves in general like the micas (cf. Table IV). For the introduction of potassium into vermiculites their change to illites creates particular conditions because of the *fixation* of this large cation.[596] The products are no more able to swell (cf. C. ¶ 298). For kaolinite the base exchange is probably restricted to the exterior basis face without hydroxyl groups, but on the crystal edges and on the basis face with hydroxyl groups the exchange of hydroxyl by fluorine anions is possible. The amount of exchangeable cations depends on the thickness of the crystals; the amount of exchangeable hydroxyl anions depends on the thickness and on the diameter of the crystals. If pH is below 10, the ≡Si—OH groups do not participate in the cation exchange. The exchange between cations with different valencies is *not* equivalent,

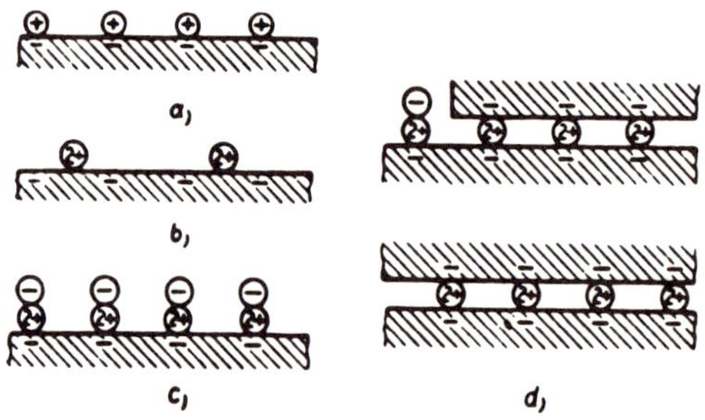

FIG. C. 146. Schematic representation of the surface of a kaolinite crystal: (*a*) with exchangeable monovalent cations; (*b*) after equivalent exchange by divalent cations; (*c*) after compensation of electrostatic charges by simultaneous binding of anions; (*d*) compensating anions, *X*, are replaced by negative charges on a second kaolinite particle (Weiss).

but approximately *equimolecular*,[597] with a compensation to electroneutrality by simultaneous binding of anions on the surface, or by a compensation on the surface of a neighboring kaolinite particle, as shown schematically in Fig. C. 146. With higher valency cations, the kaolinite may even become able to show an additional anion exchange capacity since the polyvalent cations react with the kaolinite to bind anions (Cl^-, Br^-, NO_3^-, or ClO_4^-) on the surface by a reaction of the type (kaolinite)$^-Me^+ + Me^{2+} + X^- \rightleftharpoons$ (kaolinite)$^-Me^{2+}X^- + Me^+$. This reaction also explains the different swelling behavior of montmorillonites with exchangeable monovalent and polyvalent cations. It may have a special meaning further in the flocculation

[596] Cf. A. Weiss, A. Scholz, and U. Hofmann, *Z. Naturforsch.*, **11** b, 1956, 429-430; H. W. van der Marel, *Soil Sci.*, **78**, 1954, 163-179.

[597] Cf. A. Weiss, *Kolloid-Z.*, **158**, 1958, 22-28.

TABLE IV

Correlation between Layer Charges, Swelling Properties, and Cation Exchange Capacities of Micas, Illite, Vermiculite, and mortmorillonite (Weiss)

Mineral	Charges per unit $[(Si, Al)_4O_{10}]$	Innercrystalline swelling for Me^+ ions	Cations in the natural mineral	Cation exchange capacity in meq./100 gm.: For inorganic cations	Cation exchange capacity in meq./100 gm.: For n-R.NH_3^+ ions or $[H_3N(CH_2)_xNH_3]^{2+}$ ions
Brittle Micas (Margarite)	1.8–2.0	$[H_3N—(CH_2)_x—NH_3]^{2+} > 10$	Ca^{2+}	~5 Variable with particle size	Up to 380[a]
Muscovite, Biotite	0.88–1.1	$[H_3N—(CH_2)_x—NH_3]^{2+} > 8$	K^+	~5	Up to 250
		$[R—NH_3]^+$ and $[R_2NH_2]^+$, below 30°C. also inorganic	(Na^+, Li^+)		
		Cations different from K^+, NH_4^+, Rb^+, Cs^+, Tl^+			
Illite	0.66–0.88	Same as in muscovite		Up to about 35	Up to 200
Vermiculite	0.55–0.66	With all cations but K^+, NH_4^+, Rb^+, Cs^+, Tl^+	Mg^{2+}, Ca^{2+}(Na^+)	100–170	100–170
Montmorillonite (Hectorite, Nontronite)	0.25–0.55	With all cations	Mg^{2+}, Ca^{2+}, Na^+	60–120	60–120

[a] Maximum observed after 26 months, at 65°C.

process in colloid suspensions. For kaolinite the density of exchangeable cations per surface unit is larger than for mica, vermiculite, and montmorillonite. Because of high density of the cations per surface unit a quantitative cations exchange with very large quaternary ammonium ions (e.g. tetra-n butyl ammonium ion) is no more possible. The qualitative exchange of hydroxyl by fluorine anions is represented by the accompanying simple structural schemes (I, II). For a higher concentration

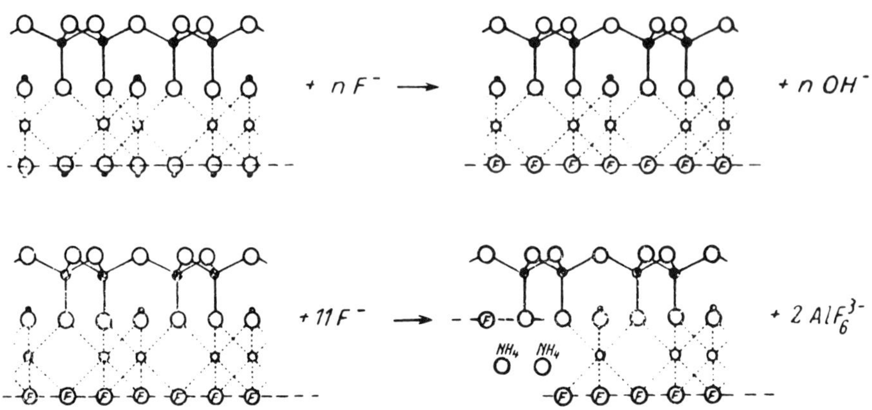

of F^- anions in the solution medium the aluminum of the octahedral layers is "leached" out, and by formation of the complex anion $(AlF_6)^{3-}$ the precipitation of ammonium cryolite is observed. At the same time, some silica is freed which is easily dissolved by esterification in a concentrated ammoniacal solution of pyrocatechol. The process of the surface decomposition of the kaolinite is excellently observed in the electron microscope, especially if the pyrocatechol is removed by treatment with H_2O_2 at elevated temperatures.

273. The acidity of hydrogen montmorillonite, in the acid-treated fresh samples, is a typically *protonic* one, according to I. Martin, R. Glaeser, and J. Méring,[598] but in this state the hydrogen ion is rapidly exchanged by aluminum. This aging effect is highly sensitive to temperature. A stabilization of the protonic montmorillonite is possible at low temperatures and with exclusion of moisture. An aluminum montmorillonite can also be treated repeatedly with 1 *N* HCl solution to "exhaust" the exchangeable aluminum contents with 80 per cent of easily exchangeable hydrogen ions. Evidently, the aluminum ions change their positions to those of the hydrogen ions, as described above, from the edges and corners of the crystals. Whereas, the common aluminum montmorillonite is similar in its reactions to calcium montmorillonite, the protonic modification shows similarities to sodium montmorillonite, with a basal distance of 12.4 to 12.7 Å. for 30 to 40 per cent relative humidity,

[598] *Silicates inds.*, 24, 1959, 131-132.

and 9.6 Å. over P_2O_5, whereas for calcium montmorillonite it is equal to 15 Å., with a double interlayer of water. Protonic montmorillonite forms a stable water suspension and swells strongly at 100 per cent moisture.

274. An interesting relation of the hydration degree with the exchangeable cations on kaolinite and bentonite was studied by E. Gutiérrez Ríos and J. L. Martin Vivaldi[599] to the polarization capacity, for variable partial vapor pressures of water (equilibrium pressure $= p$). The quantity of water, H, adsorbed is a function of the polarization, and of p, in the equation

$$H = A + \beta(Ze/r^2)$$

with $A = (bp)/i$ (i the number of ions per gram unit of substance; β and b are constants), for the alkali ions. For alkaline earth cations the equation is

$$A = (bp)/i + 108\, i$$

The term $(bp)/i$ represents water held by negative charges of the crystal lattice, the polarization term $\beta(Ze/r^2)$ the water held jointly by the layers and the exchange cations. The term $108\, i$ represents the formation of ions with six molecules of water coordinated (only for alkaline earth metals). Between lithium and the other monovalent cations there is no remarkable difference. The very small value of the constant $b = 8.1 \times 10^{-9}$ for kaolinite illustrates the low degree of accessibility to hydration of the interlaminar layers and of interior cations (for bentonite, $b = 1.0$ to 4.5×10^{-6} for different fractions). The uptake of water in clay minerals was determined by displacement of the liquid meniscus in a horizontal microburette in a very simple funnel device with a porcelain plate on which the sample is placed. The logarithmic suction curves show two break points indicating three ranges for the imbibition: first a rapid initial raise, followed by a slow raise, and then an intermediate rate of imbibition, attributed to capillary, intramicellar, and osmotic imbibition, respectively, as different functions of time and cation type.

275. A critical comparison of the different methods recommended for the quantitative determination of base exchange capacities of clay minerals was given by A. Weiss,[600] with respect to the effects of the time periods during which the sample is in contact with the exchange solution, of the pH conditions in this solution, and of the specific effects of the anions present. Particularly important is also the problem of which are the optimum conditions for the drying of samples before the analytical

[599] *Anales real soc. espan. fis. y quim.* (*Madrid*), **45** B, 1949, 291-342, 343-358; *Anales edafol. y fisiol. vegetal* (*Madrid*), **8**, 1949, 289-366.

[600] *Z. anorg. u. allgem. Chem.*, **297**, 1958, 232-256, with a very extensive literature bibliography. On *equimolecular* cation exchange for low-charged ion exchangers see A. Weiss, *Kolloid-Z.* **158**, 1958, 22-28.

procedure of base exchange. Accurate data for different methods used in this comprehensive investigation of comparison are possible only if all the cations and anions present in the exchange system are definitely included in a "chemical substance balance," as well for the exchanging solution as for the "bottom body." In the end effect, Weiss finds several rapid methods: first, the ammonium chloride method of U. Hofmann and K. Giese (1939), the ammonium acetate method,[601] and the calcium or barium chloride–triethanol amine method of A. Mehlich,[602] to be sufficiently accurate for comparison and routine determinations of base exchange capacity of natural or industrially enriched clays and kaolins. Exchangeable hydrogen ions introduced quantitatively by 0.01 *N* acid solutions, or by electrodialysis, or by a contact exchange with ion exchange columns, always bring about irreversible damage to the structure and uniformity of the clay minerals due to chemical decomposition of the silicate layers. Whereas contact exchange with exchange resins is a rapid and reliable measure for kaolinite, the same treatment for montmorillonite is a very slow and often incomplete process, also combined with an extraction of silica and alumina which contaminate the exchange products as amorphous materials. The calcium or barium triethanol amine method (with a buffered $pH = 8.2$) has the characteristic that the exchange isotherms have a different character if the exchange reaction is reversed. The ions $[NH(CH_2CH_2OH)_3]^+$ are evidently more firmly bound on montmorillonite than the Ca^{2+} and Ba^{2+} ions by the formation of hydrogen bridges to the oxygen atoms in the silicate layers.

276. A particularly useful method is that recommended by G. Frysinger and H. C. Thomas,[603] based on the exchange of radioactive isotopes such as Cs^{137} in an "equilibrium column" with rare earth cations ($R.E.^{3+}$). The equilibrium column technique was previously used by N. T. Coleman and M. E. Harward,[604] and then by G. L. Gaines and H. C. Thomas,[605] for exchange of cesium and strontium (Sr^{91}) in chloride solutions. The exchanger is brought to equilibrium with the solution, and then its composition is determined by the radioactive isotope exchange as indicator. C. N. Merriam and H. C. Thomas[606] also applied this method in studying the exchange of alkali ions on attapulgite, for which cesium ions are the most strongly adsorbed cations, followed by potassium, sodium, and lithium. The specificity of the cesium adsorption is obliterated at elevated temperatures of the exchange reaction. Gaines and Thomas used the equilibrium column method on montmorillonite with $Cs^+ \rightleftharpoons$

[601] See an interesting micromethod of this type developed by R. C. Mackenzie, *Clay Minerals Bull.*, 1, 1952, 203-205.

[602] *Soil Sci. Soc. Am. Proc.*, 3, 1938, 162-166; *Soil Sci.*, 53, 1942, 1-14; *Ibid.*, 66, 1948, 429-446.

[603] *Natl. Acad. Sci.-Natl. Research Council, Publ.*, 395, 1955, 239-245; *J. Phys. Chem.*, 64, 1960, 224-228.

[604] *J. Am. Chem. Soc.*, 75, 1953, 6045-6046.

[605] *J. Chem. Phys.*, 23, 1955, 2322-2326.

[606] *J. Chem. Phys.*, 24, 1956, 993-995.

Sr^{2+} whereas Frysinger and Thomas worked with the pairs cesium-cerium and cesium-yttrium. Typical exchange adsorption isotherms of cesium with rare earth elements are shown in Fig. C. 147, for different temperatures. From experimental data and activity coefficients of the solutions the thermodynamical parameters of free energy changes, ΔF, further ΔH and ΔS could be derived for the reactions of the general type

$$Cs_3M + (R.E.)^{3+} \rightleftharpoons 3Cs^+ + (R.E.)M$$

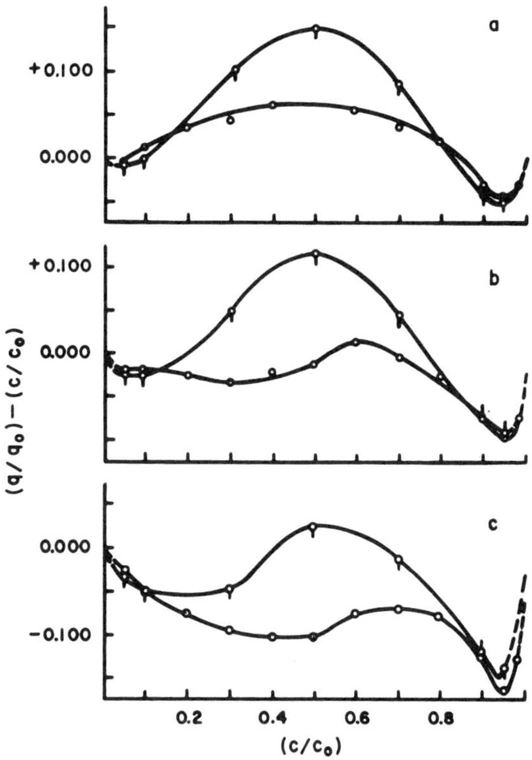

FIG. C. 147. Exchange adsorption isotherms for cesium—rare earth elements on montmorillonite in solutions of 0.02 *N* concentration (Frysinger and Thomas): (*a*) at 30°C.; (*b*) at 50°C.; (*c*) at 75°C.; open circles, for Y^{3+}; circles with vertical bars for Ce^{3+}.

277. H.-P. Boehm and K. H. Lieser[607] used the radioactive isotopes Ag^{110}, Sr^{90}, and Y^{90} as indicators (tracers) for the exchange of silver, strontium, and yttrium ions from 1 *N* salt solutions on kaolinite and illite to obtain a direct determination of the base exchange capacity. The results of this elegant and accurate method,

[607] *Z. anorg. u. allgem. Chem.*, 304, 1960, 207-217.

which is especially useful for minerals of a low capacity, are in excellent agreement with the data from classical analytical determinations. There is a very strong selectivity of the kaolinite surface for Y^{3+} ions, which are adsorbed in the ratio of about $10^3 : 1$ from mixed yttrium and strontium salt solutions. This high factor of separation may be used for enrichment of rare earth elements from solutions with alkalies or alkaline earths. The measurements must consider the time required for the establishment of the radiochemical equilibrium of Sr^{90} with its daughter element Y^{90}. Boehm and Lieser also give valuable proposals for other radioactive tracer elements with mono- and divalent cations which may be used for similar determinations of base exchange capacity of silicate minerals, either with β of γ radiation emission, and discuss advantages offered by those elements either of short or long life-times. An important practical advantage of radioactive tracer methods for clay minerals is the requirement of only small samples, in view of inferior filtration properties, and the difficult washing behavior they show.

278. In additon, A. Weiss (see above) demonstrated that hydroxyl ions influence the base exchange reactions of clay minerals above pH = 7.0. A destruction of the phyllosilicate by alkaline solutions, however, can be avoided by treating a perfectly dried sample with sodium or potassium alcoholate solution. Na^+ and K^+ ions were definitely bound on the clay mineral in this case and, following hydrolysis, the product shows in water a strongly alkaline reaction. If huminic acids are present in the clay mineral, the effects of phenolic hydroxyl groups in the humines react at pH above 8.2, whereas the carboxyl groups react at pH = 7.0. For higher pH (above 10) the methoxyl groups are saponified and additional reactive groups formed.

279. An interesting X-ray fluorescence method for the study of base exchange reactions and capacities of clay minerals, especially of montmorillonite and kaolinite in small amounts, was recommended by D. N. Hinckley and Th. F. Bates.[608] Because of the very low exchange of kaolinite with strontium salt solutions it is possible to determine the montmorillonite relatively easily by adsorption of strontium ions on this mineral, then by removing the excess Sr^{2+} by dialysis (in cellophane dialysis tubing), and analyzing the exchanged sample for Sr by the fluorescence test using a lithium fluoride crystal analyzer.

280. For a general orientation, the quantitative base exchange capacity of the most important clay minerals may vary in the following limits: kaolinite from 0 to 15 meq./100 gm.; halloysite from 5 to 50; metahalloysite from 3 to 15; montmorillonite from 60 to 130; vermiculite from 100 to 170; illite from 5 to 40; and chlorite from 3 to 40. Small amounts of contaminations by organic material (huminic acids) may strongly falsify the results. This organic material must, therefore, be carefully re-

[608] *Am. Mineralogist,* **45,** 1960, 239-241; D. Hinckley, *Ibid.* **47,** 1962, 993-996.

moved by a long-time extraction with a 1 N sodium acetate solution and finally treated with purified hydrogen peroxide solution at water bath temperatures.[609] Another important factor in varying the base exchange capacity data is the more or less disordered state of the crystal structure of the clay minerals. Especially G. F. Walker[610] studied the variability of the base exchange capacity for vermiculites of weathering soils of Aberdeenshire, Scotland, in which biotite is changed to vermiculite, with a high increase of capacity data from initially 40 to 50 meq./100 gm. to 100 to 120. The exchange positions here are always occupied by magnesium ions, the frequency of the occurrence of these cations between the basal planes surfaces being about 1.17 per unit cell, in octahedral $[Mg(H_2O)_6]^{2+}$ coordination (cf. Vol. IV, Section A).

281. Another remarkable example, in the opposite direction of variability of the base exchange capacity, is stevensite (described by G. T. Faust and K. J. Murata[611]) which is different from saponite by lacking essential aluminum, and from hectorite by lacking lithium and fluorine (cf. B. ¶ 39). Its base exchange capacity is surprisingly *low* for a mineral of the montmorillonite group, with 36 meq./100 gm., whereas this number is about 95 for average montmorillonites. The structural defects which are deficiencies in the total number of ions in positions of octahedral coordination in this example are a consequence of the very particular genetic conditions of stevensite (as pseudomorphs after pectolite in basalts). Stevensite practically contains no polyvalent cations (except silicon), the higher electrostatic charges of which would compensate for the deficiency in number; it especially lacks aluminum, the presence of which in tetrahedral positions in common montmorillonites brings about the charge deficiency required for strong base exchange capacities.

282. Lithium-containing montmorillonites (bentonites) are particularly sensitive to treatment at higher temperatures and show after exposure at only 110°C. a base exchange capacity reduced by about 50 per cent; this was noted by U. Hofmann and A. Hausdorf (1945). U. Hofmann and R. Klemen[612] studied this phenomenon by

[609] For details of these procedures, see A. Weiss, *Z. anorg. u. allgem. Chem.*, **297**, 1958, pp. 235-236.

[610] *Nature*, **163**, 1949, 726; *Clay Minerals Bull.*, **1**, 1947, 5-8; *Mineral. Mag.*, **28**, 1949, 693-703; *Ibid.*, **29**, 1950, 72-84, especially pp. 80 ff. on "clay biotite."

[611] *Am. Mineralogist*, **38**, 1953, 973-987. On the peculiar occurrence of a corresponding "magnesium silicate gel" mineral in limburgite of Sasbach in the Kaiserstuhl hills, cf. G. Lorent, *Ber. naturforsch. Ges. Freiburg Breisgau*, **32**, 1933 (2), especially pp. 33 f.

[612] *Z. anorg. u. allgem. Chem.*, **262**, 1950, 95-99, see also R. Greene-Kelly, *Clay Minerals Bull.*, **2**, 1953, 52-56. That lithium ions introduced into cation exchange sites of kaolinite become non-exchangeable (fixed) after a calcination at 200°C. is also emphasized by R. Greene-Kelly, *J. Phys. Chem.*, **59**, 1955, 1151-1152. This observation is interesting because of the structural relations existing between HAl_5O_8 and $LiAl_5O_8$. The problem of Li^+ migration in dioctahedral montmorillonite was recently investigated by R. Tettenhorst, using sensitive infrared absorp-

structural examination and observed in the X-ray diffraction diagram the appearance of an interference (002) at 10 Å. in a sample exposed to 125°C., nevertheless a mica structure of very low order, with only a few (*hk*) interferences. The swelling of the original lithium montmorillonite is still preserved. In this respect it is remarkable that the reduction of the base exchange capacity is extended over a wide temperature range, whereas the reduction of swelling properties is related to hydroxyl losses at higher temperatures, not only for lithium montmorillonite but also for montmorillonites with sodium and calcium ions. The Enslin value is also reduced to about 16 per cent of the original. (See the accompanying tabulation.)

Cations	Temperature	Base exchange in meq./100 gm.	Loss of inner swelling
Li^+	105°C.	56	At 105° to 125°C.
H^+	200°	60	At 200° to 300°C.
Ca^{2+}	300°	45	At 300° to 390°C.
Na^+	300°	80	

283. These effects are particularly pronounced for lithium. They are explained by the removal of Li^+ ions from exchange-active positions when the loose silicate layers of the original montmorillonite structure are collapsing. The cations migrate by a kind of place exchange reaction in the solid state into octahedral layers, fillling gaps in these, as it is observed in hectorite and in lithium mica. From a thoroughly modern viewpoint, U. Hofmann[613] discussed transitions from ideal crystal structures of phyllosilicates to defect types, as well as the gradual changes by a prolonged grinding (see below), in their X-ray diffraction effects and changes in base exchange capacity. For the binding of exchangeable cations on kaolinite, illite, or montmorillonite the systematic widening by an exchange with large organic ammonium cations, as with stearyl ammonium ions, is highly characteristic, combined with a transition from hydrophilic to hydrophobic qualities. In benzene suspension they can be stabilized for an application in lacquers, lubricants, and the like. Such reactions, indicated in a tremendous reduction of base exchange capacity, influence also fundamentally deflocculation effects and peptization, in which diffuse double-layers reduce binding and cohesion forces and drive the crystalline flakes and sili-

tion spectrometry; *Am. Mineralogist*, **47**, 1962, 769-773. The results are consistent with migration from interlayer positions into hexagonal holes, rather than a deeper penetration into empty sites in octahedral layers.

[613] *Silikat Tech.*, **8**, 1957, 224-230. The general swelling phenomena in phyllosilicates (montmorillonites) are in close analogy to those observed in polyphosphates, collagen, and nucleinic acids, as demonstrated by U. Hofmann, *Kolloid-Z.* **169**, 1960, 58-71.

cate layers asunder. In her equally fundamental investigation on the swelling properties of sodium montmorillonites as a consequence of cation substitution, Margaret D. Foster[614] refutes the sometimes suggested theory that swelling and base exchange capacity are directly correlated factors. There is also no relation to electrostatic charges of the tetrahedral layers although there exists such a relation with the octahedral parcel units in the direction or a systematic decrease in swelling with an increase in octahedral substitution. Swelling is an osmotic phenomenon of the colloid electrolyte as defined by G. S. Hartley,[615] related to the degree of dissociation of the structural complexes with exchangeable cations. The substitution of aluminum by iron and magnesium in montmorillonite always decreases the swelling capacity. Thus, the functional relation of swelling and octahedral substitution is established by a smooth continuous curve, whereas swelling and base exchange capacity do not show better than an irregular scattering of data, although the extremes of highest and lowest swelling coincide with highest and lowest base exchange data.

284. A very recent determination of base exchange capacities as a function of the surface area measured by the BET method was given by W. C. Ormsby and J. M. Shartsis[616] for kaolins, in comparison with those surface areas measured by water adsorption isotherms in the gas-current instrument described by L. A. Tomes, C. M. Hunt, and R. L. Blaine.[617] The determination of base exchange capacity data was made by the rapid manganese-saturation method of C. A. Bower and E. Truog,[618] from a solution of manganous chloride, and a secondary replacement of Mn^{2+} by ammonium acetate solution, with the analytical-colorimetric determination as MnO_4^-. For kaolin fractions between 10 and 0.05 μ (calculated as spherical particles) the agreement of the data of BET and base exchange measurements is highly satisfactory. In other words, the base exchange for kaolinite is a primarily surface-determined reaction and, therefore, does not depend on any ionic replacements and structural substitutions. This is the almost ideal case of a high degree of crystalline perfection in a clay mineral; isomorphic substitutions may bring about structural misfits, and even inhibit the crystal growth.

285. The effects of *grinding* over long time periods on the base exchange capacity were specifically studied by R. C. Mackenzie and A. A. Milne[619] for muscovite,

[614] *Am. Mineralogist*, **38**, 1953, 994-1006.

[615] *Trans. Faraday Soc.*, **31**, 1935, 31-50.

[616] *J. Am. Ceram. Soc.*, **43**, 1960, 44-47.

[617] *J. Research Natl. Bur. Standards*, **59**, 1957, 357-364; cf. T. C. Powers and T. L. Brownyard, *Proc. Am. Concrete Inst.*, **43**, 1947, 249-278; 469-504.

[618] *Ind. Eng. Chem. Anal. Ed.*, **12**, 1940, 411-413. A. Weiss, *Z. anorg. u. allgem. Chem.*, **297**, 1958, pp. 244 ff., criticized the manganese method because of inconsistency of the oxidation state of manganese in solutions with pH of about 7.5. See also W. C. Ormsby and L. B. Sand, *Natl. Acad. Sci-Natl. Research Council, Publ.*, **327**, 1953, 254-263.

[619] *Clay Minerals Bull.*, **2**, 1953, 57-62; *Mineral. Mag.*, **30**, 1953, 178-185.

biotite, and vermiculite. Biotite in its original state showed a rather high cation exchange, evidently due to a slight hydration. The fracturing of bonds in the crystal structure brings about a considerable increase in the analytically determined cation exchange capacity in muscovite and biotite. Vermiculite is somewhat peculiar in showing a most erratic time/grinding rate, probably because of a higher resistance to mechanical distortion as a consequence of lubricating water interlayers which impede fracturing of the crystal units. For muscovite, the base exchange capacity is especially high on dry-grinding, as a consequence of a nearly complete break-down

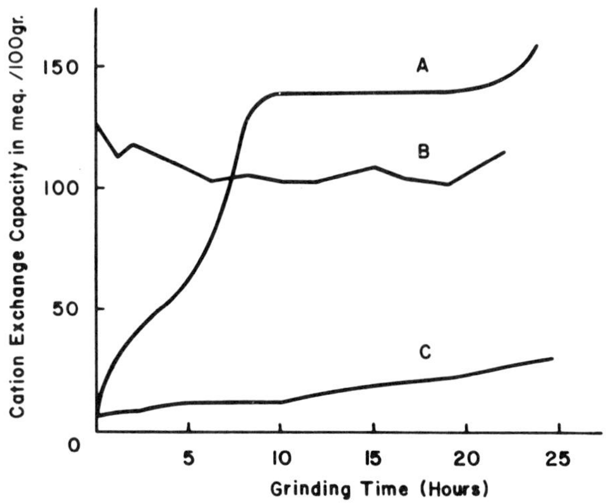

FIG. C. 148. Cation exchange capacity of muscovite and vermiculite powders as a function of grinding time: (Mackenzie and Meldau): (*A*) muscovite, dry-ground; (*B*) vermiculite, dry-ground; (*C*) muscovite, wet-ground.

of the crystal structure followed by some recrystallization which causes the exchange to reach a certain maximum limit. A more extensive investigation of grinding properties of muscovite and vermiculite by R. C. Mackenzie and R. Meldau[620] gave particularly valuable information on the difference in dry- and wet-grinding effects. Dry-ground muscovite is much more intensely distorted, its base exchange capacity very high (see Fig. C. 148). The mechanical disintegration is far more a fracturing

[620] *Ber. deut. keram. Ges.*, **33**, 1954, 222-229. An excellent recent investigation of E. Köhler, U. Hofmann, E. Scharrer, and K. Frühauf, *Ibid.*, **37**, 1960, 493-503, on the effects of grinding on kaolinite and bentonite, confirms the relatively mild action of wet-grinding but the deep-going destruction by dry-grinding. Plasticity and dry bending strength are considerably increased by wet-grinding, whereas a prolonged dry-grinding causes a considerable reduction. B. M. Bishui, N. R. Dhar, and S. S. Mandal, *Central Glass & Ceram. Research Inst. Bull.* (*India*), **8**, 1961, 15-22, studied the changes in the differential-thermal effects on green and ruby mica (India) by grinding.

process in all possible directions in this case, not the perhaps expected basal cleavage effect which, however, in wet-grinding is manifest as an intense shearing along the basis planes. The analytical results make evident that the total exchange capacity of the ground material cannot exclusively be explained by the reaction with potassium cations of the interlayers. An approximative calculation shows that 10 to 15 per cent of the observed cation exchange is to be attributed to broken bonds on edges and corners. When kaolinite was dry-ground on a pebble mill for up to 1000 hours, progressive structural distortions were also observed by S. J. Gregg, T. W. Parker, K. J. Hill, and M. J. Stephens,[621] with an increase of the specific surface by four times up to 600 hours, followed by a slight reduction in surface at 1000 hours. The base exchange capacity is continuously increased with the time of grinding.

286. Base exchange reactions with the cations of basic organic dyestuffs, especially with methylene blue, were observed already in the staining reaction studies of N. E. Vedeneeva (C. ¶ 193–¶ 198). W. Worrall[622] observed that methylene blue cations are firmly fixed on the clay minerals, but there is a definite limit beyond which a physical adsorption of the dyestuff molecules ensues, indicated by an inflection point on the curve representing the amount of dyestuff adsorbed as a function of the concentration in the solution. The linear function of cation exchange is changed beyond this limit to a curve which follows a Freundlich adsorption isotherm equation. The concentration at the break point, however, is *not* the base exchange capacity value but a lower value. For this reason it is not possible to develop a simple colorimetric method for the determination of capacity based on adsorption behavior of methylene blue. The limit of the irreversible adsorption of the dyestuff, as on fireclays, is in every case *not* determined by the base exchange capacity. In the irreversible cation exchange range, however, there is a linear correlation between adsorption and specific area, and it is possible to approximate the area by a simple sorption measurement.

287. From the practical viewpoint, the rapid ion exchange of Wyoming bentonites in strongly acidic exchange resin beds or columns is important for a complete conversion to the hydrogen form of the clay mineral, as D. R. Lewis[623] demonstrated. This preferably rapid exchange is a consequence of very high mobility of the H^+ ions in solution. In equilibrium with solutions containing both H^+ and NH_4^+ ions, the clay mineral exhibits an important preference to H^+. Another very selective adsorption of cations on clay mineral particles is that of UO_2^{2+} and of radioactive isotopes of Ca, Sr, Ba; Co, Ni, Ag, Zr, and others from waste solutions of the isolation of U^{235}, and of the nuclear disintegration process in atomic piles. For these purposes

[621] *J. Appl. Chem. London*, **4**, 1954, 631-632; 666-674.

[622] *Trans. Brit. Ceram. Soc.*, **57**, 1958, 210-217.

[623] *Ind. Eng. Chem.*, **45**, 1953, 1782-1783; cf. L. Wiklander, *Kgl. Lantbruks-Högskol. Ann.*, **16**, 1951, 154-162.

organic exchange resins cannot be used because of their thermal sensitivity to intense γ-radiation at 300°C., as described by R. Platzer and R. Bittel.[624] Such adsorption measures are especially necessary for removing the cations from cooling water in reactor loops of the piles. H-montmorillonite pellets, precipitated by oriented deposition from a suspension and then compacted to tablets, are for these purposes sufficiently stable and act promptly and completely, as for the fixation of UO_2^{2+} in the column.

288. The agriculturally important phenomenon of fixation of potassium and magnesium ions in soils by base exchange was studied by A. Malquori and L. Wiklander[625] under conditions of alternate wetting and drying, in comparison with the base exchange capacity of synthetic precipitated aluminum and iron silicate gels of variable composition. For these compounds the irreversible fixation of K^+, and particularly of Mg^{2+}, is most characteristic. The degree of fixation reaches a practically constant value after a few days. Saturated with K^+ and Mg^{2+} ions, the silicates showed an increase in exchange capacity whereas unsaturated silicates show a considerable decrease with alternate wetting and drying. Even after a treatment for 18 days the precipitated silicates did not indicate anything other than X-ray amorphous material. The studies of Malquori and Wiklander demonstrate that ion fixation in soils is not exclusively a property of crystalline silicates but may also be found in the same way with amorphous constituents of the so-called "soil permutites." The fixation is caused by formation of an insoluble potassium or magnesium alumino- and ferri-silicate, but because of the lack of crystallinity it cannot be determined as to whether or not the Mg^{2+} ions enter the octahedral coordination layers in phyllosilicate complexes, whereas it may be accepted for certain that the K^+ ions remain in external bonding. We may adopt here the definition of this fixed potassium as it was given by R. F. Reitemeier,[626] namely as "potassium which is not immediately replaceable by the normal cation exchange reagents as ammonium acetate."

289. These results are in good agreement with those of Ch. E. Weaver[627] on the structural conditions of ion adsorption on 2 : 1 clay minerals in which the irreversible fixation is connected with a collapse, as from the basal distances of 14 Å. in soil vermiculite to 10 Å., after a simple treatment of the original mineral with a

[624] *Silicates inds.*, **24**, 1959, 550-553; cf. R. Platzer, R. Bittel, J. Huré, J. Fourré, and R. Wey, *Proc. 2nd Intern. Conf. on Peaceful Uses of Atomic Energy, Geneva*, **28**, 1958, 31-47. See also R. E. Grim, *Am. Mineralogist*, **45**, 1960, especially p. 268 on the disposal of atomic waste.

[625] *Trans. Intern. Congr. Soil Sci., 4th Congr., Amsterdam*, 1950, Vol. I, 5 pp. Cf. *Kgl. Lantbruks-Högskol. Ann.*, **14**, 1946, 1-171, and G. W. Volk, *Soil Sci.*, **45**, 1938, 263-76.

[626] *Advances in Agronomy*, **3**, 1951, 113-164.

[627] *Am. Mineralogist*, **38**, 1953, 698-706; cf. G. W. Kunze, Ph. D. Thesis, Pennsylvania State Univ., University Park, Pennsylvania, 1952.

boiling solution of potassium acetate in not more than 10 minutes. Weaver[628] further observed that vermiculite and montmorillonite derived from original muscovite have "inherited" much of the high interlayer charges of the micas, therefore, and show a similar contraction to 10 Å. after treatment with potassium hydroxide solution, or even by curing in sea water. Vermiculites which were derived from low-charged minerals different from micas (hornblende, chlorite, or volcanic material), do not show this remarkable phenomenon of contraction. The great significance of these observations for geological and sediment-petrological conclusions is most obvious. The character of recent marine sediments, as those of the Gulf of Mexico, is of the latter type, whereas sediments from the Atlantic Coast behave as montmorillonitic clays derived from muscovite. The fixation of potassium may sometimes bring about adsorption products which contain more potash than even common illitic clays formed by the weathering and mechanical breakdown of muscovite.

290. Analogous to the fixation of potassium, the irreversible adsorption of ammonium ions on clay minerals (kaolinite, illite, bentonite, vermiculite) was investigated by A. D. Scott, J. J. Hanway, and G. Stanford,[629] especially in thermal decomposition as related to the nature of the exchange sites and/or the entrapment of NH_4^+ ions in the lattice. On kaolinite and illite the fixation is rather loose and ammonia is evolved off below 100°C., the loss being complete at 400°C. Vermiculite, however, holds the NH_4^+ ions much more firmly, ammonia loss starting at 400° and being complete at 600°C. On bentonite a surprising change from exchangeable to fixed ammonium sites is observed when it is heated to 300°-350°C., and then the loss of ammonia does not start below 400°C., with complete destruction at 600°C. Fixed ammonium ions are therefore much more thermally stable than exchangeable ammonium ions and the differential-thermal curve shows convincing reaction effects. It is, however, practically not possible to maintain a sharp distinction between fractions of exchangeable ions held on different sites and fixed ions. The difficulty is that, in both cases, hydrogen ions remain after the thermal decomposition of the ammonium ion. Normally not decomposed NH_4^+ ions, however, undergo a place exchange with the H^+ ions to sites where they are subsequently decomposed. In ammonium bentonite, an exothermic peak on the differential-thermal curve at 550°C. is due to the loss of ammonia from fixed ammonium ions (formed at 300° to 350°C.). In vermiculite no exothermic effects of this kind are observed. It is apparent that in bentonite the fixed ammonium ions do not have water associated with them, whereas in vermiculite this is the case. F. J. Stevenson[630] gave convincing evidence by micro-Kjeldahl determinations that the nitrogen content of granitic rocks and of

[628] *Am. Mineralogist*, **43**, 1958, 839-862, especially pp. 843 ff.

[629] *Am. Mineralogist*, **41**, 1956, 701-721.

[630] *Science*, **130**, 1959, 221-222; cf. A. P. S. Dhariwal and F. J. Stevenson, *Soil Sci.*, **85**, 1958, 42-46; *Ibid.*, **86**, 1958, 343-349.

Palaeozoic shales is for a great portion present in the form of ammonium ions fixed on the crystal structure of silicates, more in illitic than in montmorillonitic soils. Quantitatively these results are to a high degree interesting, since a calculation shows that the fixed ammonium in the earth's crust would make up almost 20 times more nitrogen than is present as the element in the atmosphere.

291. A basic review of the correlations between the cation exchange properties of bentonites and their swelling behavior was given by E. Gruner and R. E. Vogel,[631] based on plastometric consistency measurements for six different bentonite occurrences. The pulp-like slips resulting from alkali or alkaline earth hydroxide containing slips show characteristic curves (Fig. C. 149) as a function of the concentration

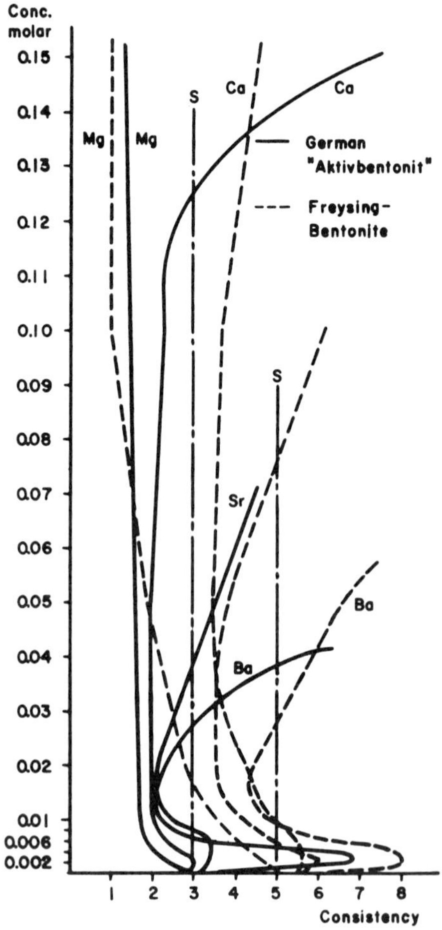

FIG. C. 149. Plastometric consistency of two bentonite slip systems as a function of molar concentration in alkali and alkaline earth hydroxide solutions (Gruner and Vogel).

in hydroxyl ions. The curves are interpreted by two contrary processes: (1) deflocculation and liquefaction which is well known in ceramic slip-casting; and (2) swelling as an adsorption phenomenon of water on layers of the montmorillonite structure. This latter property is chiefly dependent on the size and the electrostatic charge of the cations and their degree of hydration, which also influences swelling by the introduction of additional water molecules into the interlayers. Increasing basicity of the hydroxides increases chemisorption and screening of the siloxan sheets, and reduces swelling. A comparison with previous data of A. Lottermoser and E. Schmidl (1931) of viscosity curves of kaolin-alkali hydroxide slips shows that these conclusions are in good agreement for analogous bentonite–hydroxide systems. In contrast with this typical reaction mechanism of hydroxide solutions which form typical slips, alkali of alkaline earth salt solutions are not able to exert the same effects. It is also not true that under every condition sodium bentonites must have a higher swelling capacity than calcium or magnesium bentonites. There are, because of the retrograde course of the consistency curves in Fig. C. 149, even concentrations of the hydroxides for which calcium and magnesium bentonites show a stronger swelling than those with sodium. Important aging effects distinguish the swelling properties of bentonites of recent volcanic origin from those of geologically ancient occurrences. The dissociation of the colloid-electrolytic systems with cations having different hydration degrees is equally important, ruling the "screening" of the structural sheet planes. Aging effects are particularly pronounced for calcium bentonites and finally lead to the formation of nonswelling aluminosilicate hydrates. In Gruner's opinion the calcium bentonites are intermediate between aluminum silicate hydrates without any cations (pyrophyllite as the extreme example) and calcium zeolites as aluminosilicate hydrates.

292. K. Norrish[632] studied swelling montmorillonites in different salt solutions with X-ray low-angle diffraction methods. The initial swelling is in every case a crystalline interlayer phenomenon, dependent on the hydration energy of the cations in the interlayers. With monovalent cations the swelling proceeds to the "macroscopic" state of liquid crystallinity. Calculating the attraction forces between neighboring sheets, one finds that this force decreases at a lower rate with respect to the distance than would be expected if pure van der Waals forces were in action. Concerning basal spacings during swelling, Norrish finds a two-step process: below 22 Å. spacing the interlayer expansion depends strongly on the exchangeable cation; it then takes a "jump" to 35 Å. with a further continuous expansion up to 130 Å. which is for monovalent cations independent of their nature. In a third step the swollen sodium montmorillonite system shows a gradual transition to the typical gel.

[631] *Kolloid-Z.*, **116**, 1950, 88-99; *Ibid.*, **118**, 1950, 100-110.
[632] *Nature*, **173**, 1954, 255-256; 256-257; *Discussions Faraday Soc.*, 1954 (18), 120-134.

293. For bi-ionic montmorillonites, such as those of mixed calcium-sodium type, the fraction, f, of the exchange capacity related to calcium, and $(1 - f)$ for sodium saturation, R. Glaeser and J. Méring[633] observed that there is a limit value $f = 0.3$ below which the system behaves as if it were only sodium montmorillonite. Above $f = 0.3$ there is a separation effect called "demixing" for which the interlayer spaces of the clay mineral structure are shared by both cations side-by-side, independent of the liquid present (water or alcohol). X-ray diffraction measurements confirm this theoretical concept, if the water inputs are plotted as a function of the fractions f for variable moistures, but below a relative humidity of 75 per cent. Also statistical calculations of the probability for an incidental occurrence of limited numbers of continuous double-charged sites as spots of silicon replaced by aluminum are in agreement with this concept. The demixing is therefore a redistribution which is characteristic of the montmorillonite mixed phases.

ADSORPTION OF ORGANIC CATIONS; ORGANIC DERIVATIVES OF CLAY MINERALS

294. It is not always self-evident how one is to classify what are generally called organophilic clay minerals, especially montmorillonite, by distinguishing them from their normal hydrophilic state after the introduction of aliphatic or aromatic groups. As shown by the fundamental investigations of J. E. Gieseking (1939), however, it is possible to distinguish between the products of exchange reations of organic cations for inorganic ions and the real introduction of molecular radicals into the clay minerals, modifying their constitutional properties. In the latter case typically organic-synthetic methods are used, such as the Friedel-Crafts condensation method, reactions with diazomethane, and related specifically organic reactions. We may call products of the latter group of compounds "organic derivatives" of the clay minerals.

295. The first group is represented, in continuation of Gieseking's early studies, by the introduction of very large cations of aliphatic quaternary amine ions, as dimethyldodecylammonium, from the acetate solution as described by F. W. Jordan,[634] with new and highly interesting specific effects, such as swelling properties in organic solvents, particularly pronounced in solvents with increased dielectric

[633] *Clay Minerals Bull.*, **2**, 1954, 188-183.

[634] *J. Phys. Chem.*, **53**, 1949, 294-306; with F. J. Williams, *Kolloid-Z.*, **137**, 1954, 40-48. The change from hydrophobic complexes by exchange with large organic cations coating clay minerals is what R. E. Grim, *Am. Mineralogist*, **45**, 1960, especially p. 267, called "organic-clad" clay minerals or "tailor-made" products. These fit the particular organic materials with which they are to be used, e.g., in conditioning of soils for agricultural purposes, or in stabilizing of roads. The water repellency of modified clay minerals with normal long-chain alkyl substitutions may be compared to the widely variable behavior of water-repellent dyestuffs on cotton and wool yarn as described by C. C. Dewitt and P. D. Shroff, *Ind. Eng. Chem.*, **45**, 1953, 302-307.

constants. These effects excel even those previously described by H. Erbring and H. Lehmann (1944) with laurylpyridinium ions. The coatings of bentonite with such unusually large cations are very smooth and technologically interesting for the reinforcement of rubber or for emulsifying water in oils, whereas normal (hydrophilic) bentonites emulsify oils in water. The mechanism of the exchange of organic and inorganic cations on montmorillonite was especially studied by J. L. McAtee,[635] by stepwise additions of halogenide solutions of primary aliphatic amines, as the chloride of octadecylamine reacting with sodium and/or calcium montmorillonite. In mixed montmorillonites, half or more of the sodium ions are replaced before any of the calcium was removed in a normal base exchange reaction (cf. Fig. C. 150).

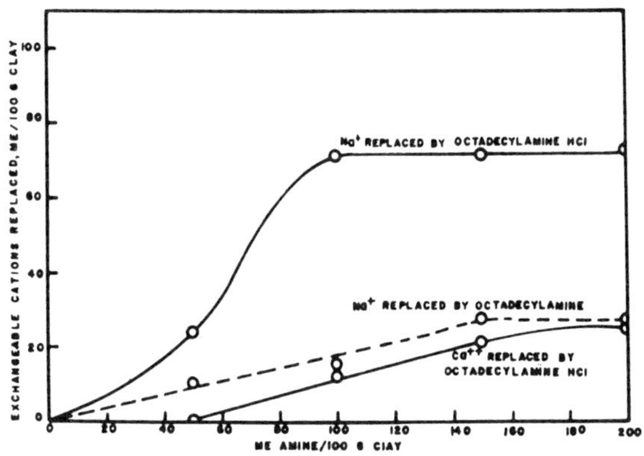

FIG. C. 150. Base exchange of sodium and calcium ions from montmorillonite by the cation of octadecylamine (McAtee).

But even the octadecylamine itself has such a high basicity that it replaces 27 meq. of sodium when 150 meq. of the amine are added. As a quaternary amine salt, McAtee used dimethyldioctadecylammonium chloride, or dimethylbenzyllaurylammonium chloride. All of the sodium ions are replaced by the organic cation before any calcium is removed (Fig. C. 151). For pure sodium bentonite the replacement is almost exactly 1 : 1. In hectorite, first the sodium ions, then those of calcium, and finally those of magnesium are exchanged by the last-mentioned quaternary agent with marked speed.

296. Very interesting thermal effects in the resulting organic (dimethyldioctadecylammonium montmorillonite) complexes described by J. L. McAtee and Ch. B. Concilio.[636] After suspension in methylethyl ketone and drying these complexes form an excellently oriented film with characteristic changes in the basal spacings as

[635] *Am. Mineralogist*, 44, 1959, 1230-1236.
[636] *Am. Mineralogist*, 44, 1959, 1219-1229.

a function of temperature [measured by the heating X-ray oscillation diffraction method (cf. Vol. IV, Section B)]. A reversible espansion from 28 to 32 Å. is the first observed effect followed by a sudden collapse at about 180°C. to 14.5 Å., at higher temperatures to 13.0 Å., and finally to 9.8 Å. The room temperature complex is a four-layer adsorbate as shown by a structural Fourier analysis. The collapse mentioned is not an oxidation phenomenon but is caused by the rupture (pyrolysis) of bonds (desorption of molecules), whereas the thermal decomposition leaves a residue of a mixture of methylene groups, carbon, and short chains with nitrogen.

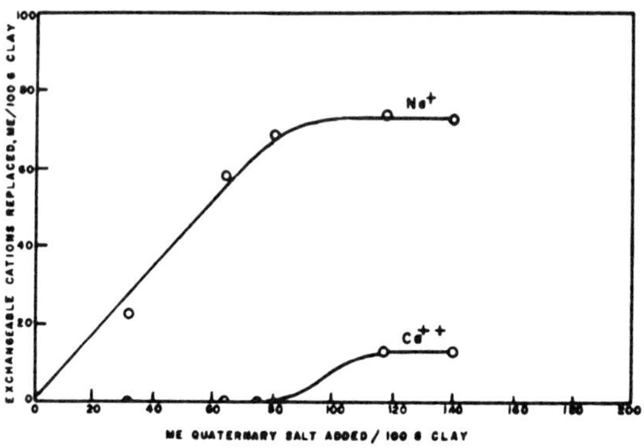

FIG. C. 151. Base exchange of sodium and calcium ions from montmorillonite by dimethyldioctadecylammonium (McAtee).

297. P. Franzen[637] investigated the structure of adsorption layers of Lissolamine (cetyltrimethylammonium bromide) ions on montmorillonite which comprises up to 2.5 times the base exchange capacity, with a basal spacing of 20.5 Å. Combining these with Jordan's results, one comes to the conclusion that for aliphatic chains with more than 12 carbon atoms in the ions the basal spacing is widened by 2 to 4 Å. The halogen anion remains statistically distributed over the available space in the complexes. The kinetic (mass action) and thermodynamic reaction conditions for the exchange of aliphatic amine ions for inorganic cations were examined by W. H. Slabaugh[638] for Wyoming bentonite. The equilibrium constant

$$K = \frac{\gamma\, C_{RNH_3^+} \times a_{Na^+}}{\gamma\, C_{Na^+} \times a_{RNH_3^+}}$$

[637] *Clay Minerals Bull.*, 2, 1955, 223-235. On the use of the acetates of aliphatic amines with carbon chain numbers 12 to 14 as an optimum of their collector efficiency in flotation, e.g., of silica from phosphate ore concentrates, see Th. H. Lentz, D. E. Teery, and H. Wittcoff, *Ind. Eng. Chem.*, 47, 1955, 468-471.

[638] *J. Phys. Chem.*, 58, 1954, 162-165.

is for $R = CH_3$ on the average equal to 0.77 at 25°C. For higher amines with C_n chains (where n is more than 10) the quantities of the amine cations adsorbed on the montmorillonite by exchange rapidly increase with concentrations of the amine in the solution, with K values of 26 for $n = 10$; 64 for $n = 14$; and 102 for $n = 18$. For $C_4H_9NH_3^+$ the enthalpy of exchange and adsorption was directly measured in the calorimeter, with $\Delta H_r^{25} = -244 \pm 10$ cal./mole. The free energy changes, ΔF, and other parameters are tabulated in Table V.

TABLE V

THERMODYNAMIC DATA FOR THE REACTIONS Na-BENTONITE + $RNH_3^+ \rightleftharpoons Na^+ + RNH_3$-BENTONITE (SLABAUGH)

RHN_3^+ for n carbons	K	ΔF (cal.)	ΔH (cal.)	ΔS (e. u.)
n = 1	0.77 ± 0.12	+ 150	150	0.0
4	1.8 ± 0.5	— 350	— 240	0.4
8	6.9 ± 2.0	— 1140	— 900	0.8
10	26 ± 5.3	— 1920	— 1780	0.5
14	64 ± 5.9	— 2450	— 2500	— 0.2
18	102 ± 5.3	— 2740	—[a]	—[a]

[a] Not determined.

298. The change from hydrophilic normal clay minerals to hydrophobic organophilic substances is not restricted to montmorillonite. A. Weiss, A. Scholz, and U. Hofmann[639] observed the same phenomenon also for illite, with a base exchange of potassium ions with alkylammonium ions, especially of stearylammonium and oleylammonium ions from the chloride solution. The basal spacing of the oleylammonium complex (after drying at 90°C.) is 23.4 Å.; under water it is 28.8 Å.; under ethyl alcohol, 28.9 Å.; under nitrobenzene, 38.6 Å. The originally brittle crystals become very soft, fatty to touch, and go into benzene-water mixture. A. Weiss, W. Mehler, and U. Hofmann[640] extended these studies on alkyl pyridinium ions exchanged for alkalies on vermiculite and batavite (cf. B. ¶ 94), combined with a considerable increase in the basal spacings. The voluminous complex formed in this way is further able to take up reversibly, in the interlayers, noncharged but polar compounds to

[639] *Z. Naturforsch.* **11** b, 1956, 429-430.

[640] *Z. Naturforsch.*, **11** b, 1956, 431-434. Hexadecylpyridinium ions, in comparison with hexadecyltrimethylammonium, ethyldimethylammonium, tetramethylammonium, and phenyltrimethylammonium ions (from bromide or iodide solutions) were studied in regard to their adsorption exchange with kaolinite and montmorillonite by S. K. Chakravarti, *J. Indian Soc. Soil Sci.*, **5**, 1957, 85-90.

form one-dimensionally swollen products, as with aliphatic alcohols, ketones, aldehydes, nitriles, ethers, esters, but not paraffin hydrocarbons, whereas toluene and nitrobenzene are imbibed. The X-ray diffraction analysis showed a characteristic orientation of the normal-alkyl ammonium ions with their elongated chain under an angle to the basal planes of the clay mineral. This angle is varied if a shearing pressure is applied. Evidently, the sheets are gliding on one another, with a decrease in the spacing as long as stress is applied. Similar conditions are also observed in montmorillonite with alkylammonium ions. For normal hexylammonium and octylammonium bentonite the angle is about 56°, but reduced by lateral stress to 50° and even 42°. For *n*-butylammonium, dodecylammonium, and stearylammonium complexes the angle is about 90° (cf. Fig. C. 152 a, b). In the imbibition products

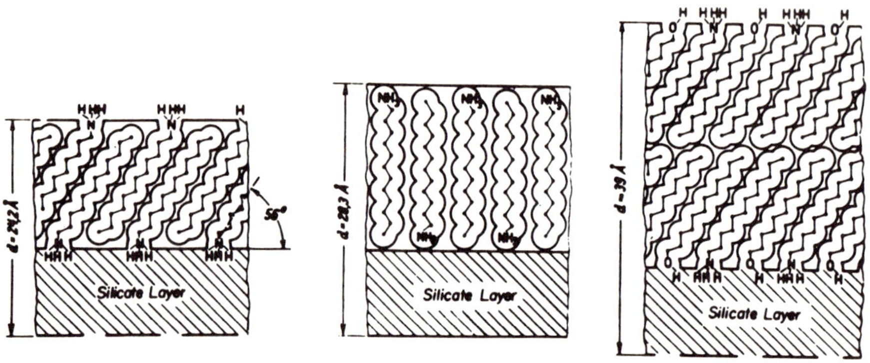

(*a*) (*b*) (*c*)

FIG. C. 152. Orientation of alkylammonium ions between silicate layers of batavite and vermiculite (Weiss, Mehler, and Hofmann): (*a* and *b*) in the original state for hexyl- and octylammonium ion (Angle of 56°), and butyl-, dodecyl-, and stearylammonium ion (angle of 90°); (*c*) in swollen state with dodecyl alcohol.

mentioned above the aliphatic chains of aliphatic alcohols are inserted between those of the alkyl ammonium ions (Fig. C. 152 c). The thickness of these squeezable interlayers may be up to 40 Å. and interesting speculations may be made, from the biochemical viewpoint, concerning these extreme swelling phenomena. Amines originating from the putrefaction of proteins are probably adsorbed on montmorillonite and may play a role in highly swellable clay minerals observed in petroleum occurrences.

299. A. Weiss, W. Mehler, and U. Hofmann[641] further extended their studies on micas in which a quantitative exchange of potassium ions by alkylammonium ions

[641] *Z. Naturforsch.*, 11 b, 1956, 435-438. In this way it was even possible indirectly to synthesize micas in which K^+ cations were replaced by Al^{3+} and Si^{4+}; cf. A. Weiss and G. Thamerus,

is possible, combined with a one-dimensional innercrystalline swelling with many organic liquids, as mentioned above for batavite and vermiculite but now with basal spacings up to more than 50 Å. (e.g., in *n*-octylamine). The alkylammonium ions can be reversibly replaced by sodium in sodium alcoholate solution; the sodium can then be replaced by calcium and magnesium in aqueous solution to form swellable mica products which, however, lose this property after a mild thermal treatment. The basal spacings measured for the alkylammonium micas and of their swelling imbibition products are very similar to those observed in the vermiculite and motmorillonite group. For lithium micas (lepidolite), the exchange of lithium ions, probably from octahedral positions, by alkylammonium ions is particularly remarkable.

300. The exchange of cations in montmorillonite by proteinic materials and amino acids was observed by O. Talibudeen.[642] Globular protein molecules may have been subdivided in the reaction to polypeptide chains with a flat orientation of the amino acid radicals into the montmorillonite structure, as U. Hofmann[643] remarked. This phenomenon may illustrate why montmorillonite flocculates their solutions near the isolectric point.

301. On kaolinite, A. Weiss[644] studied the cation exchange with divalent organic ions, such as those of hydrazine- and alkylene(α, ω)diammonium type ions, with a preferably equivalent exchange for the special condition that the distance between neighboring positive charges in the ion is about the same as that of the fixed negative charges on the basis plane of kaolinite, with 5.14 or 10.28 Å. for [100], and 8.9 Å. for [010]. The number n of the carbon atoms in straight $(CH_2)_n$ chains definitely influences the exchange effect, if one considers that propylene-(1, 3-diammonium, and heptamethylene-1, 7-diammonium ions with 4.9 and 10.0 Å. for the distance of N—N atoms show a higher exchange than even that with monovalent cations on kaolinite. The orientation of the diammonium ions shows possibilities (as shown in Fig. C. 153) either for an equivalent or an equimolecular exchange, including the formation of hydrogen bridge bonds to the oxygen atoms on the kaolinite layers.

Naturwissenschaften, **48**, 1961, 70 (cf. B ¶ 92 f.). It is very interesting that the so-called uranium micas of the general formula $Me^+[UO_2.XO_4]^-(H_2O)_n$, with X = P, As, V, also show the same alkylammonium and alkylpyridinium cation exchange reactions, one-dimensional swelling, and organophilic properties as in the silicate micas; cf. a Weiss, K. Hartl, and U. Hofmann, *Z. Naturforsch.*, **12** b, 1957, 351-355. See also on the influence of hydrogen bondings on one- and two-dimensional innercrystalline swelling phenomena, A. Weiss, E. Michel, and Al. Weiss, *in* "Hydrogen Bonding: Proceedings of the Ljubljana Symposium, 1957" (D. Hadzi and H. W. Thompson, ed.), Pergamon Press, London 1959, pp. 495-508. Innercrystalline swelling as a model of swelling processes in general are discussed by A. Weiss, *Chem. Ber.*, **91**, 1958, 487-502.

[642] *Nature*, **166**, 1950, 236.

[643] *Angew. Chem.*, **68**, 1956, especially pp. 56 f.

[644] *Z. anorg. u. allgem. Chem.*, **299**, 1959, 92-120, especially pp. 115-117.

A. Weiss and E. Michel[645] observed the same principle of cation exchange verified by large organic ammonium (e.g., trimethylcetyl ammonium) and inorganic cations in polyphosphates of chain-shaped configuration. E. Thilo and K.-H. Rattay[646] discussed this for use in water softening agents by a true cation exchange reaction. The results of Weiss *et al.* are remarkable in that the X-ray study of the alkylammo-

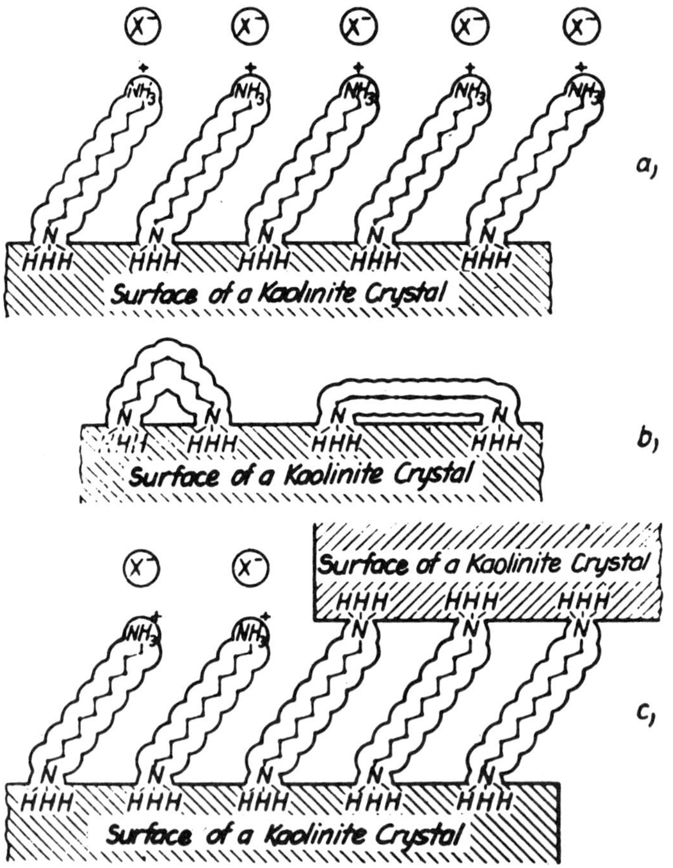

FIG. C. 153. Possible orientation of alkylene-(α, ω)-diammonium ions on the surface of kaolinite: (a) for an equimolecular exchange; (b) for an equivalent exchange, with characteristic fitting distances of fixed negative charges; (c) for a partially equivalent exchange between two neighboring crystals (Weiss).

[645] *Z. anorg. u. allgem. Chem.*, **296**, 1958, 313-332. On the structure of potassium-polyphosphate see K.-H. Jost, *Naturwissenschaften*, **49**, 1962, 229-230.

[646] *J. prakt. Chem.*, [4] **1**, 1954, 15; cf. E. Thilo, G. Schulz, and E. M. Wichmann, *Z. anorg. u. allgem. Chem.*, **272**, 1953, 182-200; E. Thilo, Anneliese Sonntag, and K.-H. Rattay, *Ibid.*, **283**, 1956, 365-371.

nium phosphate complexes showed these as *layer* compounds, and not, as might have been expected, as chain-shaped configurations. The fiber axis of the original polyphosphate chains, however, remains unchanged. The same imbibition phenomena of the ammonium-exchanged products with organic liquids (alcohols, ketones, esters, etc.) are described like those characteristic of montmorillonite or micas. They swell with these liquids to highly viscous jellies or solutions.

302. Simple base exchange reactions are not only possible with aliphatic cations on montmorillonite, but also with the ions of aromatic amines, as was especially studied by D. White and C. T. Cowan[647] for the hydrochlorides of aniline, α-naphthylamine, quinoline, and finally for methylene blue. When the base exchange capacity of the clay mineral is exceeded, hydrochloric acid is liberated, indicating that the free aromatic amine base, and not the hydrochloride complex was irreversibly adsorbed. By using an organic solvent, the tendency for the organic ions to go onto the clay mineral and of the inorganic ions to go into the solution phase is reversed, hereby facilitating the removal of the aniline, and the like, from the clay mineral, with a quantitative output. If, however, the adsorption product is aging, the material becomes pinkish to deep-red due to an irreversible adsorption of the aniline, and even steam distillation in the presence of magnesia is not able to remove the aromatic base. With α-naphthylamine the aging color is dark-blue. Quinoline hydrochloride and methylene blue are especially chemisorbed in agreement with the results of W. E. Worrall (C. ¶ 286), with an apparent surface area of 700 m.^{2}/gm. for montmorillonite. However, these results must be interpreted with caution since the position of the molecules of the dyestuff is in doubt in the critical range between 26 and 10 Å.^{2} per exchange site.

303. We should emphasize finally the interesting saturated adsorption complexes with heterocyclic and hydroaromatic ring compounds on montmorillonite, described by R. Greene-Kelly,[648] in two different groups, one with a (001) spacing of 13.3 $\pm$ 0.01, the other with 14.6 $\pm$ 0.01 KX. To the first group belong the adsorbates of tetrahydropyrrole, α-methylcyclohexane, and the hydrochlorides of piperidine and α-methylpiperidine; to the second group, those of tetrahydrofuran, piperidine, cyclohexanol, and cyclohexane. In the first type the heterocyclic or hydroaromatic molecules are intercalated with the rings parallel to the layers of the phyllosilicate; in the second type, they are approximately perpendicular to the sheets, as a one-dimensional Fourier analysis proved (Fig. C. 154). The observed basal spacings, however, are by more than 1 Å. unit smaller than those calculated from the steric conditions of a random arrangement of the molecules between the silicate layers.

[647] *Trans. Brit. Ceram. Soc.*, **59**, 1960, 16-21.

[648] *J. Phys. Chem.*, **60**, 1956, 808-809. The calculation method is described in *Trans. Faraday Soc.*, **51**, 1955, 412-424. On the packing of pyridine molecules cf. Ibid., pp. 425-430.

The projected van der Waals contact distances assumed between the oxygen atoms and the organic molecules between layers are evidently loo large.

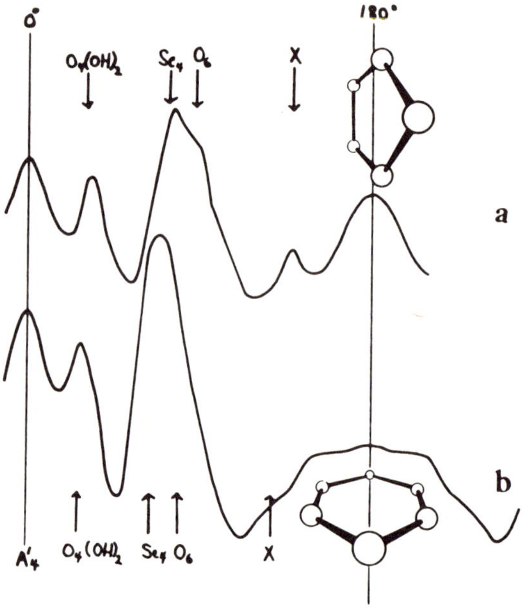

Fig. C. 154. One-dimensional Fourier analysis of: (*a*) the tetrahydropyrrole complex with montmorillonite; (*b*) the piperidine complex (Green-Kelly). Orientation of the intercalated molecules.

304. True organic *derivatives* of clay minerals were especially investigated by H. Deuel[649] in experiments starting from a treatment of the hydroxyl groups in silanol sheets with thionyl chloride, $SOCl_2$, to change the OH groups to chlorine atoms, and by treatment of the chlorinated product with methanol to methoxyl groups, $—O.CH_3$. These reactions are related to the reactions with trimethylchlorosilane (cf. C. ¶ 147, ¶ 204) and with G. Berger's diazomethane reactions (1941). H. Deuel, G. Huber, and R. Iberg[650] observed a very interesting reaction of montmorillonite, kaolinite, attapulgite, and the like, with mustard gas, $Cl(CH_2)_2S(CH_2)_2Cl$ (in alcoholic solution) to form rather stable hydrophobic compounds with the bond type $\equiv Si—O—(CH_2)_2S(CH_2)_2OH$, which swell in alcohol. Propylene oxide,

$$\underbrace{CH_2—CH}_{O}.CH_3$$

[649] Especially *Ber. deut. keram. Ges.*, **31**, 1954, 1-7, with extensive literature. See also J. Uytterhoeven, *Silicates inds.*, **25**, 1960, 403-409, on organosilicates of this type.

[650] *Helv. Chim. Acta*, 33, 1950, 1229-1232; see also *Kolloid-Z.*, **124**, 1951, 164-169; *Clay Minerals Bull.*, **1**, 1952, 205-211.

reacts with bentonite to form oxypropylates of montmorillonite which are decomposed with water. Real clay mineral chlorides with the bonding ≡Si—Cl are prepared by the reaction of the silicates with $SOCl_2$[651]. They develop *direct* bonds Si—Cl by interaction with Grignard's reagent, e.g. C_2H_5MgI and form the ethyl-silicon bond. This reaction type modified for aromatic compounds by Friedel-Crafts synthesis forms phenyl montmorillonite with the bond $\equiv Si—C_6H_5$ which shows in the infrared absorption spectrum a characteristic decrease in intensity of the hydroxyl bands in the range from 3200 to 3600 cm.$^{-1}$ (2.8 to 3.1 μ) and near 1645 cm.$^{-1}$ (6 μ), and new bonds characteristic of the phenyl group at 1500, 1522, 1562, 1577, and 1700 cm.$^{-1}$. The differential-thermal curves of montmorillonite modified by the introduction of phenyl groups, in place of hydroxyls, show additional exothermic peaks at 450°, 570°, and 700°C. Even after treatment with hydrogen peroxide a peak at 600°C. corresponds to strongly fixed phenyl radicals in covalent binding. Phenyl montmorillonite forms in benzene swelling and highly viscous systems.

305. Conclusions of Deuel *et al.* were not accepted without some rather sharp criticisms, notably by G. Brown, R. Greene-Kelly, and K. Norrish,[652] which especially concern also G. Berger's diazomethane reaction. W. F. Spencer and J. E. Gieseking[653] confirmed, however, the existence of phenyl derivatives, with phenyl groups inserted in the montmorillonite structure in a regular manner between the layers. R. M. Barrer and J. S. Reay[654] investigated benzene sorption in phenyl montmorillonite prepared by Deuel, with BET surface areas of 4 to 5 m.^{2}/gm. The benzene sorption is very slow to reach equilibrium; 10 mg./gm. are sorbed at 50°C. under a relative pressure $p/p_s = 0.3$. Water is sorbed somewhat more and with greater speed. The fact that benzene is sorbed with an amount six times larger than that required by the BET-determined surface suggests a relation between excess benzene and the interlamellar phenyl contents. In every case, Deuel's phenyl montmorillonite is distinctly different from initial hydroxyl montmorillonite in its general behavior, or from alkyl-ion exchange forms and so-called clay esters of the aliphatic

[651] H. Deuel and G. Huber, *Helv. Chim. Acta*, **34**, 1951, 1697-1701; *Ibid.*, **35**, 1952, 1799-1802; *Ibid.*, **36**, 1953, 808-810; *Agrochimica*, **1**, 1957, 248-267. Decomposition of phenyl montmorillonite was studied by H. Deuel and R. Gentili, *Helv. Chim. Acta*, **40**, 1957, 106-113, with conclusions concerning the constitution of silicon containing fractions by chromatography, e.g., with $Si—C_6H_5$ bonds, cyclosiloxanes, and acyclic siloxanes, but with subordinate Si—OH groups. Over the montmorillonite chloride, H. Mukerjee, *Naturwissenschaften*, **42**, 1955, 412-413, prepared acetyl, benzoyl, palmityl, and cetyl derivatives, also by reaction of the acidyl chlorides with the hydrogen clay mineral by refluxing in dry dioxane at 110° to 120°C.

[652] *Nature*, **169**, 1952, 756-757; *Clay Minerals Bull.*, **1**, 1952, 214-220.

[653] *J. Phys. Chem.*, **56**, 1952, 751-753.

[654] *Clay Minerals Bull.*, **3**, 1958, 214-220. A thorough thermodynamic treatment of interlamellar complexes of the type of alkylammonium montmorillonites, and their intercalation of hydrocarbons in sorption reactions, was developed by R. M. Barrer and Mrs. K. E. Kelsey, *Transact. Farad. Soc.*, **57**, 1961, 452-462.

type. Concerning the organic "filler" substance in the interlayers, it is not clearly decided whether this material is chemically bonded or only strongly adsorbed in a physical mechanism. Nitrogen cannot penetrate into the interlamellar spaces. There is also no change in the basal spacings of the montmorillonite by sorption of bases. The polar water molecule also does not have easy access into the phenyl-filled interspaces.

306. An examination of the problem of how organic dispersion media (different normal alcohols, glycols, ketones, benzene, diethyl phtalate) influence 1 per cent suspensions of kaolinite was made by Ch. L. Folkers and A. R. Welch,[655] especially for the presence of complex compounds between the organic molecules and the clay mineral. These questions are particularly important for the formation of deflocculated systems when sodium salts of certain alcohols (sodium alcoholates) are used as deflocculants. Besides the base exchange reactions discussed before, methyl and ethyl alcohol form two-layer complexes, higher alcohols only one layer with montmorillonite (cf. MacEwan, 1948) by hydrogen bonding. Kaolinite is inert, say to glycols. The van der Waals forces of attraction exerted by several alcohols are seen in Fig. C. 155 and 156, which also indicate decreasing polarity of the alcohols with increasing sizes. The specific action of higher sodium alcoholates is evident by reducing the association. The substitution of a shorter alkyl chain makes the defloc-

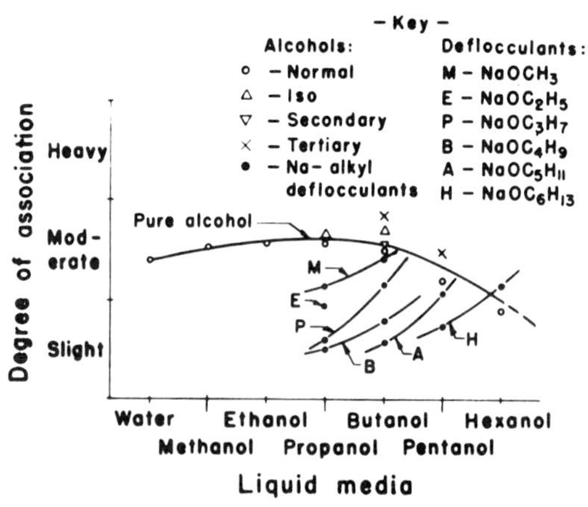

FIG. C. 155. Association of kaolinite suspensions in different aliphatic suspension media, with and without sodium alcoholates as deflocculants (Folkers and Welch).

[655] *J. Am. Ceram. Soc.*, **38**, 1955, 454-461. Polyelectrolytes, e.g., sodium polyacrylate or polyacrylamide, in their adsorption and flocculation properties with important hydrogen-bonding on kaolinite suspensions, were examined by A. S. Michaels and O. Morelos, *Ind. Eng. Chem.*, **47**, 1955, 1801-1809. The polyacrylamide is a strong flocculant in the pH range from 6.3 to 7.5; sodium polyacrylate, on the other hand, is a deflocculant for pH = 5 to 6, and a weak flocculant for 6.5 to 7.5.

culation products more soluble and more active to be adsorbed to the clay particles. For aromatic compounds the association-flocculation degree is equally evident (Fig. C. 157). Highly refined infrared techniques were used by R. O. French, M. E. Wadsworth, M. A. Cook, and I. B. Cutler[656] for detection of small amounts of or-

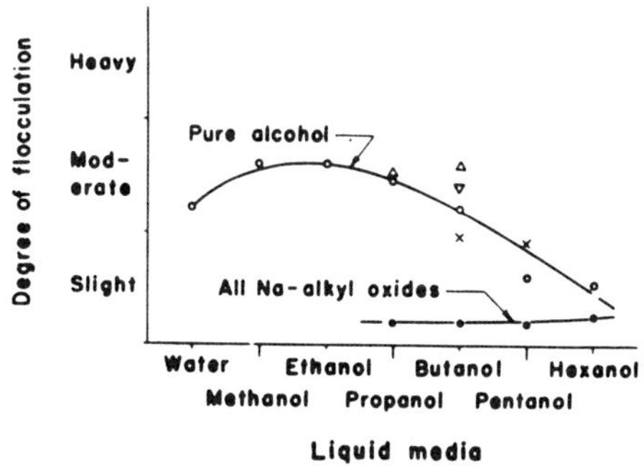

FIG. C. 156. Degree of flocculation in aliphatic alcohols, with and without sodium alcoholate deflocculants (Folkers and Welch).

ganic flocculants on Wyoming bentonite particles, or for that of oleic acid on fluorite in flotation systems or of ammonia on silica-alumina catalysts. As commercial organic flocculants were used polyanionic electrolytes of the "Lytron 886" type (a heteropolymer of maleic anhydride), or "Galactasol" and "Guartec," of similar characteristics.

307. The problems of complex formation of clay minerals with organic solvents was also investigated for the system montmorillonite-acetone by R. Glaeser.[657]

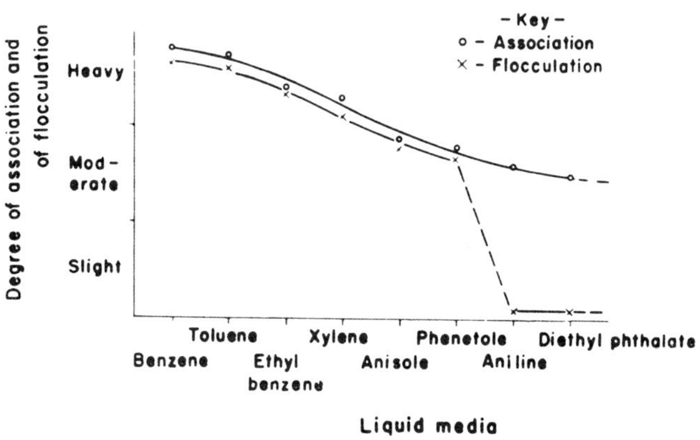

FIG. C. 157. Association and flocculation in aromatic suspension media (Folkers and Welch).

[656] *J. Phys. Chem.*, **58**, 1954, 805-811.
[657] *Clay Minerals Bull.*, **1**, 1949, 88-90.

The adsorption curve for acetone vapors shows the typical S-shape, and widening of the structure of the montmorillonite by 3.7 Å. for sodium montmorillonite and by 3.9 Å. for calcium montmorillonite.[658] The same end results were observed by immersion in acetone for a long period. The widened structure corresponds to a one-molecular layer of acetone molecules similar to that in halloysite, whereas in a 1 : 1 mixture of water and acetone vapor two-molecular layers occur. It is, therefore, possible that a mixed acetone-water complex is formed, similar to that of water and glycerin. The acetone is able to penetrate into water interlayers of montmorillonite and to widen the basal spacings by formation of the mixed complex. At the same time, however, acetone acts as a dehydration agent reducing the number of liquid molecular interlayers in the mixed complex. This dehydration effect, however, is not observed in the vapor phases.

308. G. W. Brindley and M. Rustom[659] especially studied the adsorption of the polyethylene glycol ester of oleic acid (in the commercial wetting agent "Nonisol 250") on montmorillonite, in relation to the water content of the system. The long-chained molecule of the ester:

$$OH.CH_2.CH_2(OCH_2.CH_2)_x.OCH_2{-}CH_2O \,\Big|\, CO.(CH_2)_7.CH = CH.(CH_2)_7.CH_3$$

is able to form *ordered* complexes with one or two layers between the silicate sheets, and in the appropriate complex ranges, mixtures of one and two layers are observed rather than mixed sequences. If the organic "filler" molecule is gradually removed by repeated washings and leachings, the two-layer type can be systematically reduced to one-layer sequences, but the single layer type between the silicate sheets is firmly held and cannot be removed by a Soxhlet extraction. The X-ray analysis of different stages of adsorption for the organic material gave a spacing $d(001)$ of 18.5 to 19.0 Å., rather irrespective of the exchangeable cations (Na^+, Ca^{2+}, Mg^{2+}) in the amount of adsorbed organic substance. Only for sodium montmorillonite the spacing could range up to 22 Å. for small quantities of adsorbed material, prior to drying. Very remarkable is the evidence concluded from the X-ray data, for formation of *regular* structures as the main features of these organic complexes and only a subordinate irregular stratification. Characteristic, therefore, is the formation of an integral series of $(00l)$ Bragg reflections, demonstrated for the calcium montmorillonite complexes in Fig. C. 158. In a similar way, R. W. Hoffmann and G. W.

[658] Polar liquids in their typical imbibition and solvation behavior with montmorillonite were also discussed by W. D. Johns and R. T. Tettenhorst, *Am. Mineralogist*, **44**, 1959, 894-896; cf. B. ¶ 108, especially concerning glycerin and ethylene glycol.

[659] *Am. Mineralogist,* **43**, 1958, 627-640.

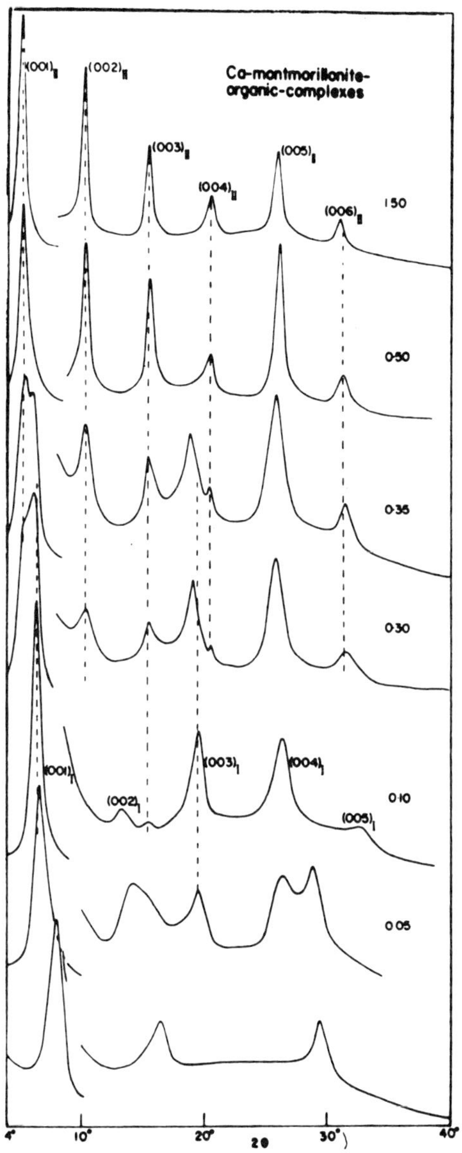

FIG. C. 158. X-ray recording of calcium montmorillonite complexes with the polyethylene glycol ester of oleic acid (Brindley and Rustom). Numbers at the right side indicate ratio of total organic material to amount of clay in the system.

Brindley[660] studied the adsorption of *nonionic* organic compounds, especially more or less water-soluble 1,3- and 1,4-diketones, esters of polyethylene glycols, polyethylene glycols, and others, not obstructed by exchange reactions. With increasing concentration, the diketones showed progressively increasing adsorption, whereas the polyethylene glycols and their esters reached a saturation limit of adsorption. In the same way as in the studies of Brindley and Rustom, the tendency to form one and two layers in preference to mixed-layer formations was most evident.

309. In this connection we may also emphasize the great importance of so-called *plasticity aids* (plastifiers) for ceramic bodies. They are based on polar, long-chained organic molecules. The classical substance of this kind is the lignin sulfonate group of products from the sulfite cellulose waste brines and from paper pulp manufacturing. They find a wide application in ceramics, but also for making cement pastes of higher workability (cf. Vol. V, Section C), and "lubricating" extrusion mixes. G. C. Robinson and J. J. Keilen[661] gave a detailed description of the specific action of commercial substances, of the sodium lignin sulfonate type of "Polyfon T" and related materials, for ceramic shaping purposes in extrusion processes. The surface-active agents (surfactants) of this class especially influence the optimum water content of the bodies, reducing it markedly in satisfactory extrusion. Equally, they increase the speed of extrusion for a constant power consumption: when used in a 0.4 per cent concentration, they increase the dry strength by 77 per cent; with a 0.2 percent addition, they increase the dry strength by 63 per cent. Similar specific actions as plasticity aids were described by J. Stawitz[662] for methylcellulose (MC) and carboxymethylcellulose (CMC). The latter is not a well defined chemical compound but includes a group of anion-active colloid electrolytes of the general type:

[660] *Bull. Geol. Soc. Am.*, **70**, 1959, 1618-1619. See also *Am. Mineralogist*, **46**, 1961, 450-452; *Geochim. et Cosmochim. Acta*, **20**, 1960, 15-29; *J. Phys. Chem.*, **64**, 1960, 1655-1662 (with L. G. Tensmeyer); *Ibid.*, **65**, 1961, 443-448, especially on clay-organic complex formation and infrared absorption spectra.

[661] *Am. Ceram. Soc. Bull.*, **36**, 1957, 422-430, with extensive literature references. A very complete discussion of characteristic physicochemical properties of surfactants was given by A. M. Mankovich, *Ind. Eng. Chem.*, **45**, 1953, 2759-2766. Another practically important effect of surfactants is the hydrophobicity of hydraulic binders developed by additions of such agents to grinding cements, as theoretically discussed by M. I. Khigerovich, *Silikattechnik*, **8**, 1957, 534-536 (cf. Vol. V. C. ¶ 218).

[662] *Ber. deut. keram. Ges.*, **32**, 1955, 304-305; *Ibid.*, **35**, 1958, 56-59. Chemical purity makes MC and CMC products superior to common plastifiers like gum arabic, starch, tragacanth, or gelatine, with their higher resistance to decomposition by bacteria and fungi. Especially in modern oxide ceramics, the slurry bodies are commonly plastified by additions of starch or amyloid starch; cf. A. Novikov, *Ogneupory*, **22**, 1957 (1), 12-21.

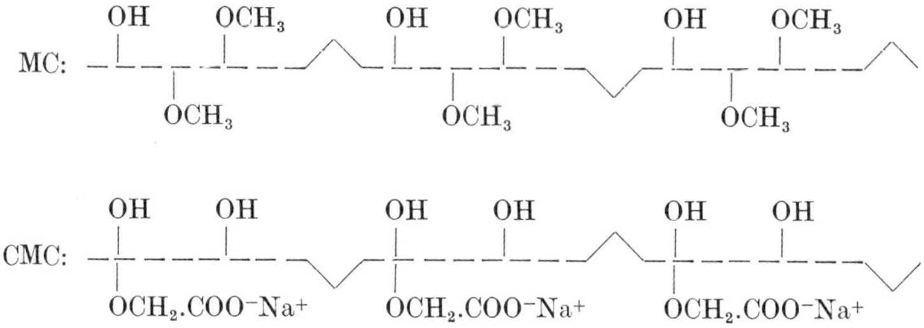

(/\ and \/ mean bridges between glucose rings)

The strong adsorption action for cations exerted by the surface of a clay mineral particle builds up a layer of sodium ions around it, which attracts, in its turn, the complex anions of the surfactant. The increase in the electrophoretic migration rate of such clay particles with adsorbed CMC complexes is evidence of this mechanism. The migration speed, U, was measured for the particles of a 1 per cent bentonite suspension. It is equal to 3.25×10^{-4}. For an addition of 0.005 per cent CMC, $U = 3.54 \times 10^{-4}$ and for a 0.05 per cent addition $U = 4.45 \times 10^{-4}$ cm./sec.

310. W. C. Ormsby, R. M. Witucki, and W. A. Weyl[663] developed elaborate experimental techniques for measurement of the deformation of kaolinite-water systems under the action of different wetting agents of the anionic, cationic, and nonionic reagent classes. We omit here a detailed enumeration of characteristic chemical compounds active in the commercial products here concerned. But we emphasize how important results of the determination of deformation behavior are for the problems of plasticity in ceramic bodies, chiefly if one critically discusses the theory of "stretched membranes" of F. H. Norton, and yield point determinations by W. D. Kingery and J. Francl (cf. C. ¶ 240). We may further emphasize how important it is to consider the effects of preferred orientation of clay particles gliding under the "lubricant" effects of organic additions. Stress-strain diagrams obtained in an elementary plastometer test, and filtration experiments, with kaolinite-water systems containing wetting agents, also play an important role in the correct interpretation of the plasticity-workability improving actions of the surfactants. We finally recall how different are the actions of cationic and nonionic reagents, on the one hand, and of anionic agents, on the other, for the decreasing and increasing phenomena in the deformability of those systems, respectively (cf. C. ¶ 241, and Fig. C. 126).

311. In all the interactions of clay-water systems, particularly those with organic additions, the role of the dielectric properties of the liquid phase, first water and then aqueous solutions, is of fundamental significance. The dielectric constant of water in wet clays was determined systematically by L. S. Palmer[664] The dielectric constant of mixed clay-water systems cannot be simply identified with that of a common

[663] *Natl. Acad. Sci.-Natl. Research Council, Publ.*, **456**, 1956, 251-272.
[664] *Proc. Roy. Soc.* (*London*), **65** B, 1952, 674-678.

mixture and it does not follow the elementary additivity rules. Relatively dry clay consists of closely packed, water-enveloped particles in a matrix of air. Wet clay consists of the particles uniformly distributed in a matrix of water. Thus the effective dielectric constant of the water associated with clay particles varies exponentially from that of free water (about 80) to that of bound water (about 3), as a function of the decreasing water content. In every case, the water films in clay-water systems have a very low dielectric constant in comparison with that of water in bulk when measured with high-frequency methods.

NEGATIVE ADSORPTION ON CLAY MINERALS; HYDRATION AND PERMEABILITY PHENOMENA

312. The particular phenomena of the change from positive to negative adsorption as of chloride anions on kaolinite, were studied by R. K. Schofield and H. R. Samson,[665] and explained by the positive charges developed on the edge faces of the crystallites under acid conditions. The original negative charges on kaolinite are evident by the adsorption of exchangeable sodium ions if the kaolinite is treated with a 1 *M* solution of sodium chloride containing a 0.001 *M* solution of HCl. The adsorption is then combined with an intense flocculation. If the clay mineral is further shaken in a 0.005 *M* NaCl solution, it adsorbs particularly the chloride anions, making evident a positive charge on the edges. The attraction of the positive charges on the edge faces for the negative charges of the "body" of the kaolinite crystals causes the flocculation in the absence of any salt. A small but sufficient amount of sodium hydroxide deflocculates the salt-free suspension. Deflocculation is also caused by small amounts of sodium oxalate, pyrophosphate, polymetaphosphate, alginate, and by sodium bentonite. Regardless of the agent used, when kaolinite is deflocculated in the absence of salt, the crystals repel chlorine anions when placed in a dilute sodium chloride solution. Natural kaolins sometimes contain small amounts of montmorillonite with its characteristic negatively charged tabular crystallites. If these are lying against the positively charged edge faces of the much larger kaolinite crystals, they must profoundly affect the working properties of the kaolin. G. H. Bolt and B. P. Warkentin[666] studied the negative adsorption of chlorine anions in sodium montmorillonite over a wide range of electrolyte concentrations. If the clay mineral was purified by ultrafiltration, a good agreement was observed between the measured and calculated data, the latter being derived from a general equation based on the Gouy theory of diffuse double layers and on derivations given by R. K. Schofield.[667]

[665] *Clay Minerals Bull.* 2, 1953, 45-51; especially in *Discussions Faraday Soc.*, No. 18, 1954, 135-145.

[666] *Kolloid-Z.*, 156, 1958, 41-46.

[667] *Nature*, 160, 1947, 408.

If the electrolyte concentration is very low, the measured negative adsorption is found smaller than the calculated amounts, probably because of the presence of positive charges. Polyanions, especially those of polymetaphosphoric and acrylic acids, increased the negative adsorption in this range to the theoretical values.

313. The adsorption processes under the action of dipole liquids on the surface of highly dispersed materials (especially silica gels, clay minerals, but also on the products of the hydration of hydraulic substances; cf. Vol. V, Section C) were discussed from the physical viewpoint by B. V. Il'in,[668] based on the rich experimental material collected in the last decade in the school of A. V. Kiselev, V. F. Kiselev, and S. P. Zhdanov, and other Russian authors. An approximative relation for calculating the energy of adsorption, U, was developed by Il'in in the form $U = \mu\, e_1/r_0 S_n$ with μ the dipole moment of the adsorbed molecule, e_1 the electrostatic charge of the touched ion of the adsorbent solid (crystal), r_0 the distance between ion and dipole, and S_n the surface required for the adsorbed molecule. The equation given here was examined in its application for water, phenol, monochlorobenzene, nitrobenzene, and a homologous series of alcohols, on strontium, barium, and lead sulfate.

314. The commonly used experimental method for determination of *adsorption energies* is identical with that of heats of immersion, of wetting by the classical calorimetric instruments, specifically the sensitive twin calorimeters. V. F. Kiselev[669] discussed modern calorimeters for this purpose with highly sensitive temperature indicators and electronic self-controlling systems, especially for the adiabatic twin calorimeter principle. The results of determinations with such instruments are accurate to $\pm$ 1 to 2.10^{-3} cal./gm.; microcalorimeters are accurate even to 4.10^{-4}cal./gm. For practical purposes, as for the investigation of hydraulic binders and common industrial materials of high dispersity, less sensitive calorimeters are often used, especially in the determination of *heats of wetting.*

315. Kiselev[670] gave a very characteristic example of the interrelation of specific surface for highly dispersed adsorbents (silica, and mixed silica-alumina gels) with absolute isotherms (see C. ¶ 344) and heats of adsorption. M. M. Egorov, K. G. Krasil'nikov, and E. A. Sysoev[671] determined the heat of wetting for well-defined silica gels (of the specific surface data 695, 400, and 275 m.^{2}/gm., determined by the BET method) as a function of heating temperatures. At 300°C. there is a distinct maximum of the heat of wetting of such gels. Above 300°C. heats of wetting

[668] *Trudy Soveshchaniya Khim. Tsementa, Nauch.-Tekh. Obshchestvo Prom. Stroitel. Materialov,* 1956, 235-242.

[669] *Trudy Soveshchaniya Khim. Tsementa, Nauch.-Tekh. Obshchestvo Prom. Stroitel. Materialov,* 1956, 243-263.

[670] *Metody Isledovaniya Struktury vysokodispersnykh i poristykh Tel, Trudy Soveshchanya,* Acad. Sci. U.S.S.R., Moscow, 1951 (publ. 1953), 86-113.

[671] *Doklady Akad. Nauk S.S.S.R.,* **108**, 1956, 103-106.

are decreased in parallel with decreasing water contents, but both parameters are rather unchanged over the range of 700° and 1000°C. There is no direct proportionality between the water content and the heat of wetting; the latter, Q, for 1 m.^{2} surface, is equal to $q_1S_1 + q_2S_2$, in which q_1 is the heat of wetting for the completely dehydrated surface, S_1, and q_2 that of the completely hydrated surface, S_2, with $S_1 + S_2 = 1$. The value for q_1 was found by extrapolation to be 50 ergs/cm.^{2} and Q was found to be between 110 and 200 ergs/cm.^{2} for gels of different degrees of hydration. In any case it is not correct to speak of an "absolute" heat of wetting for any silica gel or related material. G. I. Aleksandrova, V. F. Kiselev, K. G. Krasil'nikov, V. V. Murina, and E. A. Sysoev[672] determined with the same silica gels the heats of wetting for water, methyl alcohol, propyl alcohol, and heptane.

316. The formula $Q = (q_2 - q_1)S_2 + q_1$ is also valid here, but the slopes of the straight lines $dQ/dS = (q_2 - q_1)$ differ for different liquids, indicating individual features of adsorption in the interaction of liquid molecules with active groups of the adsorbent. The heats of wetting of the organic liquids are determined for differently porous gels in the same state of hydration, Q, as a linear function of the water content. For heptane and propyl alcohol the lines for different porosities are parallel. In totally dehydrated gels, for methyl alcohol $Q = 75$ ergs/cm.^{2}, and for water $Q = 50$ ergs/cm.^{2} (see above). For a gel with 4.5 μmoles water, $Q = 230$ ergs/cm.^{2}; for heptane adsorbed on two gels of different porosity, $Q = 75$ and 50 ergs/cm.^{2}. Even for hydrocarbons, heats of wetting cannot be regarded as characteristic constants of gels in general. A. A. Isirikyan and A. V. Kiselev[673] studied the adsorption of methyl alcohol, butane, hydrocarbons, and nitrogen near the state of complete capillary condensation, in which the differential heat of adsorption is a maximum (see Fig. C. 159), and by high-precision calorimetry (for n-pentane). The caloric effects are a maximum for definite values of the relative pressure, p/p_s. For a fine-porous gel the maximum is at $p/p_s = 0.4$ to 0.7; for a wide-porous gel, the maximum is at 0.95. By a reduction of pore diameter from 100 to 20 Å., the differential heat of wetting is increased by about 40 per cent and the difference $Q - L$ (heat of condensation) increased by 300 to 400 per cent.

317. The investigations of the Russian authors are in many points analogous to other studies on clay minerals in their water adsorption behavior. We first mention the investigations of H. Balduin and P. Wieden[674] with activated bleaching earths of commercial origin and, therefore, also in relation to the peroxide value of tetraline (cf. C. ¶ 218). A strict separation of heats of wetting and reaction was in many cases not possible, say if polymerization reactions occur with additionally superimposed

[672] *Doklady Akad. Nauk S.S.S.R.*, **108**, 1956, 283-286.
[673] *Doklady Akad. Nauk S.S.S.R.*, **110**, 1956, 1009-1012.
[674] *Ber. deut. keram. Ges.*, **29**, 1952, 344-350.

heat evolution effects. This corresponds to early observations of L. Gurwitsch[675] that amylene adsorbed on Floridine (Florida fuller's earth) or on alumina gels, undergoes a polymerization reaction and evolves unusually high heat amounts. The same is the case with pinene adsorbed on bentonites, and the like. Representative data of Bal-

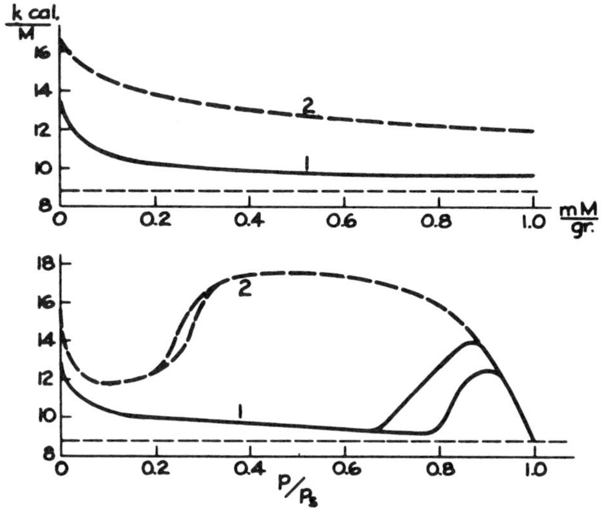

FIG. C. 159. Relation of heat of adsorption for n-pentane of a coarsely porous (curves 1), and a finely porous (curves 2) silica gel (Isirikyan and A. V. Kiselev). The upper part of the figure shows caloric effects as a function of adsorbed amounts; the lower part as a function of p/p_s, up to saturation.

duin and Wieden on the water imbibition are nevertheless given in Table VI. It is very evident that the heat of wetting is not simply related to data of the bleaching tests, although with increasing dehydration both parameters are systematically increased (see Fig. C. 160).

TABLE VI

HEAT OF WETTING OF ACTIVATED BLEACHING EARTHS, AFTER WETTING WITH 20 CM.[3] WATER/GM.

	Q (cal.)	Bleaching test data [a]
Tonsil AC	19.3 cals.	2.04
Rusmil Extra	15.3	2.41
Bleaching Earth	25.5	2.09

[a] Ratio of layer thickness of bleached oil, to that of initial oil, for light wave $\lambda = 5{,}700$ Å.

[675] *Kolloid-Z.*, 32, 1923, 80-91.

318. Heats of immersion in water were measured by A. C. Zettlemoyer, G. J. Young, and J. J. Chessick[676] for bentonite, in connection with the problems of a first and second interlayer of water molecules between the crystal sheets, which

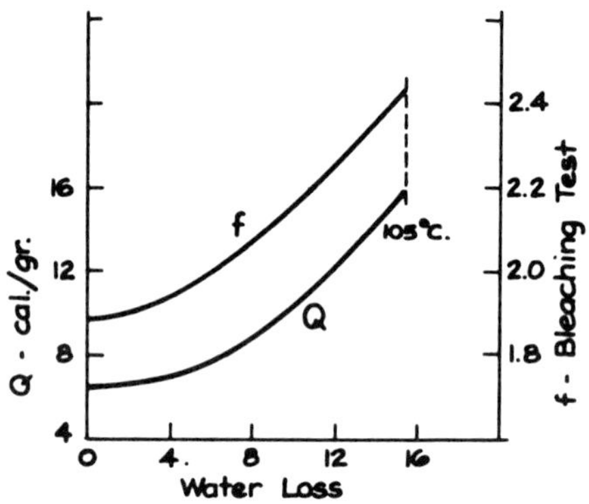

FIG. C. 160. Comparison of heat of wetting of "Rumsil Extra" (dried at 105°C.), with bleaching tests, as a function of water losses (Baldwin and Wieden).

could be well observed in the widening (swelling) effects of the basal spacings by X-ray analysis. The interrelation of caloric data with structural swelling effects are seen in Fig. C. 161. Also the heat of desorption of water from bentonite (activated at 100°C. for 48 hours) as a function of p/p_s shows a characteristic curve (Fig. C.

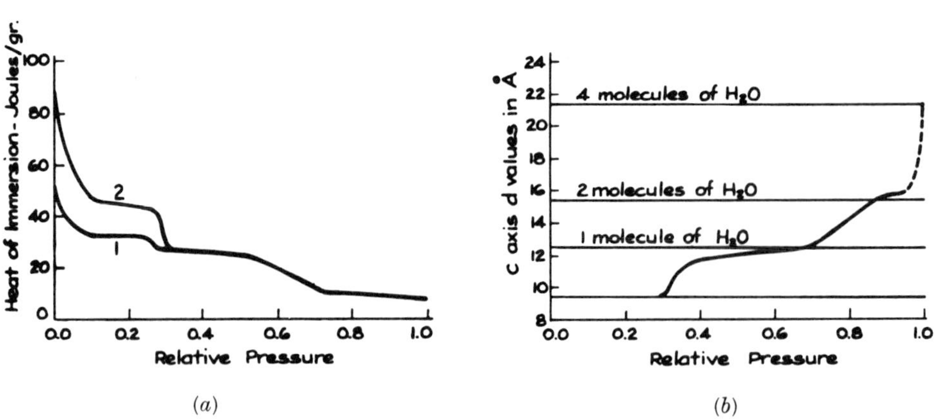

FIG. C. 161. (*a*) Heats of immersion of bentonite in water (degassed (1) at 25°C. for 48 hours; (2) at 100°C. for 24 hours), and (*b*) widening of the structural units in bentonite, both as a function of the relative pressure (p/p_s) (Zettlemoyer, Young, and Chessick).

[676] *J. Phys. Chem.*, **59**, 1955, 962-966; *Ibid.*, **60**, 1956, 1181-1184, on attapulgite adsorption. On the Thermistor calorimeter used see *Ibid.*, **57**, 1953, 649-652, applied for determination of heat of wetting on asbestos.

162), obtained from heats of immersion in water and energy of vaporization of the adsorbed water at 25°C. The activation at 100°C. not only produces highly energetic sites on the external surface, but also between the silicate sheets. These sites are apparently exchange ion sites. W. H. Slabaugh[677] extended to measurements of the heats of wetting on sodium and calcium montmorillonite in water, butylamine,

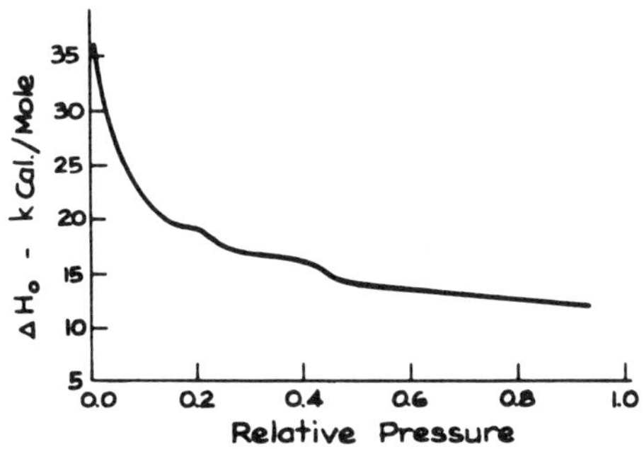

FIG. C. 162. Heats of desorption of water from bentonite at 25°C. (bentonite activated at 100°C. for 48 hours). (Zettlemoyer, Young, and Chessick).

butylamine hydrochloride, and alkyl($C_{12}H_{25}$)dimethylbenzylamine chloride (BTC) solutions. From the data a considerable difference between the reaction with the clay mineral of an amine and of an amine hydrochloride solution was calculated as seen in Table VII. The immersion of a clay mineral in a high-molecular weight amine salt solution evolves a surprisingly large amount of heat.

TABLE VII

HEATS OF REACTION BETWEEN CLAYS AND AMINES (CAL./GM. CLAY)

	With 0.1 M. solution of $C_4H_9NH_2$	With 0.1 M. solution of $C_4H_9NH_2.HCl$	With 0.1 M. solution of BTC
Centrifuged Bentonite	1.63	—0.26	5.96
Sodium Bentonite	0.87	0.07	5.28
Calcium Bentonite	0.08	—1.22	3.85
Hydrogen Bentonite	4.82	0.24	

319. A particularly sensitive vacuum flask microcalorimeter for determination of heats of wetting, and other data, for clays was developed by R. Greene-Kelly

[677] *J. Phys. Chem.*, **59**, 1955, 1022-1024.

and R. C. Gallavan,[678] with an accuracy of $\pm$ 1 per cent in the results. For montmorillonites particular precaution must be made to prevent a delayed wetting of the samples. Rapid wetting was obtained by suspending the powdered sample in glass wool. The microcalorimeter proved to be highly useful in discriminating swellable and non-swelling clay minerals, or to detect changes in surface energy of non-swelling minerals, with changes in particle sizes.

320. The calorimetric method for determination of heats of wetting and reaction was used by O. I. Lukyanova, E. E. Segalova, and P. A. Rebinder[679] for study of the action of detergents (surfactants) in hydraulic binder pastes. The characteristic induction period caused by the organic materials in the initial setting period (cf. Vol. V, Section C) can be studied very precisely in the caloric effects they produce.

321. In the school of A. V. and V. F. Kiselev, the study of heats of wetting as a means of understanding the mechanism of adsorption of aliphatic alcohols on silica gels and related highly dispersed systems was developed to a high level. In this wide field we mention particularly the recent investigations of V. F. Kiselev, K. G. Krasil'nikov, and K. V. Murina[680] based on calorimetric measurements (cf. the instrument of Aleksandrova and Krasil'nikov, C. ¶ 315) of the heats of wetting. These systematically decrease with increasing numbers, n, of carbon atoms in the aliphatic chains, but with a practically constant value above $n = 8$ (cf. Fig. C. 163). The heat of wetting is proportional to the energy of bonding between the molecules with the adsorption centers and concentration of the adsorbed molecules on the surface. The function of increasing water contents of gels with pores of 60 to 65 Å. diameter and heat of wetting for water and methyl alcohol is linear. The same is valid for the series of normal aliphatic alcohols for a nonporous Aerosil, deviations occurring only for gels with very fine pores (44 Å.). L. G. Ganichenko, V. F. Kiselev, and K. G. Krasil'nikov[681] observed that with increasing dehydration the heat of

[678] *Clay Minerals Bull.*, **3**, 1957, 170-176.

[679] *Doklady Akad. Nauk S.S.S.R.*, **117**, 1957, 1034-1036.

[680] *Doklady Akad. Nauk S.S.S.R.*, **130**, 1960, 569-572; see also **114**, 1957, 1257-1260 (A. K. Bonetskaya and K. G. Krasil'nikov).

[681] *Doklady Akad. Nauk S.S.S.R.*, **125**, 1959, 1277-1280. See further M. M. Egorov, T. S. Egorova, V. F. Kiselev, and K. G. Krasil'nikov, *Doklady Akad. Nauk S.S.S.R.*, **114**, 1957, 579-582, on the "absolute" adsorption isotherms of water vapor, the *variable* area of H_2O molecules occupied on the surface, in analogy with S. P. Zhdanov's determinations of the "absolute" isotherms of adsorption on fine quartz powders, *Ibid.*, **120**, 1958, 103-106; *Ibid.*, **123**, 1958, 716-719. The variability of surface area occupied by one water molecule on quartz was explained by W. Stöber, *Kolloid-Z.*, **145**, 1956, 17-46, by assumption of contaminating clay mineral particles on natural quartz crystals, but Zhdanov is of the opinion that water may be present in inclusions and occluded forms, especially on finest cracks. See also L. G. Ganichenko, M. M. Egorov, V. F. Kiselev, K. G. Krasil'nikov, and G. S. Khodakov, *Doklady Akad. Nauk S.S.S.R.*, **131**, 1960, 597-600, especially on strong effects of amorphization of quartz by continued grinding. The

wetting of silica gels decreased with increasing numbers n of the carbon atoms in aliphatic alcohols, whereas that of n-pentane (in CCl_4 solution) was not changed. With increasing n the sorption properties appeared to approach those of corresponding hydrocarbons. The alcohols are sorbed on the silanol portion of the active gel surface, whereas the hydrocarbon radicals may be oriented more or less perpendicular to it (cf. the same phenomena with clay minerals as observed by A. Weiss, A. Mehler, and U. Hofmann, in C. ¶ 298 f.). This, however, is *not* confirmed by the experiments of Ganichenko, Kiselev, and Krasil'nikov, who come to the opinion that the molecules lie rather flat on the surface although more inclined to it the longer the hydrocarbon chain.

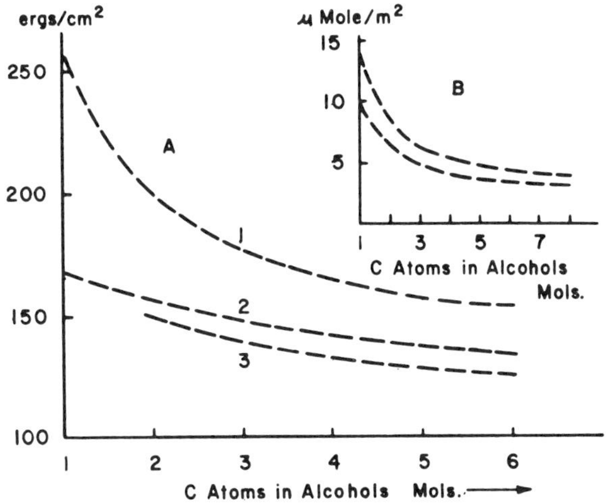

FIG. C. 163. Heat of wetting (A), and of specific adsorption (B), for the homologous series of normal alcohols, on silica gel calcined at: (1) 300°C.; (2) 500°C.; and (3) 700°C. (V. F. Kiselev Krasil'nikov, and Murina).

322. The important conditions of adsorption and heats of adsorption on aluminosilicate cracking catalysts were especially studied by M. M. Egorov, Yu. A. Zarif'ants, V. F. Kiselev, and K. G. Krasil'nikov.[682] Such commercial alumina-silica gels of 15, 30, and 50 per cent alumina were first determined by their specific surface evolution

energy of the siloxane surface is given as 259 ergs/cm.², that of silanol as 129 ergs/cm.²; cf. St. Brunauer, C. H. Weise, and D. L. Kantro, *Can. J. Chem.*, **34**, 1956, 1483-1498; *Research and Develop. Labs. Portland Cement Assoc. Research Dept. Bull.*, No. 76, 1956.

[682] *Doklady Akad. Nauk S.S.S.R.*, **120**, 1958, 326-329. Increased heats of wetting are also observed in vacuum-dehydrated silica gels, relative to air-dehydrated samples, especially in connection with *oxidizing* properties of the gels (cf. Vol. II, A. ¶ 396); see K. G. Krasil'nikov, V. F. Kiselev, and E. A. Sysoev, *Ibid.*, **116**, 1957, 990-993.

(BET method), their structural water content, and heat of wetting. The curves show the latter parameter in ergs/cm.^{2}, as a function of water content, and the characteristic maxima for those samples which were precalcined at 300°C. (see above). The heats of wetting for alumina-silica gels are considerably higher than for silica gels, that with 50 per cent alumina reaching 570 ergs/cm.^{2} At the same time, it has the highest structural water content (about 13.5 μmoles/m.^{2}), and the highest adsorption capacity for water and methyl alcohol vapors, the desorption curves being remarkably below those of adsorption. The general character of the curves for alumina-silica gels is much different from that of silica gels. The alumina-silica gels contain $[AlO_4]$ groups replacing $[SiO_4]$, and develop $\equiv$Si—OH and $\equiv$Al—OH bondings of acidity functions (cf. C. ¶ 208--¶ 215) of the active centers.

323. A mass-spectrometric analysis of the dehydration process in crystalline and amorphous forms of silica was performed by A. V. Bondarenko,[683] with a hot anode, without constrained ionization, to determine the presence of free positive ions (protons and H_3O^+) besides water molecules, as evolved from the heated silica samples, and frozen out in the high-vacuum apparatus. The samples were fine quartz powders (specific surfaces 35.8 and 8 m.^{2}/gm.), and silica gels previously used by Egorov *et al.* for adsorption studies and dehydrated between 400° and 1000°C. The sensitivity of the instrument was sufficient to detect a ratio of 1 ion : 10^5 in neutral water molecules, but only slight traces of masses 23, 19, and 41 were identified in dehydration products at 1000°C., corresponding to a thermal ionization of traces of sodium and potassium from the wolfram spiral of the anode. The absence of protons and H_3O^+ ions speaks decidedly against W. A. Weyl's theory of dehydration (cf. Vol. II, A. ¶ 396).[684] No hydrogen molecules were found, but oxygen was identified in the dehydration products of quartz and some gels with maxima for temperature range of 500° to 550°, and 900° to 950°C., most probably evolved from thermal decomposition of nitrogen oxides for the lower range, or derived from firmer adsorbed ("structural") oxygen on the silica surface for the higher temperature range.

324. Of the greatest theoretical interest is the change of surface qualities and of heats of wetting on fine-dispersed silica gels which, by a treatment with thionyl chloride, or trimethylchlorosilane, were "modified" and made hydrophobic in the same way as described for silica gels and clay minerals in C. ¶ 206 ff. A. V. Kiselev, A. Ya. Korolev, N. V. Kovaleva, and K. D. Shcherbakova[685] discussed such reactions

[683] *Doklady Akad. Nauk S.S.S.R.*, **125**, 1959, 573-576; cf. M. M. Egorov, T. S. Egorova, K. G. Krasil'nikov, and V. F. Kiselev, *Zhur. Fiz. Khim.*, **32**, 1958, 2624-2633.

[684] *Research (London)*, **3**, 1950, 230-235. Also W. Stöber, *Kolloid-Z.*, **145**, 1956, especially p. 30, considers Weyl's theories on the dehydroxylation mechanism as rather problematic from the energetic viewpoint.

[685] *Doklady Akad. Nauk S.S.S.R.*, **124**, 1959, 617-620.

in a wider scope for materials of molecular sieve type (cf. Vol. IV, Section B), the methoxylation, alkoxylation, chlorination, fluorination of graphite, or the H. W. Kohlschütter reactions with chlorosilanes (cf. C. ¶ 206 ff.). Kiselev *et al.* used silica gels and soots in parallel changing them to hydrophobic materials, especially with trimethyl chlorosilane (as vapor, or in benzene solution at 20°C. for 8 days), replacing hydroxyl groups by —O—Si≡$(CH_3)_3$. The adsorption isotherms of the initial and modified gels are highly different for water (Fig. C. 164). The initial gel has a normal S-shaped isotherm with capillary condensation; the modified gel has a reversible character, not changed by aging. Even near $p/p_s = 1$, no densely packed

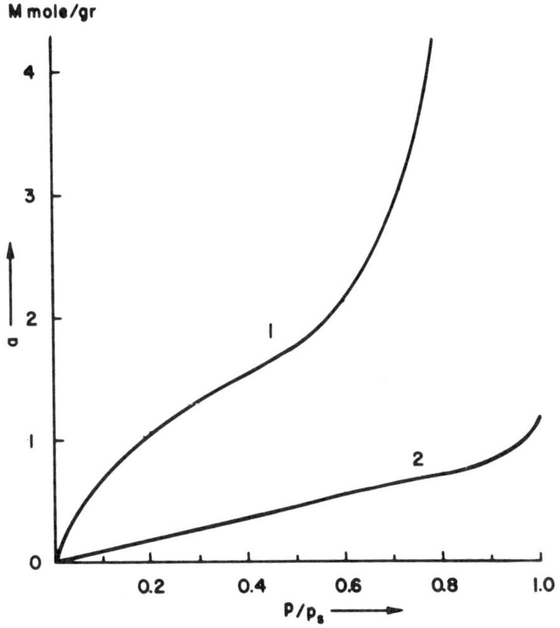

FIG. C. 164. Adsorption and desorption isotherms of water vapor on Aerosil: (1) for the hydroxylated surface; (2) for surface modified by reaction with trimethylchlorosilane (A. V. Kiselev, Kovaleva, Korolev, and Shcherbakova).

monolayer of water molecules is formed, capillary condensation is entirely suppressed and water does not wet the surface. For carbon black samples calcined at 1500°C., the general reaction type is entirely analogous, with a high degree of hydrophobicity, and reduced adsorption even for nonpolar reagents. Very contrary results, namely increased hydrophilic properties, are observed if a soot suspension is oxidized by sodium chlorite at 75°C. (pH = 3.75), to form carboxylic and phenolic groups. The adsorption isotherms have in this case the common S-shaped type with capillary condensation and a hysteresis.

325. I. Yu. Babkin, V. S. Vasil'eva, I. V. Drogaleva, A. V. Kiselev, A. Ya. Korolev, and K. D. Shcherbakova[686] calculated for trimethylchlorosilane-treated silica gels, from microanalytical data, the average degree of coating, Θ, (the area of one molecule $= 42$ Å^2, or the concentration $\alpha_m = 2.35$ groups $Si(CH_3)_3$ on 100 Å^2) with $\Theta = \alpha/2.35$. The protected surface has a mosaic structure, with $Si(HC_3)_3$ groups and "open gaps" which remain active for sorption reactions with water, methyl alcohol, hexane, and benzene. For the modified gel, $\Theta = 0.58$; for the same gel hydroxylated in the autoclave, $\Theta = 0.93$. The adsorption capacity with increasing modification, for hexane at $p/p_s = 0.1$, is reduced fifteen times in the autoclaved modified gel; for benzene, it is reduced sixty-five times. Benzene is adsorbed only on the $Si(CH_3)_3$ groups, water and methyl alcohol on the "gaps," and their slightly concave isotherms are rapidly decreased, with increased screening, by the modifying agent. These results are important for gas-chromatographic separations of higher hydrocarbons on silica gels. Babkin and Kiselev[687] further developed a model of the modified gel structure based on a tridymite skeleton, with the (0001) face covered by the methyl groups. For the model, the *energy of adsorption* can be calculated using a theory of N. N. Avgul, A. A. Isirikyan, A. V. Kiselev, I. A. Lygina, and D. P. Poshkus[688] which correlates this energy to the dipole moments, polarizabilities, and magnetic susceptibilities of the molecules in the adsorbent and the adsorbate. The equilibrium distance, z_0, of the adsorbed molecules from the modified layer is the sum of the van

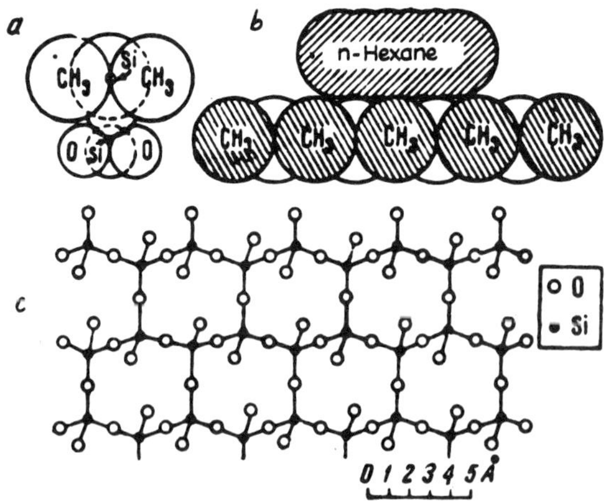

FIG. C. 165. Model showing the combination of the $Si(CH_3)_3$ group with silica layers (*a*); relation of the methyl groups with an adsorbed molecule of *n*-hexane (*b*); and tridymite-like structure of the adsorbent (*c*) (Babkin and A. V. Kiselev).

[686] *Doklady Akad. Nauk S.S.S.R.*, **129**, 1959, 131-134.
[687] *Doklady Akad. Nauk S.S.S.R.*, **129**, 1959, 357-360.
[688] *Izvest. Akad. Nauk S.S.S.R., Otdel. Khim. Nauk*, 1957, 1314-1327.

der Waals radii of the methyl groups and of the half-layer molecules adsorbed. For the tridymite substrate the attraction potential, Φ, of the adsorbed molecules is calculated from the distance 5.5 Å., which is the sum of the half-layer thickness of *n*-hexane molecules, and the distance of the modified layer surface to the plane of centers in the first basal plane of tridymite (Fig. C. 165). The curve Φ/z shows the characteristic deep minimum (Fig. C. 166). Because of the preponderance of disper-

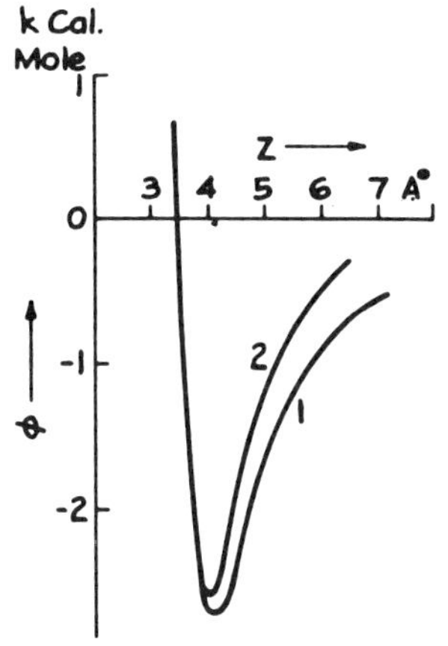

FIG. C. 166. Potential curve for adsorption (1) of *n*-hexane and (2) of benzene, on the surface of a modified silica gel covered with methyl groups (Babkin and A. V. Kiselev).

sion forces for the adsorption process, there is no great distinction between partial and total energies of adsorption for *n*-hexane and benzene (4,300 and 4,000 cal./mole, respectively). These are only half the heat of condensation, L, and the differences $\Phi - L = -3{,}300$ and $-4{,}100$ cal./mole are the true heats of adsorption, respectively.

326. A certain specificity of adsorption for copper, zinc, cobalt, and others on silica gels, by the introduction of organic compounds with primary, secondary, or tertiary amino groups is possible by a modification of the surface. The metals in such cases are able to form ammoniate-like complexes on the gels. I. E. Neĭmark, V. M. Chertov, R. Yu. Sheĭnfaĭn, and N. S. Kruglikova[689] demonstrated such modifications by a

[689] *Doklady Akad. Nauk S.S.S.R.*, **132**, 1960, 1356-1359. Effects here described must be distinguished from mere "*conditioning*" of silica gels studied by Z. Z. Vysotskiĭ and M. V. Polyakov, *Ibid.*, **129**, 1959, 831-834, as caused by changes in porosity, due to drying on H_2SO_4, and treatment with vapors of benzene, toluene, and *o*-xylene. The action of conditioning agents is that of a replacement of intermicellar liquid by aromatic hydrocarbons, which have, respectively, the following dipole moments: 0.0, 0.39, and 0.62 debyes; vapor pressures at 20°C.: 74.1, 22.5, and 10.1 mm.; parachors: 207.1, 246.1, and 258.1. Surface tension data, however, do not fit the regularities observed with those parameters. I. E. Neĭmark and V. M. Chertov,

reaction of amino alcohols with the hydroxyl groups of the silica gel at 100° to 160° on the refluxing condenser, or at 160° to 250°C. in the autoclave, and a subsequent vacuum drying at 100° to 180°C. Steric hindrance effects are observed with secondary and particularly with tertiary amines. Adsorption curves for carbon dioxide and acetic acid (but not for methyl alcohol and dimethylamine) show characteristic differences in influences exerted by the modifying groups on the amounts adsorbed, and the isotherms as functions of p/p_s (Figs. C. 167 and 168). Carbon dioxide and acetic acids are better adsorbed on ammoniated gels than on hydroxylated surfaces. The chemosorption is stronger with the tertiary amino-gel than with that modified by the primary amine.

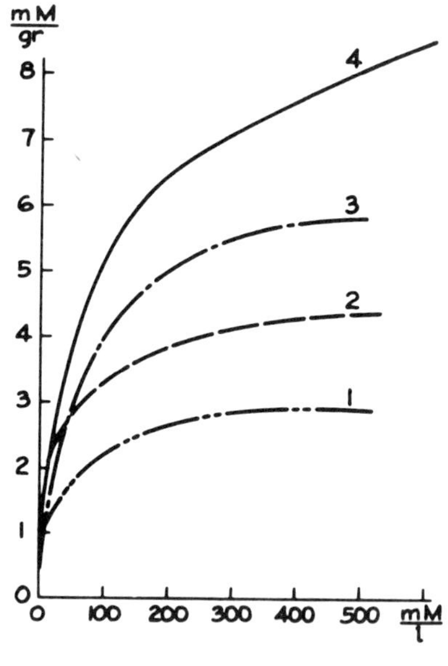

FIG. C. 167. Adsorption isotherms of methyl alcohol (from CCl_4 solution) on silica gel: (1) in the original state; (2) modified by introduction of 1.75 meq./gm. $C_2H_4NH_2$; (3) of 1.75 meq./gm. $(C_2H_4)_3N$; (4) of 3.3 meq./gm. $C_2H_4NH_2$ (Neĭmark, Chertov, Sheĭnfaĭn, and Kruglikova).

327. W. Stöber and his collaborators worked, in the same direction as the Russian authors mentioned above, on surface properties of silica in different forms, with special reference to silicosis problems. A monograph-like investigation on adsorption properties of quartz powders by W. Stöber[690] must here particularly be emphasized;

Ibid., **138**, 1961, 877-879, describe the synthesis of silica gels with introduction of *sulfonic acid* radicals (by reaction with diphenyl dichlorosilane, then with sulfonation with $H_2SO_4 + SO_3$). Such modified silica gels have interesting properties in reactions with amines and many become important ionic exchange material in strongly acid media.

[690] *Kolloid-Z.*, **145**, 1956, 17-46, with extensive bibliography. See further on chemisorption

this author examined the constitution of the hydroxylated surface in relation to the crystallographic-structural orientation of quartz.[691] The experimental identification of surface-active hydroxyl groups was made with semiquantitative fluorescence tests using the adsorbed dyestuff ions of Rhodamine B. Stöber gives an

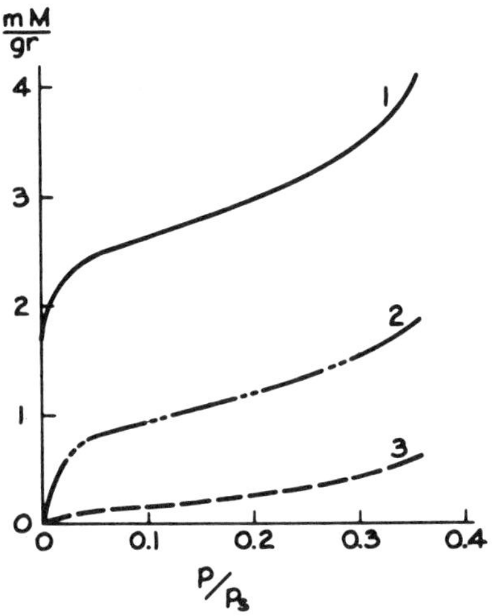

FIG. C. 168. Adsorption isotherm of diethylamine vapors on silica gel; (1) in the initial state; (2) modified by monoethanolamine; (3) by triethanolamine, both in 1.75 meq./grm. concentration (Neĭmark, Chertov, Sheĭnfain, and Kruglikova).

exhaustive demonstration of thermal decomposition of the silanol units, as schematized in Fig. C. 169. The data observed are in agreement with an average surface covering concentration of 6 μmoles/m.². The regeneration of the silanol groups by water and methyl alcohol vapors has to overcome an activation threshold of more

and esterification of alcohols on amorphous silica (Aerosil) W. Stöber, G. Bauer, and K. Thomas, *Ann. Chem. Liebigs*, **604**, 1957, 104-110. The esterificated surfaces are hydrophobic; the esterification reaction occurs more promptly on surfaces of type *d* in Figure 169, than on those of type *a*. For the investigation of *silicosis* problems, the studies of A. Weiss, G. Reiff, and G. Hofmann, Die Staublungenerkrankungen, III, *Wiss. Forschungsber. Naturw. Reihe*, **66**, 1958, 79-91, on water stable esters of silicic acid, are of particular importance in connection with the analysis of dusts from silicotic lungs (cf. Vol. II, ¶ 334 ff.). See also *Beitr. Silikose-Forsch.*, Sonderband "Grundfragen der Silikoseforschung." III, 1960, 45-68. We owe to Mrs. Fr. Oberlies, Würzburg, the following important additional information on investigations to these problems, by H. Lüchtrath and K. G. Schmidt, *Beitr. Silikose-Forsch.*, 1955. No. 37,43 pp.; 1956, No. 44, 32 pp.; 1959, No. 58, 37 pp.; 1959, No. 61, 60 pp., on intratracheal animal experiments on silicosis-releasing silicates.

[691] Cf. on this important point: H. Seifert, R. Buhl, and K.-F. Seifert, *Kolloid.Z.*, **141**, 1955, 146-159, and C. ¶ 187.

than 11,000 cal./mole. With aliphatic alcohols a partial esterification occurs, according to the scheme shown in Fig. C. 170. The reversible share in adsorption is explained by hydrogen bridge bonding to OH or to the surface ester groups. Ammonia and sulfur dioxide undergo only a slow and reversible specific adsorption by hydrogen bonding. Hydrogen sulfide and carbon dioxide are even nonspecific in their

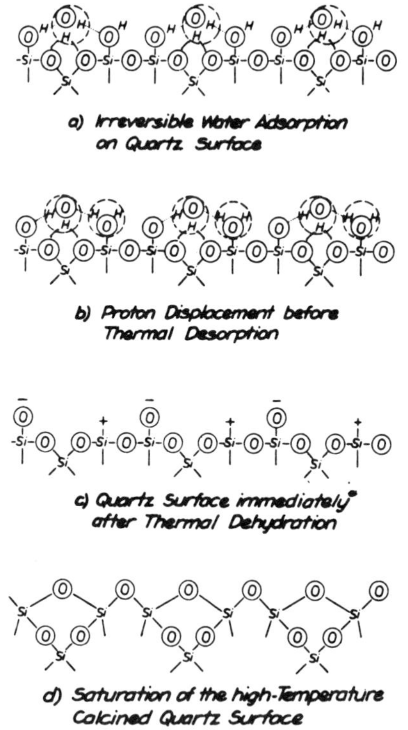

FIG. C. 169. Scheme of thermal transitions on the quartz surface (Stöber).

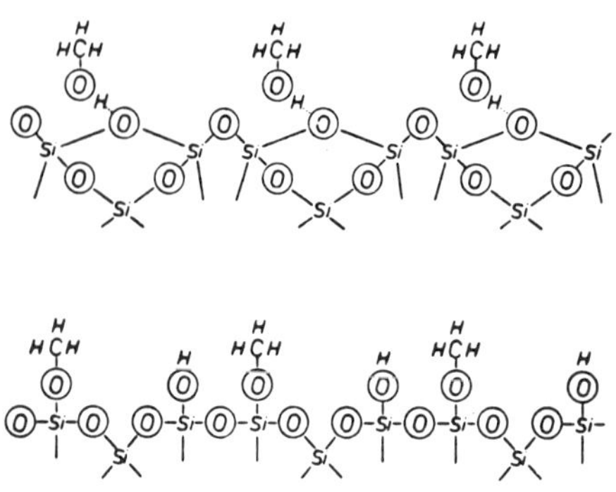

FIG. C. 170. Esterification of methyl alcohol on a calcined surface of quartz (Stöber).

mere van der Waals surface sorption. Also of interest is the hydrogen exchange catalysis on quartz surfaces, and the decomposition of propyl alcohol into water and propylene, as shown by Fig. C. 171.[692].

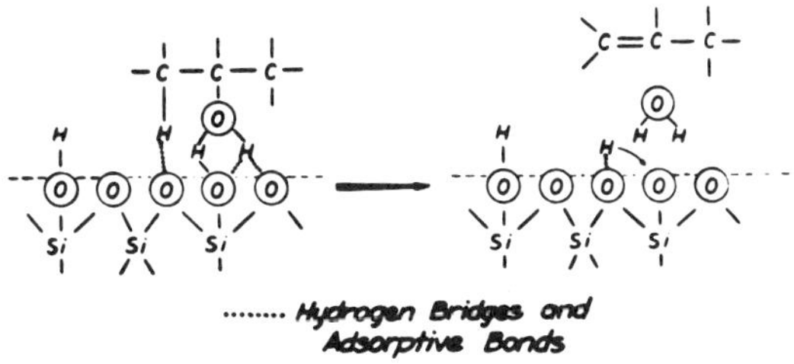

FIG. C. 171. Dehydration of higher alcohols on oxide surfaces (Stöber).

328. For mixtures of aliphatic alcohols and water, G. Bauer and W. Stöber[693] found a considerable reduction of surface esterification, even if small amounts of water are present (Fig. C. 172). For ethyl amines these authors[694] calculated the maximum surface concentration in monomolecular covering for diethylamine $n_{\infty D} = 5.4$

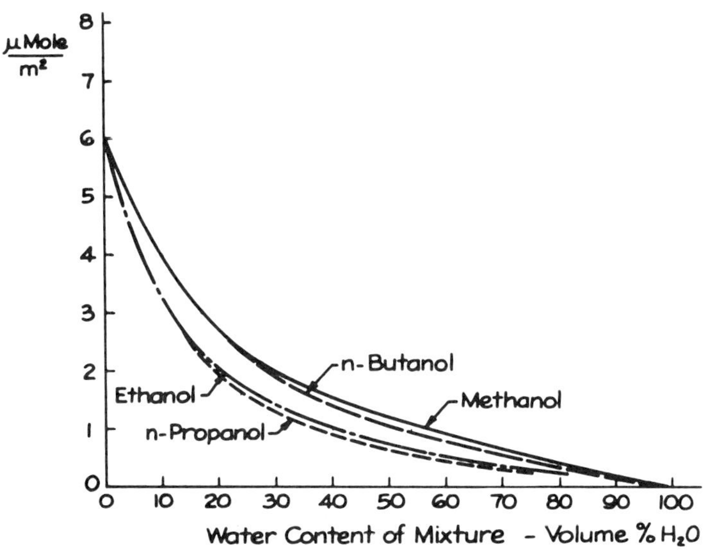

FIG. C. 172. Occupation of the surface of Aerosil by alcohol-water mixes (Bauer and Stöber).

[692] Cf. A. Eucken, *Naturwissenschaften*, **36**, 1949, 48-53, on the theory of this dehydration catalysis.

[693] *Kolloid-Z.*, **160**, 1958, 142-146.

[694] *Kolloid-Z.*, **167**, 1959, 27-30.

μmoles/m.^{2}, and for triethylamine = 4.4 μmoles/m.^{2}. The heats of adsorption for Aerosil at 25°C. and after calcination at 800°C. are

$$W_a = 2.303 \cdot RT \cdot \log b \cdot \frac{n_\infty RT}{2 r_a}$$

with r_a the average radius of the adsorbed molecule, and the constant b in two correlated Langmuir isotherms for

$$n = \frac{n_\infty bp}{1 + bp}$$

as tabulated in Table VIII.

TABLE VIII

HEATS OF ADSORPTION OF ETHYL AMINES ON AEROSIL (in CAL./MOLE)

	Monoethylamine	Diethylamine	Triethylamine
For 25°-Aerosil:			
W_{a1}	13,800	12,400	10,900
W_{a2}	9,700	9,600	9,800
For 800°-Aerosil:			
W_{a1}	11,600	11,600	10,900
W_{a2}	9,800	9,500	10,100

An analogous examination of chemisorption of methylchlorosilanes on Aerosil, quartz, and silica glass by W. Stöber[695] gives a surface concentration of 2.5 μmoles/m.^{2}, if the hydroxyl occupation of the surface was intact before the chlorosilane was acting. If this condition is not fulfilled, the concentration is reduced to 0.65 μmoles/m^2. It is important to observe that the hydrolysis of hydrophobic trimethyl chlorosilane-treated Aerosil cannot start until after an induction period of about 20 hours, whereas nontreated Aerosil sends molybdate-active silica into solution immediately after wetting.

329. The formation of water hulls around colloid materials, especially around clay particles, was discussed from a general physical viewpoint by E. Gruner[696] because of its correlation with plasticity and workability in ceramic bodies. Gruner gives special attention to the property of plastic systems that enables them to undergo in the solid phase a chemisorptive bonding, which excels the trivial wetting

[695] *Kolloid-Z.*, **149**, 1956, 39-46.

[696] *Ber. deut. keram. Ges.*, **31**, 1954, 135-142.

and formation of a liquid hull. This decisive property rules out all the more specific factors influencing modifications in the quantitative constants of the material, such as crystallographic constants, particle size, and the like. Liquid hulls and their volume properties play a further decisive role in the setting phenomena of Portland cement pastes. Especially R. Dittrich[697] emphasized the necessity of a clear distinction of volume changes of liquid hulls around the setting particles and their total effects in pastes and mortars. For this purpose, R. Dittrich developed a quantitative testing method based on a "volumenometer" adapted to these special needs for hydraulic binders, to determine the volume changes in setting and hardening (Vol. V, Section C) combined with phenomena of shrinkage. For clay suspensions, E. Plötze, A. Rubin, G. Huckert, and J. Kiefer[698] determined the density of the adsorbed water

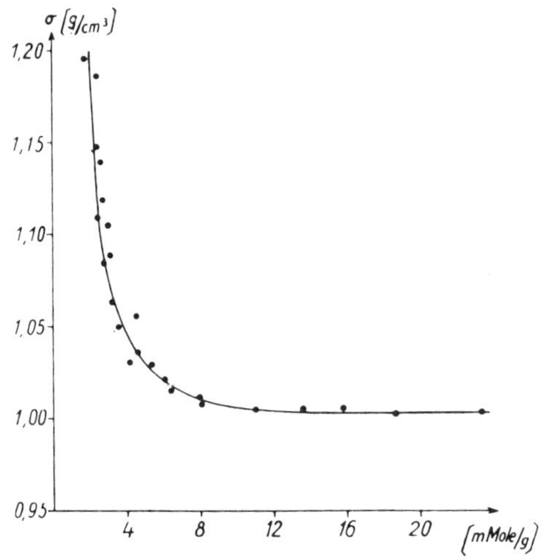

FIG. C. 173. Density of water adsorbed on kaolinite, as a function of water content of clay-water mix (Plötze, Rubin, Huckert, and Kiefer).

from volume changes, measured either by direct determinations of density and vapor pressure of the liquid phase or by electron microscopy, as well as indirectly from light diffraction and spectroscopic or surface tension measurements, electric resistance, dielectric constants, and pH ion concentrations. The adsorption state of water on clay particles is interpreted by Plötze *et al.* as highly dense packing with oriented dipoles in the adsorption layers. The curve in Fig. C. 173 shows that with decreasing water contents in kaolinite-water mixtures the density of water grows rapidly to 1.10 and even 1.20 if the water concentration is below 4 μmoles/gm.

[697] *Tagungsber. Zementind.*, No. 10, 1954, 51-54; No. 12, 1956, 158-60.
[698] *Kolloid-Z.*, **147**, 1956, 62-65.

330. The problems here involved are in a close interrelation with the technologically important properties of clay-water systems in *permeation* by liquids. The permeability of kaolinite was especially investigated by A. S. Michaels and C. S. Lin[699]; it decreases markedly with increasing polarity of the permeating liquid or gas. For the void ratio, e, and the specific surface, S, of the porous kaolinite aggregate, the permeability coefficient, K, is equal to $e^3/k_0S^2(1 + e)$, with k_0 as a function of the pore shape and the ratio of length of the actual flow path to bed thickness. For the additional "slip flow" correction given with the "slip factor," Z, is

$$K' = \frac{e^3}{k_0 S^2(1 + e)} + \frac{ZFe^2}{k_0 S(1 + e)}$$

with F the mean free path. In this way an experimental method was developed which made possible the comparison of different gas and liquid flow phenomena in a kaolinite bed. The results demonstrate that the classical Kozény-Carman equation (cf. C. ¶ 14) is *not* fulfilled. The specific surface of the kaolinite as calculated, however, is a linear function for all permanent fluids for $\varrho > 1$. Replacement of a polar by a nonpolar fluid by gradual desolvation resulted only in a small increase in permeability, for K and $\sqrt{k_0} \cdot S$. Significant increases occurred when water is replaced by a 10 to 30 per cent solution of sodium versenate (ethylenediaminetetraacetic acid salt). The most important factor controlling permeability is apparently the degree of dispersion or of disaggregation taking place in the suspension medium, colloid effects, such as adsorbed liquid films, being relatively subordinate.

331. On the role of electroosmotic counterflow, see C. ¶ 56. An increase in sodium ion content in the kaolinite introduced by a permeating solution of NaCl brings about a 200 to 300 per cent (irreversible) reduction in permeability caused by increased dispersion. Increasing the pH of a permeating fluid containing a fixed NaCl concentration is accompanied by a considerable and irreversible decrease in permeability of hydrogen-kaolinite. In part, this decrease is attributed to improved particle dispersion by trivial base exchange. Hydrogen kaolinite in a 0.5 N NaCl solution, acidified to pH = 2, has a permeability ($K = 2.32$) which is about half that of distilled water ($K = 4.25 \cdot 10^{-12}$ cm.). An interpretation of this remarkable phenomenon is not yet possible. Whether this particular effect is caused by a normal ion exchange and adsorption of H^+ ions by the clay, an increase in the dielectric constant of the solution, or by other factors remains to be investigated.

332. Drilling muds for mineral oil wells have a distinct effect on the permeability of sandstone cores as was discussed by H. T. Kennedy and R. A. Pfile.[700] Permeability

[699] *Ind. Eng. Chem.*, **46**, 1954, 1239-1246; on the counter-electroosmosis phenomena cf. *Ibid.*, **47**, 1957, 1249-1253. More recent studies on permeability of ceramic clay bodies see by W. Deen, *Ber. deut. keram. Ges.*, **37**, 1960, 489-492.

[700] *J. Phys. Chem.*, **59**, 1955, 870-873.

to aqueous solutions is considerably lower than air permeability of the dried cores. A very great effect is exerted by the swelling of clay mineral particles in the cores, evolved by the minerals coming in contact with alkaline solutions, as in filtrates from a lime-base mud with pH = 11.9, or a sodium silicate solution with pH = 10.9, with a practically complete suppression of permeability. An increased reduction of permeability is also observed if distilled water or highly dilute salt solutions are used. The susceptibility of the cores to plugging effects of the same solutions varied widely. There is apparently no clear interrelation between the action of different solutions and distilled water on one and the same core. Especially for bentonite, U. Hofmann[701] used a Baroid filter press for measuring the quantity of water pressed from a bentonite suspension under an air pressure of 7 atm. after 30 minutes, with a determination of the thickness of the filter cake.

THIXOTROPY PHENOMENA IN CLAY-WATER SYSTEMS

333. The characteristic framework structure of the clay mineral particles in bentonite-water mixtures, which show a high degree of thixotropy, was made most plausible in an elegant experimental method used by A. Weiss, R. Fahn, and U. Hofmann.[702] The stiffening from the shaken suspension is nothing but the reorganization of such a framework under the action of Brownian movements and of mutual attractions. In the gravity field or by centrifuging, the frameworks settle and collapse but they can be isolated by deep-freezing the gel and by a subsequent subliming of the solvent in a high vacuum. Very loose but distinctly coherent frameworks of the exterior shape of the original gels were thus isolated from calcium kaolinite, sodium bentonite, and of halloysite, that of kaolinite having the greatest mechanical strength, that of halloysite the lowest, and bentonite intermediate. The bentonite framework shows a remarkable elastic deformation like a sponge (compacting by 25 per cent of the original height), illustrating the lamellar type of the "house-of-cards" framework of montmorillonite crystallites (Fig. C. 174). The volume weight of such a bentonite framework is only 0.055. In general, the ratio of weights or volumes of the liquid and solid components of a thixotropic gel is related to the time required for stiffening of the system shaken in a glass tube, this time being reproducible and a characteristic of the system. Kaolinite, in thixotropic systems takes up 6 times its own volume. Sodium bentonite even takes up 35 times its volume because of the highly anisotropic shape of its particles. Prolonged centrifuging (90 minutes)

[701] *Angew. Chem.*, **68**, 1956, especially page 60; cf. "Standard Field Procedure of Testing Drilling Fluids," Am. Petroleum Inst., New York, July, 1942.

[702] *Naturwissenschaften*, **39**, 1952, 351 ff.; *Ber. deut. keram. Ges.*, **30**, 1953, especially pp. 24 f.; see also *Angew. Chem.*, **68**, 1956, pp. 59 f. See also the report of U. Hofmann, *Proc. Third Internation. Congr. Surface-Activity, Cologne, 1960* (publ. 1961), Vol. II. B-IV-4. No. 93, 658-673.

decreases the volume of thixotropic gels almost irreversibly. If the particles are isotropic (as in soots) a sphere packing of more or less loose cohesion takes place.

334. In a more extensive investigation[703] the authors mentioned above studied especially the thixotropy of kaolinite gels using H. G. F. Winkler's (1938, 1943) elementary method for determination of "thixotropy volume," for water and hydrocarbon liquids (hexane, benzene, toluene, and xylene). The thixotropy volume of kaolinite is generally the greater the less the liquid wets the surface of the clay mineral particles. The crystalline plates have possibly a maximum contact between one another. They are attracted by the action of mutual forces, and the lower the wetting, the better is the resulting cohesion, whereas with better wetting the local contacts between the particles are impeded and the gel becomes more dense and less voluminous, forming at the same time a framework of distinct solidity. The sedimentation volume (cf. C. ¶ 257 ff.) shows a symbatic analogy with thixotropic properties of kaolinites. For montmorillonite the volumes in nonaqueous media are very low, whereas an extremely high volume is reached for sodium bentonite in water. For different aliphatic alcohols as liquid media the thixotropy volume shows some analogy with the data of exterior friction of glass on glass in the same media, previously determined by R. Ramsauer.[704] One may indirectly understand the gliding of tabular kaolinite particles on one another on contacts in the framework. The thixotropy volumes for calcium kaolinite in alcohols of increasing chain lengths are distinctly lowered. Probably, the alcohols act like a protecting colloid orienting with the polar hydroxyl groups in the surface of the plates of kaolinite, whereas the hydrocarbon radicals are directed outwards. In consequence of this property the mutual "sticking" in a framework is the more impeded the longer the aliphatic chain radicals. Small additions of octanol to organic media are sufficient to suppress entirely the thixotropy of kaolinite suspensions.

335. In aqueous electrolyte solutions Hofmann, Fahn, and Weiss found for all alkali kaolinites an important increase in the sedimentation volume and of the liquid contents of equally rapidly consolidating thixotropic gels with the concentration of the electrolyte solutions, whereas for alkaline earth kaolinites they are little changed. Also low is the effect exerted by the anions. For alkaline earth kaolinites the sedimentation volume increases with temperature. Large distances in the structural layers correspond to low thixotropy volumina, and vice-versa. Innercrystalline swelling does not bring about any marked dispersion of montmorillonite crystals, even in highly dilute solutions, if alkaline earth ions are introduced, whereas sodium ions cause a strong increase in thixotropy volume, with increasing concentration of the salt solutions, and the X-ray diffraction data show an intense dispersion of the crys-

[703] *Kolloid-Z.*, **151**, 1957, 97-115.
[704] *Glastech. Ber*, **22**, 1948, 205-211.

tallites to finest lamellae, although the thixotropy volume indicates a tendency to build up a rather consistent framework. In general, there is a surprising analogy between the innercrystalline swelling of montmorillonite and the thixotropy of kaolinite in electrolyte solutions, caused by similar forces of cohesion and analogous thickness (of 5 to 10 Å.) of the water layers between the particles on their contacts, which is plausible from the house-of-cards structure.

336. Concerning the manner in which the contacting of the single particles takes place, it is not necessary only that they attract each other face-to-face; as shown in Fig. C. 174, it is also possible that face-to-edge contacts are equally suitable for

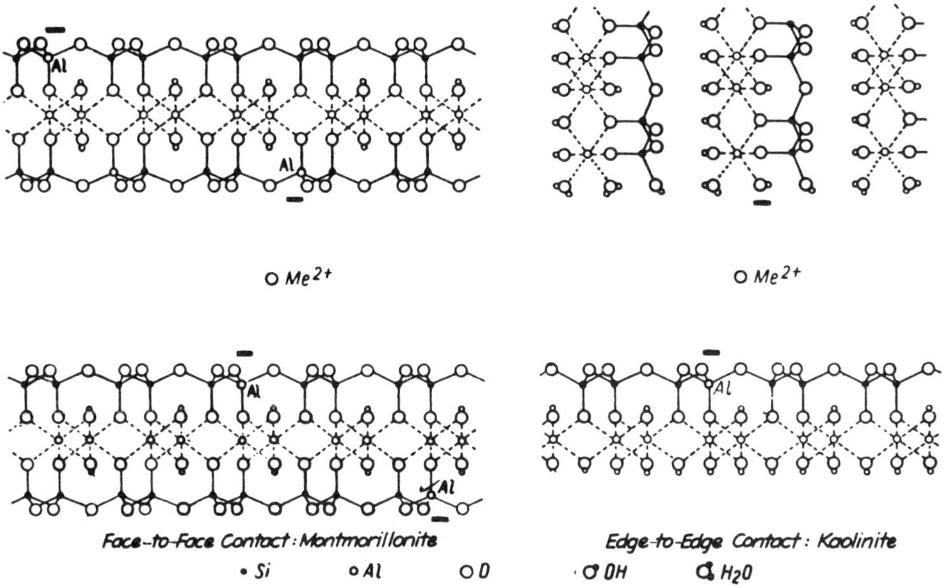

FIG. C. 174. Contact positions of kaolinite and montmorillonite particles in a consolidated thixotropic gel, either face to face, or face to edge (Hofmann, Fahn, and Weiss).

"flocculation" effects (cf. R. K. Schofield and H. R. Samson, C. ¶ 46, ¶ 312). This is especially true if substitution of $[AlO_4]$ tetrahedral groups occurs in the silica sheets, and a compensation of electrostatic charges by an Al—OH bonding takes place to explain the analogy of thixotropy of a kaolinite gel and the innercrystalline swelling of montmorillonite suspensions. Total potential curves between parallel plates, according to the theories of E. J. W. Verwey,[705] as given in Fig. C. 175, illustrate the

[705] *Kolloid-Z.*, **136**, 1954, 46-52. See also the fascinating discussion of surface tension and its role in mineral suspensions, by M. Papadakis, *Bulles et gouttes*, Symposium Paris June 1962, Question VI. E. (reprint, 5 pp.).

quantitative conditions for building up potential moulds by van der Waals-London forces in the case of the presence of electrolyte solutions. For the condition of monovalent electrolytes (as NaCl), a double layer potential $\psi = 100$ mv. (repulsion), and the attraction potential $A = 10^{-12.5}$. The practical importance of the results of Hofmann, Fahn, and Weiss consists not only in the analogy between innercrystalline swelling and thixotropy, but also in the possibility of understanding the deflocculation of ceramic slips (cf. C. ¶ 367), as Hofmann previously indicated.[706]

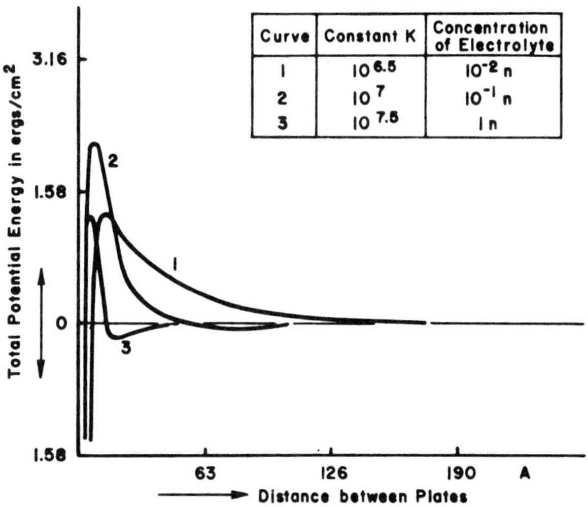

FIG. C. 175. Total potential curves between two parallel plates (Verwey).

337. The experimental determination of thixotropy, as in bentonite suspensions, was often criticized in its basic physical conditions. T. Alfrey and C. W. Rodewald[707] emphasized the necessity of sufficient control of temperature and well-defined agitation in manipulation of the tests. If the initial rates of restoration are well defined there is a characteristic functional relation with the rate of shear in the thixotropic system, and no influence of the measurement acts on the results caused by an alteration of the structure during the manipulation. Immediately after stirring the bentonite suspensions behave essentially as Newtonian fluids, but the gelation brings about a yield point phenomenon and reduces the slope of the flow curve. The gelation process is reversible, and a thorough stirring returns the system to a reproducible state as a sol, and in the same way the gelation is reversible and reproducible. There is, however, superimposed an irreversible aging process of consolidation, in

[706] *Ber. deut. keram. Ges.*, **26**, 1949. 21-32, especially pp. 28-30.
[707] *J. Colloid Sci.*, **4**, 1949, 283-298.

other words the gelation properties are different for gels of different ages. Temperature is the chief factor ruling the aging process. Alfrey and Rodewald further observed a specific effect of dioxane which, even in small amounts, reduces the tendency toward gelation in bentonite suspensions.

338. N. N. Serb-Serbina and T. M. Polonskiĭ[708] also measured the thixotropy of bentonite suspensions by determining the tangential shift of a platinum sheet (with an indentation to avoid sticking during gliding) the strains of which are plotted versus the relative deformation at a constant rate of increasing loads. The structural viscosity, η^*, was measured by a common Ubbelohde viscosimeter at temperatures from 0° to 80°C. for bentonite suspensions of 16 to 18 per cent concentration, gelating by thixotropy in 2 hours and more. With increasing temperature the solidity of the structures is markedly increased and sodium hydroxide considerably accelerates consolidation. The "plastic viscosity," η^*, increases with decreasing temperature. Particularly strong is the effect of temperature in the dispersity of "Askangel," but a reversal above 50° to 60°C. is indicated in the presence of small amounts of lithium chloride or sodium carbonate. Evidently, the bondings between particles of the consolidated thixotropic system are weakened at elevated temperature, and the surface hydration is reduced. Intensified Brownian movement accelerates coagulation and structure formation.

339. R. Dallendörfer[709] constructed an apparatus called "thixotrometer" based on the principle of an ultraoscillator forming a part in a Höppler falling-sphere viscosimeter. The oscillator, in tube form is made of barium titanate ceramic material of a practically homogeneous distribution of intensity over the cross-section. An increase in temperature during the oscillations cannot be avoided; therefore, the whole is built into an ultrathermostat, and the operation of oscillation measurements completed in a few seconds. The fall time of the spheres is determined by changes in capacity of inductivity in a resonance circuit with two coils in definite positions around the tube at $\nu = 998$ kc. The relaxation time of the suspensions must be in the order of magnitude of the falling time and no "instantaneous" viscosities are measured. If the thixotropy, however, is moderate, accuracy reached is $\pm$ 0.3 per cent. Dallendörfer also described a rotation viscosimeter in which a barium titanate oscillating tube is used over a frequency range from 200 kc. to 2.5 Mc. The coupling of movable parts is made by permanent magnets to eliminate vagabonding oscillations. This viscosimeter is, in principle, similar to that of A. Wesp,[710] which was developed for the study of polymerization processes in organic liquids, the torsion moment of the rotating cylinder being compensated by an astatic electromagnetic

[708] *Doklady Akad. Nauk S.S.S.R.*, **96**, 1954, 103-106.
[709] *Silikat Tech.*, **6**, 1955, 519-524.
[710] *Kunstoffe*, **41**, 1951, 213-219.

countermoment and the anode current of the circuit used as the measure of apparent viscosity, η^*.

340. The structural yielding of thixotropic clay bodies over a range of shearing stresses of 3.10^4 to 2.10^6 dynes/cm.^{2} was observed by S. P. Nichiporenko[711] in a complete "creeping diagram" given by a Volarovich capillary viscosimeter and a Tolstoĭ plastometer with parallel plates. The observed visocity values η^* are of the order of magnitude of 10^{10} to 10^{11} poises for shears in the range indicated above. At higher shears, however, the structure breaks down with sudden decreases in apparent viscosity η^* of 10^9, then 10^8, and even only 10^4 poises. The importance of these observations for processes such as ceramic machine-shaping and extrusion is evident. Likewise they show the formation of "knots" by an inhomogeneous distribution of shears in the ceramic bodies. The process of the yield value formation in thixotropic gels was further studied for vanadium pentoxide and bentonite suspensions by T. Watanabe and Y. Takasaki.[712] The rate of gel formation follows first order kinetic with the constant $k = \ln (A_\infty / A_\infty - 1) \cdot 1/t$, if A_∞ is the maximum yield value, using a DuNouy torsion balance. Two processes of this kind, however, must be distinguished when superimposed on each other if higher concentrations are used, with two different values $A_{1\infty}$ and $A_{2\infty}$; these are proportional to the fourth and second power of the concentration, respectively, for Wyoming bentonite, but proportional to the second and the first power for a Japanese bentonite of higher dispersibility. During the first process a structure may be developed in which the dispersed particles are linearly associated, with a relatively small number of cross-linkages or entanglements between the chains. In the second process another framework structure is built up due to a considerable increase in cross-linkage. Both structural characteristics correspond to formation of edge-to-edge and edge-to-surface associations of units in the shape of corrugated ribbons, and a "house-of-cards" structure as U. Hofmann had demonstrated.

341. The kinetics of formation of thixotropic structures were studied by E. M. Tarasova[713] with bentonites, using the "elastometer" of S. Ya. Veĭler and P. A. Rebinder,[714] with tangential shearing forces exerted on the elastic material. The suspensions were varied in concentration from 10 to 50 per cent. The consolidation process as a function of time shows minimum and maximum shearing strains and mechanical strengths as a function of the concentration in solid material. The maxima

[711] *Doklady Akad. Nauk S.S.S.R.*, **118**, 1958, 785-787. On the plastimeter of D. M. Tolstoĭ, cf. *Kolloid Zhur.*, **10**, 1948, 133-147.

[712] *Naturwissenschaften*, **45**, 1958, 334-335.

[713] *Izvest. Turkmensk. Filiala Akad. Nauk S.S.S.R.*, 1951, 77-83.

[714] *Compt. rend. acad. sci. U.R.S.S.*, **49**, 1945, 345-348. On the cone-dipping method for quantitative determination of structure-mechanical properties of plastic-viscous materials, see P.A. Rebinder and N. A. Semenenko, *Doklady Akad. Nauk S.S.S.R.*, **64**, 1949, 835-838.

of the curves correspond to a linear function of the montmorillonite and hydromica content of the clays. The framework structure is built up from the tabular particles between which dissolved surface films are inserted. It offers an explanation of the thixotropy by the orientation including the specific effects of electrolytes added. The aggregation could be treated mathematically under the assumption that the equations for bimolecular reactions can be applied to interaction of the particles. The time function of maximum limit shearing strains could also be formally described.

342. The intimate correlations of thixotropy and plasticity phenomena of size-fractionated clay minerals were experimentally investigated by H. von Platen and H. G. F. Winkler,[715] extending the comparative measurements over different kaolinites, halloysites, illites, montmorillonites, feldspar and quartz powders. The specific surfaces were determined by the BET method. They are essential for development of both fundamental properties of those materials, and a relatively simple linear functional relation was established with the surface area. Nevertheless, there is an important factor of individuality of the silicates in question, concerning the base exchange capacity, particle shape, swelling properties, and so on. It is also particularly important that the monodisperse suspensions studied are free from aggregations so as to clearly develop the linear proportionality to the specific surface. If this condition, however, is fulfilled, all of the kaolinites and others of the same specific surface, are equal in their thixotropic limit data (Winkler's method of 1938 was used), but different minerals have in distilled water different thixotropic characteristics, even if their specific surface is the same (Fig. C. 176). In a 3.6 N NaCl solution, on the other hand, the thixotropy parameter depends even for different (aggregation-free) minerals exclusively on the specific surface, independent of the mineral composition or the mass and shape of the particles; in other words, thixotropic gels retain a definite liquid volume per surface unit of the solid phase. These remarkable results are not compatible with the theory of thixotropy developed by W. von Engelhardt (1943) on attraction forces in the suspensions. They are also in some contradiction with U. Hofmann's ideas on the "house-of-cards" structures, which should be expected to have a different unwieldiness if shapes of the particles are different. This was *not* observed by von Platen and Winkler. Concerning correlations of thixotropy to viscosity, S. Thornton[716] determined the absolute viscosity of instantaneously thixotropic fluids with a modified Poiseuille tube device. The investigated thixotropic suspensions of titania particles in liquid paraffin, or the concentric rotation viscosimeter gave analogous results, although the Poiseuille tube raised experimental difficulties (cf. C. ¶ 349).

[715] *Kolloid-Z.*, **158**, 1958, 3-22. On dilatancy of lean ceramic mixes on Th. Plaul, *Silikattechnik*, **13**, 1962, 139-140.

[716] *Proc. Phys. Soc.* (*London*), **66**, 1953, 115-119; 120-128 (with D. Rae).

343. In extension of W. von Engelhardt's investigations (1942) on the *grinding hardness* of silicates in different suspension liquids, recent studies on the surface factors ruling grindability show according to R. D. Azelitskaya and P. P. Gaïdzhurov[717] the great effect of wet grinding, such as Portland cement clinker minerals,

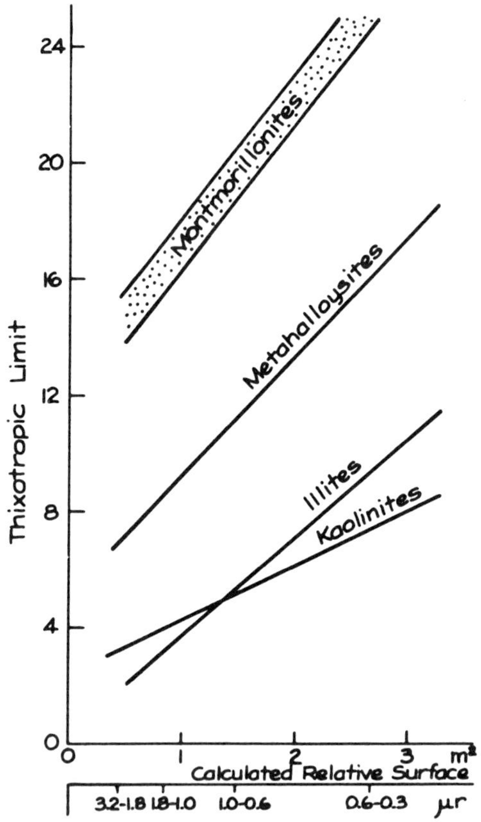

FIG. C. 176. Thixotropic limit values of aggregate-free minerals in distilled water, as a function of calculated relative specific surface (von Platen and Winkler).

either in pure water, or with additions of surfactants (chiefly lignin sulfonates) which are adsorbed on the material. The solutions of alkali carbonates or sulfates if added to the grinding liquid were also considered for accelerating the grinding process, especially sodium carbonate and sulfate even in concentrations of only

[717] *Byull. Stroitel. Tekh.*, **14**, 1957 (3), 12-15. On application of surfactants as grinding aids, see especially extensive experiments of E. Ziegler, *Schriftenreihe Zementind.*, No., 19, 1956, 80 pp., chiefly pp. 51 ff.; also, A. Götte, *Tagungsber. Zementind.*, **13**, 1956, 28-32; *VDI Zzitschrift*, **98**, 1956, 373-376.

0.01 per cent, with an initial increase up to 450 per cent of the mechanical strength of the product. In their investigations on the mechanochemical properties of glass (cf. Vol. II, A. ¶ 394), W. A. Weyl and E. Ch. Marboe[718] discuss the abrasion hardness and the rate-determining role for effect of water and aqueous solutions on quartz, as a consequence of the addition of hydroxyl ions and not of proton penetration. The abrasion of glass is strongly decreased if liquid paraffin is used as a grinding liquid. If one sets the volume of glass removed in paraffin equal to one, the volume removed in dry-grinding is six times larger, and seven times that if ground in water. The technologically important effects of increased speed of abrasion are here omitted. G. S. Khodakov and P. A. Rebinder[719] examined the same problems using water, ethyl alcohol, and benzene as grinding liquids for quartz in comparison with the effects of dry-grinding. Only 4 per cent water is sufficient to raise the specific surface of the products considerably in the first 10 to 30 seconds, whereas the organic liquids are rather ineffective. These phenomena are explained by the formation of dense aggregates of particles between which the interstices are densely packed with much finer material. A great portion of the inner surface even remains inaccessible for nitrogen in the BET test. Such aggregations are indicated by the majority of common dispersion-analytical methods only as "coarse" particles; the amorphous particles are not accessible at all. Water alone is able to disintegrate them, but on drying the dispersion effect is again nullified, the dried material becoming dense again. In acetone with a small addition of water, the same occurs. After drying the surface appears to be considerably reduced, but if the aggregations are then ground anew with water, a much higher dispersity is reached.

344. The primary accelerating effect by water is noted if only 0.04 per cent water is added, but it reaches a maximum at 2 per cent, and comes to a minimum at higher percentages, whereas at 50 to 80 per cent another optimum is observed (cf. Fig. C. 177). The minimum is caused by formation of a structure in the particles (of quartz) with liquid meniscus and water films impeding the grinding process. If organic liquids are added, even in small amounts, an increase in specific surface is observed as a consequence of the decrease in cohesion strength by the adsorbed liquid which lowers the interphase surface tension.[720] Triethanolamine has a distinct grinding-accelerating effect; oleic acid and paraffin have a very low effect on quartz, but for calcite these agents excel even water, because calcite is much less hydrophilic than quartz or corundum. V. F. Kiselev, K. G. Krasil'nikov, and G. S. Kho-

[718] *Glastech. Ber.*, **32** K, 1959 (VI), 1-18, especially pp. 11 f. See also recent contributions by W. von Engelhardt and S. Haussühl, *Kolloid-Z.*, **173**, 1960, 20-35, on grinding hardness and specific free boundary surface energy of alkali halides (NaCl structure type).

[719] *Doklady Akad. Nauk S.S.S.R.*, **127**, 1959, 1070-1073.

[720] Cf. P. A. Rebinder, V. I. Likhtman, and L. A. Kochanova, *Doklady Akad. Nauk S.S.S.R.*, **111**, 1956, 1278-1281, for such effects in metallic systems.

dakov[721] explained by systematic grinding experiments on the same basis the differences in adsorption amounts of water and nitrogen on quartz observed by S. P. Zhdanov,[722] and therefore the inconsistencies in the BET results of specific surface determinations. The specific surface of the air-dry ground powder in their experiments was 35.4 m.^{2}/gm., but in the water-ground powder 53.0 m.^{2}/gm., with 5.4 and 14.8 μmoles/cm.^{2} of structural water content. The adsorption qualities of quartz are distinctly changed by grinding as soon as the water-ground powder is exposed to air in the milling process and the particles aggregate. Heats of wetting determined

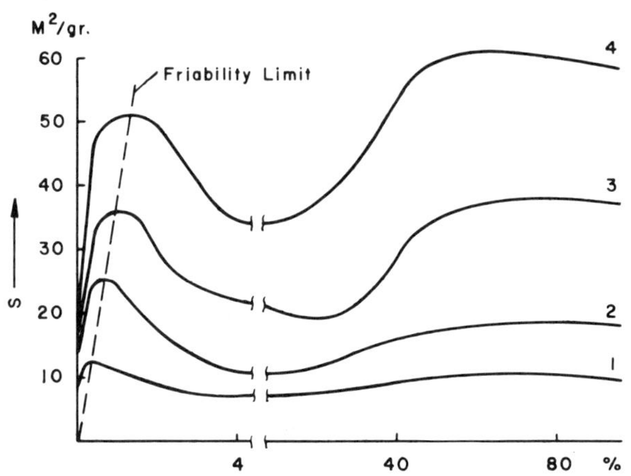

FIG. C. 177. Efficiency of grinding quartz sand, as a function of addition of water (Khodakov and Rebinder). Grinding time is: for (1) 2 minutes; for (2) 4 minutes; for (3) 8 minutes; for (4) 16 minutes.

from such powders give inconsistent results.[723] These consequences are important for correct interpreation of Kiselev's "absolute" adsorption isotherms. The fact that up to $p/p_s = 0.1$ the isotherms for water coincide for both kinds of quartz powders indicates that water spreads onto the whole surface and "peptizes" it by surface-active forces. The aggregation in dry air, however, is not exclusively characteristic of quartz, it also occurs with corundum, calcite, and silica glass powders with a one-tenth reduction of apparent specific surface.

345. The pronounced reduction of the *microhardness* of minerals by action of surfactants, especially of aluminum chloride and sodium carbonate added to the water required in "dry"-pressing of ceramic refractories, was studied for quartz, magnesite,

[721] *Doklady Akad. Nauk S.S.S.R.*, **130**, 1960, 1273-1276.
[722] *Doklady Akad. Nauk S.S.S.R.*, **120**, 1958, 103-106.
[723] Cf. A. C. Makrides and N. Hackerman, *J. Phys. Chem.*, **63**, 1959, 594-598.

periclase, and fireclay grogs by I. S. Kaĭnarskiĭ and I. Ya. Piven'.[724] The favorable effect of these agents is interpreted as an adsorption phenomenon. Analytical data show that aluminum ions are more intensely adsorbed than chlorine anions. On periclase a hydration to $Mg(OH)_2$ occurs. The measured reduction of microhardness was about 20 per cent for aluminum chloride and about 25 per cent for soda. In magnesite bodies Na_2CO_3 acts particularly well for a reduction of porosity of the pressed bodies, whereas $AlCl_3$ is better in fireclay grog mixtures.

346. Often the very specific action of organic materials on the thixotropy and viscosity properties of clay-water systems is particularly evident in the industrial use of Quebracho extracts, hydrolyzable tannin, or other polyphenolic complexes as discussed by Th. White[725] for casting slips with illite and montmorillonite as characteristic ingredients, or for drilling muds, for Portland cement mixes, and others (cf. C. ¶ 216, ¶ 332). In a much more comprehensive way, L. Wolf and R. Kurtz[726] discussed the interrelation of thixotropy, sedimentation volume, and other characteristic properties of fine-dispersed minerals and commercial inorganic products (chiefly used as pigments), and specifically the action of polar and nonpolar liquids on these, of which we can here only mention a small number. Rapid settling is characteristic in nonpolar CCl_4, benzene, cyclohexane, tetrahydrofuran, *n*-hexane, to form fluffy sediments of great volume, whereas in polar liquids, say in water, primary alcohols or amines, chiefly in formamide, $HCONH_2$, the contrary is observed. In liquids of intermediate polarity, like ethyl ether, methyl acetate and dioxane the sedimentation volumes are large to fairly large and the structure is fluffy. Fine-granular sediments of low volume are characteristic in the strongly polar acetonitrile, butylonitrile, and acetone. In nonpolar liquids the sedimentation products are generally thixotropic, rarely in polar liquids. In nonpolar liquids, settling was often complete in fraction of one minute, in polar liquids it took hours and days. Milky supernatant liquids indicate the tendency to form stable suspensions. The sedimentation volume generally depends on the dielectric constant of the liquid medium but it is more dependent on the presence of polar molecules with accessible polar groups. Thixotropy is in every case *not* restricted to flat-tabular particle shapes, as was previously often assumed; it is equally observed, in suitable liquids, with isometric crystalline particles.

[724] *Ogneupory*, **22**, 1957 (1), 27-29.

[725] *Ber. deut. keram. Ges.*, **29**, 1952, 302-309; *Tonind.-Ztg. u. Keram. Rundschau*, **77**, 1953, 193-197; *Ceram. Age*, **62**, Oct. 1953, 31-34.

[726] *Farbe u. Lack*, **60**, 1954, 483-490; 528-530. In his excellent report on actual problems of colloid chemistry, *Vestnik Akad. Nauk S.S.S.R.*, 1955 (2) 8-17, P. A. Rebinder demonstrates on page 12 the high importance of the use of surfactants in chemical technology by instructive examples.

APPARENT AND ABSOLUTE VISCOSITY OF CLAY-WATER MIXTURES

347. The fundamental factors controlling viscosity of kaolin suspensions were re-examined by L. Žagar and C. Schumann[727] with the Ostwald capillary viscosimeter, especially in regard to the Bingham formula for the viscosity of flowing clay slips, and Einstein's equation (1906). The specific hydrodynamic volume, $\varphi_s = v/V$, is for most of these systems rather constant, and on the average equals 1. But one is not permitted to conclude from this fact that the particles did not enlarge if suspended in water. The calculated parameter φ_0 for the hydrodynamic volume at the fluidity equal to 0 in the classical Bingham equation for the calculation of viscosity applied in the elementary form

$$1/\eta^2 = 1 - \varphi/\varphi_0$$

cannot be separated from the ratio φ, which is that of the volume of the suspended system, v, to the volume of the suspension, V. Nevertheless, the Newtonian type of flow can be maintained up to a volume concentration of 12.5 per cent, but for higher concentrations, or for higher shearing stresses than $\tau = 100$, anomalies are observed. Einstein's formula

$$\eta^* = 1 + 2.5\,\varphi$$

gives inconsistent results for the specific hydrodynamic volume. The assumption of spherical particles flowing, and other physical variables for the state of the particles, are not in agreement with the nature of clay-water systems. Moreover, a better agreement with experimental data is here found if the extended formula

$$\eta^* = 1 + 2.5\,\varphi + 14.1\,\varphi^2$$

is used, at least up to weight concentrations of 15 per cent, with $\varphi_s = 1$. For higher concentrations anomalies again occur.

348. The particles are *solvated* and have hydrodynamic volumina larger φ than the volume of dry material. In the range of applicability of the extended Einstein equation the volume of the suspended kaolin particles is 2.8 times larger than in the dry state. The linear relation between reciprocal viscosity $1/\eta^*$ as a function of the quotient v/V is seen in Fig. C. 178. Corresponding studies of E. W. J. Mardles[728] of kaolinite and bentonite suspensions in organic liquids showed interesting effects. For example, if ethyl alcohol is added to a montmorillonite suspension in water, namely an increase is indicated in viscosity and thixotropy for bentonite, whereas in kaolinite suspensions the dispersion is enhanced. Specific viscosity is a complex function of the nature of the dispersion medium and of the time of orientation of the solid par-

[727] *Ziegelindustrie,* 10, 1957, 585-590.
[728] *Clay Minerals Bull.,* 1, 1948, 41-43.

ticles. For a constant rate of shear its ratio to the volume percentage of the solid is a useful criterion as well as the point of pasty-plastic consolidation. The viscosity of a kaolinite suspension in a highly viscous mineral oil is very sensitive to the action of oleic acid and linseed oil added. The resulting viscosity in the mixture and the sedimentation volume is then similar to those of kaolinite in oleic acid and linseed oil alone. The suspension immediately thickens if a few drops of water are added to the mixture after addition of oleic acid.

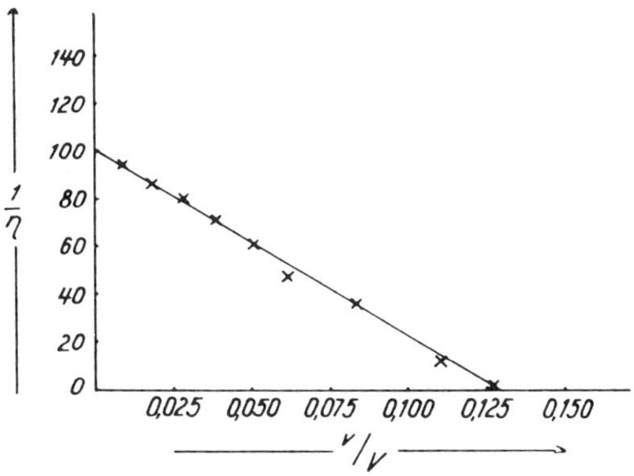

FIG. C. 178. Linear relation of reciprocal viscosity, $1/\eta^*$, as a function of the ratio of the volumina of suspended kaolinite particles to the volumina of the suspension, v/V (Žagar and Schumann).

349. For the flow characteristics of non-Newtonian fluids, I. M. Krieger and S. H. Maron[729] proposed a schema for the systematic handling of η^* data using the variables apparent fluidity, Φ^*, and shearing stress, F. The rate of shear, $g(F)$, is given in the equation

$$g(F)/F = \eta^*[1 + \Delta(F)]$$

with the correction term $\Delta(F)$ determined experimentally from viscosity measurements either with the rotation or with the capillary flow viscosimeter. For the latter purpose, S. H. Maron, I. M. Krieger, and A. W. Sisko[730] developed a precision flow viscosimeter (shown in Fig. C. 179), with a wide range of shearing stresses in a single series of determinations. The principle of the device is the flow under pressure (from the syringe of 20 cm.³ contents, G) of mercury in the column tube m (or of a liquid of lower density in tube w) which drives the liquid sample or suspension through the capillary A (in a water jacket F, and with the sidearm tube B). The height difference in the column, before and after the capillary flow of the sample,

[729] *J. Appl. Phys.*, **25**, 1954, 72-75.
[730] *J. Appl. Phys.*, **25**, 1954, 971-976.

gives both the primary drop and the flow rate as a function of time. The density of the sample must not be accurately known but that of the driving liquid must be known. The apparent viscosity is calculated in an elementary formula from a modification of Poiseuille's law. The device is very suitable for the direct determination of absolute viscosities of Newtonian media and the flow curves of non-Newtonian fluids like latex rubber or colloid suspensions.

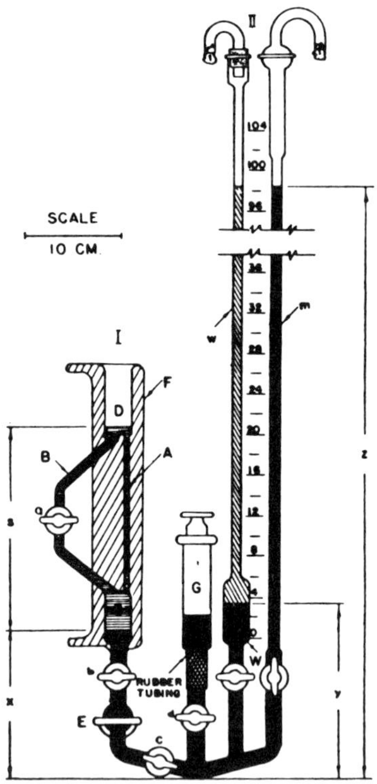

FIG. C. 179. Precision capillary viscosimeter for Newtonian, and non-Newtonian fluids (Maron, Krieger, and Sisko).

350. Flow curves, $g(R)$, characterizing 6 per cent suspensions of different bentonites were given by J. Barna[731] as derived from a rotation viscosimeter, calibrated with liquids of Newtonian type. The suspensions show yield points for the lowest rotation speeds, and a maximum at low R values, as an indication of a thixotropic

[731] *Silicates inds.*, **24**, 1959, 554-560. Ou the viscosimeter used see B. Marschalkó, *Acta Tech. Acad. Sci. Hung.*, **15**, 1955, 77-97, working in the range from 0.0001 to 1200 r.p.m., and with experimental shearing stresses extending over six orders of magnitude for $R = 30\,\Omega/\pi$. On the calculation methods, see F. Schultz-Grunow and H. Weymann, *Kolloid-Z.*, **131**, 1953, 61-66, and M. Reiner and R. Riwlin, *Ibid.*, **43**, 1927, 1-5, and 72-78.

disruption of the solid bonds between the interior and the exterior cylinder surface contacting the bentonite suspension. The appearance of an inflection point (see Fig. C. 180) on the curves for η^* versus shearing velocity in flow is of the same type as observed for greases.[732] The solvated hulls around the suspension particles begin being sheared at 0.1 to 1.0 r.p.m. The process is completed at very high speeds (50 to 1200 r.p.m.) and at highest R values the suspension is completely fluidized to an ideal Bingham medium with a constant η^*. The maximum mentioned corresponds to the thixotropic strength; the minimum, to the gel strength (rigidity).

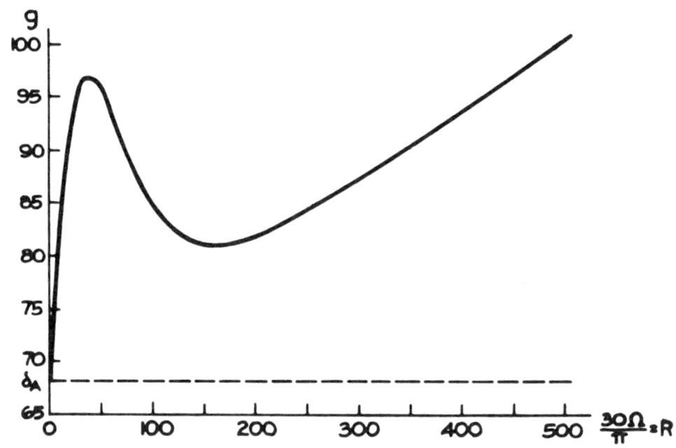

FIG. C. 180. Flow curve, $g(R)$, of an Hungarian bentonite, enriched in montmorillonite (Barna).

351. The unavoidable orientation effects in flowing clay suspensions were already previously discussed in C. ¶ 234, chiefly based on the recent investigations of H. Detaille.[733] We may here add how this author studied the extrusion process in flow-model experiments with alternately colored and colorless disks of clay under the action of an unidirectional pressure. The rheological characteristics of the extrusion bodies can be evaluated to flow diagrams; especially characteristic are the shear planes oriented under 45° to the pressure axis. E. E. Segalova, P. A. Rebinder and O. I. Luk'yanova[734] also found a tendency to form structures in cement-water mixes

[732] Cf. D. S. Velikovskiĭ, *Kolloidn. Zhur.*, **16**, 1954, 406-411, and G. V. Vinogradov and V. P. Pavlov, *Ibid.*, **20**, 1958, 248-253.

[733] *Silicates inds.*, **25**, 1959, 370-375. We mention further practical studies of structure problems (e. g., the formation of "ice-lenses") in extrusion-shaped clay bodies by W. Hagen, *Ber. deut. keram. Ges.*, **32**, 1955, 114-119, who introduced vibrating grits into helix presses which destroy the thixotropic consolidation of the clay-water mixes and reduce the tendency to form orientations in the mass.

[734] *Vestnik Moskov. Univ. Ser. Fiz. Mat. i Estestven. Nauk*, **9**, 1954 (1), 17-32; see also E. E. Segalova and E. S. Solov'eva, *Trudy Soveshchaniya Khim. Tsementa, Nauch.-Tekh. Obshchestvo*

with or without addition of lignin sulfonates. The decrease of plastic strengths by the action of lignin sulfonates, as measured in a conical plastometer, is very distinct, followed by a rapid increase in strength.

352. Absolute viscosity determinations in the Ostwald viscosimeter were made by E. Plötze, G. Huckert, and R. Plötze[735] for highly dilute kaolinite suspensions with particles 1.5 μ in diameter. As a function of the concentration in solids the η^*

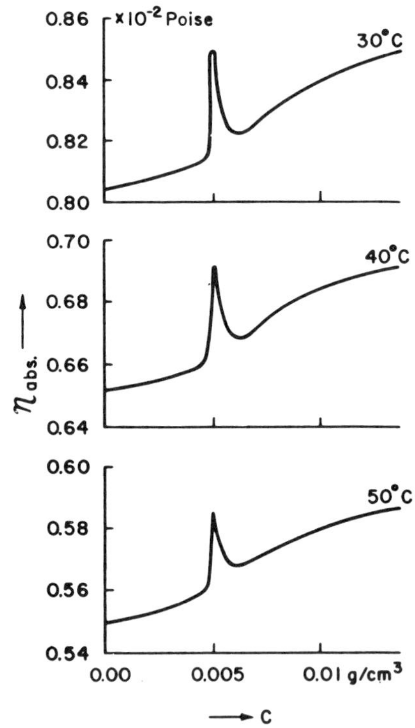

FIG C. 181. Absolute viscosity of highly dilute kaolinite suspensions in water, with charactristic maxima corresponding to solvation, and liquid structure formation (Plötze, Huckert, and Plötze).

data show distinct maxima for 0.005 gm./cm.^{3} suspensions with a half-width Δc, of less than 0.001 gm./cm.^{3}, whereas for concentrations of 0.043 and 0.084 gm./cm.^{3} two other maxima have distinctly larger half-width values (Fig. C. 181). The maxima at 0.005 gm./cm.^{3} are not explained only by the formation of solvation hulls around

Prom. Stroitel. Materialov, 1956, 138-153. Cf. Vol. V, Section C. See also P. A. Rebinder and E. E. Segalova, *Proc. Second Internat. Congr. Surface Activity*, London 1957, 492-505, on electrostatic phenomena and solid-liquid interfaces (cf. Vol. V. C. ¶ 117).

[735] *Naturwissenschaften*, **45**, 1958, 36-37; *Kolloid-Z.*, **147**, 1956, especially pp. 63-64.

the particles, nor by intraionic actions as B. Weblus[736] assumed. The striking sharpness of the maxima is better understood as indicating characteristic structures developed in the liquid, for well-defined intercrystalline distances in the suspensions, by a far-reaching (over about 8 μ distance) molecule bridging, over many intermediate water molecules. The sheet structure of the kaolinite crystals acts to induce on a liquid network structure different mesomeric packings of the water molecules. Structural viscosity effects in such suspensions are evident, as they are present in H. Staudinger's organic molecules of "linear colloids" in macromolecular homoeopolar compounds.[737]

353. For absolute viscosimetry, R. Schwaben and H. Umstätter[738] gave valuable instructions for the suitable construction of rotation viscosimeters, based on the theory of viscous flow under the action of a rotating body and exerted stress and torque. For a constant gap width between conical rotating parts and stators, the regular flow velocity is expressed in Bessel functions. For the constant apex angle of 70.5° as one limit and the cylindrical rotor system as the other limit, the general equations referring hereto were evolved; particularly simple relations were found for conical viscosimeter elements with a double-conical taper, and for spherical types of rotors and stators, with consideration of wall effects introduced at the bottom surface of the container. The practical evaluation was given for the consistometer of Umstätter and the so-called thixometer device (cf. Vol. II, A. ¶ 18). Of the cylindrical types, the Stormer viscosimeter is a commonly used instrument which was more recently improved by E. R. Ballantyne[739] by application of two different closed-cylindrical rotors of the same diameter, but with different heights, and a cylindrical cup without baffles. The theory of this modification in the use of the original Stormer viscosimeter is equally suitable for Newtonian liquids (such as mineral oils), and non-Newtonian systems. Ballantyne claims to have obtained more reliable results in using the two different rotors and calculating viscosity directly from the relevant dimensions than by using a single rotor and calculating viscosity indirectly from the calibration with different liquids of known viscosity. If high rates of shear are used, the kinetic energy $q = \eta \cdot D^2$ (with $D = dv/dr = r \cdot d\omega/dr$ per second) imparted to the liquid must be taken into consideration. Prolonged shearing in the Ballantyne apparatus may cause temperature rises which would cause, in their turn, appreciable errors in results. The viscosimeter method was equally used for different mineral oils as for bentonite sols and concrete-sealing linings.[740]

[736] *Kolloid-Z.*, **132**, 1953, 16-21.

[737] Cf. H. Staudinger, "Organische Kolloidchemie," Vieweg, Braunschweig, 1950, 308 pp.

[738] *Konstrukt. Maschinen App. u. Gerätebau*, **5**, 1953, 79-95.

[739] *Australian J. Instr. Technol.*, **10**, 1954, 102-113.

[740] *Australia, Commonwealth Sci. Ind. Research Organization. Div. Bldg. Research, Tech. Paper*, Section of Tribophysics, Report A 159, 1948; *Ibid., Tech. Paper* No. 2, 1954.

354. The use of rotation viscosimeters for determination of the apparent viscosity of Portland cement raw muds (slips, slurries) was especially demonstrated by E. P. Tkachev[741]; these are typical non-Newtonian fluids, with a Bingham yield value and a slight thixotropy. An automatization of the control of consistency of the slips is possible by the use of two photoelectric cells, on the margins of which an optical device projects the parabola of the flowing slip jet. Anomalies of the flow will more or less screen the photocells which then, over an electronic circuit, regulate automatically the feeding rate from the slip container into the calcinator of the cement kiln. F. Becker and W. Schrämli[742] gave an extensive theoretical discussion of the rheological behavior of Portland cement slips, based on the Bingham theory and its integral equation of the flow process, and for evaluation of the flow curves according to J. Pawlowski.[743] A recent discussion of the rheology of the cement raw slurry with a particular application of deflocculation effects (cf. Vol. V, Section C), especially of these in kaolinite clays, was given by V. R. Pfrunder and W. Gromes.[744]

355. A very valuable general discussion of Portland cement slips as rheological systems was also given by M. Papadakis,[745] treating them as plastic non-Newtonian fluids with a yield point. Thixotropy is also a phenomenon observed in Portland cement suspensions, as a function of the water/cement (W/C) ratio, with a hysteresis of breaking and reorganization of inner bonding forces in the structure of pastes down to $W/C = 0.35$, which determines the castibility of mortars and pastes. Additions of clay (1 to 5 per cent) to a suspension with $W/C = 0.35$ considerably increases apparent viscosity, whereas for mortars with $W/C = 0.40$ the effect is much less pronounced even when 10 per cent clay is added. The chief practical interest in changes of rheological properties, however, is connected with the use of organic surfactants with hydrophobic aliphatic chains. Five per cent of tannic acid added brings

[741] *Tsement*, **19**, 1953, No. 2, 8-14; *Silikat Tech.*, **5**, 1954, 177-178. See also an excellent critical report on rotational viscosimeter methods, by W. K. Asbeck, *J. Soc. Cosmetic Chem.*, **8**, 1957, 282-300.

[742] *Zement-Kalk-Gips*, **8**, 1955, 345-352.

[743] *Kolloid-Z.*, **130**, 1953, 129-131, based on a graphic differentiation of the integral flow curve.

[744] *Zement-Kalk-Gips*, **12**, 1959, 501-504. See also M. Papadakis, *Rev. matériaux construct. et trav. publ.*, **531**, 1959, 48 pp., especially pp. 5-18, and *Ibid.*, **537**, 1960, 157-159, on a self-controlling apparatus based on rheological properties of slurries in kiln operation. The use of lignin sulfonates as an addition promoting the fluidity of Portland cement raw mixes was discussed also by J. Magasrévy, *Zement-Kalk-Gips*, **10**, 1957, 12-16; especially the calcium salt is recommended, applied in a mill-circulation-flotation process.

[745] *Silicates inds.*, **22**, 1957, 612-615; 675-681. On apparatus for the continuous rheological control of Portland cement raw muds, see M. Papadakis and J.-P. Bombled, *Rev. matériaux construct. et trav. publ.*, **537**, 1960, 157-159. We emphasize the necessity of correct definitions of thixotropy, dilatancy, and rheopexy effects in rheological esperiments, and recommend a clear explanation of these phenomena by B. C. Weber, *Keram. Zeitschr.*, **14**, 1962, particularly pp. 462 ff., in slip-casting processes.

about a marked reduction in apparent viscosity and permits a considerable reduction of water content which is necessary for the regular manipulation of raw slips of limestone and clay. Also 0.1 to 0.2 per cent of sodium pyrophosphate is sufficient in this case, but the suspension then shows a dilatancy, not a thixotropy. The stiffening in flow may bring about much trouble in kiln operation.

356. The general use of phosphates in different forms as an addition to clay-water suspensions and related systems was recommended by J. R. Van Wazer and E. Besmertnuk.[746] They studied sodium pyrophosphate, which is preferably used in ceramic practice for slip-casting improvement, as a deflocculant (Vol. V, Section B) and also metaphosphates in their polymerized forms. The latter phenomena are especially described by A. S. Michaels[747] with "sequestering" or "chelating" action exerted on the polyvalent metal ions, especially of aluminum in the suspensions. The dimer pyrophosphate is relatively inefficient, the triphosphate much better, and the tetramer highest in activity. Beyond the tetramer there is little increase in effectiveness with increase in chain length (cf. Vol. IV, Section B, especially on anion adsorption phenomena). For the purposes of preparation of homogeneous dispersions of clays, as for particle size analysis, the use of sodium polyphosphates is of greatest importance, according to G. W. Phelps and S. G. Maguire,[748] because of their excellent action as deflocculants. The percentage of electrolyte to be used for an optimum dispersion effect must be determined from curves representing fluidity as a function of the concentration in the electrolyte. As an example, Phelps and Maguire (see Fig. C. 182) noted a 50 per cent Florida kaolin suspension, slips of which were prepared with an addition of sodium hexametaphosphate, sodium silicate (waterglass), and sodium lignin sulfonate as the most characteristic commercial agents of this kind. The hexametaphosphate and the silicate solution deflocculate more rapidly whereas the lignin sulfonate more gradually approaches the minimum apparent viscosity. The points of maximum deflocculation marked in Fig. C. 182 are very close to the percentage of electrolytes causing maximum fineness.

357. The application of organic agents for adjusting the rheological properties of clay-water suspensions to ceramic purposes is rather common and not free from a certain commercial empirism. The valuable general discussion on characteristics of natural tannin products and especially of Quebracho extracts (commonly used with slight additions of sodium hydroxide) as given by Th. White[749] was already

[746] *J. Phys. Chem.*, **54**, 1950, 89-106.

[747] *Ind. Eng. Chem.*, **50**, 1958, 951-958.

[748] *J. Am. Ceram. Soc.*, **40**, 1957, 399-402. On the corresponding use of sodium lignin sulfonate in comparison with other organic salts and agents like methyl cellulose for slip casting, see also D. G. Burgess, *Am. Ceram. Soc. Bull.*, **36**, 1957, 168-171.

[749] *Ber. deut. keram. Ges.*, **29**, 1952, 302-307; *Tonind.-Ztg. u. Keram. Rundschau*, **77**, 1953, 193-197; *Ceram. Age*, **62**, 1953, 31-34. See also H. Lehmann and G. Marx, *Tonind.-Ztg. u. Keram.*

mentioned in C. ¶ 346, and may here be emphasized again as an important source of information. Figure C. 183 gives an example of the characteristic effects in the non-Newtonian flow of a rotation viscosimeter experiment for a suspension with and without increasing additions of the extract. For practical purposes the addition of 0.5 per cent is evidently sufficient to reduce apparent viscosity and yield point. Also important is the fact that apparent viscosity, usually determined in a

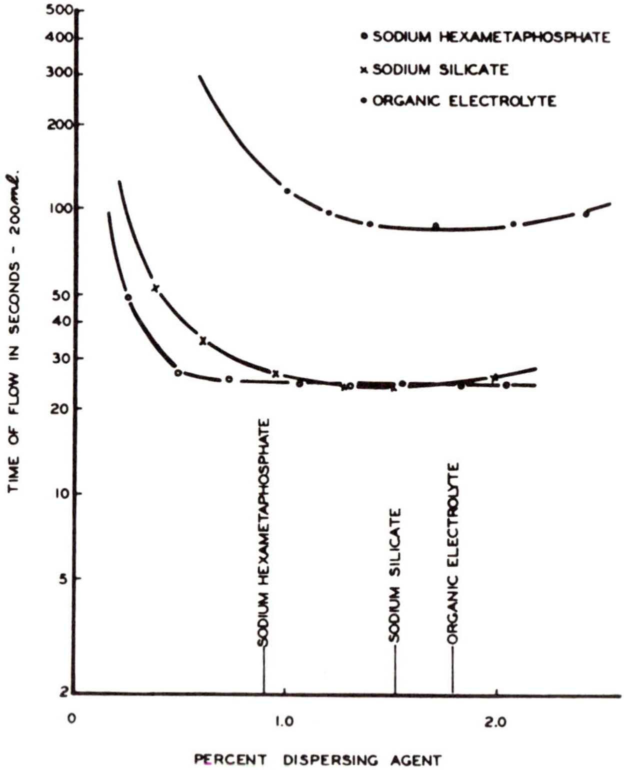

FIG. C. 182. Fluidity of a 50 per cent Florida kaolin suspension, as a function of the concentration of commonly used ceramic deflocculants (Phelps and Maguire).

Stormer viscosimeter with a constant rotation speed, may become misleading by the "gliding" phenomenon as a consequence of preferred orientation of the clay particles (especially in illitic clays) on the rotor and stator. It is, therefore, indispensable to determine apparent viscosity not only for one rotation speed (the commonly used 600 r.p.m.) but for a widely varying range of speeds to avoid results which are

Rundschau, **78**, 1954, 323-325, especially on the combined action of additions of sodium carbonate and Quebracho extract to reduce apparent viscosity of slips, and to improve their stability.

contradictory to what practice will show, as in the slip casting process. A more recent investigation of the action of tannin on clay suspensions by V. V. Khresina and V. A. Chernov[750] emphasizes the interaction of phenolic and carboxylic groups with surface hydroxyl groups of clay minerals, whereas the nonpolar glucoside radical in tannin is directed outwards. The surface thus becomes less hydrophilic and

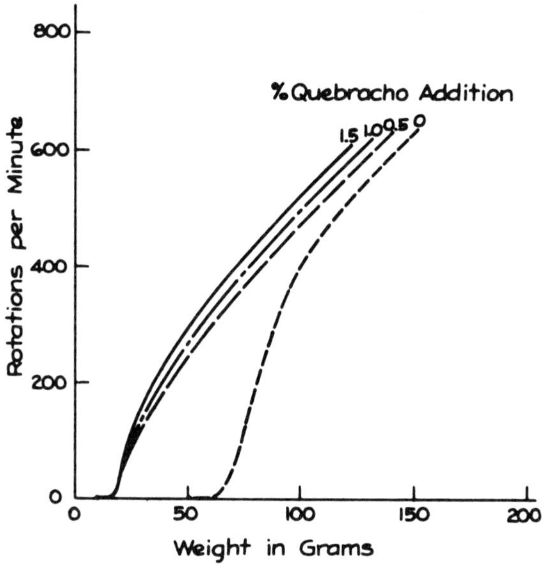

FIG. C. 183. Change of apparent viscosity, determined in rotation viscosimeter, for variable rotation speeds, as a function of varying driving weights, by addition of Quebracho extract W hite).

brings about a lower apparent viscosity and lower deflocculation effects. A combined action of tannin and waterglass added to clay-water suspensions increases the filtration speed of the slips. It promotes the rate at which ceramic cast bodies can be released from plaster moulds.

358. Besides the Stormer viscosimeter often used in industry, some other instruments of similar type were recommended, of which we mention here the Gallenkamp torsion viscosimeter used by P. Coraducci,[751] that constructed by H. H. Greger and M. Berg[752] and represented in its principal parts in Fig. C. 184, or that recommended by F. Moorc and L. J. Davies[753] particularly because of a rapid change of the shear rate over a wide range provided in this construction to measure thixotropic

[750] *Trudy Nauch. Issledovatel. Inst. Stroitel. Keram.*, No. 13, 1958, 14-29.
[751] *Ceramica* (*Milan*), **8**, 1957 (2), 50-55, especially pp. 53 f.
[752] *J. Am. Ceram. Soc.*, **39**, 1956, 98-103.
[753] *Trans. Brit. Ceram. Soc.*, **55**, 1956, 313-318.

effects in their progressive build-up (Fig. C. 185) and reversibility (Fig. C. 186) as in casting slips. J. A. Nelson and A. I. Andrews[754] measured the torque exerted in rotation by a steel rod on the clay-water mixture within its own structure and in a closed system, to bring about slippages, or a combination of slippages and shear.

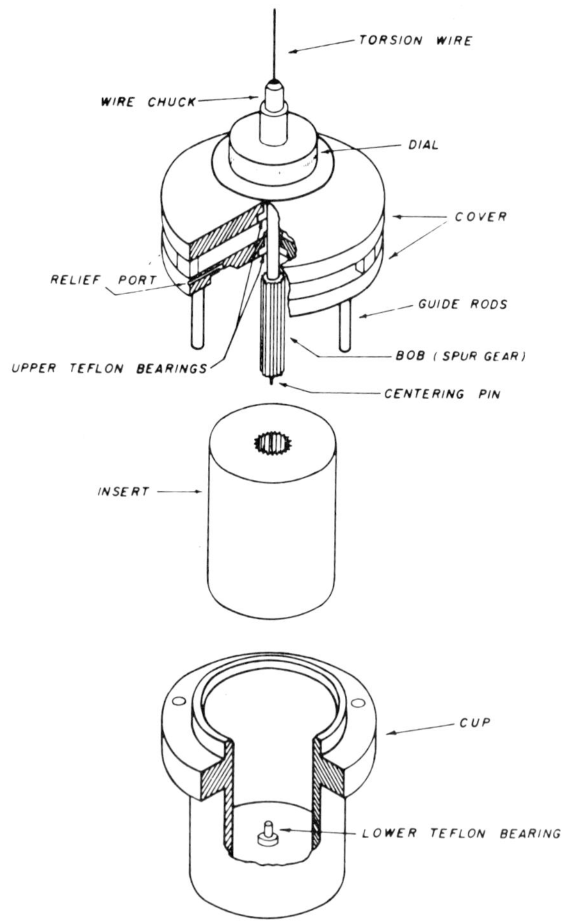

FIG. C. 184. Rotation viscosimeter for determination of workability of clay-water systems (Greger and Berg).

Equally the pressure which is exerted by a plastic clay can be measured when it is acted upon by an externally applied force. Characteristic torque-pressure curves, with a rather constant torque reached at certain limit pressures, are given as a function of the water content. Such an instrument is suitable also for the determi-

[754] *Am. Ceram. Soc. Bull.*, **38**, 1959, 447-455.

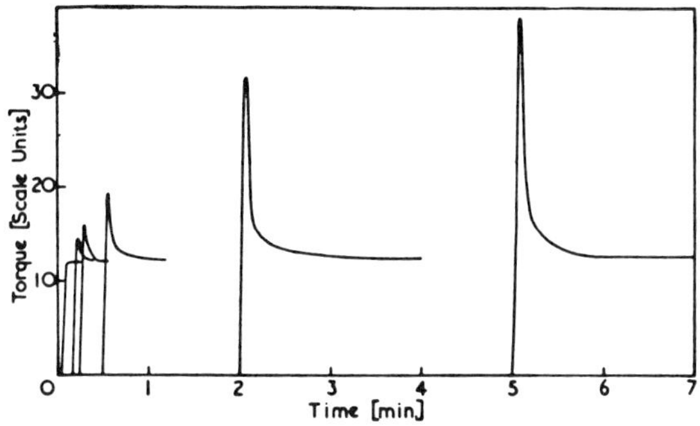

FIG. C. 185. Progressive thixotropic build-up of a casting slip (Moore and Davies).

nation of clay-water systems combined with round-grain blasting sands. The (external) friction between the steel rod and the clay is in this case a distinctly linear function of the sand addition if this is above 25 per cent. As a "mobilometer" D. H. De Clerck, R. E. Heeks, and Gouq-Jen Su[755] described a relatively simple instrument based on the falling "bob" principle which has the shape either of a plunger or of a tear-drop, or better of a double cone (about the shape of the dipping bob in Casagrande's hydrometer of optimum hydrodynamic conditions) useful for the rheological interpretation of non-Newtonian fluids like aqueous suspensions of glass frits and enamels, of moderate and low apparent viscosity, under the action of variable pH, and of additions of electrolytes to the solvent.

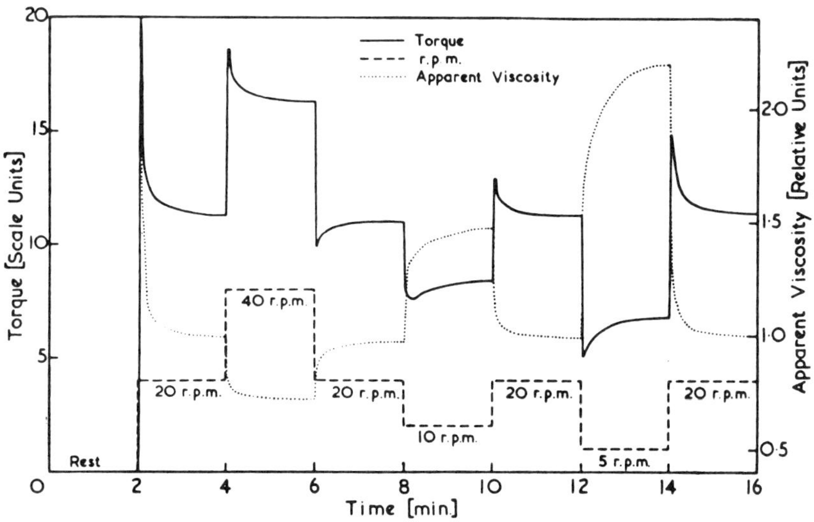

FIG. C. 186. Reversibility of thixotropic changes in casting slips (Moore and Davies).

[755] *Am. Ceram. Soc. Bull.*, 38, 1959, 459-464.

ELECTROLYTES IN VISCOUS CLAY-WATER SYSTEMS; DEFLOCCULATION PHENOMENA

359. In the center of practical interest is actually the important role of flocculants and deflocculants, on the basis of inorganic *polyanionic* compounds on the one hand and of organic surfactants on the other. For flocculation by polyanionic agents, Th. L. Mackay, I. B. Cutler, and M. E. Wadsworth[756] compared both series of possibilities and came to the general conclusion that organic materials are superior to their inorganic predecessors, say solutions of sodium chloride, copper sulfate, or aluminum sulfate, and bring about flocs of increased density. Taking the apparent viscosity of electrodialyzed Putnam Kaolin as a measure for the changes by different flocculant agents, they compare the action of polyelectrolytes of both classes, at variable pH conditions, especially for "Lytron"[757] as a model of the organic group. Charac-

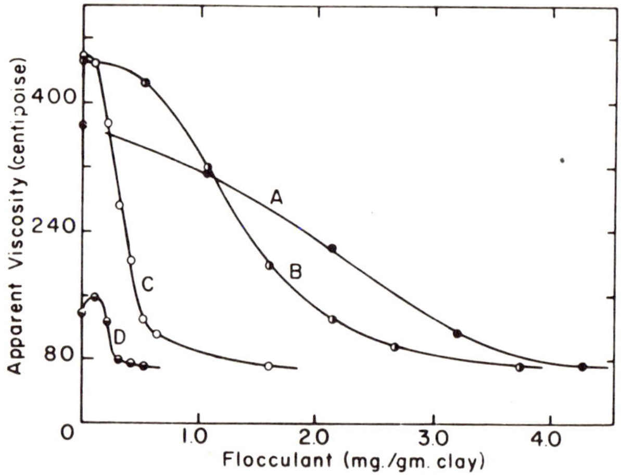

FIG. C. 187. Apparent viscosity as a function of the flocculant concentration for "Lytron 889," in a 10 per cent suspension of Putnam kaolin. The curves (*A*), (*B*), (*C*), and (*D*) refer to pH = 2, 4, 7, and 10, respectively (Mackay, Cutler, and Wadsworth).

teristic curves are given in Fig. C. 187, showing apparent viscosity as a function of the concentration in the agent at different pH. Flocculation resulting in floc growth, caused by swelling or the addition of more mineral particles, increases viscosity; this is evident for inorganic and organic flocculants, but the floc density is greater for organic agents, combined with a much greater decrease in apparent viscosity. The tight bonding is attributed to nuclear bridging between the particles. W. E. Worrall and C. V. Green[758] made an extensive investigation of the role of *natural*

[756] *Am. Ceram. Soc. Bull.*, **35**, 1956, 11-14.

[757] "Lytron" 886 and 887 are polymers of vinyl acetate and maleic anhydride, "Lytron" 889 is a sodium methacrylate polymer.

[758] *Trans. Brit. Ceram. Soc.*, **52**, 1953, 528-553.

organic constitutents of British (South Devonshire) ballclays which are genetically closely related to lignite (soft coal) deposits. The fractionation of such lignites in different solvents makes evident the complicated association (besides cellulose, resins, and waxes) of tpycal lignin-humus fractions and the so-called humic acids: the former has a strong effect on the physical properties of chinaclays; the latter act as protective colloids, enabling chinaclays to be suspended to pint weights of the slips even higher than are commonly in use, and to increase the green strength of ceramic bodies.

360. Very interesting laboratory and semi-industrial experiments were performed to apply *electrophoresis* directly for production of ceramic bodies from slips onto molds, such as for the production of crucibles as described by I. S. Kaĭnarskiĭ and K. B. Malinovskiĭ,[759] in a device derived from the classical Burton electrophoresis tube. More advanced methods for production of ceramic high-voltage insulators and the like were developed by R. Bahn,[760] who especially studied the electrophoretic deposition of oxides on conducting surfaces, i.e., wolfram heating elements in electronic tubes. The deposition of alumina from an organic suspension medium was an industrial success. Experimental data are given by Bahn on the determination in the Burton transference tube for varying dielectric potentials and increasing conductance. Aluminum or calcium citrate additions regulate in the methyl alcohol or acetone plus amyl acetate suspension the electrostatic potential of the particles. To avoid disturbing thixotropic effects, the addition of some collodion is necessary. It also avoids the spontaneous formation of gel skeletons in the suspension. The elimination of water is essential in this process, but is not easily possible in view of the high hygroscopicity of the fine-dispersed oxide. The deposition can be conducted either cathodic or anodic. The distribution of the electrostatic field lines strongly affects the anaphoretic deposition, and only very slightly affects cataphoretic deposition.

361. Concerning the classical O. Ruff method of slip-casting of refractory, nonplastic oxides to ceramic bodies from a highly acidic slurry suspension P. D. S. Saint Pierre[761] studied the important system zirconia–water–hydrochloric acid, with additions of polyvinyl alcohol, in its viscosity relationships to the pH conditions. It is very remarkable that the polymeric alcohol (addition mostly 1 per cent) has no great effect on apparent viscosity, whereas it definitely improves green strength of the oxide casts resulting from the slips of 70 to 80 per cent concentration in solids.

[759] *Steklo i Keram.*, **15**, 1958 (4), 26-30. On the electrophoresis of fine-ceramic slips see also I. S. Kaĭnarskiĭ and K. B. Malinovskiĭ, *Trudy Khar'kovsk. Politekhnich. Inst, im, V. I. Lenina, Ser. Khim. - Tekhnol.*, **8**, 1956, (3) 181-189., **31**. 1959 (1) 59-69.

[760] *Silikat Tech.*, **9**, 1958, 299-303.

[761] *Trans. Brit. Ceram. Soc.*, **51**, 1952, 260-268. B. C. Weber, *Keram. Zeitschr.*, **14**, 1962, 454-465, demonstrated the particular suitability of slip casting methods for production of cermet shaped articles (cf. Vol. V. A. ¶¶ 25, 66).

The pH is in every case the governing factor controlling viscosity characteristics. The optimum for industrial casting is for pH = 1.5 to 2.0, or in the minimum of apparent viscosity (determined by a Stormer viscosimeter) as a function of acidity of the slips.

362. A much detailed analysis of the slip casting process as a problem of *diffusion* of water in the slips (diffusion coefficient D_s) and in the mold (D_m) was given by E. Deeg.[762] In general, the diffusion theory of slip-casting is analogous to the process of heat conduction in a laminated medium. The diffusion coefficients D_s and D_m probably remain constant during the casting. Under this assumption the distribution coefficient, L, through the phase boundaries can be calculated. Deeg estimated the numerical data for these three coefficients. He calculated also the concentration displacements for variable times in the linear diffusion system (assumed to extend infinitely), or for one with limits on both sides of the diffusion current. For determination of the diffusion coefficients as a function of the concentration and the concentration displacement, the method of L. Boltzmann[763] could be applied, for different water diffusion coefficients of the slip and the cast, and different resorption capacities. The law stating that layer thickness equals $K \cdot \sqrt{t}$ is valid for progressive thickening of the cast in the slip-filled space. K is a practical measure of the slip quality, to be empirically determined[764] as a function of apparent viscosity and cast growth.

363. The water content of ceramic pressed bodies can be determined by elementary measurements of electric resistance, as a function of the weight per volume and of the load imposed on the sample. E. Plötze, A. Rubin, and A. Zwetsch[765] based upon this principle a method for control of the ion content in pressed bodies; the characteristic resistance curves for two examples are given in Fig. C. 188 a, b. The conductivity is typically ionic and the curves indicate adsorption and hydration effects in rapidly increasing curves at high weights per volume. If, however, a load, say of 2000 kg./cm.^{2}, is applied to the clay-water body (see curve II on Fig. 188 b) during the resistance measurement, an elastic deformation of the adsorption and hydration hulls occurs, freeing the ions for the conductance effect observed. The squeezing of water from clay-water bodies under a very high pressure was more recently described by P. A. Kryukov and N. A. Komarova,[766] stressing the practical significance of this phenomenon for soils, and the natural pressure effects of rocks in mines, or during tectonic displacements, or for the natural water household in epigenetic sedimentation

[762] *Ber. deut. keram. Ges.*, **30**, 1953, 129-138. In an Appendix are given the full vectorial-analytical derivations and a numerical example of the calculations.

[763] *Ann. Physik*, **53**, 1894, 959.

[764] Cf. J. G. Woodward, *Am. Ceram. Soc. Bull.*, **31**, 1952, 389-391; cf. also C. ¶ 364.

[765] *Naturwissenschaften*, **40**, 1953, 408-409.

[766] *Doklady Akad. Nauk S.S.S.R.*, **99**, 1954, 617-619.

systems. Experiments with pressures up to 20 tons/cm.^{2} applied to samples of 5 mm. thickness showed that the diffusion of squeezed water is very slow. The experiments had to be extended over two months to reach an equilibrium state, such as for bentonite, Askangel, silica gel, and the like. There is a linear relation between the residual

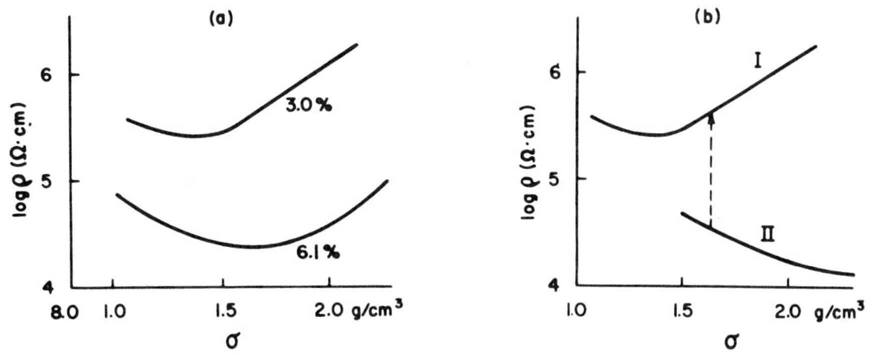

FIG. C. 188. Specific resistance (ϱ) of a clay-water system, as a function (a) of volume per weight, and (b) of a load of 2,000 kg./cm^2., applied to the system [Curve II in (b)] (Plötze, Rubin, and Zwetsch).

(equilibrium) water content and the log p, whereas for organic materials (starch, peach, and others) an approximately hyperbolic relation was observed, which shows above 2000 kg./cm.^{2} a transition to a constant water content. The dehydration curve with increasing pressure is particularly steep for Askangel. No break point appears on the curves, which might indicate a "molecular moisture capacity" effect.

364. A very helpful instrument for examination of rheological properties of clay-water systems, especially for casting slips in the casting process itself, proved to be the *vibration plate viscosimeter* developed by J. G. Woodward.[767] It is characterized by the action of an electromagnetic resonance vibrator to which a thin circular metal plate is fixed which oscillates in its plane laterally with about 800 c.p.s. in air. The reduction of the amplitude of the oscillations when immersed in the liquid or suspension to be tested is a function of the density and viscosity of the medium. The amplitude is measured by an electronic piezoelectric device which gives direct readings. The amplitude of the oscillation in the plate is an inverse function of the mechanical impedance, Z, of the medium, its damping effect (mechanical resistance), R, the frequency, ν, the mass effect, M, of the medium, on its stiffness, K, by the equation

$$Z = R + (2\pi\nu M - K/2\pi\nu)\sqrt{-1}$$

[767] *J. Colloid Sci.*, **6**, 1951, 481-491; *Am. Ceram. Soc. Bull.*, **31**, 1952, 389-391; *Electronics*, **25**, 1952 (2), 98-100. The range of apparent viscosity is from 0.1 to 10^5 centipoises, with a single instrument.

R is the factor most sensitive to the viscosity, and can be adjusted by manual adjustment of the radio-frequency oscillator so that during the measurements $2\,\pi\nu M = K/2\,\pi\nu$, hence that $Z = R$. As G. W. Phelps, J. G. Woodward, and S. G. Maguire[768] demonstrated, each clay-water slip must first be tested for fluidity by observing the flow from a common tube consistometer, and time of flow plotted against the per cent electrolyte content. Experimental results with the vibration plate viscosimeter are given for a high in talc and ballclay artware body (AW, of density $\varrho = 1.720$) and for a sanitary ware slip (SW of $\varrho = 1.880$), deflocculated with soda-sodium silicate electrolyte simply in a series (1), and in another series (2) with an addition of 0.02 per cent gypsum, as seen in Fig. C. 189, from which the typ-

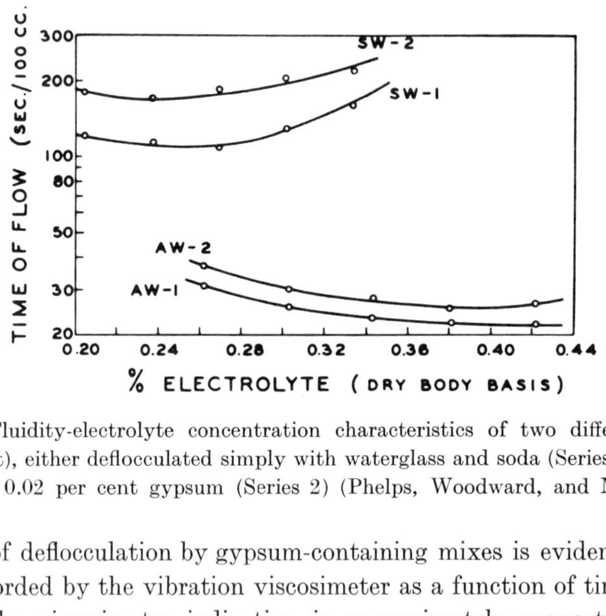

FIG. C. 189. Fluidity-electrolyte concentration characteristics of two different slips (SW and AW; see text), either deflocculated simply with waterglass and soda (Series 1), or with further addition of 0.02 per cent gypsum (Series 2) (Phelps, Woodward, and Maguire).

ical reduction of deflocculation by gypsum-containing mixes is evident. The casting curve type recorded by the vibration viscosimeter as a function of time is shown in Fig. C. 190; the viscosimeter indication is approximately proportional to $1/\eta\varrho$. The gradual thickening of the slip during the casting process is indicated on the slope of the curve which reaches at time T_c the wall of the cast, and the blade at point C. Water loss is the essential reason of gradual change in consistency, from the initially rather constant amount in the interval A–B. The water is sucked into the plaster mold from the interior of the slip as this is thickening, and gelling occurs without wall formation in a thixotropic slip. The more gradual the transition from the fluid slip in the first state to the plastic cast, the softer the resulting cast.

365. The complex ion exchange reactions occurring in the cast layer on plaster molds during their growth from the slip are particularly illustrated in Fig. C. 191,

[768] *J. Am. Ceram. Soc.*, **36**, 1953, 62-65. For flow time determinations see also W. L. Shearer, *Ibid.*, **11**, 1928, 542-559.

given by V. D. Fréchette,[769] and showing the conversion of the calcium sulfate of the mold surface to sodium sulfate which migrates with the water from the cast into the interior of the mold, leaving the surface pitted. The calcium ions going into the clay body stiffen the cast but promote the important act of releasing it from the mold while it is still quite wet. The application of flocculating salts, especially of calcium and aluminum chloride painted onto the mold surfaces, may be a means

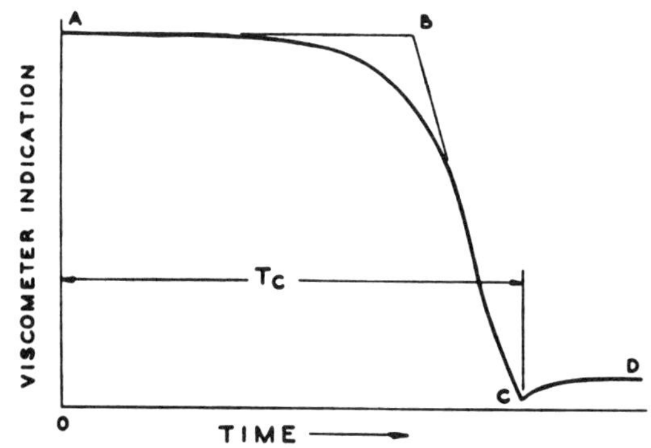

FIG. C. 190. Typical curve of slip casting as it is used in evaluating the indications of the vibration-plate viscosimeter (Phelps, Woodward, and Maguire).

of prolonging the service life of the molds by preventing disaggregation of the plaster by loss of calcium onto the clay. It also promises a faster casting by providing the strongly flocculating ions. The corrosion of the plaster molds, however, is considerable if deflocculants of the mono- and dibasic polyanionic type are used. This corrosion is strongest if a highly soluble or a highly insoluble salt is formed between the gypsum and the deflocculant, as R. Gauglitz, H. E. Schwiete, and L. Be-

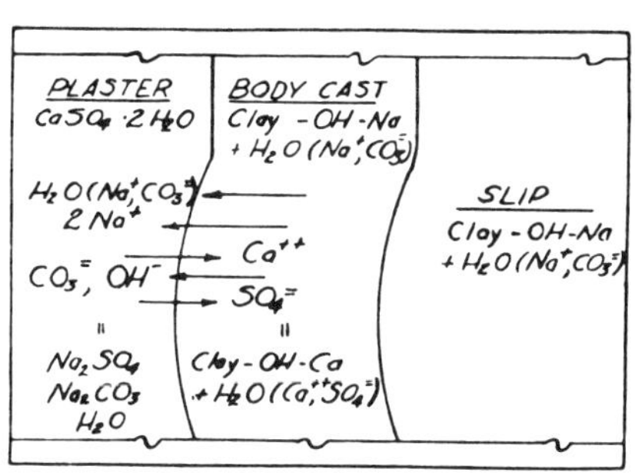

FIG. C. 191. Relationships during casting of a clay slip in plaster molds (Fréchette).

[769] *Proc. Intern. Symposium on Reactivity of Solids, Gothenburg*, 1952, Vol. II, 1954, especially p. 685. On the use of calcium and aluminum chloride paintings on mold surface, see also A. L. Johnson and F. H. Norton, *J. Am. Ceram. Soc.*, **25**, 1942, especially pp. 343 f.

rens[770] demonstrated. The least corrosive agent was the polyvinyl sulfonate but sodium tripolyphosphate is highly unsuitable and causes serious trouble by adhesion of the cast on the mold surface, and easily soluble calcium-sodium polyphosphates, such as $CaNa_3P_3O_{10}$ or $Ca(Na_4P_3O_{10})_2$ are found, besides the sparingly soluble $Ca_5(P_3O_{10})_2$ and calcium carbonate.

366. Practical details on the slip-casting process were recently given by G. W. Phelps[771] concerning the role of lignite and alkali-soluble humic acid as natural contamination of dark ballclays from Devonshire, England. In direct preparation of casting slips, the quantity of inorganic deflocculating electrolytes is normally greater than that needed for filtercake bodies, by removal of lignite dissolved in the alkaline solution during pressing. Sodium hydroxide dissolves lignite more readily than soda, and this in turn is by far more effective than the ordinary high silica waterglass solutions in the release of organic colloids. Experimental series showed that lignite-free slips tend to pass rapidly from poor casting through the range of good casting, then into "hard" casting. Lignite-free casting slips are usually poor casters, especially in the presence of appreciable amounts of soluble sulfates. English chinaclays are characterized in their casting properties by the much higher mica (muscovite) content occurring in these than in normal kaolins, especially in the coarser grain sizes (cf. C. ¶ 232). In their fluidity, after an addition of different defloccultans, synthetic mixes of kaolins with mica, talc, and pyrophyllite powders can be well adjusted to that of natural chinaclays.

367. The deflocculation effects as a consequence of the cation adsorption phenomena were extensively studied by E. Gruner and R. E. Vogel,[772] especially for bentonite suspensions, as well as by U. Hofmann[773] and by W. Czerch, K. Frühauf, and U. Hofmann.[774] It is a most remarkable result of these investigations that, in contrast to many previous interpretations of deflocculation in slips by action of cation exchange in the suspension, an alkaline reaction of the agents is *not* the essential point. Explicitly, it is not necessary that hydroxyl ions be present. Their action, doubtless, is of advantage if the kaolin has exchangeable hydrogen ions. Kaolins elutriated and flocculated with calcium hydroxide do not contain such hydrogen ions which, however, may occur in osmosis-prepared kaolins and in humic

[770] *Sprechsaal*, **92**, 1959 (19), 490-496.

[771] *Am. Ceram. Soc. Bull.*, **38**, 1959, 246-250; 411-414. See also K. O. Jiri, *Asahi Garasu Kenkyu Hokoku*, **9**, 1959, 128-136.

[772] *Kolloid-Z.*, **116**, 1950, 88-99; *Ibid.*, **118**, 1950, 100-110.

[773] *Ber. deut. keram. Ges.*, **26**, 1949, 21-32, especially pp. 28-30.

[774] *Ber. deut. keram. Ges.*, **37**, 1960, 255-265; cf. also U. Hofmann, R. Fahn, and Armin Weiss, *Kolloid-Z.*, **151**, 1957, 97-115, and *Silikat Tech.*, **8**, 1957, especially pp. 227-230. See also Kl. Frühauf, W. Burck, and U. Hofmann, *Kolloid-Z.*, **180**, 1962, 150-160, with extensive viscosity measurements and a correlation of the rheological properties of kaolinite with exchangeable cations with the grain size distribution characteristics.

acid-reacted raw kaolins and clays. The exchange of divalent cations, chiefly of calcium by sodium ions, however, is the most important factor if one studies the swelling and plasticity properties of clay-water systems. It is especially emphasized by A. Weiss[775] that deflocculation has many analogies with biochemically essential peptization and swelling phenomena.

368. Experiments with carefully purified kaolins uniformly treated with electrolytes containing sodium, potassium, magnesium, calcium, and barium cations show that there is a definite optimum if the alkaline earths are precipitated as sparingly soluble compounds by equivalent cationic exchange with solutions of soda, sodium sulfate, oxalate, fluoride, and partially also with sodium hydroxide; the resulting kaolin is exclusively the sodium clay substance. The deflocculation effect was extensively studied by determinations of apparent viscosity, thixotropy, water content required for the standard consistency, the mechanical tensile strength, and deformation pressure of moist extrusion rods (cf. C. ¶ 244). If in the same way pure calcium montmorillonite is treated with soda solutions in equivalent amounts, innercrystalline swelling causes a complete exfoliation of the silicate lamellae layers to highest dispersibility. Potassium ions act, in contrast with sodium ions, only as deflocculants if the kaolin-water suspensions are very dilute; they stiffen the suspensions with higher contents of solids. The alkaline earth ions act exclusively as stiffening agents. Evidently, the sodium ions can better screen the negative charges of the free crystal surfaces if two kaolinite crystals are separated from each other in the suspension, than can the alkaline earth cations, the number of which is only half of the alkali ions of equal total positive charges. Sodium ions counteract the mutual cohesion of the crystals, whereas alkaline earth cations favor cohesion to aggregates.

369. The deflocculating action of potassium ions in highly dilute kaolinite suspensions is caused by their hydration, whereas in more concentrated clay-water mixtures the same potassium ions are easier dehydrated and are then able to become fixed in coordination with twelve surrounding oxygen anions (cf. C. ¶ 290). The manner in which cations and water layers are arranged for sodium and calcium ions on the cohesion spots of two kaolinite crystals or on two silicate layers of montmorillonite is shown in Fig. C. 192 a; the corresponding arrangement on the surface of a kaolinite crystal or on a silicate layer alone is shown in Fig. C. 192 b. The screening actions mentioned above are best illustrated by the potential distribution on the surface of a silicate layer or of a kaolinite crystal, on the left side of Fig. C. 193 for Na^+, on the right side for Ca^{2+} ions. Similar to Hofmann's model of the "house-of-cards" structure of thixotropic montmorillonite suspensions, two schematic arrangements are given for kaolinite particles in a plastic body (Fig. C. 194 *a*) and in a

[775] *Chem. Ber.*, **91**, 1958, 487-502, especially concerning layer structure silicates of the mica and montmorillonite group.

casting slip (Fig. C. 194 *b*). The formation of diffuse double layers between the particles in the dilute solution of a sodium salt in Fig. C. 192 *a*, and the quasi-crystalline state of the introduction of calcium ions between two silicate layers in Fig. C. 192 *b* is to a certain degree related to previous schematic discussions of deflocculation effects based on the Gouy-Freundlich theory. Coming back to the attraction-repulsion theory as formulated by E. A. Hauser and S. Hirshon (1939), W. H. East,[776]

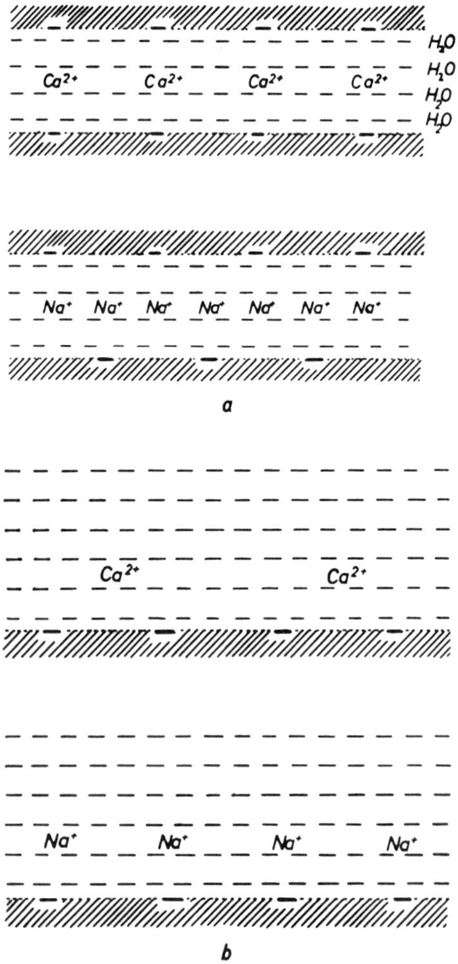

FIG. C. 192. Cations and water layers: (*a*) at contacts of two kaolinite crystals, or of two silicate layers; (*b*) on surface of a kaolinite crystal, or on one silicate layer (Czerch, Frühauf, and Hofmann).

[776] *J. Am. Ceram. Soc.*, 41, 1958, 136-140; see also E. Fusarini, *Ceramica (Milan)*, 13, 1958 (12) 61-65.

and especially F. H. Norton and A. L. Johnson (1941) and Norton and L. W. Coughanor (1948), W. G. Lawrence[777] calculated the bond energies for cation adsorption on kaolinite crystals on a crystallochemical basis, also considering the degree of cation hydration (cf. C. ¶ 251 f.) not only for the system kaolinite-water, but also for kaolinite-NaOH, NaCl, HCl, and NH_4Cl. Lawrence in discussing bond energies in the replacing power of cations in the order $Li^+ < Na^+ < K^+ < NH_4^+ < Mg^{2+} <$

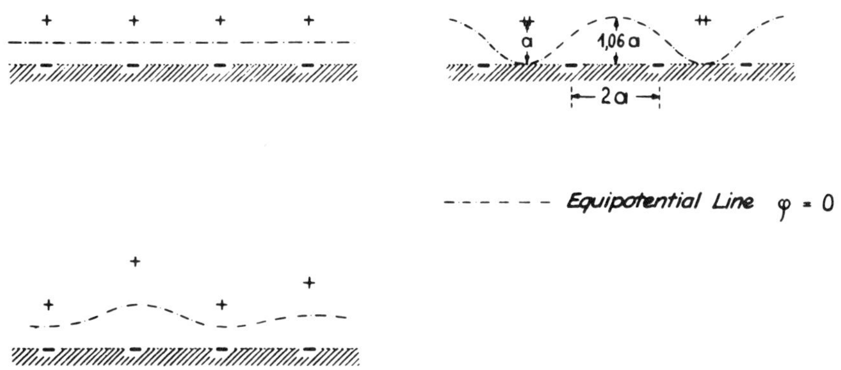

FIG. C. 193. Equipotential line for $\varphi = 0$, and potential distribution on the surface of a silicate layer, or of a kaolinite crystal, for monovalent cations (left side), and divalent ions (right side) (Czerch, Frühauf, and Hofmann). Twodimensional projection.

$Ca^{2+} < Al^{3+} < H_3O^+$ includes in his theory the effects of concentrations of the electrolyte on ion replacement and gelation, and emphasizes agreement of results with the theory of Johnson and Norton, although the proposed mechanism from which he started was quite different. The bond energies of the monovalent ions make these the most effective in the development of charges and they are also the most affected by thermal energies.

370. In spite of all attempts to describe exhaustively the properties of casting slips by ionic hydration theories, H. Salmang[778] emphasizes that these are unsatisfactory chiefly to explain many practical phenomena related to the surface tension effects in the complex phase boundary relations of a slip and casting mold. Certain anomalies, such as the much feared casting spots and striae, have their origin in surface tension effects on the slip-to-air phase boundaries, by formation of a typical tension membrane (skin). By specific staining experiments Salmang could demonstrate that nonporous, anomalously compact curved skins and striae in slip-cast ceramic bodies are caused by different slip currents meeting one another, as in double-walled

[777] *J. Am. Ceram. Soc.*, **33**, 1950, 211-218.

[778] *Naturwissenschaften*, **41**, 1954, 115; *Bull. soc. franç. céram.*, No. 34, 1957, 3-10; with many details in photographs chiefly in *Ber. deut. keram., Ges.*, **32**, 1955, 251-256.

molds, and bring about dark lines of demarcation in the unfired ceramic products and a less vitrified zone in the fired ceramic products. How great relatively small changes in surface tension in the liquid phase of clay-water systems can become, especially if wetting agents are used, was particularly evident in studies of W. C. Ormsby, R. M. Witucki, and W. A. Weyl,[779] discussed in C. ¶ 310.

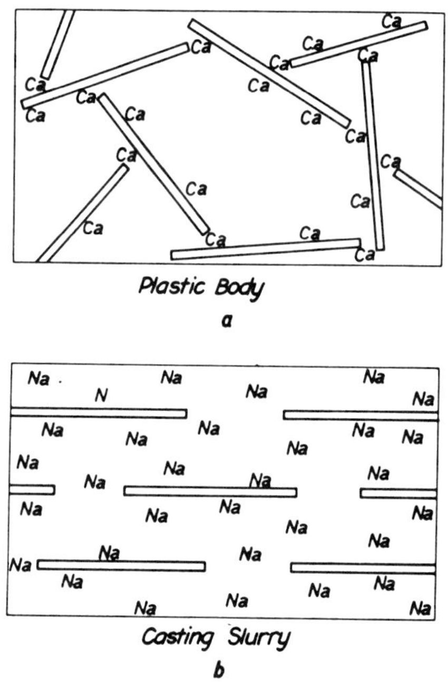

FIG. C. 194. Model of a plastic body (*a*), and of a casting slip (*b*), of kaolinite, occupied by calcium and sodium ions (simplified in two-dimensional section) (Czerch, Frühauf, and Hofmann).

371. The special use of pronounced surfactants in deflocculation of clay slips was investigated in extensive experiments of H. E. Schwiete with K. Dietz and R. Gauglitz,[780] and R. Gauglitz.[781] The flow curves studied were observed with a self-

[779] *Natl. Acad. Sci.-Natl. Research Council, Publ.*, **456**, 1956, 251-272.

[780] *Angew. Chem.*, **69**, 1957, 712; *Ziegelindustrie*, **10**, 1957, 353-359. Practical viewpoints on the liquification of a casting slip under optimum conditions of a low water content, combined with a subsequent plastification by chemical agents ("Adsorba"), were given by K. Zimmermann and W. Cremer, *Ber. deut. keram. Ges.*, **37**, 1960, 316-323.

[781] *Forschungsber. Wirtsch. u. Verkehrsministeriums Nordrhein-Westfalen*, No. 559, 1958, 65 pp., with bibliography; see also R. Gauglitz, Dissertation, Techn. Hochschule Aachen, 1958; *Veröffentl. Inst. Gesteinshüttenkunde Rhein.-Westfäl. Tech. Hochschule Aachen*, **3**, 1960, 93-148.

recording structure viscosimeter, developed by E. Helmes,[782] in shear stress versus speed slope diagrams. They show typical yield phenomena as a measure of the relative liquefaction and stiffening characteristics, with maximum flow limits determined from the diagrams. The deflocculants were added in 1 *N* aqueous solutions, in milliequivalents per 100 gm. of the solid, for kaolins, different montmorillonites, illites, muscovite, and others. Particularly important in the results is the comparison of the action of common inorganic deflocculants like soda, waterglass, sodium hydroxide, and polyacid salts (say sodium tripolyphosphate). Carboxymethyl cellulose (cf. C. ¶ 309), polyvinyl sulfonates and the sodium salts of mellitic ($C_6(COOH)_6$) and mellaphanic acid (1,2,3,5), and pyromellitic acid (1,2,4,5-benzenetetracarboxylic acid, $C_6H_2(COOH)_4$) are representatives of the organic type. Polymers of monomeric acids act increasingly, deflocculating up to a degree of polymerization of 170, but at higher polymerizations a stiffening effect is observed. The role of hydroxyl

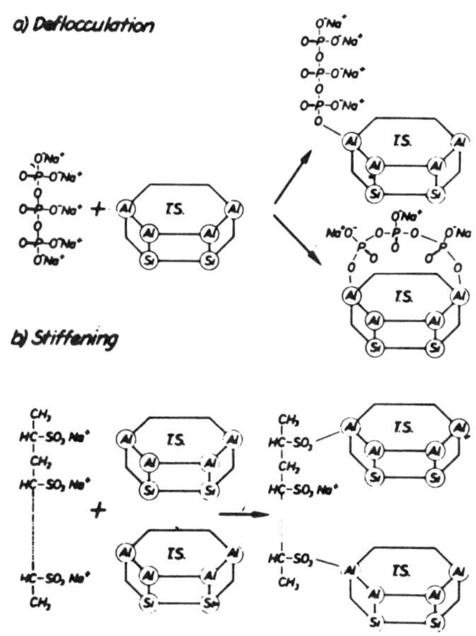

FIG. C. 195. Model of liquefaction (deflocculation) and stiffening (flocculation) of clay mineral suspensions by addition of polyanionic agents (Gauglitz and Schwiete).

anions is subordinate in comparison with these effects, of which Fig. C. 195 gives a good general impression. Hydrargillite is easily deflocculated by sodium citrate, which is adsorbed and its anion even chemically bound just as is the polyvinyl sulfonate anion with the aluminum ions of the lattice. Calcium is bound also in such complexes; particular deflocculation effects in limestone powders are thus explained.

[782] *Chem. Ingr. Tech.*, 7, 1953, 390-394. The instrument is similar to the Stormer viscosimeter.

The acid groups increase negative charges of the particles, and, therefore, increase their repulsion, but decrease the adhesion forces and the formation of micellae. An often committed error is the assumption that capillary activity and deflocculation action are closely interrelated properties of organic compounds. This is generally *not* the case. However, they do show a reduction in apparent viscosity if the chemical constitution of the acids is similar. An important reduction of slip viscosity by about 25 per cent and an increase in rate of water suction on the plaster mold is possible if the temperature of casting is increased from 20° to 50°C. Concerning the constitution of polyacidic acids, Schwiete *et al.* postulate that for good efficiency of the organic deflocculants at least three carboxyl or sulfonic acid groups must be present in the molecule. Another frequent error was the assumption that the deflocculants would act only on particles of colloidal size, since they also do it with considerably larger ones. Finally, it is interesting that hydrargillite was found to behave like a true clay mineral, because of its layer structure type.

372. Repeatedly we have mentioned how often soluble sulfates may interfere with the deflocculation process, mostly in an unfavorable manner. If soluble sulfates are present in the raw materials used for the slip casting process or if the mixing water is contaminated by dissolved sulfates, serious defects in the ceramic manufacturing are observed. It is, therefore, most important to recognize very early how to correct such effects by adequate addition of soluble barium salts to precipitate the sulfate anion. A very convenient and simple instrument for this purpose, based on rapid determination of electric conductance, the "Solu-Bridge," is recommended by W. E. Brownell,[783] with an analysis nomogram for different soluble sulfates (also including alum) to be precipitated by additions of barium carbonate. A rapid turbidimetric test for soluble sulfates was recommended by L. G. Tubbs,[784] specifically to reduce troublesome efflorescences ("dry-house scum") on ceramic products after firing.

[783] *Am. Ceram. Soc. Bull.*, **32**, 1953, 45-46.
[784] *Am. Ceram. Soc. Bull.*, **32**, 1953, 181-182.

AUTHOR INDEX

C

H

M

N

O

P

R

U

V

W

SUBJECT INDEX

A

B

C

D

E

F

G

H

I

K

L

M

N

O

Q

R

U

V

W

X

Y

Z

COMPOUND INDEX

Inorganic Compounds

Q

S

T

U

V

W

Y

Z

Organic Compounds

A

B

C

D

E

F

G

H

R

S

T

U

V

W

X

MINERAL INDEX

D

E

F

G

H

I

J

K

L